ADMISSIBILITY OF LOGICAL
INFERENCE RULES

STUDIES IN LOGIC

AND

THE FOUNDATIONS OF MATHEMATICS

VOLUME 136

Honorary Editor:

P. SUPPES

Editors:

S. ABRAMSKY, *London*
S. ARTEMOV, *Moscow*
R.A. SHORE, *Ithaca*
A.S. TROELSTRA, *Amsterdam*

ELSEVIER
AMSTERDAM • LAUSANNE • NEW YORK • OXFORD • SHANNON • TOKYO

ADMISSIBILITY OF LOGICAL INFERENCE RULES

Vladimir V. RYBAKOV
Mathematics Department
Krasnoyarsk University
Krasnoyarsk, Russia

1997

ELSEVIER
AMSTERDAM • LAUSANNE • NEW YORK • OXFORD • SHANNON • TOKYO

ELSEVIER SCIENCE B.V.
Sara Burgerhartstraat 25
P.O. Box 211, 1000 AE Amsterdam, The Netherlands

ISBN: 0 444 89505 1

Contents

QA9
R986
1997
MATH

1

0.1 Preface and Acknowledgments

This book is intended to develop the advanced theory of inference rules for general logical consequence relations, classical and non-standard logics. The study of logical consequence relations and derivability in formal deductive systems for different logics can be considered as an interdisciplinary subject lying somewhere between pure Mathematical Logic (as it is commonly understood), Universal Algebra, Algebraic Logic, and Model Theory. Working as it does with different non-classical logics (such as modal, temporal and intuitionistic logics) this field also holds considerable intersect for those concerned with Philosophical Logic, Computer Science, and Information Science. Any formal logical deductive system (in particular, any logic) presupposes a choice of axioms and inference rules, or can be generated by such a choice.

It is a popular practice to analyze the axioms of a logic, and to generate different logics by varying them. However inference rules are more resistant to study of this sort since the semantic theory for inference rules was considerably much less developed. Nevertheless, potentially this field is very attractive and important, because inference rules influence the derivability power and the effectiveness of deductive systems to a greater degree than axioms. Inference rules play an active role in derivations while axioms are in some sense static. Of course inference rules form the basic subject matter of Proof Theory. However, the general theory of inference rules and their effective applications is far less developed than, say, model theory or the aspects of equational logic in non-classical logics. The initial research concerning the nature and effectiveness of inference rules was connected with the work of formalizing logic as a whole.

This investigation could be dated from the beginning of our century. The early pioneers in the development of this field include Hilbert (1920), Lukasiewicz (1920), Lewis (1920), Gödel(1930), Tarski (1931), and Gentzen (1934). The important contribution to this field, consisting of introduction and development of algebraic semantics, was made by McKinsey and Tarski (1944). And the *possible worlds* or *relational model* semantics for different non-classical logics (which we will often employ in this book) was introduced by S.Kripke (1966). Then Lemmon (1966) and Scott (1968) developed algebraic and model-theoretic semantics further. Up to this time the field was fragmented and many subdisciplines were treated as separate subjects. The canonization of logical deductive systems for the classical propositional and predicate logics was an important step in the development of mathematical logic. Classical logic forms a powerful field which from its early days up to today has had a very important influence on mathematics and especially on set theory. Within classical mathematical logic, proof theory is connected with the foundations of mathematics and has become an important discipline in its own right.

Research concerned with formalizing logical systems which differ from classical systems is conventionally referred to as non-classical or non-standard logic.

The first investigations directly connected with clarifying the role of inference rules are works of Los (1955), Suszko (1958), Tarski (1956), and Wojcicki (1970). In fact, inference rules or derivation rules have always been among the main objects of mathematical logic. And nowadays inference rules are in active studying for different reasons and by people from different backgrounds ranging from philosophical logic through mathematics to computer science. Central question concerns which inference rules we can apply in derivations for a given logic while preserving the original set of provable formulas. Equivalently, how we can alter the inference rules of an axiomatic system while preserving the original set of its theorems (as an essence of given axiomatic system)?

This question naturally leads us to consider the class of admissible inference rules, the class of exactly those rules which satisfy this requirement. The admissible inference rules for a given logic λ can be defined as follows: they are all those rules with respect to which the logic λ is closed. They form the greatest class of rules which can be applied in proofs within the logic λ while preserving the set of theorems of λ (P.Lorenzen, 1955). In particular, these rules allow us to simplify derivations within the deductive systems of given logics. It is easy to see that the notion of admissibility is an invariant insofar that this notion does not depend on the choice of a particular axiomatic system for any given logic.

At the same time admissible rules are an unusual phenomena. Sometimes they are not derivable in a system for which they are admissible, i.e. the conclusions of these rules cannot be proved from their premises by the deductive tools of the logical system. Therefore such rules are in a sense *invalid* from viewpoint of the deductive system itself, but are *valid* nevertheless in reality, they can be applied in derivations, although their consistency with the deductive system cannot be shown inside of this system. Note that the study of even the most basic logical systems (the classical propositional calculus, predicate logic etc.) generally begins by collecting of certain necessary sets of admissible rules. Nevertheless the problem of giving a complete description of the admissible rules for many specific logics and important classes of non-standard logics, as well as the question of whether there exist algorithms for determining the admissibility of inference rules in many logical systems where open problems.

Not to anticipate the detailed content of this book (which will be outlined in the following introductory section) note that this book is intended to constructing of general theory of application inference rules, although the primary attention is paid to admissible inference rules. In particular, this book grew out of three problems concerning the admissibility of inference rules. The first problem comes from Harvey Friedman ([48]): whether there exists an algorithm for recognizing the admissible inference rules of the intuitionistic propositional logic H. The interest in H comes from the important role which this logic plays in the foundations of mathematics as a basis of constructive reasoning.

Quite a while ago R.Harrop (1960) pointed out certain examples of inference rules which are admissible but not derivable in H. Therefore the question of determining the inference rules admissible in H was an important open problem. G.Mints [104] has proved that if a rule r is admissible in H and r does not contain occurrences of even one logical connective from the list $\rightarrow$, $\vee$ then r is derivable in H. Hence H is almost structurally complete, and its admissible non-derivable rules must include enough many logical connectives, i.e. to be complicated. The second problem was set by A.Kuznetsov's (1973): is there a finite basis for the rules admissible in H? An affirmative answer to A.Kuznetsov's question would also give the positive solution to H.Friedman's problem. And the third problem is: if there exists an algorithm determining solvability of logical equations for intuitionistic logic H. A particular case of this problem is the substitution problem for H, which goes back to P.S.Novikov (1950). These problems can be reduced to the question whether we can recognize the admissibility in H of inference rules with meta-variables. The reader will find answers to all these questions in this book and also to the analogous questions for many other non-standard propositional logics and classical first-order theories via general theorems describing classes of logics for which these questions have affirmative or negative answers.

This book is intended for students in pure or applied mathematics and in computer science, who have mastered the material which is ordinarily covered by a first basic course in logic. The aim of the book is to present the fundamental theoretical results concerning inference rules in deductive formal systems. Primary attention is focused on *admissible* or *permissible* inference rules, the derivability of the admissible inference rules, the structural completeness of logics and the bases for admissible and valid inference rules. Stress is laid on propositional non-standard logics (primary, superintuitionistic and modal logics), however we also consider general logical consequence relations and classical first-order theories. The content of the book can also be considered as dealing with a branch in universal algebra. From the algebraic viewpoint we will study special quasi-varieties of algebras, primary the quasi-varieties generated by free algebras. Nowadays there are many different results concerning inference rules, and many different trends in research. Of course this book can not cover all of them. Rather, learning from those who have had a lot of experience in writing of books, we decided not to stray too far away from our research interests. Special attention is paid to make the material convenient for the reader. The presentation of material is almost self-contained and we include proofs of many results which are not readily obtainable elsewhere. We have tried to make our exposition as readable as was compatible with the presentation of complete proofs, using the most elegant arguments we knew of, employing standard notation, and reducing *hair* (as it is technically known).

The book is written at a level appropriate to first-year graduate students in mathematics or computer science. The only prerequisite is some exposure to elementary logic and universal algebra. All the results from universal algebra and all the logic that the reader needs are briefly presented in the first two introductory sections of the first chapter. The book can be used as a textbook for graduate students in mathematics and computer science. The first two chapters are suitable for a two-quarter introductory course in non-classical logic. All the material in these chapters is presented in detail, and there are many examples and exercises. The chapters from 3 to 6 contain the main results of this book concerning inference rules. They proceed more rapidly and require more sophistication on the part of the student. Their content could be the basis for a four quarter course for graduate students who are familiar with the basic tools of non-classical logic from first two introductory chapters.

Acknowledgments. The research that this book is based upon and the writing of this book were supported in many ways by many people. This book grew, in particular, out of a number of advanced courses on non-standard logics at the Krasnoyarsk State University in 1986-1991 and 1993-1995, where I found support and understanding during all this time. I would like to express my gratitude to the Alexander-Von-Humboldt Foundation (Bonn) for supporting my research program while my say at the Free University of Berlin during 1991-1993. Thanks are also due to the Science University of Tokyo and the Japan Advanced Institute of Science and Technology (Kanazava) where I spent several fruitful months in 1993. Finally I am grateful to the Fulbright Scholar Program (Washington, D.C.) which supported my research during my stay at Stanford University from September 1995 to March 1996 and at Vanderbilt University from March to June 1996.

It is my pleasant duty to thank many those who supported and inspired my work. My special thanks go to L.L.Maksimova for teaching me non-classical logic and giving me scientific advice at Novosibirsk State University, and for collaborating with me for many years, and also to D.M.Smirnov who taught me universal algebra. Thanks are also due to all the participants in the Algebra and Logic seminar at Novosibirsk State University where by giving talks and listening to others I developed up as a mathematician taking advantage of its inspiring atmosphere. I am grateful to many colleagues who listened to my lectures and talks for their helpful criticism and advice. Particular thanks are due to J.van Benthem, D.de Jongh and A.Visser for their attending my lectures at Amsterdam University and their helpful discussions, as well as to D.Gabbay for his useful advice during my lectures at Imperial College London. My thanks also go to W.Rautenberg, M.Kracht and F.Wolter for the collaboration and help during my stay at Berlin. I also wish to thank T.Hosoi and H.Ono for inspiring my work, and for their long term support.

This book was completed during my stay at Stanford University during my collaboration with G.Mints, and then during my stay at Vanderbilt University during my collaboration with R.McKenzie to whom I wish to express my gratitude for their support and for creating good conditions for me to complete my work. My thanks also go to Ph.Kremer for all his help in dealing with Stanford University life. Space does not permit me to list the names of all the colleagues who supported and inspired my work in their own way. Nevertheless I must still mention some names. I was significantly influenced by my meetings with A.Kuznetsov (who sadly deceased much too early) and colleagues in his research group, the results of one of them - A.Citkin - are well represented in this book.

I must mention my long time contact with S.Artemov, A.Chagrov, V.Shehtman and M.Zakharyaschev, representatives of Moscow's school of mathematical logic. My private conversations and discussions with them filled me with new ideas and with the desire to work. My conversations with representatives of the Polish Logical School of the 1970's have influenced my understanding of the nature of mathematical logic and given me many new ideas. My special thanks are due to W.Dziobiak, J.Czelakowski and P.Wojtylak for many useful discussions. I am grateful to D.Pigozzi for giving me useful advice in our mail correspondence. I learned many of the technical tools which are used in this book by reading the papers of K.Fine. In spite of never having discussions with M.Fitting or K.Segerberg, I wish to thank them since their books *Intuitionistic Logic, Model Theory and Forcing* and *An Essay in Classical Modal Logic* were my textbooks when I studied non-standard logics. All my colleagues in the department of Algebra and Mathematical Logic at Krasnoyarsk University always inspired my work constantly asking me when it would finally be finished. I wish to specially thank M.Golovanov for his many patient corrections of my lecture notes for graduate students.

During my stay in Stanford University I was assisted by A.J. Everett in proof-reading the present volume and I thank him for the tremendous number of improvements he suggested. Without the help of all these colleagues this book would never have been written.

Stanford University, March 1996
Vanderbilt University, May 1966
Krasnoyarsk University, August 1996

Vladimir V. Rybakov

0.2 Introduction

Any logical research is connected with a certain logical deductive system or a class of such systems. Usually we are concerned with certain logical axiomatic systems where each such system is determined by a set of axioms and a set of inference rules. A more general approach would be to deal with logical consequence relations. However, by the Los-Suszko theorem, any structural logical consequence relation can be generated by a set of axioms and a set of inference rules. Thus the axiomatic approach to logic consisting of choosing certain axioms and inference rules has strong substantiation.

By choosing a logical language, axioms, and inference rules we can introduce various logical systems (or simply logics). A special role is played by classical propositional logic (classical propositional calculus PC) and by the classical predicate calculus (classical first order logic). These logical systems are wide acknowledged to be the logical basis of all the non-logical parts of mathematics, and were developed by the founders of mathematical logic: D.Hilbert, K.Gödel, S.Kleene, and G.Gentzen from 1920 to 1940. Logics which differ from classical logic, and which are conventionally called *non-classical* or *non-standard* logics, appeared practically at the same time, but they where developed for different purposes. They played approximately those role with respect to classical logic which Non-Euclidian geometries played in the study of real space with respect to Euclidian geometry. Logics are only models of human reasoning and nothing more (of course this includes the classical logical systems themselves also). Intuitionistic logic, which was formalized by A.Heyting in 1930, goes back to the work of Brouwer from 1920 to 1930 in developing intuitionistic mathematics which was intended to reduce nonconstructive procedures in mathematical proofs. Many valued logics come from the work of Lukasiewicz (1920) who introduced the idea of a truth value other than *true* or *false*. Modal logic, which in a non-mathematical form is already present in Aristotle was axiomatized by C.Lewis (1920-1930). The motivation for considering these modal logics was based upon the idea of introducing new logical connectives, such as the connectives *necessarily* and *possibly*, in order to express the notion of logical truth more precisely. This was really an important step since it opened the way to the introduction of new logical connectives and reconsideration the meaning of usual classical logical connectives. Then came temporal logic, relevant logics (whose history goes back to Belnap), programming dynamic logics and many others.

It is a remarkable fact that this logical research, which initially might be considered as merely a branch of discreet mathematics (looking only at its syntactic aspect), also formed an important field in universal algebra. This is due to the fact that semantic tools for this logical research employed various algebraic systems and models. Historically the first semantic apparatus for logics were boolean algebras employed in connection with classical propositional logic (which are called after G.Boole, their inventor). The early pioneers of the semantic analysis of first-order logic were A.Tarski and A.Malt'sev who created the basis of model theory and established many of the fundamental results of this field. The treatment of propositional formulas as algebraic functions in certain algebras, which is merely a generalization of the truth-table approach, can be found of in the formulations of many-valued logics by Lukasiewicz and Post. And Stone and Tarski established the existence of fundamental connections between the intuitionistic propositional calculus H and the algebras of open subsets of topological spaces, which can algebraically be understood as pseudo-boolean algebras. Similar connections between modal propositional calculus and the algebras of subsets of topological spaces, which algebraically form the variety of topoboolean algebras, were found by Mckinsey. These results were later presented in an algebraic form by McKinsey and Tarski in [100]. Later on many logicians were involved in research concerning the semantics for non-classical logics, and the field became well established (see, in particular, the works of Dummett and Lemmon [34], and Rasiowa [120]). The second-order relational semantics for non-classical logics was initially presented in a various forms by several logicians (Hintikka, for instance), but the systematic use of this semantics really first appears in the works of S.Kripke in the 1960's.

It is very natural that in the beginning researchers paid primary attention to the choice of axioms for logical deductive systems and studied their theorems, i.e. the derivable formulas of given axiomatic systems. And the above mentioned semantic analyses concentrated upon the study of theorems, clarifying their nature and describing the semantic objects they involve. Thus a diversity of logical axiomatic systems (logics) was established. It was then observed that we can also vary the set of inference rules we postulate, and in this way too we can generate various different logics. Or if we need the same logic we can increase its derivation apparatus by adding new inference rules. Clearly inference rules play a more active role in derivations than axioms, and they prescribe the dynamic character of such derivations. This naturally gives rise to the question of what inference rules are consistent with given logical axiomatic systems.

First the derivable inference rules have attracted attention of researchers. Derivable rules of a logical system $\mathcal{AS}$ are those the conclusions of which can be derived from the premises of that rules in $\mathcal{AS}$, i.e. by using axioms and derivation apparatus of $\mathcal{AS}$. But it was observed that other possibly non derivable but stronger rules could also be consistently employed while derivations. This observation traced to P.Lorenzen [84], who introduced the idea of *admissible* or *permissible* inference rules. These rules are exactly all the rules with respect to which the logics are closed. By definition the admissible inference rules form the greatest class of inference rules which can be consistently employed in derivations in a given logic, i.e. while preserving the set of provable theorems. It turned out that even for well known important logics there are inference rules which are admissible but not derivable. For instance, P.Harrop discovered in [64] that the intuitionistic propositional logic H possesses admissible but non derivable rules.

Consequently the structurally complete logics: those logics for which the classes of admissible and derivable rules coincide, became important since these logics are in some sense self-contained: their axiomatic systems generate all inference rules consistent with them.

Research concerning the structural completeness has a long history. For instance, the structural completeness of some Lukasiewicz's logics was established by M.Tokarz [182], the structural completeness of Medvedev's logic of finite tasks was proved by T.Prucnal [117], and P.Wojtylak [204] showed that certain classes of many-valued logics are structurally complete. W.Dziobiak [35] found a complete description of locally finite structurally complete modal logics extending modal system $K4$. A detailed investigation took place for the admissible rules of the intuitionistic propositional logic H, which plays a special role in the foundation of mathematics. For instance, G.Mints [104] precisely specified the rules admissible but not-derivable in H, he showed that a rule r is admissible but not derivable in H only if r necessary includes formulas containing all the logical connectives of the list $\vee, \rightarrow$.

Nevertheless right up until 1980's precise criteria of the consistency of inference rules with logics had not been formulated. The primary question was to give (if possible) algorithms which would determine the admissible inference rules for a number established axiomatic systems. It seems that, structurally incomplete logic, for which the algorithm recognizing the admissibility of rules has been constructed, was quite unknown. For instance, the problem of whether there is an algorithm which determines the admissible rules of the intuitionistic logic H was posed by Friedman ([48], problem 40). The second related open problem connected with this field was the problem of description of bases for rules admissible in H. For example, in 1973 A.Kuznetsov posed the problem: whether the intuitionistic logic H has a finite basis for its admissible inference rules. The problem of recognizing the solvability of logical equations in H, and the related substitution problem in intuitionistic logic (which goes back to P.S.Novikov (1950's)) can also be regarded as questions concerning our ability to decide the admissibility of inference rules with meta-variables.

During the years 1980 to 1996 a solid theory for admissible, derivable and valid inference rules was established. All the questions mentioned and many others were answered. The theory now looks like a fusion of logic and theory of quasi-varieties of algebraic systems. In any case the results obtained were spread in many papers and were sometimes not easy accessible. It became evident that it was necessary to write a book describing this theory and in writing this book we hope to do this.

This book contains a systematic and fundamentally self-contained description of the modern theory of inference rules for formal systems. We develop a theory concerning the applicability of inference rules in general. But we pay primary attention to admissible inference rules. They are really an interesting phenomena, they are consistent with the logics for which they are admissible, form the greatest class of consistent rules, do not depend on the choice of axiomatization, but sometimes are not derivable. This means that there is no explanation of their consistency within of the logical systems in question. In this respect there is an analogy to *conscious* and *subconscious* reasoning. There is a sense in which a derivation inside a logical system using its derivable rules corresponds to conscious reasoning. In contrast a derivation using admissible rules (the strongest possible conception of consistent reasoning) corresponds, in a sense, to subconscious reasoning. For although such a derivation will be valid there is no explanation inside of the system itself of why it is.

DESCRIPTION OF THE CONTENT OF THIS BOOK

This book is structured as follows: it consists of six chapters, the first two are an introduction to the syntax and semantics of axiomatic logical systems for classical and non-standard logics. Chapters 3 - 6 then develop a fundamental theory of inference rules, and are the central part of this book.

In the first two sections of Chapter 1 we include some preliminary information concerning the axiomatization of classical and non-standard logical systems, and some basic results concerning classical first-order logic and model theory. A general approach to semantics for arbitrary propositional logical systems via universal algebra is presented in Section 3 of Chapter 1. Here the reader will find various simple but important results which where primary established by members of the Polish Logical School. Section 4 of this chapter includes a general description of admissible inference rules by employing valid quasi-identities of free algebras, and a study of the bases for admissible rules of algebraic logics. The properties of general logical consequence relations are the subject of Section 5, where we give theorems describing logical consequence relations (including the Los-Suszko theorem, which says that any structural consequence relation can be axiomatized by a set of axioms and a set of inference rules). An approach to giving a more general algebraic semantics rather than simply varieties of algebraic systems is presented in Section 6. Here we develop semantics via quasi-varieties of algebraic systems in a manner due to Blok and Pigozzi [18], and present general theorems describing logics which are algebraizable in this way. The next section, Section 7, gives theorems describing the admissible inference rules of logics which are algebraizable in the Blok-Pigozzi manner, and the bases of the admissible rules in such logics. Section 8 gives a general introductory information concerning the properties of lattices of logical consequence relations.

Chapter 2 is devoted to giving a detailed description of the semantic theory for nonstandard logics, which is necessary to understand this book. Sections 1 and 2 present algebraic semantics for superintuitionistic, modal, and temporal logics via pseudo-boolean, modal, and temporal algebras. The relational second-order Kripke semantics for these logics (in volume which we need for our purposes) is covered in Sections 3 and 4. Section 5 presents a development of Stone's theory for presentation of algebras using Kripke-style models, a description of the duality between subalgebras, homomorphic images, and direct products of algebras and the corresponding operations over Kripke models. Sections 6 and 8 present various results concerning the finite model property, i.e. the property of approximation of logics by finite algebras and models. Using algebraic semantic techniques, in Section 7 we describe a fundamental connection between modal and superintuitionistic logics, which goes back to K.Gödel [57]. Section 9 contains examples from van Benthem [190] of Kripke incomplete logics with simple axiomatic systems, and Section 10 includes K.Fine's theorems (cf. [46]) concerning the Kripke completeness of modal logics extending $K4$ with finite width.

We generally do not mean to claim that our presentation of semantic theory in this chapter is complete, rather it merely provides information which we will need in the following chapters of the book.

Chapter 3 is, in a sense, the heart of this book. Here we begin to examine admissibility using advanced methods and we give various criteria for recognizing admissibility. In Section 1 we present theorems describing reduced forms of inference rules in modal and temporal logics. These theorems, in a sense, provide us with certain canonical forms for inference rules. In this sense, they have a resemblance to the canonical formulas in classical logic (such as conjunctive and disjunctive normal form), that theorems allow us to transform rules into equivalent rules that have a very simple and uniform structure. Section 2 contains theorems which extend Gödel-McKinsey-Tarski translation T of intuitionistic formulas into modal formulas to cover the case of inference rules. We study here also the relations between the admissible, valid, and derivable rules of superintuitionistic logics and the related corresponding properties of their T-translations in the modal counterparts of superintuitionistic logics. A major step towards determining the admissibility of rules is maid in Section 3, where we give semantic criteria for recognizing admissibility of inference rules in modal and superintuitionistic.

Section 4 consists of the proof of several crucial technical lemmas which allow us to obtain in Section 5 the main theorems concerning effective determining the admissibility of inference rules for certain classes of non-classical logics. Here we give general theorems which describe the wide classes of modal and superintuitionistic logics for which the admissibility of inference rules is decidable. In particular, we give a positive answer to H.Friedman problem [48] concerning determining admissible rules of intuitionistic logic, and also to the analogous question for many other superintuitionistic and modal logics. Using the technique developed we investigate particular examples of inference rules and compare their admissibility and derivability for various logics. These results entail, in particular, that the quasi-equational theories of free algebras from many varieties of algebras corresponding to non-classical logics are decidable. In Section 6 we investigate the elementary theories of such free algebras and show, that, quite to the contrary, almost always these theories are hereditarily undecidable.

Section 7 is devoted to study of admissible inference rules for classical first order theories. We do this via schema logics of first order theories, which are merely special poly-modal logics extending the poly-modal analog of Lewis's system $S5$. It surprisingly turns out that these logics have high complexity and are almost always undecidable. Using this observation we show that the problem: whether an inference rule is admissible, is almost always undecidable for first order theories, it is undecidable even for decidable theories with very simple models. More precisely, we give here a complete description of all classical first order theories which are decidable with respect to admissibility. Section 8 presents certain examples of decidable propositional logics which are undecidable with respect to admissibility, which were first given by A.Chagrov [26]. Finally, Section 9 contains certain criteria for determining admissibility of inference rules employing particular properties of the reduced forms of inference rules.

Chapter 4 is devoted primarily to the study of the bases for admissible rules. Section 1 includes some initial algebraic results concerning the possession or lack of finite bases for quasi-identities of arbitrary algebraic systems (which we need for our exposition). Section 2 contains main results concerning the lack of finite bases for admissible inference rules. We give here general theorems showing that any logic λ from various broad classes of non-standard logics which satisfy certain specific conditions cannot have bases for admissible rules in finitely many variables, and, in particular, all such λ have no finite basis. This provides a negative solution to A.Kuznetsov's problem concerning whether intuitionistic logic H has a finite basis for its admissible rules, as well as a negative answer to the analogous question for many other superintuitionistic and modal logics. However some representative classes of logics sometimes have finite bases for admissible rules. In particular, in Section 3 we investigate such classes and prove a finite bases theorem. In particular, we settle the question about bases of admissible rules in pretabular modal and logics. Section 4 is intended to bases for valid inference rules. We show that although there are finite simple modal and pseudo-boolean algebras which have no finite bases for their quasi-identities, tabular logics always have finite bases for their valid inference rules. The subjects of Section 5 are the bases for admissible inference rules of tabular logics. We show that even tabular logics sometimes have no finite, and even no independent bases for admissible inference rules.

The property of the structural completeness is the main topic of Chapter 5. Section 1 gives some simple and general observations concerning structural completeness and establishes general theorems describing structurally complete logics. Section 2 develops the technique of quasi-characteristic inference rules which allows us to give the another proof for the description of structurally complete superintuitionistic and modal logics over system $K4$. And we present also a technique of determining admissibility for quasi-characteristic rules. Sections 3 and 4 give the main result of this chapter: a complete description of hereditarily structurally complete superintuitionistic logics and modal logic extending system $K4$. Section 5 includes an investigation concerning structurally complete fragments of logics. In particular, it contains the theorem of G.Mints [104], which shows that any fragment of intuitionistic logic not including a logical connective from the list $\vee, \rightarrow$ is structurally complete.

Chapter 6 contains some advanced topics using techniques developed previously and considers several related problems. Section 1 includes considerations connected with the problem of recognizing solvability of logical equations (i.e. with the problem: whether exists an algorithm which determines if logical equation have a solution for any given logical equation in a given logic), and the substitution problem. It turns out these problems can be reduced to a problem concerning the admissibility of inference rules in a generalized form: that is to say inference rules with meta-variables. We present here general theorems which show that logics belonging to certain broad classes non-standard logics, which satisfy certain specific conditions, have algorithms determining the admissibility of inference rules with meta-variables. By means of this technique we are able to give a positive answer to the problems mentioned above. The results of this section could be considered as a generalization of the main results of Chapter 3. Section 2 describes those modal logics which preserve inference rules admissible in the modal Lewis system $S4$, and Section 3 contains analogous theorems concerning the preservation of admissibility for rules admissible in the intuitionistic logic. The remaining three sections include examples of Kripke non-compact modal and superintuitionistic logics. The effect of Kripke non-compactness, which implies the impossibility of making a logic Kripke complete by adding inference rules, was first observed by S.K.Thomason [178]. [178] contains as an example of Kripke non-compact modal logic a logic extending modal system T of von Wright and contained in modal system $S4$. Section 4 presents examples of Kripke non-compact modal logics above logic $S4$. In Section 5 we show that there is Kripke non-compact (in particular, Kripke incomplete, and non-finitely axiomatizable) but decidable modal logic extending modal system $S4$. And Section 6 contains an example of Kripke non-compact superintuitionistic logic which was constructed by V.Shehtman [173].

Of course this book cannot cover all the interesting results concerning inference rules bearing in mind their diversity and how wide spread they are in the literature. It is also clear that the theory is far from complete and there are many challenging non-answered problems (see the open problems mentioned in this book). However we present here many important new results and supply the reader with the basic knowledge concerning this domain. Therefore this book could be considered as an invitation for further research in this field.

How to Read this Book

As we have already noted this book is a mathematical monograph devoted to modern trends in the study of inference rules. It could be used as a textbook for graduate students in Mathematics or Computer Science. The first to chapters are suitable for a two-quarter course giving an introduction to the semantic analysis of non-classical logics. The reader will find a detailed, self-contained presentation of the basic techniques together with examples and exercises which will be helpful for further study. Chapters 3 - 5 could be a basis for a full-year advanced course for students already experienced in the basic techniques of non-classical logic and model theory and interested in the study of inference rules. Some results of Chapter 6 could also be included by the choice of the lecturer.

All the definitions, propositions, examples, exercises, lemmas and theorems have the following numeration throughout: $x.y.z$ where x is the number of chapter, y is the number of the section in Chapter x, and z is the number of the definition (proposition, theorem, etc.) in Section y. Displayed equations have numeration $x.y$ where x is the number of chapter and y is the number of the equation. Authorship of all the significant results in the book is marked either by a direct reference in the formulation of the result or by the preliminary (or following) comment. If a reference is not given this is either because the author himself established the result or because he considers the result to be easy.

For convenience we briefly describe below the dependence of the chapters and sections upon one another. For students having no prior contact with non-standard logic we recommend beginning with Section 1 of first chapter, through until the last section of the second chapter. The experienced readers should only look briefly at Sections 4, 6 and 7 of the first chapter and then go straight to Chapter 3. From this point everyone have to read every section of Chapter 3 and in sequence. Readers have to study Chapter 4 in sequence, this chapter uses many results from Chapter 3. Chapter 5 can be read through, although we do not refer here actively to results from Chapter 3 or 4. Sections 1, 2 and 3 of Chapter 6 use the results of Chapter 3 extensively, but the other sections of this chapter are rather independent.

Chapter 1

Syntaxes and Semantics

1.1   Syntax of Formal Logic Systems

This section contains a brief summary of syntax and axiomatic systems for certain popular logical deductive systems, i.e. logical calculus. We also discuss here the general notions *axiomatic system, derivation, provability, propositional logic, first order theory*, etc. We start with some very simple common definitions and facts.

PROPOSITIONAL LOGICS

The language $\mathcal{L}$ for any given propositional logic λ consists of two special sets $C_{\mathcal{L}}$ and $P_{\mathcal{L}}$ and brackets (,). $C_{\mathcal{L}}$ is a collection of functional symbols which are called *logical connectives* or *truth-functors*. $P_{\mathcal{L}}$ is an infinite (but countable) collection of letters which are called *propositional letters (or variables)*. For any given logic, the logical connectives have fixed, specified meanings. For instance, the classical propositional logic PC has as logical connectives the following truth-functors $\wedge, \vee, \rightarrow, \neg$, the first three of which are binary (two arguments) functions, and last connective is a unary (one argument) function. They correspond (in a sense) to the meaning of logical expressions (or compositions) of the following form *(... and ...), (... or...), (if ... then ...)* and *(not ...)*; they are called *conjunction, disjunction, implication* and *negation* respectively.

Propositional variables from $P_{\mathcal{L}}$ can be interpreted as *variables for arbitrary propositions, variables for arbitrary logical expressions*. Therefore, these letters have no fixed meaning and are primitive (or undefined) symbols of propositional logic. It is presupposed only (informally) that they can take some truth values. Usually propositional letters are designated by small Roman letters (with or without numerical subscripts). We will not fix any special notation for propositional variables, so any letter standing alone (possibly with numerical subscripts) can be understood as propositional letter.

The formulas (or expressions) of a given language $\mathcal{L}$, more precisely, *well-formed formulas* (we will use the abbreviation *wff*), are constructed from propositional letters by means of logical connectives. That is,

Definition 1.1.1 *The set $WFF_{\mathcal{L}}$ of well-formed formulas in the language $\mathcal{L}$ is the smallest set of expressions containing the set $P_{\mathcal{L}}$ (propositional letters of $\mathcal{L}$) and closed with respect to all logical connectives from $C_{\mathcal{L}}$. That is,*

- *A propositional letter is a wff.*

- *If $\alpha_1, ..., \alpha_n$ are wffs and f is a n-ary logical connective then the expression $f(\alpha_1, ..., \alpha_n)$ is wff as well.*

Let us pause briefly to recall that often there are some special conventions governing writing formulas in order to make formulas more readable. For instance, binary logical connectives $\vee, \wedge$ and $\rightarrow$ are usually placed between arguments (the outermost brackets round any complete formula in this case are usually omitted) and unary logical connectives often omit brackets (in languages including these connectives like classical propositional calculus PC etc.). So, instead of such formally correct but artificially looking formulas as $\vee(\alpha, \beta)$, $\wedge(\alpha, \beta)$, $\rightarrow (\alpha, \beta)$, and $\neg(\alpha)$, we write more natural expressions $\alpha \vee \beta$, $\alpha \wedge \beta$, $\alpha \rightarrow \beta$ and $\neg\alpha$ respectively.

Under these conventions, any corresponding formula can be easily read and just the sound gives the sense of expression. And conversely, we can naturally write down expressions corresponding to logical laws, for example, *the law of excluded middle* in classical logic can be expressed by means of the following formula: $p \vee \neg p$. As a one more example consider the following formula concerning implication $(p \rightarrow (q \rightarrow p))$. For convenience, the outermost brackets round any complete formula can usually be omitted. However sometimes additional brackets convenient for reading are acceptable as well. Notations can be further simplified by conventions on power of connectives. So it is presupposed that all unary connectives are weaker than binary ones and that $\rightarrow$ is stronger than $\vee$ and $\wedge$. This allows us to omit some brackets. Thus

$$\neg\alpha \wedge \beta \text{ means } ((\neg\alpha) \wedge \beta); \ \neg\alpha \rightarrow \beta \text{ means } ((\neg\alpha) \rightarrow (\beta))$$
$$\alpha \wedge \beta \rightarrow \alpha \vee \beta \text{ means } ((\alpha \wedge \beta) \rightarrow (\alpha \vee \beta))) \text{ etc.}$$

A general definition of *axiomatic systems for propositional logics* in given logical language $\mathcal{L}$ can be introduced as follows. An *axiomatic system* $\mathcal{AS}$ (or synonymously *formal deductive system*) consists of a collection $\mathcal{Ax}$ of axioms and a collection $\mathcal{Ir}$ of inference rules.

Definition 1.1.2 *An* axiom *is an expression of the form* $\varphi(x_1, \ldots, x_n)$ *obtained from some formula* $\varphi(p_1, \ldots, p_n)$ *(which is built up of propositional letters* $p_1, \ldots, p_n$*) of* $WFF_\mathcal{L}$ *by means of replacement of the letters* $p_1, \ldots, p_n$ *on variables* $x_1, \ldots, x_n$ *for formulas.*

Thus informally, any formula of $WFF_\mathcal{L}$ can be considered as an axiom if we consider their propositional letters as being variables for formulas.

Definition 1.1.3 *An (structural) inference rule r is an expression of the form*

$$\frac{\varphi_1(x_1, \ldots, x_n), \ldots, \varphi_m(x_1, \ldots, x_n)}{\psi(x_1, \ldots, x_n)},$$

where $\psi(p_1, \ldots, p_n)$ *and all* $\varphi_i(p_1, \ldots, p_n)$ *are formulas from* $WFF_\mathcal{L}$ *built up of propositional letters* $p_1, \ldots, p_n$, *which are replaced in r by the new formulas variables* $x_1, \ldots, x_n$.

Hence, informally speaking, axioms can be understood as inference rules, which have the empty set of premises. That is to say, as inference rules without any premise. The notion of *derivability* (or *provability*) in a given $\mathcal{AS}$ looks as follows:

Definition 1.1.4 *A finite sequence* $\alpha_1, ..., \alpha_n$ *of formulas from* $WFF_{\mathcal{L}}$ *is said to be a derivation in* $\mathcal{AS}$ *if and only if, for each member* α_i, *at least one from the below conditions holds:*

- α_i *is obtained from some* $\alpha \in \mathcal{A}x$ *(that is from some axiom of* $\mathcal{AS}$*) by substitution formulas from* $WFF_{\mathcal{L}}$ *in place of the variables of* α.

- α_i *is obtained from some formulas* $\mu_1, ..., \mu_k$ *of our sequence which occur before* α_i *by means of some inference rule* r *from* $\mathcal{I}r$. *This means that* r *has the form*

$$\frac{\beta_1(x_1, ..., x_m), ..., \beta_k(x_1, ..., x_m)}{\delta(x_1, ..., x_m)}$$

and there exists a tuple $\rho_1, ..., \rho_m$ *of formulas from* $WFF_{\mathcal{L}}$ *such that* $\beta_j(\rho_1, ..., \rho_m) = \mu_j$ *for every* $1 \leq j \leq k$, *and* $\delta(\rho_1, ..., \rho_m) = \alpha_i$.

A *wff* α is called *derivable* (or *provable* or *deducible*) in $\mathcal{AS}$ if there exists a derivation in $\mathcal{AS}$ which has α as the final concluding formula. The derivable formulas of any $\mathcal{AS}$ are called *theorems* of $\mathcal{AS}$. We will denote by $Th(\mathcal{AS})$ the set of all theorems of $\mathcal{AS}$, $Th(\mathcal{AS})$ is called the *theory* of the system $\mathcal{AS}$.

Proposition 1.1.5 *(Substitution Property). If* α *is a formula derivable in* $\mathcal{AS}$ *and* $\alpha_1 \in WFF_{\mathcal{L}}$ *is some formula obtained from* α *by means of a substitution of certain formulas from* $WFF_{\mathcal{L}}$ *in place of some arbitrary chosen variables of* α *then* α_1 *is also derivable in* $\mathcal{AS}$. *That is if* $\beta_1, ... \beta_m$ *are wffs of* $\mathcal{AS}$ *then*

$$\alpha(p_1, ..., p_n, q_1, ..., q_m) \in Th(\mathcal{AS}) \Rightarrow \alpha(p_1, ..., p_n, \beta_1, ..., \beta) \in Th(\mathcal{AS}).$$

Proof. This follows immediately from our definition. Indeed it is sufficient to take a derivation of α in $\mathcal{AS}$ and to make the appropriate substitutions in all formulas of this derivation. The resulting sequence will be a derivation as well and will have α_1 as the concluding formula. ∎

An axiomatic system which has finite number of the axioms is said to be *logical calculus*. We now recall the notion of *propositional logic* which is related to the notion of axiomatic system. Let a logical language $\mathcal{L}$ be fixed. We say that a subset X of $WFF_{\mathcal{L}}$ is closed with respect to an inference rule

$$r := \frac{\varphi_1(x_1, ..., x_n), ..., \varphi_m(x_1, ..., x_n)}{\psi(x_1, ..., x_n)},$$

if, for every tuple $\alpha_1, ..., \alpha_n$ of formulas from $WFF_{\mathcal{L}}$, $\psi(\alpha_1, ..., \alpha_n) \in X$ holds provided $\varphi_j(\alpha_1, ..., \alpha_n) \in X$ for all φ_j. For the definition of a propositional logic, we have to fix some collection $\mathcal{I}r$ of inference rules in $\mathcal{L}$.

Definition 1.1.6 *A propositional logic λ in the language $\mathcal{L}$ with respect to inference rules $\mathcal{I}r$ is a subset of $WFF_{\mathcal{L}}$ such that*

(i) λ is closed with respect to each inference rule from $\mathcal{I}r$

(ii) λ is closed with respect to substitution: if $\alpha \in \lambda$ and β is obtained from α by substitution of formulas from $WFF_{\mathcal{L}}$ in place of the propositional letters of α then $\beta \in \lambda$.

Propositional logics are counterparts of axiomatic systems. Indeed, from Proposition 1.1.5 and definitions we immediately obtain

Proposition 1.1.7 *The set $\lambda(\mathcal{AS}) := Th(\mathcal{AS})$, i.e. the set of all theorems of $\mathcal{AS}$, where $\mathcal{AS} := \{Ax, \mathcal{I}r\}$, is a propositional logic in $\mathcal{L}$ with respect to $\mathcal{I}r$.*

If the language $\mathcal{L}$ and the set of inference rules $\mathcal{I}r$ are known and fixed we refer to λ as simply *logic*. Sometimes $\lambda(\mathcal{AS})$ is said to be the *logic* of the axiomatic system $\mathcal{AS}$. In a similar way we refer to formulas from any propositional logic λ as *theorems* of λ. On the other hand, any propositional logic λ can be considered as an axiomatic propositional system $\mathcal{AS}(\lambda)$ in the same language, with the same set of (structural) inference rules and, with the set of axioms which coincides with λ. Now we turn to presenting a collection of axiomatic systems for different propositional logics. This presentation is suitable not only for providing further references but for providing first acquaintance with these systems as well (though we cannot discuss below in detail the issues of introduction of these systems). Certain *superintuitionistic (intermediate), modal* and *temporal* logics will be presented.

EXAMPLES OF AXIOMATIC SYSTEMS FOR PROPOSITIONAL LOGICS

First of all, we recall that there are different axiomatic systems having the same logics (the sets of provable formulas, theorems). Moreover, even languages of axiomatic systems can be different but the systems are *expressible* each in other, in a such way, that the axiomatic systems are in a sense *equivalent*. That is to say there exist translations of formulas of one language into another such that these translations preserve derivability.

For instance, for classical propositional logic PC we can choose logical connectives to be the usual complete list of connectives $C_1 := \{ \wedge, \vee, \rightarrow, \neg \}$; we can also choose for the same purposes the following restricted list $C_2 := \{ \rightarrow, \neg \}$. The axiomatic system $\mathcal{AS}_1$ for PC in the first language (Hilbert style axiomatization for PC) looks as follows: it has as axioms

(A1) $(x \rightarrow (y \rightarrow x))$
(A2) $(x \rightarrow (y \rightarrow z)) \rightarrow ((x \rightarrow y) \rightarrow (x \rightarrow z))$
(A3) $x \rightarrow (x \vee y)$
(A4) $y \rightarrow (x \vee y)$
(A5) $(x \rightarrow z) \rightarrow ((y \rightarrow z) \rightarrow ((x \vee y) \rightarrow z))$
(A6) $(x \wedge y) \rightarrow x$
(A7) $(x \wedge y) \rightarrow y$
(A8) $(x \rightarrow y) \rightarrow ((x \rightarrow z) \rightarrow (x \rightarrow (y \wedge z)))$
(A9) $(x \rightarrow y) \rightarrow ((x \rightarrow \neg y) \rightarrow \neg x))$
(A10) $\neg\neg x \rightarrow x$

and single inference rule - *modus ponens*:

$$\frac{x, x \rightarrow y}{y}.$$

The axiomatic system $\mathcal{AS}_2$ for PC in the second language consists of axioms (A1), (A2), (A9), (A10) and the same inference rule - *modus ponens*. The translation t of propositional formulas in the language of the axiomatic system $\mathcal{AS}_1$ into formulas in the language of $\mathcal{AS}_2$ is the following:

$$t(p) := p, \quad t(\alpha \rightarrow \beta) := (t(\alpha) \rightarrow t(\beta)), \quad t(\neg\alpha) := \neg T(\alpha)$$
$$t(A \vee B) := (\neg t(A) \rightarrow t(B)), \quad t(\alpha \wedge \beta) := \neg(t(\alpha) \rightarrow \neg t(\beta)).$$

The converse translation is immediate: $\varphi \mapsto \varphi$. It is well known (see any textbook in mathematical logic, for example, E.Mendelson [102] that the above mentioned translations preserve derivability in $\mathcal{AS}_1$ and $\mathcal{AS}_2$. Thus these two different axiomatic systems give us, in a sense, the same logic. Because there are different axiomatic systems which give the same logics, axiomatic systems for propositional logics given below are just some (probably the most well known) variants from possible alternatives.

The axiomatic system for *intuitionistic propositional logic* H can be obtained from the axiomatic system $\mathcal{AS}_1$ for PC in the following way. We simply exchange the axiom (A10) $\neg\neg x \rightarrow x$ (which, in particular, expresses the law of excluded middle) for the axiom (A11) $x \rightarrow (\neg x \rightarrow y)$. This axiomatic system is called *the Heyting calculus* and denoted by H (after name of A.Heyting, who first gave an axiomatization for intuitionistic proportional logic).

If we have an axiomatic system $\mathcal{AS}$ in a language $\mathcal{L}$ and $\mathcal{X}$ is a set of axioms in the same language $\mathcal{L}$ (say $\mathcal{X} = \{\alpha_1, \ldots, \alpha_n\}$) then

$$\mathcal{AS} \oplus \mathcal{X} \quad (\mathcal{AS} \oplus \{\alpha_1, \ldots, \alpha_n\}, \text{ respectively})$$

denotes the axiomatic system which is obtained from $\mathcal{AS}$ by adding the set $\mathcal{X}$ of new axioms (the inference rules of $\mathcal{AS} \oplus \mathcal{X}$ are the rules of $\mathcal{AS}$). We call $\mathcal{AS} \oplus \mathcal{X}$ the axiomatic extension of $\mathcal{AS}$.

Definition 1.1.8 *A superintuitionistic axiomatic system is an axiomatic sys-tem of form $H \oplus \mathcal{X}$. Any system $H \oplus \{\alpha_1, \ldots, \alpha_n\}$ (that is superintuitionistic system with a finite number of axioms) is called superintuitionistic calculus.*

Certain examples of superintuitionistic axiomatic systems are given in the Ta-ble 1.1. Let For_p be the set of all *wffs* of the axiomatic system $\mathcal{AS}_1$ for classical

Table 1.1: A list of superintuitionistic axiomatic systems

For_p	$=$	$H \oplus p$
PC	$=$	$H \oplus p \vee \neg p$
KC	$=$	$H \oplus \neg p \vee \neg\neg p$
LC	$=$	$H \oplus (p \rightarrow q) \vee (q \rightarrow p)$
KP	$=$	$H \oplus (\neg p \rightarrow q \vee r) \rightarrow (\neg p \rightarrow q) \vee (\neg p \rightarrow r)$
SL	$=$	$H \oplus ((\neg\neg p \rightarrow p) \rightarrow \neg p \vee p) \rightarrow \neg p \vee \neg\neg p$

propositional calculus PC.

Definition 1.1.9 *A superintuitionistic logic (or an intermediate logic) is a subset of For_p which contains all the derivable formulas of H and is closed with respect to modus ponens and substitutions of formulas in place of variables.*

Thus for any superintuitionistic axiomatic system $\mathcal{AS}$ the logic $\lambda(\mathcal{AS})$ is a su-perintuitionistic logic. In particular, the logics $\lambda(\text{PC})$, $\lambda(\text{KC})$, $\lambda(\text{LC})$, $\lambda(\text{KP})$ and $\lambda(\text{SL})$ are superintuitionistic (intermediate) logics (see Table 1.1).

Hereafter we shall often identify an *axiomatic system* and its *logic* (the set of theorems), of course keeping in mind that different axiomatic system can gen-erate the same logic. We follow the common tradition of introducing axiomatic systems and calling them logics. Therefore we also will use this terminology more freely if this gives no misunderstands. In any case, when our terminology has to be clarified we will give the corresponding clarification.

Now we turn to the language $\mathcal{L}_m$ of *modal propositional logic*. In accordance with early pioneers of modal logic (Lewis, Langford), we say modal logic is de-signed to formalize discourse about the nature of *necessity, possibility* and *strict implication*. Modal logic is naturally based on language of classical proposition-al calculus. Therefore its language includes a complete (or restricted) list of the logical connectives of PC. We put as the basis for $\mathcal{L}_m$ the complete list: $\wedge, \vee, \rightarrow, \neg$ of the logical connectives of PC. The language $\mathcal{L}_m$ is enriched (in comparison to PC) with a new unary logical connective $\square$ which can be read as *necessary*. The set of all formulas in our modal propositional language is also closed with respect to the formation rule: if α is a formula, so is $\square\alpha$. The second dual modal connective $\Diamond$ (which can be read as *possible*) can be introduced as $\Diamond a := \neg\square\neg a$.

Remark. Of course, in the case when classical propositional logic is the basis of the modal logic, we can choose restricted list of the connectives, say, $\neg$, $\vee$, and the other connectives can be introduced as abbreviations: $\alpha \to \beta := \neg \alpha \vee \beta$, $\alpha \wedge \beta := \neg(\neg \alpha \vee \neg \beta)$. Also, it is possible to take $\Diamond$ as our basic modal logical connective instead of $\Box$, then we can express $\Box$ by means of $\Diamond$: $\Box \alpha := \neg \Diamond \neg \alpha$. We merely fixed above the language $\mathcal{L}_m$ as a certain appropriate variant.

With respect to axiomatic modal systems we start description with smallest normal modal logic K (for S.Kripke, who in fact never actually studied this system). Its axiomatic system is the following. The axioms include all axioms of PC (say, the Hilbert-style axioms of $\mathcal{AS}_1$) and an additional axiom connected with the modal connective $\Box$:

$$\Box(x \to y) \to (\Box x \to \Box y).$$

The inference rules for K are

$$(R1) \ \frac{x, x \to y}{y} \ \text{(modus ponens)} \quad (R3) \ \frac{x}{\Box x} \ \text{(necessitating rule)}.$$

(other names for (R3) are the *normalization rule*, or the *Gödel rule*). Below we list some modal formulas with the names for them that are more or less standard in literature.

$$
\begin{array}{lll}
T & := & \Box p \to p \\
4 & := & \Box p \to \Box\Box p \\
D & := & \Box p \to \Diamond p \\
E & := & \Diamond p \to \Box\Diamond p \\
B & := & p \to \Box\Diamond p \\
Tr & := & (\Box p \to p) \wedge (p \to \Box p) \\
V & := & \Box p \\
M & := & \Box\Diamond p \to \Diamond\Box p \\
G & := & \Diamond\Box P \to \Box\Diamond p \\
Crz & := & \Box(\Box(p \to \Box p) \to p) \to p \\
Dum & := & \Box(\Box(p \to p) \to (\Diamond\Box p \to p) \\
W & := & \Box(\Box p \to p) \to \Box p
\end{array}
$$

Any formula above reflects a certain *modal principle* presupposing some specific property of modality. Very often axiomatic systems for modal logics have been introduced by adding some of these formulas or their combinations to the system K as new axioms (see Table 1.2).

The major difference between non-normal modal logics and normal ones consists of in the fact that non-normal logics reject *the Gödel normalization rule*. Usually they have some additional weaker rules which replace the Gödel normalization Rule (R3). Well known rules of this kind are

$$(R2) \ \frac{x \to y}{\Box x \to \Box y}, \quad (R3)' \ \frac{\Box(x \to y)}{\Box(\Box x \to \Box y)}.$$

Table 1.2: Some popular normal modal calculi

For_m	=	$K \oplus \{p\}$
T	=	$K \oplus \{\Box p \to p\}$
B	=	$T \oplus \{p \to \Box \Diamond p\}$
D	=	$K \oplus \{\Box p \to \Diamond p\}$
K4	=	$K \oplus \{\Box p \to \Box \Box p\}$
S4	=	$K4 \oplus \{\Box p \to p\}$
S4.1	=	$S4 \oplus \{\Box \Diamond p \to \Diamond \Box p\}$
S4.2	=	$S4 \oplus \{\Diamond \Box p \to \Box \Diamond p\}$
S4.3	=	$S4 \oplus \{\Box(\Box p \to \Box q) \vee \Box(\Box q \to \Box p)\}$
S5	=	$S4 \oplus \{\Diamond p \to \Box \Diamond p\}$
Grz	=	$S4 \oplus \{\Box(\Box(p \to \Box p) \to p) \to p\}$
GL	=	$K4 \oplus \{\Box(\Box p \to p) \to \Box p\}$
K4Dum	=	$K4 \oplus \{\Box(\Box(p \to p) \to (\Diamond \Box p \to p))\}$
K4.1	=	$K4 \oplus \{\Box \Diamond p \to \Diamond \Box p\}$
K4.2	=	$K4 \oplus \{\Diamond \Box p \to \Box \Diamond p\}$

A list of non-normal modal logics (including some of Lewis's weak logics, family E-logics, etc.) is given in the Table 1.3. Remind that $Ax(\mathcal{AS})$ is the set of axioms of the axiomatic system $\mathcal{AS}$, and $Ir(\mathcal{AS})$ denotes the set of inference rules for $\mathcal{AS}$.

Table 1.3: Some non-normal modal logics

System	Axioms	Rules
C2	Ax(K)	R1, R2
D2	$Ax(K) \cup \{\Box x \to \Diamond x\}$	R1,R2
E2	$Ax(K) \cup \{\Box x \to x\}$	R1,R2
E3	$Ax(E2) \cup \{\Box x \to \Box(\Box y \to \Box x)\}$	R1, R2
ET	$Ax(E2) \cup \{\Box x \to \Box \Box x\}$	R1, R2
E3	$Ax(E3) \cup \{\Box x \to \Box \Box x)\}$	R1, R2
S2	$\{\Box A : A \in Th(K)\} \cup \{\Box(\Box x \to x)\}$	R1, R3'
S3	$Ax(S2) \cup \{\Box x \to \Box(\Box y \to \Box x)\}$	R1

Temporal propositional logic is (in a sense) closely related to modal logic and can be considered as a special case of *bimodal logic*. The language $\mathcal{L}_t$ is based also on the language of PC and includes it. $\mathcal{L}_t$ has only two additional unary logical temporal primitive connectives: F and P. The rule of formation for *wffs* is enriched with: if α is a wff, so is $F\alpha$ and $P\alpha$. The meaning of these connectives is as follows: $F\alpha$ - *in a certain time in future α will be*; $P\alpha$ - *in a some time in past α was*. The other temporal logical truth functions: G (*always will be*) and H (*always was*), can be expressed by means of our basic ones as follows

$$G\alpha := \neg F\neg\alpha, \quad H\alpha := \neg P\neg\alpha.$$

Minimal, in a sense, (or smallest, weakest) temporal propositional logic T_0 is defined as follows. It has the axioms of PC together with the axioms

$$G(x \to y) \to (Gx \to Gy)$$
$$H(x \to y) \to (Hx \to Hy)$$
$$p \to HFp, \quad p \to GPp$$

The inference rules of T_0 are following:

$$\frac{x, x \to y}{y}, \quad \frac{x}{Gx}, \quad \frac{x}{Hx}$$

An axiomatic system of temporal logic is an axiomatic system in the language $\mathcal{L}_t$ containing all axioms of T_0 and having the same inference rules. *Temporal logic*, respectively, is a set of formulas in the language $\mathcal{L}_t$ containing T_0 and closed with respect to the uniform substitution and all the inference rules of the logic T_0.

THE LANGUAGE OF FIRST ORDER LOGIC

The language of the first order logic is more expressive and (compared to the languages of propositional logic) is richer into two respects:

(i) The meaning of *propositional variables* is specified (in a sense), *atomic formulas*, which can express particular specific properties of objects, play the role of propositional variables;

(ii) *Quantification over variables* is introduced into the language.

In more detail, a first-order predicate language L includes a collection P_L of *predicate letters*, a collection F_L of *function symbols*, and a collection C_L of *constant symbols*. Informally, the meaning of these symbols is as follows. The members of P_L denote relations between objects, symbols from F_L mark functions on objects, and symbols of C_L denote certain specific members of the set of objects. Any set from the above mentioned sets P_L, F_L and C_L can be empty or finite, or infinite. Each functional symbol $f \in L$ (every predicate symbol $R \in P$) has a positive integer $\sharp(f)$ ($\sharp(R)$) assigned to it; if $n = \sharp(f)$ ($n = \sharp(R)$) then f is the n-ary functional symbol (respectively, R is said to be an n-ary relational symbol). Symbols from C_L can be understood as 0-ary functional symbols.

The set V of *variables* is a fixed countable (or enumerated) set of certain letters $x, y, z, \ldots$ (possibly with numerical subscripts). The set of *terms* of L is the smallest set of expressions containing V and C_L and closed under the formation rule: if $t_1, ..., t_n$ are terms and $f \in F_L$ is a certain n-ary functional symbol, then the expression $f(t_1, ..., t_n)$ is a term.

Definition 1.1.10 *If t and g are terms then $t = g$ is an atomic formula; if $t_1, ..., t_n$ are terms and $P \in P_L$ is and n-ary predicate symbol hen $P(t_1, ..., t_n)$ is an atomic formula.*

The logical connectives of a first-order predicate language L include the usual connectives of PC, say $\vee, \wedge, \rightarrow, \neg$; and furthermore L has the *quantifier* symbols $\forall$ and $\exists$. The definition of well-formed formulas (wffs) in L is as follows.

Definition 1.1.11 *The wffs form the smallest set of expressions containing the atomic formulas and closed with respect to the following formation rules:*

- *If A, B are wffs then the expressions $A \vee B$, $A \wedge B$, $A \rightarrow B$, $\neg A$ are also wffs.*

- *If A is a wffs and $x \in V$ (x is variable) then $\forall x A$, $\exists x A$ are also wffs.*

So the following expressions

$$\forall x \forall y((P(x, y) \rightarrow P(y, x)), \quad \forall x \exists y(f(x) = y),$$
$$(\exists x P(x, y)) \vee (\exists x P(y, x)), \quad (\exists x \forall y P(x, y)) \rightarrow (\forall y \exists x P(x, y))$$

are wffs of L (provided P and f are a predicate letter and a function symbol, respectively). The set $SF(A)$ of all *subformulas* of a wff A can be defined by induction on the length of A as follows.

$$SF(A \circ B) := SF(A) \cup SF(B) \cup \{A \circ B\} \text{ where } \circ \in \{\vee, \wedge, \rightarrow\},$$
$$SF(\neg A) := SF(A) \cup \{\neg A\},$$
$$SF(QxA) := SF(A) \cup \{QxA\} \text{ where } Q \in \{\forall, \exists\},$$
$$SF(A) := \{A\} \text{ if } A \text{ is an atomic formula.}$$

For instance the formula $\forall x \forall y((P(x, y) \rightarrow P(y, x))$ has subformulas

$$\forall x \forall y((P(x, y) \rightarrow P(y, x)) \quad \forall y((P(x, y) \rightarrow P(y, x)),$$
$$P(x, y) \rightarrow P(y, x) \qquad\qquad P(x, y)$$
$$P(y, x)$$

An occurrence of variable x in an wff A is called *free* if this occurrence does not occur in some subformula of A of the form $\forall x B$ or $\exists x B$. If an occurrence of a variable x in A is not free then we say that this occurrence is *bound*. The set of all variables in a wff A which have a free occurrence is denoted by $FV(A)$. A wff A is said to be a *sentence* if A has no free occurrences of variables.

An axiomatic system in a first-order language usually (likewise to the case of propositional logic) consists of a set of axioms and a set of inference rules. But there is the following difference from propositional logic. With propositional logic, as we have seen above, axioms are just formulas and inference rules are just constructed out of formulas and nothing more. However in the case of first-order logic some axioms or inference rules have special restrictions on free or bound occurrences of variables.

We describe here a basic axiomatic system CPC (in any L) which is called the *classical predicate calculus*. This system is a Hilbert-style formal system (of course, there are lots of equivalent similar Hilbert-stile systems, there are also Gentzen-type formal systems which are equivalent to CPC.).

Definition 1.1.12 *A tautology is a wff in L which is obtained from a theorem of PC by substitutions some wffs of L in place of all the propositional letters of this theorem.*

If $A(x)$ is a wff with a free variable x and t is a term then $A(x/t)$ (or A_t^x, or, more simply, $A(t)$) denotes the result of replacing of all free occurrences of x by the term t throughout $A(x)$. When using this notation we always assume that none of the variables in t occur as bound variables in the formula $A(t)$.

The axioms of CPC are as follows (where x, y, z, x_i, y_i denote arbitrary variables, n is an arbitrary natural number, P is an arbitrary n-ary predicate symbol from L, and t is an arbitrary n-ary term)

(i) All tautologies,

(ii) All equality axioms:

    (1) $x = x$,

    (2) $(x = y) \rightarrow (y = x)$,

    (3) $(x = y) \wedge (y = z) \rightarrow (x = z)$,

    (4) $(x_1 = y_1 \wedge \ldots \wedge (x_n = y_n) \rightarrow (P(x_1 \ldots x_n) \rightarrow (P(y_1 \ldots y_n))$,

    (5) $(x_1 = y_1) \wedge \ldots \wedge (x_n = y_n) \rightarrow (t(x_1 \ldots x_n) = t(y_1 \ldots y_n))$

(iii) All formulas of either of the forms
    $(\forall x A(x)) \rightarrow A(t)$, $A(t) \rightarrow \exists x A(x)$

The rules of inference for CPC are as follows, where A and B are arbitrary formulas, x and y are arbitrary variables but x has no free occurrences in A,

(i) $\frac{A \rightarrow B, A}{B}$ (modus ponens)

(ii) $\frac{A \rightarrow B(x)}{A \rightarrow \forall y B(y)}$, $\frac{B(x) \rightarrow A}{\exists y B(y) \rightarrow A}$.

Thus, generally speaking, certain inference rules and axioms of CPC can have restrictions on variables or substituted terms (hence they are not *structural*). A *derivation* (or *proof*) of a formula A from a set of sentences X in CPC is a finite sequence $B_1 \ldots B_n$ of formulas, with $B_n = A$, where every B_i is either an axiom of CPC, a member of X, or else B_i follows from earlier B_j by one of the above inference rules. A formula A is *derivable* in CPC (or is a *theorem* of CPC, or is *provable* in CPC) if there is a derivation of A from $\emptyset$. If there is a proof of A from a set of hypotheses X we write $X \vdash_{CPC} A$ and say that A is provable from X in CPC.

Any other Hilbert-style axiomatic system for any L can be introduced in similar way, it is merely that the sets of axioms and inference rules could be altered. Also the language of a given system can be restricted. An important axiomatic subsystem of CPC is the so-called *pure classical predicate calculus* PCPC which can be obtained from CPC by omitting every formula which contains the equality symbol $=$. More precisely, we remove symbol $=$ from the language, all formulas of the form $f = g$, and all equality axioms from CPC. This gives the axiomatic system PCPC. The systems CPC and PCPC are the smallest classical axiomatic first order (Hilbert-style) systems with equality and without equality respectively.

Definition 1.1.13 *The classical axiomatic first-order (Hilbert style) system with equality (without equality) is an axiomatic system S such that the set of all theorems in $\mathcal{AS}$*

(i) contains all theorems of CPC (PCPC),

(ii) is closed with respect to the inference rules of CPC (PCPC).

For any classical axiomatic first order system $\mathcal{AS}$, the set $Th(\mathcal{AS})$ of all the theorems of $\mathcal{AS}$ is a (classical) first order theory. Of course the classical axiomatic first-order systems are not the only first-order axiomatic systems. For example, first order logic based upon intuitionistic predicate calculus is very well studied and there are many elegant results concerning it, and also other non-classical first order systems play an important role in mathematics and Computer Science. But the classical first order logic is very important because it has a lot of applications in familiar areas of mathematics (which have no direct relation with logic). Even the semantic basis of non-classical logics usually uses the original semantics of CPC. Formal axiomatic systems are only one aspect of the logic theory, an other (perhaps the primary one) is the semantics for axiomatic systems. A description of semantic tools of first order logic is the main topic of the next section.

1.2 First-Order Semantics and Universal Algebra

In this section we rehearse some of the fundamental notions and theorems concerning abstract algebra and model theory. We describe all necessary notation,

definitions and theorems which we will use in our research further. Majority of
results are given without proofs and we refer readers interested to learn proofs to
monographs [25, 61, 24, 98]. The necessity to recall these knowledge lie in a fact
that the basic semantics for first-order logic can be described via universal algebra or model theory. We begin with opening definition of an algebraic structure
or a model. Let L be a first-order language.

Definition 1.2.1 *A model for L (or, synonymously, algebraic system, or an
algebraic structure for L, or just a L-structure) is a pair $\mathcal{A} := \langle A, I \rangle$, where
A is a nonempty set, which is called the universe (or the basis) of $\mathcal{A}$, and I
is a function with the domain L such that*

- *if $P \in L$ is an n-ary relation symbol, then $I(P) \subseteq A^n$ (that is $I(P)$ is an
  n-ary relation on A),*

- *if $f \in L$ is an n-ary function symbol, then $I(f) : A^n \mapsto A$,*

- *if $c \in L$ is a constant symbol then $I(m) \in A$ (that is $I(c)$ is an element
  of the set A).*

This mapping I (which is called an *interpretation*) gives a correspondence between predicate, functional and constant symbols from L on the one hand and
relations, functions (which are defined on A and have corresponding arity), and
elements of A on the other. Of course in a given universe A there are many different permissible interpretation of symbols from L; every interpretation I just
fixes some possible meaning of symbols of L in A. Because the meaning of symbols of L under any interpretation is fixed, when we dealt with a given language
L of the form $\{P_1, \ldots P_n, F_1 \ldots F_m, c_1 \ldots c_k\}$, we can write $\langle A, I \rangle$ in displayed
form as

$$\langle A, \ P_1, \ldots P_n, \ F_1 \ldots F_m, \ c_1 \ldots c_k \rangle,$$

bearing in mind the prescribed meaning of the languages symbols. When the
symbols of L are familiar, we shall agree to write $\langle A, L \rangle$ instead of $\langle A, I \rangle$. For
instance, we write $\mathcal{A} = \langle A, \wedge, \vee \rangle$ for models on the language L= $\{\wedge, \vee\}$.

We may also write $\mathcal{A} = \langle A, \wedge_A, \vee_A \rangle$, $\mathcal{B} = \langle B, \wedge_B, \vee_B \rangle$ etc. if the context
of a discussion requires it. In particular, we will use the following notation: if
$\mathcal{M} = \langle M, L \rangle$ is a model and P, f, c are certain predicate, function and constant
symbols respectively then P_M, f_M, c_M are corresponding relation, function and
constant symbol value from $\mathcal{M}$. The set of all predicate, functional and constant
symbols from L is said to be the *signature* of the model $\mathcal{M}$ It should be clear that
the definition of $\mathcal{M}$ depends on only the meaning of the symbols from the signature. Therefore instead of *model in language* ... we often (or sometimes) say
model of the signature ...). For an algebraic system (or a model) $\mathcal{M} = \langle M, L \rangle$,

$|\mathcal{M}|$ denotes the basis set M of the model $\mathcal{M}$. However following the usual custom, sometimes we will use the symbol $\mathcal{M}$ to denote both model $\mathcal{M}$ itself and its basis set (provided that the meaning of symbols from the signature is clear from context). For a set X, $||X||$ is the number of elements of X or cardinality of X if X is infinite.

Let $\mathcal{M} = \langle M, L \rangle$ be a model for a language L. An *assignment* (or a *valuation*) in $\mathcal{M}$ is a function s taking as its domain the set of variables for L and as its range a subset of M. We understand s as assigning a meaning $s(x)$ to the variable x. We extend s to the set of all terms of L by induction on the length of terms in the following way.

Definition 1.2.2 *For t a term of L define $s(t)$ as follows:*

- *if $t = c$ where c is a constant symbol then $s(t) = c_M$;*

- *if $t = f(t_1, \ldots, t_n)$ where f is an n-ary functional symbol then*
  $v(t) := f_M(s(t_1), \ldots, s(t_n))$.

If $t(x_1, \ldots, t_n)$ is a term of L with variables $x_1 \ldots, x_n$, and $a_1, \ldots n$ are some elements of $\mathcal{M}$ then we mean $t(a_1, \ldots, a_n)$ to be the value of the assignment $x_1 \mapsto a_1, \ldots, x_n \mapsto a_n$, on term $t(x_1, \ldots, t_n)$.

Let $\mathcal{M} = \langle M, L \rangle$ be a model for L, Φ be a first order formula of L, and s be an assignment of variables in $\mathcal{M}$. The basic notion of *truth relation* of Φ in $\mathcal{M}$ with respect to s is given below by induction on the length of the formula:

Definition 1.2.3 *For any assignment s,*

- $\mathcal{M} \models_s (t = q)$ *iff $s(t) = s(q)$,*

- $\mathcal{M} \models_s P(t_1, \ldots, t_n)$ *iff $(s(t_1), \ldots, s(t_n)) \in P_M$,*

- $\mathcal{M} \models_s \neg\Phi$ *iff $\mathcal{M} \models_s \Phi$ does not hold,*

- $\mathcal{M} \models_s \Phi \vee \Psi$ *iff $\mathcal{M} \models_s \Phi$ or $\mathcal{M} \models_s \Psi$,*

- $\mathcal{M} \models_s \Phi \wedge \Psi$ *iff $\mathcal{M} \models_s \Psi$ and $\mathcal{M} \models_s \Psi$,*

- $\mathcal{M} \models_s \Phi \rightarrow \Psi$ *iff $\mathcal{M} \models_s \Phi$ is fail or $\mathcal{M} \models_s \Psi$,*

- $\mathcal{M} \models_s \forall x\Phi$ *iff for all $a \in M$ $\mathcal{M} \models_v \Phi$ where*
  $(\forall y \neq x)v(y) = s(y)$ *and $v(x) := a$,*

- $\mathcal{M} \models_s \exists x\Phi$ *iff there exists an $a \in M$ such that $\mathcal{M} \models_v \Phi$ where*
  $(\forall y \neq x)v(y) = s(y)$ *and $v(x) := a$.*

The reader can observe that the truth or falsity of $\mathcal{M} \models_s \Phi$ depends only on the values of $s(x)$ for variables x which are actually free in Φ. Let $\Phi(x_1 \ldots, x_n)$ be a first-order formula of L with free variables $x_1, \ldots, x_n$. Then $\mathcal{M} \models \Phi(a_1, \ldots a_n)$ precisely means

$$\mathcal{M} \models_s \Phi \text{ where } s(x_i) := a_i, \forall i (1 \leq i \leq n),$$
$$(\forall y \neq x_1, \ldots, x_n)(s(y) := a_1)).$$

It is clear that this definition does not depend on the value of s for variables y differing from $x_1, \ldots, x_n$. In particular, if Φ is a sentence we can define $\mathcal{M} \models \Phi$ as $\mathcal{M} \models_s \Phi$ with respect to any valuation s (that is $\mathcal{M} \models \Phi$ can be defined without any reference to an assignment). If Φ is a formula (not a sentence) then $\mathcal{M} \models \Phi$ means that $\mathcal{M} \models_s \Phi$ for any assignment s. It is easy to see that in this case $\mathcal{M} \models \Phi$ is equivalent to $\mathcal{M} \models \forall x_1, ..., \forall x_n \Phi$, where $x_1, ..., x_n$ are all the free variables of Φ. If $\mathcal{M} \models \Phi$ ($\mathcal{M} \models_s \Phi$) we say Φ is *true* (or *valid*) in the model $\mathcal{M}$ (with respect to the assignment s). The following fundamental theorem connects validity and provability in CPC.

Theorem 1.2.4 GÖDEL COMPLETENESS THEOREM. *A first-order formula Φ in a language L is a theorem of CPC if and only if Φ is valid (true) in every model $\mathcal{M}$ for L.*

Thus this completeness theorem says that classical *provability (or derivability)* in CPC is equivalent to the *truth* in all models. Let us pause briefly to recall the completeness theorem for classical propositional logic PC. Of course the case of PC is considerably more simple than the case of CPC. In particular, the completeness theorem for PC can, in a sense, be extracted from completeness theorem for CPC. Indeed we can arrange the translation π from the set For_{PC} of all formulas of PC into the set For_{CPC} of all formulas of the CPC in the following way: for every propositional letter p_i,

$$\pi(p_i) := P_i(x), \ \pi(A \wedge B) := \pi(A) \wedge \pi(B),$$

$$\pi(A \vee B) := \pi(A) \vee \pi(B), \ \pi(A \to B) := \pi(A) \to \pi(B),$$

$$\pi(\neg(A)) := \neg(\pi(A)),$$

where P_i is an unary predicate letter, x is a (fixed) variable. This transformation preserves *derivability* and *non-derivability* in PC. Indeed if a $\pi(A)$ is derivable in CPC then we can transform a given derivation $\mathcal{D}$ of $\pi(A)$ into a derivation $\mathcal{D}'$ of A in PC as follows. First, we replace all the free variables of the formulas in $\mathcal{D}$ by the variable x. Next, we replace all atomic subformulas, which are not of the form $P_i(x)$, from the resulting sequence of formulas, with the formula $P_1(x) \to P_1(x)$. Finally, we remove all quantifiers from all formulas, and replace all $P_i(x)$ by p_i . It is clear that the resulting sequence of formulas $\mathcal{D}'$ can be completed till a derivation of A in PC. Thus

$$\vdash_{PC} A \Leftrightarrow \vdash_{CPC} \pi(A).$$

Therefore by the Completeness Theorem for CPC we get $\vdash_{PC} A$ iff $\pi(A)$ is valid in every model with a single-element basis set. Of course we can not consider predicates $P_i(X)$ but merely consider the truth value of every propositional letter p_i in the given model. Thus for us a model now is just a valuation of the propositional letters p_i in the set $\{\mathbf{true}, \mathbf{false}\}$. Hence A is derivable in PC iff A is valid with respect to any of the above mentioned valuations of its propositional letters. Formulas with the latter property belonging For_{PC} are said to be *tautologies*. Thus we have

Theorem 1.2.5 COMPLETENESS THEOREM FOR PC. *For any formula α, $\vdash_{PC} \alpha$ if and only if α is a tautology.*

We intend to supply the reader by a collection of theorems which will be involved lately in our proofs. Let $\mathcal{M}$ be a model of L. The *elementary theory* of $\mathcal{M}$ is the set $Th(\mathcal{M}) := \{\Phi | \mathcal{M} \models \Phi, \Phi \in For_L\}$ (i.e. $Th(\mathcal{M})$ denotes the set of all formulas which are true in $\mathcal{M}$). If $\mathcal{K}$ is a collection (a class) of models then $Th(\mathcal{K}) := \bigcap_{\mathcal{M} \in \mathcal{K}} Th(\mathcal{M})$; this class of formulas is the *elementary theory* of the class $\mathcal{K}$. A model $\mathcal{M}$ is a *model* for a set of formulas S if there is a valuation v of all free variables from formulas of S in $\mathcal{M}$ such that $\mathcal{M} \models_v \Phi$ for all $\Phi \in S$. In this case we say that S is (semantically) *consistent*, that it has a model $\mathcal{M}$. S is called *locally consistent* if for every finite $X \subseteq S$, X is consistent.

Theorem 1.2.6 COMPACTNESS THEOREM (MALT'SEV 1938). *A set of formulas S is consistent if and only if S is locally consistent.*

Theorem 1.2.7 *(Löwenheim-Skolem) If a set of formulas S is consistent then S has some model of cardinality κ, where $\kappa \leq \chi + \omega$, where $\chi = \|S\|$.*

Now we list a number of popular algebraic constructions involving models (algebraic systems) and recall results concerning truth of first order formulas in resulting models.

Definition 1.2.8 *Given models (algebraic systems) $\mathcal{M}_1$ and $\mathcal{M}_2$ in the language L, we say $\mathcal{M}_1$ is a submodel (subsystem) of $\mathcal{M}_2$ if $|\mathcal{M}_1| \subseteq |\mathcal{M}_2|$ and*

(i) *For any n-ary predicate's relation P from L and every $a_1, ..., a_n$ from $|\mathcal{M}_1|$, $\mathcal{M}_1 \models P(a_1, ..., a_n) \Leftrightarrow \mathcal{M}_2 \models P(a_1, ..., a_n)$*

(ii) *For any n-ary functional symbol f from L and every $a_1, ..., a_n, b$ from $|\mathcal{M}_1|$, $[\mathcal{M}_1 \models f(a_1, ..., a_n) = b] \Leftrightarrow [\mathcal{M}_2 \models f(a_1, ..., a_n) = b]$*

(iii) The interpretations of all constants from L in both models coincide.

We will use the abbreviation $\mathcal{M}_1 \preceq \mathcal{M}_2$ for $\mathcal{M}_1$ *is a submodel of* $\mathcal{M}_2$.

Definition 1.2.9 *Suppose $\mathcal{M}_1$ and $\mathcal{M}_2$ are algebraic systems (models) in the language L and f is a mapping from $|\mathcal{M}_1|$ into $|\mathcal{M}_2|$. The mapping f is called a* homomorphism *if*

(i) For any n-ary predicate relation P from L and every $a_1, ..., a_n$ from $|\mathcal{M}_1|$,
$$[\mathcal{M}_1 \models P(a_1, ..., a_n)] \Rightarrow [\mathcal{M}_2 \models P(f(a_1), ..., f(a_n))]$$

(ii) For every n-ary functional symbol g from L and every $a_1, ..., a_n$ from $|\mathcal{M}_1|$, $[\mathcal{M}_1 \models g(a_1, ..., a_n) = b] \Rightarrow [\mathcal{M}_2 \models g(f(a_1), ..., f(a_n)) = f(b)]$

(iii) For every constant symbol c from L, $c_{\mathcal{M}_2} = f(c_{\mathcal{M}_1})$.

The model $f(\mathcal{M}_1)$ is called a *homomorphic image* of the model $\mathcal{M}_1$. If f is a mapping onto $|\mathcal{M}_2|$ then f is called *endomorphism* (homomorphism *on* or *onto*). If f is a one-to-one mapping then f is called an *isomorphism* of $\mathcal{M}_1$ into $\mathcal{M}_2$. In this case the inverse mapping f^{-1} from $f(\mathcal{M}_1)$ is also an isomorphism from $f(\mathcal{M}_1)$ onto $\mathcal{M}_1$. If there is an isomorphism from a model $\mathcal{M}_1$ onto a model $\mathcal{M}_2$ we say $\mathcal{M}_1$ and $\mathcal{M}_2$ are *isomorphic*. We also use the denotation $\mathcal{M}_1 \cong \mathcal{M}_2$ to indicate that $\mathcal{M}_1$ and $\mathcal{M}_2$ are isomorphic. If there is an isomorphism from a model $\mathcal{M}_1$ into a model $\mathcal{M}_2$ we say $\mathcal{M}_1$ is *isomorphically embedded* in $\mathcal{M}_2$. Hence, in this case, $\mathcal{M}_1$ can be considered (by modulo of an isomorphism) as a subsystem (a submodel) of $\mathcal{M}_2$. A homomorphism f is said to be *strong* if

For any *n*-ary predicate relation P from L and every $a_1, ..., a_n$ from $|\mathcal{M}_1|, [\mathcal{M}_1 \models P(a_1, ..., a_n)] \Leftrightarrow [\mathcal{M}_2 \models P(f(a_1), ..., f(a_n))]$

An useful thing is the representation theorem for homomorphic images by means of the quotient system by the kernel congruence relation, which we rehearse now. For any set A, an *equivalence relation on A* is a binary relation R such that

(i) For every $a \in A$, $R(a, a)$ *(reflexivity)*;

(ii) For any $a, b \in A$, $R(a, b) \Rightarrow R(b, a)$ *(symmetricity)*;

(iii) For any $a, b, c \in A$, $R(a, b) \& R(b, c) \Rightarrow R(a, c)$ *(transitivity)* .

Definition 1.2.10 *A congruence relation on an algebraic system $\mathcal{M} = \langle M, \Sigma \rangle$ (in a signature Σ) is an equivalence relation $\approx$ defined on M satisfying the following condition:*

For every n-ary functional symbol $f \in \Sigma$, if $a_1 \approx b_1, ..., a_n \approx b_n$, then $f(a_1, ..., a_n) \approx f(b_1, ..., b_n)$;

Suppose we are given an algebraic system $\mathcal{M} = \langle M, \Sigma \rangle$ and a congruence relation $\approx$ on $\mathcal{M}$. The *quotient algebraic system* $\mathcal{M}/\approx$ on $\mathcal{M}$ by $\approx$ is the algebraic system defined as follows.

(i) The base set of this system is the quotient set of M by $\approx$:

$$M/\approx \;:= \{[a]_\approx \,|a \in M\},$$

where $[a]_\approx := \{b | b \in A, a \approx b\}$, that is the elements of this quotient set are all equivalence classes by $\approx$;

(ii) For every n-ary functional symbol $f \in \Sigma$ and any
$[a_1]_\approx, ..., [a_n]_\approx \in M/\approx$, $f([a_1]_\approx, ..., [a_n]_\approx) := [f(a_1, ..., a_n)]_\approx$;

(iii) For every n-ary predicate symbol $P \in \Sigma$ and any
$[a_1]_\approx, ..., [a_n]_\approx \in M/\approx$, $P([a_1]_\approx, ..., [a_n]_\approx) :=$ true $\Leftrightarrow$
$\exists b_1 \in [a_1]_\approx ... \exists b_1 \in [a_1]_\approx (P(b_1, ..., b_n) =$ true$)$

(iv) For any constant symbol $c \in \Sigma$, $c_{\mathcal{M}/\approx} := [c]_\approx$.

All above mentioned functions and predicates are well-defined in the quotient algebraic system according to properties of any congruence relation (i.e. according to consistency with operations).

Theorem 1.2.11 *The mapping $a \mapsto [a]_\approx$ is a homomorphism onto from $\mathcal{M}$ on $\mathcal{M}/\approx$ (*natural epimorphism*).*

Suppose h is an epimorphism (i.e. a homomorphism onto) of an algebraic system $\mathcal{M}$ onto a similar system $\mathcal{M}_1$. We define the kernel equivalence relation $\approx_h$ on $\mathcal{M}$ as follows:

$$a \approx_h b \;\Leftrightarrow\; h(a) = h(b)$$

Theorem 1.2.12 *Homomorphism Theorem. The following hold:*

*(i) $\approx_h$ is a congruence relation on $\mathcal{M}$ (*kernel congruence relation*);*

(ii) The canonical mapping $[a]_{\approx_h} \mapsto h(a)$ is a homomorphism from $\mathcal{M}/\approx_h$ onto $\mathcal{M}_2$;

(iii) If the homomorphism h is strong then the canonical mapping is an isomorphism, i.e. $\mathcal{M}/\approx_h \cong \mathcal{M}_2$.

Now we briefly rehearse the construction of direct products for algebraic systems.

Definition 1.2.13 *Let $\mathcal{M}_i$, $i \in I$ be a family of models in a language L. The direct product $\prod_{i \in I} \mathcal{M}_i$ of the models $\mathcal{M}_i$ is the model which is defined as follows.*

(i) $|\prod_{i \in I} \mathcal{M}_i| := \{(a_i | i \in I) | a_i \in \mathcal{M}_i, i \in I\}$

(ii) for every n-ary functional symbol f from L and every
$(a_i^1 | i \in I), ..., (a_i^n | i \in I)$ *from* $\prod_{i \in I} \mathcal{M}_i$, $f((a_i^1 | i \in I), ..., (a_i^n | i \in I)) :=$
$(f(a_i^1, ..., a_i^n) | i \in I)$

(ii) for every n-ary predicate symbol P from L and every tuple
$(a_i^1 | i \in I), ..., (a_i^n | i \in I)$ *from* $\prod_{i \in I} \mathcal{M}_i$, $[\prod_{i \in I} \mathcal{M}_i \models$
$P((a_i^1 | i \in I), ..., (a_i^n | i \in I))] \Leftrightarrow \forall i \, [\mathcal{M}_i \models P((a_i^1, ..., a_i^n)]$

(iii) For every constant symbol c from L, $c_{\prod_{i \in I} \mathcal{M}_i} := (c_{\mathcal{M}_i} | i \in I)$.

If $I = \{1, ..., n\}$ then sometimes we will write $\mathcal{M}_1 \times ... \times \mathcal{M}_n$ instead of $\prod_{i \in I} \mathcal{M}_i$. If all $\mathcal{M}_i = \mathcal{M}$ we will also use the brief denotation $\mathcal{M}^I$ for $\prod_{i \in I} \mathcal{M}_i$. Finally if μ is a cardinal number we denote by $\mathcal{M}^\mu$ the product $\prod_{i \in I} \mathcal{M}_i$, where $\mathcal{M}_i = \mathcal{M}, I := \{i \mid i \leq \mu\}$.

We say a formula Φ from For_{CPC} has the *normal prenex form* if

$$\Phi = Q_1 x_1 ... Q_n x_n \Psi(x_1, ..., x_m),$$

where $Q_i \in \{\forall, \exists\}$, $\Psi(x_1, ..., x_n)$ is a formula having no occurrences of quantifier symbols, and $\Psi(x_1, ..., x_n) = \bigvee_{i \in I} \bigwedge_{j \in J} \Theta_{i,j}$, where all $\Theta_{i,j}$ are certain atomic formulas. The sequence $Q_1 x_1, ..., Q_n x_n$ is said to be the *(quantifier) prefix* of Φ; $\Psi(x_1, ..., x_n)$ is the *matrix* of Φ.

Theorem 1.2.14 *For any formula* Θ *from* For_{CPC}, *there is a formula* Φ *which is in a prenex normal form and* $\vdash_{CPC} (\Theta \rightarrow \Phi) \wedge (\Phi \rightarrow \Theta)$, *i.e.* Φ *is equivalent to* Θ *in CPC.*

We say a formula Φ has the *prenex* form if

$$\Phi = Q_1 x_1 ... Q_n x_n \Psi(x_1, ..., x_m),$$

where as above Q_i are some quantifier symbols (i.e. $Q_i \in \{\forall, \exists\}$), but Ψ is merely a formula without occurrences of quantifier. A formula Φ is called *universal* if Φ has a prenex form and all the quantifiers in its quantifier prefix are universal (of the form $\forall x_i$). A formula Φ is said to be *positive* if Φ has no occurrences of the logical connective $\neg$.

Theorem 1.2.15 *(Preservation of truth with respect to submodels).* *Suppose* $\mathcal{M}_1$ *and* $\mathcal{M}_2$ *are models and* $\mathcal{M}_1 \preceq \mathcal{M}_2$, *and* Φ *is an universal formula. When* $\mathcal{M}_2 \models \Phi$ *is valid* $\mathcal{M}_1 \models \Phi$ *also holds.*

Theorem 1.2.16 *(Preservation of truth with respect to homomorphisms).* *Suppose* $\mathcal{M}_1$ *and* $\mathcal{M}_2$ *are models and there is a homomorphism from* $\mathcal{M}_1$ *onto* $\mathcal{M}_2$. *If* Φ *is a positive first order formula and* $\mathcal{M}_1 \models \Phi$ *then* $\mathcal{M}_2 \models \Phi$.

We will need some specific properties of universal formulas in the special form. First we consider the quasi-identities.

Definition 1.2.17 *A quasi-identity is an universal formula* Φ *in the form*

$$\forall x_1...\forall x_n[(\Psi_1 \wedge ... \wedge \Psi_m) \rightarrow \Theta],$$

where all Ψ_i *and* Θ *are some atomic formulas.*

Definition 1.2.18 *A first order formula is said to be an identity if it has form*

$$\forall x_1...\forall x_n\Psi,$$

where Ψ *is a certain atomic formula.*

Thus every identity can be considered as a quasi-identity with the empty set of premises. A model (algebraic system, algebraic structure) $\mathcal{M}$ is said to be an *algebra* indexalgebra if its signature (language) has no predicate's symbols. In a such language identities and quasi-identities have a more customary form: a quasi-identity is a formula in the form

$$\forall x_1...\forall x_n(f_1 = g_1 \wedge ... \wedge f_n = g_n \rightarrow f = g),$$

an identity, in this case, is a formula

$$\forall x_1...\forall x_n(f = g),$$

where all f_i, g_i and f, g are certain terms. In writing of identities and quasi-identities we can omit the quantifier's prefix and just write

$$f = g, \text{ or } f_1 = g_1 \wedge ... \wedge f_n = g_n \rightarrow f = g.$$

Because any quasi-identity q is an universal formula the truth of q will be preserved under submodels (Theorem 1.2.15) and the truth of identities must be preserved under homomorphic images.

Theorem 1.2.19 *(Preservation truth with respect to direct products). Suppose that* $\mathcal{M}_i, i \in I$ *is a family of models, and* Φ *is a quasi-identity which is valid in every model* $\mathcal{M}_i$. *Then* $\Pi_{i \in I}\mathcal{M}_i \models \Phi$.

Recall that if Γ is a family of formulas in some language L then

$$Mod(\Gamma) := \{\mathcal{M} | \forall \Phi \in \Gamma(\mathcal{M} \models \Phi)\}$$

Definition 1.2.20 *Suppose that Γ is a set of identities (quasi-identities, universal formulas) in a language L. The class of models $Mod(\Gamma)$ is the variety (quasivariety, universal class, respectively) generated by the identities (quasi-identities, universal formulas) of Γ.*

We call Γ the axiomatization of $Mod(\Gamma)$, or the set of axioms for $Mod(\Gamma)$. If we are not interested in a particular description of the axiomatizing formulas we say merely variety *(or* quasi-variety, universal class*) without mentioning an axiomatizing set of formulas.*

We now turn to the description of specific properties of such classes. First we need a certain additional denotation. If L is a language then

$\mathrm{For}_{I,L}$ is the set of all identities from For_L;

$\mathrm{For}_{Q,L}$ is the set of all quasi-identities from For_L;

$\mathrm{For}_{U,L}$ is the set of all universal formulas from For_L;

If the meaning of symbols in L is clear from a context the symbol L can be omitted from the notation above. For any class of models (algebraic systems) $\mathcal{K}$ in a language L,

$$Th(\mathcal{K}) := \{\Phi | \Phi \in \mathrm{For}_L, \forall \mathcal{M} \in \mathcal{K}(\mathcal{M} \models \Phi)\}$$
$$Th_U(\mathcal{K}) := Th(\mathcal{K}) \cap \mathrm{For}_{U,L}$$
$$Th_Q(\mathcal{K}) := Th(\mathcal{K}) \cap \mathrm{For}_{Q,L}$$
$$Th_I(\mathcal{K}) := Th(\mathcal{K}) \cap \mathrm{For}_{I,L}$$

Classes of formulas $Th(\mathcal{K}), Th_U(\mathcal{K}), Th_Q(\mathcal{K})$ and $Th_I(\mathcal{K})$ are called the *elementary, universal, quasi-equational, and equational* theories, respectively, of the class of models (algebraic systems) $\mathcal{K}$.

Definition 1.2.21 *Let $\mathcal{K}$ be a class of models in a signature L.*

(i) The variety generated by $\mathcal{K}$ is the variety $\mathcal{K}^V := \bigcap\{\mathcal{K}_i | \mathcal{K} \subseteq \mathcal{K}_i, \mathcal{K}_i$ is a variety $\}$ (that is $\mathcal{K}^V$ is the smallest variety containing $\mathcal{K}$);

(ii) The quasi-variety generated by $\mathcal{K}$ is the quasi-variety $\mathcal{K}^Q := \bigcap\{\mathcal{K}_i | \mathcal{K} \subseteq \mathcal{K}_i, \mathcal{K}_i$ is a quasi-variety $\}$ (that is $\mathcal{K}^Q$ is the smallest quasi-variety containing the class of models $\mathcal{K}$);

(iii) The universal class generated by $\mathcal{K}$ is the class $\mathcal{K}^U := \bigcap\{\mathcal{K}_i | \mathcal{K} \subseteq \mathcal{K}_i, \mathcal{K}_i$ is a universal class $\}$ (that is $\mathcal{K}^U$ is the smallest universal class containing the class $\mathcal{K}$);

(iv) The elementary class generated by $\mathcal{K}$ is the class $\mathcal{K}^E := \bigcap\{\mathcal{K}_i | \mathcal{K} \subseteq \mathcal{K}_i, ck_i$ is an elementary class$\}$ (i.e. $\mathcal{K}^E$ is the smallest elementary class containing the class $\mathcal{K}$);

Sometimes denotations $Var(\mathcal{K})$, $Q(\mathcal{K})$, $Un(\mathcal{K})$ and $E(\mathcal{K})$ can be used instead of $\mathcal{K}^V$, $\mathcal{K}^Q$, $\mathcal{K}^U$ and $\mathcal{K}^E$ respectively. An evident theorem concerning these above mentioned classes is

Theorem 1.2.22 *For any class $\mathcal{K}$ of signature L,*

(i) $\mathcal{K}^V := Mod(\{\Phi \mid \Phi \in For_{I,L},\ \forall \mathcal{M} \in \mathcal{K}(\mathcal{M} \models \Phi)\})$;

(ii) $\mathcal{K}^Q := Mod(\{\Phi \mid \Phi \in For_{Q,L},\ \forall \mathcal{M} \in \mathcal{K}(\mathcal{M} \models \Phi)\})$;

(iii) $\mathcal{K}^U := Mod(\{\Phi \mid \Phi \in For_{Q,L},\ \forall \mathcal{M} \in \mathcal{K}(\mathcal{M} \models \Phi)\})$;

(iv) $\mathcal{K}^E := Mod(\{\Phi \mid \Phi \in For_L,\ \forall \mathcal{M} \in \mathcal{K}(\mathcal{M} \models \Phi)\})$;

A much more difficult task is to find an algebraic description of such classes (without any using of the notion of *truth*). Such descriptions are known and will be reviewed below.

If $\mathcal{K}$ is a class of algebraic systems (models) then $H\mathcal{K}$ denotes the class of all homomorphic images of algebraic systems from $\mathcal{K}$; $S\mathcal{K}$ is the class of all subsystems of algebraic systems from $\mathcal{K}$; $\Pi\mathcal{K}$ is the class of all direct products of all families of algebraic systems from $\mathcal{K}$.

Theorem 1.2.23 *(Birkhoff) In order for a class $\mathcal{K}$ to be a variety it is necessary and sufficient that $\mathcal{K}$ be closed with respect to*

(i) homomorphic images,

(ii) taking subsystems,

(iii) direct products

Theorem 1.2.24 *(Birkhoff) Representation Theorem for Varieties. For every given class $\mathcal{K}$ of algebraic systems, $Var(\mathcal{K}) = HS\Pi\mathcal{K}$.*

Let $\mathcal{M}$ be an algebraic system. We introduce the set of variables $V_{\mathcal{M}} := \{x_a\ a \in |\mathcal{M}|\}$ indexed by elements of $|\mathcal{M}|$ and the set $CD_{\mathcal{M}}$ of all first order formulas of the kind:

(i) $P(x_{a_1}, ..., x_{a_n})$, where P is a predicate letter and
$\mathcal{M} \models P(a_1, ..., a_n)$;

(ii) $f(x_{a_1}, ..., x_{a_n}) = x_b$, where f is a functional letter and
$\mathcal{M} \models f(a_1, ..., a_n) = b$;

(iii) $c = x_c$, where c is a constant's symbol.

We call the set $CD_{\mathcal{M}}$ the *complete diagram* of the algebraic system (model) $\mathcal{M}$.
complete diagram

Proposition 1.2.25 *If there is an interpretation v of all variables from $CD_{\mathcal{M}}$ in an algebraic system $\mathcal{M}_1$ such that*

$$(\forall \Phi \in CD_{\mathcal{M}})[\mathcal{M}_1 \models_v \Phi]$$

then $\mathcal{M} \preceq \mathcal{M}_1$ (more precisely $x \mapsto v(x)$ is an isomorphic embedding of the system $\mathcal{M}$ into the system $\mathcal{M}_1$). If $\mathcal{M} \preceq \mathcal{M}_1$ then there is an interpretation v with described above property.

If $\mathcal{M}$ is a finite system of a finite signature, $|\mathcal{M}| = \{a_1, ..., a_n\}$ then instead of $CD_{\mathcal{M}}$ we can consider the formula

$$\mathrm{FD}_{\mathcal{M}} := \exists x_{a_1} ... \exists x_{a_n} [\bigwedge \{\Phi | \Phi \in CD_{\mathcal{M}}\}].$$

The formula $\mathrm{FD}_{\mathcal{M}}$ is called the *finite diagram* of the model $\mathcal{M}$. finite diagram

Proposition 1.2.26 *For any algebraic system $\mathcal{M}_1$, $\mathcal{M} \preceq \mathcal{M}_1$ if and only if $\mathcal{M}_1 \models FD_{\mathcal{M}}$.*

That is, the property of a finite model *being a submodel* is expressible by an *existential* formula.

Suppose $\mathcal{K}$ is a class of algebraic systems in a signature Σ, $\mathcal{M}$ is an algebraic system in the same language. System $\mathcal{M}$ is said to be *locally embedded* in $\mathcal{K}$ if the following holds. For every finite $\Sigma_1 \subseteq \Sigma$ and every finite $A \subseteq |\mathcal{M}|$, there is a system $\mathcal{M}_1 = \langle M_1, \Sigma \rangle$ from $\mathcal{K}$ such that

$$\langle A, \Sigma_{1,p} \rangle \preceq \langle M_1, \Sigma_{1,p} \rangle,$$

where $\Sigma_{1,p}$ is the signature obtained from Σ_1 by replacing every function symbol f with the corresponding predicate symbol R_f; the interpretation of P_f in the models above is the following: $(P_f(a_1, ..., a_n, b) = \mathrm{true}) \Leftrightarrow f(a_1, ..., a_n) = b$.

Definition 1.2.27 *A class $\mathcal{K}$ of algebraic systems is called* locally closed *if any system $\mathcal{M}$ which is locally embeddable in $\mathcal{K}$ is a member of $\mathcal{K}$.*

If L is a language then E_L is the algebraic system of L, where $|E_L|$ has only one element and where all the predicates relations in L are *true* in E_L. For any class $\mathcal{K}$ of algebraic systems $\mathcal{K}$ in the language L, $\mathcal{K}_e$ is the class consisting of $\mathcal{K}$ and E_L.

Theorem 1.2.28 *(Malt'sev)[98]. In order for a class $\mathcal{K}$ in a language L to be a quasi-variety it is necessary and sufficient that $\mathcal{K}$*

(i) be locally closed,

(ii) be closed with respect to direct products,

(iii) contain E_L.

Definition 1.2.29 *A class $\mathcal{K}$ of algebraic systems (models) in a language L is called axiomatizable iff there is a set Γ of first-order formulas in the language L such that $\mathcal{K} = Mod(\Gamma)$.*

Theorem 1.2.30 *(Malt'sev, Tarski) ([98], Corollary 9, p.233) If $\mathcal{K}$ is an axiomatizable class of algebraic systems then $\mathcal{K}^Q = S\Pi\mathcal{K}_e$.*

Definition 1.2.31 *Suppose $\mathcal{M}$ is an algebraic system in a signature Σ and $X \subseteq |\mathcal{M}|$. The subsystem of $\mathcal{M}$ generated by X is the smallest subsystem of $\mathcal{M}$ containing X. We denote this subsystem by $\mathcal{M}[X]$ and call X the set of generators of $\mathcal{M}[X]$; if $\mathcal{M}[X] = \mathcal{M}$ then we call X the set of generators of the algebraic system $\mathcal{M}$.*

Proposition 1.2.32 $\mathcal{M}[X] = \langle\{t(a_1, ..., a_n) \mid a_i \in X, t \text{ is a term in } \Sigma\}, \Sigma\rangle.$

Definition 1.2.33 *A free algebraic system in a class $\mathcal{K}$ of similar systems (in a same language) is a system $\mathcal{F}_\mathcal{K}$ from $\mathcal{K}$ with the following properties. System $\mathcal{F}_\mathcal{K}$ has a set X of generators (which are called $\mathcal{K}$-free generators, or just free generators) such that, for any $\mathcal{M} \in \mathcal{K}$ and for every mapping f from X into $\mathcal{M}$, f can be extended to a homomorphism h of $\mathcal{F}_\mathcal{K}$ into $\mathcal{M}$.*

Let $\mathcal{M}[X]$ be an algebraic system from a class $\mathcal{K}$ and X be a set of generators for this algebra.

Lemma 1.2.34 *Algebraic system $\mathcal{M}[X]$ is free in the class $\mathcal{K}$ and X is a set of free generators iff for every elements $a_1, ... , a_n$ from X and any terms f and g the following holds*

$$[\mathcal{M}[X] \models f(a_1, ..., a_n) = g(a_1, ..., a_n)] \Leftrightarrow \Leftrightarrow (\forall \mathcal{M}_s \in \mathcal{K})[\mathcal{M}_s \models \forall x_1...\forall x_n(f(x_1, ..., x_n) = g(x_1, ..., x_n))].$$

If $\mathcal{M}[X]$ is a free system in $\mathcal{K}$ with the set of free generators X and X has cardinality κ then we call $\mathcal{M}[X]$ a *free system of rank κ.*

Theorem 1.2.35 *If a class $\mathcal{K}$ of algebraic systems in a signature Σ is closed with respect to subalgebras and direct products then $\mathcal{K}$ has free algebraic systems of any rank.*

Moreover we can clarify the structure of the free algebraic systems as follows.

Theorem 1.2.36 *Suppose $\mathcal{K}$ is the quasi-variety (or variety) generated by a class $\{\mathcal{M}_i \mid i \in I\}$. Then the free algebraic system $\mathcal{F}_\mathcal{K}(r)$ of rank r from $\mathcal{K}$*

*is isomorphic to a subalgebra of a certain direct product of systems from the
family* $\{\mathcal{M}_i \mid i \in I\}$. *More precisely,*

$$\mathcal{F}_{\mathcal{K}}(r) \preceq \prod_{i \in I}(\mathcal{M}_i^{|\mathcal{M}_i|^r})$$

holds by modulo of some isomorphism.

We will follow in this book non-formal conventions concerning notions from
recursive theory, these notions will be recalled as soon as they will be necessary.
All they can be found, for example, in Rogers [129], but we will need very few
ones. For instance, a set $\mathcal{S}$ is *decidable (or recursive)* if there is an algorithm
which can determine by any element whether this element belong to $\mathcal{S}$. Similarly, a set $\mathcal{S}$ is *recursively enumerable* if there is an algorithm which can effectively
enumerate by natural numbers all elements of $\mathcal{S}$.

The semantics for non-standard (non-classical) logical systems also directly involves the notions and constructions of universal algebra. In this case the
logical connectives are usually interpreted as functions (operations) in certain
algebraic systems. We will study in detail this kind of semantics in the following sections. Here we merely briefly consider some popular algebraic systems
which are actively used for mentioned above purposes and some their algebraic
properties. The connectives $\wedge$ and $\vee$ play an important role and very often are
included in the logical language. The semantic tools for logical systems of this
kind usually include certain varieties of lattices.

Definition 1.2.37 *An algebra* $\mathcal{A} := \langle A, \wedge, \vee \rangle$ *with two binary functions is
said to be a* lattice *provided the following equations hold on* $\mathcal{M}$:

(i) $x \wedge y = y \wedge x, \quad x \vee y = y \vee x,$

(ii) $x \wedge (y \wedge z) = (x \wedge y) \wedge z, \quad x \vee (y \vee z) = (x \vee y) \vee z,$

(iii) $(x \wedge y) \vee y = y, \quad (x \vee y) \wedge y = y.$

It is possible to define lattices using special properties of partially ordered sets.

Definition 1.2.38 *Suppose* $\mathcal{M} := \langle M, R \rangle$ *is a model, where R is a binary
relation. We call $\mathcal{M}$ a* quasi-ordered set *if*

$\forall x, y, z[R(x, y)\&R(y, z) \Rightarrow R(x, z)]$ *(transitivity)*
$\forall x R(x, x),$ *(reflexivity)*

hold for $\mathcal{M}$. *In this case the relation R is called a* quasi-ordering *on* M.

If a quasi-ordering R has the following additional property:

$\forall x, y[(R(x, y) \vee R(y, z)]$

then R is said to be a *linear (or connected) quasi-ordering*.

Definition 1.2.39 *Suppose* $\mathcal{A} = \langle A, \leq \rangle$ *is a quasi-ordered set and the following holds in* $\mathcal{A}$:

$$\forall a, b \in A[(a \leq b)\&(b \leq a) \;\Rightarrow\; (a = b)] \quad (anti\text{-}symmetry),$$

then we call $\mathcal{A}$ *a* partially ordered *set (correspondingly,* $\leq$ *is a* partial order *on* A*)* .

A partially ordered set $\mathcal{M}$ is called *linearly ordered* if $\leq$ is a linear quasi-ordering. Suppose that $\mathcal{A}$ is a partially ordered set (we call it briefly a *poset*) and X is a certain subset of $|\mathcal{A}|$. An element $a \in |\mathcal{A}|$ is called an *upper (lower) bond* for X in $\mathcal{A}$ if

$$\forall x \in X(x \leq a) \quad (\text{or } \forall x \in X(a \leq x), \text{ respectively}).$$

If $X \subseteq |\mathcal{A}|$ and $a \in X$, then we call a *maximal (minimal)* in X if

$$\forall y \in X[a \leq y \Rightarrow a = b] \quad (\text{or } \forall y \in X[y \leq a \Rightarrow a = b] \text{ , respectively}).$$

An element $a \in X$ is said to be *greatest (least)* in X if $\forall y \in X[y \leq a]$ (or $\forall y \in X[a \leq y]$, respectively). A greatest lower bound of a set X (if there is any) is denoted by $\inf(X)$, we denote a least upper bound of X by $\sup(X)$. We call $\sup(X)$ and $\inf(X)$ the *supremum* and the *infimum* of X respectively. We will often use the following classical fact of the set theory

Lemma 1.2.40 Zorn Lemma. *Let* $\mathcal{A} := \langle A, \leq \rangle$ *be a partially ordered set. If, for any* $X \subset A$ *such that* $\langle X, \leq \rangle$ *is linearly ordered set, there is an upper bound for* X *in* A *then, for every element* $x \in A$*,* x *has an upper bound in* $\mathcal{A}$ *which is a maximal element of* $\mathcal{A}$*.*

Recall also the notions of cover and co-cover which we will need further. Let Q be a subset of the basis set of a model $\mathcal{A} := \langle A, R \rangle$ with transitive binary relation R, i.e. $\forall x, y, z[(xRy)\&(yRz) \Rightarrow (xRz)]$ holds in $\mathcal{A}$. An element a from $|\mathcal{A}|$ is a cover for Q if (i) $\forall b \in Q(bRa)$ and (ii) $\forall c \in Q[(\forall b \in Q)(bRc)\&(cRa)\Rightarrow(aRc)]$. An element a is a co-cover for Q if (i) $\forall b \in Q(aRb)$ and (ii) $\forall c \in Q[(\forall b \in Q)(cRb)\&(aRc)\Rightarrow(cRa)]$.

Let $\mathcal{A} := \langle A, \wedge, \vee \rangle$ be a lattice. For all $a, b \in |\mathcal{A}|$, we call $a \vee b$ the *join (or union)* of a and b; $a \wedge b$ is the *meet* (or intersection) of a and b. We note here without proof the following theorem:

Theorem 1.2.41 *If* $\mathcal{A} = \langle A, \wedge, \vee \rangle$ *is a lattice, then for every* $a, b \in A$

$$a \wedge b = a \;\Leftrightarrow\; a \vee b = b.$$

The binary relation $\leq$*, where* $a \leq b \Leftrightarrow a \wedge b = a$ *(or, equivalently,* $a \vee b = b$*), is a partial ordering on* A *(we call* $\leq$ *the* lattice ordering *on* A*). For a partially ordered set* $\langle A, \leq \rangle$*,* $a \wedge b$ *is the greatest lower bound of* $\{a, b\}$*, i.e.* $a \wedge b$ *is* $\inf\{a, b\}$*;* $a \vee b$ *is the least upper bound of* $\{a, b\}$*, i.e.* $a \vee b$ *is* $\sup\{a, b\}$*.*

Conversely, it is possible to define lattices taking by prime notion the notion of partial ordering, as it is shown below.

Theorem 1.2.42 *Suppose an partially ordered set $\langle A, \leq \rangle$ is given and, for all $a, b \in A$, $sup\{a, b\}$ and $inf\{a, b\}$ exist. Then the algebra $\langle A, \wedge, \vee \rangle$, where*

$$a \wedge b := inf\{a, b\}, \quad a \vee b := sup\{a, b\},$$

is a lattice and $\leq$ is its lattice ordering.

Using definitions of $\wedge$ and $\vee$ from Theorem 1.2.42 it is not difficult to confirm directly that

$$a \wedge a = a, \quad a \vee a = a;$$
$$a \leq a \vee b), \quad a \wedge b \leq a;$$
$$b \leq a \vee b, \quad a \wedge b \leq b;$$
$$[(a \leq c)\&(b \leq c)] \Rightarrow (a \vee b \leq c);$$
$$[(c \leq a)\&(c \leq b)] \Rightarrow (c \leq a \wedge b);$$
$$[(a \leq c)\&(b \leq d)] \Rightarrow [(a \vee b) \leq (c \vee d)];$$
$$[(a \leq c)\&(b \leq d)] \Rightarrow [(a \wedge b) \leq (c \wedge d)].$$

Let $\mathcal{A}$ be a lattice. If there exists the greatest element a in $|A|$ then we say $\mathcal{A}$ has the *unit element* a and denote this element by $\top$ or 1, provided $\mathcal{A}$ has the smallest element we denote this element by $\bot$ or 0 and say $\mathcal{A}$ has the *zero element*.

Definition 1.2.43 *Suppose that $\mathcal{A}$ is a lattice and ∇ is a subset of $|\mathcal{A}|$. We call ∇ a* filter *if*

$$\forall a, b \in |\mathcal{A}|[(a \wedge b) \in \nabla \Leftrightarrow ((a \in \nabla)\&(b \in \nabla))].$$

If ∇ is such that

$$\forall a, b \in |\mathcal{A}|[(a \vee b) \in \nabla \Leftrightarrow ((a \in \nabla)\&(b \in \nabla))],$$

then we call ∇ an ideal *in $\mathcal{A}$.*

Definition 1.2.44 *If $\mathcal{A}$ is a lattice and $X \subseteq \mathcal{A}$ then the filter (ideal) generated by X is the smallest filter (ideal) in $\mathcal{A}$ containing X.*

Proposition 1.2.45 *The filter $X^{\leq}$ generated by X coincides with the set*

$$\{a \mid a \in |\mathcal{A}|, \exists a_1 \in X, ..., \exists a_n \in X(a_1 \wedge ... \wedge a_n \leq a), \}.$$

The ideal $X^{\geq}$ generated by X is the set

$$\{a \mid a \in |\mathcal{A}|, \exists a_1 \in X, ..., \exists a_n \in X(a \leq a_1 \vee ... \vee a_n), \}.$$

A filter (ideal) ∇ is said to be *proper* provided there is an element $a \in \mathcal{A}$ which is not a member of ∇. A filter (ideal) ∇ in $\mathcal{A}$ is called *maximal* if ∇ is a proper filter (ideal) and ∇ is not a proper subset of any proper filter (ideal).

Definition 1.2.46 *A filter (ideal) ∇ is said to be* prime *if it is proper filter (ideal) and $(a \vee b) \in \nabla$ implies that either $a \in \nabla$ or $b \in \nabla$ (correspondingly, $(a \wedge b) \in \nabla$ implies that either $a \in \nabla$ or $b \in \nabla$).*

Theorem 1.2.47 *If a lattice $\mathcal{A}$ has a zero element $\bot$ (a unit element $\top$) then any proper filter (ideal, respectively) in $\mathcal{A}$ can be included in a maximal filter (ideal); any element $a \neq \bot$ ($a \neq \top$) is a member of a maximal filter (ideal).*

Definition 1.2.48 *A lattice $\mathcal{A}$ is called* distributive *if the following identities*

$$x \wedge (y \vee z) = (x \wedge y) \vee (x \wedge z), \quad x \vee (y \wedge z) = (x \vee y) \wedge (x \vee z)$$

are valid in $\mathcal{A}$.

An evident example of a distributive lattice is the lattice $L(X) := \langle 2^X, \cup, \cap \rangle$ of all subsets of a given set X; the set-theoretical union $\cup$ and intersection $\cap$ play the role of $\wedge$ and $\vee$. This simple example plays more important role than it seems at first glance. It turned out that any distribute lattice can be represented as a sublattice of a lattice of such kind, this is described below in Stone representation theorem. First we recall the following

Proposition 1.2.49 *If $\mathcal{A}$ is a distributive lattice and $a, b \in |\mathcal{A}|$ but $a \not\leq b$ then there is a prime filter ∇ (a prime ideal ∇) such that $a \in \nabla$, $b \notin \nabla$ (such that $a \notin \nabla, b \in \nabla$).*

Definition 1.2.50 *Given a distributive lattice $\mathcal{A}$, let $S(\mathcal{A})$ be the set of all prime filters in $\mathcal{A}$. We define the mapping h from $\mathcal{A}$ in $2^{S(\mathcal{A})}$ as follows. For any $a \in |\mathcal{A}|$, we set*

$$h(a) := \{ \nabla \mid \nabla \in S(\mathcal{A}), \ a \in \nabla \}.$$

The topological space on $2^{S(\mathcal{A})}$ with the base of topology $\{ h(a) \mid a \in \mathcal{A} \}$ is called Stone space of $\mathcal{A}$.

Theorem 1.2.51 *(Stone Representation Theorem) For every distributive lattice $\mathcal{A}$, the mapping h is the isomorphism from $\mathcal{A}$ in the lattice of all subsets of the set $S(\mathcal{A})$.*

Note that the similar description of distributive lattices by the set of all prime ideals can be given using symmetricity between filters and ideals. We also describe here an particular important case of distributive lattices, the so called *boolean algebras*. We will consider them in an extended signature. We assume the signature of boolean algebras to be the set $\Sigma := \{ \wedge, \vee, \neg, \bot, \top \}$.

Definition 1.2.52 *An algebra* $\mathcal{A} := \langle A, \wedge, \vee, \neg, \perp, \top \rangle$ *is said to be a* boolean algebra *if the following holds.*

(i) $\langle A, \wedge, \vee, \perp, \top \rangle$ *is a distributive lattice with the unit element* $\top$ *and the zero element* $\perp$;

(ii) *the identity* $\neg\neg x = x$ *is valid in* $\mathcal{A}$;

(iii) *the identities* $\neg(x \vee y) = (\neg x) \wedge (\neg y)$, $\neg(x \wedge y) = (\neg x) \vee (\neg y)$ *(de Morgan laws) are valid in* $\mathcal{A}$;

(iv) $\mathcal{A} \models (x \vee \neg x) = \top$, $\mathcal{A} \models (x \wedge \neg x) = \perp$.

The most well known example of a boolean algebra is the boolean algebra of all subsets of a given set. More precisely, suppose that X is a set and $\mathcal{B}(X) := \langle 2^X, \cup, \cap, \neg, X, \emptyset, X \rangle$, where $\neg Y := \{a \mid a \in X, a \notin Y\}$. It is not hard to check directly that $\mathcal{B}(X)$ is a boolean algebra which we call the *boolean on* X. The set $\neg Y$ is said to be the complement of Y. Boolean algebras of this kind play an important role because all others are subalgebras of boolean algebras of this type as we will see below. Let $\mathcal{A}$ be a boolean algebra. The Stone space of $\mathcal{A}$ is the set $S(\mathcal{A})$ of all maximal filters on $\mathcal{A}$. The mapping $h : \mathcal{A} \mapsto \mathcal{B}(S(\mathcal{A}))$ is defined as follows:

$$h(a) := \{\nabla \mid \nabla \in S(\mathcal{A}), a \in \nabla\}.$$

Theorem 1.2.53 . *The mapping h is an isomorphic embedding of the boolean algebra $\mathcal{A}$ into the boolean algebra $\mathcal{B}(S(\mathcal{A}))$. If $\mathcal{A}$ is a finite algebra then h is isomorphism onto, i.e. $\mathcal{A} \cong \mathcal{B}(S(\mathcal{A}))$.*

Finally we note the important and useful notion of sub-direct product and sub-directly irreducible algebras. Let $\prod_{i \in I} \mathcal{A}_i$ be a direct product of algebras. The projection π_i, $i \in I$, is the mapping of $\prod_{i \in I} \mathcal{A}_i$ onto the algebra $\mathcal{A}_i$, where for any $[(a_i)_{i \in I}] \in \prod_{i \in I} \mathcal{A}_i$,

$$\pi_i([(a_i)_{i \in I}]) := a_i.$$

It is not hard to see that any projection π_i is a homomorphism onto $\mathcal{A}_i$, i.e. is an endomorphism.

Definition 1.2.54 *Let $\mathcal{A}_i, i \in I$ be a family of algebras of the same signature. Let $\mathcal{A}$ be a subalgebra of $\prod_{i \in I} \mathcal{A}_i$. Algebra $\mathcal{A}$ is a subdirect product of algebras $\mathcal{A}_i, i \in I$ iff the projection π_i maps $\mathcal{A}$ onto $\mathcal{A}_i$ for every $i \in I$.*

Theorem 1.2.55 *(G.Birkhoff, see [61, 24]). Let $\mathcal{A}$ be an algebra. Let θ_i, $i \in I$ be a family of congruence relations such that $\bigwedge_{i \in I} \theta_i = \perp$, where $\perp$ is the smallest congruence relation, i.e. the trivial congruence relation, the* *equality. Then $\mathcal{A}$ is isomorphic to a certain subdirect product of the quotient-algebras $\mathcal{A}/\theta_i$, $i \in I$.*

For each $a \in |\mathcal{A}|$ we define a $f_a \in |\prod_{i \in I} \mathcal{A}/\theta_i|$ in the following manner:

$$f_a(i) := [a]_{\theta_i} \in |\mathcal{A}/\theta_i|.$$

Let

$$|\mathcal{A}_1| := \{f_a \mid a \in |\mathcal{A}|\} \subseteq |\prod_{i \in I} \mathcal{A}/\theta_i|.$$

Then $\mathcal{A}_1$ is a subalgebra of $\prod_{i \in I} \mathcal{A}/\theta_i$ and the mapping

$$\varphi(a) := f_a$$

is an isomorphism between $\mathcal{A}$ and $\mathcal{A}_1$. Furthermore, $\mathcal{A}_1$ is a subdirect product of algebras $\mathcal{A}/\theta_i$, $i \in I$.

Definition 1.2.56 *An algebra $\mathcal{A}$ is called subdirectly irreducible if for any family θ_i, $i \in I$ of congruence relations on $\mathcal{A}$, the relation*

$$[\bigwedge_{i \in I} \theta_i] = \bot = \{(a, a) \mid a \in |\mathcal{A}|\}$$

implies the existence of some $i \in I$ such that $\theta_i = \bot$.

Theorem 1.2.57 *(G.Birkhoff, see [61, 24]). Any algebra $\mathcal{A}$ is isomorphic to a subdirect product of certain subdirectly irreducible algebras.*

We conclude this section by recalling some facts connected with congruence distributive algebras.

Definition 1.2.58 *An algebra $\mathcal{A}$ is said to be congruence distributive if the lattice of all congruence relations on $\mathcal{A}$ is distributive.*

Lemma 1.2.59 *(Jonsson, [70]) Suppose that $\mathcal{A}$ is an algebra which has a ternary term $t(x, y, z)$, compound from some signature's functions such that*

$$t(x, y, x) = x, \ t(x, z, z) = z,$$

Then $\mathcal{A}$ is congruence distributive.

For instance for any lattice $\mathcal{A}$, the term $t(x, y, z) := (x \wedge y) \vee (x \wedge z) \vee (y \wedge z)$ satisfies the condition above. Therefore any algebra having a presentation by terms of lattice structure is congruence distributive. In particular such are all boolean algebras.

Theorem 1.2.60 *(K.Baker [5]). If $\mathcal{A}$ is a finite congruence distributive algebra then the variety $\mathcal{A}^V$ generated by $\mathcal{A}$ is finitely based, i.e. there is a finite set of identities $\mathcal{I}$ such that $Mod(\mathcal{I}) := \mathcal{A}^V$.*

Immediately from this theorem and the comment placed immediately before this theorem, we derive

Corollary 1.2.61 *If $\mathcal{A}$ is a finite algebra which is a lattice with respect to certain operations generated by terms in the original signature of $\mathcal{A}$ then the variety $\mathcal{A}^V$ generated by $\mathcal{A}$ is finitely based.*

Lemma 1.2.62 *(Jonsson) If a variety V consists of congruence distributive algebras then the lattice of all subvarieties of V is distributive.*

1.3 Algebraic Semantics for Propositional Logics

Semantic methods will play very important role through this book. In this section we introduce a general algebraic semantics for propositional logics. The application of algebraic methods to the study of propositional logic can be said to have begun in the 1920s and 1930s with the pioneering papers by Tarski and Lukasiewicz. Nowadays algebraic methods are very well developed and their use is wide spread in the research of different logical systems. The introduction of algebraic semantics is based on the very general, simple and fruitful idea of introducing the characterizing matrix. We now turn to description of this idea in detail. First we note that in Section 6 of this chapter we will develop the algebraic semantics for logical consequence relations, and all results of the present section will be simple consequences of those. However presentation there is comparatively more complicated for first reading. Therefore we prefer to start with a more simple explanation. Experienced readers can omit this section, but, for the less experienced reading this section would be useful and desirable.

Definition 1.3.1 *A logical matrix for a logic λ in a language $\mathcal{L}$ is an algebraic system $\mathfrak{M} := \langle M, Con_{\mathcal{L}}, D \rangle$ where M is a nonempty set, called the universe of $\mathfrak{M}$, $Con_{\mathcal{L}}$ are functions (operations) on M corresponding to all logical connectives from $\mathcal{L}$ (the rank of any such function correspond to the arity of the corresponding logical connective), and D is a subset of M called the set of designated elements.*

Less formally, M is a set of all possible *truth values* of this matrix and D consists of the set of all *true (or designated)* values. The set D can be understood as an unary predicate relation, thus this definition exactly corresponds to the usual definition of an algebraic system with a set of operations and a single unary predicate relation. If $\alpha(p_1, ..., p_k)$ is a formula from $\mathcal{F}or_\lambda$ built up of propositional letters $\{p_1, ..., p_n\}$ and V is a mapping of $\{p_1, ..., p_k\}$ into elements $a_1, ..., a_n$ of M, then V is called a valuation (or an interpretation) of the propositional letters of the formula α in $\mathfrak{M}$.

Definition 1.3.2 *Let α be a formula from $\mathcal{F}or_\lambda$.*

- *We say that α is* valid *with respect to a valuation (or under a valuation) V of its propositional letters in $\mathfrak{M}$ if and only if the value of $\alpha(a_1, ..., a_k)$ in $\mathfrak{M}$ is a member of D (takes a designated truth value), that is $\alpha(a_1, ..., a_k) \in D$.*

- *α is said to be valid in $\mathfrak{M}$ if this formula is valid in $\mathfrak{M}$ with respect to every valuation of its propositional letters.*

We use $\mathfrak{M} \models \alpha$ to express that a formula α is valid in a matrix $\mathfrak{M}$, and, if $\mathfrak{M} \models \alpha$ we also say that the formula α is *valid* in $\mathfrak{M}$.

Definition 1.3.3 *A matrix $\mathfrak{M}$ is called* adequate *for a logic λ if, for any formula $\alpha \in \lambda$, $\mathfrak{M} \models \alpha$ holds.*

Definition 1.3.4 *A matrix $\mathfrak{M}$ is called* characterizing *for a logic λ if, for any formula $\alpha \in \mathcal{F}or_\lambda$, α is a theorem of λ (that is $\alpha \in \lambda$) if and only if $\mathfrak{M} \models \alpha$.*

Let λ be a propositional logic. The Lindenbaum-Tarski matrix for λ is the matrix $\mathfrak{M}_L(\lambda) := \langle \mathcal{F}or_\lambda, Con_\lambda, \lambda \rangle$, where Con_λ consists of merely the logical connectives from the language of λ.

Theorem 1.3.5 *The matrix $\mathfrak{M}_L(\lambda)$ is a characterizing matrix for the logic λ.*

Proof. This is evident because by the definition every logic λ is closed with respect to substitution.

The calculation of the truth values of formulas in $\mathfrak{M}_L(\lambda)$ is hampered if we have no knowledge beforehand about whether the formulas are actually theorems of λ. The Lindenbaum-Tarski matrix is very simple by definition but this matrix is little suited for real evaluations. Therefore it is desirable to find a more convenient and effective algebraic characterization of propositional logics. There are many approaches for finding an effective algebraic semantics, every of which consists of some restriction of the class of logics considered by some additional requirements on provability. We will also consider only logics having certain properties necessary for an effective algebraic semantics (like theorem of the replacement of equivalents etc.). As we will see later all these properties are also natural and important for the case of the logical consequence relation.

Definition 1.3.6 *We say that a formula α is derivable from a set of formulas X in a logic λ (abbreviation $X \vdash_\lambda \alpha$) if there is a sequence S of formulas terminating in α, every formula of which is either a theorem of λ, or a formula in X, or obtained from formulas preceding it by an inference rule for λ. For sets of formulas X, Y, we write $X \vdash_\lambda Y$ if for every $\alpha \in Y$ the relation $X \vdash_\lambda \alpha$ holds.*

The following definition, generally speaking, is similar to the definition of algebraic logic due to J.Czelakowski [32].

Definition 1.3.7 *We say that a logic λ is algebraic if there is a formula $\top$ and a collection of formulas $\phi_1(x, y), ..., \phi_n(x, y)$ each of which is build up from two propositional letters x, y such that the following hold. We will further abbreviate the tuple $\phi_1(A, B), \ldots , \phi_n(A, B)$ by $A \equiv B$, and refer to formulas $\phi_1(x, y), ..., \phi_n(x, y)$ as for $\equiv$. For every formulas α, β, γ, $\alpha_1, ..., \alpha_n$, $\beta_1, ..., \beta_n$ and every n-placed logical connective δ from the language of λ,*

> *a)* $\vdash_\lambda \alpha \equiv \alpha$,
>
> *b)* $\alpha \equiv \beta \vdash_\lambda \beta \equiv \alpha$,
>
> *c)* $\alpha \equiv \beta, \beta \equiv \gamma \vdash_\lambda \alpha \equiv \gamma$
>
> *d)* $\alpha_1 \equiv \beta_1, ..., \alpha_n \equiv \beta_n \vdash_\lambda \delta(\alpha_1, ..., \alpha_n) \equiv \delta(\beta_1, ..., \beta_n)$
>
> *e)* $\alpha \equiv \beta, \beta \vdash_\lambda \alpha$
>
> *f)* $\alpha \dashv\vdash_\lambda \alpha \equiv \top$
>
> *g)* $\vdash_\lambda \top$.

We call the formulas $\equiv, \top$ an algebraic presentation for λ.

Thus, the formulas of the list $\equiv$ give an equivalence relation on formulas which is consistent in λ with logical connectives. The formula $\top$ is a formula expressing, in a sense, the *truth* in λ. It is also easy to see that d) above is nothing but the *theorem of the replacement of equivalents*. Without exaggeration it can be said, that the majority of propositional logics which are actively studied are algebraic ones (at the same time, not all propositional logics are algebraic or even algebraizable, see Section 6 for examples). Before to describe the advantage for semantics of algebraic logics, it is natural to point out some examples of algebraic logics.

Theorem 1.3.8 *Any superintuitionistic logic λ is an algebraic logic with algebraic presentation: $(\equiv) := \{x \rightarrow y, \ y \rightarrow x\}$, $\top := (x \rightarrow (x \rightarrow x))$. Also we can take as an algebraic presentation for λ formulas $(\equiv) := \{(x \rightarrow y) \wedge (y \rightarrow x)\}$, $\top := (x \rightarrow (x \rightarrow x))$.*

Proof. In the proof of this theorem, it is convenient to use the following deduction theorem.

Theorem 1.3.9 *(Deduction Theorem) Let λ be a propositional logic in a language with the logical connective $\rightarrow$, and with modus ponens as the single inference rule, and which has the axioms (A1) and (A2) of PC as its theorems. For any set of formulas X and any formulas α and β the following holds:*

$$X, \alpha \vdash_\lambda \beta \Leftrightarrow X \vdash_\lambda (\alpha \rightarrow \beta).$$

Proof. The direction $\Leftarrow$ follows trivially by modus ponens. The converse implication simply follows by induction on the length of the derivation. Indeed, if this proof has length 1 then either (1) $\beta = \alpha$, or (2) $\beta \in X$, or (3) β is a theorem of λ. Case (1): it suffices to note that, for any formula ξ,

$$\xi \to (\xi \to \xi) \ (\text{Axiom (A1)})$$
$$\xi \to ((\xi \to \xi) \to \xi) \ (\text{Axiom (2A1)})$$
$$(\xi \to ((\xi \to \xi) \to \xi)) \to ((\xi \to (\xi \to \xi)) \to (\xi \to \xi)) \ (\text{Axiom (A2)})$$
$$(\xi \to (\xi \to \xi)) \to (\xi \to \xi)$$
$$(\xi \to \xi)$$

is a derivation of the formula $\xi \to \xi$ from the empty set of premises in λ. Thus

$$\vdash_\lambda (\xi \to \xi) \tag{1.1}$$

Hence $\vdash_\lambda \alpha \to \alpha$ holds. If case (2) or (3) holds then $\beta \to (\alpha \to \beta)$ (Axiom (A1)), $\alpha \to \beta$ is a derivation of $\alpha \to \beta$ in λ from the premise set X. Suppose we have proved already that, for derivations of formulas β_1 from sets X_1, α_1 with length of not more than n, our conclusion holds. Let there is a derivation of β from X, α with length $n + 1$. Then if β is α or is a member of X, or is a theorem of λ, our conclusion follows by the same arguments as above. If β is the result of an application of the inference rule, then some formulas of the form $\gamma \to \beta$, γ precede β in the derivation. Therefore these formulas have a shorter than β derivation from the same premises. Applying the inductive hypothesis,

$$X \vdash_\lambda \alpha \to (\gamma \to \beta), \quad X \vdash_\lambda \alpha \to \gamma.$$

By concatenating these two derivations, writing to the tile of this concatenation the axiom $(\alpha \to (\gamma \to \beta)) \to ((\alpha \to \gamma) \to (\alpha \to \beta))$ (Axiom (A2)), and applying modus ponens to this axiom twice, we arrive at the formula $\alpha \to \beta$. Thus the arrow $\Rightarrow$ holds for derivations of arbitrary length. ∎

Let as turn to the proof of Theorem 1.3.8. By definition relation b) from the definition of algebraic logics holds, and a) holds also by (1.1). If we have hypothesis $\alpha \to \beta$ and $\beta \to \gamma$ then

$$(\beta \to \gamma) \to (\alpha \to (\beta \to \gamma))$$
$$\alpha \to (\beta \to \gamma)$$
$$(\alpha \to (\beta \to \gamma)) \to ((\alpha \to \beta) \to (\alpha \to \gamma))$$
$$(\alpha \to \beta) \to (\alpha \to \gamma)$$
$$\alpha \to \gamma$$

is a derivation of $\alpha \to \gamma$ in H and moreover in any superintuitionistic logic. Thus, for any superintuitionistic logic λ,

$$\alpha \to \beta, \ \beta \to \gamma \vdash_\lambda \alpha \to \gamma \tag{1.2}$$

Therefore c) holds also. Because $\top$ is a particular case of Axiom (A1), we have that g) is valid as well. If we have hypothesis $\alpha \equiv \beta$ and β then we derive in $\lambda\,\alpha$ by modus ponens, hence e) also holds. Now, let a formula α be given as a hypothesis. Then $\top \to \alpha$ follows by modus ponens from $\alpha \to (\top \to \alpha)$ and $(\alpha \to \top)$ follows by modus ponens from $\top \to (\alpha \to \top)$. Thus f) also holds. It remains only to show d). Suppose $\alpha \to \beta$, $\beta \to \alpha$, $\mu \to \rho$, $\rho \to \mu$ are given as hypotheses. By Theorem 1.3.9 in order to show that $\alpha \equiv \beta$, $\mu \equiv \rho \vdash_\lambda \alpha \wedge \mu \equiv \beta \wedge \rho$, it is sufficient to prove that $\alpha \equiv \beta$, $\mu \equiv \rho$, $\alpha \wedge \mu \vdash_\lambda \beta \wedge \rho$. The required derivation is given below:

$$\alpha \equiv \beta,\ \mu \equiv \rho,\ \alpha \wedge \mu$$
$$\alpha \wedge \mu \to \alpha \ \ (\text{Axiom (A6)})$$
$$\alpha \wedge \mu \to \mu \ \ (\text{Axiom (A7)})$$
$$(\alpha \to \beta) \to ((\alpha \to \rho) \to (\alpha \to \beta \wedge \rho)) \ \ (\text{Axiom (A8)}$$
$$(\alpha \to \rho) \to (\alpha \to \beta \wedge \rho)$$
$$\rho \to (\alpha \to \rho) \ \ (\text{Axiom (A1)})$$
$$\alpha,\ \mu,\ \rho, \alpha \to \rho, \alpha \to \beta \wedge \rho, \beta \wedge \rho.$$

To show that $\alpha \equiv \beta$, $\mu \equiv \rho \vdash_\lambda \alpha \vee \mu \equiv \beta \vee \rho$, it is sufficient to find a derivation $\alpha \equiv \beta$, $\mu \equiv \rho$, $\alpha \vee \mu \vdash_\lambda \beta \vee \rho$ (Theorem 1.3.9). A derivation of this kind is given below:

$$\alpha \equiv \beta,\ \mu \equiv \rho,\ \alpha \vee \beta$$
$$(\alpha \to \beta \vee \rho) \to ((\mu \to \beta \vee \rho) \to (\alpha \vee \mu \to \beta \vee \rho)) \ \ \text{Axiom (A5)}$$
$$(\alpha \to (\beta \to \beta \vee \rho)) \to ((\alpha \to \beta) \to (\alpha \to \beta \vee \rho)) \ \ (\text{Axiom (A2)})$$
$$(\beta \to \beta \vee \rho) \to (\alpha \to (\beta \to \beta \vee \rho)) \ \ (\text{Axiom (A1)})$$
$$\beta \to \beta \vee \rho \ \ \text{Axiom (A3)}),\ \alpha \to (\beta \to \beta \vee \rho)$$
$$(\alpha \to \beta) \to (\alpha \to \beta \vee \rho),\ \alpha \to \beta \vee \rho$$
$$(\mu \to \beta \vee \rho) \to (\alpha \vee \mu \to \beta \vee \rho),\ \rho \to \beta \vee \rho \ \ (\text{Axiom (A4)}$$
$$(\rho \to \beta \vee \rho) \to (\mu \to (\rho \to \beta \vee \rho)) \ \ (\text{Axiom (A1)})$$
$$\mu \to (\rho \to \beta \vee \rho)$$
$$(\mu \to (\rho \to \beta \vee \rho)) \to ((\mu \to \rho) \to (\mu \to \beta \vee \rho)) \ \ \text{Axiom (A2)})$$
$$(\mu \to \rho) \to (\mu \to \beta \vee \rho),\ \mu \to \rho,\ \mu \to \beta \vee \rho$$
$$\alpha \vee \mu \beta \vee \rho$$
$$\beta \vee \rho.$$

According to Theorem 1.3.9, for the proof that $\alpha \equiv \beta, \mu \equiv \rho \vdash_\lambda ((\alpha \to \mu) \to (\beta \to \rho))$, it is sufficient to show that $\alpha \equiv \beta, \mu \equiv \rho, (\alpha \to \mu), \beta \vdash_\lambda \rho))$. Indeed, from β and $\alpha \equiv \beta$ we can derive α, then from α and $\alpha \to \mu$ we infer μ, and, finally, from μ and $\mu \equiv \rho$ we derive ρ. Thus $\equiv$ is consistent with $\to$. Now we consider the connective $\neg$. Again, in order to show that $\alpha \equiv \beta \vdash_\lambda (\neg\alpha \equiv \neg\beta)$ it is sufficient (by Theorem 1.3.9) to prove that $\alpha \equiv \beta$, $\neg\alpha \vdash_\lambda \neg\beta$. This is proved as follows.

$$\alpha \equiv \beta, \ \neg\alpha, \ (\beta \to \alpha) \to ((\beta \to \neg\alpha) \to \neg\beta) \ \text{(Axiom (A9))}$$
$$(\beta \to \neg\alpha) \to \neg\beta, \ \neg\alpha \to (\beta \to \neg\alpha) \ \text{(Axiom (A1))}$$
$$\beta \to \neg\alpha, \ \neg\beta.$$

Thus d) holds also. Therefore any superintuitionistic logic is algebraic. To show the last claim of our theorem it is sufficient to note that $x \to y, y \to x \ \dashv\vdash_\lambda (x \to y) \wedge (y \to x)$ in any superintuitionistic logic λ. ∎

Now having at our disposal the Deduction Theorem, it is rather natural to show it at work. Let us pause briefly to note some basic properties of the logical connectives $\neg, \to, \wedge, \vee$, which can be easily proved using Theorem 1.3.9, i.e. the Deduction Theorem.

Exercise 1.3.10 *For any superintuitionistic logic* λ,

    a) $\vdash_\lambda \alpha \equiv \alpha \wedge \alpha, \ \vdash_\lambda \alpha \equiv \alpha \vee \alpha$ *(idempotency);*

    b) $\vdash_\lambda \alpha \wedge \beta \equiv \beta \wedge \alpha, \ \vdash_\lambda \alpha \vee \beta \equiv \beta \vee \alpha$ *(commutativity);*

    c) $\vdash_\lambda (\alpha \wedge \beta) \wedge \gamma \equiv \alpha \wedge (\beta \wedge \gamma), \ \vdash_\lambda (\alpha \vee \beta) \vee \gamma \equiv \alpha \vee (\beta \vee \gamma)$
       (associativity);

    d) $\vdash_\lambda \alpha \wedge (\alpha \vee \beta) \equiv \alpha, \ \vdash_\lambda \alpha \vee (\alpha \wedge \beta) \equiv \alpha$ *(absorption*
       equivalence);

    e) $\vdash_\lambda \alpha \vee (\beta \wedge \gamma) \equiv (\alpha \vee \beta) \wedge (\alpha \vee \gamma),$
       $\vdash_\lambda \alpha \wedge (\beta \vee \gamma) \equiv (\alpha \wedge \beta) \vee (\alpha \wedge \gamma),$ *(distributive law);*

    f) $\vdash_\lambda \alpha \to (\beta \to \gamma) \equiv (\alpha \wedge \beta) \to \gamma$;

    g) $\vdash_\lambda \beta \wedge (\alpha \to \beta) \equiv \beta;$

    h) $\vdash_\lambda \alpha \wedge (\alpha \to \beta) \equiv \alpha \wedge \beta$;

    i) $\vdash_\lambda (\alpha \to \beta) \wedge (\alpha \to \gamma) \equiv (\alpha \to (\beta \wedge \gamma));$

    j) $\vdash_\lambda (\alpha \to \gamma) \wedge (\beta \to \gamma) \equiv (\alpha \vee \beta) \to \gamma.$

For PC the following hold:

    k) $\vdash_{PC} (\alpha \to \beta) \equiv (\neg\alpha \vee \beta);$

    l) $\vdash_{PC} \neg\neg\alpha \equiv \alpha$;

    m) $\vdash_{PC} \neg(\alpha \wedge \beta) \equiv (\neg\alpha \vee \neg\beta);$

    n) $\vdash_{PC} \neg(\alpha \vee \beta) \equiv (\neg\alpha \wedge \neg\beta);$

    o) $\vdash_{PC} \neg\alpha \vee \alpha;$

Method: use the axioms of H and PC, the derivation rule modus ponens, Theorem 1.3.9, and, if necessary, d) from the definition of algebraic logic (the theorem of the replacement of equivalents).

Definition 1.3.11 *Let g be a subset of the set $\{\rightarrow, \neg, \wedge, \vee\}$. For any superintuitionistic logic λ, the g-fragment of λ is the set of all formulas which are contained in λ and which are built up only using connectives from g. We will denote this fragment by λ_g.*

If g contains $\rightarrow$ then, as it is easy to see, λ_g forms a propositional logic with the modus ponens inference rule (in the corresponding language).

Theorem 1.3.12 *The $\{\rightarrow\}$, $\{\rightarrow, \neg\}$, $\{\rightarrow, \wedge\}$, $\{\rightarrow, \vee\}$, $\{\rightarrow, \wedge, \vee\}$, $\{\rightarrow \wedge, \neg\}$, and $\{\rightarrow \vee, \neg\}$ fragments of superintuitionistic logics are algebraic logics. More precisely, for any superintuitionistic logic λ and any g, where*

$$\{\rightarrow\} \subseteq g \subseteq \{\rightarrow, \neg, \vee, \wedge\},$$

the g-fragment λ_g of λ, is an algebraic propositional logic with the algebraic presentation $(\equiv) := \{x \rightarrow y, \ y \rightarrow x\}$, $\top := (x \rightarrow (y \rightarrow x))$.

Proof. This can be easily extracted from the proof of Theorem 1.3.8. Indeed, for all the above mentioned logics, Theorem 1.3.9 holds, and, in each particular case, the derivations in the proof of Theorem 1.3.8 do not involve more logical connectives than postulated. ∎

The following definition corresponds to our conventions from Section 1 concerning definition of propositional logics.

Definition 1.3.13 *A normal modal propositional logic is a set of formulas in the language of the modal system K containing all theorems of K and closed with respect to substitution and the inference rules:*

$$(R1)\frac{x, x \rightarrow y}{y}, \qquad (R3)\frac{x}{\Box x}$$

Theorem 1.3.14 *Any normal modal logic λ is an algebraic logic with the algebraic presentation $(\equiv) := \{x \rightarrow y, y \rightarrow x\}$, $\top := \{(x \rightarrow (x \rightarrow x))\}$, still we can stipulate that $(\equiv) := \{(x \rightarrow y) \wedge (y \rightarrow x)\}$.*

Proof. We simply follow the proof of Theorem 1.3.8. The properties a), b), c), e,) f) and g) have the same proof because λ contains all tautologies (formulas obtained from theorems of PC by substitution of arbitrary modal formulas in place of letters). In the proof of d) in Theorem 1.3.8 we have used the Deduction Theorem (Theorem 1.3.9) which does not holds for modal logics in the form that it was stated. However we have shown there that, for any superintuitionistic logic λ_1, in particular for PC, the following holds

$$\alpha \equiv \beta, \ \mu \equiv \rho \vdash_{\lambda_1} \alpha \wedge \mu \equiv \beta \wedge \rho, \ \alpha \vee \mu \equiv \beta \vee \rho,$$

$$\alpha \equiv \beta, \ \mu \equiv \rho \vdash_{\lambda_1} \alpha \rightarrow \mu \equiv \beta \rightarrow \rho, \ \alpha \equiv \beta \vdash_{\lambda_1} \neg\alpha \equiv \neg\beta.$$

We can transform the derivation in PC into a derivation in λ because λ contains all tautologies and modus ponens is a postulated inference rule for λ. Therefore all above mentioned relations hold for λ. It remains only to show $\alpha \equiv \beta \vdash_\lambda \Box\alpha \equiv \Box\beta$. The derivation

$$\alpha \to \beta, \ \Box(\alpha \to \beta) \ \text{(by Rule (R3))}$$
$$\Box(\alpha \to \beta) \to (\Box\alpha \to \Box\beta) \ \text{(Axiom of K)}, \ \Box\alpha \to \Box\beta \ \text{(by (R1))}$$

gives us what was required. Thus d) holds for λ, that is λ is an algebraic logic. The last assertion of our theorem evidently follows from the proved part. ∎

We mentioned above that the Deduction Theorem in the form it was stated above for superintuitionistic logics does not hold for modal logics. Since this theorem is important and useful, in passing, we give a variant of this theorem for modal logics. We introduce the notation: $\Box^0\gamma := \gamma$, $\Box^{n+1}\gamma := \Box\Box^n\gamma$.

Theorem 1.3.15 *(Deduction Theorem for Modal Logics) Let λ be a normal modal logic, X a set of formulas in the language of λ. Suppose α and β are modal propositional formulas. Then*

$$X, \alpha \vdash_\lambda \beta \Leftrightarrow (\exists n_1, ..., n_k \in N) \ X \vdash_\lambda \Box^{n_1}\alpha \wedge ... \wedge \Box^{n_k}\alpha \to \beta.$$

If α is a modal logic extending K4 then $X, \alpha \vdash_\lambda \beta \Leftrightarrow X \vdash_\lambda \Box\alpha \wedge \alpha \to \beta$. If α is a modal logic extending S4 then $X, \alpha \vdash_\lambda \beta \Leftrightarrow X \vdash_\lambda \Box\alpha \to \beta$.

Proof. The direction $\Leftarrow$ of the main claim and of the other claims simply follows from $\gamma, \delta \vdash_{PC} \gamma \wedge \delta$ and the presence of the rules (R1), (R3). The converse direction $\Rightarrow$ of the first claim can be proved by induction on the length of the derivation. The basis of the induction has the same proof as in Theorem 1.3.9 and we turn to the inductive step. Assume β is obtained by (R1) from $\gamma \to \beta$ and γ which occur in the derivation before β. Then by the inductive hypothesis we have

$$X \vdash_\lambda \Box^{n_1}\alpha \wedge ... \wedge \Box^{n_k}\alpha \to \gamma, \ X \vdash_\lambda \Box^{m_1}\alpha \wedge ... \wedge \Box^{m_d}\alpha \to (\gamma \to \beta)$$

for some $n_1, ..., n_k, m_1, ..., m_d$. Then because $\mu \to \gamma, \delta \to (\gamma \to \beta) \vdash_{PC} \mu \wedge \delta \to \beta$ (to show this use Theorem 1.3.9) we have

$$X \vdash_\lambda \Box^{n_1}\alpha \wedge ... \wedge \Box^{n_k}\alpha \wedge \Box^{m_1}\alpha \wedge ... \wedge \Box^{m_d}\alpha \to \beta,$$

which is what we need. Suppose β is obtained by (R3), that is $\beta = \Box\gamma$, where γ occurs in the derivation before β. Then by the inductive hypothesis

$$X \vdash_\lambda \Box^{n_1}\alpha \wedge ... \wedge \Box^{n_k}\alpha \to \gamma.$$

Now we continue the derivation as follows

$$\Box^{n_1}\alpha \wedge ... \wedge \Box^{n_k}\alpha \to \gamma, \ \Box(\Box^{n_1}\alpha \wedge ... \wedge \Box^{n_k}\alpha \to \gamma)$$
$$\Box(\Box^{n_1}\alpha \wedge ... \wedge \Box^{n_k}\alpha \to \gamma) \to$$
$$(\Box(\Box^{n_1}\alpha \wedge ... \wedge \Box^{n_k}\alpha) \to \Box\gamma) \ (\text{Axiom of K})$$
$$\Box(\Box^{n_1}\alpha \wedge ... \wedge \Box^{n_k}\alpha) \to \Box\gamma \ (\text{by R1})$$

and arrive at the concluding formula. Now we need the following

Proposition 1.3.16 *For every formulas α and β,*

$$\vdash_K (\Box\alpha \wedge \Box\beta) \to \Box(\alpha \wedge \beta), \ and \ \vdash_K \Box(\alpha \wedge \beta) \to (\Box\alpha \wedge \Box\beta).$$

Proof. The following derivation

$$\alpha \wedge \beta \to \alpha \ (\text{Axiom}), \ \alpha \wedge \beta \to \beta \ (\text{Axiom})$$
$$\Box(\alpha \wedge \beta \to \alpha) \ (\text{by (R3)}), \ \Box(\alpha \wedge \beta \to \beta) \ (\text{by R3})$$
$$\Box(\alpha \wedge \beta \to \alpha) \to (\Box(\alpha \wedge \beta) \to \Box\alpha) \ (\text{Axiom of K})$$
$$\Box(\alpha \wedge \beta \to \beta) \to (\Box(\alpha \wedge \beta) \to \Box\beta) \ (\text{Axiom of K})$$
$$\Box(\alpha \wedge \beta) \to \Box\alpha \ (\text{by R1}), \ \Box(\alpha \wedge \beta) \to \Box\beta \ (\text{by R1})$$
$$\Box(\alpha \wedge \beta) \to \Box\alpha \wedge \Box\beta \ (\text{by Axiom (A8) and (R1)})$$

gives us $\Box(\alpha \wedge \beta) \to \Box\alpha \wedge \Box\beta$. Conversely, by Theorem 1.3.9, for every formulas α and β, we have $\vdash_{PC} \alpha \to (\beta \to \alpha \wedge \beta)$, therefore $\vdash_K \alpha \to (\beta \to \alpha \wedge \beta)$. Then the following sequence:

$$\alpha \to (\beta \to \alpha \wedge \beta), \ \Box(\alpha \to (\beta \to \alpha \wedge \beta)) \ (\text{by R3})$$
$$\Box(\alpha \to (\beta \to \alpha \wedge \beta)) \to (\Box\alpha \to \Box(\beta \to \alpha \wedge \beta)) \ (\text{Axiom of K})$$
$$\Box\alpha \to \Box(\beta \to \alpha \wedge \beta) \ (\text{by R1})$$
$$\Box(\beta \to \alpha \wedge \beta) \to (\Box\beta \to \Box(\alpha \wedge \beta)) \ (\text{Axiom of K})$$
$$\Box\alpha \to (\Box\beta \to \Box(\alpha \wedge \beta)) \ (\text{by (1.2)}), \ \Box\alpha \wedge \Box\beta \to \Box\alpha \ (\text{Axiom (A6)})$$
$$\Box\alpha \wedge \Box\beta \to (\Box\beta \to \Box(\alpha \wedge \beta)) \ (\text{by (1.2)})$$
$$(\Box\alpha \wedge \Box\beta \to (\Box\beta \to \Box(\alpha \wedge \beta))) \to ((\Box\alpha \wedge \Box\beta \to \Box\beta) \to$$
$$\to (\Box\alpha \wedge \Box\beta \to \Box(\alpha \wedge \beta))) \ (\text{Axiom (A1)})$$
$$(\Box\alpha \wedge \Box\beta \to \Box\beta) \to (\Box\alpha \wedge \Box\beta \to \Box(\alpha \wedge \beta)) \ (\text{by R1})$$
$$\Box\alpha \wedge \Box\beta \to \Box\beta \ (\text{Axiom (A7)})$$
$$\Box\alpha \wedge \Box B \to \Box(\alpha \wedge \beta) \ (\text{by (R1)})$$

can be completed in the way indicated above until we obtain a derivation of the concluding formula in modal logic K. ∎

Continuing the proof of our theorem, we apply Proposition 1.3.16 to the formula $\Box(\Box^{n_1}\alpha \wedge ... \wedge \Box^{n_k}\alpha)$ several times together with (1.2), d) from the definition of algebraic logic and Theorem 1.3.8 for PC, we obtain

$$\vdash_K \Box^{n_1+1}\alpha \wedge ... \wedge \Box^{n_k+1}\alpha \to \Box(\Box^{n_1}\alpha \wedge ... \wedge \Box^{n_k}\alpha)$$

From this and $\vdash_K \Box(\Box^{n_1}\alpha \wedge ... \wedge \Box^{n_k}\alpha) \to \Box\gamma$ by (1.2) we infer $\vdash_K \Box^{n_1+1}\alpha \wedge$ $... \wedge \Box^{n_k+1}\alpha \to \Box\gamma$, which is what we needed. This completes the inductive step, and the central claim of the theorem is proved. To extract corollaries for particular cases, note that if $K4 \subseteq \lambda$ then $\Box\gamma \to \Box\Box\gamma \in \lambda$, therefore by (1.2), and d) from the definition of algebraic logic we have $X \vdash_\lambda \Box\alpha \wedge \alpha \to \beta$. The case $S4 \subseteq \lambda$ is similar, noting that $\Box\alpha \to \alpha \in S4$. ∎

Now we briefly consider temporal propositional logic.

Definition 1.3.17 *A temporal propositional logic is a set of formulas in the language of temporal logic T_0 containing all the theorems of T_0 and closed with respect to substitution and the rules*

$$\frac{x, x \to y}{y}, \quad \frac{x}{Gx}, \quad \frac{x}{Hx}.$$

Theorem 1.3.18 *Any temporal logic λ is an algebraic logic with algebraic presentation* $(\equiv) := \{x \to y, \ y \to x\}$, $\top := (x \to (x \to x))$. *In particular we can set* $(\equiv) := \{(x \to y) \wedge (y \to x)\}$.

Proof. It is easy to see that the connectives G and H of T_0 satisfy the single modal axiom of the logic K, if we interpret G or H as $\Box$. Therefore we can simply follow the proof of Theorem 1.3.14 repeating twice the inductive step for $\Box$, once for G and once for H. ∎

We have now given series examples of algebraic logics, and we are now going to describe an algebraic semantics which is much more effective than Lindenbaum-Tarski matrixes for algebraic logics. Consider an arbitrary algebraic logic λ with certain algebraic presentation $\equiv$, $\top$. We define the following relation $\sim$ on its Lindenbaum-Tarski matrix $\mathfrak{M}_L(\lambda)$:

$$g \sim h \iff \vdash_\lambda (g \equiv h).$$

From properties a) - d) of the definition of algebraic logics, it immediately follows that this relation is a congruence relation on the Lindenbaum-Tarski matrix. Moreover, by e) this congruence relation is consistent with the property being designated, i.e.:

$$g \in \lambda, \ h \sim g \Rightarrow h \in \lambda, \text{ and } g \sim \top \iff \vdash_\lambda g$$

Thus the quotient algebraic system $\mathfrak{M}_L(\lambda)/_\sim$ of $\mathfrak{M}_L(\lambda)$ by this congruence relation will have only a single designated element - the class of all formulas which are equivalent to $\top$. We will denote the designated element of the quotient algebra also by $\top$.

Theorem 1.3.19 *For every algebraic logic λ, the algebraic system $\mathfrak{M}_L(\lambda)/_\sim$ is a characterizing logical matrix for λ with a single designated element $\top$.*

Proof. This follows directly from the relation $g \sim \top \iff g \in \lambda$ which holds by e), g), and f) from the definition of algebraic logics.

Now we are in a position to introduce algebraic semantics for every algebraic propositional logic λ as a variety of algebras (not algebraic systems). Let $\mathcal{A}(\lambda)$ be the class of all algebras with a signature consisting of all operations of $\mathfrak{M}_L(\lambda)$, and a single constant symbol $\top$. Let $Var(\lambda)$ be the following variety of algebras:

$$Var(\lambda) := \{\mathfrak{A} \mid (\mathfrak{A} \in \mathcal{A}(\lambda)) \& (\forall \alpha \in \lambda(\mathfrak{A} \models \alpha = \top)),$$

$$(\forall \alpha)(\forall \beta)(\vdash_\lambda \alpha \equiv \beta \Rightarrow (\mathfrak{A} \models \alpha = \beta))\}.$$

It is clear that $\mathfrak{M}_L(\lambda)/_\sim$ is a member of $Var(\lambda)$. From this fact we immediately derive

Theorem 1.3.20 *(Algebraic Completeness Theorem.) A formula α is a theorem of an algebraic logic λ iff the identity $\alpha = \top$ is valid in the variety $Var(\lambda)$.*

We will often abbreviate $\mathfrak{A} \models \alpha = \top$ by $\mathfrak{A} \models \alpha$ and say in this case that α is valid in algebra $\mathfrak{A}$. Note that the variety $Var(\lambda)$ characterizing λ is constructed uniquely by λ modulo of fixing the formulas expressing $\equiv$ and $\top$. Thus when talking of $Var(\lambda)$, we always have to remember which choice of algebraic presentation we have made.

Theorem 1.3.21 *The algebra $\mathfrak{M}_L(\lambda)/_\sim$ is a free algebra of countable rank from the variety $Var(\lambda)$ with free generators $[p_i]_\sim$, where p_i are all propositional letters.*

Proof.. Let, for some formulas $\alpha(p_1, ..., p_n)$ and $\beta(p_1, ..., p_n)$,

$$\mathfrak{M}_L(\lambda)/_\sim \models \alpha([p_1]_\sim, ..., [p_n]_\sim) = \beta([p_1]_\sim, ..., [p_n]_\sim)$$

holds. Then $\mathfrak{M}_L(\lambda)/_\sim \models [\alpha(p_1, ..., p_n]_\sim = [\beta(p_1, ..., p_n)]_\sim$, that is $\vdash_\lambda \alpha \equiv \beta$. By definition of $Var(\lambda)$ we conclude that, for every $\mathfrak{A} \in Var(\lambda)$, $\mathfrak{A} \models \alpha = \beta$. Thus we can continue every mapping $f([p_i]_\sim) \to a_i \in \mathfrak{A}$ from $\mathfrak{M}_L(\lambda)/_\sim$ into an arbitrary algebra $\mathfrak{A}$ from $Var(\lambda)$ up to a homomorphism defining the value of every term $\alpha([p_1]_\sim, ..., [p_n]_\sim)$ as follows $f(\alpha([p_1]_\sim, ..., [p_n]_\sim)) := \alpha(a_1, ..., a_n)$. ∎

Thus if a logic λ is algebraic then we can construct an algebraic semantics for λ which consists of algebras, not algebraic systems (but which can be considered as logical matrixes with a single designated element $\top$). Further in this book since this section every logic λ, if not otherwise specified, is an algebraic logic, $Var(\lambda)$ is the above mentioned variety which gives an algebraic semantics for λ, and $\mathfrak{F}_\omega(\lambda)$ is the free algebra of countable rank $\mathfrak{M}_L(\lambda)/_s im$ from this variety.

In order to show algebraic completeness theorem in work we can consider the axiomatizations of tabular logics.

Definition 1.3.22 *We say that an algebraic logic λ is tabular if there is a finite algebra $\mathfrak{A}$ from $Var(\lambda)$ which generates the variety $Var(\lambda)$. In particular,*
$$\lambda(\mathfrak{A}) := \{\alpha \mid \mathfrak{A} \models \alpha = \top\} = \lambda.$$

Theorem 1.3.23 *Let λ be a tabular algebraic logic with the set of inference rules $\mathcal{R}$ and $Var(\lambda)$ be generated by a finite algebra $\mathfrak{A}$. Let there be a finite set of formulas $\alpha_1, ..., \alpha_n \in \lambda$ such that any logic having the set of inference rules $\mathcal{R}$ and any set of axioms including formulas $\alpha_1, ..., \alpha_n$ is an algebraic logic. Then if $\mathfrak{A}$ is a lattice with respect to certain operations generated by its terms then λ can be axiomatized by a finite set of axioms and the set inference rules $\mathcal{R}$.*

Proof. By Corollary 1.2.61 the variety $\mathfrak{A}^V$ generated by $\mathfrak{A}$ is finitely axiomatizable and has, say, axioms $\gamma_1 = \delta_1, ..., \gamma_k = \delta_k$. Then formulas of the list $\mathcal{L} :=$ $\gamma_1 \equiv \delta_1, ..., \gamma_k \equiv \delta_k, \alpha_1, ... , \alpha_n$ belong to λ and the axiomatic system $\mathcal{L} \cup \mathcal{R}$ generates an algebraic logic λ_1 which is sublogic of λ. But by Theorem 1.3.20 $\lambda_1 = \lambda_2$. ∎

Thus, in particular, any tabular modal, temporal or superintuitionistic logic is finitely axiomatizable, i.e. has an axiomatic system with a finite set of axioms and a finite set of inference rules.

Let λ be a logic, $\mathcal{A}, \mathcal{B}, \mathcal{G}_i, i \in I$ be a collection of algebras in the signature of $Var(\lambda)$, and α be a formula in the language of λ.

Theorem 1.3.24 *The following hold:*

(i) If $\mathcal{B}$ is a homomorphic image of $\mathcal{A}$ and $\mathcal{A} \models \alpha$ then $\mathcal{B} \models \alpha$,

(ii) If $\mathcal{B}$ is a subalgebra of $\mathcal{A}$ and $\mathcal{A} \models \alpha$ then $\mathcal{B} \models \alpha$,

(iii) If $(\forall i)$ $\mathcal{G}_i \models \alpha$ then $\prod_{i \in I} \mathcal{G}_i \models \alpha$.

Proof. This is trivial because the validity of the formula is equivalent to the validity of the corresponding identity. And truth of identities is preserved under taking homomorphic images, subalgebras and direct products (see Theorems 1.2.15, 1.2.16, 1.2.19). ∎

Let λ be a logic, and $\mathcal{K}$ be a class of algebras from $Var(\lambda)$.

Definition 1.3.25 *We say that an inference rule r is valid in $\mathcal{K}$ if, for every algebra $\mathcal{A} \in \mathcal{K}$, for every valuation V of the variables of the rule r in $\mathcal{A}$ the following holds. If V maps all the premises of r to $\top$ then V also maps the conclusion of r to $\top$.*

Note that it is easy to see that the value of the term $\top$ in $\mathcal{A}$ is unique and equal to the constant $\top$, according to the existence of the free algebra $\mathfrak{F}_\omega(\lambda)$ in $Var(\lambda)$

(in fact, the value of $\top$ is unique in this algebra). For any class $\mathcal{K}$ of algebras in an appropriate signature, we let

$$\lambda(\mathcal{K}) := \{\alpha \mid \alpha \in \mathcal{F}or_\lambda, \ \forall \mathcal{A} \in \mathcal{K}, \ \mathcal{A} \models \alpha\}.$$

It is not hard to see that $\lambda(\mathcal{K})$ is a propositional logic with respect to any set of inference rules which are valid in $\mathcal{K}$. Let λ be a logic, and $\mathcal{A}, \mathcal{B}, \mathcal{G}_i, i \in I$ be algebras from $Var(\lambda)$. From Theorem 1.3.24 we immediately derive

Corollary 1.3.26 *The following hold:*

(i) If φ is a homomorphism from $\mathcal{A}$ then $\lambda(\varphi(\mathcal{A})) \subseteq \lambda(\mathcal{A})$;

(ii) If $\mathcal{B}$ is a subalgebra of $\mathcal{A}$ then $\lambda(\mathcal{B}) \subseteq \lambda(\mathcal{B})$;

(iii) $\lambda(\prod_{i \in I} \mathcal{G}_i) = \lambda(\bigcap_{i \in I} \mathcal{G}_i)$.

Generally speaking the algebras of the variety $Var(\lambda)$ are not necessary closed with respect to inference rules of the logic λ. If for any $\mathcal{A} \in Var(\lambda)$, $\mathcal{A}$ is closed with respect to any postulated inference rule of λ then we call $Var(\lambda)$ *regular* for the logic λ.

1.4 Admissible Rules in Algebraic Logics

An axiomatic system for a given propositional logic λ (normally) consists of axioms and inference rules. The theorems of a logic are all the formulas which can be derived from the axioms by means of the inference rules. Keeping the set of all theorems of λ constant, we can vary the set of axioms and inference rules in order to make the axiomatic system more powerful and convenient. In the case of axioms it is clear how this should be done: set of new axioms must always consist of theorems of λ. However it is a long standing problem to provide a general characterization of the inference rules which can consistently (with respect to preserving theorems) be added to the postulated inference rules of λ. Admissible inference rules, all those with respect to which the logic λ is closed, form the greatest class of such rules (P.Lorenzen, 1955, [84]).

Definition 1.4.1 *Given a logic λ in the language $\mathcal{L}$ and an inference rule*

$$r := \frac{\alpha_1(x_1, ..., x_n), ..., \alpha_m(x_1, ..., x_n)}{\beta(x_1, ..., x_n)},$$

r is said to be admissible in the logic λ if and only if for every substitution instance $x_i \mapsto \gamma_i \in \mathcal{F}or_{\mathcal{L}}$ it is always the case that $\beta(\gamma_1, ..., \gamma_n) \in \lambda$ when $\forall j \ [\ \alpha_j(\gamma_1, ..., \gamma_n) \in \lambda\ ]$.

It is easy to see that the notion of admissibility is invariant insofar that it does not depend on the choice of particular axiomatic system for any given logic. Admissible rules allow us to simplify derivations, and even the study of basic logical systems (classical propositional calculus, predicate logics etc.) starts as a rule with collecting the necessary sets of admissible rules. Important particular cases of admissible rules are the so-called derivable rules.

Definition 1.4.2 *An inference rule* $r := \alpha_1, ..., \alpha_m / \beta$ *is called* derivable *in a propositional logic* λ *if* $\alpha_1, ..., \alpha_m \vdash_\lambda \beta$.

Proposition 1.4.3 *Any inference rule derivable in* λ *is admissible for* λ.

Proof. This is almost self evident. Suppose that

$$\alpha_1(x_1, ..., x_n), ..., \alpha_m(x_1, ..., x_n) \vdash_\lambda \beta(x_1, ..., x_n)$$

holds. Consider a substitution v, $v(x_i) = \gamma_i \in \mathcal{F}or_\mathcal{L}$, such that for every j, the inclusion $\alpha_j(\gamma_1, ..., \gamma_n) \in \lambda$ holds. We take an arbitrary derivation $\mathcal{S}$ of β from $\alpha_1, ..., \alpha_n$ in λ. Furthermore, we choose the substitution w which coincides with v on the domain $Dom(v)$ of v and maps any letter lying not in $Dom(v)$ onto β, say. The sequence $\mathcal{S}^w$, obtained from $\mathcal{S}$ by applying w to each their members, will be a derivation in λ from the empty set of hypothesis. Indeed, under substitution w all hypothesis will turn into theorems of λ, the set of theorems of λ is closed with respect to substitutions, and all inference rules are structural (consistent with substitutions). Thus $\vdash_\lambda \beta^v$, that is $\beta(\gamma_1, ..., \gamma_n) \in \lambda$. ∎

Using the deduction theorems, which are stated in Section 3, we can characterize the derivable rules for particular axiomatic systems more precisely.

Theorem 1.4.4 *Let* $r := \alpha_1, ..., \alpha_m / \beta$ *be a rule in a language of a logic* λ.

(i) If λ is an superintuitionistic logic then
r is derivable in λ iff $\vdash_\lambda \alpha_1 \wedge ... \wedge \alpha_n \to \beta$;

(ii) If λ is a modal propositional logic over $K4$ then
r is derivable in λ iff $\vdash_\lambda \alpha_1 \wedge ... \wedge \alpha_n \wedge \Box\alpha_1 \wedge ... \wedge \Box\alpha_n \to \beta$;

(iii) If λ is a modal propositional logic containing $S4$ then
r is derivable in λ iff $\vdash_\lambda \Box\alpha_1 \wedge ... \wedge \Box\alpha_n \to \beta$;

Proof. The assertion of (i) follows directly from Theorem 1.3.9, and (f) of Exercise 1.3.10. (ii) and (iii) are also corollaries of Theorem 1.3.15 and (f) from Exercise 1.3.10.

So, any derivable rule is admissible, but even for some logics considered in Theorem 1.4.4, there are admissible rules that are not derivable. We will pay more attention to this fact later. At the same time, it is easy to see that derivability is not an invariant and is quite sensitive to the choice of the axiomatic system. A very simple and general description of admissible inference rules is given in the following theorem. For simplicity, we will denote the free algebra of countable rank described in Theorem 1.3.21 by $\mathfrak{F}(\lambda)$.

Theorem 1.4.5 *An inference rule*

$$r := \frac{\alpha_1(x_1, ..., x_n), ..., \alpha_m(x_1, ..., x_n)}{\alpha(x_1, ..., x_n)},$$

is admissible in an algebraic logic λ iff the quasi-identity

$$q(r) := \alpha_1 = \top, ..., \alpha_m = \top \;\Rightarrow\; \alpha = \top$$

is valid in the free algebra of countable rank $\mathfrak{F}(\lambda)$ from the variety $Var(\lambda)$.

Proof. Suppose

$$r := \frac{\alpha_1(x_1, ..., x_n), ..., \alpha_m(x_1, ..., x_n)}{\alpha(x_1, ..., x_n)},$$

is an admissible rule in λ. Let $x_i \mapsto [\delta_i]_\sim$ be an interpretation of the variables of $q(r)$ in $\mathfrak{F}(\lambda)$. Suppose that

$$\mathfrak{F}(\lambda) \models \alpha_1([\delta_i]_\sim) = \top, \;\; ... \;\;, \;\; \mathfrak{F}(\lambda) \models \alpha_m([\delta_i]_\sim) = \top.$$

By definition of $\mathfrak{F}(\lambda)$ this entails

$$\mathfrak{F}(\lambda) \models [\alpha_1(\delta_i)]_\sim = \top, \;\; ... \;\;, \;\; \mathfrak{F}(\lambda) \models [\alpha_m(\delta_i)]_\sim = \top,$$

or, in another words, for all j, $\vdash_\lambda \alpha_j(\delta_i) \equiv \top$. The logic λ is algebraic, therefore $\alpha_1(\delta_i) \in \lambda, ..., \alpha_m(\delta_i) \in \lambda$. By assumption r is admissible, and our last observation yields $\alpha(\delta_i) \in \lambda$. Then, again since λ is an algebraic logic, it follows that $\vdash_\lambda \alpha(\delta_i) \equiv \top$ and

$$\mathfrak{F}(\lambda) \models \alpha([\delta_i]_\sim) = \top.$$

That is, the quasi-identity $q(r)$ is valid in $\mathfrak{F}(\lambda)$.

Conversely, let $\mathfrak{F}(\lambda) \models q(r)$. Suppose that $\alpha_1(\delta_i) \in \lambda, \; ... \;, \alpha_m(\delta_i) \in \lambda$ for some tuple of formulas δ_i. Then by definition of $\mathfrak{F}(\lambda)$

$$\mathfrak{F}(\lambda) \models [\alpha_1(\delta_i)]_\sim = \top, \;\; ... \;\;, \mathfrak{F}(\lambda) \models [\alpha_m(\delta_i)]_\sim = \top.$$

Therefore again $\mathfrak{F}(\lambda) \models \alpha_1([\delta_i]_\sim) = \top, \; ... \;, \mathfrak{F}(\lambda) \models \alpha_m([\delta_i]_\sim) = \top$. Because $\mathfrak{F}(\lambda) \models q(r)$, we have $\mathfrak{F}(\lambda) \models \alpha([\delta_i]_\sim) = \top$ which yields $\alpha(\delta_i) \in \lambda$. ∎

It is possible also to describe the *truth* of quasi-identities in the free algebra $\mathfrak{F}(\lambda)$ through the admissibility of the corresponding inference rules in λ.

Theorem 1.4.6 *Let λ be an algebraic logic. A quasi-identity*

$$q := [g_1 = f_1, \quad \dots \quad , g_m = f_m \Rightarrow g = f]$$

is valid in $\mathfrak{F}(\lambda)$ iff all rules in the collection of inference rules (we mean any formula in the conclusion of the rule below produces the single separated rule)

$$r(q) := \frac{(g_1 \equiv f_1), \dots, (g_m \equiv f_m)}{g \equiv f}$$

are admissible in λ.

Proof. Let $q := [g_1 = f_1, \dots, g_m = f_m \Rightarrow g = f]$ be a quasi-identity valid in $\mathfrak{F}(\lambda)$. Suppose, for some tuple of formulas δ_i, the following inclusions

$$(g_1 \equiv f_1)(\delta_i) \subseteq \lambda, \quad \dots \quad , (g_m \equiv f_m)(\delta_i) \subseteq \lambda$$

hold. Then we directly obtain from the definitions of $\mathfrak{F}(\lambda)$ and $\sim$ that

$$\mathfrak{F}(\lambda) \models g_1([\delta_i]_\sim) = f_1([\delta_i]_\sim), \quad \dots \quad , \mathfrak{F}(\lambda) \models g_m([\delta_i]_\sim) = f_m([\delta_i]_\sim).$$

Because q is valid in $\mathfrak{F}(\lambda)$ we conclude $\mathfrak{F}(\lambda) \models g([\delta_i]_\sim) = f([\delta_i]_\sim)$, hence

$$(g \equiv f)(\delta_i) \subseteq \lambda.$$

That is, we proved all rules from $r(q)$ are admissible in λ. For the converse, suppose all inference rules from $r(q)$ are admissible in λ and for some tuple of formulas δ_i,

$$\mathfrak{F}(\lambda) \models g_1([\delta_i]_\sim) = f_1([\delta_i]_\sim), \quad \dots \quad , \mathfrak{F}(\lambda) \models g_m([\delta_i]_\sim) = f_m([\delta_i]_\sim).$$

Then, by definition of $\mathfrak{F}(\lambda)$, $(g_1(\delta_i) \equiv f_1(\delta_i)) \subseteq \lambda, \dots, (g_m(\delta_i) \equiv f_m(\delta_i)) \subseteq \lambda$. And, since all rules from $r(q)$ are admissible in λ, we have $g(\delta_i) \equiv f(\delta_i) \subseteq \lambda$. This implies

$$\mathfrak{F}(\lambda) \models g([\delta_i]_\sim) = f([\delta_i]_\sim).$$

Thus we have proved that q is valid in $\mathfrak{F}(\lambda)$. ∎

So we now have a description of admissible rules in algebraic logics in terms of *truth* of quasi-identities in free algebras and vice versa. Since from now onward in this section every logic λ, if not otherwise specified, is an algebraic logic, $Var(\lambda)$ is the corresponding variety which gives an algebraic semantics for λ, $\mathfrak{F}_\omega(\lambda)$ is the free algebra of countable rank from this variety.

We are interested to have at our disposal all admissible inference rules of a logic λ. However the collection all of them can be infinite and too complicated for direct description. Therefore it is natural to select a collection which consists of all the inference rules which we actually need. More precisely, it would be desirable to select a collection of admissible rules such that all the admissible rules are consequences of the selected ones. For this purpose, we need an understanding for *what it means for an inference rule to be a consequence of a collection of rules.*

Definition 1.4.7 *A sequence $\mathcal{S}$ of formulas in the language of a logic λ is said to be a* derivation *of a formula α in λ from a set of hypothesis $\mathcal{H}$ by means of inference rules from a set $\mathcal{R}$ of rules if the following hold. The formula α is the concluding formula of $\mathcal{S}$, and, for every formula γ from $\mathcal{S}$, at least one of the following propositions holds.*

- *γ is a theorem of λ*

- *γ is a formula from $\mathcal{H}$*

- *γ can be obtained from formulas placed in $\mathcal{S}$ before γ by using some postulated inference rule for λ or some inference rule from $\mathcal{H}$.*

If there is a derivation $\mathcal{S}$ of α in λ from $\mathcal{H}$ by means of a given set of inference rules $\mathcal{R}$ then we denote this fact by

$$\mathcal{H} \vdash_{\lambda, \mathcal{R}} \alpha$$

and say that we can derive α in λ from $\mathcal{H}$ by means of $\mathcal{R}$. If $\mathcal{R} = \emptyset$ we write $\mathcal{H} \vdash_{\lambda} \alpha$ and say that α is derivable in the logic λ from the set of hypothesis $\mathcal{H}$. We will use the abbreviation $\overline{x}$ for an ordered tuple $x_1, ..., x_n$ of variables.

Definition 1.4.8 *We say an inference rule*

$$r := \frac{\alpha_1(\overline{x}), ..., \alpha_n(\overline{x})}{\beta(\overline{x})}$$

is derivable in a logic λ from a collection of rules $\mathcal{R}$ (or that r is a consequence of $\mathcal{R}$ in λ, and abbreviate it by $\mathcal{R} \vdash_{\lambda} r$) if there is a derivation in λ for the conclusion $\beta(\overline{x})$ from the premises $\alpha_1(\overline{x}), ..., \beta_n(\overline{x})$, as a set of hypothesis, by means of rules from $\mathcal{R}$.

Before more deeply studying the bases for admissible rules, we need some additional information. As we have seen above, the definition for a rule to be a consequence of a set of inference rules is pure syntactical and is given in terms of derivability. We need to clarify the semantic content of this notion.

Definition 1.4.9 *A rule $r := \alpha_1, ..., \alpha_m \ / \ \beta$ in the language of a logic λ is said to be* valid *in an algebra $\mathfrak{A} \in Var(\lambda)$ iff, for every valuation V of variables from r in $\mathfrak{A}$, $\mathfrak{A} \models V(\beta) = \top$ provided $\forall j[\mathfrak{A} \models V(\mathfrak{A}_j) = \top]$.*

For a logic λ, $R_p(\lambda)$ are all the postulated inference rules for a given axiomatic system for λ. Further in the remaining part of this section we will suppose the axiomatic system for any given logic λ to be fixed.

Definition 1.4.10 *A rule r is a* semantic corollary *of a system of rules $\mathcal{R}$ in a logic λ (denotation $\mathcal{R} \models_\lambda r$) if, for every algebraic system $\mathfrak{A}$ from $Var(\lambda)$, if all inference rules from $\mathcal{R}$ and all rules postulated for λ are valid in $\mathfrak{A}$ then r is also valid in $\mathfrak{A}$.*

For a set of rules $\mathcal{R}$ and an algebra $\mathfrak{A}$, we write $\mathfrak{A} \models \mathcal{R}$ if all rules from $\mathcal{R}$ are valid on $\mathfrak{A}$.

Theorem 1.4.11 *Let $\mathcal{R} \cup \{r\}$ be a family of inference rules in the language of an algebraizable logic λ. Then $\mathcal{R} \vdash_\lambda r$ if and only if $\mathcal{R} \models_\lambda r$. In particular, if $\mathcal{R} \not\models_\lambda r$ then there is some algebra $\mathfrak{B} \in Var(\lambda)$ such that $\mathfrak{B} \models \mathcal{R} \cup R_p(\lambda)$ and $\beta \not\models r$ and, in addition, $\mathfrak{B}$ satisfies the quasi-identities:*

$$\bigwedge_{\delta \in (x \equiv y)} [\delta = \top] \Rightarrow (x = y), \quad (x = y) \Rightarrow (\delta = \top), \quad \forall \delta \in (x \equiv y).$$

Proof. Suppose $\mathcal{R} \vdash_\lambda r$ holds, and all rules from $\mathcal{R}$ and all rules postulated for λ are valid in $\mathfrak{A} \in Var(\lambda)$. Let V be a valuation under which all premises of the rule

$$r := \frac{\alpha_1(\overline{x}), \dots, \alpha_n(\overline{x})}{B(\overline{x})}$$

are valid in $\mathfrak{A}$. There exists a derivation $\mathcal{S}$ of $\beta(\overline{x})$ from $\alpha_1(\overline{x}), \dots, \alpha_n(\overline{x})$ in λ by means of $\mathcal{R}$. We take the valuation W of all propositional variables occurring in formulas from $\mathcal{S}$ which coincides with V on all propositional variables from premises of r and maps all other letters to an arbitrary (but fixed) element of the algebra $\mathfrak{A}$.

We claim that under W all formulas from S take the value $\top$ in $\mathfrak{A}$, that is all they are valid under W. This can be easily shown by direct induction on the length of the derivation $\mathcal{S}$. Indeed, all theorems of λ from $\mathcal{S}$ must take true designated values from $\mathfrak{A}$ because $\mathfrak{A} \in Var(\lambda)$. Every premise of r is valid under W by our supposition. Again, by supposition, using the inference rules postulated for λ or rules from $\mathcal{R}$ preserves the validity of formulas in $\mathfrak{A}$. Thus $\beta(\overline{x})$ is valid in $\mathfrak{A}$ under W. Hence r is a semantic corollary of $\mathcal{R}$ in λ.

Conversely, suppose that $\mathcal{R} \vdash_\lambda r$ does not hold, that is

$$\alpha_1, \dots, \alpha_n \not\vdash_{\lambda, \mathcal{R}} B.$$

We take the algebraic system $\mathfrak{F}(\lambda)$ from $Var(\lambda)$ and define the following relation in $\mathfrak{F}(\lambda)$ using the premises of r:

$$[f]_\sim \simeq [g]_\sim \Leftrightarrow [\alpha_1(\overline{x}), \dots, \alpha_n(\overline{x}) \vdash_{\lambda, R} f \equiv g].$$

The correctness of this definition (that is, the fact that it does not depends on the choice of the representatives from the equivalence classes) follows directly from the property c) of the definition of an algebraic logic. It is not hard to see that this relation is a congruence relation on the basis set of $\mathfrak{F}(\lambda)$ according to properties a) - d) from the definition of algebraic logic. Taking the quotient algebra of $\mathfrak{F}(\lambda)$ with respect to this congruence relation produces an algebra $\mathfrak{B}$ from $Var(\lambda)$ (because the quotient algebra is a homomorphic image of the algebra (see Theorem 1.2.11), and, since any variety is closed with respect to homomorphic images (see Theorem 1.2.23)) with a single designated element $[[\top]_\sim]_\simeq$. We intend to check that all rules from $\mathcal{R}$ and all postulated rules for λ are valid in $\mathfrak{B}$. Indeed, let

$$\rho := \frac{\gamma_1(\overline{x}), \, ..., \gamma_m(\overline{x})}{\gamma(\overline{x})}$$

be a rule from $\mathcal{R}$ or a postulated for λ rule. Suppose that the formulas of the list $\gamma_1(\overline{x}), \, ... \, , \gamma_m(\overline{x})$ take the designated value in $\mathfrak{B}$ for a certain valuation V of variables. That is, we have $V(x_i) := [[\theta_i]_\sim]_\simeq$,

$$[\alpha_1(\theta_1, ..., \theta_k)]_\sim \simeq [\top]_\sim, \, ..., \, [\alpha_m(\theta_1 S, ..., \theta_k)]_\sim \simeq [\top]_\sim.$$

By f) from the definition of algebraic logics, we conclude

$$\alpha_1, ..., \alpha_n \vdash_{\lambda, \mathcal{R}} \gamma_1(\theta_1, ..., \theta_k), \, ..., \, \alpha_1, ..., \alpha_n \vdash_{\lambda, \mathcal{R}} \gamma_m(\theta_1, ..., \theta_k).$$

Therefore $\alpha_1, ..., \alpha_n \vdash_{\lambda, \mathcal{R}} \gamma(\theta_1, ..., \theta_k)$. Once again applying f) from the definition of algebraic logics, it follows

$$\alpha_1, ..., \alpha_n \vdash_{\lambda, \mathcal{R}} \gamma(\theta_1, ..., \theta_k) \equiv T.$$

That is, $V(\alpha) = [[\top]_\sim]_\simeq$. In other words, the conclusion of ρ became a designated value. Hence we have proved ρ is valid in $\mathfrak{B}$. At the same time, the rule r is not valid in $\mathfrak{B}$ and is invalidated by the valuation V, where $x_i \mapsto [x_i]_\sim$. Indeed, for every j,

$$\alpha_1, ..., \alpha_n \vdash_{\lambda, \mathcal{R}} \alpha_j(x_1, ..., x_d), \, \alpha_1, ..., \alpha_n \vdash_{\lambda, \mathcal{R}} \alpha_j(x_1, ..., x_d) \equiv T,$$

that is, under V, all premises of r take the designated value. If the conclusion of r also takes the designated value under V then

$$\alpha_1, ..., \alpha_n \vdash_{\lambda, \mathcal{R}} \beta(x_1, ..., x_d) \equiv T,$$

and applying f) from the definition of algebraic logics again, we infer from this that $\alpha_1, ..., \alpha_n \vdash_{\lambda, \mathcal{R}} \beta(x_1, ..., x_d)$. This contradicts the assumption that r is not derivable from $\mathcal{R}$ in λ. Hence $\mathfrak{B}$ detaches r from $\mathcal{R}$, and we proved that r is not a semantic consequence of $\mathcal{R}$. It is easy to see that $\mathfrak{B}$ satisfies the additional quasi-identities in the formulation of this theorem. ∎

Lemma 1.4.12 *If $\mathcal{R}$ is a collection of admissible inference rules for a logic λ and $\mathcal{R} \vdash_\lambda r$ then r also is admissible for λ.*

Proof. Indeed, being a corollary of $\mathcal{R}$ in λ, r must also be a semantic corollary by Theorem 1.4.11. All rules in $\mathcal{R}$ are admissible in λ, hence by Theorem 1.4.5 all quasi-identities corresponding to rules in $\mathcal{R}$ are valid on $\mathfrak{F}(\lambda)$. This and f) from the definition of algebraic logics entail that all rules from $\mathcal{R}$ are valid in $\mathfrak{F}(\lambda)$, thus r is valid in $\mathfrak{F}(\lambda)$ also. Therefore the quasi-identity $q(r)$ is valid on $\mathfrak{F}(\lambda)$, again using f) and Theorem 1.4.5 we arrive at the conclusion that r is admissible for λ. ∎

We are now in a position to characterize the bases for admissible rules.

Definition 1.4.13 *A collection $\mathcal{G}$ of admissible inference rules for some logic λ is said to be a basis for all admissible rules of λ iff every rule r is admissible for $\lambda \Leftrightarrow r$ is a consequence of $\mathcal{G}$ in λ.*

A quasi-identity q is said to be a *semantic corollary (or consequence)* of a set of quasi-identities Q if for every algebraic system $\mathcal{A}$, $(\forall q_1 \in Q)[\mathcal{A} \models q_1]$ implies $\mathcal{A} \models q$.

Definition 1.4.14 *A subset B of a set of quasi-identities Q closed with respect to semantic corollaries is a basis for Q if the following holds. For any quasi-identity q, $q \in Q$ iff q is a semantic corollary of B.*

For a collection of algebraic systems $\mathcal{K}$, $Th_e(\mathcal{K})$ denotes the set of all identities valid in all systems from $\mathcal{K}$ (the equational theory of $\mathcal{K}$); $Th_q(\mathcal{K})$ is the set of all quasi-identities valid in all systems from $\mathcal{K}$ (the quasi-equational theory of $\mathcal{K}$). If $\mathcal{K}$ consists of a single algebraic system $\mathfrak{A}$ we write $\mathfrak{A}$ instead of $\{\mathfrak{A}\}$ in the above denotations.

Theorem 1.4.15 *Let λ be an algebraic logic and ε be the quasi-identity defined below*

$$[\bigwedge_{\delta \in (x \equiv y)} \delta(x, y) = \top] \Rightarrow (x = y).$$

A collection of inference rules $\mathcal{R}$ forms a basis for admissible rules of λ if and only if

$$\mathcal{R}^* := \{\varepsilon, q(r) \mid (r \in \mathcal{R}) \vee (r \in R_p(\lambda))\} \cup Th_e(Var(\lambda))$$

forms a basis of quasi-identities for $Th_q(\mathfrak{F}_\lambda(\omega))$.

Proof. Let $\mathcal{R}$ be a basis for the admissible inference rules of λ. We intend to show $\mathcal{R}^*$ is a basis for $Th_q(\mathfrak{F}(\lambda))$. Consider an arbitrary quasi-identity $q := (f_1 = g_1, ..., f_m = g_m \Rightarrow f = g)$ which is a semantic corollary of $\mathcal{R}^*$. Quasi-identity ε is valid in $\mathfrak{F}(\lambda)$ by definition of $\mathfrak{F}(\lambda)$ and by f) from the definition of algebraic logics. All other members of $\mathcal{R}^*$ are valid in $\mathfrak{F}(\lambda)$ according to Theorem 1.4.5. Therefore q must be valid in $\mathfrak{F}(\lambda)$ as well, that is $q \in Th_q(\lambda)$. Now suppose $q \in Th_q(\lambda)$, i.e. $\mathfrak{F}(\lambda) \models q$. We must now show that q is a semantic consequence of $\mathcal{R}^*$. By Theorem 1.4.6 the collection of inference rules $r(q)$ consists of rules admissible in λ. By hypothesis $\mathcal{R}$ is a basis for admissible rules, that is $\mathcal{R} \vdash_\lambda r(q)$. Then by Theorem 1.4.11 all rules from $r(q)$ are semantic corollaries of $\mathcal{R}$ in λ. Let $\mathfrak{A}$ be an algebra in the signature of λ. Suppose that all members of $\mathcal{R}^*$ are valid in $\mathfrak{A}$. Because $\mathcal{R}^*$ contains $Th_e(Var(\lambda))$, this yields $\mathfrak{A} \in Var(\lambda)$. Furthermore, all inference rules from $\mathcal{R}$ and all rules postulated for λ are valid in $\mathfrak{A}$ according to the definition of quasi-identities of kind $q(r_1)$. Therefore all rules from $r(q)$ are valid in $\mathfrak{A}$ because we have shown above that all rules from $r(q)$ are semantic consequences of $\mathcal{R}$ in λ. We will show that this implies that q itself is valid in $\mathfrak{A}$. Indeed, we have $q := \alpha_1 = \beta_1 \wedge ... \wedge \alpha_k = \beta_k \Rightarrow \alpha = \beta$,

$$r(q) := \frac{(\alpha_1 \equiv \beta_1), ..., (\alpha_k \equiv \beta_k)}{\alpha \equiv \beta}.$$

Suppose for all i, $\mathfrak{A} \models (\alpha_i(c_j) = \beta_i(c_j))$. Because $\mathfrak{A} \in Var(\lambda)$, $\vdash_\lambda (x \equiv x)$ (since λ is algebraic) and $\mathfrak{A} \models \delta(x, x) = \top$ for all $\delta \in (x \equiv y)$, we have

$$\mathfrak{A} \models \delta((\alpha_i(c_j), \beta_i(c_j)) = \top, \ \forall \delta \in (x \equiv y).$$

Since rules of $r(q)$ are valid in $\mathfrak{A}$, it follows that

$$\mathfrak{A} \models \delta(\alpha(c_j), \beta(c_j)) = \top, \forall \delta \in (x \equiv y).$$

Because $\mathfrak{A} \models \varepsilon$, this entails $\mathfrak{A} \models (\alpha(c_j) = \beta(c_j))$. Hence q is valid in $\mathfrak{A}$, and q is a semantic corollary of $\mathcal{R}^*$, and we proved that $\mathcal{R}^*$ is a basis for the quasi-identities of $\mathfrak{F}(\lambda)$.

Now let $\mathcal{R}^*$ be a basis for $Th_q(\mathfrak{F}(\lambda))$ (Recall ε is valid in $\mathfrak{F}(\lambda)$). Then by Theorem 1.4.5 all the rules in $\mathcal{R}$ are admissible for λ. By Lemma 1.4.12 all consequences of $\mathcal{R}$ in λ will be admissible for λ also. Now we need to show that all rules admissible in λ are consequences of $\mathcal{R}$ in λ. Let r be an inference rule admissible in λ.

By Theorem 1.4.5 $q(r)$ is valid in $\mathfrak{F}(\lambda)$. Therefore $q(r)$ is a semantic corollary of $\mathcal{R}^*$. Suppose that $\mathcal{R} \not\vdash_\lambda r$. Then by Theorem 1.4.11 there is an algebra $\mathfrak{B} \in Var(\lambda)$ such that all rules in $\mathcal{R}$ and all the postulated inference rules of λ are valid in $\mathfrak{B}$, ε is valid in $\mathfrak{B}$, but r is false in $\mathfrak{B}$. Then all quasi-identities from $\mathcal{R}^*$ are valid in $\mathfrak{A}$ as well. Hence $q(r)$ must be valid in $\mathfrak{B}$. At the same time $q(r)$ is false in $\mathfrak{B}$, giving us a contradiction. Hence $\mathcal{R} \vdash_\lambda r$ and $\mathcal{R}$ is a basis for rules admissible in λ. ∎

1.5 Logical Consequence Relations

One of the basic topics in logical investigation is logical derivability in axiomatic systems. As we have seen a logical axiomatic system $\mathcal{AS}$ consists of a collection of (finite) inference rules $\mathcal{AS}_r$ and a collection $\mathcal{AS}_{ax}$ of axioms. Recall that a formula α is called derivable in $\mathcal{AS}$ from the set of formulas Γ (abbreviation $\Gamma \vdash_{AS} \alpha$) if there is a derivation in $\mathcal{AS}$ for α from Γ: i.e. a sequence of formulas terminating in α in which every formula is either a member of Γ, or is a theorem of $\mathcal{AS}$, or is a consequence from formulas placed earlier by some inference rule from $\mathcal{AS}_r$. In particular, $\Gamma \vdash_{AS} \alpha$ iff α is a member of the smallest set of formulas that includes Γ and all axioms of $\mathcal{AS}$, and is closed with respect to applications of all the rules from $\mathcal{AS}_r$. Thus logical derivability relation $\vdash_{AS}$ is a binary relation between sets of formulas and individual formulas. By varying axiomatic systems $\mathcal{AS}$ we obtain various relations of this kind. It looks rather attractive to study general proprieties of such relations and to find a general characterization for all of them. In this section we intend to undertake this task.

Let $\mathcal{L}$ be a language of a propositional logic, and $\mathcal{F}or_{\mathcal{L}}$ be the corresponding set of all formulas over $\mathcal{L}$, and $P_{\mathcal{L}}$ be the set of all propositional letters from $\mathcal{F}or_{\mathcal{L}}$. Given any axiomatic system $\mathcal{AS}$, we define the logical *derivability* relation $\vdash_{AS}$ between sets of formulas and individual formulas as shown above. Clearly this relation has the following properties

$$\forall \alpha (\alpha \in \Gamma \Rightarrow \Gamma \vdash_{AS} \alpha) \ \text{(reflexivity)} \tag{1.3}$$

$$\Gamma \vdash_{AS} \alpha \ \& \ \Gamma \subseteq \Delta \Rightarrow [\Delta \vdash_{AS} \alpha] \ \text{(monotonicity)} \tag{1.4}$$

$$(\Gamma \vdash_{AS} \alpha) \& (\forall \beta \in \Gamma (\Delta \vdash_{AS} \beta)) \Rightarrow [\Delta \vdash_{AS} \alpha] \ \text{(transitivity)} \tag{1.5}$$

$$\Gamma \vdash_{AS} \alpha \Leftrightarrow \exists \Delta (\Delta \subseteq \Gamma)(\Delta \text{ is finite}) \& (\Delta \vdash_{AS} \alpha) \ \text{(finiteness)} \tag{1.6}$$

$$(\forall e : P_{\mathcal{L}} \mapsto \mathcal{F}or_{\mathcal{L}})(\Gamma \vdash_{AS} \alpha) \Rightarrow [e(\Gamma) \vdash_{AS} e(\alpha))] \ \text{(structurallity)} \tag{1.7}$$

Definition 1.5.1 *A binary relation $\vdash$ between the powerset of $\mathcal{F}or_{\mathcal{L}}$ and $\mathcal{F}or_{\mathcal{L}}$ itself is called a* logical consequence relation *on $\mathcal{F}or_{\mathcal{L}}$ if it has all the properties analogous to the properties (1.3) - (1.7). For every $\mathcal{AS}$ the relation $\vdash_{AS}$ is called a* logical derivability relation *for $\mathcal{AS}$.*

Theorem 1.5.2 *(*LOS-SUSZKO(1958)*) [85] A relation $\vdash$ is a logical consequence relation on $\mathcal{F}or_{AS}$ iff there exists an axiomatic system $\mathcal{AS}$ such that $(\vdash) = (\vdash_{AS})$ (that is, iff $\vdash$ is a logical derivability relation).*

Proof. Given $\vdash$ with (1.3) - (1.7), we have to find some axiomatic system $\mathcal{AS}$ such that $(\vdash) = (\vdash_{AS})$. We choose the set of axioms for this system as follows: $\mathcal{AS}_{ax}$ is the set of all formulas α such that $\emptyset \vdash \alpha$. The set $\mathcal{AS}_r$ of inference rules contains all rules Γ/α such that Γ is finite and $\Gamma \vdash \alpha$. Because the relation (1.6) holds for $\vdash$, in order to show $(\vdash) = (\vdash_{\mathcal{AS}})$ it is sufficient to proof

$$\Gamma \vdash \alpha \Leftrightarrow \Gamma \vdash_{\mathcal{AS}} \alpha$$

only for finite Γ. From left to right the condition is evident. Suppose that $\Gamma \vdash_{\mathcal{AS}} \alpha$ and Γ is finite. Consider some derivation $\mathcal{S}$ of α from Γ in $\mathcal{AS}$. We need to prove that for every formula β from $\mathcal{S}$, the following $\Gamma \vdash \beta$ holds. This can be easily shown by induction on the length of the derivation $\mathcal{S}$. If β is an axiom of $\mathcal{AS}$ then $\vdash \beta$ by definition and $\Gamma \vdash \beta$ by (1.4). If $\alpha \in \Gamma$ then $\Gamma \vdash \alpha$ by (1.3). Let β be a result of applying the inference rule

$$\frac{\alpha_1(x_1, ..., x_n), ..., \alpha_m(x_1, ..., x_n)}{\gamma(x_1, ..., x_n)}.$$

This means that there is some tuple of formulas $\delta_1, ..., \delta_n$ such that all formulas from the list $\alpha_1(\delta_1, ..., \delta_n), ..., \alpha_m(\delta_1, ..., \delta_n)$ are predecessors of β in $\mathcal{S}$ and that $\gamma(\delta_1, ..., \delta_n) = \beta$. By the inductive hypothesis we conclude

$$\Gamma \vdash \alpha_1(\delta_1, ..., \delta_n), ..., \Gamma \vdash \alpha_m(\delta_1, ..., \delta_n).$$

By definition, $\alpha_1, ..., \alpha_m \vdash \gamma$ and applying (1.7) we arrive at

$$\alpha_1(\delta_1, ..., \delta_n), ..., \alpha_m(\delta_1, ..., \delta_n) \vdash \gamma(\delta_1), ..., \delta_n)$$

This by (1.5) entails that $\Gamma \vdash \beta$. ∎

An axiomatic system $\mathcal{AS}$ is said to be a *basis* for a logical consequence relation $\vdash$ iff $(\vdash) = (\vdash_{\mathcal{AS}})$. Thus every logical consequence relation has a certain basis according to Theorem 1.5.2. Taking in mind the Los-Suszko theorem (Theorem 1.5.2), in what follows by a derivability relation we will mean a logical consequence relation $\vdash$ between the powerset of $\mathcal{F}or_{\mathcal{L}}$ and $\mathcal{F}or_{\mathcal{L}}$ itself that satisfies all the conditions (1.3)-(1.7) assuming there is no choice of the axiomatic deductive system (if it is not specified otherwise).

We also intend to exhibit a more general approach to logical consequence originating in the Polish and German logical schools. This approach is based on the notion of a logical consequence operator on the powerset of $\mathcal{F}or_{\mathcal{L}}$. Let $\mathcal{C}$ be a mapping from $2^{\mathcal{F}or_{\mathcal{L}}}$ into $2^{\mathcal{F}or_{\mathcal{L}}}$. For a given formula α and a set of formulas Γ, $\alpha \in \mathcal{C}(\Gamma)$ can be understood as α *is a logical consequence of* Γ; thus $\mathcal{C}(\Gamma)$ is the set of all logical consequences of Γ. Rather natural requirements on $\mathcal{C}$ are that $\mathcal{C}$ has to be absorbing, transitive, monotonic, compact, and to preserve substitution (i.e. to be structural):

Definition 1.5.3 *A logical* consequence operator *is a mapping* $\mathcal{C} : 2^{\mathcal{F}or} \mapsto 2^{\mathcal{F}or}$ *having properties:*

$$\Gamma \subseteq \mathcal{C}(\Gamma) = \mathcal{C}(\mathcal{C}(\Gamma)) \tag{1.8}$$

$$\Gamma \subseteq \Delta \Rightarrow \mathcal{C}(\Gamma) \subseteq \mathcal{C}(\Delta) \tag{1.9}$$

$$\mathcal{C}(\Gamma) = \bigcup \{\mathcal{C}(\Delta) \mid \Delta \subseteq \Gamma, \Delta \text{ is finite}\} \tag{1.10}$$

$$(\forall e : P_{\mathcal{L}} \mapsto \mathcal{F}or_{\mathcal{L}}) \;\; e(\mathcal{C}(\Gamma)) \subseteq \mathcal{C}(e(\Gamma)). \tag{1.11}$$

A reasonable question is: does this definition reflects properly all general regiments on a logical consequence operator? It is clear by definition that a logical consequence operator $\mathcal{C}$ is a closure operator in the topological space on the set of formulas $\mathcal{F}or_{\mathcal{L}}$ the topology of which is defined by the sets $\mathcal{C}(\Gamma)$ as a basis for the close sets. It can be easily seen that this topological space is compact, and moreover, according to this definition, every substitution is an continuous isomorphism of the algebra of formulas $\mathcal{F}or_{\mathcal{L}}$ into itself.

We can compare the logical consequence operators introduced above and the logical consequence relations (logical derivability relations). Let $\mathcal{AS}$ be an axiomatic system over the language $\mathcal{L}$. We define the logical consequence operator $Cn_{\mathcal{AS}}$ associated with $\mathcal{AS}$ in the following way:

Definition 1.5.4 *The axiomatic consequence operator $Cn_{\mathcal{AS}}$ is the mapping of the powerset of $\mathcal{F}or_{\mathcal{L}}$ into itself defined as follows*

$$Cn_{\mathcal{AS}}(\Gamma) := \{\alpha \mid \Gamma \vdash_{\mathcal{AS}} \alpha\}.$$

We say, for any logical consequence operator $\mathcal{C}$, $\mathcal{C}$ is axiomatizable consequence operator if there is an axiomatic system $\mathcal{AS}$ such that $\mathcal{C} = Cn_{\mathcal{AS}}$.

We now show that the definition of logical consequence operator was chosen precisely:

Theorem 1.5.5 *A mapping $\mathcal{C} : 2^{\mathcal{F}or_{\mathcal{L}}} \mapsto 2^{\mathcal{F}or_{\mathcal{L}}}$ is a logical consequence operator if and only if $\mathcal{C}$ is an axiomatizable consequence operator.*

Proof. Every axiomatic consequence relation $Cn_{\mathcal{AS}}$ is a logical consequence operator. This can be shown directly using relations (1.3)-(1.7). Indeed, if $\alpha \in \Gamma$ then by (1.3)) $\Gamma \vdash_{\mathcal{AS}} \alpha$ and $\alpha \in Sn_{\mathcal{AS}}(\Gamma)$, that is, the first part of (1.7) holds. If $\alpha \in Cn_{\mathcal{AS}}(\Gamma)$ then $\alpha \in Cn_{\mathcal{AS}}(Cn_{\mathcal{AS}}(\Gamma))$ by the first part of (1.8). If $\alpha \in Cn_{\mathcal{AS}}(Cn_{\mathcal{AS}}(\Gamma))$ then by (1.6) there exists a finite $\Delta \subseteq Cn_{\mathcal{AS}}(\Gamma)$ such that $\Delta \vdash_{\mathcal{AS}} \alpha$. It is clear then that according to (1.5) $\Gamma \vdash_{\mathcal{AS}} \alpha$, that is $\alpha \in Cn_{\mathcal{AS}}(\Gamma)$. Thus the second part of (1.8) holds as well.

If $\Gamma \subseteq \Delta$ and $\alpha \in Cn_{\mathcal{AS}}(\Gamma)$ then $\Gamma \vdash_{\mathcal{AS}} \alpha$ and by (1.4) $\Delta \vdash_{\mathcal{AS}} \alpha$. Then it follows directly by definition that $\alpha \in Cn_{\mathcal{AS}}(\Delta)$. Thus (1.9) holds. The equality (1.10) is the direct consequence of (1.8), (1.9) and (1.6). Now let $\alpha \in Cn_{\mathcal{AS}}(\Gamma)$, that is $\Gamma \vdash_{\mathcal{AS}} \alpha$. Then by (1.7) $e(\Gamma) \vdash_{\mathcal{AS}} e(\alpha)$, that is $e(\alpha) \in Cn_{\mathcal{AS}}(e(\Gamma))$. Thus the relation (1.11) holds as well.

Conversely, let $\mathcal{C}$ be a logical consequence operator. We have to find an axiomatic system $\mathcal{AS}$ such that $\mathcal{C} = Cn_{\mathcal{AS}}$. As before we let $\mathcal{AS}_{ax}$ be the set of all formulas α such that $\alpha \in \mathcal{C}(\emptyset)$ and we take $\mathcal{AS}_r$ to be the set of all Γ/α such that Γ is finite and $\alpha \in \mathcal{C}(\Gamma)$. First we will show that for every set of formulas Δ, the inclusion $Cn_{\mathcal{AS}}(\Delta) \supseteq \mathcal{C}(\Delta)$ is valid. Suppose $\alpha \in \mathcal{C}(\Delta)$. According to (1.10) there is a finite $\Gamma \subseteq \Delta$ such that $\alpha \in \mathcal{C}(\Gamma)$. Then we have $\Gamma \vdash_{\mathcal{AS}} \alpha$ using the inference rule Γ/α. That is, by (1.4) we infer that $\Delta \vdash_{\mathcal{AS}} \alpha$ and $\alpha \in Cn_{\mathcal{AS}}(\Delta)$. Now we will show that for every Δ the opposite inclusion

$$Cn_{\mathcal{AS}}(\Delta) \subseteq \mathcal{C}(\Delta)$$

holds. Suppose that $\alpha \in Cn_{\mathcal{AS}}(\Delta)$, this means $\Delta \vdash_{\mathcal{AS}} \alpha$. By (1.6) there exists a finite subset Γ of Δ such that $\Gamma \vdash_{\mathcal{AS}} \alpha$. And this means that there exists a finite derivation $\mathcal{S}$ of α from Γ in $\mathcal{AS}$. We will show by induction on the position of any formula $\beta \in \mathcal{S}$ that $\beta \in \mathcal{C}(\Gamma)$ holds. Indeed, if β is an axiom, we have $\beta \in \mathcal{C}(\emptyset)$, then $\beta \in \mathcal{C}(\Gamma)$ by (1.4). If $\beta \in \Gamma$ then $\beta \in \mathcal{C}(\Gamma)$ by (1.3). Suppose that β is obtained from formulas proceeding it by an inference rule

$$r := \frac{\alpha_1(x_1, ..., x_n), ..., \alpha_m(x_1, ..., x_n)}{\gamma(x_1, ..., x_n)}.$$

Then there are formulas $\delta_1, ..., \delta_n$ such that all the formulas from the following sequence

$$\alpha_1(\delta_1, ..., \delta_n), \ ... \ , \alpha_m(\delta_1, ..., \delta_n)$$

are predecessors of β in $\mathcal{S}$ and $\gamma(\delta_1, ..., \delta_n) = \beta$. By the inductive hypothesis we have

$$\alpha_1(\delta_1, ..., \delta_n) \in \mathcal{C}(\Gamma), ..., \alpha_m(\delta_1, ..., \delta_n) \in \mathcal{C}(\Gamma) \tag{1.12}$$

By definition, because the rule r is chosen for our axiomatic system, it follows that $\gamma(x_i) \in \mathcal{C}(\alpha_1(x_i), ..., \alpha_m(x_i))$. Then by (1.11)

$$\gamma(\delta_i) \in \mathcal{C}(\alpha_1(\delta_i), ..., \alpha_m(\delta_i)).$$

Therefore from this (1.11) and (1.9), we conclude $\beta \in \mathcal{C}(\Gamma)$. Hence by induction all β are in $\mathcal{C}(\Gamma)$, in particular, α is in $\mathcal{C}(\Gamma)$. Then according to (1.9) $\alpha \in \mathcal{C}(\Delta)$. Thus we have proved $\mathcal{C} = Cn_{\mathcal{AS}}$. ∎

MATRIX SEMANTICS FOR LOGICAL CONSEQUENCE

We have already been supplied an algebraic matrix semantics for propositional logics. In a sense, such semantics is a semantics for a logical derivability relation (logical consequence relation) of kind $\vdash_{\mathcal{AS}}$, that is for derivability from an empty set of hypotheses. Now we are going on to extend this semantics to cover derivability from hypotheses. Let $\mathcal{L}$ be the language of an axiomatic system $\mathcal{AS}$. An $\mathcal{L}$-matrix, as we know, is an algebraic system $\mathcal{A} := \langle A, F, D \rangle$, where A is an nonempty basis set; F a collection of operations on A corresponding to the logical connectives of $\mathcal{L}$; D is a subset of A (which consists of *designated elements* of $\mathcal{A}$). Of course D can be interpreted as unary predicate relation on A. For a mapping v of some set of propositional letters X into A (an *assignment or, synonymously, a valuation* of X on A), we extend v onto the set of all formulas which are built up from X in the obvious way: for every such formula $\alpha(x_i)$ we put $v(\alpha) := \alpha(v(x_i))$. Every $\mathcal{A}$ defines the relation $\models_\mathcal{A}$ between sets of formulas and individual formulas in the following way: for every Γ and α

$$\Gamma \models_\mathcal{A} \alpha \Longleftrightarrow \text{for every assignment } v \; [(v(\Gamma) \subseteq D) \Rightarrow (v(\alpha) \in D)].$$

It is not difficult to see that every $\models_\mathcal{A}$ satisfies (1.3),(1.4), (1.5), and (1.7), but not necessarily (1.6). That is, the property of finiteness could be not satisfied.

Definition 1.5.6 *A $\mathcal{L}$-matrix $\mathcal{A}$ is said to be* matrix model *for the relation* $\vdash_{\mathcal{AS}}$ *if* $\vdash_{\mathcal{AS}} \subseteq \models_\mathcal{A}$.

Definition 1.5.7 *A class K of $\mathcal{L}$-matrixes is a* matrix semantics *for* $\vdash_{\mathcal{AS}}$ *iff for every Γ and α*

$$\Gamma \vdash_{\mathcal{AS}} \alpha \Longleftrightarrow (\forall \mathcal{A} \in K) \; \Gamma \models_\mathcal{A} \alpha.$$

For every class K of matrixes we can define the relation $\models_K$ as follows:

$$\Gamma \models_K \alpha \Leftrightarrow \forall \alpha \in K(\Gamma \models_\mathcal{A} \alpha).$$

Thus K forms a matrix semantics for $\vdash_{\mathcal{AS}}$ if and only if $\models_K = \vdash_{\mathcal{AS}}$. Is it possible to find a matrix semantics for every logical derivability relation $\vdash_{\mathcal{AS}}$ (or, equivalently, for every logical consequence relation)? We are going to answer this question affirmatively. But first we note a general observation due to Bloom(1975). Let K be a matrix semantic for $\vdash_{\mathcal{AS}}$. Suppose that D denotes the unary predicate of designated elements. We introduce the quasi-variety $\mathcal{L}$-algebraic systems K^Q as follows

$$K^Q := \{\mathcal{A} \mid (\mathcal{A} \models \bigwedge_{\beta \in \Gamma} D(\beta) \Rightarrow D(\alpha)), \; \Gamma \text{ is finite}, \; \forall \mathcal{B} \in K(\Gamma \models_\mathcal{B} \alpha)\}.$$

The right part of the definition above contains the collection of quasi-identities defining this quasi-variety; we denote this set of quasi-identities by $Th_q(K)$. We say K^Q is the *quasi-variety generated by K*.

Theorem 1.5.8 *For every matrix semantics K of the relation $\vdash_{\mathcal{AS}}$, the quasi-variety K^Q generated by K forms a matrix semantics for $\vdash_{\mathcal{AS}}$. In particular, every $\vdash_{\mathcal{AS}}$ having a matrix semantics has also some quasi-variety in signature of $\mathcal{L}$ as its matrix semantics*

Proof. Let K be a matrix semantics for $\mathcal{AS}$. Because $K \subseteq K^Q$ it follows that $\models_{K^Q} \subseteq \models_K$. Suppose $\Gamma \not\models_{K^Q} \alpha$. Then for every finite subset Δ of Γ, the relation $\Delta \not\models_{K^Q} \alpha$ holds. By the definition of K^Q this means that $\Delta \not\models_K \alpha$. Because K is the matrix semantics we infer from this $\Delta \not\vdash_{\mathcal{AS}} \alpha$ for every finite subset Δ of Γ. By finiteness (1.6), the last observation gives us $\Gamma \not\vdash_{\mathcal{AS}} \alpha$. As K is a matrix semantics for $\vdash_{\mathcal{AS}}$, from this we infer $\Gamma \not\models_K \alpha$ which is what we need. ∎

According to this theorem, in what follows we can always suppose that a matrix semantics is some quasi-variety. Now we turn to prove the existence of matrix semantics for every $\vdash_{\mathcal{AS}}$.

Definition 1.5.9 *A subset T of $\mathcal{F}or_{\mathcal{L}}$ is a theory for $\vdash_{\mathcal{AS}}$ if for every formula $\alpha \in \mathcal{F}or_{\mathcal{L}}$, $T \vdash_{\mathcal{AS}} \alpha \Rightarrow \alpha \in T$.*

Definition 1.5.10 *For every subset Γ of $\mathcal{F}or_{\mathcal{L}}$, the $\vdash_{\mathcal{AS}}$-theory generated by Γ is the set $Th_{\vdash_{\mathcal{AS}}}(\Gamma) := \{\alpha \mid \alpha \in \mathcal{F}or_{\mathcal{L}}, \Gamma \vdash_{\mathcal{AS}} \alpha\}$.*

It is clear that $T_{\vdash_{\mathcal{AS}}}(\Gamma)$ is the smallest theory for $\vdash_{\mathcal{AS}}$ containing Γ. Let Γ be a subset of $\mathcal{F}or_{\mathcal{L}}$, and F be the set of all logical connectives of the logical language $\mathcal{L}$.

Definition 1.5.11 *The Lindenbaum-Tarski matrix for $\vdash_{\mathcal{AS}}$ over Γ is the matrix*

$$\mathcal{A}_L(\Gamma, \vdash_{\mathcal{AS}}) := \langle \mathcal{F}or_{\mathcal{L}}, F, Th_{\vdash_{\mathcal{AS}}}(\Gamma)\rangle.$$

Theorem 1.5.12 (Matrix Completeness Theorem) *For every logical derivability relation $\vdash_{\mathcal{AS}}$, the class of all Lindenbaum-Tarski matrixes for $\vdash_{\mathcal{AS}}$ over all sets of formulas, i.e.*

$$K_L(\vdash_{\mathcal{AS}}) := \{\mathcal{A}_L(\Gamma, \vdash_{\mathcal{AS}}) \mid \Gamma \subseteq \mathcal{F}or_{\mathcal{AS}}\}$$

forms a matrix semantic for $\vdash_{\mathcal{AS}}$.

Proof. First we will check that every $\mathcal{A}_L(\Gamma, \vdash_{\mathcal{AS}})$ is a matrix model of $\vdash_{\mathcal{AS}}$. Indeed, let $\Delta \vdash_{\mathcal{AS}} \alpha$ and, for some assignment v of variables from $\Delta \cup \alpha$ on $\mathcal{A}_L(\Gamma, \vdash_{\mathcal{AS}})$, $v(\Delta) \subseteq Th_{\vdash_{\mathcal{AS}}}(\Gamma)$ holds. Then $\forall \beta \in \Delta$ ($\Gamma \vdash_{\mathcal{AS}} v(\beta)$). Because $\Delta \vdash_{\mathcal{AS}} \alpha$, using (1.7), it follows $v(\Delta) \vdash_{\mathcal{AS}} v(\alpha)$, and applying (1.6) we infer that $\Gamma \vdash_{\mathcal{AS}} v(\alpha)$. Thus $v(\alpha) \in Th_{\vdash_{\mathcal{AS}}}(\Gamma)$ which is what we needed. Hence $\mathcal{A}_L(\Gamma, \vdash_{\mathcal{AS}})$ is a matrix model for $\vdash_{\mathcal{AS}}$. Conversely, if $\Gamma \not\vdash_{\mathcal{AS}} \alpha$ then $\alpha \notin Th_{\vdash_{\mathcal{AS}}}(\Gamma)$. That is

$$\Gamma \not\models_{\mathcal{A}_L(\Gamma, \vdash_{\mathcal{AS}})} \alpha$$

Hence we have showed $K_L(\vdash_{\mathcal{AS}})$ is the matrix semantics for $\vdash_{\mathcal{A}} \mathcal{S}$. ∎

As a short comment, we note here that the content of this section belongs, in a sense, to the folklore of algebraic logic and goes back to the pioneering work of Tarski, Lindenbaum, Lukasiewicz and Los in the 1930s and 1940s. We presented here only small number of results concerning the theory of logical consequence relations (which however from the basis of this theory). Readers interested to study the theory more deeply can use the papers Bloom [19], Blok and Pigozzi [18], Czelakowski [32], Los and Suszko [85], Rautenberg [124], Suszko [176], Wojcicki [201] and Zygmunt [215]. The fundamental summarizing monograph concerning this subject is the book Wojcicki [203]. Also the reader can find a good motivation for study inference rules and various information on interaction inference rules with problematic of computer science in Fagin, Halpern and Vardi [41].

1.6 Algebraizable Consequence Relations

In Section 5 we investigated the properties of logical consequence relation $\vdash$, and, in particular, we demonstrated that every $\vdash$ has a matrix semantics. The matrix semantics is easy to obtain, but this semantics is not convenient for really dealing with more deep investigating derivability in axiomatic systems. As we have seen, every matrix semantics can be understood as a quasi-variety of algebraic systems, not algebras. The single predicate relation D is concerned with picking out designated elements simulating the *true values.*

More convenient and widespread algebraic semantics for $\vdash$ consist of quasi-varieties of algebras, not algebraic systems (that is the predicate *to be designated* is omitted there by some manipulations). Till up recent time this semantics was developed only for consequence relations $\vdash_{\mathcal{AS}}$ of particular axiomatic systems $\mathcal{AS}$ and there was no general theory. But at the present time, several theories of this kind (cf., for instance, Czelakowski [32]) have been suggested. The most universal and complete approach has been developed by W.Blok and D.Pigozzi (1989). We are going to present this theory below. Hence, the majority of this section is based on Block and Pigozzi [18]. The main idea of this approach consists of expressing derivability using a collection of equations and quasi-equations which are obtained from certain composed terms in the signature of algebras.

Let $\mathcal{L}$ be a propositional language with a collection of logical connectives and certain propositional variables. $\mathcal{F}or_{\mathcal{L}}$ is the set of all formulas of this language, and $Eq_{\mathcal{L}}$ denotes the set of all $\mathcal{L}$-equations, that is all expressions of the form $\alpha = \beta$, where $\alpha, \beta \in \mathcal{F}or_{\mathcal{L}}$. Let K be a class of algebras in the signature Σ consisting of functional symbols corresponding to logical connectives from $\mathcal{L}$. We define the relation $\models_K$ between the powerset of $Eq_{\mathcal{L}}$ and the set $Eq_{\mathcal{L}}$ itself as follows:

$$\Gamma \models_K \alpha = \beta \Longleftrightarrow (\forall \mathcal{A} \in K)(\forall v : Var(\mathcal{L}) \mapsto \mathcal{A})$$

$$[(\forall (\delta = \gamma) \in \Gamma)(\mathcal{A} \models (v(\delta) = v(\gamma)) \Rightarrow (\mathcal{A} \models (v(\alpha) = v(\beta))]$$

If $\Gamma \models_K \alpha = \beta$ then $\alpha = \beta$ is said to be a K-consequence of Γ. The relation $\models$ is called the (semantic) equational consequence relation determined by K. As with the case of the semantic consequence relations determined by matrices, this relation has all the properties analogous to (1.3) - (1.7) in Section 5, except finiteness. If Δ is finite then $\Delta \models_K \alpha = \beta$ if and only if K satisfies the quasi-identity

$$\forall \overline{x} [\bigwedge_{\delta = \gamma \in \Delta} (\delta = \gamma) \Longrightarrow (\alpha = \beta)]$$

Let $\vdash$ be a logical consequence relation and K be a class of algebras in the signature corresponding to the language $\mathcal{L}$ for $\vdash$.

Definition 1.6.1 K is an algebraic semantics for $\vdash \Longleftrightarrow$ there exists a finite system of $\mathcal{L}$-equations $\delta_i = \gamma_i$, $i < n$ depending on a single variable p such that for all $\Gamma \cup \alpha \subseteq \mathcal{F}or_{\mathcal{L}}$ and every $j < n$

(i) $\Gamma \vdash \alpha \Leftrightarrow \{\delta_i(\beta) = \gamma_i(\beta) \mid i < n, \beta \in \Gamma\} \models_K \delta_j(\alpha) = \gamma_j(\alpha)$.

The equations $\delta_i = \gamma_i$ are called defining equations for $\vdash$ and K.

For simplicity we will abbreviate every system of defining equations $\delta_i = \gamma_i$ by $\delta = \gamma$. Say, the relation (i) from Definition 1.6.1 can be written in the following more compact form

(i)$\Gamma \vdash \alpha \Longleftrightarrow \{\delta(\beta) = \gamma(\beta) \mid \beta \in \Gamma\} \models_K \delta(\alpha) = \gamma(\alpha)$.

The right hand side says that the all equations from $\delta = \gamma$ are satisfied. Because every $\vdash$ is finitary (cf. (1.6)), we can also assume that Γ from the definition of the algebraic semantics is always finite. From this observation, we have that $\{\delta(\beta) = \gamma(\beta) \mid \beta \in \Gamma\} \models_K \delta(\alpha) = \gamma(\alpha)$ holds if and only if K satisfies the finite system of quasi-identities

$$\bigwedge_{\beta \in \Gamma} (\delta(\beta) = \gamma(\beta)) \Longrightarrow (\delta(\alpha) = \gamma(\beta))$$

(here we mean the collection of quasi-identities corresponding to every equation in the conclusion $\delta(\alpha) = \gamma(\alpha)$). Therefore (i) from Definition 1.6.1 holds for $\models_K$ iff it holds for $\models_{K^Q}$, where K^Q is the quasi-variety generated by K. Therefore from the definition of the algebraic semantics we immediately derive

Corollary 1.6.2 *Every ⊢ having an algebraic semantics K has a certain algebraic semantics which is a quasi-variety, in particular, the quasi-variety K^Q is an algebraic semantics for ⊢.*

Note that the definition of algebraic semantics can also be interpreted, in some way, through having a special matrix semantics (cf. Definition 1.5.7). Indeed, let K be an algebraic semantics for ⊢ with the system of defining equations $\delta = \gamma$. We define the matrix $\langle A, F, D_A^{\delta=\gamma} \rangle$ by any given $\mathcal{A} \in K$, where $\mathcal{A} = \langle A, F \rangle$, and where

$$D_A^{\delta=\gamma} := \{a \mid a \in A, \mathcal{A} \models \delta(a) = \gamma(a)\}.$$

Theorem 1.6.3 *If K is a quasi-variety of $\mathcal{L}$-algebras and $\delta(p) = \gamma(p)$ is a finite collection of equations from $Eq_{\mathcal{L}}$ with single variable then the following are equivalent*

  (i) K is an algebraic semantics for ⊢ with defining equations $\delta(p) = \gamma(p)$.

  (ii) The class $M := \{\langle A, F, D_A^{\delta=\gamma} \rangle \mid \mathcal{A} = \langle A, F \rangle \in K\}$ forms a matrix semantics for ⊢.

Proof. Directly by our definitions, we conclude that

$$\Gamma \models_M \alpha \Longleftrightarrow \{\delta(\beta) = \gamma(\beta) \mid \beta \in \Gamma\} \models_K \delta(\alpha) = \gamma(\alpha),$$

and our theorem follows immediately. ∎

It is not so easy to give a straightforward answer to whether an algebraic semantics exists for a given logical consequence relation ⊢. We will pay more attention to this question below in this section. Now note that, even if such a kind of semantics exists, it needs not to be unique (cf. [18], page 15). Nevertheless, perhaps the definition of the algebraic semantics given above is, it seems, the strongest one known to present time. In a sense, all canonical logical consequence relations have semantics of this kind. Moreover, very often when some another pure algebraic semantics is constructed, the corresponding logical consequence relation also has an algebraic semantics due to Block-Pigozzi. Of course any algebraic logic has an algebraic semantics due to Blok-Pigozzi (we will show this lately). But now we prefer to specify this semantics more, the matter is we see that the definition of algebraic semantics allows us to express derivability by means of equations and quasi-equations. An especially desirable additional property is the ability to express equality through the defining equations.

Definition 1.6.4 *Let K be an algebraic semantics for some $\vdash$ with defining equations $\delta_i(p) = \gamma_i(p)$, $i < n$ that is*

$$(i) \quad \Gamma \vdash \alpha \Leftrightarrow \{\delta(\beta) = \gamma(\beta) \mid \beta \in \Gamma\} \models_K \delta(\alpha) = \gamma(\alpha).$$

K *is said to be* equivalent *to $\vdash$ iff there exists a finite system $t_j(p,q)$, $i < m$ of terms in $\mathcal{L}$ which are build up of two variables such that, for every equation $\alpha = \beta \in Eq_{\mathcal{L}}$ the following holds (we again abbreviate $t_j(p,q), i < m$ by $t(p,q)$, etc.)*

$$(ii) \quad \alpha = \beta \models_K \delta(t(\alpha,\beta)) = \gamma((t(\alpha,\beta)),$$

$$\delta(t(\alpha,\beta)) = \gamma((t(\alpha,\beta)) \models_K \alpha = \beta.$$

The terms $t_j(p,q)$ $j < m$ satisfying (ii) are called a system of **equivalence formulas** *for $\vdash$ and K.*

As we have seen in Corollary 1.6.2, for every algebraic semantics K of $\vdash$, the quasi-variety K^Q also gives an algebraic semantics for $\vdash$. The condition (ii) above is equivalent to a finite system of quasi-identities, therefore we have

Corollary 1.6.5 *Let $\vdash$ be a logical consequence relation. If a class of algebras K forms an algebraic semantics equivalent to $\vdash$ with some set of defining equations $\delta_i = \gamma_i$ and some set of equivalence formulas $t_j(p,q)$ then the quasi-variety K^Q is a semantics equivalent to $\vdash$ with respect to $\delta_i = \gamma_i$ and $t_j(p,q)$.*

Definition 1.6.6 *A logical consequence relation $\vdash$ is called* algebraizable *if there is an algebraic semantics equivalent to $\vdash$.*

The existence of an equivalent algebraic semantics produces some special requirements on consequence relations. First we give an equivalent form of the definition to having an equivalent algebraic semantics.

Theorem 1.6.7 *Let K be a class of $\mathcal{L}$-algebras, $\delta_i = \gamma_i$ be a finite set of $\mathcal{L}$-equations and $t_j(p,q)$ be a finite collection of $\mathcal{L}$-terms. Then K forms an algebraic semantics equivalent to $\vdash$ with defining equations $\delta = \gamma$ and equivalence formulas $t(p,q)$ iff for every $\Gamma \cup \{\varphi = \psi\} \subseteq E_{\mathcal{L}}$ and each $\eta \in \mathcal{F}or_{\mathcal{L}}$ the following hold:*

$$(i) \quad \Gamma \models_K \varphi = \psi \text{ iff } \{t(\mu,\nu) \mid (\mu = \nu) \in \Gamma\} \vdash t(\varphi,\psi)$$

$$(ii) \quad (\eta \vdash t(\delta(\eta), \gamma(\eta)), \quad t(\delta(\eta), \gamma(\eta)) \vdash \eta.$$

Proof. Suppose that K is an algebraic semantics equivalent to $\vdash$ with defining equations $\delta = \gamma$ and equivalence formulas $t(p,q)$. Then the following are equivalent

$$\{t(\mu,\nu) \mid (\mu = \nu) \in \Gamma\} \vdash t(\varphi,\psi) \text{ (by (i) of Def. 1.6.4)}$$
$$\{\delta(t(\mu,\nu)) = \gamma(t(\mu,\nu))\} \models_K \delta(t(\varphi,\psi)) = \gamma((\varphi,\psi)) \text{ ((ii) of Def. 1.6.4)}$$
$$\{\mu = \nu \mid \mu = \nu \in \Gamma\} \models_K \varphi = \psi$$
$$\Gamma \models_K \varphi = \psi$$

Hence (i) holds. Now let $\eta \in \mathcal{F}or_\mathcal{L}$. Then the following are equivalent

$$\eta \dashv\vdash t(\delta(\eta),\gamma(\eta)) \text{ (by (i) of Def. 1.6.4)}$$
$$\delta(\eta) = \gamma(\eta) =\!\models_K \delta(t(\delta(\eta),\gamma(\eta))) = \gamma(t(\delta(\eta),\gamma(\eta)))$$
$$\text{(by (ii) of Def. 1.6.4)}$$
$$\delta(\eta) = \gamma(\eta) =\!\models_K \delta(\eta) = \gamma(\eta).$$

Thus (ii) also holds. Now we turn to the proving the converse.

$$\{\delta(\psi) = \gamma(\psi) \mid \psi \in \Gamma\} \models_K \delta(\phi) = \gamma(\phi) \Leftrightarrow$$
$$\text{(by (i))} \{t(\delta(\psi),\gamma(\psi)) \mid \psi \in \Gamma\} \vdash t(\delta(\phi),\gamma(\phi)) \Leftrightarrow$$
$$\text{(by (ii) and transitivity of } \vdash) \{\psi \mid \psi \in \Gamma\} \vdash (\phi).$$

Thus (i) from Definition 1.6.4 holds. Furthermore,

$$\phi = \psi =\!\models_K \delta(t(\phi,\psi)) = \gamma(t(\phi,\psi)) \Leftrightarrow \text{(by (i))}$$
$$t(\phi,\psi) \dashv\vdash t(\delta(t(\phi,\psi)),\gamma(t(\phi,\psi))) \Leftrightarrow$$
$$\text{(by (ii))} \, t(\phi,\psi) \dashv\vdash t(\phi,\psi)).$$

Hence (ii) from Definition 1.6.4 also holds. ∎

Now we provide certain additional properties of those logical consequence relations $\vdash$ which have algebraic semantics equivalent to $\vdash$, i.e. are algebraizable.

Lemma 1.6.8 *Let $\vdash$ be a logical consequence relation. Suppose there is an algebraic semantics K equivalent to $\vdash$ with defining equations $\delta_i = \gamma_i$ and equivalence formulas $t_j(p,g)$. Then for all formulas $\alpha,\beta,\theta \in \mathcal{F}or_\mathcal{L}$ the following hold.*

(i) $\vdash t(\alpha,\alpha)$

(ii) $t(\alpha,\beta) \vdash t(\beta,\alpha)$

(iii) $t(\alpha,\beta), t(\beta,\theta) \vdash t(\alpha,\theta)$

(iv) $(\forall p \in Var(\alpha))t(\beta,\theta) \vdash t(\alpha(^p_\beta), \alpha(^p_\theta))$

(v) $\alpha, t(\alpha,\beta) \vdash \beta$

Proof. Indeed, from the relation (i) of Theorem 1.6.7 it immediately follows that (i), (ii), (iii) and (iv) hold. Furthermore, it is evident that

$$\delta(\alpha) = \gamma(\alpha), \alpha = \beta \models_K \delta(\beta) = \gamma(\beta)$$

According to (ii) from the definition of an algebraic semantics equivalent to $\vdash$, $\alpha = \beta$ is equivalent in K to $\delta(t(\alpha, \beta)) = \gamma(t(\alpha, \beta))$. From this observation, using substitution of equivalent components in the relation above, it follows that

$$\delta(\alpha) = \gamma(\alpha), \delta(t(\alpha, \beta)) = \gamma(t(\alpha, \beta)) \models_K \delta(\beta) = \gamma(\beta)$$

Using this and (i) of Definition 1.6.4, we conclude $\alpha, t(\alpha, \beta) \vdash \beta$. ■

Theorem 1.6.9 *(Uniqueness Theorem.) Assume that K and K' are algebraic semantics equivalent to $\vdash$ with sets of defining equations $\delta = \gamma$, $\delta' = \gamma'$ and set of equivalence formulas t, t' respectively. Then the following hold: $K^Q = K'^Q$, $t \dashv\vdash t'$, and $\delta = \gamma \models\dashv_K \delta' = \gamma'$*

Proof. From (iv) of Lemma 1.6.8 taking $\alpha := t^{prime}(\phi, p)$, we immediately obtain

$$t(\phi, \psi) \vdash t(t'(\phi, \phi)), t'(\phi, \psi)))$$

Because by (i) from Lemma 1.6.8 $\vdash t'(\phi, \phi)$, using (v) from Lemma 1.6.8, we conclude $t(\phi, \psi) \vdash t'(\phi, \psi)$. The converse consequence can be shown in a symmetrical way. Hence, $t(\phi, \psi) \dashv\vdash t'(\phi, \psi)$ for any ϕ, ψ. In particular, this holds for $\phi = p, \psi = q$, that is $t \dashv\vdash t'$. Let $\Gamma \cup (\phi = \psi) \subseteq Eq_{\mathcal{L}}$. Then

$\Gamma \models_K \phi = \psi \Leftrightarrow \{t(\mu, \nu) \mid \mu = \nu \in \Gamma\} \vdash t(\phi, \psi)$
(by (i) from Theorem 1.6.7)$\Leftrightarrow$
$\{t'(\mu, \nu) \mid \mu = \nu \in \Gamma\} \vdash' t'(\phi, \psi)$ (because $t \dashv\vdash t'$)$\Leftrightarrow$
$\Gamma \models_{K'} (\phi = \psi)$(again by (i) from Theorem 1.6.7).

Thus $(\models_K) = (\models_{K'})$. Suppose that $(\wedge_{i<n}[\mu_i = \nu_i]) \Rightarrow [\phi = \psi]$ is a quasi-identity valid in K. This means $\{\mu_i = \nu_i) \mid i < n\} \models_K \phi = \psi$. Then

$$\{\mu_i = \nu_i) \mid i < n\} \models_{K'} \phi = \psi.$$

Therefore the mentioned above quasi-identity is also valid in K'. The converse can be shown analogously. Using these facts we immediately obtain $K^Q = K'^Q$. Now we turn to prove that $\delta = \gamma \models\dashv_K \delta' = \gamma'$.

$\delta(p) = \gamma(p) \models\dashv_K \delta'(p) = \gamma'(p) \Leftrightarrow$
$t(\delta(p), \gamma(p)) \dashv\vdash t(\delta'(p), \gamma'(p))$(by (i) of Theorem 1.6.7)$\Leftrightarrow$
$t(\delta(p), \gamma(p)) \dashv\vdash t'(\delta'(p), \gamma'(p))$(because $t \dashv\vdash t'$)$\Leftrightarrow p \dashv\vdash p$
(by (ii) of Theorem 1.6.7).

This completes the proof of our theorem. ∎

Now we are in a position to give a final description of algebraizable logical consequence relations.

Theorem 1.6.10 *(Block,Pigozzi [18]) A logical consequence relation ⊢ is algebraizable iff there exist a finite system of formulas in two variables $t_j(p,q)$ and a finite system $\delta(p) = \gamma(p)$ of equations in a single variable such that the following conditions (i) - (v) hold for all $\phi, \psi, \eta \in \mathcal{F}or_{\mathcal{L}}$:*

(i) $\vdash t(\phi, \phi)$

(ii) $t(\phi, \psi) \vdash t(\phi, \psi)$

(iii) $t(\phi, \psi), t(\psi, \eta) \vdash t(\phi, \eta)$

(iv) for all formulas α, which are formed just by a single applying of some logical connective to variables, and any propositional letter p from α,
$t(\phi, \psi) \vdash t(\alpha(^p_\phi), \alpha(^p_\psi))$

(v) $(\forall \alpha \in \mathcal{F}or_{\mathcal{L}})[\alpha \dashv\vdash t(\delta(\alpha), \gamma(\alpha))]$

In this event t and $\delta = \gamma$ are systems of equivalence formulas and defining formulas for ⊢.

Proof. Let ⊢ be an algebraizable, in this case we have the sets t and $\delta = \gamma$ are defined. Then the conditions (i) - (iv) correspond to conditions (i) - (iv) from Lemma 1.6.8. The condition (v) is a consequence of (ii) from Theorem 1.6.7. Suppose now that all (i) - (v) hold. For every given $\Gamma \subseteq \mathcal{F}or_{\mathcal{L}}$ we define the algebra $\mathcal{A}(\Gamma)$ as follows. We introduce the binary relation $\sim$ on the algebra of formulas $\langle \mathcal{F}or_{\mathcal{L}}, F \rangle$:

$$f \sim g \Leftrightarrow \Gamma \vdash t(f, g).$$

By (i) - (iii) the relation $\sim$ is an equivalence relation, and by (iv) and (iii) ⊢ is structural and transitive, therefore $\sim$ is a congruence relation. Let $\mathcal{A}(\Gamma)$ be the quotient-algebra of $\mathcal{F}or_{\mathcal{L}}$ by this congruence relation.

We intend to show that the class K of all algebras $\mathcal{A}(\Gamma)$ is an algebraic semantics equivalent to ⊢ with the defining equations and equivalence formulas employed in the formulation of this theorem. Suppose $\Gamma \nvdash \alpha$, then according to (v) it follows $\Gamma \nvdash t(\delta(\alpha), \gamma(\alpha))$. We take the algebra $\mathcal{A}(\Gamma)$ and the assignment $p_i \mapsto [p_i]_\sim$ for variables of Γ and α. For every $\beta \in \Gamma$, $\Gamma \vdash \beta$, since ⊢ is reflexive. Therefore by (v) $[\beta \in \Gamma] \Rightarrow [\Gamma \vdash t(\delta(\beta), \gamma(\beta))]$. Hence $\delta(\beta) \sim \gamma(\beta)$ for every $\beta \in \Gamma$ (that is, more precisely, this holds for all pairs δ_i, γ_i). At the same time, $\delta(\alpha) \nsim \gamma(\alpha)$. Thus

$$\{\delta(\beta) = \gamma(\beta) \mid \beta \in \Gamma\} \nvDash_K \alpha.$$

Conversely, suppose $\Gamma \vdash \alpha$ and $\mathcal{A}(\Delta)$ is an algebra from K. Suppose $p_i \mapsto [\mu_i]_\sim$ is an assignment of the variables from Γ and α on this algebra. Assume that for every $\beta \in \Gamma$ the following holds $\delta(\beta(\begin{smallmatrix}p_i\\\mu_i\end{smallmatrix})) \sim \gamma(\beta(\begin{smallmatrix}p_i\\\mu_i\end{smallmatrix}))$. This implies

$$\Delta \vdash t(\delta(\beta(\begin{smallmatrix}p_i\\\mu_i\end{smallmatrix})), \gamma(\beta(\begin{smallmatrix}p_i\\\mu_i\end{smallmatrix}))).$$

Then by (v) we infer $\Delta \vdash \beta(\begin{smallmatrix}p_i\\\mu_i\end{smallmatrix})$ for every β from Γ. According to the structurality and transitivity of $\vdash$, using this observation we conclude $\Delta \vdash \alpha(\begin{smallmatrix}p_i\\\mu_i\end{smallmatrix})$. By (v) again we derive that the relation $\Delta \vdash t(\delta(\alpha(\begin{smallmatrix}p_i\\\mu_i\end{smallmatrix})), \gamma(\alpha(\begin{smallmatrix}p_i\\\mu_i\end{smallmatrix})))$ is valid, which entails

$$\delta(\alpha(\begin{smallmatrix}p_i\\\mu_i\end{smallmatrix})) \sim \gamma(\alpha(\begin{smallmatrix}p_i\\\mu_i\end{smallmatrix})).$$

Thus $\{\delta(\beta) = \gamma(\beta) \mid \beta \in \Gamma\} \models_K \alpha$. Hence condition (i) from the definition of algebraic semantics equivalent to $\vdash$ holds. To prove (ii) from that definition, let $\mathcal{A}(\Gamma)$ be an algebra from K and let $\phi, \psi \in \mathcal{F}or_\mathcal{L}$. Suppose $p_i \mapsto [\alpha_i]_\sim$ is an assignment of variables on this algebra. Then

$$\phi(\begin{smallmatrix}p_i\\\alpha_i\end{smallmatrix}) \sim \psi(\begin{smallmatrix}p_i\\\alpha_i\end{smallmatrix}) \Leftrightarrow \delta(t(\phi(\begin{smallmatrix}p_i\\\alpha_i\end{smallmatrix}), \psi(\begin{smallmatrix}p_i\\\alpha_i\end{smallmatrix}))) \sim \gamma(t(\phi(\begin{smallmatrix}p_i\\\alpha_i\end{smallmatrix}), \psi(\begin{smallmatrix}p_i\\\alpha_i\end{smallmatrix})))$$

which holds according to (v) and definition of $\sim$. Thus

$$\phi = \psi = \|\models_K \delta(t(\phi, \psi)) = \gamma(t(\phi, \psi)),$$

and (ii) holds, which completes the proof of our theorem. ∎

Corollary 1.6.11 *A sufficient condition for $\vdash$ to be algebraizable is that there exists a system $t_j(p, q)$ of formulas in two variables satisfying (i) - (iv) from Theorem 1.6.10 together with the following properties.*

(v) $\phi, t(\psi, \phi) \vdash \psi$ (detachment)

(vi) $\phi, \psi \vdash t(\phi, \psi)$ (G-rule).

In this case $t_j(p, q)$ and $p = t(p, p)$ are the equivalence formulas and defining equations for $\vdash$.

Proof. Let $\delta(p) := p$, $\gamma(p) := t(p, p)$. For every $\alpha \in \mathcal{F}or_\mathcal{L}$, it follows immediately that $t(\delta(\alpha), \gamma(\alpha)) = t(\alpha, t(\alpha, \alpha))$. According to (vi) we have $\alpha, t(\alpha, \alpha) \vdash t(\delta(\alpha), \gamma(\alpha))$. Thus $\alpha \vdash t(\delta(\alpha), \gamma(\alpha))$, because $\vdash t(\alpha, \alpha)$. Conversely, by detachment (v) $t(\delta(\alpha), \gamma(\alpha)) \vdash_S \alpha$, therefore (v) from Theorem 1.6.10 holds. ∎

Now we are in a position easily to show that the consequence relations $\vdash_\lambda$ for algebraic logics λ are algebraizable.

Theorem 1.6.12 *If λ is an algebraic logic then $\vdash_\lambda$ is an algebraizable consequence relation with equivalence formulas $t(x, y) := (x \equiv y)$ and defining equations $x = \top$.*

Proof. In fact, $t(x, y)$ has properties (i) - (iv) from Theorem 1.6.10 by properties (a) - (d) from the definition for algebraic logics. That $x \dashv\vdash_\lambda$ ($x \equiv \top$) follows from property (f) of that definition. Hence by Theorem 1.6.10 $\vdash_\lambda$ is algebraizable. ∎

Thus all normal modal logics, all superintuitionistic logics and all temporal logics generate algebraizable logical consequence relation. The criteria given above are convenient tools for showing that a given derivation system is algebraizable. At the same time, as yet we do not have at our disposal any tool for determining that a system is not algebraizable. The above given descriptions are difficult to work with. A convenient tool for this purpose is so-called Leibniz operator invented by W.Blok and D.Pigozzi [18]. The basic idea goes back to the definition of equality in Leibniz's philosophy, which can be expressed as: x and y are identical just in case every property of x is a property of y and vice versa.

Definition 1.6.13 *Let $\mathcal{L}$ be a language and $\mathcal{A}$ be a $\mathcal{L}$-matrix. An arbitrary n-placed predicate relation P in $\mathcal{A}$ is said to be* definable *over $\mathcal{A}$ if there exists a first-order formula $\Psi(x_1, ..., x_n, y_1, ..., y_m)$ without equality (in the signature of $\mathcal{A}$) and some elements $c_1, ..., c_m$ from A such that:*

$$\forall a_1, ..., a_n \in A[P(a_1, ..., a_n) \Longleftrightarrow \mathcal{A} \models \Psi(a_1, ..., a_n, c_1, ..., c_m)].$$

Definition 1.6.14 *Let A be an algebra and let F be a subset of A. Then $\Omega_A(F) := \{\langle a, b \rangle \mid P(a) \Leftrightarrow P(b)$ for every P definable over $\langle A, F \rangle\}$. The function Ω_A from 2^A into A is called the* Leibniz operator *on A.*

If the meaning of an algebra $\mathcal{A}$ is known we omit subscript $\mathcal{A}$ in the writing of operation Ω_A and write merely Ω. It is possible to specify more the definition of the Leibniz operator and to give another convenient description.

Lemma 1.6.15 *For any algebra A and any $F \subseteq A$,*

$$\Omega(F) := \{\langle a, b \rangle \mid \varphi(a, c_1, ..., c_n) \in F \Leftrightarrow \varphi(b, c_1, ..., c_n) \in F,$$
$$\forall \varphi(y, x_1, ..., x_n) \in \mathcal{F}or_{\mathcal{L}}, \forall c_1, ..., c_n \in A\}$$

Proof. This can be shown by straightforward induction on the length of the formulas which describe the definable predicates from the definition of the Leibniz operator. The basis of induction follows from the conditions assumed in our Lemma. The inductive steps are obvious. ∎

A binary relation $\sim$ on an algebra $\mathcal{A}$ is said to be *compatible* with a subset F of the set $|\mathcal{A}|$ iff for any $a, b \in |\mathcal{A}|$ $[(a \sim b) \Rightarrow (a \in F \Leftrightarrow b \in F)]$.

Theorem 1.6.16 *Let $\mathcal{A}$ be an algebra and let F be a subset of $|\mathcal{A}|$. Then $\Omega_A(F)$ is the largest congruence relation of $\mathcal{A}$ compatible with F.*

Proof. It follows directly from Lemma 1.6.15 that $\Omega_A(F)$ is an equivalence relation compatible with F. With respect to property of being a congruence relation, suppose that $\langle a_1, b_1 \rangle, ..., \langle a_m, b_m \rangle \in \Omega_A(F)$. Let θ be an m-placed operation on $\mathcal{A}$. Then according to Lemma 1.6.15, for any $\phi \in \mathcal{F}or_{\mathcal{L}}$,

$$\phi(\theta(a_1, ..., a_m), c_1, ..., c_n) \in F \Leftrightarrow \phi(\theta(b_1, a_2 ..., a_m), c_1, ..., c_n) \in F.$$

Furthermore, we replace the remaining $a_2, .., a_n$ on $b_2, ..., b_m$ step by step (letter by letter in step). This reasoning gives us

$$\langle \theta(a_1, ..., a_m), \theta(b_1, ..., b_m) \rangle \in \Omega_A(F).$$

Hence it remains only to show that $\Omega_A(F)$ is the largest congruence relation of $\mathcal{A}$ compatible with F. Suppose that μ is a congruence relation compatible with F and $\alpha(x, y_1, ..., y_n)$ is a formula from $\mathcal{F}or_{\mathcal{L}}$. Then, for each $\langle a, b \rangle \in \mu$ and any $c_1, ..., c_n \in A$, $\langle \alpha(a, c_1, .., c_n), \alpha(b, c_1, .., c_n) \rangle \in \mu$. From this and the compatibility of μ with F we infer $\alpha(a, c_1, .., c_n) \in F \Leftrightarrow \alpha(b, c_1, .., c_n) \in F$. Thus $\langle a, b \rangle \in \Omega_A(F)$ and hence $\mu \subseteq \Omega_A(F)$. ∎

Theorem 1.6.17 *Let $\mathcal{A}$ be an algebra and F be a subset of A. Let θ be a binary relation on A which is definable over the matrix $\langle A, F \rangle$ by a first-order formula with parameters and without equality.*

(i) If θ is reflexive, then $\Omega_A(F) \subseteq \theta$.

(ii) If θ is a congruence on A and is compatible with F then $\Omega_A(F) = \theta$.

Proof. Suppose that θ is defined by a first order formula $\Psi(x, y, r_1, ..., r_k)$. Let $c_1, ..., c_k \in A$. Suppose $\langle a, b \rangle \in \Omega_A(F)$. Because θ is reflexive we obtain that $\langle A, F \rangle \models \Psi(b, b, c_1, ..., c_k)$. Using this and the definition of the Leibniz operator, we conclude that $\langle A, F \rangle \models \Psi(a, b, c_1, ..., c_k)$, i.e. $\langle a, b \rangle \in \theta$ and (i) holds. The claim (ii) follows directly from (i) applying Theorem 1.6.16. ∎

Recall that for any class K of algebras, a congruence relation θ on any algebra $\mathcal{A}$ is called a K-congruence iff $\mathcal{A}/_\theta \in K$. For any logical consequence relation $\vdash$ and any algebra $\mathcal{A}$ in the signature of $\vdash$, a subset F of $|\mathcal{A}|$ is called an $\vdash$-filter if the following holds: $\Gamma \vdash \alpha$ implies that for any valuation h of variables from $\Gamma \cup \alpha$ on the algebra $\mathcal{A}$, the truth of $(\forall \beta \in \Gamma)(h(\beta) \in F))$ implies $h(\alpha) \in F$.

Theorem 1.6.18 *(Blok, Pigozzi [18]) Let $\vdash$ be an algebraizable logical consequence relation. Let K be the quasi-variety forming an algebraic semantics equivalent to $\vdash$. Then for any algebra $\mathcal{A}$ in the signature corresponding to $\vdash$, the Leibniz operator Ω_A is an isomorphism between the lattices of all $\vdash$-filters on $\mathcal{A}$ and K-congruences of $\mathcal{A}$.*

Proof. First we need the following

Lemma 1.6.19 *Let* $\vdash$ *be an algebraizable consequence relation in a language* $\mathcal{L}$ *and let* $t(x, y)$ *be a system of equivalence formulas. Then, for every* $\mathcal{L}$-*algebra* $\mathcal{A}$ *and every* $\vdash$-*filter* F *on* $\mathcal{A}$, $\Omega_A(F) = \{\langle a, b \rangle \mid t(a, b) \in F\}$.

Proof. Suppose $\Theta := \{\langle a, b \rangle \mid t(a, b) \in F\}$. We need to show that Θ is a congruence relation. Indeed, by Theorem 1.6.10 for the system $t(x, y)$, the following holds: $\vdash t(x, x), t(x, y) \vdash t(y, x), \{t(x, y), t(y, z)\} \vdash t(x, z)$. Therefore, because F is a $\vdash$-filter, it follows that Θ is an equivalence relation. According to (iv) from Theorem 1.6.10 (the property of substitution of t-equivalent formulas) and the transitivity of Θ we obtain that Θ is a congruence relation. Moreover, Θ is compatible with F. Indeed, $x, t(x, y) \vdash y$ holds because of (v) from Lemma 1.6.8 and because F is a $\vdash$-filter. Therefore if $a \in F$ and $t(a, b) \in F$ then $b \in F$. By the definition of Θ, this relation is definable by first order formulas (even without parameters). Hence according to (ii) from Theorem 1.6.17 $\Omega_A(F) = \Theta$. ∎

Let $\mathcal{A}$ be any $\mathcal{L}$-algebra, and let F be a subset of $|\mathcal{A}|$ which is $\vdash$-filter. First we show that $\Omega_A(F)$ is a K-congruence. To prove this we have to show that any quasi-identity that is valid in K must be valid in the quotient algebra $A/_{\Omega_A(F)}$. Suppose $E \Rightarrow (\varphi = \psi)$ is a valid in K quasi-identity, that is $E \models_K \varphi = \psi$, where E is a finite system of equalities. Suppose that

$$\langle \rho(\overline{a}) = \mu(\overline{a}) \rangle \in \Omega_A(F)$$

for all $\rho = \mu \in E$. By Lemma 1.6.19 then

$$t(\rho(\overline{a}), \mu(\overline{a})) \in F \tag{1.13}$$

for every $\rho = \mu \in E$. According to Theorem 1.6.7 $E \models_K \varphi = \psi$ is equivalent to $\{t(\rho, \mu) \mid \rho = \mu \in E\} \vdash t(\varphi, \psi)$. Using the last relation, (1.13) and that F is a $\vdash$-filter we conclude $t(\varphi(\overline{a}), \psi(\overline{a})) \in F$. Using Lemma 1.6.19 we derive $\langle \varphi(\overline{a}), \psi(\overline{a}) \rangle \in \Omega_A(F)$. Hence any quasi-identity which is valid in K is valid in $A/_{\Omega_A(F)}$ as well, and consequently $\Omega_A(F)$ is a K-congruence.

Suppose Θ is a K-congruence on $\mathcal{A}$. Let $H_{A,\Theta}$ be the subset of F defined as follows:

$$H_{A,\Theta} := \{a \mid a \in |\mathcal{A}|, \langle \delta(a), \gamma(a) \rangle \in \Theta\}.$$

That $H_{A,\Theta}$ is a $\vdash$-filter follows from the fact that Θ is a K-congruence and from (i) of the Definition 1.6.4 to algebraizable consequence relations:

$$\Gamma \vdash \alpha \Leftrightarrow \{\delta(\beta) = \gamma(\beta) \mid \beta \in \Gamma\} \models_K \delta(\alpha) = \gamma(\alpha).$$

Now we turn to showing $\Omega_A(H_{A,\Theta}) = \Theta$. By our definitions and Lemma 1.6.19 it follows

$$\langle a, b \rangle \in \Omega_A(H_{A,\Theta}) \Leftrightarrow \langle \delta(t(a, b)), \gamma(t(a, b)) \rangle \in \Theta.$$

Using (ii) from Definition 1.6.4, we infer $\langle \delta(t(a,b)), \gamma(t(a,b)) \rangle \in \Theta$ iff $\langle a, b \rangle \in \Theta$. Hence $\Omega_A(H_{A,\Theta}) = \Theta$. Thus Ω_A maps the set of all $\vdash$-filters of $\mathcal{A}$ onto the set of all K-congruences of $\mathcal{A}$.

According to (ii) of Theorem 1.6.7, $x \dashv\vdash t(\delta(x), \gamma(x))$ holds. Therefore for any $\vdash$-filter F on $|\mathcal{A}|$, $a \in F$ iff $t(\delta(a), \gamma(a)) \in F$. Furthermore,

$$t(\delta(a), \gamma(a)) \in F \Longleftrightarrow \langle \delta(a), \gamma(a) \rangle \in \Omega_A(F)$$

by Lemma 1.6.19. Thus $a \in F$ iff $\langle \delta(a), \gamma(a) \rangle \in \Omega_A(F)$. Hence Ω_A is a one-to-one mapping on the set of all $\vdash$-filters which preserves the order, i.e. Ω_A is a one-to-one correspondence between the set of $\vdash$-filters and the set of K-congruences. The mapping Ω_A evidently preserves the order on the set of all $\vdash$-filters (i.e. $F_1 \subseteq F_2 \Rightarrow \Omega_A(F_1) \subseteq \Omega_A(F_2)$). Hence Ω_A is a lattice isomorphism. ∎

Using this result we are able to show that certain logical consequence relations are not algebraizable. Showing that certain consequence relations are not algebraizable, it is very convenient that we can make evaluations in arbitrary algebras in the given signature. For instance, we can easily describe some non-algebraizable axiomatic systems in modal logic. Recall that Weisberg has offered a modal axiomatic system $S5^W$ which has the modal logic $S5$ as the set of its theorems (see J.Porte [116]). This system has axioms:

(1)$\Box \psi$ (for all classical tautologies ψ),
(2)$\Box(\Box p \rightarrow p)$,
(3)$\Box(\Diamond p \rightarrow \Box \Diamond p)$,
(4)$\Box(\Box(p \rightarrow q) \rightarrow \Box(\Box p \rightarrow q))$,

and $S5^W$ has the following inference rule: $p, \Box(p \rightarrow q)/q$. Another related axiomatic system is Carnap's modal axiomatic system $S5^C$. Axioms of this system are

(1)$\Box \psi$ (for all classical tautologies ψ),
(2)$\Box(\Box(p \rightarrow q) \rightarrow (\Box p \rightarrow \Box q))$,
(3)$\Box(\Box p \rightarrow p)$,
(4)$\Box(\Diamond p \rightarrow \Box \Diamond p)$,

the single inference rule of this system is $p, p \rightarrow q/q$. J.Porte [116] showed that the set of theorems for $S5^C$ also coincides with modal logic $S5$.

Proposition 1.6.20 *[18] Deductive systems $S5^W$ and $S5^C$ are not algebraizable.*

Proof. Let A be an 4-element boolean algebra, that is the basis set of $\mathcal{A}$ is $\{\bot, a, b, \top\}$. We take $\Box$ as a unary operation on A which maps all elements except $\top$ into $\bot$ and $\Box \top = \top$. Thus A is a modal algebra. Moreover it is easy to see A is an $S5$-algebra, that is, all theorems of $S5^C$ (as well as $S5^W$ and $S5$) take the value $\top$ with respect to every valuation of their variables. This follows from the fact that all the axioms of these systems are such, and all their inference rules preserve having the value $\top$. Let

$$F_a := \{x \mid x \in A, a \leq x\}, \quad F_b := \{x \mid x \in A, b \leq x\}.$$

It is not hard to see that F_a and F_b are $\vdash_{S5W}$ and $\vdash_{S5C}$ filters. In order to show this it is sufficient to verify that all the axioms of the corresponding systems are always evaluated to $\top$ and F_a and F_b are closed with respect to the inference rules of $S5^W$ and $S5^C$. At the same time A is a simple algebra, i.e. it has no congruence relations other than the identity relation $\{\langle x, x \rangle \mid x \in A\}$ and the universal relation $A \times A$. Therefore the images of F_a and F_b under Ω_A must coincide or one of them is a subset of the other. Therefore $S5^W$ and $S5^C$ cannot be algebraizable by Theorem 1.6.18 ∎

An axiomatic system $\mathcal{S}$ in the language of propositional modal logic is called a *quasi-normal (semi-normal)* modal system if (i) the postulated inference rules of $\mathcal{S}$ are $x, x \rightarrow y/y$ ($x, \square(x \rightarrow y/y$, respectively) but only these, and (ii) the set of all theorems for $\mathcal{S}$ contains all the theorems of the minimal normal propositional modal logic K. Thus $S5^C$, for instance, is quasi-normal, while $S5^W$ is semi-normal. The modal system K' obtained from the modal axiomatic system K by omitting the rule $x/\square x$ is also quasi-normal.

Corollary 1.6.21 *Every quasi-normal subsystem of $S5^C$ and every semi-normal subsystem of $S5^W$ fail to be algebraizable. In particular, K' is not algebraizable.*

To end this section we present the second characterization (which was given by Blok, Pigozzi [18]) of algebraizable consequence relations $\vdash$ employing only properties of the Leibniz operator Ω. For this we consider the operation of the Leibniz operator Ω on the algebra of all formulas $\mathcal{F}or_{\mathcal{L}}$. Consider an equational theory T_e in the language of $\vdash$ (recall that any equational theory T_e is a set of identities closed with respect to the relation $\models_K$ for some class of algebras K). It is easy to see that the set $c(T_e) := \{\langle \varphi, \psi \rangle \mid \varphi = \psi \in T_e\}$ is a congruence on $\mathcal{F}or_{\mathcal{L}}$. Conversely, any congruence relation θ on $\mathcal{F}or_{\mathcal{L}}$ is equal to $c(T_e)$ for some T_e (it is sufficient to take the set of all identities valid on the quotient-algebra $\mathcal{F}or_{\mathcal{L}}/\theta$ as T_e). Thus there is a one-to-one correspondence between the lattice of all equational theories Th_e and the lattice of congruence relations on $\mathcal{F}or_{\mathcal{L}}$.

We know, for any $T \subseteq \mathcal{F}or_{\mathcal{L}}$, ΩT is a congruence relation on $\mathcal{F}or_{\mathcal{L}}$ (see Theorem 1.6.16). Thus ΩT is associated with a unique equational theory. Therefore for any logical consequence relation $\vdash$, there is the mapping of the lattice of all $\vdash$-theories $Th(\vdash)$ into a sublattice of the lattice of all equational theories. We fix this mapping. Hence we do not bother to distinguish between the Leibniz relation ΩT and the unique equational theory T_e such that $c(T_e) = \Omega T$. Note that a directed (up) subset D of a lattice $\mathcal{L}$ is a subset D such that, for any finite subset S of D, the union of S belongs to D. Now we are in a position to state without proof the following strong theorem

Theorem 1.6.22 *(Blok, Pigozzi [18], Theorem 4.2) A logical consequence relation $\vdash$ is algebraizable iff the Leibniz operator satisfies the following two conditions on $\mathcal{F}or_{\mathcal{L}}$:*

(i) *Ω is injective (i.e. is a one-to-one mapping in) and order preserving on the lattice $Th(\vdash)$ of all $\vdash$-theories.*

(ii) *Ω preserves unions of directed subsets of $Th(\vdash)$.*

In applying this theorem to show a deductive system $\mathcal{S}$ is algebraizable it is sufficient to verify that Ω is injective and preserves unions since the latter condition implies Ω is order preserving.

1.7 Admissibility for Consequence Relations

We already have some general information concerning admissible inference rules for algebraic logics, their algebraic descriptions and their general properties (see Section 4 of this chapter). With respect to logical consequence relations of an arbitrary nature we know that any consequence relation is generated by a collection of finite inference rules (Los-Suszko Theorem, Theorem 1.5.2), therefore derivation rules play a basic role for arbitrary consequence relations also. We know that the logical consequence relation $\vdash_\lambda$ of any algebraic logic λ is algebraizable (Theorem 1.6.12). A reasonable question is whether is it possible to extend the notion of admissibility to logical consequence relations, and if so which properties the rules admissible for logical consequence relations have? This section is devoted to consideration this question.

Definition 1.7.1 *Let Γ be a set of formulas in the language $\mathcal{L}$ of a consequence relation $\vdash$, and $\Gamma \subseteq \mathcal{F}or_{\mathcal{L}}$. The $\vdash$-theory generated by Γ is the set $\{\alpha \mid \Gamma \vdash \alpha\}$. We denote this theory by $Th_\vdash(\Gamma)$.*

The generalization of admissibility inference rules to the case of consequence relations can be maid very naturally in the following way. The admissibility of a rule r for an algebraic logic λ means that λ is closed with respect to r, but $\lambda := Th_{\vdash_\lambda}(\emptyset)$. Therefore we can just immediately transfer this property to consequence relations.

Definition 1.7.2 *Let $\vdash$ be a logical consequence relation and $r := \Gamma/\alpha$ be an inference rule. We say that r is* admissible *for $\vdash$ iff $Th_\vdash(\emptyset)$ is closed with respect to r, and the set of all admissible inference rules for $\vdash$ is denoted by $Ad(\vdash)$. A rule r is said to be* derivable *in $\vdash$ if $\Gamma \vdash \alpha$.*

Let $\vdash$ be a consequence relation. By the Los-Suszko theorem (Theorem 1.5.2) there is always an axiomatic system $\mathcal{AS}$ such that $\vdash = \vdash_{\mathcal{AS}}$.

Definition 1.7.3 *If $\mathcal{AS}$ is an axiomatic system then the admissible closure of $\mathcal{AS}$ is the system $\mathcal{AS}^{Ad}$, obtained from $\mathcal{AS}$ by adding of all inference rules admissible for $\mathcal{AS}$. Let $\mathcal{AS}$ be a certain axiomatic system that $(\vdash_{\mathcal{AS}}) = (\vdash)$ for a given consequence relation $\vdash$. The* admissible closure *of $\vdash$ is the consequence relation $\vdash^{Ad} := \vdash_{\mathcal{AS}^{Ad}}$.*

It is clear that the notion of admissible closure is well-defined, that is, the definition of the admissible closure for $\vdash$ does not depend on the choice of axiomatic system $\mathcal{AS}$. Now we note some simple properties of the notions introduced above in the following two theorems which follow directly from our definitions.

Theorem 1.7.4 *Let r be an inference rule and $\mathcal{AS}$ be an axiomatic system. Then the following properties hold.*

(i) *r is admissible for $\mathcal{AS}$ iff r is admissible for $\vdash_{\mathcal{AS}}$;*
(ii) *r is admissible for $\mathcal{AS}$ iff $Th_{\vdash_{\mathcal{AS}}}(\emptyset) = Th_{\vdash_{\mathcal{AS} \cup \{r\}}})(\emptyset)$;*
(iii) *$Th_{\vdash}(\emptyset) = Th_{\vdash^{Ad}}(\emptyset)$.*

Theorem 1.7.5 *The following hold*

(i) *For any $\vdash$, $\vdash \subseteq \vdash^{Ad}$,*

(ii) *If $\vdash_1$ and $\vdash_2$ are certain consequence relations and $\vdash_1 \subseteq \vdash_2$ then $\vdash_1^{Ad} \subseteq \vdash_2^{Ad}$. That is the operation of taking the admissible closure of logical consequence relations is monotonic.*

Lemma 1.7.6 *Let $\vdash$ be a consequence relation, and $\Gamma \cup \{\alpha\}$ be a finite set of formulas. Then $\Gamma \vdash^{Ad} \alpha$ iff Γ/α is admissible for $\vdash$.*

Proof. In fact, the proof of the right to left direction is evident. To proof the opposite direction, let $\Gamma \vdash^{Ad} \alpha$. Then there is a derivation $\mathcal{S}$ of α from Γ by means of the theorems of $\vdash$ and inference rules admissible for $\vdash$. If V is a valuation which turns all formulas from Γ into theorems from $Th_{\vdash}(\emptyset)$ then we apply this valuation to $\mathcal{S}$. The resulting $V(\mathcal{S})$ will be again a derivation from Γ^V by means of inference rules admissible for $\vdash$. Because $V(\Gamma)$ consists of theorems of $\vdash$, the conclusion $V(\alpha)$ will be a theorem of $\vdash$. Thus Γ/α is admissible for $\vdash$. ∎

Definition 1.7.7 *A logical consequence relation $\vdash$ is* structurally complete *if every inference rule admissible for $\vdash$ is derivable for $\vdash$.*

Theorem 1.7.8 *For any $\vdash$, $\vdash^{Ad}$ is structurally complete. Any given $\vdash$ is structurally complete iff it coincides with its admissible closure $\vdash^{Ad}$.*

Proof. The relation $\vdash^{Ad}$ is structurally complete. Indeed, if r is an admissible rule for $\vdash^{Ad}$ then $Th_{\vdash_{Ad}}(\emptyset)$ is closed with respect to r. But this set coincides with $Th_{\vdash}(\emptyset)$ (by (iii) from Theorem 1.7.4). Hence r is admissible for $\vdash$, therefore r is derivable in $\vdash^{Ad}$. Conversely, if $\vdash$ is structurally complete then $(\vdash) = (\vdash^{Ad})$ by definition. ∎

Thus taking the admissible closure of a given logical consequence relation $\vdash$ can be considered as taking the structural completion of $\vdash$. Indeed,

Theorem 1.7.9 *Relation* $\vdash^{Ad}$ *is the minimal structurally complete consequence relation containing* $\vdash$.

Proof. Relation $\vdash^{Ad}$ is structurally complete by Theorem 1.7.8. Let $\vdash_1$ be a structurally complete consequence relation and $\vdash\subseteq\vdash_1\subseteq\vdash^{Ad}$. Then any admissible for $\vdash$ rule r is also admissible for $\vdash_1$ (otherwise r would be not admissible in $\vdash$ by $Th_{\vdash}(\emptyset) = Th_{\vdash_{Ad}}(\emptyset)$ (see (iii) of Theorem 1.7.4)). Since $\vdash_1$ is structurally complete this rule r is derivable in $\vdash_1$, consequently $\vdash^{Ad}\subseteq\vdash_1$. ∎

Now having these simple initial observations concerning admissible rules of consequence relations, we are in a position to start a deeper study of the subject. We know that there is a complete algebraic description for admissible inference rules of every algebraic logic λ (Theorem 1.4.5), as well as that there exists a matrix semantics for every consequence relation $\vdash$ (Theorem 1.5.12), and, that, finally, there is a complete description of $\vdash$ which have certain algebraizable semantics (Theorem 1.6.10). Whether it is possible to develop also an algebraic semantics for admissible rules of consequence relations? Answering this question, first we note that, being a partial case of consequence relations, every admissible closure of a logical consequence relation has a matrix semantics. But it is possible to give a simpler matrix semantics for the admissible closure of consequence relations consisting of a quasi-variety generated by a single matrix.

Theorem 1.7.10 Matrix Completeness Theorem for Admissibility *For every logical consequence relation* $\vdash$ *the quasi-variety Q generated by the single Lindenbaum-Tarski matrix $\mathcal{A}_L(\emptyset,\vdash)$ for $\vdash$ over $\emptyset$ forms a matrix semantics for the admissible closure $\vdash^{Ad}$. The single algebra $\mathcal{A}_L(\emptyset,\vdash)$ forms a matrix semantics for the set of finite sequents $\{\langle\Gamma,\alpha\rangle \mid \Gamma\vdash^{Ad}\alpha, ||\Gamma|| < \omega\}$.*

Proof. By definition $\mathcal{A}_L(\emptyset,\vdash) := \langle\mathcal{F}or_{\mathcal{A}}, F, Th_{\vdash}(\emptyset)\rangle$. Therefore for every Γ and α such that $||\Gamma|| < \omega$ and $\Gamma\vdash^{Ad}\alpha$, $\forall\mathcal{A}\in Q$ $\Gamma\models_{\mathcal{A}}\alpha$ holds. Assume that Δ is an infinite set of formulas and $\Delta\vdash^{Ad}\alpha$. Then by definition of $\vdash^{Ad}$ there exists a finite subset Γ of Δ such that $\Gamma\vdash^{Ad}\alpha$. Then as we have showed above the last sequent is valid on all matrixes of Q. This implies $\forall\mathcal{A}\in Q$ $\Delta\models_{\mathcal{A}}\alpha$, thus $\Gamma\vdash^{Ad}\alpha$ entails $(\forall\mathcal{A}\in Q)\models_{\mathcal{A}}\alpha$.

Conversely, first suppose that, for a certain Γ and α, the following two properties $||\Gamma|| < \omega$ and $\Gamma\nvdash^{Ad}\alpha$ hold. Then the sequent $\langle\Gamma,\alpha\rangle$ is disprovable in the matrix $\langle\mathcal{F}or_{\mathcal{A}}, F, Th_{\vdash}(\emptyset)\rangle$ itself.

Now suppose that $\Gamma \nvdash^{Ad} \alpha$ and that Γ is infinite. This, in particular, means that for every finite subset Δ of Γ, $\Delta \nvdash^{Ad} \alpha$ holds. As we have seen above this implies every finite sequent $\langle \Delta, \alpha \rangle$ is disprovable on $\langle \mathcal{F}or_A, F, Th_\vdash(\emptyset) \rangle$. Therefore the set of first-order formulas

$$Th_q(Q) \cup \bigcup_{\Delta \subset \Gamma, |\Delta| < \omega} \{ \bigwedge_{\beta \in \Delta} (D(\beta) \wedge (\neg D(\alpha)))(^{p_i}_{a_i}) \},$$

(where D is the predicate distinguishing the designated elements, and all a_i are new constant's letters replacing all the variables from the corresponding formulas) is locally consistent. By the compactness theorem of first-order logic, this set is consistent, that is there is a model $\mathcal{G}$ for this set. Then $\mathcal{G} \in Q$ and it is easy to see that $\Gamma \nvDash_{\mathcal{G}} \alpha$. Hence $\Gamma \vdash^{Ad} \alpha \Leftrightarrow (\forall \mathcal{A} \in Q) \vDash_{\mathcal{A}} \alpha$. The last part of the theorem is just a corollary of our proof above. ∎

Now let us clarify what can be said about algebraic semantics for admissible closure of logical consequence relations, how can we characterize algebraically admissible rules for algebraizable consequence relations.

Theorem 1.7.11 Algebraic Completeness Theorem *Suppose $\vdash$ is a consequence relation which has an algebraic semantics K with defining equations $\delta = \gamma$. Then*

(i) *The quasi-variety $\mathfrak{F}_\lambda(\omega)^Q$ which is generated by the free algebra of countable rank from K^Q forms an algebraic semantics for $\vdash^{Ad}$ which has defining equations identical with K*

(ii) *An inference rule Γ/α is admissible for $\vdash$ iff*

$$\mathfrak{F}_\lambda(\omega) \vDash [\{\delta(\beta) = \gamma(\beta) \mid \beta \in \Gamma\} \Rightarrow \delta(\beta) = \gamma(\beta)].$$

(iii) *If $\vdash$ is algebraizable (that is, if there is an algebraic semantics K equivalent to $\vdash$) then so is $\vdash^{Ad}$. In this event $\mathfrak{F}_\lambda(\omega)^Q$ forms an algebraic semantics equivalent to $\vdash^{Ad}$ with defining equations and equivalence formulas identical to those of K.*

Proof. Suppose that K is an algebraic semantics for $\vdash$. Then K^Q is an algebraic semantics for $\vdash$ as well. Every quasi-variety must have free algebras of arbitrary rank (Theorem 1.2.35). Let $\mathfrak{F}_\lambda(\omega)$ be the free algebra of countable rank from K^Q. Let $\delta_i = \gamma_i, i < n$ be the defining equations for $\vdash$ in K. We fix this set as a candidate for the defining equations for $\vdash^{Ad}$ in the class consisting of the single algebra $\mathfrak{F}_\lambda(\omega)$. Suppose a set of formulas Γ and a formula α are such that $|\Gamma| < \omega$ and $\Gamma \vdash^{Ad} \alpha$. Suppose that

$$\{\delta(\beta) = \gamma(\beta) \mid \beta \in \Gamma\} \nvDash_{\mathcal{A}} \delta(\beta) = \gamma(\beta)$$

for some algebra $\mathcal{A} \in \mathfrak{F}_\lambda(\omega)^Q$. Because the relation above means that the corresponding quasi-identity is false on $\mathcal{A}$ this quasi-identity is false on $\mathfrak{F}_\lambda(\omega)$ itself as well. That is we can assume that $\mathcal{A} = \mathfrak{F}_\lambda(\omega)$. Then there are formulas μ_j such that for all $\beta \in \Gamma$,

$$\mathfrak{F}_\lambda(\omega) \models (\delta(\beta))(\mu_j) = (\gamma(\beta))(\mu_j), \ \mathfrak{F}_\lambda(\omega) \models (\delta(\alpha))(\mu_j) \neq (\gamma(\alpha))(\mu_j).$$

Because the algebra from the left hand side of the relations above is free, and other countable algebras of the quasi-variety $\mathfrak{F}_\lambda(\omega)^Q$ are homomorphic images of this free algebra, we conclude

$$\forall \mathcal{A} \in K^Q(\mathcal{A} \models (\delta(\beta)(\mu_j)) = (\gamma(\beta))(\mu_j).$$

The quasivariety K^Q is the algebraic semantics for $\vdash$. Therefore all formulas $\beta(\mu_j)$ are theorems of $\vdash$, that is $\beta(\mu_j) \in Th_\vdash(\emptyset)$. Simultaneously we have by the same reasoning, $\alpha(\mu_j) \notin Th_\vdash(\emptyset)$. Hence the rule Γ/α is not an admissible rule, therefore $\Gamma \nvdash^{Ad} \alpha$ (see Theorem 1.7.4) giving a contradiction. That is

$$\{\delta(\beta) = \gamma(\beta) \mid \beta \in \Gamma\} \models_{\mathfrak{F}_\lambda(\omega)^Q} \delta(\beta) = \gamma(\beta). \tag{1.14}$$

If a set of formulas Ψ is infinite and a formula α is such that $\Psi \vdash^{Ad} \alpha$ then for some finite subset Δ of Ψ we have $\Delta \vdash^{Ad} \alpha$. Then (1.14) holds for Δ and therefore this relation is valid for Ψ as well.

Conversely, assume that for some Γ and α, the relation

$$\{\delta(\beta) = \gamma(\beta) \mid \beta \in \Gamma\} \models_{(\mathfrak{F}_\lambda(\omega))^Q} \delta(\beta) = \gamma(\beta) \tag{1.15}$$

holds. Suppose that $\Gamma \nvdash^{Ad} \alpha$. This means that for every finite $\Delta \subset \Gamma, \Delta \nvdash^{Ad} \alpha$ holds. Applying Theorem 1.7.4 we obtain that the rule Δ/α is not admissible for $\vdash$. Therefore there are formulas μ_j such that for every $\beta \in \Delta$, the relation $\beta(\mu_j) \in Th_\vdash(\emptyset)$ is true but $\alpha(\mu_j) \notin Th_\vdash(\emptyset)$. Then, in K, and consequently in $\mathfrak{F}_\lambda(\omega)$, for all β and α

$$\delta(\beta(\mu_j)) = \gamma(\beta(\mu_j)), \ \delta(\alpha(\mu_j)) \neq \gamma(\alpha(\mu_j)).$$

Hence, if Γ is finite then

$$\{\delta(\beta) = \gamma(\beta) \mid \beta \in \Gamma\} \nvDash_{\mathfrak{F}_\lambda(\omega)^Q} \delta(\beta) = \gamma(\beta),$$

yielding a contradiction. Suppose that Γ is infinite. Then the set of first-order formulas

$$Th_q(\mathfrak{F}_\lambda(\omega)) \cup (\bigcup_{\Delta \subset \Gamma, |\Delta| < \omega} \{(\delta(\beta) = \gamma(\beta))\binom{p_i}{a_i}) \mid \beta \in \Delta\}) \cup$$

$$\cup (\delta(\alpha) \neq \gamma(\alpha))\binom{p_i}{a_i})),$$

where a_i are new constant's symbols replacing all the variables p_i, is locally consistent. The compactness theorem for first-order logic entails that all this set is consistent, i.e. this set has a model $\mathcal{G}$. Then $\mathcal{G} \in \mathfrak{F}_\lambda(\omega)^Q$ and we have a contradiction with (1.15). Hence we can conclude that $\Gamma \not\vdash^{Ad(\vdash)} \alpha$. Thus we proved

$$\Gamma \vdash^{Ad} \alpha \Longleftrightarrow \{\delta(\beta) = \gamma(\beta) \mid \beta \in \Gamma\} \models_{(\mathfrak{F}_\lambda(\omega))^Q} \delta(\beta) = \gamma(\beta),$$

i.e. we have proved that the our choice gives an algebraic semantics for $\vdash^{Ad}$ and so (i) is proved. A rule Γ/α is admissible for $\vdash \Leftrightarrow \Delta \vdash^{Ad} \alpha$ (Theorem 1.7.4) $\Leftrightarrow$ (by (i))

$$\mathfrak{F}_\lambda(\omega) \models [\{\delta(\beta) = \gamma(\beta) \mid \beta \in \Gamma\} \Rightarrow \delta(\beta) = \gamma(\beta)].$$

Hence (ii) holds. To show (iii), it is sufficient to note that the property (ii) from the definition of algebraic semantics equivalent to $\vdash$ does not concern $\vdash$ but is concerned only with the truth of quasi-identities on K. ∎

Now we recall that an axiomatic system $\mathcal{AS}$ is said to be a basis for a logical consequence relation $\vdash$ if $(\vdash) = (\vdash_{\mathcal{AS}})$. We intend to find a description of the bases of logical consequences in matrix and algebraic semantics. First we note that according to Theorem 1.2.16 (by the Los-Suszko theorem we can consider every logical consequence operation as an axiomatic consequence relation) every $\vdash$ has the matrix semantics $K_L(\vdash)$. By Theorem 1.2.30 it follows that the quasivariety $K_L(\vdash)^Q$ generated by $K_L(\vdash)$ forms a matrix semantics for the consequence $\vdash$ as well.

For any quasi-variety Q, $Th_{pq}(Q)$ denotes the *pure quasi-equational theory* of Q, that is the set of all quasi-identities without occurrences of $=$ (which we call by pure quasi-identities) which are valid in all members of Q.

Definition 1.7.12 *A pure quasi-equational basis for Q is a set S of pure quasi-identities from $Th_{pq}(Q)$ such that, for any pure quasi-identity q from $Th_{pq}(Q)$, q is a consequence of S, which means that for any algebraic system $\mathcal{A}$, if $(\forall q_1 \in S)\mathcal{A} \models q_1$ then $\mathcal{A} \models q$.*

Theorem 1.7.13 *An axiomatic system $\mathcal{AS}$ with the set of axioms $\mathcal{Ax}$ and the set of inference rules $\mathcal{R}$ forms a basis for a logical consequence relation $\vdash$ if and only if the set $\mathcal{Ax}^e \cup \mathcal{R}^q$, where*

$$\mathcal{Ax}^e := \{D(\alpha) \mid \alpha \in \mathcal{Ax}\},$$

$$\mathcal{R}^q := \{(\wedge_{\beta \in \Gamma} D(\beta)) \Rightarrow D(\alpha) \mid (\Gamma \vdash \alpha) \in \mathcal{R}\},$$

forms the pure quasi-equational basis for the quasivariety $K_L(\vdash)^Q$.

Proof. Suppose $\mathcal{AS}$ is a basis for $\vdash$. Then all formulas from $\mathcal{Ax}^e \cup \mathcal{R}^q$ are quasi-identities which are valid in $K_L(\vdash)^Q$ because $K_L(\vdash)$ is a matrix semantics for $\vdash$. Assume that a pure quasi-identity $(\wedge_{\beta \in \Delta} D(\beta)) \Rightarrow D(\alpha)$ is valid in $K_L(\vdash)^Q$. In particular, this quasi-identity must be valid in $\langle \mathcal{F}or_{\mathcal{A}}, F, Th_\vdash(\Delta)\rangle$. Therefore

$\Delta \vdash_{AS} \alpha$. Thus there exists a derivation $\mathcal{S}$ of α from Δ in $\mathcal{AS}$. By induction on the occurrence of formulas δ in $\mathcal{S}$ it is not difficult to show that the quasi-identity $(\wedge_{\beta \in \Delta} D(\beta)) \Rightarrow D(\delta)$ is a consequence of $\mathcal{A}x^e \cup \mathcal{R}^q$. Then $(\wedge_{\beta \in \Delta} D(\beta)) \Rightarrow D(\alpha)$, in particular, is a consequence of $\mathcal{A}x^e \cup \mathcal{R}^q$. Thus $\mathcal{A}x^e \cup \mathcal{R}^q$ is a pure quasi-equational basis for the quasivariety $K_L(\vdash)^Q$.

Conversely, suppose $\mathcal{A}x^e \cup \mathcal{R}^q$, is a pure quasi-equational basis for $K_L(\vdash)^Q$. Then every quasi-identity $(\wedge_{\beta \in \Gamma} D(\beta)) \Rightarrow D(\alpha)$ from $\mathcal{A}x^e \cup \mathcal{R}^q$ must be valid in $\langle \mathcal{F}or_A, F, Th_\vdash(\Gamma) \rangle$. Therefore $\Gamma \vdash \alpha$ holds. Similarly we have $\vdash \alpha$ for $D(\alpha) \in \mathcal{A}x^e$. From this and Thereon 1.5.2 (Los-Suszko Theorem) we infer that for every Δ and β if $\Delta \vdash_{AS} \beta$ then $\Delta \vdash \beta$. Suppose that for some Δ and α, $\Delta \nvdash_{AS} \alpha$. Then for every finite $\Gamma \subseteq \Delta$ $\Gamma \nvdash_{AS} \alpha$. In particular, the quasi-identity $(\wedge_{\beta \in \Gamma} D(\beta)) \Rightarrow D(\alpha)$ is not valid in the matrix $\langle \mathcal{F}or_{\mathcal{L}}, F, Th_{\vdash_{AS}}(\Gamma) \rangle$ which belongs to $K_L(\vdash)^Q$ (this holds because $\mathcal{A}x^e \cup \mathcal{R}^q$ is a basis for this quasivariety). Therefore $\Gamma \nvdash \alpha$ since $K_L(\vdash)^Q$ is a matrix semantics for $\vdash$. From this and the finiteness of $\vdash$ (cf. (1.6)) we conclude $\Delta \nvdash \alpha$. Hence $\vdash = \vdash_{AS}$, that is $\mathcal{A}x^e \cup \mathcal{R}^q$ is a basis for $\vdash$. ∎

As we know the algebraic semantics is a very useful and convenient tool for the study of derivability in formal systems. Therefore we now intend to extend the description above for bases of logical consequence relations $\vdash$ to the case where there is an algebraic semantics equivalent to $\vdash$.

Definition 1.7.14 *If $\vdash$ is a logical consequence relation and Γ/α is an inference rule, then we call Γ/α derivable in $\vdash$ if $\Gamma \vdash \alpha$ holds.*

Suppose that $\vdash$ is an algebraizable logical consequence relation. Then there is an equivalent to $\vdash$ algebraic semantics K. This in particular means, that the class K consists of $\mathcal{L}$-algebras, such that there is a fixed system $\delta_i = \gamma_i, i < n$ of defining equations, and that there is a finite system of terms (or *equivalence formulas*) $t_j(p, q), j < m$ such that the following quasi-identities

$$x = y \Rightarrow \delta_i(t_j(x, y)) = \gamma_i(t_j(x, y)),$$

$$(\bigwedge_{i,j} \delta_i(t_j(x, y)) = \gamma_i(t_j(x, y))) \Rightarrow (x = y) \tag{1.16}$$

are valid in K. We denote this family of quasi-identities by K_A. Moreover, according to (ii) of Theorem 1.6.7 and Lemma 1.6.8, the set of rules $D^\vdash$ consisting of

$$\frac{x}{t(\delta(x), \gamma(x))}, \quad \frac{t(\delta(x), \gamma(x))}{x}, \quad \frac{z}{t(x, x)}, \quad \frac{t(x, y)}{t(y, x)},$$

$$\frac{t(x, y), t(y, z)}{t(x, z)}, \quad \frac{x, t(x, y)}{y}, \quad (\forall x \in Var(\alpha)) \; \frac{t(y, z)}{t(\alpha(^x_y), \alpha(^x_z))},$$

where α is any term of the kind **a single application of a logical connective to variables**, includes only rules derivable for $\vdash$. Thus we can assume that every basis for $\vdash$, when there is an algebraic semantics equivalent to $\vdash$, includes the rules of $D^\vdash$.

Theorem 1.7.15 *An axiomatic system $\mathcal{AS}$ consisting of axioms $\mathcal{Ax}$ and inference rules $\mathcal{R}$ (where, $D^\vdash \subseteq \mathcal{R}$) is a basis for an algebraizable consequence relation $\vdash$ if and only if the set of quasi-identities $\mathcal{Ax}^e \cup \mathcal{R}^q \cup K_A$, where*

$$\mathcal{Ax}^e := \{\delta(\alpha) = \gamma(\alpha) \mid \alpha \in \mathcal{Ax}\}$$
$$\mathcal{R}^q := \{(\bigwedge_{\beta \in \Gamma, i} \delta_i(\beta) = \gamma_i(\beta)) \Rightarrow \delta_j(\alpha) = \gamma_j(\alpha) \mid (\Gamma \vdash \alpha) \in R, \ j\},$$

forms a basis for the quasivariety K^Q.

Proof. For the left to right direction, let $\vdash = \vdash_{\mathcal{AS}}$. Then every identity and quasi-identity from $\mathcal{Ax}^e \cup \mathcal{R}^q \cup K_A$ will be valid in K^Q because this quasi-variety forms an algebraic semantics equivalent to $\vdash$. Assume that a quasi-identity $(\bigwedge_k f_k = g_k) \Rightarrow f = g$ is valid in K^Q. By our assumption about the algebraizability of $\vdash$ the quasi-identity

$$(\bigwedge_{k,i,j} \delta_i(t_j(f_k, g_k)) = \gamma_i(t_j(f_k, g_k))) \Rightarrow \delta_m(t_d(f, g))) = \gamma_m(t_d(f, g)))$$

is valid in K^Q for every m, d. Because K^Q is an algebraic semantics, we have $\{(t_j(f_k, g_k)) \mid j, k\} \vdash t_d(f, g)$ for every d. Therefore there exists a derivation $\mathcal{S}$ of every $t_d(f, g)$ from $\{(t_j(f_k, g_k)) \mid j, k\}$ in $\mathcal{AS}$. By induction on the occurrence of every formula β in $\mathcal{S}$ it is not hard to show that the quasi-identity

$$(\bigwedge_{k,i,j} \delta_i(t_j(f_k, g_k)) = \gamma_i(t_j(f_k, g_k))) \Rightarrow \delta_m(\beta) = \gamma_m(\beta))$$

is a corollary of $\mathcal{Ax}^e \cup \mathcal{R}^q \cup K_A$ for every m. Then, in particular, the latter holds for $\beta = t_d(f, g)$ and every d. From this and property (1.16) of *equivalence formulas* $t_j(p, q)$ we have that $(\bigwedge_k f_k = g_k) \Rightarrow f = g$ is a corollary of $\mathcal{Ax}^e \cup \mathcal{R}^q \cup K_A$. Thus every quasi-identity q is a corollary of $\mathcal{Ax}^e \cup \mathcal{R}^q \cup K_A$ iff q is valid on K^Q. Thus $\mathcal{Ax}^e \cup \mathcal{R}^q \cup K_A$ forms a basis for quasivariety K^Q.

Conversely, suppose now that $\mathcal{Ax}^e \cup \mathcal{R}^q \cup K_A$ is a basis for K^Q. Then every quasi-identity and identity from this basis is valid in K^Q. Therefore, because K^Q is an algebraic semantics for $\vdash$, every rule Γ/α from $\mathcal{R}$ is an $\vdash$-sequent, that is $\Gamma \vdash \alpha$, and every axiom α from $\mathcal{Ax}$ is an $\vdash$-axiom, i.e. $\vdash \alpha$. This and Theorem 1.5.2 (Los-Suszko Theorem) yield $\vdash_{\mathcal{AS}} \subseteq \vdash$. Suppose now that for some Δ and μ, $\Delta \vdash_{\mathcal{AS}} \alpha$ does not hold. We take the free algebra $\mathfrak{F}$ from K^Q with free generators denoted by the same letters as propositional variables from the set of formulas $\Delta \cup \{\alpha\}$. We now consider formulas from this set as elements of the free algebra, taking all propositional letters as free generators. We define the binary relation $\sim$ on this algebra as follows:

$$f \sim g \Leftrightarrow \forall j (\Delta \vdash_{AS} t_j(f, g))$$

According to $D^{\vdash} \subseteq \mathcal{R}$ we have this relation is a congruence on $\mathfrak{F}$. Next we consider the quotient algebra $\mathfrak{N} := \mathfrak{F}/\sim$ of $\mathfrak{F}$ by this congruence relation.

First we have to check that this algebra is from K^Q. In fact, all identities from Ax_e are valid in $\mathfrak{N}$ because this algebra is a homomorphic image of $\mathfrak{F}$. All quasi-identities from $\mathcal{R}^q$ corresponding to inference rules from $\mathcal{R}$ are also valid in $\mathfrak{N}$. Indeed, let $\Gamma \vdash \alpha$ be from $\mathcal{R}$ and the equality

$$\bigwedge_{\beta \in \Gamma, i} [\delta_i(\beta^s)]_\sim = [\gamma_i(\beta^s)]_\sim$$

holds in $\mathfrak{N}$. This means for every j, $\Delta \vdash_{AS} t_j(\delta_i(\beta^s), \gamma_i(\beta^s))$ which implies by (ii) of Theorem 1.6.7 that $\Delta \vdash_{AS} \beta^s$ for all $\beta \in \Gamma$. Then $\Delta \vdash_{AS} \alpha^s$. By (ii) from Theorem 1.6.7 again, we have $\Delta \vdash_{AS} t(\delta(\alpha^s), \gamma(\alpha^s)))$. This entails $[\delta(\alpha^s)]_\sim = [\gamma(\alpha^s)]_\sim$, which is what we needed. Hence $\mathfrak{N} \in K^Q$.

The quasi-identities from K_A will be also valid in $\mathfrak{N}$. Indeed, let $[\alpha]_\sim = [\beta]_\sim$, i.e. for all j, $\Delta \vdash_{AS} t_j(\alpha, \beta)$. Because $x/t(\delta(x), \gamma(x))$ belongs to $D^{\vdash}$ we have for every r, m $\Gamma \vdash_{AS} t_r(\delta_m(t_j(\alpha, \beta)), \gamma_m(t_j(\alpha, \beta))$. That is

$$[\delta_m(t_j(\alpha, \beta))]_\sim = [\gamma_m(t_j(\alpha, \beta))]_\sim$$

for all m and j. Thus the quasi-identities of first kind in (1.16) are valid in $\mathfrak{N}$. Let us now verify the truth of the remaining part of K_A. Assume that for all i, j, $[\delta_i(t_j(\alpha, \beta))]_\sim = [\gamma_i(t_j(\alpha, \beta))]_\sim$. This means for every j,

$$\Gamma \vdash_{AS} t_j(\delta_i(t_j(\alpha, \beta)), \gamma_i(t_j(\alpha, \beta))).$$

Because $D^{\vdash} \subseteq \mathcal{R}$ we infer from this $\Delta \vdash_{AS} t_j(\alpha, \beta)$. That is $[\alpha]_\sim = [\beta]_\sim$. Hence algebra $\mathfrak{N}$ belongs to K^Q.

Note that the relation $\Delta \nvdash_{AS} \mu$ entails implies that $\Delta \nvdash_{AS} t(\delta(\mu), \gamma(\mu))$. That is, we have $[\delta(\mu)]_\sim \neq [\gamma(\mu)]_\sim$. Simultaneously, for every $\beta \in \Delta$, $\Delta \vdash_{AS} \beta$ holds, which implies that $\Delta \vdash_{AS} t(\delta(\beta), \gamma(\beta))$. Then $[\delta(\beta)]_\sim = [\gamma(\beta)]_\sim$ for every β. Hence in view of K^Q is an algebraic semantics equivalent to $\vdash$, it follows that $\Delta \nvdash \mu$. Thus, $(\vdash) = (\vdash_{AS})$ and our theorem is proved. $\blacksquare$

Now we are in a position to give a description of the bases for those logical consequence relations which are the admissible closures of consequence relations.

Theorem 1.7.16 *An axiomatic system AS with axioms Ax and inference rules $\mathcal{R}$ forms a basis for the admissible closure $\vdash^{Ad(\vdash)}$ of a logical consequence relation $\vdash$ iff the set of quasi-identities $Ax_e \cup \mathcal{R}^q$ where $Ax^e := \{D(\alpha) \mid \alpha \in Ax\}$, $\mathcal{R}^q := \{(\bigwedge_{\beta \in \Gamma} D(\beta)) \Rightarrow D(\alpha) \mid (\Gamma \mid \alpha) \in \mathcal{R}\}$ forms the pure quasi-equational basis for the quasivariety Q generated by single Lindenbaum-Tarski matrix $\mathcal{A}_L(\emptyset, \vdash)$ for $\vdash$ over empty set of formulas.*

Proof. Let $\mathcal{AS}$ be a basis for the admissible closure. Then according to Theorem 1.7.10 all quasi-identities from $\mathcal{A}x^e \cup \mathcal{R}^q$ are valid on Q. If a quasi-identity $(\bigwedge_{\beta\Gamma} D(\beta)) \Rightarrow D(\alpha)$ is valid in Q then it is valid in $A_L(\emptyset, \vdash)$ also. Then, directly from our definition, Γ/α is admissible for $\vdash$. Because $\mathcal{AS}$ is basis for admissible closure, we have $\Gamma \vdash_{\mathcal{AS}} \alpha$. This means there exists a derivation $\mathcal{S}$ of α from Γ in $\mathcal{AS}$. By induction on the occurrence of a formula δ in $\mathcal{S}$ it is easy to see that $(\bigwedge_{\beta\in\Gamma} D(\beta)) \Rightarrow D(\delta)$ is a semantic corollary of $\mathcal{A}x^e \cup \mathcal{R}^q$. That is $(\bigwedge_{\beta\in\Gamma} D(\beta)) \Rightarrow D(\alpha)$ is a semantic corollary of $\mathcal{A}x^e \cup \mathcal{R}^q$. Thus $\mathcal{A}x^e \cup \mathcal{R}^q$ is the pure quasi-equational basis.

Conversely, suppose $\mathcal{A}x^e \cup \mathcal{R}^q$ is a pure quasi-equational basis for Q. Then all the rules (and axioms) from $\mathcal{A}x \cup \mathcal{R}$ are admissible rules for $\vdash$, again by Theorem 1.7.10. Thus $\vdash_{\mathcal{AS}} \subseteq \vdash^{Ad(\vdash)}$ by Los-Suszko Theorem. Suppose that $\Delta \nvdash_{\mathcal{AS}} \alpha$ but that nevertheless $\Delta \vdash^{Ad(\vdash)} \alpha$. Then there is a finite $\Gamma \subseteq \Delta$ such that $\Gamma \vdash^{Ad(\vdash)} \alpha$. In particular, the quasi-identity $(\bigwedge_{(\beta\in\Gamma)} D(\beta)) \Rightarrow D(\alpha))$ is valid in Q (cf. Theorem 1.7.10). The matrix $\langle \mathcal{F}or_{\mathcal{L}}, F, Th_{\vdash_{\mathcal{AS}}}(\Delta) \rangle$ is a member of Q. Indeed, this is not difficult to check using the fact that $\mathcal{A}x^e \cup \mathcal{R}^q$ is a basis of Q. Then $(\bigwedge_{\beta\in\Gamma} D(\beta)) \Rightarrow D(\alpha)$ must be valid in this matrix which contradicts the relation $\Delta \nvdash_{\mathcal{AS}} \alpha$. Thus if $\Delta \nvdash_{\mathcal{AS}} \alpha$ then Δ/α is not admissible for $\vdash$, and $\mathcal{AS}$ is really a basis for the admissible closure $\vdash^{Ad(\vdash)}$. ∎

Assume now that there is an algebraic semantics equivalent to $\vdash$, that is, $\vdash$ is algebraizable. What can be said in this case about the bases for the admissible closure of $\vdash$. According to the Theorem 1.7.11 the quasi-variety $\mathfrak{F}(\omega)^Q$ forms the algebraic semantics equivalent to $\vdash^{Ad(\vdash)}$ with the same sets of defining and equivalence formulas. From this and Theorem 1.7.15 we extract

Corollary 1.7.17 *An axiomatic system $\mathcal{AS}$ consisting of axioms $\mathcal{A}x$ and inference rules $\mathcal{R}$ forms a basis for the admissible closure $\vdash^{Ad(\vdash)}$ of $\vdash$ (when $D_{\vdash} \subseteq \mathcal{R}$) iff the set of quasi-identities $\mathcal{A}x^e \cup \mathcal{R}^q \cup K^A$ is a basis for the quasivariety $\mathfrak{F}(\omega)^Q$.*

1.8 Lattices of Logical Consequences

The aim of this section is to give a brief overview and introductory knowledge concerning the structure of the family of all logical consequence operators and relations over given language $\mathcal{L}$, which will remain constant throughout this section. We begin with an description of certain simple properties of the set of all consequence operations over $\mathcal{L}$, which is denoted by $Con(\mathcal{L})$. First we introduce a natural order on $Con(\mathcal{L})$.

Definition 1.8.1 *Given $\mathcal{C}_1$ and $\mathcal{C}_2$ from $Con(\mathcal{L})$, we put $\mathcal{C}_1 \leq \mathcal{C}_2$ if for every set $X \subseteq \mathcal{F}or_c l$, $\mathcal{C}_1(X) \subseteq \mathcal{C}_2(X)$. In this event we say $\mathcal{C}_1$ is weaker than $\mathcal{C}_2$ ($\mathcal{C}_2$ is stronger than $\mathcal{C}_1$).*

It is clear that $\langle Con(\mathcal{L}), \leq \rangle$ is a partially ordered set (poset) (i.e. $\leq$ is reflexive antisymmetric and transitive) having the smallest element $(0(X):=X)$ and greatest element $(1(X) := \mathcal{F}or_{\mathcal{L}})$. Now it would be very natural to clarify whether $Con(\mathcal{L})$ forms a lattice; that is, if there are least upper and greatest lower bounds for any finite set of elements from this set.

According to Theorem 1.5.5, for any $\mathcal{C} \in Con(\mathcal{L}), \mathcal{C} = Cn_{\mathcal{AS}}$, where $\mathcal{AS}$ is the special axiomatic system having the set of axioms $Ax(\mathcal{C})$ and the set of inference rules $Rule(\mathcal{C})$ pointed out in that theorem. Moreover, because $Ax(\mathcal{C}) \subseteq Rule(\mathcal{C})$ (more precisely, for every axiom α of $\mathcal{C}$ the rule $\emptyset/\alpha$ is a rule from $Rule(\mathcal{C})$), we have $\mathcal{C} = Cn_{Rule(\mathcal{C})}$.

Proposition 1.8.2 *Given an arbitrary family $\mathcal{C}_i = Cn_{\mathcal{AS}_i}$, $i \in I$. The least upper bound $sup(\mathcal{C}_i, i \in I)$ and the greatest lower bound $inf(\mathcal{C}_i, i \in I)$ are as follows:*

(i) $sup(\mathcal{C}_i, i \in I) = Cn_{R^+}$, where $R^+ := \bigcup\{Rule(\mathcal{C}_i) \mid i \in I\}$;
(ii) $inf(\mathcal{C}_i, i \in I) = Cn_{R^-}$, where $R^- := \bigcap\{Rule(\mathcal{C}_i) \mid i \in I\}$.

Proof. It is clear that for each operator $\mathcal{C}_i$, $\mathcal{C}_i \leq Cn_{R^+}$. Suppose the same holds for a consequence operator $\mathcal{C}$. We know, as we have noted above, that $\mathcal{C} = Cn_R$, where $R = Rule(\mathcal{C})$. We have to show that $Cn_{R^+} \leq Cn_R$. By our assumption $Rule(\mathcal{C}_i) \subseteq R$ for each i. Therefore $R^+ \subseteq R$. This immediately yields $Cn_{R^+} \leq Cn_R$. That is, Cn_{R^+} is really the least upper bound for the family $\mathcal{C}_i, i \in I$. Clearly then $Cn_{R^-} \leq \mathcal{C}_i = Cn_{Rule(\mathcal{C}_i)}$ for each i. Suppose now that $\mathcal{C} = Cn_{Rule(\mathcal{C})}$ is a consequence operator and $\mathcal{C} \leq \mathcal{C}_i$ for every i. Then for all i, $Rule(\mathcal{C}) \subseteq Rule(\mathcal{C}_i)$ which implies $Rule(\mathcal{C}) \subseteq R^-$ and $\mathcal{C} \leq Cn_{R^-}$. Hence the latter consequence operator is the greatest lower bound for the family $\mathcal{C}_i, i \in I$.
∎

According to this lemma $Con(\mathcal{L})$ is a complete lattice because it possesses least upper and greatest lower bounds for every family of elements and not merely for finite ones. Now we turn to the set of all logical consequence relations over a language $\mathcal{L}$ which is denoted by $Cr(\mathcal{L})$. What can be said about the set of all logical consequence relations? The set $Cr(\mathcal{L})$ forms a partially ordered set with respect to the usual set-theoretic inclusion $\subseteq$. It is clear that the smallest element of $Cr(\mathcal{L})$ is the relation $\vdash_0$, where $\Gamma \vdash_0 \alpha \Leftrightarrow \alpha \in \Gamma$. The greatest element of $Cr(\mathcal{L})$ is the relation $\vdash_1$, where $\forall \alpha \ (\emptyset \vdash_1 \alpha)$. We intend to show that $Cr(\mathcal{L})$ forms a lattice. For this we need some simple properties of consequence relations. Recall that, for a given a consequence relation $\vdash$, an inference rule Γ/α is said to be derivable in $\vdash$ if $\Gamma \vdash \alpha$. According to Theorem 1.5.2 every relation $\vdash$ from $Cr(\mathcal{L})$ has the representation $\vdash_{\mathcal{AS}}$, that is $(\vdash) = (\vdash_{\mathcal{AS}})$, where $\mathcal{AS}$ is an axiomatic system with axioms $Ax(\vdash)$ and inference rules $Rule(\vdash)$.

Lemma 1.8.3 *Let $\vdash$ be a consequence relation and $\mathcal{AS}$ be an axiomatic system. If for any rule Γ/α of $\mathcal{AS}$ (in particular, for any rule $\emptyset/\alpha$, where α is* *an axiom)* Γ/α *is derivable in* $\vdash$ *then* $\vdash_{AS}\subseteq\vdash$. *If* $\vdash_1$ *and* $\vdash_2$ *are consequence relations then every rule from* $Rule(\vdash_1)$ *is derivable in* $\vdash_2$ *iff* $\vdash_1\subseteq\vdash_2$.

Proof. To prove the first part of this lemma, we note $(\vdash) = (\vdash_{AS_1})$ for some axiomatic system AS_1. Suppose that $\Gamma \vdash_{AS} \alpha$. Then there is a derivation of α from Γ in AS which is a finite sequence S with the required properties. We can prove by induction on the occurrence of formula β in S that $\Gamma \vdash_1 \beta$. Indeed the initial formulas of S have to be either axioms of AS or formulas from Γ. Because all the axioms from AS are derivable in AS_1 we have $\Gamma \vdash_1 \beta$ for all such formulas. Inductive step: assume that β is obtained by a rule Δ/γ of AS from formulas occurring earlier than β and that for all such formulas δ, $\Gamma \vdash \delta$ holds. Then there is a substitution e such that all formulas from $e(\Delta)$ occur in S before β and $\beta = e(\gamma)$. Because Δ/γ is a rule from AS, $\Delta \vdash_1 \gamma$ holds, and by the structurallity of $\vdash_1$ we infer $e(\Delta) \vdash_1 e(\gamma)$. Hence for all formulas δ from $e(\Delta)$, $\Gamma \vdash_1 \delta$ hold, and $e(\Delta) \vdash_1 \beta$. Thus $\Gamma \vdash_1 \beta$. Carrying on the inductive procedure we obtain $\Gamma \vdash_1 \alpha$ and the first part of our lemma is proved.

Turning to the second part, suppose that every rule from $Rule(\vdash_1)$ is derivable in $\vdash_2$. Then, by the first part of this lemma, $\vdash_{AS}\subseteq\vdash_2$, where AS is the axiomatic system having $Rule(\vdash_1)$ as the set of inference rules (and axioms are implicitly presented there). Because, according to Theorem 1.5.2 $(\vdash_1) = (\vdash_{AS})$, it follows $\vdash_1\subseteq\vdash_2$. The converse implication is evident. ∎

Proposition 1.8.4 *The partially ordered set* $Cr(\mathcal{L})$ *forms a lattice and for every finite family* $\vdash_i, i \in I$, $\vdash_i\in Cr(\mathcal{L})$,

(i) $inf(\vdash_i, i \in I) = \bigcap_{i\in I} \vdash_i$;
(ii) $sup(\vdash_i, i \in I) =\vdash_{AS}$, *were* $AS = \bigcup\{AS_i \mid \vdash_i=\vdash_{AS_i}, i \in I\}$.

Proof. The right hand side of (i) is greatest lower bound for the family $\vdash_i, i \in I$ in the poset of all subsets of the set $2^{\mathcal{F}or(\mathcal{L})} \times \mathcal{F}or(\mathcal{L})$ being the set-theoretic intersection of the relations $\vdash_i, i \in I$. It remains only to show that $\bigcap_{i\in I} \vdash_i$ is a logical consequence relation. For this, it is sufficient to show that the relations (1.3) - (1.7) from the definition of logical consequence relation hold. The truth of the relations (1.3) - (1.5) and (1.7) follows directly from our definitions. If for all i, $\Gamma \vdash_i \alpha$ then because (1.6) holds for all $\vdash_i$ there are some finite $\Delta_i \subseteq \Gamma$ such that $\Delta_i \vdash_i \alpha$. Then $\bigcup\{\Delta_i \mid i \in I\} \vdash_i \alpha$ for all i, and the set $\bigcup\{\Delta_i \mid i \in I\}$ is finite because the set I is finite. Thus $\bigcup\{\Delta_i \mid i \in I\} \subseteq \Gamma$ and $\bigcup\{\Delta_i \mid i \in I\} \vdash \alpha$, where $\vdash= \bigcap_{i\in I} \vdash_i$. Hence (1.6) also holds and (i) is established.

It is evidently seen that the right hand side of (ii) is an upper bound for the family $\vdash_i, i \in I$. Suppose $\vdash_{AS_1}$ is also an upper bound for this family. Then for every i, Γ and α, $\Gamma \vdash_{AS_i} \alpha$ implies $\Gamma \vdash_{AS_1} \alpha$. Suppose $\Gamma \vdash_{AS} \alpha$, then there is a finite $\Delta \subset \Gamma$ such that $\Delta \vdash_{AS} \alpha$. This show that, there is a derivation S of α from Δ in AS. This derivation involves only a finite number of axioms and inference rules from $\mathcal{AS}$, in particular,. these axioms and rules are an axioms and rules of some systems $\mathcal{AS}_i$, $i \in I$. Therefore they all have to be axioms and derivable inference rules for axiomatic system $\mathcal{AS}_1$. Therefore by Lemma 1.8.3 there exists a derivation of α from Δ in $\mathcal{AS}_1$, which yields $\Gamma \vdash_1 \alpha$. Hence $\vdash_{\mathcal{AS}}$ is the least upper bound. ∎

Recall that for any logical consequence relation $\vdash$, a set of formulas T is said to be a $\vdash$-theory if T is closed with respect to all rules from $Rule(\vdash)$. The set of all $\vdash$-theories will be denoted by $Th(\vdash)$.

It is evident that $Th(\vdash)$ is closed under any set-theoretic intersection. Therefore $Th(\vdash)$ forms a complete lattice with respect to the set-theoretic inclusion. This lattice has the greatest element, namely it is the set of all formulas. The smallest element of $Th(\vdash)$ is the set $Th(\vdash) := \{\alpha \mid \emptyset \vdash \alpha\}$. According to the definition we have, for every family of $\vdash$-theories T_i, $i \in I$,

$$sup(T_i, i \in I) := \bigcap\{T \mid T \in Th(\vdash), \cup\{T_i \mid i \in I\} \subseteq T\}.$$

It is possible to give another description for the least upper bound:

Lemma 1.8.5 *Given a family of $\vdash$ theories T_i, $i \in I$,*

$$sup(T_i, i \in I) = Cn_\vdash(\bigcup\{T_i \mid i \in I\}).$$

Proof. In fact, the right part from the equality above is a $\vdash$-theory, as it is easy to see. Therefore it is an upper bound for the family. Suppose T is some other upper bound, that is $T_i \subseteq T$ for every i and $\bigcup\{T_i \mid i \in T\} \subseteq T$. Because $Cn_\vdash$ have to be monotonic, $Cn_\vdash(\bigcup\{T_i \mid i \in I\}) \subseteq Cn_\vdash(t)$. T is an $\vdash$-theory, therefore $Cn_\vdash(T) = T$. Hence $Cn_\vdash(\bigcup\{T_i \mid i \in I\})$ is the least upper bound. ∎

Definition 1.8.6 *A $\vdash$-theory is said to be* finitely generated *if there is a finite set X of formulas such that $Cn_\vdash(X) = T$*

It is clear that for every finite X, $Cn_\vdash(X)$ is a finitely generated $\vdash$-theory. In view of the finiteness of every $\vdash$ it is possible simply to characterize finitely generated $\vdash$-theories by means of pure lattice-theoretic terms. First we need the following

Definition 1.8.7 *An element e of a complete lattice L is said to be* compact *if $e \leq \bigvee\{e_i \mid i \in I\}$ implies $e \leq \bigvee\{e_i \mid i \in J\}$ for some finite $J \subseteq I$. A lattice L is* algebraic *if L is complete and every element of L is some (possibly infinite) union of compact elements.*

Lemma 1.8.8 *[18] The following hold*

(i) Any $\vdash$-theory is finitely generated iff T is compact in $Th(\vdash)$;

(ii) $Th(\vdash)$ is closed under set-theoretic unions of chains of $\vdash$-theories. I.e. if $(T_i, i \in I, T_i \in Th(\vdash))$ is a chain (that is, for every $i, j \in I$, $T_i \subseteq T_j$ or $T_j \subseteq T_i$) then $\bigcup\{T_i \mid i \in I\} \in Th(\vdash)$;

(iii) The lattice $Th(\vdash)$ is algebraic.

Proof. Let T be a finitely generated $\vdash$-theory, that is $T = Cn_\vdash(X)$, where $\|X\| < \omega$. Suppose that $T \subseteq \bigvee\{T_i \mid i \in I\}$. By Lemma 1.8.5

$$\bigvee\{T_i \mid i \in I\} = Cn_\vdash(\bigcup\{T_i \mid i \in I\}).$$

Therefore by finiteness of $Cn_\vdash$ and $\vdash$ itself it follows that

$$X \subseteq Cn_\vdash(\bigcup\{T_i \mid i \in J\}),$$

for some finite $J \subseteq I$. Then $T = Cn_\vdash(X) \subseteq Cn_\vdash(\bigcup\{T_i \mid i \in J\})$. The right hand side of this relation according to Lemma 1.8.5 is $\bigvee\{T_i \mid i \in J\}$. Hence T is compact. Conversely, let T be a compact element. It is clear that

$$T \subseteq \bigvee\{Cn_\vdash(\{\alpha\}) \mid \alpha \in T\}.$$

By compactness of T and Lemma 1.8.5 we conclude

$$T \subseteq Cn_\vdash(\bigcup\{Cn_\vdash(\{\alpha_i\}) \mid \alpha_i \in T, i \in I, \|I\| < \omega\}).$$

On the other hand, the right hand side of the relation above coincides with the set $Cn_\vdash\{\alpha_i \mid i \in I\}$ and this set is contained in T because T is $\vdash$-theory. Hence $T = Cn_\vdash\{\alpha_i \mid i \in I\}$ holds, and T is a finitely generated $\vdash$-theory. Thus (i) is proved. The proof of (ii) is evident and follows directly from our definitions. For any T from $Th(\vdash)$,

$$T = Cn_\vdash(\bigcup\{Cn_\vdash(\{\alpha_i\}) \mid \alpha_i \in T\}). \tag{1.17}$$

Every $Cn_\vdash(\{\alpha_i\})$ is compact by part (i) of our lemma, and using Lemma 1.8.5 we establish that the right side of (1.17) is the lattice union of elements $Cn_\vdash(\{\alpha_i\})$. Hence $Th(\vdash)$ is a compact lattice. ∎

It is easy to see that a $\vdash$-theory can not be closed with respect to some substitution, as well as for any given substitution e and any $\vdash$-theory T, $e(T)$ does not have to be a $\vdash$-theory. Let us clarify the behavior of $\vdash$-theories with respect to substitutions. For any given substitution e and $\vdash$-theory T, we put

$$e(\vdash)(T) := Cn_\vdash(e(T)).$$

For any set of formulas Γ, $e^{-1}(\Gamma) := \{\alpha \mid e(\alpha) \in \Gamma\}$.

Lemma 1.8.9 *For any logical consequence relation* $\vdash$ *and every substitution* *e, the following hold*

(i) For every $T \in Th(\vdash)$ *and any substitution* e, $e^{-1}(T) \in Th(\vdash)$, *i.e.* $Th(\vdash)$ *is closed with respect to inverse substitutions;*

(ii) For every substitution e *and every* $\Gamma \subseteq \mathcal{F}or(\mathcal{L})$, $e(\vdash)(\mathcal{C}n_\vdash(\Gamma)) = \mathcal{C}n_\vdash(e(\Gamma))$;

(iii) The mapping $e(\vdash)$ *is permutable with arbitrary (in particular, infinite) unions in* $Th(\vdash)$.

Proof. By the structurallity of $\mathcal{C}n_\vdash$, for every set of formulas Γ, we have that

$$e(\mathcal{C}n_\vdash(e^{-1}(\Gamma))) \subseteq \mathcal{C}n_\vdash(e(e^{-1}(\Gamma))).$$

Therefore if T is an $\vdash$-theory then

$$e(\mathcal{C}n_\vdash(e^{-1}(T))) \subseteq \mathcal{C}n_\vdash(e(e^{-1}(T))) \subseteq \mathcal{C}n_\vdash(T) = T.$$

Hence $\mathcal{C}n_\vdash(e^{-1}(T)) \subseteq e^{-1}(T)$, the converse inclusion is evident, and (i) holds. As to (ii), by structurallity of $\mathcal{C}n_\vdash$

$$e(\vdash)(\mathcal{C}n_\vdash(\Gamma) = \mathcal{C}n_\vdash(e(\mathcal{C}n_\vdash(\Gamma))) = \mathcal{C}n_\vdash(e(\Gamma)).$$

To verify (iii) we merely derive consequently

$$e(\vdash)(\bigvee\{T_i \mid i \in I\}) = e(\vdash)(\mathcal{C}n_\vdash(\bigcup\{T_i \mid i \in I\}))(\text{by Lemma 1.8.5}) =$$
$$\mathcal{C}n_\vdash(e(\bigcup\{T_i \mid i \in I\}))(\text{by (ii)}) = \mathcal{C}n_\vdash(\bigcup\{e(T_i) \mid i \in I\})) =$$
$$\mathcal{C}n_\vdash(\bigcup\{e(\vdash)(T_i) \mid i \in I\})) = \bigvee\{e(\vdash)(T_i) \mid i \in I\}(\text{by Lemma 1.8.5}).$$

∎

As a brief historical comment, note that peculiar important contribution to the subject of current section has been made by Polish Logical School (Los, 1955, Suszko 1958, Rasiowa, Sikorski 1968, Suszko 1961, and Wojcicki 1970). A comprehensive presentation of this field is given in the book of R.Wojcicki [203].

Chapter 2

Semantics for Non-Standard Logics

Having set up the basic semantic machinery for algebraic logics, in this chapter we turn to specify algebraic semantics for some particular important non-standard logics. Algebraic semantics for modal and intuitionistic logics has its origins in the works of McKinsey and Tarski [100, 101] and has been further developed by Lemmon [83] and others. In a sense, algebraic semantics did not gain widespread popularity compared to relational Kripke semantics (beginning 1960s). Although Kanger (1975) and Hintikka (1961) had similar ideas independently, relational semantics are commonly linked with the name of Saul Kripke and his writings have been more widely read and influential then theirs. We consider in this chapter some basic properties and theorems concerning Kripke's relational semantics and its connections with algebraic semantics through Stone theory. The presentation concentrates around questions of the effective use of the semantics and its applications to the study of logical systems through their having the finite model property, completeness by Kripke etc. In hole this chapter in intended to developing peculiar semantics technique for special classes of non-standard logic.

2.1 Algebraic Semantics for Intuitionistic Logic

As we have seen in Section 1.3 (Theorem 1.3.8) any superintuitionistic (intermediate) logic is an algebraic logic with the following algebraic presentation:

$$(\equiv) := \{x \rightarrow y, y \rightarrow x\}, \top := \{(x \rightarrow (x \rightarrow x))\}.$$

Therefore according to the Algebraic Completeness Theorem (Theorem 1.3.20), for any formula α and any superintuitionistic logic (abbreviation - s.l.) λ, $\alpha \in \lambda$ if and only if the identity $\alpha = \top$ is valid on any algebra $\mathfrak{A}$ from the variety $Var(\lambda)$. This, generally speaking, allows us to describe the set of theorems of λ. Nevertheless up to now we have had no precise information about the structure of algebras from $Var(\lambda)$. Therefore the use of Algebraic Completeness Theorem for the above mentioned purpose would be hampered. However in this section we intend to study in detail the algebraic semantics for superintuitionistic logics which will allow us to use algebraic semantics more effectively. We start with the introduction of the variety of all pseudo-boolean algebras which, as it will be shown, coincides with the variety $Var(H)$ characterizing the intuitionistic logic H.

Definition 2.1.1 *A pseudo-boolean algebra is an algebra*

$$\mathfrak{A} := \langle A, \wedge, \vee, \rightarrow, \neg, 0 \rangle$$

in which the following identities hold

(1) $a \vee b = b \vee a$, $a \wedge b = b \wedge a$;

(2) $a \vee (b \vee c) = (a \vee b) \vee c$, $a \wedge (b \wedge c) = (a \wedge b) \wedge c$;

(3) $(a \wedge b) \vee b = b$, $a \wedge (a \vee b) = a$;

(4) $a \wedge (a \rightarrow b) = a \wedge b$;

(5) $(a \rightarrow b) \wedge b = b$;

(6) $(a \rightarrow b) \wedge (a \rightarrow c) = a \rightarrow (b \wedge c)$;

(7) $(a \rightarrow a) \wedge b = b$;

(8) $(0 \wedge a) = a$;

According to this definition, any pseudo-boolean algebra forms a lattice.

Definition 2.1.2 *Let $\mathcal{L}$ be a lattice, $a, b, c \in \mathcal{L}$. Element c is said to be a pseudo-complement of element a to b if c is greatest element among elements x with the property $x \wedge a \leq b$.*

Lemma 2.1.3 *For any pseudo-boolean algebra $\mathfrak{A}$ and elements $a, b \in \mathfrak{A}$, the element $a \rightarrow b$ is the pseudo-complement of a to b.*

Proof. In fact, suppose that $c \leq (a \rightarrow b)$ then

$$a \wedge c \leq a \wedge (a \rightarrow b) = a \wedge b \leq b$$

by (4) from the definition of pseudo-boolean algebras. Suppose that $a \wedge c \leq b$. This yields $c \wedge a = c \wedge a \wedge b$. Therefore

$c \leq (a \rightarrow c)$ by (5)
$(a \rightarrow c) = (a \rightarrow c) \wedge (a \rightarrow a)$ by (7)
$(a \rightarrow c) \wedge (a \rightarrow a) = (a \rightarrow (c \wedge a))$ by (6)
$(a \rightarrow (c \wedge a)) = (a \rightarrow (c \wedge a \wedge b))$
$(a \rightarrow (c \wedge a \wedge b)) = (a \rightarrow (c \wedge a)) \wedge (a \rightarrow b)$ by (6)
$(a \rightarrow (c \wedge a)) \wedge (a \rightarrow b) \leq (a \rightarrow b)$.

Thus $a \rightarrow b$ is the greatest element in $\{c \mid a \wedge c \leq b\}$. $\blacksquare$

Lemma 2.1.4 *If for all elements a, b, c of a lattice $\mathcal{L}$ there is a pseudo-complement of a to $(a \wedge b) \vee (a \wedge c)$ then $\mathcal{L}$ is a distributive lattice. In particular, any pseudo-boolean algebra forms a distributive lattice.*

Proof. Let $d := (a \wedge b) \vee (a \wedge c)$, and e be the pseudo-complement of a to d. Seeing that $a \wedge b \leq d$ and $a \wedge c \leq d$, using the fact that d is the pseudo-complement of a to d, we infer $b \leq e$, $c \leq e$. Hence $b \vee c \leq e$. Again using the fact that d is the pseudo-complement of a to d, we have $a \wedge (b \vee c) \leq d$. On the other hand,

$$a \wedge b \leq a \wedge (b \vee c), \quad a \wedge c \leq a \wedge (b \vee c),$$

therefore $d \leq a \wedge (b \vee c)$. So $a \wedge (b \vee c) \leq d \leq a \wedge (b \vee c) \leq d$, i.e. $a \wedge (b \vee c) = d$. This demonstrates the first distributive law. As it is well known, the second (dual) distributive law follows from the first one. Indeed, we have

$$(a \vee b) \wedge (a \vee c) = ((a \vee b) \wedge a) \vee ((a \vee b) \wedge c) =$$
$$(a \vee (((a \wedge c) \vee (b \wedge c)) = (a \vee (a \wedge c)) \vee (b \wedge c) = a \vee (b \wedge c).$$

Thus L is a distributive lattice. ∎

Lemma 2.1.5 *Any pseudo-boolean algebra $\mathfrak{A}$ has a greatest element (unit element) which is equal to $a \to a$ for every $a \in \mathfrak{A}$, and a smallest element which is equal to 0.*

Proof. In fact, $(a \to a)$ is the greatest element by (7) from the definition of pseudo-boolean algebras; 0 is the smallest element according to (8).

Some basic simple laws which hold in pseudo-boolean algebras are summarized in the following three lemmas.

Lemma 2.1.6 *Let $\mathfrak{A}$ be a pseudo-boolean algebra, and $\top$ be the unit element of $\mathfrak{A}$. Elements of $\mathfrak{A}$ satisfy the following properties.*

(1) $(a \to b) = \top \Leftrightarrow a \leq b$;

(2) $a = b \Leftrightarrow ((a \to b) = \top$ & $(b \to a) = \top)$;

(3) $(a \to \top) = \top$;

(4) $(\top \to b) = b$;

(5) $(a_1 \leq a_2) \Rightarrow (a_2 \to b) \leq (a_1 \to b)$;

(6) $(b_1 \leq b_2) \Rightarrow (a \to b_1) \leq (a \to b_2)$;

(7) $b \leq (a \to b)$;

Proof. By Lemma 2.1.3 for arbitrary elements $x, y \in \mathfrak{A}$, $x \to y$ is the pseudo-complement of x to y. We will use this fact in relation to different elements of $\mathfrak{A}$ to show all above mentioned properties hold together with those stated in next two lemmas. So (1) follows directly from $a \wedge \top = a$ and $a \to b$ is the pseudo-complement. (2) follows from (1) and the definition of pseudo-complements. (3) and (4) are direct consequences of the definitions for the *pseudo-complement to* and for the *unit element*. As for (5), we have $a_1 \wedge (a_2 \to b) \leq a_2 \wedge (a_2 \to b) \leq b$. Thus $(a_2 \to b) \leq (a_1 \to b)$. To show (6) it is sufficient to note that $(a \to b_1) \wedge a \leq b_1 \leq b_2$ which implies $(a \to b_1) \leq (a \to b_2)$, also $a \wedge b \leq b$ which yields $b \leq (a \to b)$ and (7) holds. ∎

Lemma 2.1.7 *If $\mathfrak{A}$ is a pseudo-boolean algebra and a, b, c are any elements of $\mathfrak{A}$ then*

(8) $(a \to c) \wedge (b \to c) = (a \vee b) \to c;$

(9) $(a \to (b \to c)) = (a \wedge b) \to c = b \to (a \to c);$

(10) $(c \to a) \le ((c \to (a \to b)) \to (c \to b));$

(11) $((a \to b) \wedge (b \to c)) \le (a \to c);$

(12) $(a \to b) \le ((b \to c) \to (a \to c));$

(13) $a \le (b \to (a \wedge b));$

(14) $(a \to (b \to c)) \le ((a \to b) \to (a \to c));$

(15) $c \wedge ((c \wedge a) \to (c \wedge b)) = c \wedge (a \to b).$

Proof. To prove (8) we note first that $((a \vee b) \to c) \le (a \to c)$ and $((a \vee b) \to c) \le (b \to c)$ by (5) from Lemma 2.1.6. Therefore $((a \vee b) \to c) \le d$, where $d := (a \to c) \wedge (b \to c)$. On other hand, by (4) from the definition of pseudo-boolean algebra and Lemma 2.1.4 we derive

$$(a \vee b) \wedge d = (a \wedge d) \vee (b \wedge d) =$$

$$((a \wedge c \wedge (b \to c)) \vee ((a \to c) \wedge b \wedge c)) \le c.$$

Taking into account our definition of pseudo-complements we extract from this $d \le (a \vee b) \to c$. Hence $d = ((a \vee b) \to c)$ and (8) is established. Now turning to (9), we first note that for every element x, the following relations are pairwise equivalent:

$$x \le (a \to (b \to c)), \ (a \wedge x \le (b \to c)),$$

$$(a \wedge b) \wedge x \le c, \ x \le ((a \wedge b) \to c).$$

(9) follows immediately from the equivalence of the first and the last relations above.

Because according to (6) from Lemma 2.1.6 $(c \to (a \wedge b)) \le (c \to b)$, using (4) and then (6) from the definition of pseudo-boolean algebras, we sequentially derive in turn $c \to (a \wedge (a \to b)) \le (c \to b)$, and $(c \to a) \wedge (c \to (a \to b)) \le (c \to b)$. This immediately yields (10). (11) immediately follows from the definition of *pseudo complement to* because

$$((a \to b) \wedge (b \to c)) \wedge a = a \wedge b \wedge (b \to c) = a \wedge b \wedge c \le c.$$

(12) is a direct consequence of (11) and the definition of pseudo-complements. Turning to prove (13), we note $b \wedge a \le a \wedge b$. By the definition of *pseudo-complement to* we obtain $a \le (b \to (a \wedge b))$, thus (13) holds. To prove (14), we first see that

$$(a \wedge (a \to b) \wedge (a \to (b \to c)) = a \wedge b \wedge (b \to c) = a \wedge b \wedge c \le c$$

(by (4) from the definition of pseudo-boolean algebras). Directly using the definition of pseudo-complement, we obtain

$$(a \to b) \land (a \to (b \to c)) \leq (a \to c)$$
$$(a \to (b \to c)) \leq ((a \to b) \to (a \to c)).$$

It now remains only to prove (15). By (6) from the definition of pseudo-boolean algebras, (1) from Lemma 2.1.6, and (5) from Lemma 2.1.6, we conclude

$$(c \land a) \to (c \land b) = ((c \land a) \to c) \land ((c \land a) \to b)$$
$$= (c \land a) \to b \geq (a \to b).$$

This gives us $c \land ((c \land a)) \to (c \land b)) \geq c \land (a \to b)$. For the other side, by (4) from the definition of pseudo-boolean algebras, we infer

$$a \land c \land ((c \land a) \to (c \land b)) = (c \land a) \land (c \land b) \leq b,$$

which by the definition of pseudo-complement gives us $c \land ((c \land a) \to (c \land b)) \leq c \land (a \to b)$ and proves (15). ∎

Definition 2.1.8 *For every element a of a certain pseudo-boolean algebra $\mathfrak{A}$, the pseudo-complement of a is the element $a \to 0$. We will denote this element by $\neg a$.*

Lemma 2.1.9 *For every pseudo-boolean algebra $\mathfrak{A}$, the following relations hold*

(16) $(a \leq b) \Rightarrow \neg b \leq \neg a$;

(17) $\neg \top = \bot$, $\neg \bot = \top$;

(18) $a \land \neg a = \bot$;

(19) $\neg(a \land \neg a) = \top$;

(20) $a \leq \neg\neg a$;

(21) $\neg\neg\neg a = \neg a$;

(22) $\neg(a \lor b) = \neg a \land \neg b$;

(23) $\neg a \lor \neg b \leq \neg(a \land b)$;

(24) $\neg a \lor \neg b \leq (a \to b)$;

(25) $a \to b \leq \neg b \to \neg a$;

(26) $(a \to \neg b) = \neg(a \land b) = (b \to \neg a)$;

(27) $(a \to \neg b) = \neg\neg(a \to \neg b)$;

(28) $\neg\neg(a \to b) \leq (a \to \neg\neg b)$;

(29) $(\bot \to a) = \top$;

Proof. Suppose $a \leq b$. Then $a \wedge \neg b \leq b \wedge \neg b = \bot$. Thus $\neg b \leq \neg a$ and (16) is established. (17) and (18) follows directly from our definitions. (19) is a corollary of (18) and (17). Because $a \wedge \neg a = \bot$ we immediately infer that $a \leq \neg \neg a$, i.e. (20) holds. From (20) and (16) we have $\neg \neg \neg a \leq \neg a$. Making the substitution in (20), replacing a by $\neg a$, we get $\neg a \leq \neg \neg \neg a$, thus (21) is valid. Furthermore, $\neg(a \vee b) = (a \vee b) \rightarrow \bot = (a \rightarrow \bot) \wedge (b \rightarrow \bot) = \neg a \wedge \neg b$ according to definition of pseudo-complements and according to (8) from Lemma 2.1.7. Hence (22) holds. Because, by the distributive law $(\neg a \vee \neg b) \wedge (a \wedge b) \leq \bot$ is valid, we get $\neg a \vee \neg b \leq (a \wedge b) \rightarrow \bot = \neg(a \wedge b)$, and (23) is established. Again by the distributive law and (18) we have $(\neg a \vee b) \wedge a \leq b$ which yields (24). (25) follows from our definitions and (11) from Lemma 2.1.7 when $c = \bot$. (26) is a consequence of our definitions and (9) from Lemma 2.1.7 when $c = \bot$. To prove (27) we note that by (26) and (21)

$$\neg\neg(a \rightarrow \neg b) = \neg\neg\neg(a \wedge b) = \neg(a \wedge b) = a \rightarrow \neg b$$

by (26) and (21). Turning to (28), we use (20) and (6) from Lemma 2.1.6, and we obtain $a \rightarrow b \leq (a \rightarrow \neg\neg b)$. From this, using (16) and (27), it follows that

$$\neg\neg(a \rightarrow b) \leq \neg\neg(a \rightarrow \neg\neg b) = a \rightarrow \neg\neg b,$$

i.e. (28) holds. (29) is a directly consequence of (1) from Lemma 2.1.6. ∎

Theorem 2.1.10 (Algebraic Completeness Theorem for S.I. Logics)

(1) The variety of algebras $Var(H)$ corresponding to the intuitionistic propositional logic λ coincides with the variety V_{pb} of all pseudo-boolean algebras (with $\neg$ interpreted in V_{pb} as the pseudo-complement, and $\bot$ interpreted in $Var(H)$ as $\neg\top$). Moreover, in $Var(H)$, and V_{pb} the following identity $\top = (a \rightarrow a) = (a \rightarrow (a \rightarrow a))$ holds;

(2) The variety $Var(\lambda)$ corresponding to a superintuitionistic logic λ is a variety of pseudo-boolean algebras.

(3) For any family $\mathcal{F}$ of pseudo-boolean algebras, the set $\lambda(\mathcal{F})$ defined as follows $\lambda(\mathcal{F}) := \{\alpha \mid (\forall \mathfrak{A} \in \mathcal{F})(\mathfrak{A} \models \alpha)\}$ is a superintuitionistic logic.

(4) For every superintuitionistic logic λ, there is a pseudo-boolean algebra $\mathfrak{A} \in Var(\lambda)$ such that $\lambda(\mathfrak{A}) = \lambda = \lambda(Var(\lambda))$.

Proof. We see that the signatures of $Var(H)$ and V_{pq} are slightly different, namely algebras from $Var(H)$ have the operation $\neg$ which is omitted in the definition of pseudo-boolean algebras, in contrast, pseudo-boolean algebras have the constant $\bot$ designating the smallest element. We define $\neg a$ in the algebras from V_{pb} as before : $\neg a := a \rightarrow \bot$, and also we define the zero element in the algebras from $Var(H)$ as follows: $\bot := \neg(a \rightarrow (a \rightarrow a))$ (for any fixed element a).

According to Theorem 1.3.8 any superintuitionistic logic λ is an algebraic logic with algebraic presentation $(\equiv) := \{x \to y, y \to x\}$, $\top := \{(x \to (x \to x))\}$.

To show that all the algebras from $Var(\lambda)$ are pseudo-boolean it is sufficient to prove that the free algebra $\mathfrak{F}(\lambda)$ of countable rank (cf. Theorem 1.3.21) from $Var(\lambda)$ is a pseudo-boolean algebra. Indeed, it is sufficient to prove that all countable subalgebras of algebras from $Var(\lambda)$ are pseudo-boolean, and all such algebras are homomorphic images of $\mathfrak{F}(\lambda)$ and the truths of any identity have to be preserved under homomorphism. The algebra $\mathfrak{F}(\lambda)$ is presented as the quotient-algebra on the matrix of formulas $\mathfrak{M}_L(\lambda)$ with respect to the congruence relation $g \sim f \Leftrightarrow \vdash_\lambda (g \equiv f)$. Therefore in order to show that $\mathfrak{F}(\lambda)$ is a pseudo-boolean algebra it is sufficient to prove the following lemma

Lemma 2.1.11 *For every superintuitionistic logic λ the following hold*

(r) $\vdash_\lambda c \to c$ (reflexivity of $\to$);

(t) $\alpha \to \beta$, $\beta \to \gamma \vdash_\lambda \alpha \to \beta$ (transitivity of $\to$ or the **cut rule**);

(i) $\vdash_\lambda (a \to (a \to a)) \to (a \to a)$, $\vdash_\lambda (a \to a) \to (a \to (a \to a))$;

(a) $\vdash_\lambda [a \vee b] \equiv (b \vee a)]$, $\vdash_\lambda [(a \wedge b) \equiv (b \wedge a)]$;

(b) $\vdash_\lambda [a \vee (b \vee c) \equiv (a \vee b) \vee c]$, $\vdash_\lambda [a \wedge (b \wedge c) \equiv (a \wedge b) \wedge c]$;

(c) $\vdash_\lambda [(a \wedge b) \vee b \equiv b]$, $\vdash_\lambda [a \wedge (a \vee b) \equiv a]$;

(d) $\vdash_\lambda [a \wedge (a \to b) \equiv a \wedge b]$;

(e) $\vdash_\lambda [(a \to b) \wedge b \equiv b]$;

(f) $\vdash_\lambda [(a \to b) \wedge (a \to c) \equiv a \to (b \wedge c)]$;

(g) $\vdash_\lambda [(a \to a) \wedge b \equiv b]$;

(h) $\vdash_\lambda [(\bot \wedge a) \equiv a]$;

Proof. We recall that every superintuitionistic logic λ contains among its theorems the axioms (A1) - (A9) of the classical propositional calculus and the axiom (A11) : $x \to (\neg x \to y)$ of H. To show (r) we derive:

$c \to (c \to c)$ (A1), $c \to ((c \to c) \to c)$ (A1)
$(c \to ((c \to c) \to c)) \to ((c \to (c \to c)) \to (c \to c))$ (A2)
$(c \to (c \to c)) \to (c \to c)$ (use (R1))
$(c \to c)$ (use (R1).

Hence (r) holds.

$\alpha \to \beta$, $\beta \to \gamma$
$(\alpha \to (\beta \to \gamma)) \to ((\alpha \to \beta)) \to (\alpha \to \gamma))$ (A2)
$(\beta \to \gamma) \to (\alpha \to (\beta \to \gamma))$ (A1)
$\alpha \to (\beta \to \gamma)$ (R1)
$\alpha \to \gamma$ (using (R1) twice.)

So (t) is also valid. With respect to (i), (i) is a direct consequence of (r) and (A1). Each of the other above mentioned propositions has the form $\vdash_\lambda \alpha \equiv \beta$. Bearing in mind the meaning of $\equiv$ and the Deduction Theorem for superintuitionistic logics (Theorem 1.3.9), in each case it is sufficient to show that $\alpha \vdash_\lambda \beta$ and $\beta \vdash_\lambda \beta$. Now we turn to (a). The proof of the relation $a \vee b \vdash_\lambda b \vee a$ is given below: $a \vee b$, $(a \to b \vee a) \to ((b \to b \vee a) \to ((a \vee b) \to (b \vee a)))$ (A5)

$\quad a \to b \vee a$ (A4), $b \to b \vee a$ (A3)

$\quad (b \to b \vee a) \to ((a \vee b) \to (b \vee a))$ (by (R1))

$\quad (a \vee b) \to (b \vee a)$ (by (R1)), $b \vee a$ (by (R1)).

The converse entailment can be shown in a similar way. Now we consider the second part of (a). $a \wedge b$, $a \wedge b \to a$ (A6), $a \wedge b \to b$ (A7), $(a \wedge b \to b) \to ((a \wedge b \to a) \to ((a \wedge b) \to (b \wedge a)))$ (see (A8)), $b \wedge a$ (using (R1) three times).

Now we turn to (b).

$\quad a \vee (b \vee c)$, $(a \to ((a \vee b) \vee c)) \to (((b \vee c) \to ((a \vee b) \vee c)) \to$

$\quad ((a \vee (b \vee c)) \to ((a \vee b) \vee c)))$ (A5), $a \to a \vee b$ (A3)

$\quad (a \vee b) \to ((a \vee b) \vee c)$ (A3), $((a \vee b) \to ((a \vee b) \vee c)) \to (a \to$

$\quad ((a \vee b) \to ((a \vee b) \vee c)))$ (A1)

$\quad a \to ((a \vee b) \to ((a \vee b) \vee c))$ (by (R1))

$\quad (a \to ((a \vee b) \to ((a \vee b) \vee c))) \to$

$\quad \to ((a \to a \vee b) \to (a \to ((a \vee b) \vee c)))$ (A2)

$\quad ((a \to a \vee b) \to (a \to ((a \vee b) \vee c)))$ (by (R1)),

$\quad a \to ((a \vee b) \vee c)$ (by (R1)), $((b \vee c) \to ((a \vee b) \vee c)) \to$

$\quad ((a \vee (b \vee c)) \to$

$\quad ((a \vee b) \vee c))$ (by (R1))

$\quad (b \to a \vee b) \to ((c \to (a \vee) \vee c) \to ((b \vee c) \to ((a \vee b) \vee c)))$ (A5)

$\quad b \to a \vee b$ (A4), $c \to (a \vee) \vee c$ (A4)

$\quad (b \vee c) \to ((a \vee b) \vee c)$ (by (R1) twice)

$\quad (a \vee (b \vee c)) \to ((a \vee b) \vee c)$ (by (R1)).

The converse entailment can be shown in the same way and is left to the reader as an exercise. Now we examine the second part of (b).

$\quad [(a \wedge (b \wedge c)) \to (a \wedge b)] \to [((a \wedge (b \wedge c)) \to \wedge c) \to$

$\quad \to ((a \wedge (b \wedge c)) \to ((a \wedge b) \wedge c))]$ (A8)

$\quad (a \wedge (b \wedge c)) \to (b \wedge c)$ (A7)

$\quad (b \wedge c) \to c$ (A7)

$\quad (a \wedge (b \wedge c)) \to c$ (by (t))

$\quad ((a \wedge (b \wedge c)) \to a) \to (((a \wedge (b \wedge c)) \to b) \to$

$\quad \to ((a \wedge (b \wedge c)) \to (a \wedge b)))$ (A8)

$\quad (a \wedge (b \wedge c)) \to a$ (A6)

$\quad (b \wedge c) \to b$ (A6)

$\quad (a \wedge (b \wedge c)) \to b$ (by (t))

$\quad (a \wedge (b \wedge c)) \to ((a \wedge b) \wedge c)$ (to use (R1) twice)

The proof of the converse can be obtained in a similar way and is left to the reader as an exercise. We turn now to (c).

$(a \wedge b) \vee b$
$((a \wedge b) \to b) \to ((b \to b) \to (((a \wedge b) \vee b) \to b))$ (A5)
$(a \wedge b) \to b$ (A7), $b \to b$ (by (r))
$((a \wedge b) \vee b) \to b$ (to use (R1) twice), b (by (R1))

Conversely: $b, b \to (a \wedge b) \vee b$ (A4), $(a \wedge b) \vee b$ (to use (R1)) is the proof of the converse entailment. Thus the first part of (c) is established. Let us consider the second one. The sequence: $a \wedge (a \vee b), a \wedge (a \vee b) \to a$ (A6), a (by (R1)) entails that $a \wedge (a \vee b) \vdash_\lambda b$. Furthermore,

$a, \ a \to (a \vee b)$ (A3), $a \to a$ (by (r))
$(a \to a) \to ((a \to (a \vee b)) \to (a \to (a \wedge (a \vee b))))$ (A8)
$a \to (a \wedge (a \vee b))$ (to apply (R1) twice)
$a \wedge (a \vee b)$ (by (R1))

and (c) is proved. Now we are going to establish (d).

$a \wedge (a \to b) \vdash_\lambda a, \ (a \to b)$ (by (A6), (A7), (R1))
$a, a \to b \vdash_\lambda b$ (by (R1))
$a, b \vdash_\lambda a \wedge b$ (by (A8), (r), (A1), (R1))
$a \wedge (a \to b) \vdash_\lambda a \wedge b$

Conversely,

$a \wedge b \vdash_\lambda a, b$ (by (A6), (A7), (R1))
$a, b \vdash_\lambda (a \to b)$ (by (A1), (R1))
$a, (a \to b) \vdash_\lambda a \wedge (a \to b)$ (by (A8), (r), (A1), (R1))
$a \wedge b \vdash_\lambda a \wedge (a \to b)$

Hence (d) holds. To show (e) we calculate: $(a \to b) \wedge b \vdash_\lambda b$ (by (A7), (R1)). Conversely,

$\vdash_\lambda b \to (a \to b)$ (A1)
$\vdash_\lambda (b \to b)$ (by (r))
$b \vdash_\lambda b \to (a \to b) \wedge b$ (by (R1), (A8), (r), (A1), (R1))

and (e) is proved. We now turn to (f). In order to show $(a \to b) \wedge (a \to c) \vdash_\lambda a \to (b \wedge c)$ by the Deduction Theorem for superintuitionistic logics (Theorem 1.3.9), it is sufficient to show that $(a \to b) \wedge (a \to c), a \vdash_\lambda (b \wedge c)$. We show this below:

$(a \to b) \wedge (a \to c), a \vdash_\lambda b, c$ (by (A7), (A8), (R1))
$b, c \vdash_\lambda b \wedge c$ (by (A8), (r), (A1), (R1)),

which gives us what we need. Thus $(a \to b) \wedge (a \to c) \vdash_\lambda a \to (b \wedge c)$ holds. Conversely,

$$\vdash_\lambda (b \wedge c) \to b, \ (b \wedge c) \to c \ ((A6), (A7))$$
$$a \to (b \wedge c) \vdash_\lambda a \to b, \ a \to c \ (\text{by (t)})$$
$$a \to b, \ a \to c \vdash_\lambda (a \to b) \wedge (a \to c) \ (\text{by (A8), (r), (A1), (R1)})$$
$$a \to (b \wedge c) \vdash_\lambda (a \to b) \wedge (a \to c)$$

Since (f) holds. As for (g), the relation $(a \to a) \wedge b \vdash_\lambda b$ asserts the provability a conjunct from the conjunction which has already been proved in one of the ways discussed above. By (r) the relation $\vdash_\lambda (a \to a)$ is valid. Therefore $b \vdash_\lambda (a \to a) \wedge b$ (again using (A8), (r), (A1), (R1)). Thus (g) holds. It remains only to verify (h). Again, $\bot \wedge a \vdash_\lambda \bot$ and $\vdash_\lambda (\bot \to \bot)$ by (r). We intend to show that $\bot \vdash_\lambda a$. In fact, it is sufficient to show that $\vdash_\lambda \neg(\top) \to a$. We have $\vdash_\lambda \top \to (\neg\top \to a)$, it is (A11). Furthermore, $\vdash_\lambda \top$, hence $\vdash_\lambda \neg\top \to a$. Thus $\bot \vdash_\lambda \bot, a$ and consequently $\bot \vdash_\lambda \bot \wedge a$. Now (h) is established and our lemma is proved. ∎

From this lemma it follows that every variety $Var(\lambda)$ as well as $Var(H)$ itself are varieties of certain pseudo-boolean algebras and that in this varieties $\top = (a \to a) = (a \to (a \to a))$. If $\mathfrak{A}$ is a pseudo-boolean algebra then $a \to a$ is the unit element of $\mathfrak{A}$ (Lemma 2.1.5). Conversely, $a \to (a \to a)$ is a pseudo-complement of a to the unit element, hence we get $a \to (a \to a)$ is equal to the unit element also. Hence $\top = (a \to a) = (a \to (a \to a))$. Thus (2) is proved. In order to prove (1), it is sufficient to show $Var(H)$ coincides with V_{pb}. To establish this we have only to show that every pseudo-boolean algebra $\mathfrak{A}$ belongs to $Var(H)$. According to the definition of $Var(H)$,

$$Var(H) := \{\mathfrak{B} \mid \forall \alpha \in H(\mathfrak{B} \models \alpha = \top)\}.$$

We have to show that $\mathfrak{A} \in Var(H)$. For every $\alpha \in H$, $\vdash_{AS(H)} \alpha$, where $AS(H)$ is the axiomatic systems with the set of axioms (A1) - (A9) (for classical propositional calculus) and (A11), and single inference rule (R1) - modus ponens. Therefore it is sufficient to show that under every interpretation all the above mentioned axioms take the value $\top$ in $\mathfrak{A}$, and that if $\mathfrak{A} \models \alpha = \top$ and $\mathfrak{A} \models \alpha \to \beta = \top$ then $\mathfrak{A} \models \beta = \top$. We first intend to verify the last conjecture. By (1) from Lemma 2.1.6 if $\mathfrak{A} \models (\alpha \to \beta) = \top$ then $\mathfrak{A} \models \alpha \leq \beta$. Therefore $\mathfrak{A} \models \alpha = \top$ implies $\mathfrak{A} \models \beta = \top$. Thus (R1) preserves the value of the designated element $\top$. Now we will check the truth of the axioms in $\mathfrak{A}$. By (1) from Lemma 2.1.6 in order to show $\alpha \to \beta = \top$ in $\mathfrak{A}$ it suffices to prove that $\alpha \leq \beta$. We will use this observation and the fact that $a \to b$ is the pseudo-complement of a to b (Lemma 2.1.3) in our proofs below.

Hence $\mathfrak{A} \models x \to (y \to x) = \top$ (axiom (A1)) is equivalent to $\mathfrak{A} \models x \leq (y \to x)$ which directly follows from the definition of a pseudo-complement. Now (A2):

$(x \to (y \to z)) \wedge (x \to y) = x \to ((y \to z) \wedge y)$
(by (6) of Definition 2.1.1)
$x \to ((y \to z) \wedge y) = x \to (y \wedge z)$ (by (4) of Definition 2.1.1)
$x \to (y \wedge z) \leq (x \to z)$ (by (6) of Lemma 2.1.6).

Hence (A2) is shown. That (A3) and (A4) always take the value $\top$ follows directly from (1) of Lemma 2.1.6.

$(x \to z) \wedge (y \to z) = ((x \vee y) \to z))$ (by (8) from Lemma 2.1.7).

Thus axiom (A5) is valid also. The validity of axioms (A7) and (A8) is evident. With the axiom (A8), we derive

$(x \to y) \wedge (x \to z) = (x \to (y \wedge z))$ (by (6) from Definition 2.1.1)

therefore the value of (A8) is always equal to $\top$. Now we turn to (A9),

$(x \to y) \wedge (x \to \neg y) = (x \to (y \wedge \neg y))$ (by (6) of Definition 2.1.1)
$(x \to (y \wedge \neg y)) = (x \to \bot)$ (by (18) of Lemma 2.1.9)
$(x \to \bot) = \neg x$,

therefore the case of axiom (A9) is also established. It remains only to verify the validness of the axiom (A11).

$x \wedge \neg x = x \wedge (x \to \bot) = x \wedge \bot$ (by (4) of Definition 2.1.1)
$x \wedge \bot = \bot$ (by (8) of Definition 2.1.1),

thus (A11) take the value $\top$, this completes the proof of (1). Now we turn to (3). Actually, we have shown that any pseudo-boolean algebra $\mathfrak{A}$ is a member of $Var(H)$. Therefore $\lambda(\mathcal{F}) = \bigcup_{\mathfrak{A} \in \mathcal{F}} \lambda(\mathfrak{A}) \supseteq H$. That $\lambda(\mathcal{F})$ is closed with respect to substitutions is evident, and we have showed above that, for any pseudo-boolean algebra $\mathfrak{A}$, if $\mathfrak{A} \models \alpha = \top$ and $\mathfrak{A} \models (\alpha \to \beta) = \top$ then $\mathfrak{A} \models \beta = \top$. Therefore the set $\lambda(\mathcal{F})$ is also closed with respect to modus ponens: $(R1) := \{x, x \to y/y\}$. Thus $\lambda(\mathcal{F})$ is a superintuitionistic logic, i.e. (3) is valid. It remains only to prove (4). According to Theorems 1.3.19 and 1.3.21 the free algebra of countable rank $\mathfrak{F}(\lambda)$ from the variety $V := Var(\lambda)$ has properties: $\lambda(\mathfrak{F}(\lambda)) = \lambda = \lambda(Var(\lambda))$. By (2) $\mathfrak{F}(\lambda)$ is a pseudo-boolean algebra and has all the properties required in (4) for $\mathfrak{A}$. $\blacksquare$

Using this theorem we can show directly that some formulas which are theorems of the classical propositional logic PC are non theorems of the Heyting intuitionistic logic H. For instance, a typical example is the formula which correspond to the law of the excluded middle. Before to illustrate this we pause briefly to introduce some other definition for pseudo-boolean algebras which will be sometimes more convenient for our applications.

Definition 2.1.12 *Pseudo-boolean algebra is a lattice L with smallest element $\perp$ such that for every elements $a, b \in L$ there is the pseudo-complement of a to b.*

Lemma 2.1.13 *This definition is equivalent to the first one.*

Proof. Indeed any pseudo-boolean algebra according to the first definition is a lattice with smallest element $\perp$ and by Lemma 2.1.3 $a \to b$ is the pseudo-complement a to b. Conversely, any lattice satisfies the laws (1) - (3) from the first definition of pseudo-boolean algebras. We denote the pseudo-complement a to b by $a \to b$. That $a \wedge (a \to b) \leq b$ follows immediately by the definition. Because $a \wedge (a \to b) \leq a$ we have $a \wedge (a \to b) \leq a \wedge b$. Since $a \wedge b \leq a$ and $a \wedge b \leq (a \to b)$, we obtain $a \wedge (a \to b) = a \wedge b$). Thus the law (4) also holds. $b \leq (a \to b)$ by definition, since $(a \to b) \wedge b = b$ and (5) is true. $a \wedge (a \to b) \leq b$, $a \wedge (a \to c) \leq c$ hence $(a \to b) \wedge (a \to c) \leq (a \to (b \wedge c))$. Conversely, $a \wedge (a \to (b \wedge c)) \leq (b \wedge c)$. Therefore $(a \to (b \wedge c)) \leq (a \to b) \wedge (a \to c)$. Thus (6) holds. (7) and (8) follow directly from our definitions. ∎

A measure on how convenient semantic tools are usually consists of possibility to recognize non-theorems of a given logic. We illustrate this for the case of intuitionistic logic H using pseudo-boolean algebras in several examples below. In doing this it will be convenient to apply the following

Lemma 2.1.14 *Any finite distributive lattice L is a pseudo-boolean algebra.*

Proof. Being finite, L must have the smallest element $\perp$ and the greatest element $\top$. By Lemma 2.1.13 it is sufficient to show that there is the pseudo-complement $a \to b$ of a to b for all elements a, b from L. We take the family

$$F := \{c \mid c \in L, c \wedge a \leq b\}.$$

F is not empty, in particular $b \in F$. We intend to show that $a \to b = \bigvee_{c \in F} c$. Indeed, by the fact that L is distributive we have

$$a \wedge F = \bigvee_{c \in F} a \wedge c \leq b,$$

which yields that $\bigvee_{c \in F} c$ is the greatest element in F, and $\bigvee_{c \in F} c = a \to b$. ∎

Example 2.1.15 *The law of excluded middle $p \vee \neg p$ is not a theorem of the intuitionistic propositional logic H.*

In fact, we take the lattice $\mathfrak{A}_3$ with three elements basis set $\{\perp, a, \top\}$, whose diagram is placed in Fig 2.1. This lattice by Lemma 2.1.14 $\mathfrak{A}_3$ is a pseudo-boolean algebra. Therefore using Theorem 2.1.10 it follows $\mathfrak{A} \in Var(H)$. It is easy to see that $\neg a = \perp$, therefore $a \vee \neg a = a \neq \top$. By the completeness theorem (Theorem 2.1.10) we conclude $p \vee \neg p \notin H$, hence the law of excluded middle is not a theorem of the intuitionistic logic.

Figure 2.1:

Exercise 2.1.16 *Show that the axiom (A10): $\neg\neg p \rightarrow p$ of CPC is not a theorem of H.*

Hint: use $\mathfrak{A}_3$ and a for the evaluation of (A10).

Example 2.1.17 *The formula $(p \rightarrow q) \vee (q \rightarrow a)$ is not a theorem of H.*

This case we take the lattice $\mathfrak{A}_5$ whose diagram is placed in Fig. 2.2. Again $\mathfrak{A}_5$, is

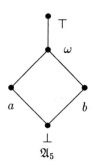

Figure 2.2:

distributive lattice, therefore by Lemma 2.1.14 $\mathfrak{A}_5$ is a pseudo-boolean algebra and $\mathfrak{A}_5 \in Var(H)$ (according to Theorem 2.1.10). Also we have

$$(a \rightarrow b) = b, (b \rightarrow a) = a, (a \rightarrow b) \vee (b \rightarrow a) = \omega \neq \top.$$

Hence $\mathfrak{A}_5 \not\models (p \rightarrow q) \vee (q \rightarrow a) = \top$. Thus by the Completeness Theorem (Theorem 2.1.10) we derive $(p \rightarrow q) \vee (q \rightarrow p) \notin H$. ∎

Exercise 2.1.18 *Show that $\mathfrak{A}_3 \models p \vee \neg p = \top$ but $\mathfrak{A}_5 \not\models \neg p \vee \neg\neg p = \top$, in particular the weak law of excluded middle $\neg p \vee \neg\neg p$ is not a theorem of H.*

These examples show us that algebraic semantics are convenient for rejecting formulas. To do this we have to evaluate formulas in algebras from varieties $Var(\lambda)$ which are varieties of pseudo-boolean algebras in the case of superintuitionistic logics. In general these varieties have infinitely many algebras and to observe all of them would sometimes be difficult. Therefore it would be desirable to reduce the family of algebras from $Var(\lambda)$ in which we have to check the truth of formulas.

Of course we can reduce this family downto a single free algebra of countable rank from $Var(\lambda)$. But sometimes this gives us too little information. Generally speaking, in a sense, all the algebras have somehow to be presented in this family. This kind of presentation can be given according to Birkhoff's Representation Theorem (cf. Theorem 1.2.57). Indeed every algebra $\mathfrak{B}$ from a variety V can be presented as a subdirect product of certain subdirectly irreducible algebras $\mathfrak{B}_i$, $i \in I$ from V. Hence $\mathfrak{B}$ is a subalgebra of $\prod_{i \in I} \mathfrak{B}_i$. Thus any identity is valid in $\mathfrak{B}$ only if this identity is valid on all $\mathfrak{B}_i$, $i \in I$. Therefore an identity e is valid in V if and only if e is valid in all subdirectly irreducible algebras from V. Consequently the description of subdirectly irreducible algebras is really important.

We now proceed to describe subdirectly irreducible pseudo-boolean algebras. First we will develop a theory of congruence relations on pseudo-boolean algebras.

Lemma 2.1.19 *Let* $\mathfrak{A}$ *be a pseudo-boolean algebra and* Δ *be a non-empty filter on* $\mathfrak{A}$. *Then the relation* $\sim_\Delta$, *where* $a \sim_\Delta b \Leftrightarrow (a \to b) \wedge (b \to a) \in \Delta$ *is a congruence relation on* $\mathfrak{A}$. *Moreover* $\forall a \in \mathfrak{A}$, $a \in \Delta$ *iff* $a \sim_\Delta \top$.

Proof. In fact, that $\sim_\Delta$ is reflexive and symmetric follows directly from the definition. Suppose $(a \to b) \in \Delta$ and $(b \to c) \in \Delta$. Then $(a \to b) \wedge (b \to c) \in \Delta$ and $(a \to b) \wedge (b \to c) \leq ((a \to c)$ by (11) from Lemma 2.1.7. Hence we get $(a \to c) \in \Delta$. Therefore $\sim_\Delta$ is transitive. In order to show $\sim_\Delta$ is a congruence we have now to check that $\sim_\Delta$ is consistent with the operations. Let $(a_1 \to b_1) \wedge (b_1 \to a_1) \in \Delta$ and $(a_2 \to b_2) \wedge (b_2 \to a_2) \in \Delta$. Then $a_1 \wedge a_2 \to a_1 = \top \in \Delta$, $a_1 \wedge a_2 \to a_2 = \top \in \Delta$ by (1) from Lemma 2.1.6. Therefore we get

$$(a_1 \wedge a_2 \to a_1) \wedge (a_1 \to b_1) \in \Delta$$
$$(a_1 \wedge a_2 \to a_2) \wedge (a_1 \to b_2) \in \Delta$$

Because Δ is a filter, using (11) from Lemma 2.1.7, we infer from this that

$$a_1 \wedge a_2 \to b_1) \in \Delta, \quad (a_1 \wedge a_2 \to b_2) \in \Delta,$$
$$(a_1 \wedge a_2 \to b_1) \wedge (a_1 \wedge a_2 \to b_2) \in \Delta.$$

Now applying (6) from Definition 2.1.1 we obtain from above memberships that

$$(a_1 \wedge a_2 \to b_1 \wedge b_2 \in \Delta),$$

what we needed. In a similar way we can show that the converse implication belongs to the filter. Thus $\equiv_\Delta$ is consistent with $\wedge$. Now we turn to the case of $\vee$. Using the fact that Δ is a filter, (1) from Lemma 2.1.6, and then (11) from Lemma 2.1.7, we derive

$$(a_1 \to b_1) \wedge (b_1 \to b_1 \vee b_2) \in \Delta, \quad (a_2 \to b_2) \wedge (b_2 \to b_1 \vee b_2) \in \Delta,$$
$$(a_1 \to b_1) \wedge (b_1 \to b_1 \vee b_2) \leq (a_1 \to b_1 \vee b_2),$$
$$(a_2 \to b_2) \wedge (b_2 \to b_1 \vee b_2) \leq (a_2 \to b_1 \vee b_2.$$

Since $(a_1 \to b_1 \vee b_2) \wedge (a_2 \to b_1 \vee b_2) \in \Delta$, applying (8) from Lemma 2.1.7, from this it follows that $(a_1 \vee a_2 \to b_1 \vee b_2) \in \Delta$.

The case of the reverse implication can be thought of in the similar way and the consistency of $\sim_\Delta$ with $\vee$ is demonstrated. Now it is turn of $\to$. We calculate:

$$(a_2 \to b_2) \leq (a_1 \wedge a_2 \to b_2) \quad \text{(by (5) of Lemma 2.1.6)},$$
$$(a_1 \wedge a_2 \to b_2) = (a_1 \to a_2) \wedge a_1 \to b_2 \quad \text{(by (4) of Def. 2.1.1)},$$
$$(a_1 \to a_2) \wedge a_1 \to b_2 = a_1 \to ((a_1 \to a_2) \to b_2) \quad \text{((9) of Lemma 2.1.7)},$$

Thus $a_1 \to ((a_1 \to a_2) \to b_2) \in \Delta$. Because $b_1 \to a_1 \in \Delta$ we conclude

$$(b_1 \to a_1) \wedge (a_1 \to ((a_1 \to a_2) \to b_2) \in \Delta.$$

Using (11) from Lemma 2.1.7 we infer from the membership relation above that $b_1 \to ((a_1 \to a_2) \to b_2) \in \Delta$ holds, and applying (9) from those lemma, we obtain that $(a_1 \to a_2) \to (b_1 \to b_2) \in \Delta$, which is what we needed.

The inclusion of the reverse implication can be shown in a similar way. Thus the consistency of $\sim_\Delta$ with $\to$ is established. The case of $\neg$ is a corollary of the consistency of $\to$ with $\sim_\Delta$ because $\neg a := a \to \bot$. If $a \in \mathfrak{A}$ then $\top \to a = a$ and $\top \to a \in \Delta$; furthermore $a \to \top = \top \in \Delta$. Thus $a \sim_\Delta \top$. Conversely, if $a \sim_\Delta \top$ then $\top \to a \in \Delta$ and hence $\top \to a = a \in \Delta$. ∎

Lemma 2.1.20 *If $\sim$ is a congruence relation on some pseudo-boolean algebra $\mathfrak{A}$ then $\Delta_\sim := \{a \mid a \in \mathfrak{A}, a \sim \top\}$ is a filter on $\mathfrak{A}$ and $\sim = \sim_{\Delta_\sim}$.*

Proof. Let $a_1, a_2 \in \Delta_\sim$, i.e. $a_1 \sim \top, a_2 \sim \top$. Since $\sim$ is a congruence relation, it follows that $a_1 \wedge a_2 \sim \top \wedge \top = \top$. Hence $a_1 \wedge a_2$ is an element from $\Delta_\sim$. Suppose now that $a \in \Delta_\sim$, i.e. that $a \sim \top$, and let $a \leq b$. Then $a \vee b \sim \top \vee b = \top$. Because $a \leq b$ we have $a \vee b = b$. Thus $b \sim \top$ and $\Delta_\sim$ is really a filter. From the lemma above we know that $\sim_{\Delta_\sim}$ is a congruence relation on $\mathfrak{A}$. If $a \sim b$ then $a \to b \sim a \to a = \top, b \to a \sim b \to b = \top$, so $a \sim_{\Delta_\sim} b$. Conversely, if $a \sim_{\Delta_\sim} b$, which entails that $a \to b \sim \top$ and $b \to a \sim \top$, then

$$\top \wedge b \sim (a \to b) \wedge b \sim \top \wedge b \sim b, \quad \top \wedge b \sim (b \to a) \wedge b = (a \wedge b)$$

by (4) from Definition 2.1.1. Thus $b \sim a \wedge b$, similarly we show that $a \sim a \wedge b$, thus we conclude $a \sim b$. ∎

Lemma 2.1.21 *Let $\mathfrak{A}$ be a pseudo-boolean algebra, Δ be a filter on $\mathfrak{A}$, and $\sim$ be a congruence relation on $\mathfrak{A}$. Then $\sim_{\Delta_\sim} = \sim$ and $\Delta_{\sim_\Delta} = \Delta$.*

Proof. The first claim of this lemma follows from Lemma 2.1.6. If $a \in \Delta$ then by Lemma 2.1.20 $a \sim_\Delta \top$ i.e. $a \in \Delta_{\sim_\Delta}$. Conversely, if $a \in \Delta_{\sim_\Delta}$ then by definition it follows $a \sim_\Delta \top$ which again by Lemma 2.1.20 entails $a \in \Delta$. ∎

The set of all filters of a pseudo-boolean algebra $\mathfrak{A}$ forms a partially ordered set $F(\mathfrak{A})$ with respect to usual set-theoretic inclusion $\subseteq$. Analogously the set of all congruence relations on an algebra $\mathfrak{A}$ forms a partially ordered set $Con(\mathfrak{A})$ with respect $\subseteq$. It is easy to see that both these posets are closed with respect to set-theoretic intersections of arbitrary families of elements. Therefore both these posets form certain complete lattices.

Theorem 2.1.22 *For every pseudo-boolean algebra $\mathfrak{A}$, the lattices $Con(\mathfrak{A})$ and $F(\mathfrak{A})$ are isomorphic. In particular the mapping $\sim \mapsto \Delta_\sim$ is an isomorphism.*

Proof. This follows directly from Lemmas 2.1.20 and 2.1.21.

We are now in a position to prove the following result useful for various applications.

Theorem 2.1.23 *A pseudo-boolean algebra $\mathfrak{A}$ is subdirectly-irreducible if and only if there is the greatest element ω in the set $\{a \mid a \leq \top, a \neq \top\}$.*

Proof. In fact, by the definition an algebra $\mathfrak{A}$ is subdirectly-irreducible if and only if the intersection of all non-trivial congruence relations on $\mathfrak{A}$ is a non-trivial congruence relation. This last condition is equivalent to the claim that there exists a smallest congruence relation on $\mathfrak{A}$ among non-trivial congruences. By theorem 2.1.22 this is equivalent to the claim that there is a smallest filter in $\mathfrak{A}$ among all filters not equal to $\{\top\}$. If the element ω with the required property exists then the filter $\mathcal{F}(a)$ generated by ω is the smallest filter among all filters which differ from the single element filter $\{\top\}$. Therefore $\mathfrak{A}$ is subdirectly irreducible.

Conversely, if $\mathfrak{A}$ is subdirectly irreducible and consequently there is a smallest filter Δ in the set of all filters which differ from $\{\top\}$ then Δ has an element a such that $a \neq \top$. We can show that a has the properties required for ω. Indeed, we claim that Δ is the filter $\mathcal{F}(a)$ generated by a. If not then there is a $b \in \Delta$ such that $b \neq \top$ and $a \nleq b$. Then the filter $\mathcal{F}(b)$ generated by b is properly included Δ, a contradiction. If $\mathcal{F}(a)$ has an element b which differs from $\top$ and a then the filter $\mathcal{F}(b)$ generated by b will be included in Δ again. Therefore $\Delta = \{a, \top\}$. Let us take arbitrary element $b \in |\mathfrak{A}|$ such that $b \neq \top$. The filter $\mathcal{F}(a)$ is included in $\mathcal{F}(b)$ which means $b \leq a$. Thus a has the required properties for ω. ∎

Now we are going to clarify the structure of arbitrary homomorphisms acting on pseudo-boolean algebras.

Lemma 2.1.24 Let $\mathfrak{A}_1$ and $\mathfrak{A}_2$ be a certain pseudo-boolean algebras and let $\varphi : \mathfrak{A}_1 \mapsto \mathfrak{A}_2$ be a homomorphism into $\mathfrak{A}_2$. Then $\Delta(\varphi) := \{a \mid a \in |\mathfrak{A}_1|, \varphi(a) = \top\}$ is a filter on $\mathfrak{A}_1$.

Proof. In fact, $\varphi(\top) = \top$, because being a homomorphism φ must preserve order. Thus the set $\Delta(\varphi)$ is nonempty. If $a_1, a_2 \in \mathfrak{A}_1$, and $\varphi(a_1) = \top$ and $\varphi(a_2) = \top$, then $\varphi(a_1 \wedge a_2) = \varphi(a_1) \wedge \varphi(a_2) = \top \wedge \top = \top$. Thus $\Delta(\varphi)$ is closed with respect to intersection. If $a, b \in \mathfrak{A}_1$ and $a \leq b$, and still $\varphi(a) = \top$ then $\varphi(a \vee b) = \varphi(a) \vee \varphi(b) = \varphi(b) = \top \vee \varphi(b) = \top$. Hence $\Delta(\varphi)$ is also closed with respect to extensions. ∎

Using this observation we can reformulate the homomorphism theorem (Theorem 1.2.12) for the case of pseudo-boolean algebras as follows.

Theorem 2.1.25 Let $\mathfrak{A}_1$ and $\mathfrak{A}_2$ be some pseudo-boolean algebras, $\varphi : \mathfrak{A}_1 \mapsto \mathfrak{A}_2$ be a homomorphism onto $\mathfrak{A}_2$, and $\approx$ be the kernel congruence relation generated by φ. The following relations hold

(i) $\approx = \sim_{\Delta(\varphi)}$;

(ii) algebras $\mathfrak{A}_2$ and $\mathfrak{A}_1 / \sim_{\Delta(\varphi)}$ are isomorphic and $[a]_{\sim_{\Delta(\varphi)}} \mapsto \varphi(a)$ is the corresponding isomorphism.

Proof. If $a \approx b$ then $\varphi(a) = \varphi(b)$ holds. Thus $(\varphi(a) \to \varphi(b)) = \top$ and $(\varphi(b) \to \varphi(a)) = \top$. Hence $\varphi((a \to b) \wedge (b \to a)) = \top \wedge \top = \top$, and $(a \to b) \wedge (b \to a) \in \Delta(\varphi)$, and therefore $a \sim_{\Delta(\varphi)} b$. Conversely, if $a \sim_{\Delta(\varphi)} b$ holds then $(a \to b) \wedge (b \to a) \in \Delta(\varphi)$. This yields $\varphi((a \to b) \wedge (b \to a)) = \top$ and therefore $\varphi(a \to b) = \top$ and $\varphi(b \to a)) = \top$. φ is a homomorphism, therefore the latter means $\varphi(a) \to \varphi(b) = \top$ and $\varphi(b) \to \varphi(a) = \top$. Applying (1) from Lemma 2.1.6 we infer from this that $\varphi(a) \leq \varphi(b)$ and $\varphi(b) \leq \varphi(a)$, i.e. $\varphi(b) = \varphi(a)$ and $a \approx b$. Thus (i) holds. (ii) follows from (i) and Theorem 1.2.12. ∎

We recall (see Section 1.3) that for every $\mathfrak{A} \in Var(H)$ (which is, as we have seen, a pseudo-boolean algebra,) $\lambda(\mathfrak{A})$ is the following set of formulas $\{\alpha \mid \alpha \in \mathcal{F}or_H, \mathfrak{A} \models \alpha\}$ which is a superintuitionistic logic. Immediately from Corollary 1.3.26 we derive

Proposition 2.1.26 Let $\mathfrak{A}_1$ and $\mathfrak{A}_2$ be some pseudo-boolean algebras, $\mathfrak{A}_i, i \in I$ is a family of pseudo-boolean algebras. Then

(i) If $\mathfrak{A}_2$ is a homomorphic image of $\mathfrak{A}_1$ then $\lambda(\mathfrak{A}_1) \subseteq \lambda(\mathfrak{A}_2)$,

(ii) If $\mathfrak{A}_2$ is a subalgebra of $\mathfrak{A}_1$ then $\lambda(\mathfrak{A}_1) \subseteq \lambda(\mathfrak{A}_2)$,

(iii) $\lambda(\prod_{i \in I} \mathfrak{A}_i) = \lambda(\bigcap_{i \in I} \mathfrak{A}_i)$.

Using the semantic technique developed we can give a description of the family of all superintuitionistic logics from a semantic point of view. The family $\mathcal{L}_{sil}$ of all superintuitionistic logics as well as the class of all varieties of pseudo-boolean algebras $\mathcal{L}_{pba}$ form partially ordered sets with respect to set-theoretic inclusion. Moreover they are both closed with respect to set-theoretic intersections of arbitrary families of elements (for the case of $\mathcal{L}_{sil}$ this follows directly from definitions, in case of $\mathcal{L}_{pba}$ it follows, for instance, from Birkhoff Representation Theorem (Theorem 1.2.24). Therefore $\mathcal{L}_{sil}$ and $\mathcal{L}_{pba}$ are both certain complete lattices: they have least upper bound and greatest lower bound for any set of elements (and in particular for any infinite set).

Theorem 2.1.27 *The mapping e, where $e(\lambda) := Var(\lambda)$, of the lattice $\mathcal{L}_{sil}$ into the lattice $\mathcal{L}_{pba}$ is the dual isomorphism onto, that is*

(i) *e is one-to-one mapping onto,*
(ii) *if $\lambda_1 \subseteq g_2$ then $e(\lambda_1) \supseteq e(\lambda_2)$,*
(iii) *$e(\lambda_1 \wedge \lambda_2) = e(\lambda_1) \vee e(\lambda_2)$, $e(\lambda_1 \vee \lambda_2) = e(\lambda_1) \wedge e(\lambda_2)$.*

Moreover e is the dual isomorphism of complete lattices:

(iv) *$e(\bigwedge_{i \in I} \lambda_i) = \bigvee_{i \in I} e(\lambda_i)$,*
(v) *$e(\bigvee_{i \in I} \lambda_i) = \bigwedge_{i \in I} e(\lambda_i)$.*

Proof. According to (4) from Theorem 2.1.10 the mapping e is a one-to-one mapping into. For every variety V of pseudo-boolean algebras, $\lambda(V) := \bigcap_{\mathfrak{A} \in V} \lambda(\mathfrak{A})$ is a superintuitionistic logic. It is clear that $e(\lambda(V)) = V$, thus e is a mapping onto. The property (ii) follows directly from definitions. From (i) and (ii) we immediately get (iii) and (iv). ∎

The smallest element of $\mathcal{L}_{pba}$ is the variety U containing the single-element algebra (in which $\bot = \top$) only. Thus this variety corresponds to the superintuitionistic logic $\mathcal{F}or_{PC}$ of all formulas in the propositional language. It is clear also that U has only single immediate successor in $\mathcal{L}_{pba}$, namely the variety V_2 generated by the two-element pseudo-boolean algebra B_2, and V_2 is the smallest among the proper successors for U in $\mathcal{L}_{pba}$. Thus there is a greatest consistent superintuitionistic logic, the logic $\lambda(B_2)$, where B_2 is the two-element pseudo-boolean algebra.

Exercise 2.1.28 *Show that $\lambda(B_2)$ is the classical propositional logic.*

Hint: verify that all axioms of PC are valid on B_2 but there are formulas which are false in this algebra.

According to the definition of the variety V_b of all boolean algebras, B_2 is a boolean algebra (it can be easily verified by calculating the truth values the defining V_b identities).

Exercise 2.1.29 *Show that V_b coincides with the variety $Var(B_2)$ generated by B_2 and $Var(PC) = V_b$.*

Hint: use Exercise 2.1.28: the fact that $Var(PC) = Var(B_2)$, the fact that every boolean algebra is a certain pseudo-boolean algebra, where $x \to y = \neg x \vee y$, and the fact that a boolean algebra is subdirectly irreducible if and only if it is isomorphic to the algebra B_2.

We postpone a more detailed examination of the structure of the lattice $\mathcal{L}_{sil}$ until the following sections and we now intend to give several simple general results about the lattice in general. First we are going to describe how lattice operations interact with the axiomatizations for superintuitionistic logics. We recall that a set of formulas Γ form an *axiomatization* of a superintuitionistic logic λ (or Δ is the set of axioms for λ) if λ is the set of theorems of the axiomatic system $H \oplus \Delta$, the axioms of which are the axioms of the Heyting intuitionistic logic H and all formulas from Δ, and having only *modus ponens* as a derivation rule. If $\Delta = \{\alpha_1, ..., \alpha_n\}$ we will sometimes to write $H \oplus \alpha_1, ..., \alpha_n$ instead of $H \oplus \Delta$.

Lemma 2.1.30 *If* $\lambda_1 = H \oplus \{\alpha_i \mid i \in I\}$ *and* $\lambda_2 = H \oplus \{\beta_j \mid j \in J\}$ *then* $\lambda_1 \vee \lambda_2 = H \oplus \{\alpha_i \mid i \in I\} \cup \{\beta_j \mid j \in J\}$.

Proof. Easy. Left as an exercise to the reader.

Theorem 2.1.31 *If*

$$\lambda_1 = H \oplus \{\alpha_i \mid i \in I\} \text{ and } \lambda_2 = H \oplus \{\beta_j \mid j \in J\}$$

then $\lambda_1 \wedge \lambda_2 = H \oplus \{\alpha_i^2 \vee \beta_j^3 \mid i \in I, j \in J\}$, *where every* α_j^2 *obtained from* α_i *by the substitution* $p_k \mapsto p_{2k}$, *and each* β_j^k *is the result of the substitution* $p_k \mapsto p_{2k+1}$ *in* β_j.

Proof. It is clear that

$$H \oplus \{\alpha_i^2 \vee \beta_j^3 \mid i \in I, j \in J\} \leq \lambda_1, \ \ H \oplus \{\alpha_i^2 \vee \beta_j^3 \mid i \in I, j \in J\} \leq \lambda_2$$

because all the axioms on the left side of the above relations are derivable in λ_1 and λ_2. Let λ be a logic such that $\lambda \leq \lambda_1$ and $\lambda \leq \lambda_2$. Suppose that $\lambda \not\subseteq H \oplus \{\alpha_i^2 \vee \beta_j^3 \mid i \in I, j \in J\}$. Then there is a formula α such that $\alpha \in \lambda_1, \alpha \in \lambda_2$ but $\alpha \notin H \oplus \{\alpha_i^2 \vee \beta_j^3 \mid i \in I, j \in J\}$. By (4) of Theorem 2.1.10 there is a pseudo-boolean algebra $\mathfrak{A}_f$ from $Var(H \oplus \{\alpha_i^2 \vee \beta_j^3 \mid i \in I, j \in J\})$ such that $\mathfrak{A}_f \not\models \alpha$. Because, according to Birkhoff Representation Theorem (Theorem 1.2.57) $\mathfrak{A}_f$ is a subdirect product of subdirectly irreducible algebras, there is a subdirectly irreducible algebra $\mathfrak{A}$ from the variety $Var(H \oplus \{\alpha_i^2 \vee \beta_j^3 \mid i \in I, j \in J\})$ which disproves α also. We claim that $\mathfrak{A}$ is a member of $Var(\lambda_1)$ or $Var(\lambda_2)$. Indeed, suppose otherwise. Then $\mathfrak{A}$ is a not member of $Var(\lambda_1)$ and is not a member of $Var(\lambda_2)$. Then there are formulas α_i and β_j such that $\mathfrak{A} \not\models \alpha_i$ and $\mathfrak{A} \not\models \beta_j$. This means there are valuations $E : p_k \mapsto a_k \in |\mathfrak{A}|$ and $W : p_k \mapsto b_k \in |\mathfrak{A}|$ such that

$E(\alpha_i) \neq \top$ and $W(\beta_j) \neq \top$. $\mathfrak{A}$ is a subdirectly irreducible pseudo-boolean algebra. Therefore according to Lemma 2.1.23 there is the greatest element ω among elements which are properly less than $\top$. Therefore $E(\alpha_i) \leq \omega$ and $W(\beta_j) \leq \omega$. We take then the valuation V as follows:

$$V(p_n) = \begin{cases} E(p_k) & \text{if } n = 2k \\ W(p_k) & \text{if } n = 2k+1 \end{cases}$$

Then $V(\alpha_i^2) = E(\alpha_i) \leq \omega$ and $V(\beta_j^3) = E(\beta_j) \leq \omega$. Thus $V(\alpha_i^2 \vee \beta_j^3) \leq \omega$ which contradicts $\mathfrak{A} \in Var(H \oplus \{\alpha_i^2 \vee \beta_j^3 \mid i \in I, j \in J\})$. Hence indeed $\mathfrak{A} \in Var(\lambda_1)$ or $\mathfrak{A} \in Var(\lambda_2)$. If $\mathfrak{A} \in Var(\lambda_1)$ ($\in Var(\lambda_2)$) $\mathfrak{A} \not\models \alpha$ contradicts $\lambda \leq \lambda_1$ (λ_2). Therefore our assumption is impossible and $H \oplus \{\alpha_i^2 \vee \beta_j^3 \mid i \in I, j \in J\}$ is the greatest lower bound for λ_1 and λ_2. ∎

Recall that a superintuitionistic logic λ is called *finitely axiomatizable* if $\lambda = H \oplus \alpha_1, ..., \alpha_n$.

Corollary 2.1.32 *The family $\mathcal{L}_{sil}^{fa}$ of all finitely axiomatizable superintuitionistic logics forms a sublattice of the lattice $\mathcal{L}_{sil}$ of all superintuitionistic logics.*

We can specify the structure of $\mathcal{L}_{sil}$ more.

Theorem 2.1.33 *The lattice $\mathcal{L}_{sil}$ is a pseudo-boolean algebra.*

Proof. For the proof we will use the second definition (Definition 2.1.12) of pseudo-boolean algebras (which is equivalent to the first one (Lemma 2.1.13). Since $\mathcal{L}_{sil}$ has the smallest element H, it is sufficient to show that the pseudo-complement of any logic λ_1 to an arbitrary given logic λ_2 exists. Let $\lambda_1 = H \oplus \Gamma$ (for instance, we can take $\Gamma := \lambda_1$). We define the set Δ of formulas in the language of H as follows:

$$\Delta := \{\beta \mid (\forall \alpha \in \Delta)(\alpha^2 \vee \beta^3 \in \lambda_2\}$$

(for the definition of α^2 and β^3 see Lemma 2.1.31). We put $\lambda := H \oplus \Delta$, and will show that $(\lambda_1 \to \lambda_2) = \lambda$. Indeed, by Lemma 2.1.31 we have

$$\alpha_1 \wedge \lambda = H \oplus \{\alpha^2 \vee \beta^3 \mid \alpha \in \Gamma, \beta \in \Delta\}.$$

For every $\alpha \in \Delta$ and $\beta \in \Delta$, we have $\alpha^2 \vee \beta^3 \in \lambda_2$ holds by definition of Δ. Thus all the extra axioms of the logic $\lambda_1 \wedge \lambda$ are theorems of the logic λ_2, and therefore we obtain $\lambda_1 \wedge \lambda \subseteq \lambda_2$. Now we show that the logic λ is the greatest element of the set $\{\lambda^p \mid \lambda^p \in \mathcal{L}_{sil}, \lambda_1 \wedge \lambda^p \subseteq \lambda_2\}$. Indeed, let $\lambda_1 \wedge \lambda^p \subseteq \lambda_2$ and $\lambda^p = H \oplus \Omega$. By Lemma 2.1.31 we have $\lambda_1 \wedge \lambda^p = H \oplus \{\alpha^2 \wedge \gamma^3 \mid \alpha \in \Gamma, \gamma \in \Omega\}$. Then for every $\alpha \in \Gamma, \gamma \in \Omega$, $\alpha^2 \vee \gamma^3 \in \lambda_2$ is valid. This means that for each $\gamma \in \Omega, \gamma \in \Delta$ according to the definition of Δ. Thus $\lambda^p = H \oplus \Omega \subseteq H \oplus \Delta = \lambda$. Hence we have proved that $\lambda_1 \to \lambda_2 = \lambda$. ∎

From this theorem and Lemma 2.1.4 we immediately derive

Corollary 2.1.34 *The lattice $\mathcal{L}_{sil}$ is distributive.*

Note that this result also could be extracted as a corollary from the general Jonsson Theorem (Theorem 1.2.62) which says that, for every variety V which consists of certain congruence-distributive algebras (i.e., all the algebras have distributive lattices of congruences), the lattice of all subvarieties of V is distributive and satisfies the infinitary distributive law:

$$V_1 \vee (\bigwedge_{i \in I} V_i) = \bigwedge_{i \in I} (V_1 \vee V_i).$$

It also known that any variety of algebras such that all its algebras are distributive lattices with respect to certain terms generated by signature's operations, is congruence distributive. (see comment after Lemma 1.2.59). Using these observations we again obtain that the lattice $\mathcal{L}_{pba}$ is distributive, and hence the dual to $\mathcal{L}_{pba}$ the lattice $\mathcal{L}_{sil}$ is distributive also.

2.2 Algebraic Semantics, Modal and Tense Logics

A remarkable property of modal and temporal logics is the fact that the algebraic semantics for them are based on boolean algebras. More precisely, this semantics consists of a varieties of boolean algebras enriched with new operations corresponding to the new logical connectives. We know (cf. Theorem 1.3.14 and Theorem 1.3.18) that modal and temporal logics are algebraic logics. Also, it is known that every algebraic logic λ has an characterizing variety $Var(\lambda)$ of the corresponding algebras (see Theorem 1.3.20).

Lemma 2.2.1 *The following hold.*

(i) Let λ be an algebraic logic in a language including the logical connectives $\wedge, \vee, \rightarrow, \neg, \perp, \top$, and let $\lambda_{\wedge,\vee,\rightarrow,\neg,\perp,\top}$ be its $\{\wedge, \vee, \rightarrow, \neg, \perp, \top\}$-fragment (the set of all formulas from λ containing occurrences of logical connectives only from the set $\{\wedge, \vee, \rightarrow, \neg, \perp, \top\}$). If $PC \subseteq \lambda_{\wedge,\vee,\rightarrow,\neg,\perp,\top}$ then, for every $\mathfrak{B} \in Var(\lambda)$, the algebra

$$\mathfrak{B}_{\wedge,\vee,\rightarrow,\neg,\perp,\top} := \{|\mathfrak{B}|, \vee, \wedge, \rightarrow, \neg, \perp, \top\}$$

is a boolean algebra, where by definition $x \rightarrow y := \neg x \vee y$.

(ii) Any boolean algebra B is a pseudo-boolean algebra, with the pseudo-complement x to y defined as follows: $x \rightarrow y := \neg x \vee y$.

Proof. It is well known that the variety $Var(PC)$ is the variety Var_b of all boolean algebras (cf. Exercise 2.1.29). Therefore $\mathfrak{B}_{\wedge,\vee,\to,\neg,\perp,\top} \in Var(PC)$ is a boolean algebra and (i) holds. In order to show (ii) according to Lemma 2.1.13 it is sufficient to show that $\neg x \vee y$ is the greatest element among the elements c such that $x \wedge c \leq y$. Actually,

$$x \wedge (\neg x \vee y) = (x \wedge \neg x) \vee (x \wedge y) =$$
$$\perp \wedge (x \wedge y) = x \wedge y \leq y.$$

Suppose that an element c is such that $x \wedge c \leq y$. Then

$$c = c \wedge \top = c \wedge (x \vee \neg x) = (c \wedge x) \vee (c \wedge \neg x) \leq$$
$$y \vee (c \wedge \neg x) \leq \neg x \vee y.$$

Thus $\neg x \vee y$ is the pseudo-complement of x to y. ∎

We know (see Theorem 1.3.14 and Theorem 1.3.18) that any normal modal logic λ, as well as every temporal logic λ, is an algebraic logics. Any of these logics includes the classical propositional calculus PC as a fragment. Therefore according to Lemma 2.2.1 the boolean fragments of all the algebras from the corresponding varieties $Var(\lambda)$ are boolean algebras. In particular, every boolean algebra is a pseudo-boolean algebra, and therefore all identities and inclusions which hold for pseudo-boolean algebras are also valid for boolean algebras. We will often use this observation in what follows. Hence we already have an information on the structure of algebras from $Var(\lambda)$, but only with respect to certain fragments of them. However how they do look overall? First we are going to answer this question for normal modal logics.

Definition 2.2.2 *We call an algebra* $\langle B, \wedge, \vee, \to, \neg, \square, \perp, \top \rangle$ *a modal algebra if* $\langle B, \wedge, \vee, \to, \neg, \perp, \top \rangle$ *is a boolean algebra and the unary operation* $\square$ *satisfies the following identities:*

(1) $\square(x \to y) \to (\square x \to \square y) = \top$,
(2) $\square\top = \top$.

According to (1) of Lemma 2.1.6 relations (1) and (2) above are equivalent to

(3) $\square(x \to y) \leq (\square x \to \square y)$,
(4) $\top \leq \square\top$,

respectively. Note that the connective $\square$ commutes with $\wedge$ and the definable dual operation $\diamondsuit := \neg\square\neg$ commutes with $\vee$:

Lemma 2.2.3 *For any modal algebra* $\mathfrak{B}$ *and any* $a, b \in |\mathfrak{B}|$,

(5) $\square(a \wedge b) = \square a \wedge \square b$,
(6) $\diamondsuit(a \vee b) = \diamondsuit a \vee \diamondsuit b$.

Proof. Actually, $a \wedge b \leq a$ and $a \wedge b \leq b$. Therefore $a \wedge b \to a = \top$ and $a \wedge b \to b = \top$ (cf. (1) from Lemma2.1.6). Applying (2), (1) from Definition 2.2.2 and the fact that $x \to y = \top \Leftrightarrow x \leq y$ we derive

$$\Box(a \wedge b) \to \Box a = \top,$$
$$\Box(a \wedge b) \to \Box b = \top.$$

Therefore $\Box(a \wedge b) \leq \Box a$ and $\Box(a \wedge b) \leq \Box(b)$, i.e. $\Box(a \wedge b) \leq \Box a \wedge \Box(b)$. Conversely, $a \to (b \to a \wedge b) = \top$ and using (1) and (2) from Definition 2.2.2 we arrive at $\Box a \to \Box(b \to a \wedge b) = \top$. Thus $\Box a \leq \Box(b \to a \wedge b)$. Again by (1) of Definition 2.2.2 $\Box(b \to a \wedge b) \leq (\Box b \to \Box(a \wedge b))$, and hence $\Box a \leq (\Box b \to \Box(a \wedge b))$. Thus

$$\Box a \wedge \Box b \leq \Box b \wedge (\Box b \to \Box(a \wedge b)) =$$
$$\Box b \wedge \Box(a \wedge b) \text{ (cf. (4) from Definition 2.1.1) } \leq \Box(a \wedge b).$$

Hence we obtain that $\Box a \wedge \Box b \leq \Box(a \wedge b)$ and (5) holds. The equality (6) can be proved completely analogous to the case (5), the details are left to the readers as an exercise. ∎

Lemma 2.2.4 *An algebra $\mathfrak{B}$ in the signature of modal algebras, which is a boolean algebra, is a modal algebra iff $(\forall a, b \in |\mathfrak{B}|) \Box(a \wedge b) = \Box a \wedge \Box b$ and $\mathfrak{B} \models \Box\top = \top$ and $\Box\top = \top$.*

Proof. Necessity follows from Lemma 2.2.3. To prove sufficiency we reason using the supposition and the laws of boolean algebras,

$$\Box(a \to b) \to (\Box a \to \Box b) = \neg\Box(\neg a \vee b) \vee (\neg\Box a \vee \Box b) =$$
$$\neg(\Box(\neg a \vee b) \wedge \Box a) \vee \Box b) = \neg(\Box((\neg a \vee b) \wedge a) \vee \Box b) =$$
$$\neg(\Box((\neg a \wedge a) \vee (a \wedge b)) \vee \Box b = \neg\Box(a \wedge b) \vee \Box b =$$
$$\neg(\Box a \wedge \Box b) \vee \Box b = \neg\Box a \vee \neg\Box b \vee \Box b = \top.$$

Thus $\mathfrak{B}$ is a modal algebra. ∎

Theorem 2.2.5 (Completeness Theorem for Normal Modal Logics)

(i) *For every normal modal logic λ and $\mathfrak{B} \in Var(\lambda)$, $\mathfrak{B}$ is a modal algebra.*

(ii) *For any family $\mathcal{F}$ of modal algebras,*
  $\lambda(\mathcal{F}) := \{\alpha \mid (\forall \mathfrak{B} \in \mathcal{F})(\mathfrak{B} \models \alpha)\}$ is a normal modal logic.

(iii) *$Var(K)$, where K is the smallest normal modal logic, coincides with the variety of all modal algebras.*

(iv) *For every normal modal logic λ, there is a modal algebra $\mathfrak{B}$ and a variety of modal algebras V which have the following property $\lambda = \lambda(\mathfrak{B})$ and $\lambda(\mathfrak{B}) = \lambda(V)$, moreover $Var(\lambda) = V$.*

Proof. To show (i) suppose $\mathfrak{B} \in Var(\lambda)$. Then $\mathfrak{B}$ is a boolean algebra and because $K \subseteq \lambda$ and λ is closed with respect to the normalization rule $R_3 := \{x/\Box x\}$, we infer that the identities (1) and (2) from the definition of modal algebras hold in $\mathfrak{B}$. Thus $\mathfrak{B}$ is a modal algebra. If $\mathcal{F}$ is a family of modal algebras then all boolean fragments of these algebras are boolean algebras and are included in $Var(PC)$. Therefore all the axioms of Hilbert style axiomatizations for PC are valid in any algebra from $\mathcal{F}$. In addition, because all the algebras from $\mathcal{F}$ are modal algebras, the additional modal axiom of K is also valid in all algebras from $\mathcal{F}$. In all boolean algebras the following $(x = \top, x \rightarrow y = \top) \Rightarrow (y = \top)$ holds (cf. (1) of Lemma 2.1.6) so the rule *modus ponens*, i.e. (R1), preserves the designated value $\top$. Similarly, since all the algebras from $\mathcal{F}$ are modal, the *necessitating rule*, - (R3): $x/\Box x$, - also preserves the designated value $\top$.

Therefore all formulas derivable in the axiomatic system of K are valid in all algebras from $\mathcal{F}$. Thus $K \subseteq \lambda(\mathcal{F})$. It follows directly from the above observations that $\lambda(\mathcal{F})$ is closed with respect to (R1) and (R3). Hence $\lambda(\mathcal{F})$ is a normal modal logic, i.e. (ii) holds. From (i) we infer that $Var(K)$ is a family of modal algebras. If $\mathfrak{B}$ is a modal algebra then according to (ii) $\lambda(\mathfrak{B})$ is a normal modal logic, hence $K \subseteq \lambda(\mathfrak{B})$ and $\mathfrak{B} \in Var(K)$. Thus actually $Var(K)$ is the variety of all modal algebras, i.e. (iii) holds. According to Theorems 1.3.19, 1.3.20 and 1.3.21 the free algebra of countable rank $\mathfrak{F}(\lambda)$ from the variety $V := Var(\lambda)$ has the property: $\lambda(\mathfrak{F}(\lambda)) = \lambda = \lambda(Var(\lambda))$. By (i) is a modal algebra and $\mathfrak{F}(\lambda)$ has all properties required in (iv) for $\mathfrak{B}$. ∎

The completeness theorem establishes a close relationship between normal modal logics and varieties of modal algebras. We are going to describe this relation in terms of lattices. The family $\mathcal{L}_{nml}$ of all normal modal logics is a partially ordered set by set-theoretic inclusion $\subseteq$. The family of all varieties of modal algebras $\mathcal{L}_{ma}$ also form a partially ordered set by $\subseteq$. Posets $\mathcal{L}_{nml}$ and $\mathcal{L}_{ma}$ are closed with respect to the set-theoretic intersection of arbitrary collections of elements (in the case of $\mathcal{L}_{pba}$ this follows, for instance, from Birkhoff's Theorem (Theorem 1.2.24) or from the definition of varieties by means of sets of identities). Therefore $\mathcal{L}_{nml}$ and $\mathcal{L}_{ma}$ are complete lattices: they have least upper bounds and greatest lower bounds for every (and, in particular, for any infinite) set of elements.

Theorem 2.2.6 *There exists a dual isomorphism of the lattice $\mathcal{L}_{nml}$ onto the lattice $\mathcal{L}_{ma}$. It is the mapping e, where $e(\lambda) := Var(\lambda)$, and, in particular,*

> *(i) e is one-to-one mapping onto,*
> *(ii) if $\lambda_1 \subseteq \lambda_2$ then $e(\lambda_1) \supseteq e(\lambda_2)$,*
> *(iii) $e(\lambda_1 \wedge \lambda_2) = e(\lambda_1) \vee e(\lambda_2)$, $e(\lambda_1 \vee \lambda_2) = e(\lambda_1) \wedge e(\lambda_2)$.*

and

> *(iv) $e(\bigwedge_{i \in I} \lambda_i) = \bigvee_{i \in I} e(\lambda_i)$,*
> *(v) $e(\bigvee_{i \in I} \lambda_i) = \bigwedge_{i \in I} e(\lambda_i)$.*

Proof. By Theorem 2.2.5 the mapping e is a one-to-one mapping into. For every variety V of modal algebras, $\lambda(V) := \bigcap_{\mathfrak{A} \in V} \lambda(\mathfrak{A})$ is a normal modal logic by Theorem 2.2.5. It is clear that $e(\lambda(V)) = V$, thus e is a mapping onto. The property (ii) follows directly from our definitions. The properties (i) and (ii) as always yield (iii), (iv) and (v). ∎

The lattice $\mathcal{L}_{ma}$ has a smallest element: the variety U_m containing only the single-element modal algebra (in which $\bot = \top$). Thus this variety corresponds to the normal modal logic $\mathcal{F}or_K$ of all formulas in the language of K. As in the case of pseudo-boolean algebras, the variety U_m has immediate successors in $\mathcal{L}_{ma}$ - the variety V_2^r generated by the two-element modal algebra M_2^r where $\Box\bot = \bot$, and the variety V_2^i generated by the modal algebra M_2^i, where $\Box\bot = \top$. Therefore there exist maximal consistent normal modal logics, in particular, the logics $\lambda(M_2^r)$ and $\lambda(M_2^i)$.

Exercise 2.2.7 *Show that $\lambda(M_2^r)$ is the normal modal logic $K \oplus (\Box p \to p), (p \to \Box p)$, and $\lambda(M_2^i)$ is the normal modal logic $K \oplus (\Box\neg\top \equiv \top)$.*

Example 2.2.8 *The formulas $\Box p \to p$ and $p \to \Box p$ are not theorems of the logics K and $K4$.*

Actually we take t he four element boolean algebra B_4 which is shown in the picture Fig 2.3. Let $\mathfrak{B}_{4.1}$ be the modal algebra on the boolean algebra B_4, where

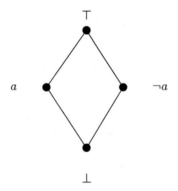

Figure 2.3:

$\Box a = \Box\neg a = \bot$ and $\Box\bot = \bot$. Then $\mathfrak{B}_{4.1} \in Var(K)$ and also $\mathfrak{B}_{4.1} \in Var(K4)$. However $a \to \Box a = \neg a \neq \top$. Therefore $p \to \Box p \notin K4$ and consequently we obtain that $p \to \Box p \notin K$. Let us consider the following modal algebra $\mathfrak{B}_{4.2}$ on the carrier B_4, where $\Box a = \Box\neg a = \Box\bot = \top$. It is not hard to check that $\mathfrak{B}_{4.2} \in Var(K4) \subseteq Var(K)$. Furthermore, in $\mathfrak{B}_{4.2}$ we have $\Box a \to a = \top \to a = a \neq \top$. Thus $\Box p \to p \notin K, \Box p \to p \notin K4$.

Example 2.2.9 *The formula* $\Diamond p \to \Box \Diamond p$ *is not a theorem of the logics* K, $K4$, $S4$, *or* $S4.3$.

We take the modal algebra $\mathfrak{B}_{4.3}$ on B_4, where $\Box a = a$, $\Box \neg a = \bot$, $\Box \bot = \bot$. We leave for the readers to show that $\mathfrak{B}_{4.3}$ is indeed a modal algebra and that $\mathfrak{B}_{4.3} \in Var(S4.3) \subseteq Var(S4) \subseteq Var(K4) \subseteq Var(K)$. Now we simply calculate: $\Diamond \neg a \to \Box \Diamond \neg a = \neg a \to \Box \neg a = \neg a \to \bot = a \neq \top$. Thus $\Diamond p \to \Box \Diamond p \notin S4.3$

Exercise 2.2.10 *The formula* $p \to \Box p$ *is not a theorem of the logic* $S5$.

Hint: Use the algebra $\mathfrak{B}_{4.1}$. ∎

As we know from Birkhoff's Theorem (Theorem 1.2.57), every variety V is generated by its own subdirectly irreducible algebraic systems. Therefore, for every modal logic λ, $\lambda = \lambda(\mathcal{K})$, where $\mathcal{K}$ is a class of subdirectly irreducible modal algebras from the variety of modal algebras $Var(\lambda)$ corresponding to λ (see Theorem 2.2.5). Thus we can restrict ourselves to only considering of subdirectly irreducible algebras in every case. Therefore we have to clarify the structure of such algebras and, in particular, we are going to give a description of congruence relations in modal algebras. For this we need to introduce a special kind filters on modal algebras.

Definition 2.2.11 *A filer* Δ *on a modal algebra* $\mathfrak{B}$ *is said to be* $\Box$-*filter if for every* $a \in \Delta$, $\Box a \in \Delta$.

Lemma 2.2.12 *For every modal algebra* $\mathfrak{B}$ *and a congruence relation* $\approx$ *on* $\mathfrak{B}$, *the set* $\Delta_{\approx} := \{a \mid a \approx \top\}$ *is a* $\Box$-*filter on* $\mathfrak{B}$.

Proof. Actually, if $a \in \mathfrak{A}$ then $a \approx \top$, therefor because $\approx$ is a congruence relation, $\Box a \approx \Box \top$. $\mathfrak{B}$ is a modal algebra, consequently $\Box \top = \top$ and $\Box a = \top$. It remains only to show that $\Delta_{\approx}$ is a filter. This can be done in a quite similar way to the case of pseudo-boolean algebras: if $a, b \in \Delta_{\approx}$ then $a \approx \top$ and $b \approx \top$. Because $\approx$ is a congruence relation we get $a \wedge b \approx \top \wedge \top = \top$. Thus $a \wedge b \in \Delta$. If $a \in \Delta_{\approx}$ and $a \leq b$ then $a \approx \top$ yields $a \vee b \approx \top \vee b = \top$. But $a \vee b = b$, therefore $b \approx \top$ and $b \in \Delta_{\approx}$. Hence $\Delta_{\approx}$ is a filter. ∎.

Lemma 2.2.13 *If* $\mathfrak{B}$ *is a modal algebra and* Δ *is a* $\Box$-*filter on* $\mathfrak{B}$ *then the relation*

$$\approx_{\Delta} := \{\langle a, b \rangle \mid (a \to b) \wedge (b \to) \in \Delta\}$$

is a congruence relation on $\mathfrak{B}$.

Proof. By definition this relation is reflexive and symmetric. Its transitivity can be shown in the following way. If $a \approx_\Delta b$ and $b \approx_\Delta c$ then $(a \to b) \in \Delta$ and $(b \to c) \in \Delta$. Then $(a \to b) \wedge (b \to c) \in \Delta$. Now we merely calculate using distributive laws, and the fact that $x \wedge \neg x = \bot$ in boolean algebras:

$$(a \to b) \wedge (b \to c) \in \Delta = (\neg a \vee b) \wedge (\neg b \vee c) =$$
$$(\neg a \wedge (\neg b \vee c)) \vee (b \wedge (\neg b \vee c)) =$$
$$([\neg a \wedge \neg b] \vee [\neg a \wedge c]) \vee ([b \wedge \neg b] \vee [b \wedge c]) =$$
$$(\neg a \wedge \neg b) \vee (b \wedge c) \leq \neg a \vee c = a \to c.$$

Thus $a \to c \in \Delta$. Similarly we can show that $c \to a \in \Delta$ and hence $a \approx_\Delta c$. We have showed that $\approx_\Delta$ is an equivalence relation. It remains to show that this relation is consistent with all operations on $\mathfrak{B}$. However every boolean algebra is a pseudo-boolean algebra. Therefore we can just reproduce the proof of Lemma 2.1.19 in the case of connectives $\wedge, \vee, \to, \neg$. Therefor it remains only to check the case of the operation $\Box$. Let $a \approx_\Delta b$, i.e. $(a \to b) \wedge (b \to a) \in \Delta$. Then $(a \to b) \in \Delta$ and $(b \to a) \in \Delta$. Because Δ is a $\Box$-filter we get $\Box(a \to b) \in \Delta$ and $\Box(b \to a) \in \Delta$. Using (1) from the definition of modal algebras we arrive at $(\Box a \to \Box b) \in \Delta$ and $(\Box b \to \Box a) \in \Delta$. Thus $\Box a \approx \Box b$ and $\approx_\Delta$ is really a congruence relation. ∎

For any modal algebra $\mathfrak{B}$ the set $\Phi(\mathfrak{B}, \Box)$ of all $\Box$-filters on $\mathfrak{B}$ and the set $\mathcal{C}(\mathfrak{B}, \approx)$ of all congruence relations on $\mathfrak{B}$ form partially ordered sets with respect to the set theoretic inclusion $\subseteq$. In is easy to see that both these posets are closed with respect to arbitrary set-theoretic intersections. Therefore $\Phi(\mathfrak{B}, \Box)$ and $\mathcal{C}(\mathfrak{B}, \approx)$ are complete lattices (with respect to $\subseteq$).

Theorem 2.2.14 *The lattices $\Phi(\mathfrak{B}, \Box)$ and $\mathcal{C}(\mathfrak{B}, \approx)$ are isomorphic. In particular, the mapping e, where $e \colon \mathcal{C}(\mathfrak{B}, \approx) \mapsto \Phi(\mathfrak{B}, \Box)$, $e(\approx) := \Delta_\approx$ is an isomorphism onto and $e^{-1}(\Delta) = \approx_\Delta$.*

Proof. According to Lemma 2.2.12 $e(\approx)$ is always a $\Box$-filter. The mapping e evidently preserves the order. Moreover, e is a one-to-one mapping. Actually, let $e(\approx_1) = e(\approx_2)$. Suppose that $a \approx_1 b$ holds . Then we have $(a \to b) \wedge (b \to a) \approx (a \to a) \wedge (a \to a) = \top$. That is $(a \to b) \wedge (b \to a) \in \Delta_{\approx_1} = \Delta_{\approx_2}$. This means $(a \to b) \wedge (b \to a) \approx_2 \top$. Then $a \wedge \top \approx_2 a \wedge (a \to b) \wedge (b \to a) = a \wedge b \wedge (b \to a) = a \wedge b$. Similarly we show that $b \approx_2 a \wedge b$. Hence $a \approx_2 b$ and we have proved that $\approx_1 \subseteq \approx_2$. The converse inclusion can be shown in an analogous way. Thus $\approx_1 = \approx_2$ and e is really a one-to-one mapping. The mapping e is onto. Indeed, let Δ be a $\Box$-filer on $\mathfrak{B}$. Then according to Lemma 2.2.13 the relation $\approx_\Delta$ is a congruence relation on $\mathfrak{B}$. Let us take $e(\approx_\Delta)$. We claim that $e(\approx_\Delta) = \Delta$. Actually, if $a \in \Delta$ then $a \approx_\Delta \top$ and consequently $a \in e(\approx_\Delta)$. Conversely, if $a \in e(\approx_\Delta)$ then $a \approx_\Delta \top$ which means $(a \to \top) \wedge (\top \to a) = a \in \Delta$. So, indeed $e(\approx_\Delta) = \Delta$ and e is a mapping onto, and $e^{-1}(\Delta) = \approx_\Delta$. Because e preserves the order it is an isomorphism of lattices. ∎

We know that an algebra is subdirectly irreducible if and only if there exists a smallest congruence among all non-trivial congruence relations (the trivial congruence relation is the relation which consists of all pairs $\langle a, a \rangle$). Applying Theorem 2.2.14 we get

Theorem 2.2.15 *A modal algebra $\mathfrak{B}$ is subdirectly irreducible if and only if there is a smallest $\square$-filter among all $\square$-filters on the algebra $\mathfrak{B}$ which differ from the filter $\{\top\}$.*

We will need some special $\square$-filters and congruence relations. To introduce them we need the following

Definition 2.2.16 *An element a of a modal algebra $\mathfrak{B}$ is called* stable *if and only if $\square a = a$.*

Lemma 2.2.17 *For every stable element a of a modal algebra $\mathfrak{B}$, the filter $\mathcal{F}(a) := \{b \mid a \leq b\}$ generated by a is a $\square$-filter.*

Proof. Indeed, if $b \in \mathcal{F}(a)$, i.e. $a \leq b$ then $a \rightarrow b = \top$. Then we obtain $\square(a \rightarrow b) = \square\top = \top$. By (1) from the definition of modal algebras we immediately derive $\square(a \rightarrow b) \leq (\square a \rightarrow \square b)$, i.e. $\square a \rightarrow \square b = \top$ consequently $\square a \leq \square b$. Still $\square a = a$ holds, therefore $a \leq \square b$ and $\square b \in \mathcal{F}(a)$. ∎

The following observations may be also of interest.

Lemma 2.2.18 *Let $\mathfrak{B}$ be a modal algebra from $Var(T)$, where T is the von Wright normal modal logic, i.e. $T := K \oplus \square p \rightarrow p$, and a is an element from $\mathfrak{B}$.*

 (i) The set $\mathcal{F}(a, \downarrow) := \{b \mid b \in |\mathfrak{B}|, (\exists n)(\square^n a \leq b)\}$ is a $\square$-filter on $\mathfrak{B}$ (where $\square^0 a := a$ and $\square^{n+1} a := \square\square^n a$).

 (ii) If Δ is a $\square$-filter from $\mathfrak{B}$ and $a \in \Delta$ then $\mathcal{F}(a, \downarrow) \subseteq \Delta$. That is $\mathcal{F}(a, \downarrow)$ is the smallest $\square$-filer containing a.

Proof. (i): First we have to verify that $\mathcal{F}(a, \downarrow)$ is a filter. Indeed, let $\square^n a \leq b$ and $\square^m a \leq c$. Then $\square^n a \wedge \square^m a \leq b \wedge c$. The elements $\square^k a, k \in N$ form a decreasing chain in $\mathfrak{B}$. Actually, $\mathfrak{B} \in Var(K)$, therefore $\square a \rightarrow a = \top$ and consequently $\square a \leq a$. In particular, $\square(\square a \rightarrow a) = \square\top = \top$. By (1) from the definition of modal algebras $\square(\square a \rightarrow a) \rightarrow (\square\square a \rightarrow \square a) = \top$ and hence $\square(\square a \rightarrow a) \leq (\square\square a \rightarrow \square a)$. Thus $(\square^2 a \rightarrow \square^1 a) = \top$ and therefore $\square^2 a \leq \square a$. Continuing this reasoning we obtain that $\square^k a, k \in N$ is really a decreasing chain.

Thus if $k = max(n, m)$ then $\square^k a \leq b \wedge c$ and the set $\mathcal{F}(a, \downarrow)$ is closed with respect to intersections. That this set is closed with respect to extension follows directly from the definition. So it is a filter. Suppose that $\square^k a \leq b$. Using (1) from the definition of modal algebras, from this, as above, it follows that $\square^{k+1} a \leq \square b$. Hence $\square b \in \mathcal{F}(a, \downarrow)$ and $\mathcal{F}(a, \downarrow)$ is a $\square$-filter. If Δ is a $\square$-filter and $a \in \Delta$ then directly by definition $\mathcal{F}(a, \downarrow) \subseteq \Delta$ and (ii) holds. ∎

Lemma 2.2.19 *If a modal algebra* $\mathfrak{A}$ *belongs to the variety* $Var(K4)$ *then the filter* $\mathcal{F}(\Box a)$ *(the filter* $\mathcal{F}(\Box_0 a)$*)generated by an arbitrary element of kind* $\Box a$ *(*$\Box_0 a := \Box a \wedge a$ *) is a* $\Box$*-filter.*

Proof. Suppose $b \in \mathcal{F}(\Box a)$ that is $\Box a \leq b$. Then it follows immediately that $\Box a \rightarrow b = \top$, and consequently $\Box(\Box a \rightarrow b) = \Box\top = \top$. Using (1) from the definition of modal algebras we have $\Box(\Box a \rightarrow b) \leq (\Box\Box a \rightarrow \Box b)$. Hence $\Box\Box a \rightarrow \Box b = \top$ and $\Box\Box a \leq \Box b$. Because $\mathfrak{B} \in Var(K4)$, $\Box a \rightarrow \Box\Box a = \top$ which means $\Box a \leq \Box\Box a$. This way we arrive at $\Box a \leq \Box b$ and $\Box b \in \mathcal{F}(\Box a)$. Thus $\mathcal{F}(\Box a)$ is in fact a $\Box$-filter. The proof of the second claim concerning $\Box_0 a$ has the same proof using Lemma 2.2.3. ∎

Thus now we know that every stable element generates a $\Box$-filter and the corresponding congruence relation. Using this observation we can give a more precise description for finite subdirectly irreducible modal algebras from the variety $Var(T)$, where T is the von Wright logic: $T := K \oplus \Box p \rightarrow p$.

Theorem 2.2.20 *Any finite modal algebra from* $Var(T)$ *is subdirectly irreducible if and only if there exists the greatest element* ω *in the set of all stable elements which differ from* $\top$.

Proof. If $\mathfrak{B}$ is subdirectly irreducible then according to Theorem 2.2.15 there is the smallest $\Box$-filter $\mathcal{F}$. Because $\mathfrak{B}$ is a finite algebra, the filter $\mathcal{F}$ has to be generated by an element a. That is $\mathcal{F} = \mathcal{F}(a) := \{b \mid a \leq b\}$. We claim that a is the biggest stable element. Indeed, suppose that a is not stable. Because the filter $\mathcal{F}(a)$ is a $\Box$-filter we have $a \leq \Box a$. We have by assumption that $\mathfrak{B} \in Var(T)$, therefore $\Box a \rightarrow a = \top$ and $\Box a \leq a$. Thus $\Box a = a$ and a is stable. Let us consider an arbitrary stable element b such that $b \neq \top$. Then the filter $\mathcal{F}(b)$ generated by b is a $\Box$-filter by Lemma 2.2.17. The fact that $\mathcal{F}$ is the smallest non-trivial $\Box$-filter implies that $\mathcal{F}(a) \subseteq \mathcal{F}(b)$ which gives us $b \leq a$. Thus a is the greatest stable element.

Conversely suppose that the greatest stable element ω exists. Let $\mathcal{F}$ be a $\Box$-filter which differs from the filter $\{\top\}$. This means that there is an element b from $\mathcal{F}$ such that $b \neq \bot$. Then because $\mathfrak{A} \in Var(T)$ we obtain $\Box b \rightarrow b = \top$, that is $\Box b \leq b$, and $\Box b \in \mathcal{F}$. Furthermore, $\Box b \rightarrow b = \top$ and $\Box(\Box b \rightarrow b) = \Box\top = \top$. Applying (1) from the definition of modal algebras we obtain $\Box(\Box \rightarrow b) \leq (\Box\Box b \rightarrow \Box b)$ which yields $\Box\Box b \rightarrow \Box b = \top$ and $\Box\Box b \leq \Box b$. Thus we get $\Box\Box b \leq \Box b$ and $\Box\Box b \in \mathcal{F}$. Continuing this reasoning we obtain that for every n, $\Box^{n+1} b \leq \Box^n b$ and $\Box^n b, \Box^{n+1} b \in \mathcal{F}$. That is the elements of kind $\Box^n b$ form a decreasing chain of elements from $\mathcal{F}(b)$. Because the algebra $\mathfrak{A}$ is finite this chain must be terminated by an element $\Box^m b$. Hence the element $\Box^m b$ has to be stable. Moreover, $\Box^m b \leq b \neq \top$. The element ω is the greatest stable element, therefore $\Box^m b \leq \omega$. Hence the filter $\mathcal{F}(\omega)$ is a subset of the filter $\mathcal{F}$. The filter $\mathcal{F}_\omega$ is a $\Box$-filter according to Lemma 2.2.17, thus we get $\mathcal{F}(\omega)$ is the smallest non-trivial $\Box$-filter. By Theorem 2.2.15 we conclude $\mathfrak{B}$ is subdirectly irreducible. ∎

We say that an element a of a modal algebra $\mathfrak{A}$ is *up-stable* if $\Box a \geq a$.

Lemma 2.2.21 *A finite modal algebra $\mathfrak{A} \in Var(K4)$ is subdirectly irreducible iff there is a greatest up-stable element ω in $(|\mathfrak{A}| - \{\top\})$.*

Proof. By definition an algebra $\mathfrak{A}$ is subdirectly irreducible iff there is a smallest non-trivial congruence relation in $\mathfrak{A}$. If θ is a congruence relation then the set $\top_\theta := \{a \mid a \in |\mathfrak{A}|, a \equiv_\theta \top\}$ is a $\Box$-filter (which means that $\top_\theta$ is closed with respect to $\Box$). The mapping $\theta \to \top_\theta$ is a one-to-one correspondence between the set of all congruence relations in $\mathfrak{A}$ and the set of all $\Box$-filters in $\mathfrak{A}$ (see Theorem 2.2.14), and by Theorem 2.2.15 $\mathfrak{A}$ is subdirectly irreducible iff there is a smallest non-trivial $\Box$-filter.

Suppose $\mathfrak{A}$ has a greatest up-stable element, ω, in $(|\mathfrak{A}| - \{\top\})$. Then $\nabla(\omega) := \{a \mid a \in \mathfrak{A}, \omega \leq a\}$ forms a $\Box$-filter. It is easy to see that $\nabla(\omega)$ is the smallest non-trivial $\Box$-filter because $\mathfrak{A}$ is finite. Conversely, suppose $\mathfrak{A}$ has a smallest non-trivial $\Box$-filter. Then, because $\mathfrak{A}$ is finite, each $\Box$-filter ∇ has a smallest element a_∇, and this element is up-stable. Every up-stable element generates a $\Box$-filter being its smallest element. Therefore the smallest element of the smallest non-trivial $\Box$-filter is the greatest up-stable element in $(|\mathfrak{A}| - \{\top\})$. ∎

Having at our disposal the description of congruence relations on modal algebras we are in a position to specify the homomorphisms of modal algebras.

Lemma 2.2.22 *For any given modal algebras $\mathfrak{B}_1$ and $\mathfrak{B}_2$ and a homomorphism $\varphi : \mathfrak{B}_1 \mapsto \mathfrak{B}_2$, the set $\Delta(\varphi) := \{a \mid a \in |\mathfrak{B}_1|, \varphi(a) = \top\}$ is a $\Box$-filter on $\mathfrak{B}_1$.*

Proof. The set $\Delta(\varphi)$ is non-empty, in fact, $\varphi(\top) = \top$ (because φ must preserve the order as it is a homomorphism). If $a_1, a_2 \in \mathfrak{B}_1$ and $\varphi(a_1) = \top$, $\varphi(a_2) = \top$ then $\varphi(a_1 \wedge a_2) = \varphi(a_1) \wedge \varphi(a_2) = \top \wedge \top = \top$. We have showed that $\Delta(\varphi)$ is closed with respect to intersection. Let $a, b \in \mathfrak{B}_1$ and $a \leq b$. Suppose $\varphi(a) = \top$. We merely calculate: $\varphi(a \vee b) = \varphi(a) \vee \varphi(b) = \varphi(b) = \top \vee \varphi(b) = \top$. Hence $\Delta(\varphi)$ is also closed with respect to extensions of elements. If $a \in \mathfrak{B}_1$ and $\varphi(a) = \top$ then, as φ is a homomorphism, $\varphi(\Box a) = \Box \varphi(a) = \Box \top = \top$. So $\Box a \in \Delta(\varphi)$ and $\Delta(\varphi)$ is a $\Box$-filter. ∎

The Homomorphism Theorem for algebraic systems (Theorem 1.2.12) can be reformulated for the case of modal algebras as follows.

Theorem 2.2.23 *If $\mathfrak{B}_1$ and $\mathfrak{B}_2$ are modal algebras and $\varphi : \mathfrak{B}_1 \mapsto \mathfrak{B}_2$ is a homomorphism onto $\mathfrak{B}_2$, and $\approx$ is the kernel congruence relation generated by φ then the following hold* *(i)* $\approx = \approx_{\Delta(\varphi)}$,

(ii) The mapping $[a]_{\approx_{\Delta(\varphi)}} \mapsto \varphi(a)$ *is the isomorphism from the algebra* $\mathfrak{A}_1 / \approx_{\Delta(\varphi)}$ *onto* $\mathfrak{B}_2$. *Thus these modal algebras are isomorphic.*

Proof. Actually, if $a \approx b$ then $\varphi(a) = \varphi(b)$ holds. Therefore we have

$$(\varphi(a) \to \varphi(b)) = \top, (\varphi(b) \to \varphi(a)) = \top.$$

Consequently it follows that $\varphi((a \to b) \wedge (b \to a)) = \top \wedge \top = \top$, consequently we have $(a \to b) \wedge (b \to a) \in \Delta(\varphi)$. Therefore we get $a \sim_{\Delta(\varphi)} b$. Conversely, let $a \sim_{\Delta(\varphi)} b$ hold. Then $(a \to b) \wedge (b \to a) \in \Delta(\varphi)$. This yields $\varphi((a \to b) \wedge (b \to a)) = \top$ and therefore we obtain $\varphi(a \to b) = \top$, $\varphi(b \to a)) = \top$. As φ is a homomorphism, these entail that $\varphi(a) \to \varphi(b) = \top$ and $\varphi(b) \to \varphi(a) = \top$. A well-known property of boolean algebras allows us to infer from this that $\varphi(a) \leq \varphi(b)$ and $\varphi(b) \leq \varphi(a)$, i.e. $\varphi(b) = \varphi(a)$ and $a \approx b$. So (i) holds. The claim of (ii) follows from (i) and Theorem 1.2.12. ∎

In accordance with the usual notation, for every $\mathfrak{B} \in Var(K)$ (which is, as we have seen, a modal algebra), $\lambda(\mathfrak{B})$ is the set of formulas $\{\alpha \mid \alpha \in \mathcal{F}or_K, \mathfrak{B} \models \alpha\}$ which is a normal modal logic. From Corollary 1.3.26 we immediately have the following

Corollary 2.2.24 *If* $\mathfrak{B}_1$ *and* $\mathfrak{B}_2$ *are modal algebras and* $\mathfrak{B}_i, i \in I$ *is a family of modal algebras then the following hold.*

(i) If $\mathfrak{B}_2$ *is a homomorphic image of* $\mathfrak{B}_1$ *then* $\lambda(\mathfrak{B}_1) \subseteq \lambda(\mathfrak{B}_2)$,
(ii) If $\mathfrak{B}_2$ *is a subalgebra of* $\mathfrak{B}_1$ *then* $\lambda(\mathfrak{B}_1) \subseteq \lambda(\mathfrak{B}_2)$,
(iii) $\lambda(\prod_{i \in I} \mathfrak{B}_i) = \lambda(\bigcap_{i \in I} \mathfrak{B}_i)$.

As we have seen the family of all normal modal logics $\mathcal{L}_{nml}$ forms a complete lattice with a smallest element K and a greatest element $\mathcal{F}or_K$ consisting of all formulas in the language of K. We now are going to clarify (as with the case of the lattice of superintuitionistic logics) how the lattice operations of $\mathcal{L}_{nml}$ interact with the axiomatizations for normal modal logics. We recall that a set of formulas Γ form an *axiomatization* of a normal modal logic λ with respect to a modal logical system $\mathcal{S}$ (or Δ is the set of axioms for λ with respect to $\mathcal{S}$) when λ is the set of all theorems of the axiomatic system $\mathcal{S} \oplus \Delta$ with axioms which are the all axioms of modal logic system $\mathcal{S}$ and still all formulas from Δ, and inference rules $(R1) : x, x \to y/y$, $R(3) : x/\Box x$. As before, if $\Delta = \{\alpha_1, ..., \alpha_n\}$ we will sometimes write $\mathcal{S} \oplus \alpha_1, ..., \alpha_n$ instead of $\mathcal{S} \oplus \Delta$.

Lemma 2.2.25 *If some normal modal logics* λ_1 *and* λ_2 *have axiomatizations* $\{\alpha_i \mid i \in I\}$ *and* $\{\beta_j \mid j \in J\}$ *with respect to* $\mathcal{S}$ *then the following holds* $\lambda_1 \vee \lambda_2 = \mathcal{S} \oplus \{\alpha_i \mid i \in I\} \cup \{\beta_j \mid j \in J\}$.

Proof. Easy, it is left as an exercise to the reader. ∎

We have accepted the following notation: $\square^0 a := a$ and $\square^{n+1} a := \square\square^n a$. Recall that $T := K \oplus (\square p \to b)$ is the von Wright normal modal logic. We denote the lattice of all normal modal logics extending the von Wright logic T by $\mathcal{L}_{nml}(T)$.

Lemma 2.2.26 *Let $\lambda_1 = T \oplus \{\alpha_i \mid i \in I\}$ and $\lambda_2 = T \oplus \{\beta_j \mid j \in J\}$. Then in the lattice $\mathcal{L}_{nml}(T)$ the following holds: $\lambda_1 \wedge \lambda_2 = T \oplus \{\square^n \alpha_i^2 \vee \square^m \beta_j^3 \mid i \in I, j \in J, 0 \le n, 0 \le m\}$, where every α_j^2 obtained from α_i by substitution $p_k \mapsto p_{2k}$, and each β_j^k is the result of the substitution $p_k \mapsto p_{2k+1}$ in β_j.*

Proof. In is not difficult to see that

$$T \oplus \{\square^n \alpha_i^2 \vee \square^m \beta_j^3 \mid i \in I, j \in J, 0 \le n, 0 \le m\} \le \lambda_1$$
$$T \oplus \{\square^n \alpha_i^2 \vee \square^m \beta_j^3 \mid i \in I, j \in J, 0 \le n, 0 \le m\} \le \lambda_2.$$

Indeed, all the extra axioms on the left hand sides of the relations above are derivable in λ_1 and λ_2. Let λ be a logic from $\mathcal{L}_{nml}(T)$ such that $\lambda \le \lambda_1$ and $\lambda \le \lambda_2$. Suppose that

$$\lambda \not\subseteq T \oplus \{\square^n \alpha_i^2 \vee \square^m \beta_j^3 \mid i \in I, j \in J, 0 \le n, 0 \le m\}.$$

Then there is a formula α such that $\alpha \in \lambda$, in particular $\alpha \in \lambda_1$, $\alpha \in \lambda_2$, but

$$\alpha \notin T \oplus \{\square^n \alpha_i^2 \vee \square^m \beta_j^3 \mid i \in I, j \in J, 0 \le n, 0 \le m\}.$$

By Theorem 2.2.26 there is a modal algebra $\mathfrak{B}_f$ from $Var(T \oplus \{\square^n \alpha_i^2 \vee \square^m \beta_j^3 \mid i \in I, j \in J, 0 \le n, 0 \le m\})$ such that $\mathfrak{B}_f \not\models \alpha$. Because according to Birkhoff Theorem (Theorem 1.2.57) $\mathfrak{B}_f$ is a subdirect product of subdirectly irreducible algebras, there is a subdirectly irreducible modal algebra $\mathfrak{B}$ from the variety $Var(T \oplus \{\square^n \alpha_i^2 \vee \square^m \beta_j^3 \mid i \in I, j \in J, 0 \le n, 0 \le m\})$ which also disproves α. We claim that $\mathfrak{B}$ is a member of $Var(\lambda_1)$ or $Var(\lambda_2)$. Indeed, suppose $\mathfrak{B}$ is neither a member of $Var(\lambda_1)$ nor of $Var(\lambda_2)$. Then there are formulas α_i and α_j such that $\mathfrak{B} \not\models \alpha_i$ and $\mathfrak{B} \not\models \beta_j$. This means there are valuations $E : p_k \mapsto a_k \in |\mathfrak{B}|$ and $W : p_k \mapsto b_k \in |\mathfrak{B}|$ such that

$$E(\alpha_i) = a \ne \top$$
$$W(\beta_j) = b \ne \top.$$

By Lemma 2.2.18 we can consider the $\square$-filters $\mathcal{F}(a, \downarrow)$ and $\mathcal{F}(b, \downarrow)$. The modal algebra $\mathfrak{B}$ is subdirectly irreducible. Therefore according to Lemma 2.2.15 there is a smallest $\square$-filter $\mathcal{F}$ among all non-trivial $\square$-filters on $\mathfrak{B}$. The filter $\mathcal{F}$ has an element $c \ne \top$. However we have $\mathcal{F} \subseteq \mathcal{F}(a, \downarrow)$, $\mathcal{F} \subseteq \mathcal{F}(b, \downarrow)$. Therefore there are $n, m \in N$ such that $\square^n a \le c$ and $\square^m b \le c$. In particular, $\square^n a \vee \square^m b \le c \ne \top$. We then take the valuation V as follows:

$$V(p_n) = \begin{cases} E(p_k) & \text{if } n = 2k \\ W(p_k) & \text{if } n = 2k+1 \end{cases}$$

Then it is easy to see $V(\alpha_i^2) = E(\alpha_i)$ and $V(\beta_j^3) = E(\beta_j)$. Thus

$$V(\Box^n \alpha_i^2 \vee \Box^m \beta_j^3) = \Box^n a \vee \Box^m b \leq c \neq \top$$

which contradicts the fact that $\mathfrak{B}$ is a member of the variety $Var(T \oplus \{\Box^n \alpha_i^2 \vee \Box^m \beta_j^3 \mid i \in I, j \in J, 0 \leq n, 0 \leq m\})$. Hence $\mathfrak{B} \in Var(\lambda_1)$ or $\mathfrak{B} \in Var(\lambda_2)$. If $\mathfrak{A} \in Var(\lambda_1)$ $(\in Var(\lambda_2))$ $\mathfrak{B} \not\models \alpha$ contradicts $\lambda \leq \lambda_1$ (λ_2). Therefore our assumption is impossible and $T \oplus \{\Box^n \alpha_i^2 \vee \Box^m \beta_j^3 \mid i \in I, j \in J, 0 \leq n, 0 \leq m\})$ is the greatest lower bound for λ_1 and λ_2. ∎

Definition 2.2.27 *We say that a normal modal logic λ is finitely axiomatizable with respect to a modal axiomatic system S if there is a finite tuple of formulas $\alpha_1, ..., \alpha_n$ such that $\lambda = S \oplus \alpha_1, ..., \alpha_n$.*

From Lemmas 2.2.25, and 2.2.26 we immediately derive

Corollary 2.2.28 *The family $\mathcal{L}_{nml}^{fa}(S4)$ of all normal modal logic extending modal logic $S4$ which are finitely axiomatizable with respect to $S4$ forms a sublattice of the lattice $\mathcal{L}_{nml}(S4)$ of all normal modal logics over $S4$.*

Theorem 2.2.29 *The lattice $\mathcal{L}_{nml}(T)$ of all normal modal logics which extend the von Wright modal logic T is a pseudo-boolean algebra.*

Proof. According to the second definition of pseudo-boolean algebras (Definition 2.1.12, which is equivalent to the first definition of *pba*, see Lemma 2.1.13) it is sufficient to show that our lattice has a smallest element and possesses pseudo-complements for all pairs of elements. It is clear that $\mathcal{L}_{nml}(T)$ has a smallest element T, and it is sufficient to show that the pseudo-complement of any logic λ_1 to an arbitrary logic λ_2 exists in this lattice. Let $\lambda_1 = T \oplus \Gamma$ (for instance, we can take $\Gamma := \lambda_1$). We define the set Δ of formulas in the language of K as follows:

$$\Delta := \{\beta \mid (\forall \alpha \in \Delta)(\forall n, m \in N)(\Box^n \alpha^2 \vee \Box^m \beta^3 \in \lambda_2\}$$

(for definition of α^2 and δ^3 see Lemma 2.2.26). We let $\lambda := T \oplus \Delta$, and will show that $(\lambda_1 \to \lambda_2) = \lambda$. Indeed, by Lemma 2.2.26 we have

$$\alpha_1 \wedge \lambda = T \oplus \{\Box^n \alpha^2 \vee \Box^3 \beta^3 \mid \alpha \in \Gamma, \beta \in \Delta, 0 \leq n, 0 \leq m\}.$$

For every $\alpha \in \Delta$, $\beta \in \Delta$, every $n \geq 0$ and $m \geq 0$, $\Box^n \alpha^2 \vee \Box^m \beta^3 \in \lambda_2$ holds by definition of Δ. Thus all the extra axioms of the logic $\lambda_1 \wedge \lambda$ are theorems of the logic λ_2 therefore we get $\lambda_1 \wedge \lambda \subseteq \lambda_2$. Now we are going to show that the logic λ is the greatest element of the set

$$\{\lambda^p \mid \lambda^p \in \mathcal{L}_{nml}(T), \lambda_1 \wedge \lambda^p \subseteq \lambda_2\}.$$

Let $\lambda_1 \wedge \lambda^p \subseteq \lambda_2$ and $\lambda_p = T \oplus \Omega$. By Lemma 2.2.26 we have $\lambda_1 \wedge \lambda^p = T \oplus \{\Box^n \alpha^2 \wedge \Box^m \gamma^3 \mid \alpha \in \Gamma, \gamma \in \Omega, 0 \leq n, 0 \leq m\}$. Then for every $\alpha \in \Gamma, \gamma \in \Omega$, $n \geq 0$ and $m \geq 0$, $\Box^n \alpha^2 \vee \Box^m \gamma^3 \in \lambda_2$ is valid. Hence, for each $\gamma \in \Omega$, $\gamma \in \Delta$ by the definition of Δ. Thus $\lambda^p = T \oplus \Omega \subseteq T \oplus \Delta = \lambda$. Hence we have showed that $\lambda_1 \rightarrow \lambda_2 = \lambda$. $\blacksquare$

From this and Lemma 2.1.4 we immediately infer

Corollary 2.2.30 *The lattice $\mathcal{L}_{nml}(T)$ is distributive.*

Note that this last corollary also is a corollary of the Jonsson's theorem (Theorem 1.2.62), which can be shown by reasoning similar to given one in the end of the previous section. Recall that by those arguments we can also show that $\mathcal{L}_{nml}(K)$ also satisfies the infinitary distributive low:

$$V_1 \vee \left(\bigwedge_{i \in I} V_i\right) = \bigwedge_{i \in I}(V_1 \vee V_i).$$

Now we turn to the case of temporal logics. We know that all temporal logics are algebraic logics (Theorem 1.3.18). Being an algebraic logic any temporal logic has a characterizing variety of appropriate algebras $Var(\lambda)$ by Theorem 1.3.20. This completeness theorem says: $\alpha \in \lambda$ iff $\alpha = \top$ is valid in $Var(\lambda)$, and now our aim is to specify the structure of algebras from $Var(\lambda)$ for temporal logics λ. We will do this by the patterns which follow from considerations concerning normal modal logics.

Definition 2.2.31 *An algebra $\mathcal{A}$ of kind $\langle A, \wedge, \vee, \rightarrow, \neg, G, H, \bot, \top \rangle$ is called a* temporal algebra *if*

    (i) the algebra $\langle A, \wedge, \vee, \rightarrow, \neg, \bot, \top \rangle$ is a boolean algebra;
    (ii) $\mathcal{A}$ satisfies the following identities (where $x \rightarrow y := \neg x \vee y$,
      $Fx := \neg G \neg x$, $Px := \neg H \neg x$):
      (1) $G(x \rightarrow y) \rightarrow (Gx \rightarrow Gy) = \top$
      (2) $H(x \rightarrow y) \rightarrow (Hx \rightarrow Hy) = \top$
      (3) $p \rightarrow HFp = \top$
      (4) $p \rightarrow GPp = \top$,
      (5) $G\top = \top$,
      (6) $H\top = \top$.

Theorem 2.2.32 (Algebraic Completeness Theorem for Temporal Logics).

    (i) If $\mathcal{F}$ is a family of temporal algebras then
      $\lambda(\mathcal{F}) := \{\alpha \mid (\forall \mathcal{A} \in \mathcal{F})(\mathcal{A} \models \alpha)\}$ *is a temporal logic.*

    (ii) For every temporal logic λ, the variety $Var(\lambda)$ consists of temporal algebras.

(iii) If T_0 is the smallest temporal logic then $Var(T_0)$ is the variety of all temporal algebras.

(iv) For any given temporal logic λ, there is a temporal algebra A from $Var(\lambda)$ such that $\lambda = \lambda(A)$.

Proof. If all algebras from $\mathcal{F}$ are temporal then they all are in particular boolean. Therefore, because $Var(PC)$ is the variety of all boolean algebras (cf. Exercise 2.1.29), all formulas in the temporal language which are obtained from formulas of PC by arbitrary substitutions of temporal formulas belong to $\lambda(\mathcal{F})$. Formulas, which are obtained from temporal axioms of T_0 by substitutions, belong to $\lambda(\mathcal{F})$ in view of (1) - (4) from the definition of temporal algebras. The set $\lambda(\mathcal{F})$ is closed with respect to modus ponens $\{x, x \to y/y\}$ because all algebras from $\mathcal{F}$ are boolean algebras. Finally, $\lambda(\mathcal{F})$ is closed with respect to the rules x/Gx, and x/Hx by (5) and (6) from the definition of temporal algebras. Thus, in particular, $\lambda(\mathcal{F})$ contains all theorems of T_0. It is evident that $\lambda(\mathcal{F})$ is closed with respect to substitutions. Hence $\lambda(\mathcal{F})$ is a temporal logic and (i) holds.

Let λ be a temporal logic. Consider an arbitrary algebra A from $Var(\lambda)$. Because $PC \subseteq \lambda$ we obtain that $\wedge, \vee, \to, \neg, \top$ - fragment of the algebra A belongs to the variety $Var(PC)$ of all boolean algebras and consequently the $\wedge, \vee, \to$ $, \neg, \top$ fragment of A is a boolean algebra. The identities (1) - (4) from the definition of temporal algebras are valid in A because the temporal axioms for T_0. It remains only to show that the identities (5) and (6) from this definition will be valid in A also. In fact, we have $\top \in \lambda$. The logic λ is closed with respect to the rules x/Gx and x/Hx. Therefore the identities $G\top = \top, H\top = \top$ are valid in the variety $Var(\lambda)$. Therefore they are also valid in A. Thus (5) and (6) are indeed valid in A and (ii) holds.

From (ii) it follows that $Var(T_0)$ is a variety of temporal algebras. Conversely, if A is a temporal algebra then A is a boolean algebra and hence all the axioms of PC are valid in A. All temporal axioms of T_0 are valid in A since the identities (1) - (4) from the definition of temporal algebra are valid in A. The set of all temporal formulas which valid in A is closed with respect to inference rules $x, x \to y/y$ and $x/Gx, x/Hx$ because $\mathcal{L}$ is a boolean algebra, and (5), (6) from the definition of temporal algebras. Therefore all theorems of T_0 are valid in A and $A \in Var(\lambda)$, which completes the proof of (iii). In order to show (iv) we take the free algebra of countable rank $\mathcal{F}(\lambda)$ from $Var(\lambda)$. According to (ii) this algebra is a temporal algebra and because it is free for $Var(\lambda)$ and since $\lambda = \lambda(Var(\lambda))$ (cf. Theorem 1.3.20), we finally conclude $\lambda(\mathcal{F}(\lambda)) = \lambda$. ∎

It is easy to see that as with normal modal logics and modal algebras the family of all temporal logics $\mathcal{L}_{tl}$ and the family of all varieties of temporal algebras $\mathcal{L}_{ta}$ form complete lattices with respect to set-theoretic inclusion $\subseteq$. As with the proof of Theorem 2.2.6, we can prove using Theorem 2.2.32

Theorem 2.2.33 *The mapping e of the lattice $\mathcal{L}_{tl}$ into the lattice $\mathcal{L}_{ta}$, where $e(\lambda) := Var(\lambda)$, is the dual lattices isomorphism onto, in particular the following hold:*

> *(i) e is one-to-one mapping onto,*
> *(ii) if $\lambda_1 \subseteq \lambda_2$ then $e(\lambda_1) \supseteq e(\lambda_2)$,*
> *(iii) $e(\lambda_1 \wedge \lambda_2) = e(\lambda_1) \vee e(\lambda_2)$, $e(\lambda_1 \vee \lambda_2) = e(\lambda_1) \wedge e(\lambda_2)$.*

Moreover e is the dual isomorphism of complete lattices:

> *(iv) $e(\bigwedge_{i \in I} \lambda_i) = \bigvee_{i \in I} e(\lambda_i)$,*
> *(v) $e(\bigvee_{i \in I} \lambda_i) = \bigwedge_{i \in I} e(\lambda_i)$.*

As we have seen, while describing an algebraic logic λ, it is important to clarify the structure of subdirectly irreducible algebras from $Var(\lambda)$. Therefore we are now interested in describing subdirectly irreducible temporal algebras. For this we need to consider the structure of congruence relations on temporal algebras.

Definition 2.2.34 *A filer Δ on a temporal algebra $\mathcal{A}$ is called a G, H-filter if for every $a \in \Delta$, $Ga \in \Delta$ and $Ha \in \Delta$.*

Lemma 2.2.35 *If $\mathcal{A}$ is a temporal algebra and $\approx$ is a congruence relation on $\mathcal{A}$ then the set $\Delta_{\approx} := \{a \mid a \approx \top\}$ is a G, H-filter.*

Proof. We show that $\Delta_{\approx}$ is a filter by the same arguments as used in the proof of Lemma 2.2.12. Let $a \in \Delta_{\approx}$. Then $a \approx \top$, and, since $\approx$ is a congruence, we get $Ga \approx G\top$. However $G\top = \top$ because $\mathcal{A}$ is a temporal algebra. Hence $Ga \approx \top$ and $Ga \in \Delta_{\approx}$. In a similar way we can show that $\Delta_{\approx}$ is closed with respect to H. Thus $\Delta_{\approx}$ is a G, H-filter. ∎

Lemma 2.2.36 *Let $\mathcal{A}$ be a temporal algebra and Δ be a G, H-filter on $\mathcal{A}$. Then the relation $\approx_{\Delta} := \{\langle a, b \rangle \mid (a \to b) \wedge (b \to a) \in \Delta\}$ is a congruence relation on $\mathcal{A}$.*

Proof. Using the same arguments as in the proof of Lemma 2.2.13 we can show that the relation $\approx_{\Delta}$ is an equivalence relation and is consistent with the boolean operations on $\mathcal{A}$. It remains to show that this relation is consistent with the operations G and H. Suppose that $a \approx_{\Delta} b$, i.e. $(a \to b) \wedge (b \to a) \in \Delta$. Then $(a \to b) \in \Delta$ and $(b \to a) \in \Delta$. Because Δ is a G, H-filter we get $G(a \to b) \in \Delta$ and $G(b \to a) \in \Delta$. Using (1) from the definition of temporal algebras we obtain $(Ga \to Gb) \in \Delta$ and $(Gb \to Ga) \in \Delta$. Thus $Ga \approx Gb$. The proof of consistency with H is analogous to the case for G. Hence the relation $\approx_{\Delta}$ is a congruence relation on $\mathcal{A}$. ∎

As with normal modal logics, for every temporal algebra $\mathcal{A}$, the set of all G, H-filters on $\mathcal{A}$, which we denote $\Phi(\mathcal{A}, G, H)$, and the set $\mathcal{C}(\mathcal{A}, \approx)$ of all congruence relations on $\mathcal{A}$ form partially ordered sets (with respect to the usual set-theoretic inclusion $\subseteq$). Again both these posets are closed with respect to arbitrary set-theoretic intersections. Therefore $\Phi(\mathcal{A}, G, H)$ and $\mathcal{C}(\mathcal{A}, \approx)$ are complete lattices.

Theorem 2.2.37 *The mapping e of the lattice $\mathcal{C}(\mathcal{A}, \approx)$ into the lattice $\Phi(\mathcal{A}, G, H)$, where $e(\approx) := \Delta_{\approx}$, is an isomorphism onto and $e^{-1}(\Delta) = \approx_{\Delta}$.*

Proof. By Lemma 2.2.35 for every congruence $\approx$, $e(\approx)$ is a G, H-filter. Obviously e preserves the order. The mapping e is one-to-one. Indeed, suppose that $e(\approx_1) = e(\approx_2)$. Let $a \approx_1 b$ hold. Then we have $(a \to b) \wedge (b \to a) \approx_1 (a \to a) \wedge (a \to a) = \top$. That is $(a \to b) \wedge (b \to a) \in \Delta_{\approx_1} = \Delta_{\approx_2}$. This yields $(a \to b) \wedge (b \to a) \approx_2 \top$. Then $a \wedge \top \approx_2 a \wedge (a \to b) \wedge (b \to a) = a \wedge b \wedge (b \to a) = a \wedge b$. Similarly we show that $b \approx_2 a \wedge b$. Hence $a \approx_2 b$ and we have proved that $\approx_1 \subseteq \approx_2$. The converse inclusion can be shown similarly. Thus $\approx_1 = \approx_2$ and e is in fact a one-to-one mapping. Now we show that the mapping e is onto. Let Δ be a G, H-filer on $\mathcal{A}$. Then according to Lemma 2.2.36 the relation $\approx_{\Delta}$ is a congruence relation on $\mathfrak{B}$. We claim that $e(\approx_{\Delta}) = \Delta$. Indeed, if $a \in \Delta$ then $a \approx_{\Delta} \top$ and therefore $a \in e(\approx_{\Delta})$. Conversely, if $a \in e(\approx_{\Delta})$ then $a \approx_{\Delta} \top$ which means $(a \to \top) \wedge (\top \to a) = a \in \Delta$. Hence, really $e(\approx_{\Delta}) = \Delta$ and e is a mapping onto, and $e^{-1}(\Delta) = \approx_{\Delta}$. Summarizing the observations e is an lattices isomorphism in view of e preserves the order. ■.

By the definition an algebra is subdirectly irreducible if and only if there exists a smallest congruence relation among all non-trivial congruence relations. Applying Theorem 2.2.37 we obtain

Theorem 2.2.38 *A temporal algebra $\mathcal{A}$ is subdirectly irreducible if and only if there is a smallest G, H-filter among all non-trivial G, H-filters on $\mathcal{A}$.*

Analogously to the case of modal algebras we can introduce some special G, H-filters and corresponding congruence relations for special temporal algebras.

Lemma 2.2.39 *If $\mathcal{A}$ is a temporal algebra from the variety $Var(T_0 \oplus \{Gp \to p, Hp \to p\}$ and a is an element of $\mathcal{A}$ then the following hold.*

(i) The set

$$\mathcal{F}(a, \downarrow) := \{b \mid b \in |\mathcal{A}|, (\exists n_1, ..., n_k)(\exists m_1, ..., m_k)$$

$$(G^{n_k} H^{m_k} ... G^{n_1} H^{m_1} a \leq b)\}$$

is a G, H-filter on $\mathcal{A}$ (where $G^0 a := a$ and $G^{n+1} a := GG^n a$, $H^0 a := a$ and $H^{n+1} a := HH^n a$).

(ii) If Δ is a G, H-filter on $\mathcal{A}$ containing a then $\mathcal{F}(a, \downarrow) \subseteq \Delta$. That is $\mathcal{F}(a, \downarrow)$ is the smallest G, H-filter containing a.

Proof. The set $\mathcal{F}(a, \downarrow)$ is obviously closed with respect to extensions of elements. Suppose that

$$G^{n_k} H^{m_k} ... G^{n_1} H^{m_1} a \leq b$$
$$G^{r_l} H^{q_l} ... G^{r_1} H^{q_1} a \leq c.$$

Then

$$G^{n_k} H^{m_k} ... G^{n_1} H^{m_1} a \wedge G^{r_l} H^{q_l} ... G^{r_1} H^{q_1} a \leq b \wedge c.$$

For every element x from $\mathcal{A}$, we have $Gx \leq x$ and $Hx \leq x$ and applying (1) and (2) from the definition of temporal algebras we obtain that the following $G^{n+1} x \leq G^n x$, $H^{n+1} x \leq H^n x$ hold. Therefore if $v = max(k, l)$ and for every i, $1 \leq i \leq v$,

$$w_i := \begin{cases} max(n_i, m_i, r_i, q_i), & \text{if } i \leq k, i \leq l \\ max(n_i, m_i), & \text{if } l < i \\ max(r_i, q_i), & \text{if } n < i \end{cases}$$

then

$$G^{w_v} H^{w_v} ... G^{w_1} H^{w_1} a \leq G^{n_k} H^{m_k} ... G^{n_1} H^{m_1} a,$$
$$G^{w_v} H^{w_v} ... G^{w_1} H^{w_1} a \leq G^{r_l} H^{q_l} ... G^{r_1} H^{q_1} a.$$

Therefore we get $G^{w_v} H^{w_v} ... G^{w_1} H^{w_1} a \leq b \wedge c$, and $b \wedge c \in \mathcal{F}(a, \downarrow)$. Thus $\mathcal{F}(a, \downarrow)$ is a filter. Let $b \in \mathcal{F}(a, \downarrow)$, that is

$$G^{n_k} H^{m_k} ... G^{n_1} H^{m_1} a \leq b.$$

Then applying (1) and (2) from the definition of temporal algebras we obtain

$$GG^{n_k} H^{m_k} ... G^{n_1} H^{m_1} a \leq Gb$$
$$HG^{n_k} H^{m_k} ... G^{n_1} H^{m_1} a \leq Hb.$$

Thus $Gb, Hb \in \mathcal{F}(a, \downarrow)$ and this set is a certain G, H-filer. If Δ is an G, H-filter then obviously $\mathcal{F}(a, \downarrow) \subseteq \Delta$ because Δ is closed with respect to G and H, hence (ii) also holds. $\blacksquare$

Definition 2.2.40 *An element a of a temporal algebra $\mathcal{A}$ is called* stable *if and only if $Ga = a$ and $Ha = a$.*

Lemma 2.2.41 *If a is a stable element of a temporal algebra $\mathcal{A}$ then the filter $\mathcal{F}(a) := \{b \mid a \leq b\}$ generated by a is a certain G, H-filter.*

Proof. The algebra $\mathcal{A}$ can be understood as a modal algebra if we understand $\square$ as G or H. Therefore applying Lemma 2.2.17 it follows that $\mathcal{F}(a)$ is a G-filter and a H-filter simultaneously. Hence $\mathcal{F}(a)$ is a G, H-filter. $\blacksquare$

Theorem 2.2.42 *Any finite temporal algebra from $Var(T_0 \oplus (Gp \to p, Hp \to p))$ is subdirectly irreducible if and only if there exists a greatest element ω in the set of all stable elements which are not equal to $\top$.*

Proof. Because $\mathcal{A}$ is subdirectly irreducible, by Theorem 2.2.38 there is the smallest non-trivial G, H-filter $\mathcal{F}$. This filter has a smallest element a in view of the fact that $\mathcal{A}$ is finite. Thus $\mathcal{F} = \mathcal{F}(a) := \{b \mid a \leq b\}$. We claim that a is the biggest stable element. In fact, because the filter $\mathcal{F}(a)$ is an G, H-filter, $a \leq Ga$ and $a \leq Ha$ hold. But $\mathcal{A} \in Var(T_0 \oplus (Gp \to p, Hp \to p))$, therefore $Ga \leq a$ and $Ha \leq a$. Hence $Ga = a$ and $Ha = a$ and a is stable.

Now we consider an arbitrary stable element b, where $b \neq \top$. Using Lemma 2.2.41 we take the filter $\mathcal{F}(b)$ generated by b which is a G, H-filter. Because $\mathcal{F}$ is the smallest non-trivial G, H-filter, we have $\mathcal{F}(a) \subseteq \mathcal{F}(b)$ which gives us $b \leq a$. So we have showed that a is the greatest stable element (among stable elements which are not equal to $\top$).

Assume that the greatest stable element ω exists. Let $\mathcal{F}$ be a G, H-filter which differs from the filter $\{\top\}$. Then there exists an element b from $\mathcal{F}$ such that $b \neq \bot$. We have that $\mathcal{A} \in Var(T_0 \oplus (Gp \to p, Hp \to p))$. Therefore we can apply Lemma 2.2.37 and then we conclude $\mathcal{F}(b, \downarrow) \subseteq \Delta$. The filter $\mathcal{F}(b, \downarrow)$ has the smallest element c because the algebra $\mathcal{A}$ is finite. That is $\mathcal{F}(b, \downarrow) = \mathcal{F}(c)$. The filter $\mathcal{F}(b, \downarrow)$ is a G, H-filter. Therefore we can prove that c is stable in quite a similar manner to the proof that a is stable in the *necessity part* of our lemma. After that we have $c \leq b \neq \top$ and consequently $c \leq \omega$. Therefore applying Lemma 2.2.41 it follows that $\mathcal{F}(c)$ and $\mathcal{F}(\omega)$ are G, H-filters and

$$\mathcal{F}(\omega) \subseteq \mathcal{F}(c) = \mathcal{F}(b, \downarrow) \subseteq \Delta.$$

Thus we have proved that $\mathcal{F}(\omega)$ is the smallest G, H-filter. Then by Theorem 2.2.36 $\mathcal{A}$ is subdirectly irreducible. ∎

We now intend to describe the action of homomorphisms on temporal algebras and to specify the Homomorphism Theorem (Theorem 1.2.12) for temporal algebras.

Lemma 2.2.43 *For any given temporal algebras $\mathcal{A}_1$ and $\mathcal{A}_2$ and a homomorphism $\varphi : \mathcal{A}_1 \mapsto \mathcal{A}_2$, the set $\Delta(\varphi) := \{a \mid a \in |\mathcal{A}_1|, \varphi(a) = \top\}$ is a G, H-filter on $\mathcal{A}_1$.*

Proof. The algebras $\mathcal{A}_1$ and $\mathcal{A}_2$ are modal algebras with respect to the $\square$ operation understood as G or H. Therefore applying Lemma 2.2.22 we conclude $\Delta(\varphi)$ is G-filter and H-filter, i.e. this filter is a G, H-filter. ∎

Theorem 2.2.44 *Suppose $\mathcal{A}_1$ and $\mathcal{A}_2$ are temporal algebras and $\varphi : \mathcal{A}_1 \mapsto \mathcal{A}_2$ is a homomorphism onto, and $\approx$ is the kernel congruence relation generated by φ. Then*

(i) $\approx = \approx_{\Delta(\varphi)}$,

(ii) The mapping $[a]_{\approx_{\Delta(\varphi)}} \mapsto \varphi(a)$ is the isomorphism from the algebra

$\mathcal{A}_1 / \approx_{\Delta(\varphi)}$ onto $\mathcal{A}_2$. Thus these temporal algebras are isomorphic.

Proof. (i): If $a \approx b$ then $\varphi(a) = \varphi(b)$ holds. Then using this we derive

$$(\varphi(a) \rightarrow \varphi(b)) = \top, \ (\varphi(b) \rightarrow \varphi(a)) = \top.$$

This yields $\varphi((a \rightarrow b) \wedge (b \rightarrow a)) = \top \wedge \top = \top$, therefore it follows that $(a \rightarrow b) \wedge (b \rightarrow a) \in \Delta(\varphi)$. Thus we have $a \sim_{\Delta(\varphi)} b$. Conversely, suppose $a \sim_{\Delta(\varphi)} b$ holds. Then $(a \rightarrow b) \wedge (b \rightarrow a) \in \Delta(\varphi)$. This yields $\varphi((a \rightarrow b) \wedge (b \rightarrow a)) = \top$ and consequently we get $\varphi(a \rightarrow b) = \top, \varphi(b \rightarrow a)) = \top$. The mapping φ is a homomorphism, consequently the last relations imply $\varphi(a) \rightarrow \varphi(b) = \top$ and $\varphi(b) \rightarrow \varphi(a) = \top$. This gives $\varphi(a) \leq \varphi(b)$ and $\varphi(b) \leq \varphi(a)$, i.e. $\varphi(b) = \varphi(a)$ and $a \approx b$. Hence (i) holds. The claim of (ii) follows from (i) and Theorem 1.2.12.
∎

As with the case of modal algebras applying Corollary 1.3.26 we immediately obtain

Corollary 2.2.45 *If $\mathcal{A}_1$ and $\mathcal{A}_2$ are temporal algebras and $\mathcal{A}_i, i \in I$ is a family of temporal algebras then the following hold.*

(i) *If $\mathcal{A}_2$ is a homomorphic image of $\mathcal{A}_1$ then $\lambda(\mathcal{A}_1) \subseteq \lambda(\mathcal{A}_2)$,*
(ii) *If $\mathcal{A}_2$ is a subalgebra of $\mathcal{A}_1$ then $\lambda(\mathcal{A}_1) \subseteq \lambda(\mathcal{A}_2)$,*
(iii) $\lambda(\prod_{i \in I} \mathcal{A}_i) = \lambda(\bigcap_{i \in I} \mathcal{A}_i)$.

The interaction of lattice operations in $\mathcal{L}_{tl}$ and the axiomatizations of temporal logics is, in a sense, standard. This can be shown on the basis of Lemma 2.2.38. The case of the union of temporal logics is evident, namely:

Lemma 2.2.46 *If temporal logics λ_1 and λ_2 have axiomatizations $\{\alpha_i \mid i \in I\}$ and $\{\beta_j \mid j \in J\}$ with respect to some temporal axiomatic system $\mathcal{S}$ then $\lambda_1 \vee \lambda_2 = \mathcal{S} \oplus \{\alpha_i \mid i \in I\} \cup \{\beta_j \mid j \in J\}$.*

The case of the intersection of temporal logics is not so evident. Let T_t be the temporal logic based on the modal logic von Wright T:

$$T_t := T_0 \oplus Gp \rightarrow p, Hp \rightarrow p.$$

Lemma 2.2.47 *Let $\mathcal{L}_{tl}(T_t)$ be the lattice of all temporal logics extending the logic T_t. Suppose λ_1 and λ_2 are temporal logics having the following presentations $\lambda_1 = T_t \oplus \{\alpha_i \mid i \in I\}$, $\lambda_2 = T_t \lambda \oplus \{\beta_j \mid j \in J\}$. In the lattice $\mathcal{L}_{tl}(T_t)$ the following holds*

$$\lambda_1 \wedge \lambda_2 = T_t \oplus \{G^{n_k} H^{m_k} ... G^{n_1} H^{m_1} \alpha_i^2 \ \vee \ G^{r_h} H^{q_h} ... G^{r_1} H^{q_1} \beta_j^3$$
$$\mid i \in I, j \in J, n_\alpha, m_\beta, r_\gamma, q_\delta \in N\}$$

where every α_i^2 obtained from α_i by substitution $p_k \mapsto p_{2k}$, and each β_j^k is the result of the substitution $p_k \mapsto p_{2k+1}$ on β_j.

Proof. It can easily be seen that

$$T_t \oplus \{ G^{n_k} H^{m_k} ... G^{n_1} H^{m_1} \alpha_i^2 \ \lor \ G^{r_h} H^{q_h} ... G^{r_1} H^{q_1} \beta_j^3 \mid i \in I,$$
$$j \in J, n_\alpha, m_\beta, r_\gamma, q_\delta \in N \} \leq \lambda_1$$
$$T_t \oplus \{ G^{n_k} H^{m_k} ... G^{n_1} H^{m_1} \alpha_i^2 \ \lor \ G^{r_h} H^{q_h} ... G^{r_1} H^{q_1} \beta_j^3 \mid i \in I,$$
$$j \in J, n_\alpha, m_\beta, r_\gamma, q_\delta \in N \} \leq \lambda_2.$$

Actually, it is not difficult to see that all the extra axioms on the left hand sides of the relations above are derivable in the logics λ_1 and λ_2 respectively.

Let λ be a temporal logic such that $T_t \leq \lambda \leq \lambda_1$ and $T_t \leq \lambda \leq \lambda_2$. Suppose that

$$\lambda \not\subseteq T_t \oplus \{ G^{n_k} H^{m_k} ... G^{n_1} H^{m_1} \alpha_i^2 \ \lor \ G^{r_h} H^{q_h} ... G^{r_1} H^{q_1} \beta_j^3 \mid i \in I,$$
$$j \in J, n_\alpha, m_\beta, r_\gamma, q_\delta \in N \}$$

Then there is a formula α such that $\alpha \in \lambda$, and consequently $\alpha \in \lambda_1$ and $\alpha \in \lambda_2$, but $\alpha \notin \lambda_0$, where

$$\lambda_0 := T_t \oplus \{ G^{n_k} H^{m_k} ... G^{n_1} H^{m_1} \alpha_i^2 \ \lor \ G^{r_h} H^{q_h} ... G^{r_1} H^{q_1} \beta_j^3 \mid i \in I,$$
$$j \in J, n_\alpha, m_\beta, r_\gamma, q_\delta \in N \}$$

According to Theorem 2.2.32 there is a temporal algebra $\mathcal{A}_f$ from $Var(\lambda_0)$ such that $\mathcal{A}_f \not\models \alpha$. By Theorem 1.2.57 $\mathcal{A}_f$ is a subdirect product of subdirectly irreducible algebras, consequently there is a subdirectly irreducible temporal algebra $\mathcal{A}$ from the variety $Var(\lambda_0)$ which also disproves α. We will show that $\mathcal{A}$ is a member of the variety $Var(\lambda_1)$ or the variety $Var(\lambda_2)$. Actually, suppose $\mathcal{A}$ nether is a member of $Var(\lambda_1)$ nor $Var(\lambda_2)$. Then there are formulas α_i and α_j such that $\mathcal{A} \not\models \alpha_i$ and $\mathcal{A} \not\models \beta_j$. This means there are valuations $\mu : p_k \mapsto a_k \in |\mathcal{A}|$ and $\nu : p_k \mapsto b_k \in |\mathcal{A}|$ such that

$$\mu(\alpha_i) = a \neq \top$$
$$\nu(\beta_j) = b \neq \top.$$

Using Lemma 2.2.39 we generate the G, H-filters $\mathcal{F}(a, \downarrow)$ and $\mathcal{F}(b, \downarrow)$. The temporal algebra $\mathcal{A}$ is a subdirectly irreducible. Therefore by Theorem 2.2.38 there is a smallest G, H-filter $\mathcal{F}$ among all non-trivial G, H-filters on the algebra $\mathcal{A}$. The filter $\mathcal{F}$ has an element $c \neq \top$. Because the filter $\mathcal{F}$ is smallest, we infer that $\mathcal{F} \subseteq \mathcal{F}(a, \downarrow)$ and $\mathcal{F} \subseteq \mathcal{F}(b, \downarrow)$. Then there is some tuple $n_k, m_k, ..., n_1, m_1,$ $r_h, q_h, ... , r_1, q_1$ of natural numbers such that

$$G^{n_k} H^{m_k} ... G^{n_1} H^{m_1} a \leq c, \quad G^{r_h} H^{q_h} ... G^{r_1} H^{q_1} b \leq c.$$

Therefore we get $G^{n_k} H^{m_k} ... G^{n_1} H^{m_1} a \ \lor \ G^{r_h} H^{q_h} ... G^{r_1} H^{q_1} b \leq c \neq \top$. We choose the valuation V on $\mathcal{A}$ as follows:

$$V(p_n) = \begin{cases} \mu(p_k) & \text{if } n = 2k \\ \nu(p_k) & \text{if } n = 2k + 1 \end{cases}$$

It can be easily seen that $V(\alpha_i^2) = \mu(\alpha_i)$ and $V(\beta_j^3) = \nu(\beta_j)$. Therefore

$$V(G^{n_k}H^{m_k}...G^{n_1}H^{m_1}\alpha_i^2 \vee G^{r_h}H^{q_h}...G^{r_1}H^{q_1}\beta_j^3) =$$
$$G^{n_k}H^{m_k}...G^{n_1}H^{m_1}a \vee G^{r_h}H^{q_h}...G^{r_1}H^{q_1}b \leq c \neq \top.$$

This contradicts $\mathcal{A} \in Var(\lambda_0)$). Hence actually $\mathcal{A} \in Var(\lambda_1)$ or $\mathcal{A} \in Var(\lambda_2)$. If $\mathcal{A} \in Var(\lambda_1)$ ($\in Var(\lambda_2)$) then $\mathcal{A} \nvDash \alpha$ which contradicts $\lambda \leq \lambda_1$ (λ_2). Therefore our assumption is contradictory and λ_0 is the greatest lower bound for λ_1 and λ_2 in the lattice $\mathcal{L}_{tl}(T_t)$. $\blacksquare$

As usually, a temporal logic λ is said to be *finitely axiomatizable* with respect to a temporal axiomatic system $\mathcal{S}$ if there is a finite tuple of temporal formulas $\alpha_1, ..., \alpha_n$ such that $\lambda = \mathcal{S} \oplus \alpha_1, ..., \alpha_n$. Let $S4_{t,c}$ be the temporal counterpart for Lewis modal system $S4$ with pairwise permutable G and H, that is

$$S4_{t,c} := T_0 \oplus \{Gp \to p, Hp \to p, Gp \to GGp, Hp \to HHp,$$
$$GHp \to HGp, HGp \to GHp\}.$$

Using Lemmas 2.2.46 and 2.2.47 we immediately derive

Corollary 2.2.48 *The family $\mathcal{L}_{tl}^{fa}(S4_{t,c})$ of all temporal logics finitely axiomatizable with respect to $S4_{t,c}$ forms a sublattice of the lattice $\mathcal{L}_{tl}(S4_{t,c})$ of all temporal logics over $S4_{t,c}$.*

Theorem 2.2.49 *The lattice $\mathcal{L}_{tl}(T_t)$ of all temporal logics which extend the temporal logic T_t is a pseudo-boolean algebra.*

Proof. The set $\mathcal{L}_{tl}(T_t)$ forms a lattice with a smallest element. Therefore it is sufficient to prove that our lattice possesses pseudo-complements for all pairs of elements. We take two arbitrary logics λ_1 and λ_2 from the lattice $\mathcal{L}_{tl}(T_t)$. Let $\lambda_1 = T_t \oplus \Gamma$ (for instance, we can take $\Gamma := \lambda_1$). Now we introduce the set Δ of formulas in temporal language for T_0 as follows:

$$\Delta := \{\beta \mid (\forall\alpha \in \Delta)(\forall n_k, m_k, ...n_1, m_1, r_h, q_h, ...r_1, q_1 \in N)$$
$$G^{n_k}H^{m_k}...G^{n_1}H^{m_1}\alpha^2 \vee G^{r_h}H^{q_h}...G^{r_1}H^{q_1}\beta^3 \in \lambda_2\}$$

(for definitions of α^2 and β^3 see Lemma 2.2.47). We set $\lambda := T_t \oplus \Delta$. Our aim is to show that $(\lambda_1 \to \lambda_2) = \lambda$. By Lemma 2.2.47 we have

$$\alpha_1 \wedge \lambda = T_t \oplus \{G^{n_k}H^{m_k}...G^{n_1}H^{m_1}\alpha^2 \vee G^{r_h}H^{q_h}...G^{r_1}H^{q_1}\beta^3 \mid \alpha \in \Gamma,$$
$$\beta \in \Delta, n_k, m_k, ...n_1, m_1, r_h, q_h, ...r_1, q_1 \in N\}.$$

All the extra axioms from the right hand side of the equality above belong to the logic λ_2 by the definition of Δ. Hence we have $\lambda_1 \wedge \lambda \subseteq \lambda_2$. It remains to prove that the logic λ is the greatest element of the set

$$\{\lambda^p \mid \lambda^p \in \mathcal{L}_{tl}(T_t), \lambda_1 \wedge \lambda^p \subseteq \lambda_2\}.$$

In fact, let us suppose that $\lambda_1 \wedge \lambda^p \subseteq \lambda_2$ and $\lambda^p = T \oplus \Omega$. By Lemma 2.2.47 we have

$$\lambda_1 \wedge \lambda^p = T_t \oplus \{G^{n_k} H^{m_k} ... G^{n_1} H^{m_1} \alpha^2 \vee G^{r_h} H^{q_h} ... G^{r_1} H^{q_1} \gamma^3$$
$$\mid \alpha \in \Gamma, \gamma \in \Omega, n_k, m_k, ... n_1, m_1, r_h, q_h, ... r_1, q_1 \in N\}.$$

Under the assumption $\lambda_1 \wedge \lambda^p \subseteq \lambda_2$, for every $\alpha \in \Gamma$, every $\gamma \in \Omega$, and every tuple $n_k, m_k, ... n_1, m_1, r_h, q_h, ... r_1, q_1$ of natural numbers,

$$G^{n_k} H^{m_k} ... G^{n_1} H^{m_1} \alpha^2 \vee G^{r_h} H^{q_h} ... G^{r_1} H^{q_1} \gamma^3 \in \lambda_2.$$

This implies that, for every $\gamma \in \Omega$, $\gamma \in \Delta$ by the definition of Δ. Thus

$$\lambda^p = T_t \oplus \Omega \subseteq T_t \oplus \Delta = \lambda.$$

Hence we have proved that $\lambda_1 \to \lambda_2 = \lambda$. ∎

This theorem and Lemma 2.1.4 immediately entail

Corollary 2.2.50 *The lattice $\mathcal{L}_{tl}(T_t)$ is distributive.*

2.3 Kripke Semantics for Modal and Temporal Logic

The initial idea for possible worlds semantics comes originally, from Leibniz who claimed that necessary truth is truth in all possible worlds. This idea was made the basis of a model theory for modal logic in Kripke [78]. (see for historical comments M.Fitting [49]). Since then possible worlds models have become widely used tools in modal investigations both because of their technical facility and because of their intuitive appeal. In general, a possible worlds model for a propositional logic is a pair, where the first element is a set of elements (possible worlds), and the second element is a binary relation on the first set simulating accessibility between worlds. We begin by developing a Kripke semantics that is suitable for normal modal logics and then we will consider the Kripke relational semantics for temporal logic and intuitionistic logic.

Definition 2.3.1 *A frame $\mathcal{F}$ is a pair $\langle F, R \rangle$, where F is a non-empty set and R is a binary relation on F, that is $R \subseteq F^2$.*

Informally, the set F can be understood as the set of all *possible worlds* or *possible states* which are simulated by elements of the given frame (recall that according to standard denotations of model theory $|\mathcal{F}| := F$), R is the accessibility relation between all possible worlds. aRb can be reformulated as b is accessible from a. The accessibility relation R itself can be interpreted as *transition* from one world (state) to another one. This transition can have different interpretations and nature such as the flow of time, communication possibilities, moving in space, etc. Involving geometrical intuition, we also accept the following conventions on placement elements in frame. We say an element b of $\mathcal{F}$ is R-seen from $a \in |\mathcal{F}|$, or a see b, if aRb. An element $a \in |\mathcal{F}|$ is R-maximal, or quasi-maximal, in $\mathcal{F}$ if $\forall b \in |\mathcal{F}|((aRb) \Rightarrow (bRa))$.

Definition 2.3.2 *Let P be a collection of the propositional letters, and let $\mathcal{F} = \langle W, R \rangle$ be a frame. A valuation of P on $\mathcal{F}$ is a mapping V which assigns to each letter $p \in P$ a subset $V(p)$ of the set W, that is $V : P \mapsto 2^W$.*

In this case the informal meaning of the interpretation V is: V assigns to the variable p the set of all possible worlds in which the claim encoded by p is valid.

Definition 2.3.3 *A Kripke model is a triple $\mathfrak{M} = \langle W, R, V \rangle$, where $\langle W, R \rangle$ is a frame and V is a valuation of a set P of propositional letters in the frame $\langle W, R \rangle$. $Dom(V) := P$ is called the domain of V. A Kripke model $\mathfrak{M}$ is weak if $Dom(V)$ is finite.*

For a modal propositional language $\mathcal{L}_m$, the truth relation $\Vdash$ in a given Kripke model $\mathfrak{M} := \langle W, R, V \rangle$ for modal formulas which are built up out of propositional variables included in the domain of the valuation V can be introduced inductively as follows:

$$(\forall p \in P)(\forall a \in W)[(a \Vdash_V p) \Leftrightarrow (a \in V(p))]$$
$$(\forall a \in W)(a \Vdash_V \alpha \wedge \beta) \Leftrightarrow (a \Vdash_V \alpha) \& (a \Vdash_V \beta)$$
$$(\forall a \in W)(a \Vdash_V \alpha \vee \beta) \Leftrightarrow (a \Vdash_V \alpha) \vee (a \Vdash_V \beta)$$
$$(\forall a \in W)(a \Vdash_V \neg \alpha) \Leftrightarrow (a \nVdash_V \alpha)$$
$$(\forall a \in W)(a \Vdash_V \alpha \rightarrow \beta) \Leftrightarrow (a \nVdash_V \alpha) \vee (a \Vdash_V \beta)$$
$$(\forall a \in W)(a \Vdash_V \Box \alpha) \Leftrightarrow (\forall b \in W[aRb \Rightarrow b \Vdash_V \alpha])$$
$$(\forall a \in W)(a \Vdash_V \Diamond \alpha) \Leftrightarrow (\exists b \in W[(aRb) \& (b \Vdash_V \alpha)])$$

So we merely extend the valuation V from propositional letters to all modal formulas which are constructed out of these letters. The introduced truth relation depends on the valuation V therefore we denote it by $\Vdash_V$. Thus the relation $a \Vdash_V \alpha$ can be spelled out as follows: the formula α is *true* (or is *valid*) on the element a in the model $\mathfrak{M}$ under the valuation V. Note that the definition of $\Vdash_V$ is consistent with the accepted abbreviation $\Diamond := \neg \Box \neg$ and the equivalence $\Box p \equiv \neg \Diamond \neg p$ which holds for modal logics:

Exercise 2.3.4 *For every Kripke model $\mathfrak{M} = \langle W, R, V \rangle$, and any element a from W, the following hold*

(i) $a \Vdash_V \Diamond p \rightarrow \neg \Box \neg p$, $a \Vdash_V \neg \Box \neg p \rightarrow \Diamond p$,
(ii) $a \Vdash_V \Box p \rightarrow \neg \Diamond \neg p$, $a \Vdash_V \neg \Diamond \neg p \rightarrow \Box p$.

Using given above definition we can introduce a truth relation for modal propositional formulas in frames:

Definition 2.3.5 *Given a frame $\mathcal{F} := \langle W, R \rangle$ and a modal propositional formula α, we say that α is true (or is valid) in the frame $\mathcal{F}$ (denotation $\mathcal{F} \Vdash \alpha$) if for every valuation V with the domain including all propositional variables from α, and for every element a from W, $a \Vdash_V \alpha$ holds.*

Definition 2.3.6 *A frame $\mathcal{F}$ is said to be an λ-frame for a logic λ if for every $\alpha \in \lambda$, $\mathcal{F} \Vdash \alpha$. A family Ω of frames is called* adequate *for a modal logic λ if $\lambda \subseteq \lambda(\Omega) := \{\alpha \mid \alpha \in \mathcal{F}or_K, (\forall \mathcal{F} \in \Omega) \mathcal{F} \Vdash \alpha\}$, that is if Ω consists of a collection of λ-frames.*

In other words, a class of frames Ω is adequate for a normal modal logic λ if and only if all theorems of λ are valid on all frames from Ω. Now we are going to present special classes of frames which are adequate for some normal modal logics. To begin, we consider the following classes.

Definition 2.3.7 *A frame $\mathcal{F} = \langle W, R \rangle$ is called* reflexive *if for every $a \in W$, aRa, that is if the relation R is reflexive. A frame $\mathcal{F} = \langle W, R \rangle$ is a transitive frame if for every $a, b, c \in |\mathcal{F}|$, aRb and bRc imply aRc, i.e. if the relation R is transitive.*

For any transitive frame $\mathcal{F} = \langle W, R \rangle$, a cluster of $\mathcal{F}$ is a subset C of W consisting of all mutually comparable by R elements, i.e.

$$(\forall x, y \in C)(xRy)\&(yRx) \text{ and}$$
$$(\forall x \in C, \forall y \in W)[(xRy)\&(yRx) \Rightarrow y \in C],$$

or $C := \{a\}$, where a is an irreflexive element, i.e. $\neg(aRa)$. For any element a, $C(a)$ is the cluster containing a. Also we introduce the following denotations:

> Fr be the class of all frames,
> Fr_r be the class of all reflexive frames,
> Fr_t be the class of all transitive frames,
> $Fr_{r,t}$ be the class of all reflexive and transitive frames,
> $Fr_e := \{\langle W, R \rangle \mid$ were R is an equivalence relation$\}$

Theorem 2.3.8 *The following hold*

> *(i) The class Fr is adequate for the normal modal logic K.*
>
> *(ii) The class Fr_r is adequate for the modal logic von Wright T.*
>
> *(iii) The class Fr_t is adequate for the modal logic $K4$.*
>
> *(iv) The class $Fr_{r,t}$ is adequate for the Lewis's modal logic $S4$.*

(v) The class Fr_e is adequate for the Lewis's modal logic S5.

Proof. The scheme of the proof for all assertions (i) - (v) is similar: for every case, we prove that all substitutions of axioms for PC as well as all substitutions of all modal axioms of the given logics are valid on corresponding classes of frames. And also we show that the set of all formulas which are valid on any class of frames is closed with respect to modus ponens and the normalization rule. This will give the adequacy of these classes frames for given logics. Actually,

Lemma 2.3.9 *If all axioms of a normal modal logic λ are valid in a frame $\mathcal{F}$ and rules (R1): $x, x \to y/y$ and (R3): $x/\Box x$ preserve the truth of formulas in $\mathcal{F}$ then all theorems of λ are valid in the frame $\mathcal{F}$.*

Proof. Given a theorem α of λ. The formula α is derivable from the schemes of axioms of λ. We fix a derivation $\mathcal{S}$ for α, and then we prove by induction on the length of the derivation $\mathcal{S}$ that all formulas of $\mathcal{S}$ are valid in $\mathcal{F}$. The first formula β of $\mathcal{S}$ must be a scheme of an axiom. Let V be a valuation in $\mathcal{F}$ of propositional letters from β. We must show β is valid in $\mathcal{F}$ under V. Indeed let β be a substitution example of an axiom γ for λ, where $\gamma = \gamma(p_i)$ (i.e. γ is build up out of letters p_i) and $\beta = \gamma(\delta_i)$. We define the valuation V_1 of all letters p_i in $\mathcal{F}$ as follows: V_1 assigns to each p_i the set $V(\delta_i)$ of all elements from $\mathcal{F}$ in which δ_i is valid under V. It is clear that the truth value of γ under V_1 coincides with the truth value of $\gamma(\delta_i)$ under V. We are given γ is valid in $\mathcal{F}$ under any valuation, therefore the formula $\gamma(\delta_i)$ (which coincides with β) is valid in $\mathcal{F}$ under V. Thus any scheme of axiom β for λ is true in $\mathcal{F}$ under any valuation. We are also given that (R1) and (R3) preserve the truth in $\mathcal{F}$. Therefore by induction on the places of the occurrences of formulas in $\mathcal{S}$ it follows that all formulas from $\mathcal{S}$ are valid in the frame $\mathcal{F}$, in particular, α is valid in $\mathcal{F}$. ∎

Thus in order to prove the theorem it is sufficient to show that the rules (R1) and (R3) preserve truth in frames and to show that all the corresponding axioms of the logics are valid in the appropriate classes of frames. First we consider inference rules.

Lemma 2.3.10 *For any class Ω of frames, the following set*

$$\{\alpha \mid \alpha \in \mathcal{F}or_K, (\forall \mathcal{F} \in \Omega)\mathcal{F} \Vdash \alpha\}$$

is closed with respect to rules (R1): $x, x \to y/y$ and (R3): $x/\Box x$.

Proof. Suppose $\alpha, \alpha \to \beta \in \lambda(\Omega)$. Let $\mathcal{F}$ be a frame of Ω. We take a valuation V of all propositional letters from the formulas β in the frame $\mathcal{F}$. If there are letters which occur in α but have no occurrences in β then we suppose that under V they all take the empty truth value in $\mathcal{F}$. Then according to the our supposition, for every element a from $\mathcal{F}$, $a \Vdash_V \alpha$ and $a \Vdash_V \alpha \to \beta$ which by definition of $\Vdash_V$

implies $a \Vdash_V \beta$. Hence $\beta \in \lambda(\Omega)$ and $\lambda(\Omega)$ is closed with respect to (R1). Let $\alpha \in \lambda(\Omega)$. Let $\mathcal{F}$ be a frame of Ω, and V is a valuation of propositional variables from α in $\mathcal{F}$. Then by our assumption, for every element a from $\mathcal{F}$, $a \Vdash_V \alpha$. Let b be an element of $\mathcal{F}$, and c is any accessible from b element, i.e. bRc. Then $c \Vdash_V \alpha$, thus we get $b \Vdash_V \Box\alpha$. Hence we have showed $\Box\alpha \in \lambda(\Omega)$ and $\lambda(\Omega)$ is closed with respect to (R3). ∎

Hence it remains only to check the truth of axioms on frames from the corresponding classes.

Lemma 2.3.11 *If a modal formula α is a substitution instance of some axiom from the list (A1) - (A10) of axioms of PC and $\mathcal{F}$ is a frame then $\mathcal{F} \Vdash \alpha$.*

Proof. The proof consists of a straightforward calculation of truth value and is left to the reader as a simple exercise.

Proving case (i), we have to show that any formula of the form $\Box(\alpha \to \beta) \to (\Box\alpha \to \Box\beta)$ is valid in any frame $\mathcal{F}$. Indeed, let $a \Vdash_V \Box(\alpha \to \beta)$ under some valuation V, where a is an element of some frame $\mathcal{F}$. Suppose $a \Vdash_V \Box\alpha$. Then for every $b \in \mathcal{F}$ such that aRb, $b \Vdash_V \alpha$. Because $a \Vdash_V \Box(\alpha \to \beta)$ we get $b \Vdash_V \alpha \to \beta$ for such b. That is, for every b such that aRb, $b \Vdash_V \beta$, which by the definition of $\Vdash_V$ yields $a \Vdash_V \Box\beta$. Hence $a \Vdash_V \Box(\alpha \to \beta) \to (\Box\alpha \to \Box\beta)$. Thus (i) holds.

Case (ii): we need only to show that the formula of kind $\Box\alpha \to \alpha$ is valid in all reflexive frames. Let $\mathcal{F}$ be a such frame, and V is a valuation of variables from α on $\mathcal{F}$. Assume that $a \in \mathcal{F}$ and $a \Vdash_V \Box\alpha$. Because the accessibility relation on $\mathcal{F}$ is reflexive we obtain $a \Vdash_V \alpha$. That is $a \Vdash_V \Box\alpha \to \alpha$ and the claim (ii) is proved.

Case (iii): Let $\mathcal{F}$ be a transitive frame, $\mathcal{F} = \langle W, V \rangle$, and let V be a valuation in $\mathcal{F}$. We take an $a \in |\mathcal{F}|$. Suppose $a \Vdash_V \Box\alpha$. We choose arbitrary $b, c \in \|\mathcal{F}\|$ such that aRb and bRc. Because R is transitive we get aRc. Therefore $a \Vdash_V \Box\alpha$ implies $c \Vdash_V \alpha$. According to the definition of $\Vdash_V$ this gives $a \Vdash_V \Box\Box\alpha$. Hence $a \Vdash \Box\alpha \to \Box\Box\alpha$ and the proof of (iii) is complete.

Case (iv): it follows directly from (ii) and (iii).

Case (v): in view of (iv), it is sufficient only to show that the formula $\Diamond\alpha \to \Box\Diamond\alpha$ is valid on all frames which are frames of the equivalence relations (because any equivalence relation is reflexive and transitive). Indeed, let $\mathcal{F} = \langle W, R \rangle$ is a frame and R is an equivalence relation. Let V be a valuation of all propositional letters from α in $\mathcal{F}$. Suppose $a \in \mathcal{F}$ and $a \Vdash_V \Diamond\alpha$. This means that there exists an element b such that aRb and $b \Vdash_V \alpha$. Let $c \in \mathcal{F}$ and aRc. Because R is an equivalence relation we get cRa, therefore cRb. Thus $c \Vdash_V \Diamond\alpha$ and we have shown $a \Vdash_V \Box\Diamond\alpha$. Hence $a \Vdash_V \Diamond\alpha \to \Box\Diamond\alpha$ and (v) is proved. ∎

The theorems about adequacy are useful for verifying whether a given formula is a theorem for a modal logic, or whether a modal logic λ_1 is a sublogic for another given logic λ_2. We give several illustrating examples below.

Examples 2.3.12 *The following hold:*

(i) $K \subset K4 \subset S4 \subset S5$, $K \subset T \subset S4$

(ii) $K4 \not\subseteq T$, $T \not\subseteq K4$.

In order to show these propositions we will use some special frames. We will depict frames and models by diagrams. We will understand the accessibility relations in transitive frames as the possibility of passing from one element to another moving along the lines which connect elements in the direction from down to top. We will depict reflexive elements of frames (those $a \in W, \mathcal{F} = \langle W, R \rangle$ that aRa) by small solid circles •, and irreflexive elements (those a such that $\neg(aRa)$) by empty circles ∘. In non-transitive frames we will designate the accessibility relation by arrows, where $a \to b$ means aRb. We introduce into considerations the following frames $\mathcal{F}_1 - \mathcal{F}_4$ depicted in the Fig. 2.4. Let us take

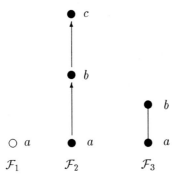

Figure 2.4:

the frame $\mathcal{F}_1$ and consider the valuation V as follows: $V(p) := \emptyset$. Then $a \Vdash_V \Box p$ because no elements b from $\mathcal{F}_1$ that aRb, and $a \nVdash_V p$. That is $a \nVdash_V \Box p \to p$. Since the class Fr is adequate to K according to Theorem 2.3.8, we get $\Box p \to p \notin K$, consequently $K \subset T$.

Now we choose the frame $\mathcal{F}_2$ and take the valuation V as follows $V(p) := \{b\}$. Then in the resulting model in $\mathcal{F}_2$ we get $a \Vdash_V \Box p$. At the same time $c \nVdash_V p$, consequently $b \nVdash_V \Box p$ and $a \nVdash_V \Box p \to \Box\Box p$. Thus $\mathcal{F}_2 \nVdash \Box p \to \Box\Box p$. By our theorem about adequacy (Theorem 2.3.8) we know Fr is adequate for K, therefore $\Box p \to \Box\Box p \notin K$ and $K \subset K4$.

Consider the frame $\mathcal{F}_2$ again. It is a reflexive frame, consequently we have $\mathcal{F}_2 \in Fr_r$. We have seen $\mathcal{F}_2 \nVdash \Box p \to \Box\Box p$. Hence according to Theorem 2.3.8, we obtain $\Box p \to \Box\Box p \notin T$, and $T \subset S4$. Because $\Box p \to \Box\Box p$ is an axiom of $K4$, we also get $K4 \not\subseteq T$. As we have already seen $\mathcal{F}_1 \nVdash \Box p \to p$, and the frame $\mathcal{F}_1$ is transitive, i.e. $\mathcal{F}_1 \in Fr_t$. Therefore, since Fr_t is adequate for $K4$, we derive $\Box p \to p \notin K4$, and consequently $K4 \subset S4$ and $T \not\subseteq K4$ hold.

We choose now the frame $\mathcal{F}_3$ and the valuation $V(p) := \{a\}$ in $\mathcal{F}_3$. Then $a \Vdash_V a$ and $a \Vdash_V \Diamond p$, but $b \nVdash_V \Diamond p$. Thus $a \nVdash_V \Box \Diamond p$ and $a \nVdash_V \Diamond p \to \Box \Diamond p$. The frame $\mathcal{F}_3$ is reflexive and transitive, therefore by our theorem concerning adequacy we infer that $\Diamond p \to \Box \Diamond p \notin S4$. Hence $S4 \subset S5$. ∎

Thus considered above normal modal logics form in the lattice $\mathcal{L}_{nml}$ the fragment designated in Fig. 2.5. Moreover, there $S4$ is the union of T and $K4$ ac-

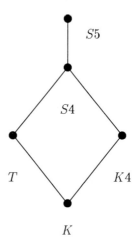

Figure 2.5:

cording to the description of the axiomatization of unions for logics.

We have already seen that the use of Kripke semantics is helpful even on the basis of the adequacy theorems. These kind of theorems allow us to recognize certain formulas as not being theorems. It is rather natural to determine if the converse is possible, i.e. whether it is possible to describe the sets of theorems for modal logics in a similar way. This question motivates our introducing the following definitions.

Definition 2.3.13 *A class of frames Ω is said to be* characterizing *for a normal modal logic λ if $\lambda = \lambda(\Omega) := \{\alpha \mid \alpha \in \mathcal{F}or_K, (\forall \mathcal{F} \in \Omega)(\mathcal{F} \Vdash \alpha)\}$.*

Definition 2.3.14 *A normal modal logic λ is said to be* Kripke complete *if there is a family Ω of Kripke frames such that*

$$\lambda = \lambda(\Omega) := \{\alpha \mid \alpha \in \mathcal{F}or_K, (\forall \mathcal{F} \in \Omega)(\mathcal{F} \Vdash \alpha)\}.$$

In other words, a modal logic λ is Kripke complete if there exists a class of frames characterizing λ. Thus the set of all theorems for any such logic coincides with the set of all formulas which are valid in all the frames of the corresponding to λ

class of frames. Our aim now is to present a simple natural tools for establishing Kripke completeness. We start with the following simple observation.

Lemma 2.3.15 *Let $\mathcal{F}$ be a frame, and Ω be a class of frames, $\mathfrak{M}$ be a Kripke model and Ω_1 be a class of Kripke models. Then*

(i) The set of formulas $\lambda(\mathcal{F}) := \{\alpha \mid \alpha \in \mathcal{F}or_K, \mathcal{F} \Vdash \alpha\}$ is a normal modal logic.

(ii) The set $\lambda(\Omega) := \bigcap_{\mathcal{F} \in \Omega} \lambda(\mathcal{F})$ is also a normal modal logic.

Proof. Indeed by (i) of Theorem 2.3.8 for any frame $\mathcal{F}$, $K \subseteq \lambda(\mathcal{F})$. By Lemma 2.3.10 the set $\lambda(\mathcal{F})$ is closed with respect to rules (R1) and (R3), and $\lambda(\mathcal{F})$ is closed under arbitrary substitutions. Indeed,

Lemma 2.3.16 *Let $\mathcal{F}$ be a frame, α be a modal formula such that $\mathcal{F} \Vdash \alpha$. Suppose that β is a substitution instance of α. Then $\mathcal{F} \Vdash \beta$.*

Proof. Suppose that the formula α is built up of propositional variables p_i and the formula β is obtained from α by substituting formulas γ_i in place of letters p_i. We will abbreviate this condition as follows: $\alpha = \alpha(p_i)$, $\beta = \alpha(\gamma_i)$. Let S be an arbitrary valuation of the propositional variables from β on $\mathcal{F}$. We introduce the new valuation of propositional letters p_i as follows:

$$V(p_i) := S(\gamma_i) := \{a \mid a \in |\mathcal{F}|, a \Vdash_S \gamma_i\}.$$

Because $\mathcal{F} \Vdash \alpha$ we get

$$(\forall a \in |\mathcal{F}|)(a \Vdash_V \alpha) \tag{2.1}$$

By simple induction on the length of subformulas $\delta(p_i)$ of the formula $\alpha(p_i)$ it can be easily shown that

$$(\forall a \in |\mathcal{F}|)[a \Vdash_V \delta(p_i) \Leftrightarrow a \Vdash_S \delta(\gamma_i)] \tag{2.2}$$

Combining (2.1) and (2.2) we arrive at $(\forall a \in |\mathcal{F}|)(a \Vdash_S \alpha(\gamma_i))$. Thus the formula β is valid in $\mathcal{F}$ under any valuation S, i.e. $\mathcal{F} \Vdash \beta$. ∎

Applying all these observations we conclude that $\lambda(\mathcal{F})$ is a normal modal logic, and (i) holds. (ii) follows directly from (i) because the family of normal modal logics is closed with respect to arbitrary intersections. ∎

We will call the set $\lambda(\mathcal{F})$ (respectively $\lambda(\Omega)$) *the modal logic* of the frame $\mathcal{F}$ (of the class Ω). logic of a class of frames $\lambda(\mathcal{F})$ Thus self-evident examples of Kripke complete normal modal logics are logics of arbitrary classes of frames. It turns out that not all normal modal logics are Kripke complete. The first examples of Kripke incomplete modal logics were independently discovered by K.Fine [45] and S.K.Thomason [178]. These examples were very intelligent but they had rather intricate axiomatizations. Then a new branch of research concerning this subject matter was founded. Lately we will describe examples of Kripke incomplete modal logics with considerably simple axiomatic systems which were offered by J. van Benthem [190], but these logics are weak logics: they positioned far down in the lattice of normal modal logics (i.e. simplicity should be paid).

Anyway possession of Kripke completeness is so useful and elegant property that a lot of papers were devoted to demonstrating that different individual logics and indeed whole classes of logics are Kripke complete. We now start with Canonical Kripke models, a very simple tool, which lies at the heart of almost all proofs of Kripke completeness. First we will give a definition of the theory of an axiomatic system for a normal modal logic, for this we need initially the following definition.

Definition 2.3.17 *A set S of modal formulas is* consistent *with an modal axiomatic system $\mathcal{AS}$ if, for any tuple $\alpha_1, ..., \alpha_n$ of formulas from S, the following holds $\nvdash_{\mathcal{AS}} \neg(\alpha_1 \wedge \alpha_2 \wedge ... \wedge \alpha_n)$.*

Let us pause briefly to give the following simple observation.

Lemma 2.3.18 *If a normal modal logic λ has a non-empty adequate class $\mathcal{C}$ of frames than λ itself is consistent with any own axiomatic system.*

Proof. Let $\mathcal{AS}$ be an axiomatic system for λ. By supposition there is a frame $\mathcal{F}$ of the class $\mathcal{C}$. Assume that there are some formulas $\alpha_1, ..., \alpha_n$ from λ such that $\vdash_{\mathcal{AS}} \neg(\alpha_1 \wedge ... \wedge \alpha_n)$. Let V be a valuation in $\mathcal{F}$ of all propositional letters having occurrences in formulas $\alpha_1, ..., \alpha_n$. Let a be an element of $\mathcal{F}$. Because all α_i are theorems of λ and $\mathcal{K}$ is adequate for λ, we get $a \Vdash_V \alpha_i$ for every i. Hence $a \Vdash_V \alpha_1 \wedge ... \wedge \alpha_n$. In view of $\vdash_{\mathcal{AS}} \neg(\alpha_1 \wedge ... \wedge \alpha_n)$, we obtain $\neg(\alpha_1 \wedge ... \wedge \alpha_n) \in \lambda$ holds. Therefore $a \Vdash_V \neg(\alpha_1 \wedge ... \wedge \alpha_n)$, a contradiction. So λ is really consistent with $\mathcal{AS}$. ∎

Definition 2.3.19 *A set of modal formulas S is said to be a* complete *modal theory for an modal axiomatic system $\mathcal{AS}$ if*

(i) *The set S is consistent with $\mathcal{AS}$;*

(ii) *For any modal formula α, either $\alpha \in S$ or $\neg\alpha \in S$.*

Lemma 2.3.20 *The following hold*

(i) *Every complete modal theory S for $\mathcal{AS}$ is a maximal consistent with $\mathcal{AS}$ set of formulas.*

(ii) *Any maximal consistent with $\mathcal{AS}$ set of formulas is a complete modal theory for $\mathcal{AS}$.*

Proof. Let S be a complete modal theory for $\mathcal{AS}$ and S_1 be a consistent with $\mathcal{AS}$ set of formulas that $S \subseteq S_1$. Suppose $\alpha \in S_1$. Since S is a complete modal theory, we get $\alpha \in S$ or $\neg \alpha \in S$. If $\neg \alpha \in S$ holds then $\neg \alpha \in S_1$. Consequently, because $\vdash_{PC} \neg(p \wedge \neg p)$, and all derivations in PC can be transformed into derivations in $\mathcal{AS}$ (because all tautologies of PC are provable in $\mathcal{AS}$, and $\mathcal{AS}$ has the postulated inference rule *modus ponens*), we derive $\vdash_{\mathcal{AS}} \neg(\alpha \wedge \neg \alpha)$ and hence S_1 is inconsistent, a contradiction. Thus $\alpha \in S$. Hence $S = S_1$ and (i) holds.

(ii): Let S be a maximal consistent with $\mathcal{AS}$ set of formulas. Suppose then that $\alpha \notin S$ but $\neg \alpha \notin S$. And suppose that $S \cup \{\alpha\}$ is inconsistent. Then for some tuple $\alpha_1, ..., \alpha_n$ of formulas from S, $\vdash_{\mathcal{AS}} \neg(\bigwedge_{1 \leq i \leq n} \alpha_i \wedge \alpha)$. Using tautologies and derivations in PC we arrive at $\vdash_{\mathcal{AS}} \bigwedge_{1 \leq i \leq n} \alpha_i \to \neg \alpha$. We claim that $S \cup \{\neg \alpha\}$ is consistent. Otherwise we would apply the similar reasoning and would arrive at $\vdash_{\mathcal{AS}} \bigwedge_{1 \leq j \leq m} \beta_j \to \alpha$ for some formulas $\beta_1, ..., \beta_m$ from S. This would give us

$$\vdash_{\mathcal{AS}} \bigwedge_{1 \leq i \leq n} \alpha_i \wedge \bigwedge_{1 \leq j \leq m} \beta_j \to (\alpha \wedge \neg \alpha).$$

Using tautologies from PC we would have $\vdash_{\mathcal{AS}} \neg(\bigwedge_{1 \leq i \leq n} \alpha_i \wedge)$ and S would be inconsistent. Thus $S \cup \{\neg \alpha\}$ is really consistent and is proper extension of S, yielding a contradiction. If we suppose now that $S \cup \{\alpha\}$ is consistent we again get a contradiction the fact that S is maximal. Thus S is a complete modal theory. ∎

Lemma 2.3.21 *For every consistent with a modal axiomatic system $\mathcal{AS}$ set of formulas S, there is a complete modal theory S^+ for $\mathcal{AS}$ containing the set S.*

Proof. Of course it is possible to give a proof on basis the Zorn Lemma. But we prefer a constructive straightforward proof. Since the set of all propositional letters is countable we can enumerate all modal formulas and to place they in a sequence: $\alpha_1, \alpha_2, ..., \alpha_n, ...$. We will construct a sequence of sets $S_n, n \in N$ consisting of formulas as follows: $S_0 := S$,

$$S_{n+1} = \begin{cases} S_n \cup \{\alpha_n\} & \text{if} \quad S_n \cap \{\alpha_n\} \text{ is consistent,} \\ S_n \cup \{\neg \alpha_n\} & \text{if} \quad S_n \cap \{\alpha_n\} \text{ is inconsistent.} \end{cases}$$

Note that all S_n are consistent with $\mathcal{AS}$. Indeed S_0 is consistent by definition. Let S_n be a consistent set. If S_{n+1} is given by first line of the definition above then by definition S_{n+1} is consistent. Suppose that $S_n \cap \{\alpha_n\}$ is inconsistent and $S_{n+1} := S_n \cup \{\neg \alpha_n\}$. Assume that this set is inconsistent with $\mathcal{AS}$. This means there are formulas $\delta_1, ..., \delta_m$ from $S_n \cup \{\neg \alpha_n\}$ such that $\vdash_{\mathcal{AS}} \neg(\delta_1 \wedge ... \wedge \delta_m)$.

Because S_n is consistent, the formula α_n is one from these formulas. We can let this formula be the last one in the our enumeration. That is

$$\vdash_{\mathcal{AS}} \neg(\delta_1 \wedge ... \wedge \delta_{m-1} \wedge \alpha_n) \tag{2.3}$$

Because all substitution examples of axioms for PC are derivable in $\mathcal{AS}$ and $\mathcal{AS}$ includes the rule modus ponens, we can use all them in derivations for $\mathcal{AS}$. Therefore from (2.3) we obtain

$$\vdash_{\mathcal{AS}} [\delta_1 \wedge ... \wedge \delta_{m-1}] \to \alpha_n.$$

Thus we can derive in $\mathcal{AS}$ the formula $\neg\alpha_n$ and by supposition the set $S_n \cup \alpha_n$ is inconsistent. Then for some formulas $\beta_1, ..., \beta_k$ from $S_n, \vdash_{\mathcal{AS}} \neg(\beta_1 \wedge ... \wedge \beta_k \wedge \alpha_n)$. Reasoning similar above (using tautologies of PC) we obtain that $\vdash_{\mathcal{AS}} [\beta_1 \wedge ... \wedge \beta_k] \to \neg\alpha_n$. Thus we have

$$\vdash_{\mathcal{AS}} \bigwedge_{1 \leq i \leq m-1} \delta_i \wedge \bigwedge_{1 \leq j \leq k} \beta_j \to \alpha_n \wedge \neg\alpha_n.$$

Then applying known equivalences of PC we get

$$\vdash_{\mathcal{AS}} \neg(\bigwedge_{1 \leq i \leq m-1} \delta_i \wedge \bigwedge_{1 \leq j \leq k} \beta_j) \vee (\alpha_n \wedge \neg\alpha_n),$$
$$\vdash_{\mathcal{AS}} \neg(\bigwedge_{1 \leq i \leq m-1} \delta_i \wedge \bigwedge_{1 \leq j \leq k} \beta_j).$$

The last yields S_n is inconsistent with $\mathcal{AS}$ which contradicts to the supposition. Thus S_{n+1} is consistent with $\mathcal{AS}$.

We put $S^+ := \bigcup_{n \in N} S_n$. The set S^+ is consistent with $\mathcal{AS}$ because $S_n, n \in N$ forms an increasing sequence of sets and every S_n is consistent. Directly by definition the set S^+ is a complete modal theory, because for every α_n we placed α_n or $\neg\alpha_n$ into S_{n+1}. ■

Lemma 2.3.22 *Let S be a complete modal theory for an axiomatic modal system $\mathcal{AS}$. Then the following hold*

(1) $(\forall\alpha)(\vdash_{\mathcal{AS}} \alpha \Rightarrow \alpha \in S)$;

(2) $(\alpha \in S \; \& \; \alpha \to \beta \in S) \Rightarrow \beta \in S$;

(3) $\alpha \wedge \beta \in S \Leftrightarrow (\alpha \in S \; \& \; \beta \in S)$;

(4) $\alpha \vee \beta \in S \Leftrightarrow (\alpha \in S \; \vee \; \beta \in S)$;

(5) $\alpha \to \beta \in S \Leftrightarrow (\alpha \notin S \; \vee \; \beta \in S)$;

(6) $\neg a \in S \Leftrightarrow \alpha \notin S$;

(7) $\forall\alpha((\alpha \in S) \vee (\neg\alpha \in S))$.

Proof. (1): Suppose that $\vdash_{AS} \alpha$ but $\alpha \notin S$. Since S is complete, using Lemma 2.3.20 we get $S \cup \{\alpha\}$ is inconsistent with $\mathcal{AS}$. This means

$$\vdash_{AS} \neg(\alpha_1 \wedge ... \wedge \alpha_n \wedge \alpha)$$

for some formulas $\alpha_1, ..., \alpha_n$ from S. Using the equivalences of PC we obtain that $\vdash_{AS} (\alpha_1 \wedge ... \wedge \alpha_n) \rightarrow \neg\alpha$. Because $\vdash_{AS} \alpha$ we infer from this relation (applying again the corresponding derivability in PC)

$$\vdash_{AS} (\alpha_1 \wedge ... \wedge \alpha_n) \rightarrow \alpha \wedge \neg\alpha$$
$$\vdash_{AS} \neg(\alpha_1 \wedge ... \wedge \alpha_n),$$

consequently we get S is inconsistent with $\mathcal{AS}$, a contradiction. Thus $\alpha \in S$.

(2): Let $\alpha \in S$ and $\alpha \rightarrow \beta \in S$. Suppose that $S \cup \{\beta\}$ is inconsistent with $\mathcal{AS}$. This, as always, means there are some formulas $\alpha_1, ..., \alpha_n$ from S such that $\vdash_{AS} \neg(\alpha_1 \wedge ... \wedge \alpha_n \wedge \beta)$. Applying known derivations in PC we get

$$\vdash_{AS} (\alpha_1 \wedge ... \wedge \alpha_n \wedge \alpha \wedge (\alpha \rightarrow \beta)) \rightarrow \neg\beta,$$
$$\vdash_{AS} (\alpha_1 \wedge ... \wedge \alpha_n \wedge \alpha \wedge (\alpha \rightarrow \beta)) \rightarrow \beta,$$
$$\vdash_{AS} (\alpha_1 \wedge ... \wedge \alpha_n \wedge \alpha \wedge (\alpha \rightarrow \beta)) \rightarrow \beta \wedge \neg\beta,$$
$$\vdash_{AS} \neg(\alpha_1 \wedge ... \wedge \alpha_n \wedge \alpha \wedge (\alpha \rightarrow \beta)).$$

The last derivability relation yields S is inconsistent with $\mathcal{AS}$, a contradiction. Hence $S \cup \{\beta\}$ is consistent with $\mathcal{AS}$ which according to Lemma 2.3.20 yields that $\beta \in S$.

(3): Let $\alpha \wedge \beta \in S$. We know that $\vdash_{PC} p \wedge q \rightarrow p, \vdash_{PC} p \wedge q \rightarrow q$. Therefore according to (1) we have $\alpha \wedge \beta \rightarrow \alpha \in S$ and $\alpha \wedge \beta \rightarrow \beta \in S$. And applying (2) we obtain $\alpha \in S$ and $\beta \in S$. Conversely, now let $\alpha, \beta \in S$. Because $\vdash_{PC} (p \rightarrow (q \rightarrow p \wedge q))$ we have $\vdash_{AS} (\alpha \rightarrow (\beta \rightarrow \alpha \wedge \beta))$, and using (1) we get $\alpha \rightarrow (\beta \rightarrow \alpha \wedge \beta) \in S$. From this, $\alpha, \beta \in S$ and (2) it follows that $\alpha \wedge \beta \in S$.

(4): Let $\alpha \vee \beta \in S$ but $\alpha \notin S$ and $\beta \notin S$. Since S is complete, we have $\neg\alpha \in S$ and $\neg\beta \in S$, and by (3) $\neg\alpha \wedge \neg\beta \in S$. Since $\vdash_{PC} \neg p \wedge \neg q \rightarrow \neg(p \vee q)$, it follows $\vdash_{AS} \neg\alpha \wedge \neg\beta \rightarrow \neg(\alpha \vee \beta)$. This by (1) yields $\neg\alpha \wedge \neg\beta \rightarrow \neg(\alpha \vee \beta) \in S$, and using (2) we obtain $\neg(\alpha \vee \beta) \in S$. Since $\vdash_{PC} \neg((p \vee q) \wedge \neg(p \vee q))$, we have $\vdash_{AS} \neg((\alpha \vee \beta) \wedge \neg(\alpha \vee \beta))$ and by (1) $\neg((\alpha \vee \beta) \wedge \neg(\alpha \vee \beta)) \in S$, which contradicts S is consistent. So, we have either $\alpha \in S$ or $\beta \in S$. Let $\alpha \in S$. Since $\vdash_{PC} p \rightarrow p \vee q$, it follows that $\vdash_{AS} \alpha \rightarrow \alpha \vee \beta$ and by (1) $\alpha \rightarrow \alpha \vee \beta \in S$. Then (2) yields $\alpha \vee \beta \in S$. The case when $\beta \in S$ can be considered similarly.

(5): Assume $\alpha \rightarrow \beta \in S$ but $\neg\alpha \notin S$. Since S is complete we conclude $\alpha \in S$. Applying (2) we obtain $\beta \in S$. Conversely, let $\neg\alpha \in \mathcal{AS}$. Because $\vdash_{PC} \neg p \rightarrow (p \rightarrow q)$ it follows that $\vdash_{AS} \neg\alpha \rightarrow (\alpha \rightarrow \beta)$ and by (1) $\neg\alpha \rightarrow (\alpha \rightarrow \beta) \in S$. This, $\neg\alpha \in \mathcal{AS}$ and (2) yield $(\alpha \rightarrow \beta) \in S$. Suppose now that $\beta \in S$. Again, $\vdash_{PC} q \rightarrow (p \rightarrow q)$, and consequently $\vdash_{AS} \beta \rightarrow (\alpha \rightarrow \beta)$, which by (1) brings $\beta \rightarrow (\alpha \rightarrow \beta) \in S$. Using $\beta \in S$ and (2) we get $(\alpha \rightarrow \beta) \in S$.

(6): Let $\neg\alpha \in S$ and $\alpha \in S$. Since $\vdash_{PC} \neg(p \wedge \neg p)$ it follows that $\vdash_{AS} \neg(\alpha \wedge \neg\alpha)$ holds, and we get a contradiction with S is consistent. Thus $\alpha \notin S$ always when $\neg\alpha \in S$. Conversely, if $\alpha \notin S$ then $\neg\alpha \in S$ because S is complete. (7) follows directly from the definition of a set of formulas being complete theory with respect to $\mathcal{AS}$. ∎

Lemma 2.3.23 *Let S be a complete modal theory for an axiomatic system $\mathcal{AS}$, and $\Box S := \{\alpha \mid \Box\alpha \in S\}$. Let β be a modal formula such that $\neg\Box\beta \in S$. Then there exists a complete modal theory $S_\Box$ such that $\Box S \subseteq S_\Box$ holds and also $\neg\alpha \in S_\Box$.*

Proof. We are going to show that the set $\Box S \cup \{\neg a\}$ is consistent with $\mathcal{AS}$. Suppose otherwise, then there are formulas $\alpha_1, ..., \alpha_n$ from the set $\Box S$ such that $\vdash_{\mathcal{AS}} \neg(\alpha_1 \wedge ... \wedge \alpha_n \wedge \neg\alpha)$. Using derivability and equivalences of PC we infer from this $\vdash_{\mathcal{AS}} (\bigwedge_{1 \leq i \leq n} \alpha_i) \to \beta$. Applying the normalization rule of $\mathcal{AS}$ we infer that $\vdash_{\mathcal{AS}} \Box((\bigwedge_{1 \leq i \leq n} \alpha_i) \to \beta)$. Because $\vdash_{\mathcal{AS}} \Box(p \to q) \to (\Box p \to \Box p)$ (note that we consider axiomatic theories for normal modal logics), we arrive at

$$\vdash_{\mathcal{AS}} \Box((\bigwedge_{1 \leq i \leq n} \alpha_i) \to \beta) \to (\Box(\bigwedge_{1 \leq i \leq n} \alpha_i) \to \Box\beta).$$

Using Proposition 1.3.16 and the fact that all normal modal logics are algebraic logics with the algebraic presentation $(\equiv) := \{x \to y, y \to x\}$, $\top$ (what gives as possibility for replacing equivalent formulas with preserving of the equivalence) we obtain from this

$$\vdash_{\mathcal{AS}} \Box((\bigwedge_{1 \leq i \leq n} \alpha_i) \to \beta) \to ((\bigwedge_{1 \leq i \leq n} \Box\alpha_i) \to \Box\beta).$$

This relation by (1) from Lemma 2.3.22 gives us

$$\Box((\bigwedge_{1 \leq i \leq n} \alpha_i) \to \beta) \to ((\bigwedge_{1 \leq i \leq n} \Box\alpha_i) \to \Box\beta) \in S. \tag{2.4}$$

Because all α_i are formulas from $\Box S$, all $\Box\alpha_i$ are enclosed in S and according to (3) from Lemma 2.3.22, $\bigwedge_{1 \leq i \leq n} \Box\alpha_i \in S$. From this, (2.4) and (2) from Lemma 2.3.22, it is follows immediately that $\Box\beta \in S$. By our supposition $\neg\Box\beta \in S$ holds. This gives us that S is inconsistent with $\mathcal{AS}$, a contradiction. Thus the set $\Box S \cup \{\neg a\}$ is consistent with $\mathcal{AS}$ and using Lemma 2.3.21 it follows that this set is included in a certain complete modal theory $S_\Box$. ∎

Now we are in a position to describe a very important construction, to introduce the canonical Kripke models.

Definition 2.3.24 *Let λ be a normal modal logic, and $\mathcal{AS}$ be an axiomatic system for λ. By $C_{\mathcal{AS}}$ we denote the family of all complete modal theories for $\mathcal{AS}$. The canonical Kripke model for $\mathcal{AS}$ is the following model*

$$\mathfrak{M}_{\mathcal{AS}} := \langle C_{\mathcal{AS}}, R_c, V_c \rangle, \text{ where}$$

(i) For any propositional letter p, and any $T_c \in C_{\mathcal{AS}}$,

$$T_c \in V_c(p) \Leftrightarrow (p \in T_c);$$

(ii) For every $T_c, H_c \in C_{\mathcal{AS}}$,

$$T_c R_c H_c \Leftrightarrow (\forall \alpha)(\Box \alpha \in T_c \Rightarrow \alpha \in H_c).$$

The use of canonical Kripke models is based on the following

Theorem 2.3.25 Canonical Kripke Models Theorem. *For any formula α and any complete modal theory T_c, in the model $\mathfrak{M}_{\mathcal{AS}}$ the following holds*

$$T_c \Vdash_{V_c} \alpha \Leftrightarrow \alpha \in T_c.$$

Proof. The proof is by induction on the length of α. The basis of induction, the case when α is a propositional letter p, follows directly from the definition of V_c. Now we check the inductive steps. The case of $\wedge$. Directly, using the definition of $\Vdash_{V_c}$ in Kripke models and the inductive assumption we calculate:

$$T_c \Vdash_{V_c} \alpha \wedge \beta \Leftrightarrow (T_c \Vdash_{V_c} \alpha)\&(T_c \Vdash_{V_c} \beta) \Leftrightarrow (\alpha \in T_c)\&(\beta \in T_c).$$

The last inclusions are equivalent by (3) from Lemma 2.3.22 the single inclusion $\alpha \wedge \beta \in T_c$. Consider now the case of $\vee$:

$$T_c \Vdash_{V_c} \alpha \vee \beta \Leftrightarrow (T_c \Vdash_{V_c} \alpha) \vee (T_c \Vdash_{V_c} \beta) \Leftrightarrow (\alpha \in T_c) \vee (\beta \in T_c).$$

The last relation by (4) from Lemma 2.3.22 is equivalent to $(\alpha \vee \beta \in T_c)$. The case of $\neg$:

$$T_c \Vdash_{V_c} \neg \beta \Leftrightarrow T_c \nVdash_{V_c} \neg \beta \Leftrightarrow (\beta \notin T_c).$$

That $\beta \notin T_c$ by (6) from Lemma 2.3.22 is equivalent to $\neg \beta \in T_c$. Consider the inductive step for $\rightarrow$:

$$T_c \Vdash_{V_c} \alpha \rightarrow \beta \Leftrightarrow (T_c \nVdash_{V_c} \alpha) \vee (T_c \Vdash_{V_c} \beta) \Leftrightarrow (\alpha \notin T_c)\&(\beta \in T_c).$$

The last conjunction of relations according to (5) from Lemma 2.3.22 is equivalent to $(\alpha \rightarrow \beta) \in T_c$. The remaining case is the case of modal logical connective $\Box$. First suppose that $T_c \Vdash_{V_c} \Box \beta$. Consider an arbitrary complete modal theory H_c such that $\langle T_c, H_c \rangle \in R_c$. According to the definition of the relation $\Vdash_{V_c}$ in Kripke models, $H_c \Vdash_{V_c} \beta$ holds. By the definition of R_c we get $(\forall \alpha)[(\Box \alpha \in T_c) \Rightarrow (\alpha \in H_c)]$. We claim that from all these observations it follows that $\Box \beta \in T_c$. Actually, if otherwise then $\neg \Box \beta \in T_c$ because T_c is complete. Then applying Lemma 2.3.23 we obtain that there exists a complete modal theory $T_{c,\Box}$ such that $\Box T_c \subseteq T_{c,\Box}$ and $\neg \beta \in T_{c,\Box}$. In particular, $\beta \notin T_{c,\Box}$ and by inductive assumption $T_{c,\Box} \nVdash_{V_c} \beta$. From $\Box T_c \subseteq T_{c,\Box}$, it immediately follows that $T_c R_c T_{c,\Box}$. Consequently, $T_c \nVdash_{V_c} \Box \beta$, a contradiction. Thus indeed $\Box \beta \in T_c$. Conversely, let $\Box \beta \in T_c$. Let H_c be a complete modal theory such that $\langle T_c, H_c \rangle \in R_c$. By the definition of R_c we have $\beta \in H_c$, and applying the inductive hypothesis we arrive at $H_c \Vdash_{V_c} \beta$. Hence we obtain that $T_c \Vdash_{V_c} \Box \beta$. $\blacksquare$

Definition 2.3.26 *Let λ be a normal modal logic. A Kripke model $\mathfrak{M}$ where $\mathfrak{M} := \langle W, R, V \rangle$ is called* characterizing *for the logic λ if for any formula α,*

$$\alpha \in \lambda \Leftrightarrow \alpha \in \lambda(\mathfrak{M}) := \{ \beta \mid \beta \in \mathcal{F}or_K, (\forall a \in W)(a \Vdash_V \beta) \}.$$

Theorem 2.3.27 *For any normal modal logic λ and any axiomatic system $\mathcal{AS}$ for λ, the canonical model $C_{\mathcal{AS}}$ is a characterizing Kripke model for λ.*

Proof. If $\alpha \in \lambda$ then $\vdash_{\mathcal{AS}} \lambda$ and by Lemma 2.3.22 $\alpha \in T_c$ for any complete modal theory T_c. Applying Theorem 2.3.25 we get $T_c \Vdash_{V_c} \alpha$ and $\alpha \in \lambda(C_{\mathcal{AS}})$. Conversely, suppose $\alpha \notin \lambda$, then $\nvdash_{\mathcal{AS}} \alpha$. We claim that $\{\neg\alpha\}$ is a consistent set for $\vdash_{\mathcal{AS}}$. Indeed, otherwise $\vdash_{\mathcal{AS}} \neg\neg\alpha$ and consequently $\vdash_{\mathcal{AS}} \alpha$. Since $\{\neg\alpha\}$ is consistent with $\mathcal{AS}$, according to Lemma 2.3.21 there is a complete modal theory T_c for $\mathcal{AS}$ containing the formula $\neg\alpha$. By Theorem 2.3.25 we obtain $T_c \Vdash_{V_c} \neg\alpha$ which yields $\alpha \notin \lambda(C_{\mathcal{AS}})$ ∎

Thus any normal modal logic possesses a characterizing Kripke model, and in order to proof Kripke completeness for a logic λ it is sufficient to find a characterizing model with the frame adequate for λ. Very often the canonical Kripke model or a model obtained from this model by some transformation will be such a model. We illustrate this in the next theorem.

Theorem 2.3.28 *The following hold:*

(i) *The minimal normal modal logic K is Kripke complete and the class Fr of all frames characterizes K.*

(ii) *The normal modal logic T is Kripke complete, the class Fr_r of all reflexive frames characterizes T and the canonical model C_T belongs to Fr_r.*

(iii) *The normal modal logic $K4$ is Kripke complete, the class Fr_t of all transitive frames characterizes $K4$ and $C_{K4} \in Fr_t$.*

(iv) *The normal modal logic $S4$ is Kripke complete, the class $Fr_{r,t}$ of all transitive and reflexive frames characterizes $S4$, and $C_{S4} \in Fr_{r,t}$.*

(v) *The normal modal logic $S5$ is Kripke complete, the class Fr_e (of all frames with equivalence relations as the accessibility relations) is characterizing for $S5$, and $C_{S5} \in Fr_e$.*

Proof. According to Theorem 2.3.8 all mentioned above classes of frames are adequate for corresponding logics. Thus in order to proof this theorem it is sufficient, for every mentioned above modal logic λ, and every $\alpha \notin \lambda$, to find a λ-frame from the pointed classes which disproves α. Theorem 2.3.28 says that every $C_{\mathcal{AS}}$ is a characterizing Kripke model for λ (where $\mathcal{AS}$ is an axiomatic system for λ). Therefore it will be sufficient to show that each $C_{\mathcal{AS}}$ (where

$\lambda(\mathcal{AS}) = \lambda$) has the frame belonging to the mentioned above corresponding to λ class of frames. Because the particular view of the axiomatic systems $\mathcal{AS}$ plays here no role any more, we will subscribe the corresponding canonical models just by the name of the logic: $C_K, C_T, C_{K4}, C_{S4}, C_{S5}$ respectively.

(i): Since the frame of the model C_K is a frame, we have $C_K \in Fr$ and K is Kripke complete. (ii): We have to show that C_T is a transitive frame. Let T_c be a complete modal theory from $|C_T|$. Because for any formula α, $\Box\alpha \to \alpha \in T$, using (1) from Lemma 2.3.22 we get $\Box\alpha \to \alpha \in T_c$. Therefore for any α, if $\Box\alpha \in T_c$ then according to (2) from Lemma 2.3.22 $\alpha \in T_c$. Hence R_c in C_T is reflexive, thus T is Kripke complete and Fr_r is the characterizing T class of frames.

(iii): We have to show that the model C_{K4} has a transitive frame. Actually, let $T_c, H_c, G_c \in |C_{K4}|$ and $T_c R_c H_c$, $H_c R_c G_c$. Consider arbitrary α such that $\Box\alpha \in T_c$. Because $\Box\alpha \to \Box\Box\alpha \in K4$, by (1) from Lemma 2.3.22 it follows that $\Box\alpha \to \Box\Box\alpha \in T_c$. Then according to (2) from Lemma 2.3.22, from this and $\Box\alpha \in T_c$ we conclude $\Box\Box\alpha \in T_c$. This and $\langle T_c, H_c \rangle \in R_c$ imply $\Box\alpha \in H_c$. From this and $\langle H_c, G_c \rangle \in R_c$ we immediately extract $\alpha \in G_c$. Thus have proved that $T_c R_c G_c$, hence R_c is transitive.

(iv): For C_{S4} we must show that the frame of this model is reflexive and transitive. All elements of C_{S4} are elements of C_T and C_{K4} simultaneously. Moreover the accessibility relations in all these models are defined identically. Thus C_{S4} is a submodel of the models C_T and C_{K4}. This yields the frame of C_{S4} is reflexive and transitive in view of the above shown facts concerning the structures of C_T and C_{K4}. In the remaining case (v), we will check that the frame of C_{S5} is a frame where the accessibility relation is an equivalence relation. Indeed, C_{S5} is a submodel of C_{S4} hence the frame of C_{S5} is reflexive and transitive. It remains to show that R_c in C_{S5} is a symmetric relation. Suppose $T_c, G_c \in |C_{S5}|$ and $T_c R_c G_c$. Suppose also that $\langle G_c, T_c \rangle \notin R_C$. This means that there is a formula α such that $\Box\alpha \in G_c$ but $\alpha \notin T_c$. Because T_c is a complete theory, it follows $\neg\alpha \in T_c$. According to Theorem 2.3.25 we derive

$$G_c \Vdash_{V_c} \Box\alpha, \quad T_c \Vdash_{V_c} \neg\alpha.$$

Since R_c is reflexive we obtain $T_c \Vdash_{V_c} \Diamond\neg\alpha$. Using Theorem 2.3.25 we conclude $\Diamond\neg\alpha \in T_c$. The fact that $\Diamond\neg\alpha \to \Box\Diamond\neg\alpha \in S5$ by (1) from Lemma 2.3.22 yields $\Diamond\neg\alpha \to \Box\Diamond\neg\alpha \in T_c$. Again using Theorem 2.3.25 we arrive at

$$T_c \Vdash_{V_c} \Diamond\neg\alpha \to \Box\Diamond\neg\alpha.$$

Therefore $T_c \Vdash_{V_c} \Diamond\neg a$ implies that $T_c \Vdash_{V_c} \Box\Diamond\neg\alpha$. This and $T_c R_c G_c$ give us $G_c \Vdash_{V_c} \Diamond\neg\alpha$ which contradicts $G_c \Vdash_{V_c} \Box\alpha$. Thus $G_c R_c T_c$, i.e. R_c is symmetric.

∎

The possession of Kripke completeness gives us a tool which can help us to recognize by pure semantic methods the formulas derivable in normal modal logic. Sometimes this is really very simple although Kripke completeness does not give us an algorithm for recognizing derivable formulas, generally speaking. We illustrate below the first part of the above assertion with several examples.

Example 2.3.29 $(\Box(\Box p \to \Box q) \vee \Box(\Box q \to \Box p)) \in S5$, *that is* $S4.3 \subseteq S5$.

Indeed, let $\langle W, R, V \rangle$ be a Kripke model, where R is an equivalence relation. Suppose $a \in W$ and $a \not\Vdash_V \Box(\Box p \to \Box q)$.

Then there is a $b \in W$ such that aRb, $b \Vdash_V \Box p$ and $b \not\Vdash_V \Box q$. The fact that aRb implies bRa, consequently $a \Vdash_V \Box p$. Therefore we obtain $a \Vdash_V \Box(\Box q \to \Box p)$. So, the formula $(\Box(\Box p \to \Box q) \vee \Box(\Box q \to \Box p))$ is valid in all frames from Fr_e and according to (v) from Theorem 2.3.28 $\Box(\Box p \to \Box q) \vee \Box(\Box q \to \Box p) \in S5$.

Exercise 2.3.30 *The following hold.*

(i) $\Diamond \Box p \to \Box \Diamond \in S5$, *that is* $S4.2 \subseteq S5$.

(ii)$\Box p \to \Diamond p \in S5$, *i.e.* $D \subseteq S5$.

(iii)$(p \to \Diamond p) \in S5$.

As an illustration of the usage of Kripke completeness we give below the semantic proof of the so-called *disjunction theorem* for normal modal logics.

Definition 2.3.31 *A normal modal logic* λ *has the disjunction property iff the following holds. For any given formulas* α, β,

$$\Box \alpha \vee \Box b \in \lambda \Leftrightarrow (\alpha \in \lambda) \vee (\beta \in \lambda).$$

Definition 2.3.32 *We call a normal modal logic* λ *down closed if the class* C *of all frames adequate for* K *is closed with respect to the following operation* $\otimes$. *Given certain frames* $\mathcal{F}_1 := \langle W_1, R_1 \rangle$ *and* $\mathcal{F}_2 := \langle W_2, R_2 \rangle$ *having no common elements, the frame* $\mathcal{F}_1 \otimes \mathcal{F}_2$ *is the frame of the following structure:*

$$\mathcal{F}_1 \otimes \mathcal{F}_2 := \langle |\mathcal{F}_1 \otimes \mathcal{F}_2|, R \rangle$$
$$|\mathcal{F}_1 \otimes \mathcal{F}_2| := |\mathcal{F}_1| \cup |\mathcal{F}_2| \cup \{\bot\}, \quad \text{where } \bot \notin |\mathcal{F}_1| \cup |\mathcal{F}_2|;$$
$$R := R_1 \cup R_2 \cup \{\langle \bot, a \rangle \mid a \in W_1 \cup W_2\} \cup \{\langle \bot, \bot \rangle\}.$$

Theorem 2.3.33 *If a normal modal logic* λ *is Kripke complete and down closed then* λ *has the disjunction property.*

Proof. Indeed, suppose that $\alpha \notin \lambda$ and $\beta \notin \lambda$. In view of Kripke completeness there are λ-frames $\mathcal{F}_1 := \langle W_1, R_1 \rangle$ and $\mathcal{F}_2 := \langle W_2, R_2 \rangle$, and some valuations V_1 and V_2 in $\mathcal{F}_1$ and $\mathcal{F}_2$ respectively such that

$$(\exists a_1 \in W_1)(a_1 \not\Vdash_{V_1} \alpha), \quad (\exists a_2 \in W_2)(a_2 \not\Vdash_{V_2} \beta).$$

Of course we can assume without lost of the generality that the frames $\mathcal{F}_1$ and $\mathcal{F}_2$ have no common elements. The logic λ is down closed, therefore the frame $\mathcal{F}_1 \otimes \mathcal{F}_2$ also is a λ-frame. We introduce in this frame the valuation V as follows:

$$(\forall p)(V(p) := V_1(p) \cup V_2(p)).$$

It is clear that the truth relation $\Vdash_V$ in the components $\mathcal{F}_1$ and $\mathcal{F}_2$ coincides with the truth relations $\Vdash_{V_1}$ and $\Vdash_{V_2}$ respectively. The general structure of the model in $\mathcal{F}_1 \otimes \mathcal{F}_2$ is depicted in the picture Fig. 2.6. Thus, in particular,

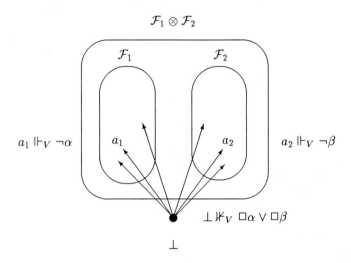

Figure 2.6:

$$(a_1 \nVdash_V \alpha), \qquad (a_2 \nVdash_V \beta).$$

Since $\perp Ra_1$ and $\perp Ra_2$ we get $\perp \nVdash_V \square\alpha$ and $\perp \nVdash_V \square\beta$, i.e. $\perp \nVdash_V (\square\alpha \vee \square\beta)$. Because $\mathcal{F}_1 \otimes \mathcal{F}_2$ is a λ-frame we conclude $\square\alpha \vee \square\beta \notin \lambda$. $\blacksquare$

Immediately from Theorem 2.3.33 and Theorem 2.3.28 we infer

Corollary 2.3.34 *Modal logics K, T, $K4$ and $S4$ have the disjunction property.*

In following sections we will develop certain advanced methods for studying Kripke completeness. Now we are going to specify Kripke semantics for the case of temporal logic. As we have seen, temporal logic can be understood as a modal logic with two $\square$-operations and certain laws governing their interaction. Therefore a Kripke semantics for temporal logic is very easily obtained from the Kripke semantics for normal modal logics. The intuition behind this semantics is very simple: the elements of the Kripke model are *states (worlds)* and the accessibility relation is just the *flow of time*. Thus any frame can be understood as some temporal frame and

Definition 2.3.35 *A temporal Kripke model is a triple* $\mathfrak{M} := \langle W, R, V \rangle$, *where* $\langle W, R \rangle$ *is a frame, and* V *is a valuation of a certain set of propositional letters* P *in* W, *i.e* $V : P \mapsto 2^W$, $\forall p \in P[V(p) \subseteq W]$.

The definition of Kripke models for temporal logic is quite similar to that for modal logics. The definition of the truth relation $\Vdash_V$ for temporal formulas can be given by induction on the length of formulas (of course, the inductive steps for logical connectives of PC are the same as for modal logics):

$$(\forall p \in P)(\forall c \in W)[(c \Vdash_V p) \Leftrightarrow (c \in V(p)]$$
$$(\forall c \in W)[(c \Vdash_V \alpha \wedge \beta) \Leftrightarrow (c \Vdash_V \alpha) \& (c \Vdash_V \beta)]$$
$$(\forall c \in W)[(c \Vdash_V \alpha \vee \beta) \Leftrightarrow (c \Vdash_V \alpha) \vee (c \Vdash_V \beta)]$$
$$(\forall c \in W)[(c \Vdash_V \neg \alpha) \Leftrightarrow (c \nVdash_V \alpha)]$$
$$(\forall c \in W)[(c \Vdash_V \alpha \rightarrow \beta) \Leftrightarrow (c \nVdash_V \alpha) \vee (c \Vdash_V \beta)]$$
$$(\forall c \in W)[(c \Vdash_V G\alpha) \Leftrightarrow (\forall b \in W[cRb \Rightarrow b \Vdash_V \alpha])]$$
$$(\forall c \in W)[(c \Vdash_V H\alpha) \Leftrightarrow (\forall b \in W[bRc \Rightarrow b \Vdash_V \alpha])]$$
$$(\forall c \in W)[(c \Vdash_V F\alpha) \Leftrightarrow (\exists b \in W[cRb \& b \Vdash_V \alpha])]$$
$$(\forall c \in W)[(c \Vdash_V P\alpha) \Leftrightarrow (\exists b \in W[bRc \& b \Vdash_V \alpha])]$$

Hence the truth of temporal formulas in Kripke models in non-formal understanding means

$$(c \Vdash_V G\alpha) \Leftrightarrow \text{in the future after } c \ \alpha \text{ always will be true}$$
$$(c \Vdash_V H\alpha) \Leftrightarrow \text{in the past before } c \ \alpha \text{ always was true}$$
$$(c \Vdash_V F\alpha) \Leftrightarrow \text{there will be in the future after } c \text{ a state when}$$
α will be true
$$(c \Vdash_V P\alpha) \Leftrightarrow \text{there was in the past before } c \text{ a state}$$
when α was true.

It is not hard to see again that these definitions are consistent with accepted abbreviations of temporal logic: $G := \neg F\neg$, $H := \neg P\neg$ and the equivalences $G\alpha \equiv \neg F\neg\alpha$, $H\alpha \equiv \neg P\neg\alpha$, $F\alpha \equiv \neg G\neg\alpha$, $P\alpha \equiv \neg H\neg\alpha$ which hold in the minimal temporal logic T_0.

Exercise 2.3.36 *Let* $\mathfrak{M}$ *be a Kripke model and* a *be an element from* $|\mathfrak{M}|$. *Then*

$$a \Vdash_V G\alpha \Leftrightarrow a \Vdash_V \neg F\neg\alpha, \ a \Vdash_V H\alpha \Leftrightarrow a \Vdash_V \neg P\neg\alpha,$$
$$a \Vdash_V F\alpha \Leftrightarrow a \Vdash_V \neg G\neg\alpha, \ a \Vdash_V P\alpha \Leftrightarrow a \Vdash_V \neg H\neg\alpha,$$

The definition of a frame or a class of frames adequate to a temporal logic is once again quite similar to the case of modal logics. We say a temporal formula α is valid (true) in a Kripke model $\mathfrak{M} := \langle W, R, V \rangle$ iff for every $c \in W$, $c \Vdash_V \alpha$ (we will sometimes abbreviate this by $\mathfrak{M} \Vdash \alpha$. A temporal formula α is valid (true) in a frame $\mathcal{F}$ (abbreviation $\mathcal{F} \Vdash \alpha$) if α is valid in any Kripke model which is based on the frame $\mathcal{F}$. A frame $\mathcal{F}$ is said to be *adequate* for a temporal logic λ if all formulas from α are valid in any Kripke model with the frame equal to $\mathcal{F}$. A class of frames $\mathcal{K}$ is adequate for a temporal logic λ if any frame from $\mathcal{K}$ is adequate for λ. To illustrate the adequacy of frames for temporal logics we introduce certain temporal logics extending T_0:

Definition 2.3.37 *Let*

$$T_{0,W} := T_0 \oplus Gp \to p, Hp \to p;$$
$$T_{0,K4} := T_0 \oplus Gp \to GGp, Hp \to HHp ;$$
$$T_{0,S4} := T_{0,W} \oplus Gp \to GGp, Hp \to HHp ;$$

Theorem 2.3.38 *The following hold:*

(i) The class Fr of all frames is adequate for the minimal tense logic T_0;

(ii) The class Fr_r of all reflexive frames is adequate for the tense logic $T_{0,W}$;

(iii) The class Fr_t of all transitive frames is adequate for the tense logic $T_{0,K4}$;

(iv) The class $Fr_{r,t}$ of all reflexive and transitive frames is adequate for the tense logic $T_{0,S4}$.

Proof. It is sufficient to show in every case that substitution instances of all the axioms for the corresponding temporal logics are valid in all frames of the corresponding classes, and that the inference rules always preserve the truth. Indeed, Lemma 2.3.11 yields that all substitution examples of axioms (A1) - (A10) of PC are valid on all frames, and Lemma 2.3.10 gives us that the inference rule (R1): $x, x \to y/y$ preserves the truth in arbitrary frames. That rules x/Gx and x/Hx preserve the truth in arbitrary frames can be shown in a quite similar way to the case of preserving the truth by $x/\Box x$ in Lemma 2.3.10. Hence it remains only to show that the special temporal axioms are valid in frames from the corresponding classes.

(i): The temporal axioms of T_0 are

$$(1) G(p \to q) \to (Gp \to Gq),$$
$$(2) H(p \to q) \to (Hp \to Hq),$$
$$(3) p \to HFp$$
$$(4) p \to GPp.$$

According to Theorem 2.3.8, Fr is adequate for modal logic T. Therefore the formula $\Box(p \to q) \to (\Box p \to \Box q)$ is valid on all frames. We can interpret G as $\Box$ in any given frame, therefore the axiom (1) is valid in all frames. For any given model $\mathfrak{M} := \langle W, R, V \rangle$, the truth of the formula Hp corresponds to the truth of the formula $\Box p$ in the model $\mathfrak{M}^{-1} := \langle W, R^{-1}, V \rangle$. Therefore the axiom (2) will be valid in $\mathfrak{M}$ also. Let $\mathfrak{M} := \langle W, R, V \rangle$ be a Kripke model and α is a temporal formula. Suppose that $a \in W$ and $a \Vdash_V \alpha$. Assume that $b \in W$ and bRa. Then $b \Vdash_V F\alpha$ and consequently $a \Vdash_V HF\alpha$. Thus for any $a \in W$, $a \Vdash_V \alpha \to HF\alpha$. i.e. the axiom (3) is valid in $\mathfrak{M}$. As to (4), if $a \Vdash_V \alpha$ and aRb, then $b \Vdash_V P\alpha$. Therefore $a \Vdash_V GP\alpha$ and the axiom (4) is also valid in every frame.

(ii): Suppose $\mathcal{F} := \langle W, R \rangle$ is a reflexive frame. By (i) all axioms of T_0 are valid on $\mathcal{F}$. According to (ii) from Theorem 2.3.8 we have that the modal formula $\Box p \to p$ is valid in the frame $\mathcal{F}$. As the truth value for $\Box$ corresponds to the truth value for G we have $Gp \to p$ is valid in $\mathcal{F}$. Consider the frame $\mathcal{F}^{-1} := \langle W, R^{-1} \rangle$. This frame is reflexive, therefore $\Box p \to p$ is valid on $\mathcal{F}^{-1}$. The truth value of $\Box$ in this frame corresponds to the truth value of H in $\mathcal{F}$, hence $Hp \to p$ is also valid in $\mathcal{F}$, i.e. (ii) holds.

(iii): Let $\mathcal{F} := \langle W, R \rangle$ be a transitive frame. Then as for the case (ii), according to Theorem 2.3.8 we obtain $Gp \to GGp$ is valid in $\mathcal{F}$. The frame $\mathcal{F}^{-1} := \langle W, R^{-1} \rangle$ will also be transitive. Therefore by Theorem 2.3.8 $\Box p \to \Box\Box p$ is valid in $\mathcal{F}^{-1}$ which implies $Hp \to HHp$ is valid in $\mathcal{F}$. (iv) follows directly from (ii) and (iii). ∎

Applying this theorem we can easily establish that certain formulas are not theorems of the above mentioned temporal logics. For this it is sufficient to find a frame adequate for a given logic which disproves the formula in question.

Exercise 2.3.39 *Prove that*

$(i) T_0 \subset T_{0,W} \subset T_{0,S4}$
$(ii) T_0 \subset T_{0,K4} \subset T_{0,S4}$
$(iii) T_{0,K4} \not\subseteq T_{0,W}, \; T_{0,W} \not\subseteq T_{0,K4},$

Hint: use the considerations from Examples 2.3.12.

As with the case of modal logics we define all the terms connected with Kripke completeness for temporal logics:

Definition 2.3.40 *Let λ be a temporal logic.*

(i) A characterizing for the logic λ class of frames is a class C such that

$$\lambda := \lambda(C) := \{\alpha \mid \alpha \in \mathcal{F}or_{K_0}, (\forall \mathcal{F} \in C)(\mathcal{F} \Vdash \alpha)\}.$$

(ii) A temporal logic λ is Kripke complete if there exists a characterizing for λ class of frames.

It is easy to see that according to Theorem 2.3.38 for any class of frames Ω, $\lambda(\Omega)$ is a temporal logic (in order to show this we can use the analog of Lemmas 2.3.15 and 2.3.16). This gives evident examples of Kripke complete temporal logics. We now intend to show that introduced above temporal logics are Kripke complete. In order to show this we need the definition of a set of temporal formulas consistent with an axiomatic temporal system, and the definition of complete temporal theory for given temporal axiomatic system. These definitions are just reformulations of corresponding definitions for modal language. The results analogous to Lemmas 2.3.20, 2.3.21 and 2.3.22 for temporal case hold also (the proofs can be reformulated directly). We summarize these properties in the following

Lemma 2.3.41 *Let $\mathcal{AS}$ be an axiomatic system for a temporal logic λ.*

(i) *If S is a complete for $\mathcal{AS}$ temporal theory then S is a maximal consistent with $\mathcal{AS}$ set of formulas.*

(ii) *If S is a maximal set of temporal formulas consistent with $\mathcal{AS}$ then S is a complete temporal theory with respect to $\mathcal{AS}$.*

(iii) *If S is a consistent with $\mathcal{AS}$ set of temporal formulas then there is a complete temporal theory S^+ containing S.*

(iv) *All relations (1) - (7) from Lemma 2.3.22 hold for every complete temporal theory S (with respect to $\mathcal{AS}$).*

Lemma 2.3.42 *Let S be a complete temporal theory for an axiomatic temporal system $\mathcal{AS}$ of a consistent temporal logic λ.*

(i) *Let $GS := \{\alpha \mid G\alpha \in S\}$ and β is a formula that $\neg G\beta \in S$. Then there is a complete temporal theory S_G such that $GS \subseteq S_G$ and $\neg\beta \in S_G$.*

(ii) *If $HS := \{\alpha \mid H\alpha \in S\}$ and β is a formula such that $\neg H\beta \in S$ then there is a complete temporal theory S_H such that $HS \subseteq S_H$ and $\neg\beta \in S_H$.*

Proof. We completely follow the proof of Lemma 2.3.23 what is possible for both cases (i) and (ii) since axioms analogous to the $\square$-axiom of K for connectives G and H are provable in $\mathcal{AS}$.

The definition of canonical Kripke models for temporal logics can also be given in a very similar way to the case of modal logics with merely a small modification:

Definition 2.3.43 *The canonical Kripke model for a temporal logic λ with an axiomatic system $\mathcal{AS}$ is the model $\mathfrak{M}_{AS} := \langle C_{AS}, R, V \rangle$, where C_{AS} is* *the set of all complete temporal theories, and V is the valuation defined as follows:*

$$V(p) := \{T_c \mid T_c \in C_{\mathcal{AS}}, p \in T_c\},$$

and the accessibility relation R is given by

$$T_c R W_c \Leftrightarrow (\forall \alpha)[(G\alpha \in T_c \Rightarrow \alpha \in W_c)\&$$
$$(\forall \beta)(H\beta \in W_c \Rightarrow \beta \in T_c)].$$

Theorem 2.3.44 (Canonical Models Theorem for Temporal Logics). *If T_c is a complete temporal theory from $\mathfrak{M}_{\mathcal{AS}}$ and α is a temporal formula then*

$$T_c \Vdash_V \alpha \Leftrightarrow \alpha \in T_c.$$

Proof. We carry out the proof by induction on the length of the formula α in a quite similar way to the proof of Lemma 2.3.25. The difference is only in the inductive steps for formulas of kind $G\beta$ and $H\beta$ which we describe below. Suppose

$$T_c \Vdash_V G\gamma.$$

Assume that $G\gamma$ does not belong to the set T_c. Since T_c is a complete temporal theory the inclusion $\neg G\beta \in T_c$ holds. Then by (i) of Lemma 2.3.42 there exists a complete temporal theory $T_{c,G}$ such that $GT_c \subseteq T_{c,G}$ and $\neg\gamma \in T_{c,G}$. We will show that $T_c R T_{c,G}$. Indeed, directly by the definition, we get $\beta \in T_{c,G}$ for all formulas β such that $G\beta \in T_c$. Let δ be a formula such that $H\delta \in T_{c,G}$. Assume that $\delta \notin T_c$. Since T_c is complete, $\neg\delta \in T_c$ holds. Because $\vdash_{\mathcal{AS}} \neg\delta \to G\neg H\neg\neg\delta$ and consequently $\vdash_{\mathcal{AS}} \neg\delta \to G\neg H\delta$, applying (iv) from Lemma 2.3.41, we arrive at $G\neg H\delta \in T_c$. Then by the definition of GT_c the inclusion $\neg H\delta \in T_{c,G}$ holds. Thus $H\delta, \neg H\delta \in T_{c,H}$ which implies $T_{c,H}$ is inconsistent, a contradiction. Hence $\delta \in T_c$ and $T_c R T_{c,G}$. In view of $\neg\gamma \in T_{c,G}$ and consistency of $T_{c,G}$, we conclude $\gamma \notin T_{c,G}$, and by inductive hypothesis we have $T_{c,G} \nVdash_V \gamma$. Since $T_c R T_{c,G}$ we arrive at $T_c \nVdash_V G\gamma$, a contradiction. Thus $G\gamma \in T_c$. Conversely, suppose that

$$G\gamma \in T_c, \text{ and } T_c R W_c.$$

By the definition of R we infer $\gamma \in W_c$ and the inductive supposition yields us that $W_c \Vdash_V \gamma$. Hence $T_c \Vdash_V G\gamma$ and the consideration of the inductive step for the case G is complete.

Now we examine the connective H. The proof (in a sense) is parallel to the case G. Actually let $T_c \Vdash_V H\gamma$. Suppose that $H\gamma \notin T_c$. Since T_c is a complete temporal theory we derive that $\neg H\beta \in T_c$ holds. Then by (ii) of Lemma 2.3.42 there is a complete temporal theory $T_{c,H}$ such that $HT_c \subseteq T_{c,H}$ and $\neg\gamma \in T_{c,H}$. Now our aim is to prove that $T_{c,H} R T_c$. In fact, by the definition, for all formulas

β that $H\beta \in T_c$, $\beta \in T_{c,H}$ holds. Let δ be a formula such that $G\delta \in T_{c,H}$. Suppose that $\delta \notin T_c$. Since T_c is complete $\neg\delta \in T_c$ holds. We also know that $\vdash_{\mathcal{AS}} \neg\delta \to H\neg G\neg\neg\delta$ and $\vdash_{\mathcal{AS}} \neg\delta \to H\neg G\delta$. Therefore applying (iv) from Lemma 2.3.41 we conclude $H\neg G\delta \in T_c$. Then by the definition of HT_c the following $\neg G\delta \in T_{c,H}$ holds. Hence we arrive at $G\delta, \neg G\delta \in T_{c,H}$ which implies $T_{c,H}$ is inconsistent, a contradiction. Hence $\delta \in T_c$ and $T_{c,H} R T_c$. Since $\neg\gamma \in T_{c,G}$ and $T_{c,H}$ is consistent, $\gamma \notin T_{c,H}$. Then by inductive hypothesis we conclude $T_{c,H} \nVdash_V \gamma$. Because $T_{c,H} R T_c$ it follows $T_c \nVdash_V H\gamma$, a contradiction. Hence $H\gamma \in T_c$. Now assume that

$$H\gamma \in T_c, \text{ and } W_c R T_c.$$

According to the definition of R it follows that $\gamma \in W_c$. Therefore by inductive hypothesis $W_c \Vdash_V \gamma$ holds. Thus $T_c \Vdash_V H\gamma$. ∎

We say a Kripke model $\mathfrak{M}$ is characterizing for a temporal logic λ if the set of all theorems for λ coincides with the set of all formulas which are valid in $\mathfrak{M}$ on all elements.

Theorem 2.3.45 *Let λ be a temporal logic with an axiomatic system $\mathcal{AS}$. The canonical model $C_{\mathcal{AS}}$ is a characterizing model for the temporal logic λ.*

Proof. Let α be a theorem of λ. Then $\vdash_{\mathcal{AS}} \alpha$ and applying (iv) from Lemma 2.3.41 we obtain $\alpha \in T_c$ for any complete temporal theory T_c. According to Theorem 2.3.44 it follows that $T_c \Vdash_V \alpha$. Thus all theorems from λ are valid on $C_{\mathcal{AS}}$. Conversely, let $\alpha \notin \lambda$. Then the set $\{\neg\alpha\}$ is consistent with $\mathcal{AS}$. Therefore according to (iii) from Lemma 2.3.41 there exists a complete temporal theory T_c containing $\neg\alpha$. Then by Theorem 2.3.44 we conclude $T_c \Vdash_V \neg\alpha$ and α is false in $C_{\mathcal{AS}}$ on the element T_c. ∎

Now we are in a position to show Kripke completeness for the individual temporal logics considered above.

Theorem 2.3.46 *The following hold:*

(i) The minimal temporal logic T_0 is Kripke complete. The class of all frames, as well as the frame of the corresponding canonical Kripke model for T_0, are characterizing classes for T_0.

(ii) The temporal logic $T_{0,W}$ is Kripke complete. The class of all reflexive frames, as well as the frame of the corresponding canonical Kripke model for $T_{0,W}$, are characterizing classes for $T_{0,W}$.

(iii) The temporal logic $T_{0,K4}$ is Kripke complete. The class of all transitive frames, as well as the frame of the corresponding canonical Kripke model for $T_{0,K4}$, are characterizing classes for $T_{0,K4}$.

(iv The temporal logic $T_{0,S4}$ is Kripke complete. The class of all reflexive and transitive frames, as well as the frame of the corresponding canonical Kripke model for $T_{0,S4}$, are characterizing classes for $T_{0,S4}$.

Proof. According to Theorem 2.3.38 and Theorem 2.3.45 it is sufficient to show that the frames of the corresponding canonical Kripke models belong to the corresponding classes of frames. (i) is evident. (ii): We must check that the frame $\mathcal{F}_{T_0,W}$ of the canonical Kripke model C_{AS} (where AS is the axiomatic system for $T_{0,W}$) is reflexive. This can be done in quite similar way to the proof of the case (ii) in Theorem 2.3.28. (iii): in order to show that the frame $\mathcal{F}_{T_0,K4}$ of the canonical Kripke model for $T_{0,K4}$ is transitive we follow the proof of the case (iii) in Theorem 2.3.28 repeating reasoning concerning $\square$ for G and H separately. (iv): again the proof is identical to the proof of case (iv) in Theorem 2.3.28 (the our case we refer on (ii) and (iii) of the present theorem which are already shown). In fact, all the complete temporal theories from $\mathfrak{M}_{T_0,S4}$ belong to $\mathfrak{M}_{T_0,W}$ and $\mathfrak{M}_{T_0,K4}$, and the accessibility relation R is defined identical in the all mentioned models. Since models $\mathfrak{M}_{T_0,W}$ and $\mathfrak{M}_{T_0,K4}$ are reflexive and transitive respectively so is $\mathfrak{M}_{T_0,S4}$. ∎

Exercise 2.3.47 *Show that*

(i) $G\top \in T_0$, $H\top \in T_0$, $F\top \notin T_{0,K4}$, $P\top \notin T_{0,K4}$;

(ii) $F\top \in T_{0,W}$, $P\top \in T_{0,W}$;

(iii) $Fp \to Pp \notin T_0$; $GFp \to Fp \notin T_{0,K4}$.

Now we are familiarized with the basis of technique concerning Kripke completeness. To end this section we present some more examples of Kripke complete modal logics which we need for further research. First we consider modal systems

$$B := T \oplus p \to \square\lozenge p,$$
$$D := K \oplus \square p \to \lozenge p.$$

Theorem 2.3.48 *The following hold:*

(i) The modal logic B is Kripke complete. The class Fr_B of all reflexive frames satisfying the condition $\forall x \forall y (x Ry \Rightarrow y Rx)$ characterizes B and the frame of the canonical model C_B belongs to this class.

(ii) The modal logic D is complete by Kripke. The class Fr_D of all frames satisfying the condition $\forall x \exists y (x Ry)$ characterizes D and the frame of the canonical model C_D belongs to Fr_D.

Proof. (i): It can be easily checked that the formula $p \rightarrow \Box\Diamond p$ is valid in frames which belong to Fr_B. We take the canonical model C_B for B. According to Theorem 2.3.27 this model is a characterizing for B. It remains only to show that this model has the frame in the required form. The model C_B is a submodel of C_T therefore C_B has a reflexive frame (cf. Theorem 2.3.28). Suppose $T_c, H_c \in C_B$ and $T_c R_c H_c$. Let α be a formula that $\Box\alpha \in H_c$. Assume that $\alpha \notin T_c$. Then $\neg\alpha \in T_c$ and according to (1), (2) from Lemma 2.3.22 $\Box\Diamond\neg\alpha \in T_c$, consequently $\Diamond\neg\alpha \in H_c$. Thus we immediately conclude $\neg\Box\alpha \in H_c$, and $\Box\alpha \in H_c$, which implies H_c is inconsistent, a contradiction. Hence $H_c R_c T_C$ and (i) holds.

(ii): That the formula $\Box p \rightarrow \Diamond p$ is valid in all frames from the class Fr_D can be easily verified directly. The canonical Kripke model C_D is characterizing for D according to Theorem 2.3.27. We must check that the frame of C_D belongs to the class Fr_D. Let T_c be a complete modal theory from C_D. The formula $\Box\top$ is provable in K and consequently also is provable in D. Therefore by (1), (2) from Lemma 2.3.22 $\Box\top \in T_c$ and $\Diamond\top \in T_c$. Then by Theorem 2.3.25 $T_c \Vdash_{V_c} \Diamond\top$. This implies that there exists some $H_c \in C_D$ such that $T_c R_c H_c$ and $H_c \Vdash_{V_c} \top$. Thus we finally conclude $T_c R_c H_c$, what we needed. ∎

Consider now the modal systems

$$K4.1 := K4 \oplus \Box\Diamond p \rightarrow \Diamond\Box p,$$
$$K4.2 := K4 \oplus \Diamond\Box p \rightarrow \Box\Diamond p,$$
$$K4.3 := K4 \oplus \Box(\Box p \rightarrow q) \vee \Box(\Box q \rightarrow p),$$
$$S4.1 := S4 \oplus \Box\Diamond p \rightarrow \Diamond\Box p,$$
$$S4.2 := S4 \oplus \Diamond\Box p \rightarrow \Box\Diamond p,$$
$$S4.3 := S4 \oplus \Box(\Box p \rightarrow q) \vee \Box(\Box q \rightarrow p).$$

Theorem 2.3.49 *The following hold:*

(i) The logic $K4.1$ is Kripke complete. The class $Fr_{K4.1}$ of all transitive frames which satisfy the condition

$$\forall x \exists y [x Ry \& yRy \& \forall z(yRz \Rightarrow z = y)]$$

characterizes $K4.1$ and the frame of the canonical Kripke model $C_{K4.1}$ belongs to $Fr_{K4.1}$.

(ii) The logic $S4.1$ is Kripke complete. The class $Fr_{S4.1}$ of all reflexive and transitive frames which satisfy the condition

$$\forall x \exists y [x Ry \& \forall z(yRz \Rightarrow z = y)]$$

characterizes $S4.1$ and the frame of $C_{S4.1}$ belongs to $Fr_{S4.1}$.

Proof. (i): In order to show that $Fr_{K4.1}$ is adequate for $K4.1$ we only need to check that the new axiom

$$\Box\Diamond p \to \Diamond\Box p$$

is valid in frames of $Fr_{K4.1}$. Suppose $\mathcal{F}$ ia a frame which meets the mentioned property, V is a valuation on $\mathcal{F}$, $a \in |\mathcal{F}|$ and $a \Vdash_V \Box\Diamond p$. Let b is an element of $\mathcal{F}$ such that aRb, bRb, $\forall c(bRc \Rightarrow c = b)$. Then $b \Vdash_V \Diamond p$, and in view of the choice of b, we get $b \Vdash_V \Box p$. Consequently $a \Vdash_V \Diamond\Box p$. Thus the class of all frames satisfying

$$\forall x \exists y[xRy \& yRy \& \forall z(yRz \Rightarrow z = y)]$$

is adequate for $K4.1$.

The canonical Kripke model $C_{K4.1}$ characterizes $K4.1$ by Lemma 2.3.27. We intend to show that the frame of $C_{K4.1}$ belongs to $Fr_{K4.1}$. The model $C_{K4.1}$ is a submodel of C_{K4}. Therefore according to Theorem 2.3.28 the frame of $C_{K4.1}$ is transitive. Also this frame has the following property

$$(\forall T_c \in C_{K4.1})(\exists H_c \in C_{K4.1})(T_c R_c H_c). \tag{2.5}$$

Actually, otherwise we get $T_c \Vdash_{V_c} \Box\Diamond\top$ and $T_c \nVdash_{V_c} \Diamond\Box\top$. This yields that $T_c \nVdash_{V_c} \Box\Diamond\top \to \Diamond\Box\top$ and $\Box\Diamond\top \to \Diamond\Box\top \notin T_c$ (by Theorem 2.3.25), what contradicts (1) from Lemma 2.3.22. For every element T_c of $C_{K4.1}$, we introduce the set $T_C^\Box$ as follows $T_c^\Box := \{\Box\alpha \mid \Box\alpha \in T_c\}$. The following holds:

$$T_c^\Box \subseteq \Box T_c. \tag{2.6}$$

Indeed, let $\Box\alpha \in T_c$. By (1) from Lemma 2.3.22 $\Box\alpha \to \Box\Box\alpha \in T_c$ and applying (2) from Lemma 2.3.22 we conclude $\Box\Box\alpha \in T_c$, and therefore it follows that $\Box\alpha \in \Box T_c$.

We fixe a T_c from $C_{K4.1}$. Consider the family $\mathcal{F}$ of sets $\mathcal{S}$ which satisfy the following conditions: (1) $\mathcal{S}$ consists of certain formulas of kind $\Box\gamma$, (2) the set $\mathcal{S} \cup \Box T_c$ is consistent. Note that the family $\mathcal{F}$ is non-empty. Actually, by (2.5) $T_c R_c H_c$ for some H_c. Then $\Box T_c \subseteq H_c$ and according to (2.6) $T_c^\Box \subseteq \Box T_c$ holds. Thus the set $T_c^\Box$ belongs to $\mathcal{F}$ because the set $\Box T_c$ being a subset of a consistent set is consistent. Applying Zorn Lemma we conclude that there is a maximal set $\mathcal{S}^*$ in the family $\mathcal{F}$. We pick some such $\mathcal{S}^*$. Since the set $\mathcal{S}^* \cup \Box T_c$ is consistent there is a complete modal theory H_c containing the set $\mathcal{S}^* \cup \Box T_c$ (according to Lemma 2.3.21). Using (2.5) we infer $H_c R_c E_s$ for some E_c. Consider the set C of all elements $E_c \in C_{K4.1}$ such that $H_c R_c E_c$. As we have seen this set is non-empty. Because $\Box T_c \subseteq H_c$ we have $T_c R_c H_c$. By transitivity of $C_{K4.1}$ it follows $T_c R_c E_c$ for every $E_c \in C$. We claim that for every $E_c \in C$,

$$E_c^\Box = H_c^\Box = \mathcal{S}^*. \tag{2.7}$$

In fact, by (2.6) $\mathcal{S}^* \subseteq H_c^\square \subseteq \square H_c$, hence $\mathcal{S}^* \subseteq H_c^\square \subseteq E_c$ and $\mathcal{S}^* \subseteq H_c^\square \subseteq E_c^\square$. Suppose that $\mathcal{S}^* \subset E_c^\square$. Then $\square T_c \cup E_c^\square \subseteq E_c$ that is $\square T_c \cup E_c^\square$ is consistent which contradicts the fact that $\mathcal{S}^*$ is maximal. Thus (2.7) holds.

$$(\forall E_c \in C)(\forall \alpha)(\square \alpha \in E_c \Rightarrow \alpha \in E_c). \tag{2.8}$$

Indeed, if $\square \alpha \in E_c$ then by (2.7) $\square \alpha \in H_c$. Then $H_c R_c E_c$ implies $\alpha \in E_c$. Now we are in a position to show that C consists of single reflexive element. Actually, any pair E_1, E_2 of elements from C by (2.7) and (2.8) consists of mutually accessible elements: $E_1 R_c E_2$ and $E_2 R_c E_1$. That is C forms a cluster in $C_{K4.1}$. Let α be a formula from E_c, where $E_c \in C$ and E_1 be a fixed element from C. Then by Theorem 2.3.25 $E_c \Vdash_{V_c} \alpha$ and $E_1 \Vdash_{V_c} \square \Diamond \alpha$, $\square \Diamond \alpha \in E_1$ correspondingly. By (1), (2) of Lemma 2.3.22 we get $\square \Diamond \alpha \to \Diamond \square \alpha \in E_1$ and $\Diamond \square \alpha \in E_1$. Theorem 2.3.25 yields $E_1 \Vdash_{V_c} \Diamond \square \alpha$. Hence for some E_2, $E_1 R_c E_2$ and $E_2 \Vdash_{V_c} \square \alpha$. But $E_2 \in C$, consequently $E_1 \Vdash_{V_c} \alpha$ and $\alpha \in E_1$. Thus $E_c = E_1$ and the single element of C has all required properties. Thus the frame of the model $C_{K4.1}$ belongs to $Fr_{K4.1}$. To prove (ii), we follow completely the line of the proof (i), it is sufficient only to note that $C_{S4.1}$ is reflexive as a submodel of C_{S4} by Theorem 2.3.28. ∎

Theorem 2.3.50 *The following hold:*

(i) The modal logic $K4.2$ is Kripke complete. The class $Fr_{K4.2}$ of all transitive frames which satisfy the condition

$$\forall x, y, z[(xRy)\&(xRy) \Rightarrow \exists w((yRw)\&(xRw))]$$

characterizes $K4.2$ and the frame of $C_{K4.2}$ belongs to $Fr_{K4.2}$.

(ii) The modal logic $S4.2$ is Kripke complete. The class $Fr_{S4.2}$ of all reflexive and transitive frames which satisfy the condition

$$\forall x, y, z[(xRy)\&(xRy) \Rightarrow \exists w((yRw)\&(xRw))]$$

characterizes $S4.2$ and the frame of $C_{S4.2}$ belongs to $Fr_{S4.2}$.

Proof. (i): To show $Fr_{K4.2}$ is adequate for $K4.2$ it is sufficient to check the truth of the new axiom of $K4.1$ (with respect to $K4$) on $Fr_{K4.2}$. Let $\mathcal{F}$ be a frame from $Fr_{K4.2}$ and let V be a valuation in $\mathcal{F}$. Suppose $a \in |\mathcal{F}|$ and $a \Vdash_V \Diamond \square p$. Then there is some b such that $b \Vdash_V \square p$. Consider arbitrary element c, where aRc. By the property of the frames from $Fr_{K.4.2}$ there is an element v from $\mathcal{F}$ such that bRv and cRv. Hence $v \Vdash_V p$ and $c \Vdash_V \Diamond p$. Thus we have proved $a \Vdash_V \square \Diamond p$. Hence $a \Vdash_V \Diamond \square p \to \square \Diamond p$. Thus the class of frames $Fr_{K4.2}$ is adequate for $K4.2$.

Now our aim is to show that the frame of $C_{K4.2}$ belongs to $Fr_{K4.2}$. In view of the fact that $C_{K4.2}$ characterizes $K4.2$, according to Lemma 2.3.27, this will complete the proof of the case (i). First note that $C_{K4.1}$ is a submodel of C_{K4}. Therefore according to Theorem 2.3.28 the frame of $C_{K4.2}$ is transitive. Let $T_c, E, H \in C_{K4.2}$ and relations $T_c R_c E$, $T_c R_c H$ hold. We intend to prove that $\Box E \cup \Box H$ is consistent. Suppose otherwise, then by (3) of Lemma 2.3.22 there are some formulas $\alpha_1, ..., \alpha_n \in \Box E$, $\beta_1, ..., \beta_m \in \Box H$ such that $\vdash_{K4.2} \neg(\alpha_1 \wedge ... \wedge \alpha_n \wedge \beta_1 \wedge ... \wedge \beta_m)$. This implies

$$\vdash_{K4.2} \bigwedge_{1 \leq i \leq n} \alpha_i \rightarrow \neg(\bigwedge_{1 \leq j \leq m} \beta_j),$$

$$\vdash_{K4.2} \Box(\bigwedge_{1 \leq i \leq n} \alpha_i) \rightarrow \Box\neg(\bigwedge_{1 \leq j \leq m} \beta_j),$$

$$\vdash_{K4.2} \Box(\bigwedge_{1 \leq i \leq n} \alpha_i) \rightarrow \neg\Diamond(\bigwedge_{1 \leq j \leq m} \beta_j),$$

$$\vdash_{K4.2} (\bigwedge_{1 \leq i \leq n} \Box\alpha_i) \rightarrow \neg\Diamond(\bigwedge_{1 \leq j \leq m} \beta_j).$$

By (3) of Lemma 2.3.22 it follows that $\bigwedge_{1 \leq i \leq n} \Box\alpha_i \in E$ and using (1), (2) of that lemma we arrive at

$$\neg\Diamond(\bigwedge_{1 \leq j \leq m} \beta_j) \in E, \tag{2.9}$$

Since every $\beta_i \in \Box H$ we infer $\bigwedge_{1 \leq j \leq m} \Box\beta_j \in H$, and using relations (1) and (2) from Lemma 2.3.22 we obtain that $\Box \bigwedge_{1 \leq j \leq m} \beta_j \in H$, then Theorem 2.3.25 yields $H \Vdash_{V_c} \Box \bigwedge_{1 \leq j \leq m} \beta_j$. Consequently

$$T_c \Vdash_{V_c} \Diamond\Box \bigwedge_{1 \leq j \leq m} \beta_j. \tag{2.10}$$

According to (1) of Lemma 2.3.22

$$\Diamond\Box \bigwedge_{1 \leq j \leq m} \beta_j \rightarrow \Box\Diamond \bigwedge_{1 \leq j \leq m} \beta_j \in T_c,$$

therefore Theorem 2.3.25 allows us to conclude

$$T_c \Vdash_{V_c} \Diamond\Box \bigwedge_{1 \leq j \leq m} \beta_j \rightarrow \Box\Diamond \bigwedge_{1 \leq j \leq m} \beta_j. \tag{2.11}$$

Relations (2.10) and (2.11) give us

$$T_c \Vdash_{V_c} \Box\Diamond \bigwedge_{1 \leq j \leq m} \beta_j.$$

Consequently the following holds $E_c \Vdash_{V_c} \Diamond \bigwedge_{1 \leq j \leq m} \beta_j$, and by Theorem 2.3.25 we obtain $\Diamond \bigwedge_{1 \leq j \leq m} \beta_j \in E_c$ which contradicts the claim of (2.9).

Thus $\Box E \cup \bar{\Box} H$ is consistent and (according to Lemma 2.3.21) this set can be enclosed in a complete modal theory M_c. Then we conclude $H R_c M_c$ and $E R_c M_c$. Hence $C_{K4.2} \in Fr_{K4.2}$ which completes the proof of case (i). The proof of case (ii) is completely the same, we only note that the model $C_{S4.2}$ is a submodel of C_{S4} and consequently (Theorem 2.3.28) the frame of the model $C_{S4.2}$ is reflexive and transitive. ∎

Recall that a frame $\mathcal{F}$ is said to be connected if $(\forall x, y \in \mathrm{F})((xRy) \vee (yRz)))$. A frame $\mathcal{F}$ is weakly connected if

$$(\forall x, y, z \in |\mathcal{F}|)[((xRy)\&(xRz))\Rightarrow((yRz) \vee (zRy))].$$

Theorem 2.3.51 *The following hold:*

> *(i) The modal logic $K4.3$ is Kripke complete. The class $Fr_{K4.3}$ of all transitive and weakly connected frames characterizes $K4.3$ and the frame of $C_{K4.3}$ belongs to $Fr_{K4.3}$.*
>
> *(ii) The modal logic $S4.3$ is Kripke complete. The class $Fr_{S4.3}$ of all reflexive transitive and weakly connected frames characterizes $S4.3$ and the frame of $C_{S4.3}$ belongs to $Fr_{S4.3}$.*

Proof. (i): In order to show that $Fr_{K4.3}$ is adequate for $K4.3$ we need to prove truth of the axiom $\Box(\Box p \to q) \vee \Box(\Box q \to p)$ in all frames of $Fr_{K4.3}$. Let $\mathcal{F}$ be a frame from $Fr_{K4.3}$, let V be a valuation on $\mathcal{F}$ and $a \in |\mathcal{F}|$. Suppose that there is some element b such that aRb and $b \Vdash_V \Box p$ but $b \nVdash_V q$. We claim that then $a \Vdash_V \Box(\Box q \to p)$. Indeed if aRc then cRb or bRc. If bRc then $c \Vdash_V \Box p \wedge p$. Consequently $c \Vdash_V \Box q \to p$. Another possibility is cRb. Then $c \nVdash_V \Box q$ and therefore $c \Vdash_V \Box q \to p$. Thus the axiom is valid in all frames from $Fr_{K4.3}$ and $K4.3$ is adequate for $K4.3$.

Now we are going to show that the class $Fr_{K4.3}$ characterizes $K4.3$. We know that the canonical Kripke model $C_{K4.3}$ characterizes $K4.3$ (cf. Lemma 2.3.27). So in order to complete the proof of (i) it is sufficient to show that the frame of $C_{K4.3}$ belongs to $Fr_{K4.3}$. The model $C_{K4.3}$ is a submodel of C_{K4}. Consequently (cf. Theorem 2.3.28) the frame of $C_{K4.3}$ is transitive. Consider any complete modal theories T_c, E_c, H_c from $C_{K4.3}$. Suppose that $T_c R_c E_c$ and $T_c R_c H$. This means $\Box T_c \subseteq E_c$ and $\Box T_c \subseteq H_c$. Assume that $\langle H_c, E_c \rangle \notin R_c$ and simultaneously $\langle E_c, H_c \rangle \notin R_c$. This yields there are formulas α, β such that $\Box \alpha \in H_c$, $\alpha \notin E_c$, $\Box \beta \in E_c$ and $\beta \notin H_c$. According to (2) from Lemma 2.3.22 this implies

$$\Box \alpha \to \beta \notin H_c$$
$$\Box \beta \to \alpha \notin E_c,$$
$$\Box(\Box \alpha \to \beta) \notin T_c,$$
$$\Box(\Box \beta \to \alpha) \notin T_c.$$

Therefore applying (4) from Lemma 2.3.22 we get

$$\Box(\Box\alpha \to \beta) \vee \Box(\Box\beta \to \alpha) \notin T_c,$$

which contradicts (1) from Lemma 2.3.22. Hence $H_c R_c E_c$ or $E_c R_c H_c$ and the frame of $C_{K4.3}$ belongs to $Fr_{K4.3}$ which complete the proof of (i). (ii) follows immediately from (i), because $C_{S4.3}$ is a submodel of $C_{K4.3}$. ∎

2.4 Kripke Semantics for Intuitionistic Logic

To specify the Kripke relational semantics for intuitionistic propositional logic we need to slightly alter the meaning a frame and the truth relation on frames. First we have to accept only partially ordered sets as frames. Second, the intuitionistic valuations of variables must be stable: the truth of a propositional variable in a world implies the truth of this variable in all worlds which are accessible from the given one. This exactly reflects the intuitionistic meaning of truth, a true proposition must be true always. And third, the truth relation of logical connectives $\to$, $\neg$ must be appropriately modified. In order to give an explicit explanation we turn to present certain precise definitions.

Definition 2.4.1 *An intuitionistic frame $\mathcal{F}$ is a partially ordered set $\langle W, \le \rangle$ that is, the relation $\le$ is reflexive transitive and antisymmetric:*

(1)$\forall x \in W(x \le x)$
(2)$\forall x, y, z \in W(x \le y \& y \le z \Rightarrow x \le z)$
(3)$\forall x, y \in W(x \le y \& y \le x \Rightarrow x = y)$.

In a similar way to the case of Kripke semantics for modal and temporal logic, the set W can be understood as the set of all *possible worlds, possible states* which are simulated by elements of the given frame. The partial order $\le$ is the accessibility relation between all possible words; $a \le b$ can be reformulated as b is accessible from a, or, less formally, a sees b.

Definition 2.4.2 *Given a set P of the propositional letters and an intuitionistic frame (poset) $\mathcal{F} = \langle W, \le \rangle$, an* intuitionistic valuation *of P in $\mathcal{F}$ is a mapping V with the following properties:*

(1)$V : P \mapsto 2^W$ *that is, $\forall p \in P(V(p) \subseteq W)$*
(2)$(\forall a, b \in W)(\forall p \in P)[a \le b \Rightarrow (a \in V(p) \Rightarrow b \in V(p))]$

As before $a \Vdash_V p$ is another way of saying that $a \in V(p)$, meaning is *the letter p is valid (true) on the element a*. The informal understanding of an intuitionistic valuation V is following: V assigns to any variable p the set of all possible worlds (states) in which the proposition encoded by p is valid (true). But in the intuitionistic case the definition of a valuation is given in the specifically intuitionistic manner: if p if true in a state a and b is a successor for a then p must be true on b also (which is encoded in relation (2) above).

Definition 2.4.3 *An intuitionistic Kripke model is a triple* $\mathfrak{M} = \langle W, \leq, V \rangle$, *where* $\langle W, \leq \rangle$ *is a partially ordered frame (poset) and* V *is an intuitionistic valuation of a set* P *of propositional letters in poset* $\langle W, \leq \rangle$.

We can extend the truth relation $\Vdash_V$ on intuitionistic Kripke models from propositional letters to all formulas which are build up of propositional letters in the following way:

(i) $\forall p \in P, \forall a \in W[(a \Vdash_V p) \Leftrightarrow (a \in V(p))]$,

(ii) $\forall a \in W[(a \Vdash_V \alpha \wedge \beta) \Leftrightarrow (a \Vdash_V \alpha) \& (a \Vdash_V \beta)]$,

(iii) $\forall a \in W[(a \Vdash_V \alpha \vee \beta) \Leftrightarrow (a \Vdash_V \alpha) \vee (a \Vdash_V \beta)]$,

(iv) $(\forall a \in W)[(a \Vdash_V \neg\alpha) \Leftrightarrow (\forall b \in W)(a \leq b \Rightarrow a \nVdash_V \alpha)]$,

(v) $\forall a \in W[(a \Vdash_V \alpha \rightarrow \beta) \Leftrightarrow$
$(\forall b \in W)(a \leq b \Rightarrow (a \Vdash_V \alpha \Rightarrow b \Vdash_V \beta))]$.

The relation $a \Vdash_V \alpha$ means: the formula α is *valid* (or *true*) on (or in) the element a in the model $\mathfrak{M}$ with respect to the valuation V. We know on the basis of the algebraic semantics for intuitionistic logic that in intuitionistic logic $\neg x$ is equivalent to the formula $x \rightarrow \bot$, where $\bot := \neg\top$. We can introduce the logical constants $\bot$ and $\top$ into Kripke semantics making $\bot$ be always false and $\top$ be always true under every valuation.

Exercise 2.4.4 *For any intuitionistic Kripke model* $\mathfrak{M} = \langle W, \leq, V \rangle$, *and each element* a *from* W,

(i) $a \Vdash_V \neg p \Leftrightarrow a \Vdash_V p \rightarrow \bot$,
(ii) $a \Vdash_V p \rightarrow \neg\neg p$,
(iii) $a \Vdash_V \neg\neg\neg p \rightarrow \neg p$.

An important and noteworthy property of truth for intuitionistic Kripke models is the *stability* of the truth: if a formula α is valid on an element a of a given intuitionistic model and another element b is accessible from a then α is also valid on b. We show this in the next lemma.

Lemma 2.4.5 *Let* $\mathfrak{M} := \langle W, \leq, V \rangle$ *be an intuitionistic model. For any formula* α *and every element* $a \in W$,

$$\forall a \in W, \forall b \in W((a \Vdash_V \alpha) \& (a \leq b) \Rightarrow b \Vdash_V \alpha).$$

Proof. It can be easily verified by induction on length of the formula α. Indeed, if α is a propositional letter then the required property follows directly from the definition of intuitionistic valuation. The inductive steps for logical connectives $\wedge$ and $\vee$ are evident. Let $\alpha = \beta \rightarrow \delta$. Suppose $a \Vdash_V \beta \rightarrow \delta$ and $a \leq b$. If $b \leq c \in W$ then $a \leq c$ and, according to the definition of truth for logical connective $\rightarrow$, if $c \Vdash_V \alpha$ then $c \Vdash_V \beta$. This, in particular, means $b \Vdash_V \alpha \rightarrow \delta$. Consider the case when $\alpha = \neg\beta$. If $a \Vdash_V \neg\beta$ then for every b such that $a \leq b$, $b \nVdash_V \beta$ holds. Therefore if $b \leq c$ then $c \nVdash_V \beta$ because $a \leq c$. Thus $b \Vdash_V \neg\beta$. ∎

If $\mathfrak{M} := \langle W, \leq, V \rangle$ is an intuitionistic Kripke model and α is a formula then we write $\mathfrak{M} \Vdash \alpha$ if $\forall a \in |\mathfrak{M}|(a \Vdash_V \alpha)$. In this event we say α is true (valid) on $\mathfrak{M}$. We can extend the notion of truth formulas in intuitionistic Kripke models for intuitionistic frames in usual manner:

Definition 2.4.6 *Let $\mathcal{F} := \langle W, \leq \rangle$ be a poset and let α be a formula in an intuitionistic propositional language. The formula α is called valid (true) on the frame $\mathcal{F}$ (denotation $\mathcal{F} \Vdash \alpha$) if for every intuitionistic valuation V, where the variables of α are from $Dom(V)$, for every element a from $|\mathcal{F}|$, $a \Vdash_V \alpha$ holds.*

Definition 2.4.7 *Let λ be a superintuitionistic logic. A poset $\mathcal{F}$ is said to be an λ-frame if for every $\alpha \in \lambda$, $\mathcal{F} \Vdash \alpha$. A family Ω of posets is adequate for λ if all frames from Ω are λ-frames.*

It is easy to see that Ω is adequate for λ iff

$$\lambda \subseteq \lambda(\Omega) := \{\alpha \mid \alpha \in \mathcal{F}or_H, (\forall \mathcal{F} \in \Omega)(\mathcal{F} \Vdash \alpha)\}.$$

Now as an example we describe some particular classes of partially ordered sets which are adequate for some particular superintuitionistic logics. We employ the following conventions:

Fr_H is the class of all posets,

Fr_{KC} is the class of all posets satisfying the condition
$\forall x, y, z[(x \leq y \& x \leq z) \Rightarrow \exists t(y \leq t \& z \leq t)]$,

Fr_{LC} is the class of all posets satisfying the condition:
$\forall x, y, z[(x \leq y \& y \leq z) \Rightarrow (y \leq z \vee z \leq y)]$; that is, the class of all posets which are disjoint unions of linearly ordered sets.

Theorem 2.4.8 *The following hold*

(i) The class Fr_H is adequate for the Heyting intuitionistic logic H;

(ii) The class Fr_{KC} is adequate for the superintuitionistic logic $KC := H \oplus \neg\neg p \vee \neg p$ (the logic of weak law of excluded middle);

(iii) The class Fr_{LC} is adequate for the superintuitionistic logic $LC := H \oplus$
$(p \to q) \vee (q \to p)$.

Proof. The proof scheme for all of cases (i) - (iii) is the same: (a) we show that all axioms of the given logical systems are valid on frames of corresponding classes, and (b) that modus ponens preserves the truth of formulas in intuitionistic Kripke models. This is sufficient to prove that the class is adequate, indeed

Lemma 2.4.9 *If all the axioms of a superintuitionistic logic λ are valid on a frame $\mathcal{F}$ and modus ponens preserves the truth in any Kripke model that is based on this frame then all theorems of λ are valid on $\mathcal{F}$.*

Proof. Indeed, any theorem α of the given logic λ is provable from schemes of axioms of this logic. We fix some derivation S for α. Consider some valuation V in $\mathcal{F}$ for all propositional letters from α. We extend V on all variables from S putting that all variables having no occurrences in α are not valid everywhere in $\mathcal{F}$ under V. All schemes of axioms β involving in S will be true under V in $\mathcal{F}$. Actually let β be a substitution instance of some axiom γ of λ, where $\gamma = \gamma(p_i)$ (γ is build up on letters p_i) and $\beta = \gamma(\delta_i)$. We introduce the valuation V_1 of all letters p_i in $\mathcal{F}$, where V_1 assigns to each p_i the set $V(\delta_i)$ of all elements from $\mathcal{F}$ on which δ_i is valid under V. This valuation is an intuitionistic valuation according to Lemma 2.4.5. Besides truth value of γ under V_1 coincides with truth value of $\gamma(\delta_i)$ under V. Since we have assumed that γ is valid on $\mathcal{F}$ under any valuation, we get the formula $\gamma(\delta_i)$ which is equal to β is valid on $\mathcal{F}$ under V. Thus all schemes of axioms β from S are true under V everywhere in $\mathcal{F}$ and by supposition modus ponens preserves truth in intuitionistic Kripke models that are built up on $\mathcal{F}$. Therefore by induction on the place of occurrences of formulas in S we arrive at α is true under V on $\mathcal{F}$. ∎

Thus, if we have at our disposal (a) and (b) proved we obtain that any corresponding class of frames described above is adequate for given logic. Hence it remains only to show that (a) and (b) hold for (i) - (iii). We start with (b): we verify that modus ponens rule preserves the truth of formulas in any Kripke model:

Lemma 2.4.10 *Let $\mathfrak{M}$ be an intuitionistic Kripke model with a valuation V. Let $L(\mathfrak{M}) := \{\alpha \mid \alpha \in \mathcal{F}or_H, \mathfrak{M} \Vdash \alpha\}$. The set $L(\mathfrak{M})$ is closed with respect to the inference rule (R1): $x, x \to y/y$.*

Proof. Suppose formulas α and $\alpha \to \beta$ are true in the model $\mathfrak{M}$ with the valuation V. Let $a \in |\mathfrak{M}|$. Then by supposition about the truth we conclude $a \Vdash_V \alpha$ and $a \Vdash_V \alpha \to \beta$ which yields $a \Vdash_V \beta$. ∎

Hence we need only to verify truth of axioms on frames from the corresponding classes. We consider below subsequently cases (i) - (iii).

(i): Let $\mathcal{F} := \langle W, \leq \rangle$ be a poset and V is a valuation on $\mathcal{F}$. H has axioms (A1) - (A9) of PC and (A11). We consider (A1): we need to show $a \Vdash_V p \rightarrow (q \rightarrow p)$ for any $a \in W$. Suppose $a \leq b$ and $b \Vdash_V p$. Now if $b \leq c$ then $c \Vdash_V p$ also since V is an intuitionistic valuation. Therefore $c \Vdash_V (q \rightarrow p)$ holds. So $a \Vdash_V p \rightarrow (q \rightarrow p)$. (A2): the formula

$$(p \rightarrow (q \rightarrow r)) \rightarrow ((p \rightarrow q) \rightarrow (p \rightarrow r))$$

is valid on any element a under V. Indeed, suppose $a \leq b$ and $b \Vdash_V (p \rightarrow (q \rightarrow r))$. Assume $b \leq c$ and $c \Vdash_V (p \rightarrow q)$. Let $c \leq d$ and $d \Vdash_V p$. Then $d \Vdash_V q$ and $d \Vdash_V (q \rightarrow r)$ since $b \leq d$. Thus $d \Vdash_V r$ what we need.

Truth of axioms (A3): $p \rightarrow p \vee q$ and (A4): $q \rightarrow p \vee q$ in $\mathcal{F}$ is evident.

(A5): We check axiom

$$(p \rightarrow r) \rightarrow ((q \rightarrow r) \rightarrow ((p \vee q) \rightarrow r)).$$

Suppose $a \in W$, $a \leq b \in W$ and $b \Vdash_V (p \rightarrow r)$. Let $b \leq c$ and $c \Vdash_V (q \rightarrow r)$. Assume that $c \leq d$ and $d \Vdash_V p \vee q$. Then $d \Vdash_V p$ or $d \Vdash_V q$. If $d \Vdash_V p$ then $d \Vdash_V r$ since $b \leq d$. If $d \Vdash_V q$ then $d \Vdash_V r$ because $c \leq d$. Thus (A5) is true on $\mathcal{F}$ under the valuation V.

Truth of axioms (A6): $p \wedge q \rightarrow p$ and (A7): $p \wedge q \rightarrow q$ on $\mathcal{F}$ follows directly from definitions.

(A8): We must check the truth of the formula

$$(p \rightarrow q) \rightarrow ((p \rightarrow r) \rightarrow (p \rightarrow q \wedge r)).$$

Let a be an element of $\mathcal{F}$. Suppose $a \leq b$ and $b \Vdash_V (p \rightarrow q)$. Assume $b \leq c$ and $c \Vdash_V (p \rightarrow r)$. Let d be an element of W such that $d \Vdash_V p$. Then since $b \leq d$ and $c \leq d$, we have $d \Vdash_V q$ and $d \Vdash_V r$, which yields $d \Vdash_V q \wedge r$.

(A9): This case we check truth of the formula

$$(p \rightarrow q) \rightarrow ((p \rightarrow \neg q) \rightarrow \neg p).$$

Let $a \in W$, $a \leq b$ and $b \Vdash_V (p \rightarrow q)$. Suppose $b \leq c$ and $c \Vdash_V (p \rightarrow \neg q)$. Then for every d such that $b \leq d$,

$$d \Vdash_V p \Rightarrow d \Vdash_V q, \qquad d \Vdash_V p \Rightarrow d \Vdash_V \neg q.$$

Therefore for all such d, $d \nVdash_V p$. That is, $b \Vdash_V \neg p$, what we needed.

(A11) We show the formula $p \rightarrow (\neg p \rightarrow p)$ is valid. Suppose that $a \in W$, $a \leq b$ and $b \Vdash_V p$. The valuation V is intuitionistic, therefore $c \Vdash_V p$ for all $c \in W$ such that $b \leq c$. Hence for every c such that $b \leq c$, $c \nVdash_V \neg p$ and consequently $c \Vdash_V (\neg p \rightarrow q)$ holds. The proof of (i) is complete.

To prove (ii) it is sufficient to show that the formula $\neg\neg p \vee \neg p$ is valid on all frames from Fr_{KC}. Let $\mathcal{F} := \langle W, \leq \rangle \in Fr_{KC}$ and let V be a valuation in $\mathcal{F}$. Suppose $a \in W$ and $a \nVdash_V \neg p$. This means there exists b such that $a \leq b$ and

$b \Vdash_V p$. Suppose c is an arbitrary element from W such that $a \leq c$. Since $\mathcal{F}$ is a frame from Fr_{KC}, there exists an element $d \in W$ such that $b \leq d$ and $c \leq d$. Since $b \Vdash_V p$ we get $d \Vdash_V p$. Therefore $c \nVdash_V \neg p$. Thus we have, for all c such that $a \leq c$, $c \nVdash_V \neg p$. Therefore $a \Vdash_V \neg\neg p$ and the proof of case (ii) is complete.

To show (iii) we only need to prove that the formula $(p \rightarrow q) \vee (q \rightarrow p)$ is valid on all frames from Fr_{LC}. Suppose $\mathcal{F} := \langle W, \leq \rangle \in Fr_{LC}$ and V is a valuation in $\mathcal{F}$. Let a be an element from W such that $a \nVdash_V (p \rightarrow q)$. This yields there is an element b such that $a \leq b$, $b \Vdash_V p$ and $b \nVdash_V q$. Suppose c is an arbitrary element from W such that $a \leq b$ and $c \Vdash_V q$. Since $\mathcal{F} \in Fr_{LC}$ we have $b \leq c$ or $c \leq b$. If $c \leq b$ then $b \nVdash_V q$ gives us a contradiction. Therefore $b \leq c$ holds and as $b \Vdash_V p$ we get $c \Vdash_V p$ and consequently $a \Vdash_V (q \rightarrow p)$, which completes the proof of (iii). ∎

Exercise 2.4.11 *Show that the following hold:*

(i) A frame $\mathcal{F}$ is adequate for the logic KC iff $\mathcal{F} \in Fr_K C$.
(ii) A frame $\mathcal{F}$ is adequate for the superintuitionistic logic LC iff $\mathcal{F} \in Fr_{LC}$.

Exercise 2.4.12 *Prove the following strong inclusions hold: $H \subset KC$ and $KC \subset LC$.*

Let Ω be a class of posets. We define $\lambda(\Omega)$ to be the set of all formulas in the language of PC such that $\forall \mathcal{F} \in \Omega(\mathcal{F} \Vdash \alpha)$.

Lemma 2.4.13 *For any class of posets Ω, $\lambda(\Omega)$ is a superintuitionistic logic.*

Proof. By Theorem 2.4.8 $\lambda(\Omega)$ includes all formulas from H. According to Lemma 2.4.10 $\lambda(\Omega)$ is closed with respect to modus ponens, and $\lambda(\Omega)$ is closed with respect to substitutions by Lemma 2.4.5.

Definition 2.4.14 *A class of frames Ω is said to be* characterizing *for a superintuitionistic logic λ if $\lambda = \lambda(\Omega)$.*

Definition 2.4.15 *A superintuitionistic logic λ is* Kripke complete *if there is a family Ω of posets such that $\lambda = \lambda(\Omega)$, that is if there is a class of posets which characterizes λ.*

We can reformulate this definition saying that λ is Kripke complete if λ is the logic $\lambda(\Omega)$ for some class of frames Ω. We now intend to show that the superintuitionistic logics considered above are Kripke complete and to develop the basic tools for investigating Kripke completeness for superintuitionistic logics. For this purpose we introduce canonical intuitionistic Kripke models below which significantly differ from canonical models for modal logics. We start with some necessary definitions.

Definition 2.4.16 *An intuitionistic theory is a pair $\langle T, F \rangle$, where T, F are sets of formulas in the language of intuitionistic propositional logic.*

A theory $\langle T, F \rangle$ is said to be *consistent* for (or with) some superintuitionistic logic λ if there are no finite subsets A, B of sets T, F respectively such that

$$\vdash_\lambda \bigwedge_{\alpha_i \in A} \alpha_i \to \bigvee_{\beta_j \in B} \beta_j .$$

Applying this definition we use the fact that, as always, the conjunction of an empty set of formulas is $\top$ and the disjunction of an empty set of formulas is $\bot$ (recall that $\bot \equiv \neg(p \to p)$). Informally the set F for the theory $\langle T, F \rangle$ means the set of all formulas which are true from the point of view of this theory, and F means the set of all formulas which are not valid in this theory. All other formulas (those which are not members of $T \cup F$) can be understood as formulas whose truth value is not determined for this theory yet.

Lemma 2.4.17 *Let $\langle T, F \rangle$ be an intuitionistic theory consistent with some superintuitionistic logic λ. For any formula α, either $\langle T \cup \{\alpha\}, F \rangle$ or $\langle T, F \cup \{\alpha\} \rangle$ is consistent with λ.*

Proof. Suppose both these theories are inconsistent with λ. Then there are finite sets of formulas A_1, A_2, B_1, B_2 such that $A_1 \subseteq T$, $A_2 \subseteq T$, $B_1 \subseteq F$, $B_2 \subseteq F$ and

$$\vdash_\lambda \bigwedge_{\alpha_i \in A_1} \alpha_i \wedge \alpha \to \bigvee_{\beta_j \in B_1} \beta_j$$
$$\vdash_\lambda \bigwedge_{\alpha_i \in A_2} \alpha_i \to \bigvee_{\beta_j \in B_2} \beta_j \vee \alpha .$$

Using this and the corresponding derivations in intuitionistic logic H we get

$$\vdash_\lambda \bigwedge_{\alpha_i \in A_1} \alpha_i \wedge \bigwedge_{\alpha_i \in A_2} \alpha_i \wedge \alpha \to \bigvee_{\beta_j \in B_1} \beta_j \vee \bigvee_{\beta_j \in B_2} \beta_j$$
$$\vdash_\lambda \bigwedge_{\alpha_i \in A_1} \alpha_i \wedge \bigwedge_{\alpha_i \in A_2} \alpha_i \to \bigvee_{\beta_j \in B_1} \beta_j \vee \bigvee_{\beta_j \in B_2} \beta_j \vee \alpha .$$

From these relations, again using the appropriate derivations in H, we arrive at

$$\vdash_\lambda \bigwedge_{\alpha_i \in A_1} \alpha_i \wedge \bigwedge_{\alpha_i \in A_2} \alpha_i \to \bigvee_{\beta_j \in B_1} \beta_j \vee \bigvee_{\beta_j \in B_2} \beta_j$$

which implies that the pair $\langle T, F \rangle$ is inconsistent with λ, a contradiction. ∎

Definition 2.4.18 *An intuitionistic theory $\langle T, F \rangle$ is said to be* complete *if $T \cup F = \mathcal{F}or_H$, that is if every formula is a member of $T \cup F$.*

We say a pair $\langle H, I \rangle$ *extends* a pair $\langle T, F \rangle$ if $T \subseteq H$, $F \subseteq I$.

Lemma 2.4.19 *For any intuitionistic theory $\langle T, F \rangle$ consistent with a superintuitionistic logic λ, there exists a complete theory $\langle T_1, F_2 \rangle$ extending $\langle T, F \rangle$ and consistent with λ.*

Proof. We can enumerate all formulas $\alpha_1, \alpha_2, \alpha_3, \ldots$ and arrange a procedure of adjoining of formulas α_i to the theory $\langle T, F \rangle$ in the following way. We put $\langle T_0, F_0 \rangle := \langle T, F \rangle$. Suppose the step $i - 1$ is already maid and we have a theory $\langle T_{i-1}, F_{i-1} \rangle$ as the result of our procedure. On step i, according to Lemma 2.4.17, we can add α_i to T_{i-1} or to F_{i-1} preserving consistency of the theory. We make the corresponding choice and enlarge the theory $\langle T_{i-1}, F_{i-1} \rangle$ till $\langle T_i, F_i \rangle$. The union of all theories $\langle T_i, F_i \rangle$ gives us a consistent theory $\langle H, L \rangle$ because, (1): on any step i theory $\langle T_i, F_i \rangle$ is consistent by choice, and (2): this theory is an extension of all theories which have been constructed before. Clearly then that $\langle H, L \rangle$ is complete and extends the theory $\langle T, F \rangle$. ∎

Lemma 2.4.20 *If $\langle T, F \rangle$ is a complete theory that is consistent with a superintuitionistic logic λ then*

$$(i) \alpha \wedge \beta \in T \Leftrightarrow (\alpha \in T) \& (\beta \in T)$$
$$(ii) \alpha \vee \beta \in T \Leftrightarrow (\alpha \in T) \vee (\beta \in T).$$

Proof. (i): Let $\alpha \wedge \beta \in T$ and $\alpha \notin T$. As the theory is complete we get $\alpha \in F$. Because $\vdash_H \alpha \wedge \beta \to \alpha$ we get the theory $\langle T, F \rangle$ is inconsistent with H and consequently with λ. Let $\alpha \in T$ and $\beta \in T$. Suppose $\alpha \wedge \beta \notin T$. Since our theory is complete we have $\alpha \wedge \beta \in F$. But $\vdash_\lambda \alpha \wedge \beta \to \alpha \wedge \beta$ which implies $\langle T, F \rangle$ is inconsistent. Thus $\alpha \wedge \beta \in T$.

(ii): Let $\alpha \vee \beta \in T$, Suppose $\alpha \notin T$ and $\beta \notin T$. Then $\alpha \in F$ and $\beta \in F$. Moreover $\Vdash_\lambda \alpha \vee \beta \to \alpha \vee \beta$ which would give $\langle T, F \rangle$ is inconsistent. Hence we get $\alpha \in T$ or $\beta \in T$. Conversely, assume that $\alpha \in T$ but $\alpha \vee \beta \notin T$. Then $\alpha \vee \beta \in F$. We know that $\vdash_\lambda \alpha \to \alpha \vee \beta$. This entails $\langle T, F \rangle$ is inconsistent, a contradiction. ∎

Lemma 2.4.21 *Let $\langle T, F \rangle$ be a complete theory that is consistent with a superintuitionistic logic λ. For any formula α the following holds:*

$$\alpha \in T \Leftrightarrow \exists A[(A \subseteq T) \& (\|A\| < \omega) \& (\vdash_\lambda \bigwedge_{\delta \in A} \delta \to \alpha)].$$

Proof. If $\alpha \in T$ then we can take $A := \{\alpha\}$ because $\Vdash_\lambda \alpha \to \alpha$ holds. Conversely, suppose $\alpha \notin T$. Then $\alpha \in F$. Since the theory $\langle T, F \rangle$ is consistent, for every finite subset A of $T, \nvdash_\lambda \bigwedge_{\delta \in A} \delta \to \alpha$. Thus the assertion of our lemma holds. ∎

Lemma 2.4.22 *If a superintuitionistic logic λ is consistent (which means $\perp \notin \lambda$) and a formula α is not a theorem of λ then there exists a complete intuitionistic theory $\langle T, F \rangle$ consistent with λ and such that $\alpha \in F$.*

Proof. The theory $\langle \emptyset, \{\alpha\} \rangle$ in consistent with λ. Indeed, otherwise either $\langle \emptyset, \emptyset \rangle$ is inconsistent and $\top \to \perp \in \lambda$, and then $\perp \in \lambda$ which gives λ is inconsistent; or $\top \to \alpha \in \lambda$, which yields $\alpha \in \lambda$. Therefore by Lemma 2.4.19 there is a complete

consistent with λ intuitionistic theory $\langle T, F \rangle$ which extends the theory $\langle \emptyset, \{\alpha\} \rangle$.
■

Now we are in a position to introduce canonical Kripke models for superintuitionistic logics. Let λ be a consistent superintuitionistic logic and W_λ be the set of all complete intuitionistic theories consistent with λ. Note that W_λ is nonempty: indeed $\langle \emptyset, \emptyset \rangle$ is a consistent theory and by Lemma 2.4.19 there is a complete theory consistent with λ and extending $\langle \emptyset, \emptyset \rangle$. We define $\leq$ to be the partial order on W_λ, where

$$(\langle T, F \rangle \leq \langle H, \mathit{I} \rangle) \Leftrightarrow (T \subseteq H \,\&\, \mathit{I} \subseteq F).$$

Definition 2.4.23 *For any consistent superintuitionistic logic λ the canonical Kripke model is the model $\mathfrak{M}_\lambda := \langle W_\lambda, \leq, V \rangle$, where the valuation V of all propositional letters in W_λ is defined as follows:*

$$(\forall p)(\forall \langle T, F \rangle \in W_\lambda) \; [(\langle T, F \rangle \in V(p)) \Leftrightarrow (p \in T)].$$

It is easy to see that the valuation V is an intuitionistic one and that $\leq$ is a partial order. The use of canonical models for superintuitionistic logics is similar to the application of canonical models of modal logics and is based on the following theorem.

Theorem 2.4.24 *(*Theorem of Canonical Models in Intuitionistic Logic.*) Let α be a formula and λ be a consistent superintuitionistic logic. For any intuitionistic theory $\langle T, F \rangle$ from W_λ,*

$$\langle T, F \rangle \Vdash_V \alpha \Leftrightarrow \alpha \in T.$$

Proof. The proof is by induction on the length of the formula α. If α is a propositional letter the claim follows directly from the definition of V. The inductive steps for connectives $\wedge$, $\vee$ follow directly from Lemma 2.4.20. It remains to consider the connectives $\neg$, $\rightarrow$.

Let $\alpha := \neg \delta$. According to Lemma 2.4.21 $\neg \delta \notin T$ iff for every finite subset A of T, the formula $\bigwedge_{\gamma \in A} \gamma \rightarrow \neg \delta$ is not provable in λ. The latter is equivalent for any finite subset A of T, the theory $\langle A \cup \{\delta\}, \emptyset \rangle$ is consistent with λ. Indeed, if $\vdash_\lambda \bigwedge_{\gamma \in A} \gamma \rightarrow \neg \delta$ then

$$\vdash_\lambda \bigwedge_{\gamma \in A} \gamma \wedge \delta \rightarrow \delta,$$
$$\vdash_\lambda \bigwedge_{\gamma \in A} \gamma \wedge \delta \rightarrow \neg \delta,$$
$$\vdash_\lambda \bigwedge_{\gamma \in A} \gamma \wedge \delta \rightarrow \bot,$$

that is, $\langle A \cup \{\delta\}, \emptyset \rangle$ is inconsistent with λ. Conversely, if this pair is inconsistent then

$$\vdash_\lambda \bigwedge_{\gamma \in A} \gamma \wedge \delta \to \bot,$$
$$\vdash_\lambda \bigwedge_{\gamma \in A} \gamma \wedge \delta \to \neg \delta,$$
$$\bigwedge_{\gamma \in A} \gamma, \delta \vdash_\lambda \neg \delta,$$
$$\bigwedge_{\gamma \in A} \gamma \vdash_\lambda \delta \to \neg \delta \text{ (cf. deduction theorem)},$$
$$\bigwedge_{\gamma \in A} \gamma \vdash_\lambda \delta \to \delta,$$
$$\bigwedge_{\gamma \in A} \gamma \vdash_\lambda (\delta \to \delta) \to ((\delta \to \neg \delta) \to \neg \delta)$$
$$\vdash_\lambda \bigwedge_{\gamma \in A} \gamma \to \neg \delta.$$

Thus the equivalence of mentioned above propositions in fact holds. So $\alpha \notin T$ iff $\langle T \cup \{\alpha\}, \emptyset \rangle$ is consistent with λ. Further, the theory $\langle T \cup \{\delta\}, \emptyset \rangle$ is consistent with λ iff there exists a complete intuitionistic theory consistent with λ and extending $\langle T \cup \{\delta\}, \emptyset \rangle$ (this holds according to Lemma 2.4.19). Hence $\neg \delta \notin T$ iff there exists a complete intuitionistic theory $\langle H, L \rangle$ consistent with λ and such that $\alpha \in H$, and $\langle T, F \rangle \leq \langle H, L \rangle$. By inductive supposition the latter is equivalent to $\langle T, F \rangle \Vdash_V \neg \delta$. Thus we have showed

$$\neg \delta \in T \Leftrightarrow \langle T, F \rangle \Vdash_V \neg \delta.$$

Consider now the step for $\to$. Let $\alpha := \delta \to \gamma$. Again according to Lemma 2.4.21 we have $\delta \to \gamma \notin T$ iff for any finite subset A of T the following formula $\bigwedge_{\phi \in A} \phi \to (\delta \to \gamma)$ is not a theorem of λ. This is equivalent to $\bigwedge_{\phi \in A} \phi \wedge \delta \to \gamma \notin \lambda$, that is, that to the theory $\langle A \cup \{\delta\}, \{\gamma\} \rangle$ is consistent. Hence $\delta \to \gamma \notin T$ iff the theory $\langle T \cup \{\delta\}, \{\gamma\} \rangle$ is consistent with λ. The latter by Lemma 2.4.19 is equivalent there is a complete intuitionistic theory consistent with λ and extending the theory $\langle T \cup \{\delta\}, \{\gamma\} \rangle$. Thus we have proved $\delta \to \gamma \notin T$ iff there is a complete intuitionistic theory $\langle H, L \rangle$ consistent with λ and extending $\langle T \cup \{\delta\}, \{\gamma\} \rangle$. In particular, by inductive hypothesis, if such $\langle H, L \rangle$ exists then

$$\langle H, L \rangle \Vdash_V \delta, \ \langle H, L \rangle \not\Vdash_V \gamma,$$
$$\langle T, F \rangle \leq \langle H, L \rangle, \ \langle T, F \rangle \not\Vdash_V \delta \to \gamma.$$

Conversely if $\delta \to \gamma$ is not valid on $\langle T, F \rangle$ then for some $\langle H, L \rangle$ all these relations hold and this, as we have seen above, implies $\delta \to \gamma \notin T$. Hence

$$\delta \to \gamma \in T \Leftrightarrow \langle T, F \rangle \Vdash_V \delta \to \gamma.$$

From the inductive procedure we infer that the conclusion of this theorem holds for any formula α. ∎

Definition 2.4.25 *Let λ be a superintuitionistic logic. An intuitionistic Kripke model $\mathfrak{M} := \langle W, R, V \rangle$ is characterizing for λ if for any formula α, $\alpha \in \lambda$ iff $\mathfrak{M} \Vdash \alpha$.*

Using the properties of intuitionistic Kripke canonical models studied above we immediately derive a general theorem describing characterizing models for superintuitionistic logics:

Theorem 2.4.26 *For any consistent superintuitionistic logic λ, the intuitionistic canonical Kripke model $\mathfrak{M}_\lambda$ characterizes λ.*

Proof. Let $\alpha \in \lambda$. Then $\vdash_\lambda \top \to \alpha$ and by Lemma 2.4.21 $\alpha \in T$ for every complete theory $\langle T, F \rangle$ consistent with λ. This according to Theorem 2.4.24 means that $\langle T, F \rangle \Vdash_V \alpha$. Thus $\mathfrak{M}_\lambda \Vdash \alpha$. Conversely suppose $\alpha \notin \lambda$. Then by Lemma 2.4.22 there is a complete intuitionistic theory $\langle T, F \rangle$ consistent with λ and such that $\alpha \in F$. Then $\alpha \notin T$ and according to Theorem 2.4.24 $\langle T, F \rangle \nVdash_V \alpha$. ∎

Theorem 2.4.27 *The following hold:*

(i) The Heyting intuitionistic logic H is Kripke complete and the class Fr_H of all posets characterizes H.

(ii) The superintuitionistic logic KC is Kripke complete and the class Fr_{KC} characterizes KC. In particular, the frame of $\mathfrak{M}_{KC}$ belongs to Fr_{KC}.

(iii) The logic LC is Kripke complete and the class Fr_{LC} characterizes LC. Moreover the frame of $\mathfrak{M}_{LC}$ belongs to Fr_{LC}.

Proof. By Theorem 2.4.8 in all the cases (i), (ii) and (iii) the respective classes of frames Fr_H, Fr_{KC}, Fr_{LC} are adequate for the logics H, KC, LC. By Theorem 2.4.26 the canonical intuitionistic Kripke model $\mathfrak{M}_\lambda$ characterizes λ for any superintuitionistic logic λ. Therefore in order to complete the proof of this theorem it is sufficient to show that the frames of canonical models are from corresponding classes of frames. For the case (i) it is evident.

(ii): Consider the frame of $\mathfrak{M}_{KC}$, suppose $\langle T, F \rangle, \langle H, L \rangle, \langle R, E \rangle$ are members of $|\mathfrak{M}_{KC}|$ and $\langle T, F \rangle \leq \langle H, L \rangle$, $\langle T, F \rangle \leq \langle R, E \rangle$. We will show that the theory $\langle H \cup R, \emptyset \rangle$ is KC-consistent. Indeed, otherwise there are some finite subsets A, B of H and R, respectively, such that $\vdash_{KC} \bigwedge_{\alpha \in A} \alpha \wedge \bigwedge_{\beta \in B} \beta \to \bot$. Using deduction theorem we infer

$$\bigwedge_{\alpha \in A} \alpha, \bigwedge_{\beta \in B} \beta \vdash_{KC} \bot$$
$$\bigwedge_{\alpha \in A} \alpha \vdash_{KC} \bigwedge_{\beta \in B} \beta \to \bot$$
$$\bigwedge_{\alpha \in A} \alpha \vdash_{KC} \neg(\bigwedge_{\beta \in B} \beta).$$

Since $\vdash_{KC} (\neg \bigwedge_{\beta \in B} \beta) \vee (\neg\neg \bigwedge_{\beta \in B} \beta)$, according to Lemma 2.4.21, we conclude that $(\neg \bigwedge_{\beta \in B} \beta) \vee (\neg\neg \bigwedge_{\beta \in B} \beta) \in T$. Therefore by Lemma 2.4.20 either $\neg \bigwedge_{\beta \in B} \beta \in T$ or $\neg\neg \bigwedge_{\beta \in B} \beta \in T$. Suppose the first inclusion holds. Then $\neg \bigwedge_{\beta \in B} \beta \in R$ (because $T \subseteq R$) which implies $\langle R, E \rangle$ is KC-inconsistent, a contradiction. Therefore $\neg\neg \bigwedge_{\beta \in B} \beta \in T \subseteq H$. Furthermore, from the fact that

$\bigwedge_{\alpha \in A} \alpha \vdash_{KC} \neg(\bigwedge_{\beta \in B} \beta)$ by Lemma 2.4.21 we derive $\neg(\bigwedge_{\beta \in B} \beta) \in H$ which again would give $\langle H, I \rangle$ is KC-inconsistent. Thus the theory $\langle H \cup R, \emptyset \rangle$ is KC-consistent. Using Lemma 2.4.19 we obtain there is an intuitionistic complete theory $\langle T^*, F^* \rangle$ extending the theory $\langle H \cup R, \emptyset \rangle$. Then

$$\langle H, I \rangle \leq \langle T^*, F^* \rangle, \; \langle R, E \rangle \leq \langle T^*, F^* \rangle,$$

what we need.

(iii): Now we consider the frame of the canonical intuitionistic model $\mathfrak{M}_{LC}$. Suppose that $\langle T, F \rangle, \langle H, I \rangle, \langle R, E \rangle$ are members of $|\mathfrak{M}_{LC}|$ and

$$\langle T, F \rangle \leq \langle H, I \rangle, \; \langle T, F \rangle \leq \langle R, E \rangle.$$

Suppose that

$$\langle H, I \rangle \not\leq \langle R, E \rangle, \; \langle R, E \rangle \not\leq \langle H, I \rangle.$$

Then there is a formula $\alpha \in H$ such that $\alpha \notin R$ and there exist a formula $\beta \in R$ such that $\beta \notin H$. At the same time $\vdash_{LC} (\alpha \rightarrow \beta) \vee (\beta \rightarrow \alpha)$. Therefore from Lemma 2.4.21 we conclude either $(\alpha \rightarrow \beta) \in T$ or $(\beta \rightarrow \alpha) \in T$. If $(\alpha \rightarrow \beta) \in T \subseteq H$ then applying Lemma 2.4.21 to $\alpha \in H$, we have $\beta \in H$, a contradiction. Similarly, if $(\beta \rightarrow \alpha) \in T \subseteq R$ then by Lemma 2.4.21 and the fact that $\beta \in R$ we have $\alpha \in R$, a contradiction again. Thus either $\langle H, I \rangle \leq \langle R, E \rangle$ or $\langle R, E \rangle \leq \langle H, I \rangle$ ∎

Exercise 2.4.28 *Show that the classical propositional calculus PC is Kripke complete and that the class of all posets which are antichains (that is posets which satisfy the condition : $\forall x, y (x \leq y \Rightarrow x = y))$ characterizes PC.*

Definition 2.4.29 *We say a superintuitionistic logic λ has the disjunction property iff for any given formulas α and β,*

$$\alpha \vee b \in \lambda \Leftrightarrow (\alpha \in \lambda) \vee (\beta \in \lambda).$$

Definition 2.4.30 *A superintuitionistic logic λ is downwards closed if the class K of all frames adequate for K is closed with respect to the operation $\otimes$, where for any frames $\mathcal{F}_1 := \langle W_1, \leq_1 \rangle$, $\mathcal{F}_2 := \langle W_2, \leq_2 \rangle$ having no common elements, $\mathcal{F}_1 \otimes \mathcal{F}_2$ is the following frame:*

$$\mathcal{F}_1 \otimes \mathcal{F}_2 := \langle |\mathcal{F}_1 \otimes \mathcal{F}_2|, \leq \rangle$$
$$|\mathcal{F}_1 \otimes \mathcal{F}_2| := |\mathcal{F}_1| \cup |\mathcal{F}_2| \cup \{\bot\}, \; where \; \bot \notin |\mathcal{F}_1| \cup |\mathcal{F}_2|;$$
$$\leq := \leq_1 \cup \leq_2 \cup \{\bot \leq a \mid a \in W_1 \cup W_2\} \cup \{\bot \leq \bot\}.$$

Theorem 2.4.31 Disjunction Theorem. *If a superintuitionistic logic λ is Kripke complete and downwards closed then λ has the disjunction property.*

Proof. Assume that $\alpha \notin \lambda$ and $\beta \notin \lambda$. Since λ is Kripke complete, there are frames $\mathcal{F}_1 := \langle W_1, \leq_1 \rangle$, and $\mathcal{F}_2 := \langle W_2, \leq_2 \rangle$ which are adequate for λ and valuations V_1, V_2 on $\mathcal{F}_1$ and $\mathcal{F}_2$ respectively such that $(\exists a_1 \in W_1)(a_1 \nVdash_{V_1} \alpha)$ and $(\exists a_2 \in W_2)(a_2 \nVdash_{V_2} \beta)$. Without loss of generality we can assume that these frames $\mathcal{F}_1$, $\mathcal{F}_2$ have no common elements. Since the logic λ is downwards closed, the frame $\mathcal{F}_1 \otimes \mathcal{F}_2$ also is a λ-frame. We introduce the valuation V on this frame as follows:

$$(\forall p)[V(p) := V_1(p) \cup V_2(p)].$$

It is clear that the truth relation $\Vdash_V$ on components $\mathcal{F}_1$ and $\mathcal{F}_2$ coincides with the truth relations $\Vdash_{V_1}$, $\Vdash_{V_2}$ respectively. The general structure of the model $\mathcal{F}_1 \otimes \mathcal{F}_2$ is depicted in the picture Fig. 2.7. In particular we have $(a_1 \nVdash_V \alpha)$ and

$$\mathcal{F}_1 \otimes \mathcal{F}_2$$

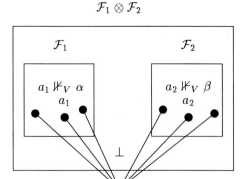

Figure 2.7:

$(a_2 \nVdash_V \beta)$. Since $\bot \leq a_1$, $\bot \leq a_2$, applying Lemma 2.4.5, we arrive at $\bot \nVdash_V \alpha$ and $\bot \nVdash_V \beta$, consequently $\bot \nVdash_V (\alpha \vee \beta)$. Because $\mathcal{F}_1 \otimes \mathcal{F}_2$ is a λ-frame we conclude $\alpha \vee \beta \notin \lambda$. ∎

From this theorem and Theorem 2.4.27 we immediately infer

Corollary 2.4.32 *Intuitionistic Heyting logic H has the disjunction property.*

Exercise 2.4.33 *Show that the logics KC and LC do not have the disjunction property.*

2.5   Stone's Theory and Kripke Semantics

The aim of this section is to develop semantic instruments for non-standard logical systems. The main topics are (i) the relationship and connections between algebraic and Kripke relational semantics, (ii) representation theorems for modal algebras, pseudo-boolean algebras, and Kripke models, (iii) the interaction between algebraic constructions on algebras and the corresponding operations over Kripke models and frames. We begin with a representation modal algebras by means of Kripke frames.

Definition 2.5.1 *Let* $\mathcal{F} = \langle F, R \rangle$ *be a frame. The algebra*

$$\mathcal{F}^+ := \langle 2^W, \wedge, \vee, \rightarrow, \neg, \Box, \bot, \top \rangle,$$

where

> *(i)* $\langle 2^W, \wedge, \vee, \rightarrow, \neg, \bot, \top \rangle$ *is the boolean algebra of all subsets of* W,
> *(ii)* $\forall X \subseteq W (\Box X := \{a \mid \forall y (\langle a, y \rangle \in R \Rightarrow (y \in X))\})$,

is called the associated (or wrapping) modal algebra for the frame $\mathcal{F}$.

Lemma 2.5.2 *For any frame* $\mathcal{F}$, *the algebra* $\mathcal{F}^+$ *is a modal algebra.*

Proof. It is clear that $\Box W = W$. Thus it only remains to check that the equality $\Box(X \rightarrow Y) \rightarrow (\Box X \rightarrow \Box Y) = W$ holds for arbitrary subsets X, Y of the set W. This equality is equivalent to the inclusion

$$\Box(X \rightarrow Y) \subseteq (\Box X \rightarrow \Box Y).$$

Suppose that $a \in \Box(\neg X \vee Y)$. This means

$$\forall y (aRy \Rightarrow (y \in Y) \vee (y \notin X). \tag{2.12}$$

If $a \in \Box X$ then $\forall y (aRy \Rightarrow (y \in X))$ and by (2.12) $\forall y (aRy \Rightarrow (y \in Y))$, that is $a \in \Box Y$. Thus $\Box(X \rightarrow Y) \subseteq (\Box X \rightarrow \Box Y)$ holds. ∎

Definition 2.5.3 *Let* $\mathfrak{M} := \langle W, R, V \rangle$ *be a Kripke model where the domain of the valuation* V *is a set of propositional letters* P. *We define the algebra* $\mathfrak{M}^+$ *to be the subalgebra of* $\langle W, V \rangle^+$ *generated by the set of elements* $\{V(p) \mid p \in P\}$. *The algebra* $\mathfrak{M}^+$ *is said to be the modal algebra associated with the Kripke model* $\mathfrak{M}$.

Note that this definition is correct since according to Lemma 2.5.2 the algebra $\mathfrak{M}^+$ is a modal algebra. Now we turn to the description of arbitrary modal algebras by means of relational Kripke semantics. This approach is based on the Stone representation of any boolean algebra as a subalgebra of the boolean algebra of all subsets of a special set.

Definition 2.5.4 *A filter* Δ *of any boolean algebra* $\mathcal{B}$ *is ultrafilter iff* Δ *is non-trivial* $(\bot \notin \Delta)$ *and* $\forall a \in |\mathcal{B}|$, *either* $a \in \Delta$ *or* $\neg a \in \Delta$.

Definition 2.5.5 *Let $\mathfrak{B}$ be a modal algebra. The Stone representation for $\mathfrak{B}$ is the Kripke model $\mathfrak{B}^+ := \langle W_{\mathfrak{B}}, R, V \rangle$, where*

> *(i) $W_{\mathfrak{B}}$ is set of all ultrafilters on $\mathfrak{B}$,*
> *(ii) $\forall \Delta_1, \Delta_2 \in W_{\mathfrak{B}}, \langle \Delta_1, \Delta_2 \rangle \in R \Leftrightarrow$*
> *$(\forall x \in |\mathfrak{B}|)(\Box x \in \Delta_1 \Rightarrow x \in \Delta_2)$,*
> *(iii) $Dom(V) := \{a \mid a \in |\mathfrak{B}|\}, \forall a \in |\mathfrak{B}|,$*
> *$V(a) := \{\Delta \mid \Delta \in W_{\mathfrak{B}}, a \in \Delta\}.$*

We will below denote the frame of any Kripke model $\mathfrak{M}$ by $\mathfrak{F}(\mathfrak{M})$. The following theorem, as well as the theorem of this section that we will call the *Stone representation theorem for temporal algebras*, are consequences of Theorems 3.10, 2.12 and 3.6 of Jonsson and Tarski [71]. These theorems describe general representations of boolean algebras with operators. However we will use a direct proof of Stone representation theorems here in order to demonstrate the work of modal and temporal operations on Stone representations, the feeling of which will be essential in our further constructions.

Theorem 2.5.6 (Stone Representation Theorem). *For every modal algebra $\mathfrak{B}$, $\mathfrak{B}$ is isomorphic to the modal algebra $\mathfrak{B}^{++}$. That is $\mathfrak{B}$ is a subalgebra of $\mathfrak{F}(\mathfrak{B}^+)^+$. In particular, if $\mathfrak{B}$ is finite then $\mathfrak{B} \cong \mathfrak{F}(\mathfrak{B}^+)^+$, and for any finite frame $\mathcal{F}$, $\mathcal{F}^{++} \cong \mathcal{F}$.*

Proof. We prove that the mapping $i : \mathfrak{B} \mapsto \mathfrak{B}^{++}$, where

$$\forall a \in |\mathfrak{B}|, \ i(a) := \{\Delta \mid \Delta \in W_{\mathfrak{B}}, a \in \Delta\},$$

is the required isomorphism. First we show that i is one-to-one mapping. Suppose $a, b \in \mathfrak{B}$ and $a \neq b$. This assumption yields $a \wedge \neg b \neq \bot$ or $b \wedge \neg a \neq \bot$. Indeed, otherwise

$$(a \wedge \neg b) \vee b = \bot \vee b = (a \vee b) \wedge (\neg b \vee b) = a \vee b = b \Rightarrow a \leq b,$$
$$(b \wedge \neg a) \vee a = \bot \vee a = (b \vee a) \wedge (\neg a \vee a) = b \vee a = a \Rightarrow b \leq a, a = b.$$

Let $a \wedge \neg b = c \neq \bot$. We will denote the set $\{x \mid x \in |\mathfrak{B}|, c \leq x\}$ by Δ_c. It is easy to see directly from our definition that Δ_c is a filter. This filter is non-trivial, $\bot \notin \Delta_c$. Furthermore,

Lemma 2.5.7 *Any non-trivial filter Δ of any boolean algebra $\mathcal{B}$ is contained in some ultrafilter Δ^* on $\mathcal{B}$.*

Proof. In order to prove this lemma it is sufficient to apply Zorn's Lemma. Indeed, the set $\mathcal{F}_{\mathcal{B}}$ of all non-trivial filters of $\mathcal{B}$ is partially ordered by the set-theoretic inclusion $\subseteq$. Suppose $\Delta_j, j \in J$ is a chain of filters from $\mathcal{F}_{\mathcal{B}}$. Then the set $\bigcup_{j \in J} \Delta_j$ is a non-trivial filter, which can be verified directly. Besides it is clear that $\bigcup_{j \in J} \Delta_j$ is an upper bound for all filters from $\Delta_j, j \in J$.

Applying Zorn's Lemma we find that Δ is contained in some maximal filter Δ^* of the poset $\mathcal{F}_\mathcal{B}$. It remains only to show that Δ^* is an ultrafilter. Suppose otherwise, then there is an element b that $b \notin \Delta^*$ and $\neg b \notin \Delta^*$. Clearly the following set

$$D := \{h \mid h \in |\mathcal{B}|, (\exists d \in \Delta^*)(b \wedge d \le h\}$$

is a filter on $\mathcal{B}$. Moreover it is non-trivial filter. Indeed, otherwise we have $b \wedge d = \bot$ for some $d \in \Delta^*$. Then

$$(b \wedge d) \vee \neg b = \neg b = (b \vee \neg b) \wedge (d \vee \neg b) = d \vee \neg b, \; d \le \neg b,$$

and $\neg b \in \Delta^*$, a contradiction. So D is a non-trivial filter, and by definition $\Delta^* \subseteq D$, moreover $b \in D$ and $b \notin \Delta^*$. Thus $\Delta^* \subset D$ which contradicts the maximality of Δ^*. Hence Δ^* is an ultrafilter. ∎

According to this lemma Δ_c is enclosed in an ultrafilter Δ on $\mathfrak{B}$. Then $c \in \Delta$ and $c \le a$, hence $a \in \Delta$ and $\Delta \in i(a)$. Suppose $b \in \Delta$. Then $c \wedge b \in \Delta$ and $c \wedge b = a \wedge \neg b \wedge b = \bot$ that is, Δ is a trivial filter, a contradiction. Thus it follows that $b \notin \Delta$ and $\Delta \notin i(b)$ and consequently $i(a) \ne i(b)$, that is the mapping i is one-to-one. Now we show that i is a homomorphism. We pick out some elements a, b from $|\mathfrak{B}|$. The operation $\vee$:

$$i(a \vee b) := \{\Delta \mid \Delta \in W_\mathfrak{B}, a \vee b \in \Delta\},$$
$$i(a) \vee i(b) := \{\Delta \mid \Delta \in W_\mathfrak{B}, (a \in \Delta) \vee (b \in \Delta)\}.$$

If $a \vee b \in \Delta$ then $(a \in \Delta) \vee (b \in \Delta)$. Indeed, otherwise $\neg a \in \Delta$ and $\neg b \in \Delta$ which implies $\neg a \wedge \neg b = \neg(a \vee b) \in \Delta$, a contradiction. Conversely, if $a \in \Delta$ $(b \in \Delta)$ then evidently $a \vee b \in \Delta$ since $a \le a \vee b$ $(b \le a \vee b,$ correspondingly$)$. Thus $i(a \vee b) = i(a) \vee i(b)$. As to the operation $\wedge$,

$$i(a \wedge b) := \{\Delta \mid \Delta \in W_\mathfrak{B}, a \wedge b \in \Delta\} =$$
$$i(a) \wedge i(b) := \{\Delta \mid \Delta \in W_\mathfrak{B}, (a \in \Delta)\&(b \in \Delta)\}.$$

Indeed, if $a \wedge b \in \Delta$ then $a \in \Delta$ and $b \in \Delta$ since $a \wedge b \le a$ and $a \wedge b \le b$. Conversely, if $a \in \Delta$ and $b \in \Delta$ then $a \wedge b \in \Delta$. Hence $i(a \wedge b) = i(a) \wedge i(b)$. We consider now the operation $\neg$:

$$i(\neg a) := \{\Delta \mid \Delta \in W_\mathfrak{B}, \neg a \in \Delta\},$$
$$\neg i(a) := \{\Delta \mid \Delta \in W_\mathfrak{B}, a \notin \Delta\}.$$

If $\neg a \in \Delta$ then $a \notin \Delta$, because otherwise Δ would be a trivial filter. If $a \notin \Delta$ then by the definition of ultrafilter we get $\neg a \in \Delta$. So $i(\neg a) = \neg i(a)$. As to operation $\rightarrow$, we know that it has in boolean algebras the following representation: $x \rightarrow y = \neg x \vee y$. Therefore from commutativity of i with $\vee$ and $\neg$ we have i is consistent with $\rightarrow$. Clearly $i(\top) = \top$, and $i(\bot) = \bot$. It remains only to check that i is permutable with $\square$. By our definitions we have directly

$$i(\square a) := \{\Delta \mid \Delta \in W_\mathfrak{B}, \square a \in \Delta\},$$

$$\square i(a) := \{\Delta \mid \Delta \in W_\mathfrak{B}, \forall \Delta_1 \in W_\mathfrak{B}(\langle \Delta, \Delta_1 \rangle \in R \Rightarrow \Delta_1 \in i(a))\}. \qquad (2.13)$$

If $\Delta \in i(\square a)$ then $\square a \in \Delta$. Suppose $\langle \Delta, \Delta_1 \rangle \in R$. Then by the definition of R we conclude $a \in \Delta_1$ that is $\Delta_1 \in i(a)$. Thus $i(\square a) \subseteq \square i(a)$. For the converse, suppose that $\Delta \notin i(\square a)$, that is $\square a \notin \Delta$ and $\neg \square a \in \Delta$. Let $\square[\Delta] := \{b \mid \square b \in \Delta\}$. We will show that the set $U := \{c \mid (\exists b \in \square[\Delta])(\neg a \wedge b \le c\}$ is non-trivial filter on $\mathfrak{B}$. That it is a filter follows directly from the definition. Suppose that this filter is trivial, that is $\neg a \wedge b = \bot$ for some $b \in \square[\Delta]$. Then

$$(\neg a \wedge b) \vee a = \bot \vee a = a = (\neg a \vee a) \wedge (b \vee a) = b \vee a.$$

Therefore $b \le a$ and $b \to a = \top$. Hence $\square(b \to a) = \top$. Since $\mathfrak{B}$ is a modal algebra, $\square(b \to a) \le (\square b \to \square a)$. Consequently $(\square b \to \square a) = \top$ and $\square b \le \square a$. But $b \in \square[\Delta]$ which means $\square b \in \Delta$, therefore we conclude $\square a \in \Delta$ which contradicts our assumption. Hence U is non-trivial filter on $\mathfrak{B}$ and according to Lemma 2.5.7 there is an ultrafilter U^* on $\mathfrak{B}$ containing U. Then $U^* \in W_\mathfrak{B}$ and $\langle \Delta, U^* \rangle \in R$. Indeed, if $\square b \in \Delta$ then $b \in \square[\Delta]$ and $b \in U \subseteq U^*$. At the same time $a \notin U^*$. Otherwise $a, \neg a \in U^*$, a contradiction. Therefore by (2.13) we get $\Delta \notin \square i(a)$. Thus $i(\square a) = \square i(a)$ and i is a homomorphism, and as the mapping i is one-to-one, i is an isomorphism into. We show now that i is a mapping onto. Note that any generating element $V(a)$, $a \in |\mathfrak{B}|$ of algebra $\mathfrak{B}^{++}$ is the image of the element a with respect to the mapping i. This, together with the fact that i is a homomorphism, entail that i maps $\mathfrak{B}$ onto $\mathfrak{B}^{++}$. Thus i is an isomorphism onto.

Consider now the case when $\mathfrak{B}$ is finite. In this case the set $W_\mathfrak{B}$ is also finite and has, say, only the elements $\Delta_1, ..., \Delta_n$. Since $\mathfrak{B}$ is finite, for any Δ_i, the intersection of all members of Δ_i gives the smallest element $g(\Delta_i)$ of Δ_i which is not equal to $\bot$. Note that $i(g(\Delta_i)) = \{\Delta_i\}$. Indeed suppose $\Delta_j \in i(g(\Delta_i))$, that is $g(\Delta_i) \in \Delta_j$ and $\Delta_i \ne \Delta_j$. Then $g(\Delta_j) < g(\Delta_i)$. Simultaneously $\neg g(\Delta_j) \notin \Delta_i$. Otherwise $\neg g(\Delta_j)$ and $g(\Delta_j) \in \Delta_j$ which contradicts the fact that Δ_j is an ultrafilter. Thus $\neg g(\Delta_j) \notin \Delta_i$ and $g(\Delta_j) \notin \Delta_i$ which again contradicts the fact that Δ_i is an ultrafilter. The single-element sets of kind $\{\Delta_i\}$ generate the modal algebra $\mathfrak{F}(\mathfrak{M}^+)^+$ and any element of $\mathfrak{F}(\mathfrak{M}^+)^+$ is a union of some finite collection of such elements. Using this observation, the fact that i is a homomorphism and $i(g(\Delta_i)) = \{\Delta_i\}$, it follows that i is an isomorphism from $\mathfrak{B}$ onto $\mathfrak{F}(\mathfrak{M}^+)^+$.

Let $\mathcal{F}$ be a finite frame. Then $\mathcal{F}^+$ is a finite modal algebra, which follows from Lemma 2.5.2, and by reasoning that similar above, any ultrafilter Δ on $\mathcal{F}^+$ has a smallest element $g(\Delta)$ which is a subset of $|\mathcal{F}|$. Since Δ is an ultrafilter, $g(\Delta)$ must be a single-element subset, i.e. $g(\Delta) = \{g\}$, where $g \in |\mathcal{F}|$. So the mapping

h of $\mathcal{F}^{++}$ into $\mathcal{F}$, where $h(\Delta) := g$, is an one-to-one embedding of $\mathcal{F}^{++}$ into the frame $\mathcal{F}$. Clearly that this embedding is a mapping onto. It remains to show that h preserves the accessibility relation R on $\mathcal{F}$. Suppose $\{g_1\}$ and $\{g_2\}$ are smallest elements of ultrafilters Δ_1, Δ_2, respectively and $\neg(g_1 R g_2)$. Then $\square\neg\{g_2\} \in \Delta_1$ but $\{g_2\} \in \Delta_2$. Thus $\langle \Delta_1, \Delta_2 \rangle \notin R^*$, where R^* is the accessibility relation on $\mathcal{F}^{++}$. Thus $\mathcal{F}^{++} \cong \mathcal{F}$. ∎

Now we will study the interaction of usual algebraic constructions on modal algebras and the corresponding operations over frames and Kripke models.

Definition 2.5.8 *Given frames* $\mathcal{F}_1 := \langle W_1, R_1 \rangle$, $\mathcal{F}_2 := \langle W_2, R_2 \rangle$, *the frame* $\mathcal{F}_2$ *is said to be an* open subframe *(or open up subframe) of* $\mathcal{F}_2$ *if* $W_1 \subseteq W_2$, $R_2 \cap W_1^2 = R_1$ *and more* $\forall a \in W_1, \forall b \in W_2 (a R_2 b \Rightarrow b \in W_1)$.

Lemma 2.5.9 *If a frame* $\mathcal{F}_1 = \langle W_1, R_1 \rangle$ *is an open subframe of a frame* $\mathcal{F}_2 = \langle W_2, R_2 \rangle$ *then there is a homomorphism* φ *from* $\mathcal{F}_2^+$ *onto* $\mathcal{F}_1^+$.

Proof. We define φ as follows: $\forall X \subseteq W_2, \varphi(X) := X \cap W_1$. It is clear that φ is a mapping onto and φ preserves $\wedge, \vee$ and $\neg$:

$$\varphi(X \cap Y) := (X \cap Y) \cap W_1 = (X \cap W_1) \cap (Y \cap W_1) = \varphi(X) \cap \varphi(Y),$$
$$\varphi(X \cup Y) := (X \cup Y) \cap W_1 = (X \cap W_1) \cup (Y \cap W_1) = \varphi(X) \cup \varphi(Y).$$
$$\varphi(\neg X) = \neg X \cap W_1 = (W_2 - X) \cap W_1 = (W_1 - X) \cap W_1 = \neg\varphi(X).$$

Now we show that φ is permutable with $\square$. Note that by our definitions

$$\varphi(\square X) := \{a \mid a \in W_2, \forall y(a R_2 y \Rightarrow y \in X)\} \cap W_1,$$
$$\square\varphi(X) := \{b \mid b \in W_1, \forall y \in W_1(b R_1 y \Rightarrow y \in X \cap W_1)\}.$$

Suppose $c \in \varphi(\square X)$. Then $c \in W_1$. Suppose $y \in W_1$ and $c R_1 y$. Then $c R_2 y$ and since $c \in \square X$ we infer $y \in X$. Because $\mathcal{F}_1$ is an open subframe of $\mathcal{F}_2$ we also obtain $y \in W_1$. Thus $y \in X \cap W_1$. Hence $c \in \square\varphi(X)$. Conversely, let c be an element of $\square\varphi(X)$. Then, in particular, $c \in W_1$. Suppose $d \in W_2$ and $c R_2 d$. Since $\mathcal{F}_1$ is an open subframe of $\mathcal{F}_2$ we conclude $d \in W_1$ and it immediately follows $d \in X \cap W_1$. Hence $d \in X$ and $c \in \varphi(\square X)$. Thus we have showed $\varphi(\square X) = \square\varphi(X)$. ∎

Definition 2.5.10 *If* $\mathfrak{M}_1 := \langle W_1, R_1, V_1 \rangle$, $\mathfrak{M}_2 := \langle W_2, R_2, V_2 \rangle$ *are Kripke models then we call* $\mathfrak{M}_1$ *an* open submodel *of* $\mathfrak{M}_2$ *if the following holds:*

(i) $\langle W_1, R_1 \rangle$ *is an open subframe of* $\langle W_2, R_2 \rangle$,
(ii) $Dom(V_1) = Dom(V_2)$ *and* $\forall p \in Dom(V_1)(V_1(p) = V_2(p) \cap W_1)$.

Lemma 2.5.11 *If a Kripke model* $\mathfrak{M}_1 := \langle W_1, R_1, V_1 \rangle$ *is an open submodel of a Kripke model* $\mathfrak{M}_2 := \langle W_2, R_2, V_2 \rangle$ *then there is a homomorphism from* $\mathfrak{M}_2^+$ *onto* $\mathfrak{M}_1^+$.

Proof. It is clear that the restriction to $\mathfrak{M}_2^+$ of the homomorphism φ from the algebra $\langle W_2, R_2 \rangle^+$ onto the algebra $\langle W_1, R_1 \rangle^+$ from Lemma 2.5.9 will give the required homomorphism. Actually, φ assigns the generators of $\mathfrak{M}_1^+$ to the generators of $\mathfrak{M}_2^+$. ■

Exercise 2.5.12 *Show that*

> (*i*) *If $\mathfrak{M}_1$ is an open submodel of a model $\mathfrak{M}_2$ then for every formula α, $\mathfrak{M}_2 \Vdash \alpha$ implies $\mathfrak{M}_1 \Vdash \alpha$.*

> (*ii*) *If a frame $\mathcal{F}_1$ is an open subframe of a frame $\mathcal{F}_2$ then $\lambda(\mathcal{F}_2) \subseteq \lambda(\mathcal{F}_1)$.*

Hint: (i) can be shown by a simple induction on the length of α, (ii) follows from (i).

Definition 2.5.13 *Let f be a mapping of a frame $\mathcal{F}_1 := \langle W_1, R_1 \rangle$ into a frame $\mathcal{F}_2 := \langle W_2, R_2 \rangle$. The mapping f is called a* p-morphism *if*

> (*i*) $(\forall a, b \in W_1)[aR_1 b \Rightarrow f(a) R_2 f(b)]$,
> (*ii*) $\forall a, b \in W_1[f(a) R_2 f(b) \Rightarrow (\exists c \in W_1)[aR_1 c \& f(c) = f(b)]]$.

Definition 2.5.14 *Let f be a mapping of the frame $\langle W_1, R_1 \rangle$ of a Kripke model $\mathfrak{M}_1 := \langle W_1, R_1, V_1 \rangle$ into the frame $\langle W_2, R_2 \rangle$ of a Kripke model $\mathfrak{M}_2 := \langle W_2, R_2, V_2 \rangle$. We say f is a* p-morphism *of the model $\mathfrak{M}_1$ into the model $\mathfrak{M}_2$ if (i) f is a p-morphism of the frame $\langle W_1, R_1 \rangle$ into the frame $\langle W_2, R_2 \rangle$, (ii) the valuations V_1, V_2 are defined on the same set of propositional variables and*

> (*iii*) $\forall p \in Dom(V_1), \forall a \in W_1[a \Vdash_{V_1} p \Leftrightarrow f(a) \Vdash_{V_2} p]$.

The primary property of p-morphisms consists of the fact that they preserve the truth of formulas:

Lemma 2.5.15 *If f is a p-morphism of a Kripke model $\mathfrak{M}_1 := \langle W_1, R_1, V_1 \rangle$ onto a Kripke model $\mathfrak{M}_2 := \langle W_2, R_2, V_2 \rangle$ then for any formula α which is built up out of letters from the domain of the valuation V_1,*

$$\forall a \in W_1 (a \Vdash_{V_1} \alpha \Leftrightarrow f(a) \Vdash_{V_2} \alpha).$$

Proof. We prove the assertion of this lemma by induction on the length of the formulas α. For $\alpha = p$, the claim follows directly from (iii) of the definition of p-morphism of Kripke models. The inductive steps for logical connectives $\wedge$, $\vee$, $\rightarrow$, $\neg$ are evident. Now let $\beta = \Box \delta$. Let $a \Vdash_{V_1} \Box \delta$. Suppose that $c \in W_2$ and $f(a) R_2 c$. Since f is a mapping onto, there exists $b \in W_1$ such that $f(b) = c$. From $f(a) R_2 f(b)$ and (ii) of the definition for p-morphism, we get there is some

$d \in W_1$ such that aR_1d and $f(d) = f(b) = c$. By $a \Vdash_{V_1} \Box\delta$ we have $d \Vdash_{V_1} \delta$, and according to the inductive assumption we arrive at $f(d) \Vdash_{V_2} \delta$. Thus $c \Vdash_{V_2} \delta$. Hence $f(a) \Vdash_{V_2} \Box\delta$. Conversely, let $f(a) \Vdash_{V_2} \Box\delta$. Suppose that aR_1b. Then by (i) from the definition of p-morphisms, $f(a)R_2f(b)$, and in view of $f(a) \Vdash_{V_2} \Box\delta$ we obtain $f(b) \Vdash_{V_2} \delta$. Applying the inductive hypothesis we derive $b \Vdash_{V_1} \delta$. Thus we have showed $a \Vdash_{V_1} \Box\delta$. ∎

Corollary 2.5.16 *If $\mathcal{F}_1$ and $\mathcal{F}_2$ are frames and there is a p-morphism f from $\mathcal{F}_1$ onto $\mathcal{F}_2$ then $\lambda(\mathcal{F}_1) \subseteq \lambda(\lambda_2)$.*

Proof. Let α be a formula from $\lambda(\mathcal{F}_1)$. Let V_2 be a valuation of letters of α in $\mathcal{F}_2$. We introduce the valuation V_1 of the same letters in $\mathcal{F}_1$ by $V_1(p) := f^{-1}(V_2(p))$. Then f is a p-morphism from the model $\langle\mathcal{F}_1, V_1\rangle$ onto the model $\langle\mathcal{F}_2, V_2\rangle$. Applying Lemma 2.5.15 it follows $\langle\mathcal{F}_2, V_2\rangle \Vdash \alpha$. Hence $\alpha \in \lambda(\mathcal{F}_2)$. ∎

We also need the following simple observation. Let $\mathfrak{M} := \langle W, R, V\rangle$ be a Kripke model. For any formula α whose variables are in the domain of V we put $V(\alpha) := \{a \mid a \in W, a \Vdash_V \alpha\}$. If $\beta(p_1, ..., p_n)$ is a formula which is built up out of letters $p_1, ..., p_n$, where all $p_1, ..., p_n$ are included in $Dom(V)$, then $\beta(V(p_1), ..., V(p_n))$ denotes the set of elements from W which is the value of the term $\beta(p_1, ..., p_n)$ in $\mathfrak{M}^+$ on elements $V(p_1), ..., V(p_n)$. In what follows below the following lemma will often be used in proofs.

Lemma 2.5.17 *For any formula $\alpha(p_1, ..., p_n)$ with propositional letters $p_1, ...$, p_n contained in the domain of V,*

$$V(\alpha(p_1, ..., p_n)) = \alpha(V(p_1), ..., V(p_n)).$$

Proof. It is easy to prove this lemma by induction on the length of the formula α. The basis of induction follows directly from our definitions. The inductive steps for non-modal logical connectives also follow easily from the definitions. Consider the step for $\Box$. Let $a \in V(\Box\beta(p_i))$. Then for all b such that aRb, $b \Vdash_V \beta(p_i)$ that is $b \in V(\beta(p_i))$ holds. By the inductive hypothesis this yields $b \in \beta(V(p_i))$. According to the definition of $\Box$ in $\mathfrak{M}^+$ this entails $a \in \Box\beta(V(p_i))$. Conversely, if $a \in \Box\beta(V(p_i))$ then for all b such that aRb, $b \in \beta(V(p_i))$ holds. Furthermore in view of inductive hypothesis this brings $b \in V(\beta(p_i))$. Thus we have $b \Vdash_V \beta(p_i)$ and, since this holds for all b such that aRb, we have $a \Vdash_V \Box\beta(p_i)$ and $a \in V(\Box\beta(p_i))$. ∎

In spirit of this lemma we can define the modal logic of a given Kripke model $\mathfrak{M} := \langle W, R, V\rangle$ as follows.

$$\lambda(\mathfrak{M}) := \{\alpha(p_1, ..., p_n) \mid \forall S \text{ where } (S(p_i) \in |\mathfrak{M}^+|)$$
$$(\langle W, R, S\rangle \Vdash \alpha(p_1, ..., p_n))\}.$$

Lemma 2.5.18 *$\lambda(\mathfrak{M})$ is a normal modal logic.*

Proof. Indeed, $\lambda(\mathfrak{M})$ contains K and is evidently closed with respect to modus ponens and necessitating rule. That $\lambda(\mathfrak{M})$ is closed with respect to substitution follows immediately from Lemma 2.5.17. ∎

Lemma 2.5.19 *Let f be a p-morphism of a Kripke model $\mathfrak{M}_1 := \langle W_1, R_1, V_1 \rangle$ onto a Kripke model $\mathfrak{M}_2 := \langle W_2, R_2, V_2 \rangle$. Then there is an isomorphism of the modal algebra $\mathfrak{M}_2^+$ into the algebra $\mathfrak{M}_1^+$.*

Proof. We define the mapping g as follows:

$$\forall X \in \mathfrak{M}_2^+, \; g(X) := f^{-1}(X).$$

Note that this mapping is well-defined that is $g(X) \in \mathfrak{M}_1^+$. Indeed, any X from $\mathfrak{M}^+$ has following representation: $X := \alpha(V_2(p_1), ..., V_2(p_n))$. By Lemma 2.5.17 then $X = V_2(\alpha(p_1, ..., p_n))$ and using Lemma 2.5.15 we conclude

$$f(V_1(\alpha(p_1, ..., p_n))) = V_2(\alpha(p_1, ..., p_n)).$$

Therefore $f^{-1}(X) = V_1(\alpha(p_1, ..., p_n))$. By Lemma 2.5.17 we derive that

$$V_1(\alpha(p_1, ..., p_n)) = \alpha(V_1(p_1), ..., V_1(p_n)).$$

Thus we conclude that $f^{-1}(X) = \alpha(V_1(p_1), ..., V_1(p_n)) \in \mathfrak{M}^+$, and g maps $\mathfrak{M}_2$ in $\mathfrak{M}_1$.

To show g is one-to-one it is sufficient to note that if $X \neq Y$, $X, Y \in \mathfrak{M}_2^+$ then $f^{-1}(X) \neq f^{-1}(Y)$ since f is a mapping onto. It remains to prove that g is a homomorphism. Equalities

$$g(X \cap Y) = f^{-1}(X \cap Y) = f^{-1}(X) \cap f^{-1}(Y) = g(X) \cap g(Y),$$
$$g(X \cup Y) = f^{-1}(X \cup Y) = f^{-1}(X) \cup f^{-1}(Y) = g(X) \cup g(Y),$$
$$g(\neg X) = f^{-1}(W_2 - X) = W_1 - f^{-1}(X) = \neg g(X)$$

hold since f is a mapping onto. Regarding $\square$, we have

$$g(\square X) := f^{-1}(\{a \mid a \in W_2, \forall b \in W_2(aR_2b \Rightarrow b \in X)\}),$$
$$\square g(X) := \{c \mid c \in W_1, \forall d \in W_1(cR_1d \Rightarrow f(d) \in X)\}.$$

Suppose $a \in g(\square X)$. This implies $\forall b \in W_2(f(a)R_2b \Rightarrow b \in X)$. Also assume $d \in W_1$ and aR_1d. Since f is a p-morphism, it follows $f(a)R_2f(d)$. Consequently $f(d) \in X$ and $a \in \square g(X)$. Conversely, let $a \in \square g(X)$. Suppose $b \in W_2$ and $f(a)R_2b$. Since f is a mapping onto, there exists some $c \in W_1$ such that $f(c) = b$. Because f is a p-morphism there is an element $d \in W_1$ which have properties: $f(d) = f(c) = b$ and aR_1d. Since $a \in \square g(X)$ we derive $d \in g(X)$ and besides $f(d) = b$. Thus we have showed $\forall b \in W_2(f(a)R_2b \Rightarrow b \in g(X))$. Consequently $a \in g(\square)$ and hence $g(\square X) = \square g(X)$. ∎

Lemma 2.5.20 *Let f be a p-morphism of a frame $\mathcal{F}_1 := \langle W_1, R_1 \rangle$ onto a frame $\mathcal{F}_2 := \langle W_2, R_2 \rangle$. Then there is an isomorphism of the modal algebra $\mathcal{F}_2^+$ into the modal algebra $\mathcal{F}_1^+$.*

Proof. It is easily seen that $\mathcal{F}_2^+$ coincides with $\mathfrak{M}_2^+$, where $\mathfrak{M}_2$ is the Kripke model obtained from $\mathcal{F}_2$ by enriching it with the valuation V_2 of the set of letters $P := \{ p_i \mid i \subseteq W_2 \}$, where $V_2(p_i) := i$. We transfer the valuation V_2 in $\mathcal{F}_1$ by introducing the valuation V_1, where $V_1(p_i) := f^{-1}(V_2(p_i))$ and obtain the model $\mathfrak{M}_1$. It is easily seen that f is a p-morphism of the model $\mathfrak{M}_1$ onto the model $\mathfrak{M}_2$. According to Lemma 2.5.19 there is an isomorphic embedding of $\mathfrak{M}_2^+$ into $\mathfrak{M}_1^+$. Besides $\mathfrak{M}_1^+$ is a subalgebra of $\mathcal{F}_1^+$, hence $\mathcal{F}_2^+$ is isomorphically embeddable in $\mathcal{F}_1^+$. ∎

Now we turn to the description of homomorphic images and subalgebras of modal algebras in term of operations on Stone representations of these algebras.

Lemma 2.5.21 *Suppose there is a homomorphism g of a modal algebra $\mathfrak{B}_1$ onto a modal algebra $\mathfrak{B}_2$. Then the Stone representation $\mathfrak{B}_2^+$ is isomorphically embeddable as an open submodel in the Kripke model $\mathfrak{B}_1^{+g}$ which is based on the frame of $\mathfrak{B}_1^+$ and has the valuation V_3, where $\forall a \in |\mathfrak{B}_2|$,*

$$V_3(a) := \{ g^{-1}(\Delta) \mid a \in \Delta, \Delta \in |\mathfrak{B}_2^+| \}.$$

Proof. Note that the valuation V_3 is well-defined, i.e. any $g_1(\Delta)$ is an ultrafilter in $\mathfrak{B}_1$. This property also is necessary for the next definition and will be proved below. We define the mapping h of $\mathfrak{B}_2^+$ into $\mathfrak{B}_1^+$ as follows:

$$\forall \Delta \in \mathfrak{B}_2^+, h(\Delta) := g^{-1}(\Delta).$$

First we need to check that h is well-defined, i.e. that $g^{-1}(\Delta)$ is an ultrafilter on $\mathfrak{B}_1$. To check it is a filter we suppose $a, b \in g^{-1}(\Delta)$ and $c \le c$. Then $g(a), g(b) \in \Delta$ and $g(a) \wedge g(b) \in \Delta$. Since g is a homomorphism we get $g(a) \wedge g(b) = g(a \wedge b)$ and $a \wedge b \in g^{-1}(\Delta)$. Furthermore, $g(a) \le g(c)$ and consequently $g(c) \in \Delta$ which implies $c \in g^{-1}(\Delta)$. So $g^{-1}(\Delta)$ is a filter. This filter is non-trivial. Indeed, if $\perp \in g^{-1}(\Delta)$ then $g(\perp) \in \Delta$ and still we know $g(\perp) = \perp$. Thus on our assumption we discover that Δ is trivial, a contradiction. It remains to show that the filter $g^{-1}(\Delta)$ is an ultrafilter. Let a be an element of $\mathfrak{B}_1$. Then $g(a) \in \mathfrak{B}_2$ and either $g(a) \in \Delta$ or $\neg g(a) = g(\neg a) \in \Delta$. Therefore either $a \in g^{-1}(\Delta)$ or $\neg a \in g^{-1}(\Delta)$. Hence $g^{-1}(\Delta)$ is an ultrafilter and h is well-defined.

To show h is one-to-one suppose $\Delta_1, \Delta_2 \in \mathfrak{B}^+$ and $\Delta_1 \ne \Delta_2$. Then also $g^{-1}(\Delta) \ne g^{-1}(\Delta)$ because g is a mapping onto. So $h(\Delta_1) \ne h(\Delta_2)$. Now we check that h preserves the accessibility relation R_2. We will show that for all $\Delta_1, \Delta_2 \in \mathfrak{B}_2$

$$\langle \Delta_1, \Delta_2 \rangle \in R_2 \Leftrightarrow \langle h(\Delta_1), h(\Delta_2) \rangle \in R_1. \tag{2.14}$$

If $\langle \Delta_1, \Delta_2 \rangle \in R_2$ and $\Box a \in h(\Delta_1)$ then $g(\Box a) = \Box g(a) \in \Delta_1$ and $g(a) \in \Delta_2$. Therefore $a \in h(\Delta_2)$. Thus $\langle h(\Delta_1), h(\Delta_2) \rangle \in R_1$. Conversely now let $\langle h(\Delta_1), h(\Delta_2) \rangle \in R_1$. Let $a \in |\mathfrak{B}_2|$ and $\Box a \in \Delta_1$. Since g is a mapping onto there is an element b from $\mathfrak{B}_1$ such that $g(b) = a$. Then $g(\Box b) = \Box g(b) = \Box a$ and $\Box b \in h(\Delta_1)$. Therefore $b \in h(\Delta_2)$ and $g(b) = a \in \Delta_2$. Thus $\langle \Delta_1, \Delta_2 \rangle \in R_2$ and (2.14) holds. Hence we have showed h preserves the accessibility relation, and h is an isomorphic embedding of the frame of the model $\mathfrak{M}_2^+$ into the frame of the model $\mathfrak{B}_2^+$. Moreover $\forall \Delta \in \mathfrak{B}_2^+, \forall a \in \mathfrak{B}_2 \ \Delta \in V_2(a) \Leftrightarrow h(\Delta) \in V_3(a)$ by definition of V_3. Hence h is an isomorphic embedding of the model $\mathfrak{B}_2^+$ in the model which is based on the frame of $\mathfrak{B}_1^+$ and has the valuation V_3.

Now it remains to show that $h(\mathfrak{B}_2^+)$ is an open submodel. Let $h(\Delta) \in |\mathfrak{B}_1^+|$. Suppose that $\langle h(\Delta), \Delta_1 \rangle \in R, \Delta_1 \in \mathfrak{B}_1^+$. Then $g^{-1}(g(\Delta_1)) = \Delta_1$. Indeed, clearly $\Delta_1 \subseteq g^{-1}(g(\Delta_1))$. Conversely, let a be an element of $g^{-1}(g(\Delta_1))$. Then $g(a) = g(b)$ for some $b \in \Delta_1$. This yields $g(b) \to g(a) = \top$ and $g(b \to a) = \top$. Consequently $\Box g(b \to a) = \top$ and $g(\Box(b \to a)) = \top$. Thus $\Box(b \to a) \in h(\Delta)$ which implies $(b \to a) \in \Delta_1$. Since $b \in \Delta_1$ we get $a \in \Delta_1$. So we have $g^{-1}(g(\Delta_1)) = \Delta_1$ holds. We now need to show $g(\Delta_1)$ is an ultrafilter on $\mathfrak{B}_2$. Clearly $g(\Delta_1)$ is a filter since g is a mapping onto. $\bot \notin g(\Delta_1)$ holds because otherwise $\bot \in g^{-1}g(\Delta_1) = \Delta_1$, a contradiction. Let $a \in |\mathfrak{B}_2^+|$. Then, as g is a mapping onto, $a = \tilde{g}(b)$ for some $b \in |\mathfrak{B}_1|$. Since Δ_1 is an ultrafilter, either $b \in \Delta_1$, or $\neg b \in \Delta_1$. Therefore we have the following either $g(b) = a \in g(\Delta_1)$ or $\neg g(b) = \neg a \in g(\Delta_1)$. Thus we proved $g(\Delta_1)$ is an ultrafilter and $h(g(\Delta_1)) = \Delta_1$. ∎

Corollary 2.5.22 *If there is a homomorphism g of a certain modal algebra $\mathfrak{B}_1$ onto a modal algebra $\mathfrak{B}_2$ then $\mathfrak{F}(\mathfrak{B}_2^+)$ is an open subframe of the frame $\mathfrak{F}(\mathfrak{B}_1^+)$.*

Lemma 2.5.23 *If a modal algebra $\mathfrak{B}_1$ is a subalgebra of a modal algebra $\mathfrak{B}_2$ then there is a p-morphism from the model $\mathfrak{B}_2^{+,'}$ onto the model $\mathfrak{B}_1^+$, where $\mathfrak{B}_2^{+,'}$ is the model obtained from $\mathfrak{B}_2^+$ by restriction of its own valuation to the set of letters from the domain of the valuation of the model $\mathfrak{B}_1^+$.*

Proof. We will prove that the mapping f, where

$$\forall \Delta \in \mathfrak{B}_2^+, \ f(\Delta) = \Delta \cap |\mathfrak{B}_1|,$$

is the required p-morphism. It is clear that f is well-defined, i.e., $f(\Delta)$ is an ultrafilter on $\mathfrak{B}_1$. We begin with verifying of properties (i), (ii) from the definition of p-morphisms.

(i): Suppose $\Delta_1, \Delta_2 \in |\mathfrak{B}_2^+|, \langle \Delta_1, \Delta_2 \rangle \in R$. Let $a \in |\mathfrak{B}_1|$ and $\Box a \in \Delta_1 \cap |\mathfrak{B}_1|$. Then $a \in \Delta_2$ and $a \in \Delta_2 \cap |\mathfrak{B}_1|$. Thus we have showed that the following holds $\langle f(\Delta_1), f(\Delta_2) \rangle \in R$.

(ii): Now suppose that $\Delta_1, \Delta_2 \in |\mathfrak{B}_2^+|$ and $\langle f(\Delta_1), f(\Delta_2) \rangle \in R$, that is $\langle \Delta_1 \cap |\mathfrak{B}_1|, \Delta_2 \cap |\mathfrak{B}_1| \rangle \in R$. This means

$$\forall a \in |\mathfrak{B}_2|, (\Box a \in \Delta_1 \cap |\mathfrak{B}_1|) \Rightarrow a \in \Delta_2 \cap |\mathfrak{B}_1|. \tag{2.15}$$

We must show that there exists an ultrafilter Δ_3 on the algebra $\mathfrak{B}_2$ such that $f(\Delta_3) = f(\Delta_2)$. In order to show this we introduce the family Σ of filters on $\mathfrak{B}_2$ and a filter Δ_0 as follows:

$$\circ(\Delta_2) := (\mathfrak{B}_1 - \Delta_2), \ \Box\Delta_1 := \{a \mid a \in |\mathfrak{B}_2|, \Box a \in \Delta_1\},$$
$$\Sigma := \{\Phi \mid \Phi \text{ is non-trivial filter on } \mathfrak{B}_2, , \Phi \cap \circ(\Delta_2) = \emptyset,$$
$$\Box\Delta_1 \cup (\Delta_2 \cap |\mathfrak{B}_1|) \subseteq \Phi\},$$
$$\Delta_0 := \{x \mid x \in |\mathfrak{B}_2|, \exists y \in \Box\Delta_2, \exists z \in \Delta_2 \cap |\mathfrak{B}_1|(y \wedge z \leq x)\}.$$

We will show that Δ_0 is a filter and $\Delta_0 \cap \circ(\Delta_2) = \emptyset$. That Δ_0 is a filter follows directly from the definition and the fact that for any elements r, q of any modal algebra, $\Box(r \wedge g) = \Box r \wedge \Box q$ (cf. Lemma 2.2.3). Suppose there is an element $x \in \Delta_0 \cap \circ(\Delta_2)$. Then there exist $y \in \Box\Delta_1$ and $z \in \Delta_2 \cap |\mathfrak{B}_1|$ such that $y \wedge z \leq x$. Using this we merely calculate:

$$y \wedge (\neg z \vee x) = (y \wedge \neg z) \vee (x \wedge y) \geq (y \wedge \neg z) \vee (y \wedge z) =$$
$$y \wedge (\neg z \vee z) = y.$$

From this we infer $y \leq (z \to x)$. This yields $\Box y \leq \Box(z \to x)$. But we have $\Box y \in \Delta_1$ consequently $\Box(z \to x) \in \Delta_1$.

Furthermore, $x \in \circ\Delta_0$ implies $x \in |\mathfrak{B}_1|$; also $z \in |\mathfrak{B}_1|$, therefore we conclude $\Box(z \to x) \in |\mathfrak{B}_1|$. Hence we obtained $\Box(z \to x) \in \Delta_1 \cap |\mathfrak{B}_1|$. Applying (2.15) we arrive at $(z \to x) \in \Delta_2 \cap |\mathfrak{B}_1|$. But $z \in \Delta_2 \cap |\mathfrak{B}_1|$ therefore the latter inclusion yields $x \in \Delta_2 \cap |\mathfrak{B}_1|$ which contradicts $x \in \circ(\Delta_2) := (\mathfrak{B}_1 - \Delta_2)$. Thus we have proved $\Delta_0 \cap \circ(\Delta_2) = \emptyset$. In particular, $\bot \notin \Delta_0$, which means the filter Δ_0 is non-trivial. Just by our definitions we have $\Box\Delta_1 \subseteq \Delta_0$ and $\Delta_2 \cap |\mathfrak{B}_1| \subseteq \Delta_0$. Hence $\Delta_0 \in \Sigma$, i.e. Σ is nonempty.

Now we show Σ includes an ultrafilter applying Zorn's Lemma. It is clear that the set-theoretic union of an arbitrary chain of filters from Σ is again some filter which belongs to Σ. Therefore by Zorn's Lemma there is a maximal filter Δ among the filters from Σ. We will show Δ is an ultrafilter. Suppose that for some a, neither $a \in \Delta$ nor $\neg a \in \Delta$. From this it follows that the set $\Theta := \{x \mid \exists z \in \Delta(a \wedge z \leq x)\}$ is a non-trivial filter. Moreover $\Delta \subset \Theta$ since $a \in \Theta$. Suppose that $\Theta \cup \circ(\Delta_2) \neq \emptyset$ and $b \in \Theta \cup \circ(\Delta_2)$. Then $b \in (\mathfrak{B}_1 - \Delta_2)$ and consequently $\neg b \in \Delta_2$ and $\neg b \in |\mathfrak{B}_1|$. This gives $\neg b \in \Delta_2 \cap |\mathfrak{B}_1|$ and therefore it follows that $\neg b \in \Theta$ which contradicts the claim that Θ is non-trivial filter. Thus Δ is an ultrafilter.

Using the definition of Σ we infer $\Delta \cap |\mathfrak{B}_1| = \Delta_2 \cap |\mathfrak{B}_1|$. Indeed, directly by definition of Σ it follows $\Delta_2 \cap |\mathfrak{B}_1| \subseteq \Delta \cap |\mathfrak{B}_1|$. Let $a \in \Delta \cap |\mathfrak{B}_1|$. Then $a \in |\mathfrak{B}_1|$ and since $(\Delta \cap |\mathfrak{B}_1| - \Delta_2) = \emptyset$, we conclude $a \in |\Delta_2|$, consequently $a \in \Delta_2 \cap |\mathfrak{B}_1|$.

Thus $\Delta \cap |\mathfrak{B}_1| = \Delta_2 \cap |\mathfrak{B}_1|$ and (ii) from the definition of p-morphisms holds. Thus f is a p-morphism of frames.

Now we turn to show that f is a p-morphism onto. Let Δ be an ultrafilter on $\mathfrak{B}_1$. The set $\Omega := \{x \mid x \in |\mathfrak{B}_2|, \exists a \in \Delta(a \le x)\}$ is a non-trivial filter on $\mathfrak{B}_2$. Therefore the set Σ consisting of all filters on $\mathfrak{B}_2$, which (a) include Δ and (b) have no elements from $(|\mathfrak{B}_1| - \Delta)$, is nonempty and contains Ω. The set-theoretic union of any chain of filters from Σ is again a filter from Σ. Hence, by Zorn Lemma, Σ has a maximal filter Δ_0. This filter is an ultrafilter. Indeed, assume for some a, neither $a \in \Delta_0$ nor $\neg \alpha \in \Delta_0$. Then the set $\Psi := \{x \mid \exists z \in \Delta_0(a \wedge z \le x)\}$ is a non-trivial filter. Indeed, otherwise $\bot = a \wedge z$ for some $z \in \Delta_0$. Then $a \le \neg z$ and $z \le \neg a$, that is $\neg a \in \Delta_0$, a contradiction. Furthermore, $\Psi \wedge (|\mathfrak{B}_1| - \Delta) = \emptyset$ holds. Really suppose $c \in \Psi$ and $c \in (|\mathfrak{B}_1| - \Delta)$. Then $\neg c \in \Delta$ and $\neg c \in \Delta_0 \subseteq \Psi$, contradicting the fact that Ψ is a non-trivial filter. Thus we showed $\Psi \in \Sigma$ and $\Delta_0 \subset \Psi$ which contradicts the fact that Δ_0 is maximal in Σ. Hence Δ_0 is an ultrafilter. Besides $\Delta \subseteq \Delta_0, (\Delta_0 \cap |\mathfrak{B}_1| - \Delta) = \emptyset$. This yields $f(\Delta_0) = \Delta_0 \cap |\mathfrak{B}_1| = \Delta)$. Thus f is a mapping onto. Finally it is easy to show that f preserves the valuation of models. Let a be an element of $\mathfrak{B}_1$. Then for any $\Delta \in \mathfrak{B}_2^+, \Delta \Vdash_{V_2} a \Leftrightarrow a \in \Delta \Leftrightarrow a \in \Delta \cap |\mathfrak{B}_1| \Leftrightarrow \Delta \cap |\mathfrak{B}_1| \Vdash_{V_1} a$.
∎

Corollary 2.5.24 *Let $\mathfrak{B}_1$ be a subalgebra of a modal algebra $\mathfrak{B}_2$. Then there exists a p-morphism from the frame $\mathfrak{F}(\mathfrak{B}_2^+)$ onto the frame $\mathfrak{F}(\mathfrak{B}_1^+)$.*

We now consider one another construction employing Kripke models and its interaction with operations on the Stone representations of modal algebras.

Definition 2.5.25 Disjoint Union.

(i) Let $\mathcal{F}_i := \langle W_i, R_i \rangle, i \in I$ be a family of pairwise disjoint frames, i.e. $W_i \cap W_j = \emptyset$, for $i \ne j \in I$. The disjoint union of this family is the frame $\bigsqcup_{i \in I} \mathcal{F}_i := \langle W, R \rangle$, where

$$W := \bigcup_{i \in I} W_i, \quad R := \bigcup_{i \in I} W_i.$$

Frames $\mathcal{F}_i$ are components of the frame $\bigsqcup_{i \in I} \mathcal{F}_i$.

(ii) If $\mathfrak{M}_i := \langle W_i, R_i, V_i \rangle, i \in I$ is a family of pairwise disjoint Kripke models which have the same domains of valuations then the disjoint union of models $\mathfrak{M}_i$ is the model $\bigsqcup_{i \in I} \mathfrak{M}_i := \langle W, R, V \rangle$, where $\langle W, R \rangle$ is the disjoint union of the frames of these models and for any propositional letter p valued under V, $V(p) := \bigcup_{i \in I} V_i(p)$. $\mathfrak{M}_i$ are components of $\bigsqcup_{i \in I} \mathfrak{M}_i$.

Lemma 2.5.26

(i) If $\bigsqcup_{i \in I} \mathfrak{M}_i$ is the disjoint union of Kripke models then for any modal formula α with propositional letters from the domain of the valuation of this model, $\bigsqcup_{i \in I} \mathfrak{M}_i \Vdash \alpha$ iff $\forall i \in I(\mathfrak{M}_i \Vdash \alpha)$.

(ii) If $\bigsqcup_{i \in I} \mathcal{F}_i$ is the disjoint union of frames then for any modal formula α, $\bigsqcup_{i \in I} \mathcal{F}_i \Vdash \alpha$ iff $\forall i \in I(\mathcal{F}_i \Vdash \alpha)$.

Proof is left as an easy exercise to the reader; (i) follows directly from our definitions and (ii) follows immediately from (i). ∎

Lemma 2.5.27 *Let $\mathcal{F}_i := \langle W_i, R_i \rangle, i \in I$ be a family of pairwise disjoint frames. Then $(\bigsqcup_{i \in I} \mathcal{F}_i)^+$ is isomorphic to $\prod_{i \in I} \mathcal{F}_i^+$*

Proof. For any $X \subseteq |\bigcup_i \mathcal{F}_i|$, we denote by $\pi_i(X)$ the set of all elements from $|\mathcal{F}_i|$ that are in X. The suggested isomorphism from $(\bigsqcup_{i \in I} \mathcal{F}_i)^+$ onto $\prod_{i \in I} \mathcal{F}_i^+$ is the mapping h:

$$\forall X \subseteq |\bigcup_i \mathcal{F}_i|, \; h(X) := (\pi_i(X) \mid i \in I).$$

It is easily seen that this mapping is an isomorphism onto. We leave to verify this to the reader as an exercise. ∎

Exercise 2.5.28 *Show that if $\mathfrak{M}_i := \langle W_i, R_i, V_i \rangle, i \in I$ is a family of pairwise disjoint Kripke models which have the same domains of valuations then $(\bigsqcup_{i \in I} \mathfrak{M}_i)^+$ is not necessarily isomorphic to $\prod_{i \in I} \mathfrak{M}_i^+$ even if I and all $\mathfrak{M}_i$ are finite.*

Hence the translation of disjoint unions of Kripke models (not frames as before) into direct products of modal algebras is not always possible. The related (in a sense converse) transition from the direct products of modal algebras to the disjoint unions of Kripke frames is also only sometimes valid. Namely

Lemma 2.5.29 *If $\prod_{i \in I} \mathfrak{B}_i$ is a product of finite modal algebras then $\prod_{i \in I} \mathfrak{B}_i \cong (\bigsqcup_{i \in I} \mathfrak{B}_i^+)^+$.*

Proof. Since any $\mathfrak{B}_i$ is finite, according to Theorem 2.5.6 $\mathfrak{B}_i \cong (\mathfrak{F}(\mathfrak{B}_i^+))^+$. Thus any $\mathfrak{B}_i$ is isomorphic to $\mathcal{F}_i^+$ for some finite frame $\mathcal{F}_i$. Applying Theorem 2.5.6 $\mathcal{F}_i \cong \mathcal{F}_i^{++}$ and the fact that $\mathcal{F}_i^+ \cong \mathfrak{B}_i$ we have $\mathcal{F}_i \cong \mathfrak{B}_i^+$. Furthermore, we have $\prod_{i \in I} \mathfrak{B}_i \cong \prod_{i \in I} \mathcal{F}_i^+$. Applying Lemma 2.5.27 it follows $\prod_{i \in I} \mathcal{F}_i^+ \cong. (\bigsqcup_{i \in I} \mathcal{F}_i)^+$. Thus $\prod_{i \in I} \mathfrak{B}_i \cong (\bigsqcup_{i \in I} \mathfrak{B}_i^+)^+$. ∎

Now we will transfer Stone representation on temporal algebras. Since any temporal algebra is a modal algebra with respect to the operator H and also with respect to the operator G, the representation here will be similar to that for the modal case and we will use facts we already know about the representation of modal algebras.

Definition 2.5.30 *Let $\mathcal{F} = \langle F, R \rangle$ be a frame. The algebra*

$$\mathcal{F}_t^+ := \langle 2^W, \wedge, \vee, \neg, G, H, \bot, \top \rangle,$$

where

(i) $\langle 2^W, \wedge, \vee, \rightarrow, \neg, \bot, \top \rangle$ *is the boolean algebra of all subsets for W,*
(ii) $\forall X \subseteq W(GX := \{a \mid \forall y(\langle a, y \rangle \in R \Rightarrow (y \in X))\}$,
(iii) $\forall X \subseteq W(HX := \{a \mid \forall y(\langle y, a \rangle \in R \Rightarrow (y \in X))\}$,

is an associated (or wrapping) temporal algebra for the frame $\mathcal{F}$.

Lemma 2.5.31 *For any frame $\mathcal{F}$, the algebra $\mathcal{F}_t^+$ is a temporal algebra.*

Proof. According to Lemma 2.5.2 it is sufficient to show the equalities $x \rightarrow GPx = \top$ and $x \rightarrow HFx = \top$ hold on $\mathcal{F}_t^+$. And this follows directly from our definitions. ∎

Definition 2.5.32 *Suppose $\mathfrak{M} := \langle W, R, V \rangle$ is a Kripke model, and the domain of the valuation V is a set of propositional letters P. We define the algebra $\mathfrak{M}_t^+$ to be the subalgebra of $\langle W, V \rangle_t^+$ generated by the set of elements $\{V(p) \mid p \in P\}$. We call $\mathfrak{M}_t^+$* the temporal algebra associated with the Kripke model $\mathfrak{M}$.

Definition 2.5.33 *Let $\mathcal{A}$ be a temporal algebra. The Stone representation for $\mathcal{A}$ is the temporal Kripke model $\mathcal{A}^+ := \langle W_{\mathcal{A}}, R, V \rangle$, where*

(i) $W_{\mathcal{A}}$ *is set of all ultrafilters on $\mathcal{A}$,*
(ii) $\forall \Delta_1, \Delta_2 \in W_{\mathcal{A}}, \langle \Delta_1, \Delta_2 \rangle \in R \Leftrightarrow$
$(\forall x \in |\mathcal{A}|)(Gx \in \Delta_1 \Rightarrow x \in \Delta_2)$,
(iii) $Dom(V) := \{a \mid a \in |\mathcal{A}|\}, \forall a \in |\mathcal{A}|$,
$V(a) := \{\Delta \mid \Delta \in W_{\mathcal{A}}, a \in \Delta\}$.

Theorem 2.5.34 Stone Representation Theorem for Temporal Algebras. *Any temporal algebra $\mathcal{A}$ is isomorphic to a temporal algebra $(\mathcal{A}^+)_t^+$, in particular $\mathcal{A}$ is a subalgebra of $\mathfrak{F}(\mathcal{A}^+)_t^+$. If $\mathcal{A}$ is finite then $\mathcal{A} \cong \mathfrak{F}(\mathcal{A}^+)_t^+$ and for any finite frame $\mathcal{F}$, $\mathcal{F}_t^{++} \cong \mathcal{F}$.*

Proof. By Theorem 2.5.6 the mapping i, where $\forall a \in |\mathcal{A}| \; i(a) := \{\Delta \mid \Delta \in W_{\mathcal{A}}, a \in \Delta\}$, is the isomorphism of the modal algebra $\mathcal{A}$ with $\square$ interpreted as G into the modal algebra $(\mathcal{A}^+)_t^+$ with the operator $\square$ given by the same interpretation. We will show that i is an isomorphism of temporal algebras. We need to show i preserves H. i is permutable with $\square$. Moreover, we have directly by our definitions

$$i(Ha) := \{\Delta \mid \Delta \in W_{\mathcal{A}}, Ha \in \Delta\},$$

$$Hi(a) := \{\Delta \mid \Delta \in W_{\mathcal{A}}, \forall \Delta_1 \in W_{\mathcal{A}}(\langle \Delta_1, \Delta \rangle \in R \Rightarrow \Delta_1 \in i(a))\}. \quad (2.16)$$

Suppose $\Delta \in i(Ha)$, then $Ha \in \Delta$. Let $\langle \Delta_1, \Delta \rangle \in R$. Since $\neg Ha \notin \Delta$, by the definition of R on $\mathcal{A}^+$ we get $G \neg Ha \notin \Delta_1$. $\mathcal{A}$ is a temporal algebra, therefore $\neg a \leq G \neg H \neg \neg a$. This implies $\neg a \notin \Delta_1$ and $a \in \Delta_1$. Thus bearing in mind (2.16) we infer $i(Ha) \subseteq Hi(a)$. Conversely, let $\Delta \notin i(Ha)$, that is $Ha \notin \Delta$ and $\neg Ha \in \Delta$. We set $H[\Delta] := \{b \mid Hb \in \Delta\}$. We need to show that the set $\Phi := \{c \mid (\exists b \in H[\Delta])(\neg a \wedge b \leq c\}$ is a non-trivial filter on $\mathcal{A}$. Clearly, it is filter. Suppose that $\neg a \wedge b = \bot$ for some $b \in H[\Delta]$. Then

$$(\neg a \wedge b) \vee a = \bot \vee a = a = (\neg a \vee a) \wedge (b \vee a) = b \vee a.$$

Hence $b \leq a$ and, since $\mathcal{A}$ is a temporal algebra, we have $Hb \leq Ha$. This and $Hb \in \Delta$ imply $Ha \in \Delta$ which contradicts our assumption. Hence Φ is a non-trivial filter on $\mathcal{A}$. Then according to Lemma 2.5.7 there is an ultrafilter U on $\mathcal{A}$ containing Φ, in particular $U \in W_{\mathcal{A}}$. We need to show $\langle U, \Delta \rangle \in R$. Indeed, let $Gb \in U$. Suppose $b \notin \Delta$. Then $\neg b \in \Delta$ and, since $\mathcal{A}$ is a temporal algebra and $\neg b \leq HF \neg b$, we conclude $H \neg G \neg \neg b \in \Delta$. Thus $H \neg Gb \in \Delta$ and $\neg Gb \in H[\Delta]$. Therefore $\neg Gb \in \Phi \subseteq U$ which contradicts $Gb \in U$. Thus $\langle U, \Delta \rangle \in R$ and using (2.16) we arrive at $\Delta \notin Hi(a)$. Hence we have showed $i(Ha) = Hi(a)$, and i is an isomorphism of temporal algebras. If $\mathcal{A}$ is finite then by Theorem 2.5.6 $\mathcal{A}$ being the modal algebra with respect to G is isomorphic to the modal algebra $\mathfrak{F}(\mathcal{A}^+)^+$ (with respect to G again), and i is the isomorphism onto. We already shown i preserves H also, so $\mathcal{A}$ will be isomorphic to $\mathfrak{F}(\mathcal{A})^+)_t^+$. If $\mathcal{F}$ is a finite frame then (cf. Theorem 2.5.6) $\mathcal{F}^{++}$ is isomorphic to $\mathcal{F}$. The frames $\mathcal{F}^{++}$, $(\mathcal{F}_t^+)^+$ have the same carrier and the accessibility relation R. Hence $(\mathcal{F}_t^+)^+ \cong \mathcal{F}$. ∎

Definition 2.5.35 *A frame* $\mathcal{F}_1 := \langle W_1, R_1 \rangle$, *is said to be a* T-open subframe *of some frame* $\mathcal{F}_2 := \langle W_2, R_2 \rangle$ *if* $W_1 \subseteq W_2$, $R_2 \cap W_1^2 = R_1$ *and*

(i) $\forall a \in W_1, \forall b \in W_2(aR_2b \Rightarrow b \in W_1)$;
(ii)$\forall a \in W_1, \forall b \in W_2(bR_2a \Rightarrow b \in W_1)$;

Definition 2.5.36 *A Kripke model* $\mathfrak{M}_1 := \langle W_1, R_1, V_1 \rangle$ *is a* T-open sub-model *of a model* $\mathfrak{M}_2 := \langle W_2, R_2, V_2 \rangle$ *if (i)* $\langle W_1, R_1 \rangle$ *is a T-open subframe of* $\langle W_2, R_2 \rangle$ *and (ii)* $Dom(V_1) = Dom(V_2)$ *and* $\forall p \in Dom(V_1)[V_1(p) = V_2(p) \cap W_1]$.

Exercise 2.5.37 *(i) Let* $\mathfrak{M}_1$ *be a T-open submodel of a model* $\mathfrak{M}_2$. *Show that, for any temporal formula* α, $\mathfrak{M}_2 \Vdash \alpha$ *implies* $\mathfrak{M}_1 \Vdash \alpha$; *(ii) Let a frame* $\mathcal{F}_1$ *be a T-open subframe of a frame* $\mathcal{F}_2$. *Show that the temporal logic* $\lambda(\mathcal{F}_1)$ *extends the temporal logic* $\lambda(\mathcal{F}_2)$.

Lemma 2.5.38 *If a frame* $\mathcal{F}_1 = \langle W_1, R_1 \rangle$ *is a T-open subframe of a frame* $\mathcal{F}_2 = \langle W_2, R_2 \rangle$ *then there is a homomorphism* φ *from the temporal algebra* $\mathcal{F}_{2,t}^+$ *onto the algebra* $\mathcal{F}_{1,t}^+$.

Proof. It is easy to see that $\mathcal{F}_1$ is an open subframe of $\mathcal{F}_2$ with respect to the direct and the converse accessibility relations. Therefore by Lemma 2.5.9 the mapping $\varphi : \forall X \subseteq W_2, \varphi(X) := X \cap W_1$ is a boolean homomorphism from $\mathcal{F}_{2,t}^{+}$ onto $\mathcal{F}_{1,t}^{+}$ which preserves G and H simultaneously. Therefore φ is temporal homomorphism onto. ∎

Lemma 2.5.39 *Suppose a Kripke model $\mathfrak{M}_1 := \langle W_1, R_1, V_1 \rangle$ is a T-open submodel of a model $\mathfrak{M}_1 := \langle W_1, R_1, V_1 \rangle$. Then there is a homomorphism from $\mathfrak{M}_{2,t}^{+}$ onto $\mathfrak{M}_{1_t}^{+}$.*

Proof. It is not hard to see that the restriction to $\mathfrak{M}_{2,t}^{+}$ of the homomorphism φ from the algebra $\langle W_2, R_2 \rangle_t^{+}$ onto the algebra $\langle W_1, R_1 \rangle_t^{+}$ from Lemma 2.5.38 will give us the required homomorphism onto. Indeed, φ maps the generators of $\mathfrak{M}_{2,t}^{+}$ onto the generators of $\mathfrak{M}_{2,t}^{+}$. ∎

Lemma 2.5.40 *Suppose there is a homomorphism g of a temporal algebra $\mathcal{A}_1$ onto a temporal algebra $\mathcal{A}_2$. Then the Stone representation $\mathcal{A}_2^{+}$ is isomorphically embeddable as a T-open submodel in the Kripke model $\mathcal{A}_1^{+g}$ which is based on the frame of $\mathcal{A}_1^{+}$ and has the valuation V_3, where $\forall a \in |\mathcal{A}_2|$, $V_3(a) := \{ g^{-1}(\Delta) \mid a \in \Delta, \Delta \in |\mathcal{A}_2^{+}| \}$.*

Proof. According to Lemma 2.5.21 the Stone representation $\mathcal{A}_2^{+}$ is isomorphically embeddable as an open submodel in the model $\mathcal{A}_1^{+g}$. In the proof of that lemma we demonstrated that the isomorphism in question is the mapping h: $\forall \Delta \in \mathcal{A}_2^{+}, h(\Delta) := g^{-1}(\Delta)$. In order to prove this lemma it is sufficient to show that $\forall \Delta \in |\mathcal{A}_2^{+}|, \forall \Delta_1 \in |\mathcal{A}_1^{+}|, \langle \Delta_1, h(\Delta) \rangle \in R$ implies $\Delta_1 \in h(\mathcal{A}_2^{+})$. Indeed, suppose that $\langle \Delta_1, h(\Delta) \rangle \in R, \Delta_1 \in \mathcal{A}_1^{+}$. We will show that $g^{-1}(g(\Delta_1)) = \Delta_1$. Actually, $\Delta_1 \subseteq g^{-1}(g(\Delta_1))$ holds evidently. Conversely, suppose that a is an element of $g^{-1}(g(\Delta_1))$. Then $g(a) = g(b)$ for some $b \in \Delta_1$. Then $g(b) \to g(a) = \top, g(b \to a) = \top$ and $Hg(b \to a) = \top$, consequently $g(H(b \to a)) = \top$. Therefore $H(b \to a) \in h(\Delta)$. We will show that this implies $(b \to a) \in \Delta_1$. Suppose otherwise, $(b \to a) \notin \Delta_1$ and $\neg(b \to a) \in \Delta_1$. Then $G\neg H\neg\neg(b \to a) \in \Delta_1$. This yields $\neg H(b \to a) \in h(\Delta)$ which contradicts $H(b \to a) \in h(\Delta)$. Thus $(b \to a) \in \Delta_1$ and since $b \in \Delta_1$ we get $a \in \Delta_1$. Thus in fact $g^{-1}(g(\Delta_1)) = \Delta_1$ holds. Using the fact that g is a homomorphism onto it is not hard to show that $g(\Delta_1)$ is an ultrafilter. Hence $g(\Delta_1) \in |\mathcal{A}_2^{+}|$ and $h(g(\Delta_1)) = \Delta_1$. ∎

Corollary 2.5.41 *If there is a homomorphism g of a temporal algebra $\mathcal{A}_1$ onto a temporal algebra $\mathcal{A}_2$ then $\mathfrak{F}(\mathcal{A}_2^{+})$ is a T-open subframe of the frame $\mathfrak{F}(\mathcal{A}_1^{+})$.*

Definition 2.5.42 *A mapping* f *of a frame* $\mathcal{F}_1 := \langle W_1, R_1 \rangle$ *into a frame* $\mathcal{F}_2 := \langle W_2, R_2 \rangle$ *is a* temporal p-morphism *if*

(*i*) $(\forall a, b \in W_1)[a R_1 b \Rightarrow f(a) R_2 f(b)]$,
(*ii*) $\forall a, b \in W_1[f(a) R_2 f(b) \Rightarrow (\exists c \in W_1)[a R_1 c \& f(c) = f(b)]$,
(*iii*) $\forall a, b \in W_1[f(b) R_2 f(a) \Rightarrow (\exists c \in W_1)[c R_1 a \& f(c) = f(b)]$.

It is easy to see that any temporal p-morphism is a certain mapping which is a p-morphism of frames for direct and converse accessibility relations. As tin the modal case, a mapping f of the domain of a Kripke model $\mathfrak{M}_1 := \langle W_1, R_1, V_1 \rangle$ into the domain of a model $\mathfrak{M}_2 := \langle W_2, R_2, V_2 \rangle$ is called a *temporal p-morphism* if the following holds: (1) f is a temporal p-morphism of the frame $\langle W_1, R_1 \rangle$ into the frame $\langle W_2, R_2 \rangle$, (2) both valuations V_1, V_2 are defined on the same set of propositional variables and $\forall p \in Dom(V_1), \forall a \in W_1[a \Vdash_{V_1} p \Leftrightarrow f(a) \Vdash_{V_2} p]$. Temporal p-morphisms also preserve the truth of temporal formulas:

Lemma 2.5.43 *If* f *is a temporal p-morphism of a Kripke model* $\mathfrak{M}_1 := \langle W_1, R_1, V_1 \rangle$ *onto a Kripke model* $\mathfrak{M}_2 := \langle W_2, R_2, V_2 \rangle$ *then for any temporal formula* α *which is built up out of letters from the domain of the valuation* V_1,

$$\forall a \in W_1(a \Vdash_{V_1} \alpha \Leftrightarrow f(a) \Vdash_{V_2} \alpha).$$

Proof. The proof of this is quite similar to the proof of Lemma 2.5.15, except that the two connectives G, H have to be considered instead of the single connective $\square$. $\square$

Corollary 2.5.44 *If* $\mathcal{F}_1$, $\mathcal{F}_2$ *are frames and there is a temporal p-morphism* f *from* $\mathcal{F}_1$ *onto* $\mathcal{F}_2$ *then the temporal logic* $\lambda(\mathcal{F}_1)$ *is included in the temporal logic* $\lambda(\lambda_2)$.

The analog of Lemma 2.5.17 for modal formulas holds for temporal formulas as well. Actually, we simply transfer our notation on temporal case: let $\mathfrak{M} := \langle W, R, V \rangle$ be a Kripke model, and let, for any temporal formula α, $V(\alpha) := \{a \mid a \in W, a \Vdash_V \alpha\}$. If $\alpha(p_1, ..., p_n)$ is a formula which is built up of letters $p_1, ..., p_n$ and all $p_1, ..., p_n$ are included in $Dom(V)$ then $\alpha(V(p_1), ..., V(p_n))$ denotes the set of elements, i.e. the element of $\mathfrak{M}_t^+$, which is the value of the term $\beta(p_1, ..., p_n)$ on the elements $V(p_1), ..., V(p_n)$.

Lemma 2.5.45 *For any temporal formula* $\alpha(p_1, ..., p_n)$ *with a list of propositional letters* $p_1, ..., p_n$ *contained in the domain of* V, $V(\alpha(p_1, ..., p_n)) = \alpha(V(p_1), ..., V(p_n))$.

Proof is left as an exercise to the reader. The proof is parallel to the proof of Lemma 2.5.17.

Note that using this lemma we can define as in the case of modal logics the notion of temporal logic of any given Kripke model (we introduced above the notion of temporal logic only for frames and classes of frames but not Kripke models).

Lemma 2.5.46 *Let f be a temporal p-morphism of a certain Kripke model $\mathfrak{M}_1 := \langle W_1, R_1, V_1 \rangle$ onto a Kripke model $\mathfrak{M}_2 := \langle W_2, R_2, V_2 \rangle$. Then there is an isomorphism of the temporal algebra $\mathfrak{M}_{2,t}^+$ into the temporal algebra $\mathfrak{M}_{1,t}^+$.*

Proof. We refer to Lemma 2.5.19 and its proof: That the mapping g, where $(\forall X \in \mathfrak{M}_{2,t}^+)(g(X) := f^{-1}(X))$ is an isomorphism from the boolean algebra of $\mathfrak{M}_{1,t}^+$ into the boolean algebra of $\mathfrak{M}_{2,t}^+$ can be shown in a quite similar way to the proof of Lemma 2.5.19, merely using Lemma 2.5.45 here instead of Lemma 2.5.17. We show that g preserves G and H in the same way as it was shown in the proof of Lemma 2.5.19 that g preserves $\square$, using the fact that any temporal p-morphism is a p-morphism for direct and converse accessibility relations. ∎

Corollary 2.5.47 *Let f be a temporal p-morphism of a frame $\mathcal{F}_1 := \langle W_1, R_1 \rangle$ onto a frame $\mathcal{F}_2 := \langle W_2, R_2 \rangle$. Then there is an isomorphism of the temporal algebra $\mathcal{F}_{2,t}^+$ into the temporal algebra $\mathcal{F}_{1,t}^+$.*

Lemma 2.5.48 *Suppose a temporal algebra $\mathcal{A}_1$ is a subalgebra of a temporal algebra $\mathcal{A}_2$. Then there is a temporal p-morphism from the model $\mathcal{A}_2^{+,s}$ onto the model $\mathcal{A}_1^+$, where $\mathcal{A}_2^{+,s}$ is the model obtained from $\mathcal{A}_2^+$ by restriction of the valuation V_2 to the set of letters from the domain of the valuation of the model $\mathcal{A}_1^+$.*

Proof. We will use Lemma 2.5.23 and its proof. $\mathcal{A}_1$ is a modal algebra with respect to G, and is a subalgebra of the modal algebra $\mathcal{A}_2$ with respect to G. Therefore in the proof of Lemma 2.5.23 we showed that the mapping f, where $\forall \Delta \in |\mathcal{A}_{2,t}^+|$, $f(\Delta) = \Delta \cap |\mathcal{A}_1|$, is a p-morphism of the frame $\mathfrak{F}(\mathcal{A}_2^+)$ onto the frame $\mathfrak{F}(\mathcal{A}_1^+)$. We also showed that f preserves the valuation V_2: if a is an element of $|\mathcal{A}_1|$ then for any $\Delta \in |\mathcal{A}_2^+|$, $\Delta \Vdash_{V_2} a \Leftrightarrow \Delta \cap |\mathcal{A}_1| \Vdash_{V_1} a$. Hence in order to prove our lemma it is sufficient to show that if $\Delta_1, \Delta_2 \in |\mathcal{A}_{2,t}^+|$ and $\langle f(\Delta_2), f(\Delta_1) \rangle \in R$ then there is an ultrafilter Δ such that $\langle \Delta, \Delta_1 \rangle \in R$ and $f(\Delta) = f(\Delta_2)$.

Let $\langle f(\Delta_2), f(\Delta_1) \rangle \in R$ that is $\langle \Delta_2 \cap |\mathcal{A}_1|, \Delta_1 \cap |\mathcal{A}_1| \rangle \in R$. This implies that

$$\forall a \in |\mathcal{A}_2|, Ha \in \Delta_1 \cap |\mathcal{A}_1| \Rightarrow a \in \Delta_2 \cap |\mathcal{A}_1|. \tag{2.17}$$

Otherwise there is an a such that $Ha \in \Delta_1 \cap |\mathcal{A}_1|$ and $\neg a \in \Delta_2 \cap |\mathcal{A}_1|$. This yields $G \neg H \neg \neg a \in \Delta_2 \cap |\mathcal{A}_1|$ and $\neg Ha \in \Delta_1 \cap |\mathcal{A}_1|$ which contradicts the fact that $Ha \in \Delta_1 \cap |\mathcal{A}_1|$. Thus (2.17) holds. Now we continue the proof as in Lemma 2.5.23 after assertion (2.15) replacing by H everywhere $\square$ and using

(2.17) instead of (2.15). As result we have there is an ultrafilter Δ on $\mathcal{A}_2$ such that $f(\Delta) = f(\Delta_2)$ and $H[\Delta_1] \subseteq \Delta$. Consider some c such that $Gc \in \Delta$. Suppose $c \notin \Delta_1$, then $\neg c \in \Delta_1$, and consequently $H \neg G \neg \neg c \in \Delta_1$, $H \neg Gc \in \Delta_1$. Since $H[\Delta_1] \subseteq \Delta$ we get $\neg Gc \in \Delta$ which contradicts $Gc \in \Delta$. Hence $\langle \Delta, \Delta_1 \rangle \in R$. So f is a temporal p-morphism. ∎

Corollary 2.5.49 *If $\mathcal{A}_1$ is a subalgebra of a temporal algebra $\mathcal{A}_2$ then there exists a temporal p-morphism from the frame $\mathfrak{F}(\mathcal{A}_2^+)$ onto the frame $\mathfrak{F}(\mathcal{A}_1^+)$.*

Now we briefly present several results concerning disjoint unions of frames and Kripke models for temporal logic.

Lemma 2.5.50
   (i) If $\bigsqcup_{i \in I} \mathfrak{M}_i$ is a disjoint union of Kripke models then, for any temporal formula α such that $Var(\alpha) \subseteq Dom(V)$, where V is the valuation of $\bigsqcup_{i \in I} \mathfrak{M}_i$, $\bigsqcup_{i \in I} \mathfrak{M}_i \Vdash \alpha$ iff $\forall i \in I(\mathfrak{M}_i \Vdash \alpha)$.
   (ii) If $\bigsqcup_{i \in I} \mathcal{F}_i$ is a disjoint union of frames then for any temporal formula α, $\bigsqcup_{i \in I} \mathcal{F}_i \Vdash \alpha$ iff $\forall i \in I(\mathcal{F}_i \Vdash \alpha)$.

Proof. Simply, the details are left as an exercise for the reader.

Lemma 2.5.51 *Let $\mathcal{F}_i := \langle W_i, R_i \rangle, i \in I$ be a family of pairwise disjoint frames. Then the temporal algebra $(\bigsqcup_{i \in I} \mathcal{F}_i)_t^+$ is isomorphic to the temporal algebra $\prod_{i \in I} \mathcal{F}_{i,t}^+$.*

Proof. Similar to the proof of Lemma 2.5.27, we leave details as an exercise to the reader.

Lemma 2.5.52 *If $\prod_{i \in I} \mathcal{A}_i$ is a product of some finite temporal algebras then $\prod_{i \in I} \mathcal{A}_i \cong (\mathfrak{F}(\bigsqcup_{i \in I} \mathcal{A}_i^+))_t^+$.*

Proof. Any $\mathcal{A}_i$ is finite, therefore according to Theorem 2.5.34 we have $\mathcal{A}_i \cong \mathfrak{F}(\mathcal{A}_i^+)_t^+$. Thus any $\mathcal{A}_i$ is isomorphic to $\mathcal{F}_{i,t}^+$ for some finite frame $\mathcal{F}_i$. By Theorem 2.5.34 $\mathcal{F}_i \cong \mathcal{F}_{i,t}^{++}$ and because $\mathcal{F}_{i,t}^+ \cong \mathcal{A}_i$ it follows that $\mathcal{F}_i \cong \mathcal{A}_i^+$. Furthermore, $\prod_{i \in I} \mathcal{A}_i \cong \prod_{i \in I} \mathcal{F}_{i,t}^+$, and applying Lemma 2.5.51 we derive that $\prod_{i \in I} \mathcal{F}_{i,t}^+ \cong (\bigsqcup_{i \in I} \mathcal{F}_i)_t^+$. Thus $\prod_{i \in I} \mathcal{A}_i \cong (\bigsqcup_{i \in I} \mathcal{A}_i^+)_t^+$. ∎

   Now we will develop the Stone theory for pseudo-boolean algebras.

Definition 2.5.53 *Let $\mathcal{F} = \langle F, \leq \rangle$ be a partially ordered set. Then a subset X of W is said to open iff $\forall x \in X, \forall y \in F \ (x \leq y \Rightarrow y \in X)$.*

Exercise 2.5.54 *The set $O(F)$ of all open subsets of F is closed with respect to $\cap$ and $\cup$.*

Definition 2.5.55 *Let* $\mathcal{F} = \langle F, \leq \rangle$ *be a partially ordered set. Then the algebra*

$$\mathcal{F}_p^+ := \langle O(F), \wedge, \vee, \rightarrow, \neg, \bot \rangle,$$

where

    $(i) O(W)$ *is the set of all open subsets of* W;
    (ii) *the operations* $\wedge$ *and* $\vee$ *are* $\cap$ *and* $\cup$ *respectively*;
    $(iii) \forall X \subseteq W, Y \subseteq W, X \rightarrow Y := \{a \mid \forall x (a \leq x \Rightarrow$
      $(x \in Y) \vee (x \notin X))\}$;
    $(iv) \forall X \subseteq W\ (\neg X := \{a \mid \forall x (a \leq x \Rightarrow x \notin X)\})$;
    $(v) \bot = \emptyset$,

is the associated (or wrapping) pseudo-boolean algebra for the frame $\mathcal{F}$.

This definition correctly uses the term *pseudo-boolean algebra* since

Lemma 2.5.56 *For any poset* $\mathcal{F}$, *the algebra* $\mathcal{F}_p^+$ *is a pseudo-boolean algebra.*

Proof. Using Exercise 2.5.54, $\mathcal{F}_p^+$ is a distributive lattice. According to Lemma 2.1.13 it is sufficient to show that $\mathcal{F}_p^+$ has a pseudo-complement of an element X to an element Y and this pseudo-complement is the element $X \rightarrow Y$. The element $X \rightarrow Y$ is an open subset of F. Indeed, let $a \in X \rightarrow Y$ and $a \leq b$. Suppose $b \leq c$. Then $a \leq c$, hence either $c \in Y$ or $c \notin X$. Thus $b \in (X \rightarrow Y)$. Consider an element $a \in X \cap (X \rightarrow Y)$, then $a \leq a$ and the definition of $X \rightarrow Y$ together imply $a \in Y$. Thus $X \cap (X \rightarrow Y) \subseteq Y$. Let Z be an open subset of F and $Z \cap X \subseteq Y$. Let $a \in Z$ and $a \leq x$. Then $x \in Z$ since Z is an open subset. If $x \in X$ then $x \in Y$ holds, thus $X \rightarrow Y$ is the pseudo-complement of X to Y. ∎

Definition 2.5.57 *Let* $\mathfrak{M} := \langle W, \leq, V \rangle$ *be an intuitionistic Kripke model with the set* P *as the domain of the valuation* V. *The pseudo-boolean algebra* $\mathfrak{M}_p^+ := \langle W, \leq, V \rangle_p^+$ *which is the subalgebra of* $\langle W, \leq \rangle_p^+$ *generated by the set of elements* $\{V(p) \mid p \in P\}$ *is the pseudo-boolean algebra associated with the Kripke model* $\mathfrak{M}$.

The theory of Stone representations for pseudo-boolean algebras is based upon prime filters on pseudo-boolean algebras (which play a role related to that played by ultrafilters in the case of modal and temporal algebras).

Definition 2.5.58 *A filter* Δ *on a lattice* $\mathcal{L}$ *is said to be* prime *iff* Δ *is non-trivial* $(|\mathcal{L}| \neq \Delta)$ *and* $\forall a, b \in |\mathcal{L}|$, *if* $a \vee b \in \Delta$ *then either* $a \in \Delta$ *or* $b \in \Delta$.

Definition 2.5.59 *Let* $\mathfrak{A}$ *be a pseudo-boolean algebra. The Stone representation for* $\mathfrak{A}$ *is the Kripke intuitionistic model* $\mathfrak{A}^+ := \langle W_{\mathfrak{A}}, \leq, V \rangle$, *where*

    (i) $W_{\mathfrak{A}}$ *is the set of all prime filters on* $\mathfrak{A}$,
    (ii) $\forall \Delta_1, \Delta_2 \in W_{\mathfrak{A}},\ (\Delta_1 \leq \Delta_2 \Leftrightarrow \Delta_1 \subseteq \Delta_2)$,
    (iii) $Dom(V) := \{a \mid a \in |\mathfrak{A}|\}$,
    $\forall a \in |\mathfrak{B}|, V(a) := \{\Delta \mid \Delta \in W_{\mathfrak{A}}, a \in \Delta\}$.

It is clear that the valuation V defined above is an intuitionistic valuation, i.e. $\mathfrak{A}^+$ is an intuitionistic Kripke model.

Theorem 2.5.60 Stone Representation Theorem. *For every pseudo-boolean algebra $\mathfrak{A}$, $\mathfrak{A}$ is isomorphic to the pseudo-boolean algebra $(\mathfrak{A}^+)_p^+$. That is, $\mathfrak{A}$ is a subalgebra of $\mathfrak{F}(\mathfrak{A}^+)_p^+$. If $\mathfrak{A}$ is finite, then $\mathfrak{B} \cong \mathfrak{F}(\mathfrak{A}^+)_p^+$ and, for any finite poset $\mathcal{F}$, $\mathcal{F}_p^{++} \cong \mathcal{F}$ holds.*

Proof. The mapping i which will play the role of the required isomorphism is defined as follows:

$$\forall a \in |\mathfrak{A}|, \; i(a) := \{\Delta \mid \Delta \in W_{\mathfrak{A}}, a \in \Delta\},$$

First we show that i is a one-to-one mapping. Suppose $a, b \in \mathfrak{A}$ and $a \neq b$. From this it follows that $a \not\leq b$ or $b \not\leq a$. Suppose $a \not\leq b$ (other case is similar).

Lemma 2.5.61 *For any element a and any filter Δ of a distributive lattice $\mathcal{L}$, if $a \notin \Delta$ then there is a prime filter Δ^* on $\mathcal{L}$ such that $a \notin \Delta^*$ and $\Delta \subseteq \Delta^*$.*

Proof. We merely apply Zorn's Lemma. Let S be the set of all filters on $\mathcal{L}$ which include the filter Δ and do not include the element a. We consider S as the poset with set-theoretic inclusion $\subseteq$. The set S is non-empty since the filter Δ belongs to S. Any chain of filters from S has the upper bound in S, it is the union of all filters of this chain. Therefore by Zorn Lemma there is a maximal filter Δ^* in S. Then $a \notin \Delta^*$ and it remains only to show that Δ^* is a prime filter. Suppose otherwise, Δ^* is not prime. Then there are elements $c_1, c_2 \in \mathcal{L}$ such that $c_1 \vee c_2 \in \Delta^*$ but $c_1 \notin \Delta^*$ and $c_2 \notin \Delta^*$. Let

$$\Delta_i := \{x \mid \exists y \in \Delta^* (c_i \wedge y \leq x\}, \; i = 1, 2.$$

Clearly each Δ_i is a filter on $\mathcal{L}$. We claim that one of Δ_i does not include a. For otherwise there are elements $y_1, y_2 \in \Delta^*$ such that $c_1 \wedge y_1 \leq a$ and $c_2 \wedge y_2 \leq a$. Then for $y := y_1 \wedge y_2$, we get $y \wedge c_1 \leq a$, $c_1 \wedge y \leq a$ and $c_2 \wedge y \leq a$. Hence

$$(c_1 \wedge y) \vee (c_2 \wedge y) = (c_1 \vee c_2) \wedge y \leq a,$$

which implies $a \in \Delta^*$, a contradiction. Thus we obtain that $a \notin \Delta_i$ for some i. We fix this i and note that Δ_i belongs to S. Besides we have $c_i \in \Delta_i$ that is $\Delta^* \subset \Delta_i$ which contradicts the fact that Δ^* is a maximal filter of S. Thus Δ^* is prime. ∎

Consider the filter $\Delta := \{x \mid a \leq x\}$. The filter Δ does not include b. Therefore according to Lemma 2.5.61 there is a prime filter Δ^* on $\mathfrak{A}$ such that $\Delta \subseteq \Delta^*$ and $b \notin \Delta^*$. Thus $\Delta^* \in i(a)$ and $\Delta^* \notin i(b)$. Hence the mapping i is one-to-one. Now we check that i is a homomorphism. We take some elements a, b from $|\mathfrak{A}|$. First we consider the operation $\vee$, we have to show $i(a \vee b) = i(a) \cup i(b)$.

If $a \vee b \in \Delta$ then $(a \in \Delta) \vee (b \in \Delta)$ because Δ is prime. Consequently $\Delta \in i(a) \cup i(b)$ holds. Conversely, if $a \in \Delta$ (or $b \in \Delta$) then $a \vee b \in \Delta$ holds since $a \leq a \vee b$ (and $b \leq a \vee b$). Thus $i(a \vee b) = i(a) \vee i(b)$.

Now we establish the permutation of i with $\wedge$. If $\Delta \in i(a \wedge b)$, i.e. $a \wedge b \in \Delta$, then $a \in \Delta$ and $b \in \Delta$ since $a \wedge b \leq a$ and $a \wedge b \leq b$. Hence it follows $\Delta \in i(a) \cap i(b)$. Conversely, if $\Delta \in i(a) \cap i(b)$, i.e. $a \in \Delta$ and $b \in \Delta$, then $a \wedge b \in \Delta$. Thus $i(a \wedge b) = i(a) \wedge i(b)$.

Now we show that $i(a \to b) = i(a) \to i(b)$. Let $a \to b \in \Delta$. In order to prove $\Delta \in (i(a) \to i(b))$ we suppose $\Delta \subseteq \Delta_1$, where $\Delta_1 \in W_{\mathfrak{A}}$ and $\Delta_1 \in i(a)$. Then $a \to b \in \Delta_1$ and $a \in \Delta_1$ which imply $a \wedge (a \to b) \in \Delta_1$. But $a \wedge (a \to b) \leq b$, therefore we have $b \in \Delta_1$, that is $\Delta_1 \in i(b)$. Thus $\Delta \in i(a) \to i(b)$. Conversely, let $\Delta \in [i(a) \to i(b)]$ and $a \to b \notin \Delta$. Then, in particular, $a \not\leq b$. We claim that the filter $\Delta_1 := \{x \mid \exists y \in \Delta (a \wedge y \leq x\}$ does not contain b. Indeed otherwise we would have $a \wedge y \leq b$ for some $y \in \Delta$ and then $y \leq (a \to b)$, and consequently $(a \to b) \in \Delta$, a contradiction. Thus indeed Δ_1 does not contain b and contains a. Besides Δ_1 includes the filter Δ. Applying Lemma 2.5.61 we conclude there is a prime filter Δ_2 containing Δ_1 and such that $b \notin \Delta_2$, $a \in \Delta_2$. Thus we arrive at $\Delta \notin i(a) \to i(b)$ which contradicts our assumption. Thus it follows that $a \to b \in \Delta$. Hence we proved $i(a \to b) = i(a) \to i(b)$. The permutation of i with the operation $\neg$ follows from this observation and the presentation $\neg a = a \to \bot$. Thus we have proved that i is an isomorphic embedding. Any generating element $V(a)$ of the algebra $\mathfrak{A}_p^{++}$ is the image of the element $a \in |\mathfrak{A}|$ with respect to i. Thus i is an isomorphism of $\mathfrak{A}$ onto $\mathfrak{A}_p^{++}$.

Suppose now that $\mathfrak{A}$ is finite. In this event the set $W_{\mathfrak{A}}$ is also finite. Since $\mathfrak{A}$ is finite, for any $\Delta \in W_{\mathfrak{A}}$, the intersections of all members of Δ gives the smallest element $g(\Delta)$ of Δ which differs with $\bot$. Any open subset of $\mathfrak{F}(\mathfrak{A}^+)_p^+$ is the union of a finite number of open subsets of kind $\Delta^{\leq} := \{\Delta_1 \mid \Delta_1 \in W_{\mathfrak{A}}, \Delta \subseteq \Delta_1\}$. Note that $i(g(\Delta)) = \Delta^{\leq}$. Using these observations and that i is a homomorphism we get that i is an isomorphism from $\mathfrak{A}$ onto $\mathfrak{F}(\mathfrak{A}^+)_p^+$.

Suppose $\mathcal{F}$ is a finite poset. Then $\mathcal{F}_p^+$ is a finite pseudo-boolean algebra (cf. Lemma 2.5.56) and any prime filter Δ on $\mathcal{F}_p^+$ has a smallest element $e(\Delta)$ which is an open subset of $|\mathcal{F}|$. It is clear that $e(\Delta)$ must be a subset of kind $g(\Delta)^{\leq} := \{x \mid x \in |\mathcal{F}|, g(\Delta) \leq x\}$, where $g(\Delta) \in |\mathcal{F}|$ (otherwise Δ would be not a prime filter). And conversely, any $g(\Delta)^{\leq}$ generates a prime filter in $\mathcal{F}_p^+$. Therefore the mapping h of $\mathcal{F}_p^{++}$ into $\mathcal{F}$, where $h(\Delta) := g(\Delta)$ is an one-to-one mapping onto. It remains only to note that h preserves the partial order. Indeed if $\Delta_1 \subseteq \Delta_2$ then $e(\Delta_2) \subseteq e(\Delta_1)$ and $g(\Delta_1) \leq g(\Delta_1)$. So $\mathcal{F} \cong \mathcal{F}_p^{++}$. ∎

Now we describe the relations between algebraic constructions on pseudo-boolean algebras and the corresponding operations over intuitionistic Kripke frames.

Lemma 2.5.62 *If a poset $\mathcal{F}_1 = \langle W_1, \leq \rangle$ is an open subframe of some poset $\mathcal{F}_2 = \langle W_2, \leq \rangle$ then there is a homomorphism φ from the pseudo-boolean alge-*

bra $\mathcal{F}_{2p}^{+}$ onto the pseudo-boolean algebra $\mathcal{F}_{1p}^{+}$.

Proof. We define φ as follows: for any open subset X of the set W_2, $\varphi(X) :=$ $X \cap W_1$. Because W_1 is an open subset of W_2 the set $\varphi(X)$ is an open subset of W_1, so φ is well-defined. It is clear that φ is a mapping onto. It also is easy to see that φ preserves $\wedge$ and $\vee$:

$$\varphi(X \cap Y) := (X \cap Y) \cap W_1 = (X \cap W_1) \cap (Y \cap W_1) = \varphi(X) \cap \varphi(Y),$$
$$\varphi(X \cup Y) := (X \cup Y) \cap W_1 = (X \cap W_1) \cup (Y \cap W_1) = \varphi(X) \cup \varphi(Y).$$

Now we show that $\varphi(X \to Y) = \varphi(X) \to \varphi(Y)$. We suppose that $a \in \varphi(X \to Y) := (X \to Y) \cap W_1$. Assume also that $a \leq b$ and $b \in X \cap W_1 = \varphi(X)$. Then $b \in W_1$ since W_1 is an open subset of W_2, and because $a \in (X \to Y) \cap W_1$ we infer $b \in Y$. Thus $b \in Y \cap W_1 = \varphi(Y)$. Hence we proved $a \in \varphi(X) \to \varphi(Y)$. Conversely, suppose $a \in \varphi(X) \to \varphi(Y) = X \cap W_1 \to Y \cap W_1$. Assume that $a \leq b$ and $b \in X$. Then $a \in W_1$ and consequently $b \in W_1$ since W_1 forms an open subframe of $\mathcal{F}_2$. Thus we conclude $b \in X \cap W_1$ therefore $a \in X \cap W_1 \to Y \cap W_1$ implies $b \in Y \cap W_1$. So $b \in Y$ and we showed $a \in (X \to Y) \cap W_1 = \varphi(X \to Y)$. That φ preserves $\neg$ follows from $\varphi(X \to Y) = \varphi(X) \to \varphi(Y)$ and the representation $\neg X = X \to \emptyset$. ∎

Lemma 2.5.63 *If an intuitionistic Kripke model $\mathfrak{M}_1 := \langle W_1, \leq, V_1 \rangle$ is an open submodel of an intuitionistic Kripke model $\mathfrak{M}_2 := \langle W_2, \leq, V_1 \rangle$ then there is a homomorphism from $\mathfrak{M}_{2p}^{+}$ onto $\mathfrak{M}_{1p}^{+}$.*

Proof. It can be easily seen that the restriction to $\mathfrak{M}_{2p}^{+}$ of the homomorphism φ of the algebra $\langle W_2, R_2 \rangle_p^{+}$ onto the algebra $\langle W_1, R_1 \rangle_p^{+}$, which we studied in Lemma 2.5.62, will give the required homomorphism onto. Indeed, φ assigns generators of $\mathfrak{M}_{1p}^{+}$ to generators of $\mathfrak{M}_{2p}^{+}$. ∎

Exercise 2.5.64 *(i) If an intuitionistic Kripke model $\mathfrak{M}_1$ is an open submodel of an intuitionistic model $\mathfrak{M}_2$ then, for any formula α in the language of H, $\mathfrak{M}_2 \Vdash \alpha$ implies $\mathfrak{M}_1 \Vdash \alpha$. (ii) If a poset $\mathcal{F}_1$ is an open subframe of a poset $\mathcal{F}_2$ then the superintuitionistic logic $\lambda(\mathcal{F}_2)$ is included in the superintuitionistic logic $\lambda(\mathcal{F}_1)$.*

Hint: (i) can be proved by simple induction on the length of α, (ii) follows directly from (i).

The definitions of p-morphism for frames which are posets and for intuitionistic Kripke models completely coincide with the definitions of p-morphisms for usual frames and usual Kripke models. Similarly to the modal and temporal cases, p-morphisms preserve the truth of formulas on intuitionistic Kripke models:

Lemma 2.5.65 *If f is a p-morphism of a given intuitionistic Kripke model $\mathfrak{M}_1 := \langle W_1, \leq, V_1 \rangle$ onto an intuitionistic Kripke model $\mathfrak{M}_2 := \langle W_2, \leq, V_2 \rangle$ then for any formula α (in the language of intuitionistic propositional logic) which is built up out of letters from the domain of the valuation V_1,*

$$\forall a \in W_1 (a \Vdash_{V_1} \alpha \Leftrightarrow f(a) \Vdash_{V_2} \alpha).$$

Proof. We proceed by induction on the length of the formula α. For $\alpha = p$, the claim follows directly from (iii) in the definition of p-morphisms of Kripke models. The inductive steps for logical connectives $\wedge$, $\vee$ are evident. Now let $\beta = \gamma \to \delta$ and let $a \Vdash_{V_1} \gamma \to \delta$. Suppose $c \in W_2$, $f(a) \leq c$ and $c \Vdash_{V_2} \gamma$. Since f is a mapping onto, there exists $b \in W_1$ such that $f(b) = c$. From $f(a) R_2 f(b)$ and (ii) of the definition of p-morphisms, we get there is some $d \in W_1$ such that $a \leq d$ and $f(d) = f(b) = c$. By $c \Vdash_{V_2} \gamma$ and the inductive hypothesis we have $d \Vdash_{V_1} \gamma$. From this and $a \Vdash_{V_1} \gamma \to \delta$ it follows that $d \Vdash_{V_1} \delta$. Again applying the inductive hypothesis we derive $c \Vdash_{V_2} \delta$. Thus we have showed $f(a) \Vdash_{V_2} \gamma \to \delta$. Conversely, let $f(a) \Vdash_{V_2} \gamma \to \delta$. Suppose $a \leq b$ and $b \Vdash_{V_1} \gamma$. By the inductive hypothesis we conclude $f(b) \Vdash_{V_2} \gamma$, and since f is a p-morphism, $f(a) \leq f(b)$. Therefore $f(b) \Vdash_{V_2} \delta$ holds, and applying the inductive assumption we arrive at $b \Vdash_{V_2} \delta$. So $a \Vdash_{V_1} \gamma \to \delta$. The inductive step for the connective $\neg$ follows directly from our consideration of the step for $\to$ and the representation $\neg \delta = \delta \to \bot$. ∎

Corollary 2.5.66 *If $\mathcal{F}_1$ and $\mathcal{F}_2$ are posets and there is a p-morphism f from $\mathcal{F}_1$ onto $\mathcal{F}_2$ then the superintuitionistic logic $\lambda(\mathcal{F}_1)$ is included in the super-intuitionistic logic $\lambda(\lambda_2)$.*

Proof. Suppose α is an intuitionistic formula from $\lambda(\mathcal{F}_1)$. Let V_2 be a valuation of the set S of letters from α in $\mathcal{F}_2$. We introduce the valuation V_1 of the letters from S in $\mathcal{F}_1$ by $V_1(p) := f^{-1}(V_2(p))$. It is easy to see that V_1 is an intuitionistic valuation and f is a p-morphism from the model $\langle \mathcal{F}_1, V_1 \rangle$ onto the intuitionistic model $\langle \mathcal{F}_2, V_2 \rangle$. Using Lemma 2.5.65 we infer $\langle \mathcal{F}_2, V_2 \rangle \Vdash \alpha$. Since it is true for every intuitionistic valuation V_2, we get $\alpha \in \lambda(\mathcal{F}_2)$. ∎

Let $\mathfrak{M} := \langle W, R, V \rangle$ be an intuitionistic Kripke model. As with the modal case, for any intuitionistic formula α with variables contained in the domain of V, we set $V(\alpha) := \{a \mid a \in W, a \Vdash_V \alpha\}$. That is, $V(\alpha)$ is the set of all elements on which the formula α is valid. If $\beta(p_1, ..., p_n)$ is a propositional formula which is built up of letters $p_1, ..., p_n$ which are included in $Dom(V)$ then $\beta(V(p_1), ..., V(p_n))$ denotes the element of $\mathfrak{M}_p^+$ which is the value of the term $\beta(p_1, ..., p_n)$ on elements $V(p_1), ..., V(p_n)$.

Lemma 2.5.67 *Let $\alpha(p_1, ..., p_n)$ be an intuitionistic propositional formula with letters $p_1, ..., p_n$ contained in the domain of V. Then*

$$V(\alpha(p_1, ..., p_n)) = \alpha(V(p_1), ..., V(p_n)).$$

Proof. We will use induction on the length of the formula p. The inductive basis ($\alpha = p_i$) follows directly from our definitions. The inductive steps for logical connectives $\vee$ and $\wedge$ also follow immediately from the definitions. Consider the step for $\rightarrow$. Let $a \in V(\gamma \rightarrow \delta)(p_i)$. Then for all b such that $a \leq b$, either $\neg(b \Vdash_V \gamma)$ or $b \Vdash_V \delta$. According to the inductive hypothesis this yields: either $b \notin \gamma(V(p_i)$ or $b \in \delta(V(p_i)$. Thus $a \in (\gamma \rightarrow \delta)(V(p_i))$. Conversely, let $a \in (\gamma \rightarrow \delta)(V(p_i))$. Suppose $a \leq b$. Then either $b \notin \gamma(V(p_i))$ or $b \in \delta(V(p_i)))$. This means (by the inductive hypothesis): either $\neg(b \Vdash_V \gamma)$ or $b \Vdash_V \delta$. Thus $a \Vdash_V \gamma \rightarrow \delta$ and $a \in V(\gamma \rightarrow \delta)$. The inductive step for $\neg$ follows from the step for $\rightarrow$ and the representation $\neg\gamma = \gamma \rightarrow \perp$. ∎

As with the case of modal logics, using this lemma we can introduce the superintuitionistic logic $\lambda(\mathfrak{M})$ of any given intuitionistic Kripke model $\mathfrak{M}$, which consists of all formulas α which are valid in all models obtained from $\mathfrak{M}$ by taking arbitrary valuations of variables from α by sets from $\mathfrak{M}_p^+$.

Lemma 2.5.68 *Let f be a p-morphism of an intuitionistic Kripke model $\mathfrak{M}_1 := \langle W_1, \leq, V_1 \rangle$ onto an intuitionistic Kripke model $\mathfrak{M}_2 := \langle W_2, \leq, V_2 \rangle$. Then there is an isomorphism of the pseudo-boolean algebra $\mathfrak{M}_{2p}^+$ into the pseudo-boolean algebra $\mathfrak{M}_{1p}^+$.*

Proof. We introduce the mapping g as follows:

$$\forall X \in \mathfrak{M}_{2p}^+, \ g(X) := f^{-1}(X).$$

First of all we note that this mapping is well-defined, i.e. $g(X) \in \mathfrak{M}_{1p}^+$. Indeed, any X from $\mathfrak{M}_p^+$ has a representation of kind: $X := \alpha(V_2(p_1), ..., V_2(p_n))$, where α is some propositional intuitionistic formula. By means of Lemma 2.5.67 we immediately derive $X = V_2(\alpha(p_1, ..., p_n))$ and according to Lemma 2.5.65 the following equality $f(V_1(\alpha(p_1, ..., p_n)) = V_2(\alpha(p_1, ..., p_n))$ holds. Hence it follows $f^{-1}(X) = V_1(\alpha(p_1, ..., p_n))$. Thus applying Lemma 2.5.67 we immediately infer $V_1(\alpha(p_1, ..., p_n)) = \alpha(V_1(p_1), ..., V_1(p_n))$. Hence we arrive at $f^{-1}(X) = \alpha(V_1(p_1), ..., V_1(p_n)) \in \mathfrak{M}_{1p}^+$. Thus g maps $\mathfrak{M}_{2p}^+$ into $\mathfrak{M}_{1p}^+$. The mapping g is one-to-one. Indeed, if $X \neq Y$ and $X, Y \in \mathfrak{M}_{2p}^+$ then $f^{-1}(X) \neq f^{-1}(Y)$ since f is a mapping onto. It remains to show that g is a homomorphism. The equalities

$$g(X \cap Y) = f^{-1}(X \cap Y) = f^{-1}(X) \cap f^{-1}(Y) = g(X) \cap g(Y),$$
$$g(X \cup Y) = f^{-1}(X \cup Y) = f^{-1}(X) \cup f^{-1}(Y) = g(X) \cup g(Y),$$

hold since f is a mapping onto. To prove f preserves $\rightarrow$ we first note:

$$g(X \rightarrow Y) := f^{-1}(\{a \mid a \in W_2, \forall b \in W_2(a \leq b \Rightarrow (b \notin X) \vee (b \in Y))\}),$$
$$g(X) \rightarrow g(Y) := \{c \mid c \in W_1, \forall d \in W_1(c \leq d \Rightarrow$$
$$(f(d) \notin X) \vee (f(d) \in Y))\}.$$

Suppose $a \in g(X \to Y)$. This implies $\forall b \in W_2[f(a) \leq b \Rightarrow (b \notin X) \vee (b \in Y)]$. Assume $d \in W_1$ and $a \leq d$. Since f is a p-morphism we get $f(a) \leq f(d)$. Consequently $f(d) \notin X$ or $f(d) \in Y$. So, $a \in g(X) \to g(Y)$. Conversely, let $a \in g(X) \to g(Y)$. Suppose $b \in W_2$, $f(a) \leq b$ and $b \in X$. Since f is a mapping onto, there exists some $c \in W_1$ such that $f(c) = b$. Because f is a p-morphism there is an element $d \in W_1$ such that $f(d) = f(c) = b$ and $a \leq d$. Then $f(d) \in X$ (and besides $f(d) = b$) which implies $f(d) \in Y$ and $b \in Y$. Thus we have proved $g(X \to Y) = g(X) \to g(Y)$. The preserving of $\neg$ under g follows from the preserving of $\to$ by g and the fact that $\neg X = X \to \emptyset$. ∎

Lemma 2.5.69 *Let f be a p-morphism of a poset $\mathcal{F}_1 := \langle W_1, \leq \rangle$ onto a poset $\mathcal{F}_2 := \langle W_2, \leq \rangle$. Then there is an isomorphism of pseudo-boolean algebra $\mathcal{F}_{2p}^+$ into pseudo-boolean algebra $\mathcal{F}_{1p}^+$.*

Proof. Note that $\mathcal{F}_{2p}^+$ coincides with $\mathfrak{M}_{2p}^+$ where $\mathfrak{M}_2$ is the intuitionistic Kripke model obtained from $\mathcal{F}_2$ by enriching it with the valuation V_2 of the set of letters $P := \{p_i \mid i \text{ is an open subset of } W_2\}$, where $V_2(p_i) := i$. We transfer the valuation V_2 onto $\mathcal{F}_1$ by introducing the valuation V_1, where $V_1(p_i) := f^{-1}(V_2(p_i))$. It is clear that the valuation V_1 is intuitionistic, so we obtain the intuitionistic model $\mathfrak{M}_1$. It is easily seen that f is a p-morphism of the model $\mathfrak{M}_1$ onto the model $\mathfrak{M}_2$. According to Lemma 2.5.68 there is an isomorphic embedding of $\mathfrak{M}_{2p}^+$ into $\mathfrak{M}_{1p}^+$. Besides $\mathfrak{M}_{1p}^+$ is a subalgebra of $\mathcal{F}_{1p}^+$; hence $\mathcal{F}_{2p}^+$ is isomorphically embeddable in $\mathcal{F}_{1p}^+$. ∎

Lemma 2.5.70 *Suppose there is a homomorphism g of a pseudo-boolean algebra $\mathfrak{A}_1$ onto a pseudo-boolean algebra $\mathfrak{A}_2$. Then the Stone representation of $\mathfrak{A}_2^+$ as an intuitionistic Kripke model is isomorphically embeddable as an open submodel into the intuitionistic Kripke model $\mathfrak{A}_1^{+g}$ which is based on the frame of $\mathfrak{A}_1^+$ and has the valuation V_3, where $\forall a \in |\mathfrak{A}_2|$,*

$$V_3(a) := \{g^{-1}(\Delta) \mid a \in \Delta, \Delta \in |\mathfrak{A}_2^+|\}.$$

Proof. We begin by noting that the valuation V_3 is well-defined, i.e. any $g^{-1}(\Delta)$ is a prime filter on $\mathfrak{A}_1$, and that the valuation V_3 is an intuitionistic valuation. Before we prove this we define the mapping h of $\mathfrak{A}_2^+$ into $\mathfrak{A}_1^+$ as follows:

$$\forall \Delta \in \mathfrak{B}_2^+, h(\Delta) := g^{-1}(\Delta).$$

Now we show that $g^{-1}(\Delta)$ is a prime filter on $\mathfrak{A}_1$. To check this we take some $a, b \in g^{-1}(\Delta)$ and some c such that $a \leq c$. Then $g(a), g(b) \in \Delta$ and $g(a) \wedge g(b) \in \Delta$. Because g is a homomorphism we infer $g(a) \wedge g(b) = g(a \wedge b)$ and $a \wedge b \in g^{-1}(\Delta)$. Furthermore, $g(a) \leq g(c)$ and consequently $g(c) \in \Delta$ which implies $c \in g^{-1}(\Delta)$. So $g^{-1}(\Delta)$ is a filer. This filter is non-trivial. Actually, if $\perp \in g^{-1}(\Delta)$ then $g(\perp) \in \Delta$ and still we know $g(\perp) = \perp$, so our assumption gives Δ is trivial, a contradiction. It remains to show that the filter $g^{-1}(\Delta)$ is prime. Suppose $a \vee b \in g^{-1}(\Delta)$. Then $g(a \vee b) = g(a) \vee g(b) \in \Delta$. Since the filter Δ is prime it follows either $g(a) \in \Delta$ or $g(b) \in \Delta$, thus either $a \in g^{-1}(\Delta)$ or $b \in g^{-1}(\Delta)$. Hence we proved $g^{-1}(\Delta)$ is a prime filter. In particular, h and V_3 are well-defined. That the valuation V_3 is an intuitionistic valuation follows directly from our definition.

The mapping h is one-to-one. Actually, suppose that $\Delta_1, \Delta_2 \in \mathfrak{A}_2^+$ and $\Delta_1 \neq \Delta_2$. Then $g^{-1}(\Delta) \neq g^{-1}(\Delta)$ because g is a mapping onto and $h(\Delta_1) \neq h(\Delta_2)$. Now we will show that h preserves the order: for all $\Delta_1, \Delta_2 \in \mathfrak{A}_2$

$$\Delta_1 \subseteq \Delta_2 \Leftrightarrow h(\Delta_1) \subseteq h(\Delta_2). \tag{2.18}$$

If $\Delta_1 \subseteq \Delta_2$ then obviously $g^{-1}(\Delta_1) \subseteq g^{-1}(\Delta_2)$, i.e. $h(\Delta_1) \subseteq h(\Delta_2)$. Conversely, let now $h(\Delta_1) \subseteq h(\Delta_2)$. Let $a \in |\mathfrak{A}_2|$ and $a \in \Delta_1$. Since g is a mapping onto there is an element b from $\mathfrak{A}_1$ such that $g(b) = a$. Hence $b \in h(\Delta_1)$ and consequently $b \in h(\Delta_2)$. This implies $a \in \Delta_2$ and we proved that $\Delta_1 \subseteq \Delta_2$ and (2.18) holds. In sum we proved that h preserves the order $\subseteq$ and h is an isomorphic embedding of the frame of the model $\mathfrak{M}_{2p}^+$ into the frame of the model $\mathfrak{B}_1^+$. Furthermore $\forall \Delta \in \mathfrak{B}_{2p}^+, \forall a \in \mathfrak{B}_2 (\Delta \in V_2(a) \Leftrightarrow h(\Delta) \in V_3(a))$ by the definition of V_3. Hence h is an isomorphic embedding of the model $\mathfrak{B}_{2p}^+$ in the model which is based on the frame of $\mathfrak{B}_{1p}^+$ and has the valuation V_3.

It remains to show that $h(\mathfrak{B}_{2p}^+)$ is an open submodel. Let $h(\Delta) \in |\mathfrak{B}_{1p}^+|$. Suppose that $h(\Delta) \subseteq \Delta_1, \Delta_1 \in \mathfrak{B}_{1p}^+$. We have to show that

$$g^{-1}(g(\Delta_1)) = \Delta_1. \tag{2.19}$$

Indeed, clearly $\Delta_1 \subseteq g^{-1}(g(\Delta_1))$. Conversely, let a be an element of the set $g^{-1}(g(\Delta_1))$. Then $g(a) = g(b)$ for some $b \in \Delta_1$. This yields that $g(b) \to g(a) = \top$ and $g(b \to a) = \top$. Thus $g(b \to a) \in \Delta$ and $(b \to a) \in h(\Delta)$ which yields $(b \to a) \in \Delta_1$. Then $b \wedge (b \to a) \in \Delta_1$ and $b \wedge (b \to a) \leq a$ which entails $a \in \Delta_1$. Hence in fact (2.19) holds. From (2.19) it is follows that $g(\Delta_1)$ does not contain $\bot$.

We now prove $g(\Delta_1)$ is a prime filter. In fact, it is clear that $g(\Delta)$ is closed with respect to intersections. Let $a \in g(\Delta_1)$ and $a \leq b$. Then $a = g(c)$, for some $c \in \Delta_1$, and since g is a mapping onto, $b = g(d)$ for some d. Then $g(d \vee c) = g(d) \vee g(c) = g(d)$, and $d \vee c \in \Delta_1$, i.e. $g(d) \in g(\Delta_1)$. Hence it is shown that $g(\Delta_1)$ is a non-trivial filter. Suppose $a \vee b \in g(\Delta_1)$, then $a = g(c)$, $b = g(d)$ for some c, d and $g(c \vee d) \in g(\Delta_1)$. By (2.19) we conclude $c \vee d \in \Delta_1$. Since Δ_1 is prime, we get either $c \in \Delta_1$ or $d \in \Delta_1$. This yields either $g(c) = a \in g(\Delta)$ or $g(d) = b \in \Delta_1$. Thus it is shown that $g(\Delta)$ is a prime filter and $h(g(\Delta_1)) = \Delta_1$.

∎

Corollary 2.5.71 *If there is a homomorphism g of a pseudo-boolean algebra $\mathfrak{A}_1$ onto a pseudo-boolean algebra $\mathfrak{A}_2$ then the poset $\mathfrak{F}(\mathfrak{A}_2^+)$ is an open subframe of the frame $\mathfrak{F}(\mathfrak{A}_1^+)$.*

Lemma 2.5.72 *If a pseudo-boolean algebra $\mathfrak{A}_1$ is a subalgebra of a pseudo-boolean algebra $\mathfrak{A}_2$ then there is a p-morphism from the intuitionistic model $\mathfrak{A}_2^{+,'}$ onto the model $\mathfrak{A}_1^+$, where $\mathfrak{A}_2^{+,'}$ is the model obtained from $\mathfrak{A}_2^+$ by restriction of the valuation to the set of letters from the domain of the valuation of the model $\mathfrak{A}_1^+$.*

Proof. We define the mapping $f\colon \forall \Delta \in \mathfrak{A}_2^+$, $f(\Delta) = \Delta \cap |\mathfrak{A}_1|$, and prove f is the required p-morphism. It is clear that for every $\Delta \in \mathfrak{A}_2^+$, $f(\Delta)$ is a prime filter on $\mathfrak{B}_1$ and f satisfies (i) from the definition of p-morphisms. Also property (ii) holds for f. In fact, suppose that $\Delta_1, \Delta_2 \in |\mathfrak{A}_2^+|$ and

$$f(\Delta_1) = \Delta_1 \cap |\mathfrak{A}_1| \subseteq f(\Delta_2) = \Delta_2 \cap |\mathfrak{A}_1|.$$

We must show that there exists a prime filter Δ_3 on $\mathfrak{A}_2$ such that $f(\Delta_3) = f(\Delta_2)$ and $\Delta_1 \subseteq \Delta_2$. In order to show this we introduce the family Σ of filters on $\mathfrak{A}_2$ and a filter Δ_0 as follows:

$$\Sigma := \{\Phi \mid \Phi \text{ is non-trivial filter on } \mathfrak{A}_2, \Delta_1 \subseteq \Phi,\ \Phi \cap \mathfrak{A}_1 = \Delta_2 \cap \mathfrak{A}_1\},$$
$$\Delta_0 := \{x \mid x \in |\mathfrak{A}_2|, \exists y \in \Delta_1, \exists z \in \Delta_2 \cap |\mathfrak{A}_1|(y \vee z \leq x)\}.$$

It is clear that Δ_0 is a filter. Besides Δ_0 is non-trivial. Indeed, if $y \in \Delta_1$ and $z \in \Delta_2 \cap |\mathfrak{A}_1|$ and $y \wedge z = \bot$ then $y \leq \neg z$ and $\neg z \in \Delta_1 \cap |\mathfrak{A}_1| \subseteq \Delta_2 \cap |\mathfrak{A}_1|$ which gives a contradiction the fact that $z \in \Delta_2$. Furthermore,

$$\Delta_2 \cap |\mathfrak{A}_1| \subseteq (\Delta_1 \cup (\Delta_2 \cap |\mathfrak{A}_1|)) \cap |\mathfrak{A}_1| \subseteq \Delta_0 \cap |\mathfrak{A}_1|.$$

Let $u \in \Delta_0 \cap |\mathfrak{A}_1|$. Then for some $y \in \Delta_1$, $z \in \Delta_2 \cap |\mathfrak{A}_1|$ and $y \wedge z \leq u$. Therefore $y \leq (z \to u)$ and consequently $(z \to u) \in \Delta_1 \cap |\mathfrak{A}_1| \subseteq \Delta_2 \cap |\mathfrak{A}_1|$. Therefore we get $(z \to u) \wedge z \in \Delta_2 \cap |\mathfrak{A}_1|$ and $(z \to u) \wedge z \leq u$, thus $u \in \Delta_2 \cap |\mathfrak{A}_1|$. So we have shown $\Delta_0 \cap |\mathfrak{A}_1| = \Delta_2 \cap |\mathfrak{A}_1|$. Hence $\Delta_0 \in \Sigma$ and Σ is non-empty.

The set Σ is ordered by the usual set-theoretic inclusion. It is clear that the union of an arbitrary chain of filters from Σ is again a filter from Σ. Therefore by Zorn Lemma there is a maximal filter Δ among the filters of Σ. The filter Δ is prime. Indeed, suppose $a \vee b \in \Delta$. $a \notin \Delta$, $b \notin \Delta$ for some a, b. Consider the filters

$$\Delta_a := \{z \mid \exists v \in \Delta(a \wedge v \leq z\}, \Delta_b := \{z \mid \exists v \in \Delta(b \wedge v \leq z\}$$

on $\mathfrak{A}_2$. Suppose that

$$\Delta_a \cap |\mathfrak{A}_1| \not\subseteq \Delta_2 \cap |\mathfrak{A}_1|,\ \Delta_b \cap |\mathfrak{A}_1| \not\subseteq \Delta_2 \cap |\mathfrak{A}_1|.$$

Then there are some $z_a \in \Delta_a \cap |\mathfrak{A}_1|$, $z_b \in \Delta_b \cap |\mathfrak{A}_1|$ such that $z_a \notin \Delta_2 \cap |\mathfrak{A}_1|$, $z_b \notin \Delta_2 \cap |\mathfrak{A}_1|$. From this it follows that $z_a \vee z_b \notin \Delta_2$. At the same time there are some $v_a, v_b \in \Delta$ such that $v_a \wedge a \leq z_a \leq z_a \vee z_b$, $v_b \wedge b \leq z_b \leq z_a \vee z_b$ (which hold by the definition of Δ_a, Δ_b). Consequently, $(v_a \wedge v_b) \wedge (a \vee b) \leq z_a \vee z_b$ which yields $z_a \vee z_b \in \Delta$. Still we have $z_a \vee z_b \in |\mathfrak{A}_1|$. So $z_a \vee z_b \in \Delta \cap |\mathfrak{A}_1|$. Furthermore we had $z_a \notin \Delta_2 \cap |\mathfrak{A}_1|$ and $z_b \notin \Delta_2 \cap |\mathfrak{A}_1|$ which yield $z_a \vee z_b \notin \Delta_2 \cap |\mathfrak{A}_1|$. Thus we have a contradiction the fact that $\Delta \cap |\mathfrak{A}_1| = \Delta_2 \cap |\mathfrak{A}_1|$. Hence either $\Delta_a \cap |\mathfrak{A}_1| = \Delta_2 \cap |\mathfrak{A}_1|$ or $\Delta_b \cap |\mathfrak{A}_1| = \Delta_2 \cap |\mathfrak{A}_1|$. Thus at least one of the filters Δ_a, Δ_b belongs to Σ and both they are proper extensions of Δ which contradicts Δ is maximal. Hence we proved Δ is a prime filter, Δ extends Δ_1 and $f(\Delta) = f(\Delta_2)$. So f is a p-morphism of frames.

We now will show that f is a mapping onto. Let Δ be a prime filter on $\mathfrak{A}_1$ then the filter $\Delta_1 := \{x \mid x \in |\mathfrak{A}_2|, \exists y \in \Delta(y \leq x)\}$ will be a non-trivial filter on $\mathfrak{A}_2$. Consider the family

$$\Sigma := \{\Phi \mid \Phi \text{ is non-trivial filter on } \mathfrak{A}_2, \Delta_1 \subseteq \Phi, \Phi \cap |\mathfrak{A}_1| = \Delta\}.$$

It is clear that $\Delta_1 \in \Sigma$. Applying Zorn Lemma there is a maximal filter Δ_0 among filters of Σ. We will show that Δ_0 is prime. Suppose $a \vee b \in \Delta_0$, $a \notin \Delta_0$ and $b \notin \Delta_0$. We define similar as in our reasoning above filters $\Delta_a := \{z \mid \exists v \in \Delta_0(a \wedge v \leq z\}$ and $\Delta_b := \{z \mid \exists v \in \Delta_0(b \wedge v \leq z\}$ on $\mathfrak{A}_2$. Suppose that $\Delta_a \cap |\mathfrak{A}_1| \not\subseteq \Delta$ and $\Delta_b \cap |\mathfrak{A}_1| \not\subseteq \Delta$. Then there are some $z_a \in \Delta_a \cap |\mathfrak{A}_1|$, $z_b \in \Delta_b \cap |\mathfrak{A}_1|$ such that $z_a \notin \Delta$, $z_b \notin \Delta$. This, in particular, implies $z_a \vee z_b \notin \Delta$. By definition of Δ_a and Δ_b there are some $v_a, v_b \in \Delta_0$ such that $v_a \wedge a \leq z_a \leq z_a \vee z_b$, $v_b \wedge b \leq z_b \leq z_a \vee z_b$. Therefore, $(v_a \wedge v_b) \wedge (a \vee b) \leq z_a \vee z_b$ which gives $z_a \vee z_b \Delta_0$. Besides we assumed that $z_a \notin \Delta$ and $z_b \notin \Delta$ which yield $z_a \vee z_b \notin \Delta$. Thus we have a contradiction the fact that $\Delta_0 \cap |\mathfrak{A}_1| = \Delta$. Consequently either $\Delta_a \cap |\mathfrak{A}_1| = \Delta$ or $\Delta_b \cap |\mathfrak{A}_1| = \Delta$. Thus at least one filter from the pair Δ_a, Δ_b belongs to Σ which contradicts Δ_0 is maximal. Thus Δ_0 is a prime filter and $f(\Delta_0) = \Delta$. Hence the mapping f is onto. It is clear that f preserves the valuation on models: Let a be an element of $\mathfrak{A}_1$ Then for any $\Delta \in \mathfrak{A}_2^+$, $\Delta \Vdash_{V_2} a \Leftrightarrow a \in \Delta \Leftrightarrow a \in \Delta \cap |\mathfrak{A}_1| \Leftrightarrow \Delta \cap |\mathfrak{A}_1| \Vdash_{V_1} a$. ∎

Corollary 2.5.73 *If $\mathfrak{A}_1$ is a subalgebra of a pseudo-boolean algebra $\mathfrak{A}_2$ then there exists a p-morphism from the frame $\mathfrak{F}(\mathfrak{A}_2^+)$ onto the frame $\mathfrak{F}(\mathfrak{A}_1^+)$.*

Lemma 2.5.74

(i) For the disjoint union $\bigsqcup_{i \in I} \mathfrak{M}_i$ of a collection of intuitionistic Kripke models $\mathfrak{M}_i$ and for any propositional formula α with propositional letters from the domain of the valuation of $\bigsqcup_{i \in I} \mathfrak{M}_i$, $\bigsqcup_{i \in I} \mathfrak{M}_i \Vdash \alpha$ iff $\forall i \in I(\mathfrak{M}_i \Vdash \alpha)$.

(ii) If $\bigsqcup_{i \in I} \mathcal{F}_i$ is the disjoint union of posets $\mathcal{F}_i$ then for any propositional formula α from $\mathcal{F}or_H$, $\bigsqcup_{i \in I} \mathcal{F}_i \Vdash \alpha$ iff $\forall i \in I(\mathcal{F}_i \Vdash \alpha)$ where $\Vdash$ is the intuitionistic truth relation.

Proof is left as an exercise: (i) follows directly from our definitions and (ii) follows immediately from (i).

Lemma 2.5.75 *Let* $\mathcal{F}_i := \langle W_i, \leq \rangle, i \in I$ *be a family of pairwise disjoint posets. Then the pseudo-boolean algebra* $(\bigsqcup_{i \in I} \mathcal{F}_i)_p^+$ *is isomorphic to* $\prod_{i \in I} \mathcal{F}_{ip}^+$

Proof. If X is an open subset of $\bigcup_i \mathcal{F}_i$ then we let $\pi_i(X)$ be the set of all elements from $|\mathcal{F}_i|$ that are in X. The isomorphism h from $(\bigsqcup_{i \in I} \mathcal{F}_i)_p^+$ onto $\prod_{i \in I} \mathcal{F}_{ip}^+$ can be defined in the following way: for every open subset X of $\bigcup_i \mathcal{F}_i$, $h(X) := (\pi_i(X) \mid i \in I)$. It is easy to see that h is an isomorphism onto. We leave simple details to the reader as an exercise. ∎

Lemma 2.5.76 *If a pseudo-boolean algebra* $\prod_{i \in I} \mathfrak{A}_i$ *is a product of a finite pseudo-boolean algebras then* $\prod_{i \in I} \mathfrak{A}_i \cong (\mathfrak{F}(\bigsqcup_{i \in I} \mathfrak{A}_i^+))_p^+$.

Proof. In fact, since any $\mathfrak{A}_i$ is finite, by Theorem 2.5.60 $\mathfrak{B}_i \cong (\mathfrak{F}(\mathfrak{B}_i^+))_p^+$. Thus any $\mathfrak{A}_i$ is isomorphic to $\mathcal{F}_i^+$ for some finite frame $\mathcal{F}_i$. Again by Theorem 2.5.60 $\mathcal{F}_i \cong \mathcal{F}_{ip}^{++}$ and because $\mathcal{F}_{ip}^+ \cong \mathfrak{A}_i$ we get $\mathcal{F}_i \cong \mathfrak{A}_{ip}^+$. Furthermore, we have $\prod_{i \in I} \mathfrak{A}_i \cong \prod_{i \in I} \mathcal{F}_i^+$, and by Lemma 2.5.75 $\prod_{i \in I} \mathcal{F}_i^+ \cong (\bigsqcup_{i \in I} \mathcal{F}_i)^+$. Thus we conclude that $\prod_{i \in I} \mathfrak{A}_i \cong$ and $(\bigsqcup_{i \in I} \mathfrak{A}_{ip}^+)^+$. ∎

2.6 The Finite Model Property

In this section we develop instruments which provide algorithms for recognizing theorems of logics by means of the *finite model property*. The possession of the *finite model property* for a given algebraic logic λ means that each non-theorem α of λ is falsified in a finite algebra $\mathfrak{A}$ from $Var(\lambda)$. More precisely,

Definition 2.6.1 *Let* λ *be an algebraic logic and*

$$Var(\lambda) := \{\mathcal{A} \mid \mathcal{A} \models \alpha, \alpha \in \lambda\}$$

be the corresponding variety of all algebras on which all theorems of λ *are valid. We say* λ *has the* finite model property *(abbreviation fmp) if, for every formula* α *which is not a theorem of* λ, *there is a certain finite algebra* $\mathcal{A}$ *from the variety* $Var(\lambda)$ *such that* $\mathcal{A} \not\models \alpha$.

We know that any algebraic logic λ is complete with respect to the variety of algebras $Var(\lambda)$. In light of this, fmp is just a strengthening of the algebraic completeness theorem to the case of finite algebras:

Lemma 2.6.2 *If* λ *has the finite model property then*

$$\lambda = \bigcap \{\lambda(\mathcal{A}) \mid \mathcal{A} \in Var(\lambda), ||\mathcal{A}|| < \omega\}.$$

Proof follows immediately from our definitions.

Not only does the fmp a special kind completeness, completeness with respect to a class of finite algebras, but the fmp gives a tool to recognize formula's derivability, in particular a tool for proving the decidability of logics. Recall that a logic λ is said to be *decidable* iff there is an algorithm which allows us effectively determine for any given formula α whether α is a theorem of λ.

Theorem 2.6.3 *(Harrop) If λ is a logic with the fmp and λ has a finite axiomatic system $\mathcal{AS}$ (i.e. with certain finite sets of axioms and inference rules) then λ is decidable.*

Proof. Since $\mathcal{AS}$ is finite we can generate an effective procedure P^+ for obtaining a sequence of all the theorems of λ (by uniform derivation of theorems by $\mathcal{AS}$). At the same time, using the fact that λ has the fmp we also can generate an effective procedure P^- which enumerates those formulas which are not theorems (by enumeration of all formulas which are falsified in finite algebras by checking of formulas in the set of all finite algebras from $Var(\lambda)$). The inclusion of finite algebras in $Var(\lambda)$ is effectively recognizable since λ has finite number of axioms. For any given formula α we start both procedures P^+ and P^-. Then either an some step of P^+ we get α or α appears at some step of P^-. In first case we conclude $\alpha \in \lambda$, in second one we have $\alpha \notin \lambda$. ∎

Clear that the decision procedure from Harrop's theorem does not stipulate a certain upper bound for the number of steps we need to recognize whether a given formula is a theorem. At the same time, very often possession of the fmp provides the effectively calculable upper bound to the number of elements from finite algebras which can disprove a given formula (this will be illustrated lately by a number examples below).

Definition 2.6.4 *Let λ be a modal, a temporal or a superintuitionistic logic. We say λ has the Kripke finite model property if*

$$\lambda = \{\alpha \mid (\forall \mathcal{F})(||\mathcal{F}|| < \omega \& \mathcal{F} \Vdash \lambda \Rightarrow \mathcal{F} \Vdash \alpha\},$$

i.e. if λ is complete with respect to some class of corresponding finite frames.

Theorem 2.6.5 *A modal, temporal or superintuitionistic logic λ has the fmp if and only if λ has the Kripke fmp.*

Proof. Let λ have the fmp. Suppose $\alpha \notin \lambda$, then there is a finite $\mathcal{A} \in Var(\lambda)$ such that $\mathcal{A} \not\models \alpha$. According to the Stone Representation Theorem (Theorem 2.5.6 for modal algebras, Theorem 2.5.34 for temporal algebras, Theorem 2.5.60 for pseudo-boolean algebras), $\mathcal{A} \cong \mathfrak{F}(\mathcal{A}^+)^+$ for the modal algebra $\mathcal{A}$, $\mathcal{A} \cong \mathfrak{F}(\mathcal{A}^+)^+_t$ for the temporal algebra $\mathcal{A}$, and $\mathcal{A} \cong \mathfrak{F}(\mathcal{A}^+)^+_p$ for the pseudo-boolean algebra $\mathcal{A}$.

We will show that $\mathfrak{F}(\mathcal{A}^+) \Vdash \lambda$. Let β be a theorem of λ and V be a (modal, temporal or intuitionistic) valuation on $\mathfrak{F}(\mathcal{A}^+)$. Then for any propositional letter p_i from β, $V(p_i) \in |\mathcal{A}|$. Using Lemma 2.5.17 if λ is a modal logic, Lemma 2.5.45 if λ is a temporal logic, and Lemma 2.5.67 when λ is a superintuitionistic logic, we have $V(\beta(p_i)) = \beta(V(p_i)) = |\mathfrak{F}|$. Thus we have showed that $\mathfrak{F}(\mathcal{A}^+) \Vdash \beta$. Because $\mathcal{A} \not\models \alpha$, there are some elements $a_j \in |\mathcal{A}|$ corresponding to propositional letters p_j from α such that $\mathcal{A} \models \alpha(a_j) \neq \top$. We define the valuation V on $\mathfrak{F}(\mathcal{A}^+)$ as follows: $V(p_i) := a_i \subseteq \mathfrak{F}(\mathcal{A}^+)$. Applying similar reasoning to the given above one we derive

$$V(\alpha(p_j)) = \alpha(V(p_j)) = \alpha(a_j) \neq \top = |\mathfrak{F}(\mathcal{A}^+)|.$$

That is, the finite frame $\mathfrak{F}(\mathcal{A}^+)$ detaches the formula α from the logic λ. Hence λ has the Kripke fmp.

Conversely, suppose λ has the Kripke finite model property. Let α be a non-theorem of λ. Then there is a finite frame $\mathcal{F}$ such that $\mathcal{F} \Vdash \lambda$ and there exists a valuation V of propositional letters from α on $\mathcal{F}$ such that $\mathcal{F} \not\Vdash_V \alpha$. We define $\mathcal{A} = \mathcal{F}^+$ for the case when λ is a modal logic, $\mathcal{A} = \mathcal{F}_t^+$ when λ is a temporal logic, and $\mathcal{A} = \mathcal{F}_p^+$ when λ is a superintuitionistic logic. First we verify that $\mathcal{A} \in Var(\lambda)$. Indeed, let V_1 be a valuation on $\mathcal{A}$ of the propositional letters p_i of some formula β which is a theorem of λ. That is $V_1(p_i) = a_i \subseteq \mathcal{F}$. Again by reasoning similar to that in the proof above of the case *necessary* we conclude $\beta(V_1(p_j)) = V_1(\beta(p_j)) = |\mathcal{F}|$. Thus $\forall \beta \in \lambda(\mathcal{A} \models \beta)$, i.e. the finite algebra $\mathcal{A}$ belongs to $Var(\lambda)$. Since $\mathcal{F} \not\Vdash_V \alpha$ we have $V(\alpha) \neq |\mathcal{F}|$. At the same time $V(\alpha(p_i)) = \alpha(V(p_i))$. Consequently $\mathcal{A} \not\models \alpha$. Hence the logic λ has the finite model property. ∎

Corollary 2.6.6 *Any modal, temporal or superintuitionistic logic with the fmp is Kripke complete.*

Having this theorem at our disposal we can prove the possession of the fmp through possession of the Kripke fmp. A very simple and popular technique for proving possession of the fmp is the so-called filtration method which goes back to D.Scott and has been further developed by K.Segerberg (cf. Segerberg [165]), and we proceed to description of this method.

Let $\mathfrak{M} := \langle W, R, V \rangle$ be a Kripke model (modal, temporal or intuitionistic). Let S be a set of formulas (modal, temporal or intuitionistic, respectively) closed with respect to subformulas, and let the propositional letters having occurrences in formulas from S be contained in $Dom(V)$. We define the equivalence relation $\equiv_S$ on W as follows:

$$a \equiv_S b \Leftrightarrow (\forall \alpha \in S)(a \Vdash_V \alpha \Leftrightarrow b \Vdash_V \alpha).$$

We take the quotient-set $W_{\equiv_S} := \{[a]_{\equiv_S} \mid a \in W\}$ of W by $\equiv_S$ and define on $W_{\equiv_S}$ a filtration as follows. Recall that $Var(S)$ is the set of all the propositional variables that occur in formulas of S.

Definition 2.6.7 *The filtration on $\mathfrak{M}$ by means of S is any Kripke model*
$\mathfrak{M}_S := \langle W_{\equiv_S}, R_1, V_S \rangle$, *where*

$(\forall p \in Var(S))(V_S(p) := \{[a]_{\equiv_S} \mid a \Vdash_V p\})$;
  if $\mathfrak{M}$ is a modal model then
    $(1)(\forall x, y \in W)(aRy \Rightarrow [x]_{\equiv_S} R_1 [y]_{\equiv_S})$,
    $(2)(\forall x, y \in W)[\ [x]_{\equiv_S} R_1 [y]_{\equiv_S} \Rightarrow (\Box\alpha \in S \& x \Vdash_V \Box\alpha \Rightarrow y \Vdash_V \alpha)\ ]$;
  if $\mathfrak{M}$ is a temporal model then (1) and
    $(3)(\forall x, y \in W)[\ [x]_{\equiv_S} R_1 [y]_{\equiv_S} \Rightarrow (G\alpha \in S \& x \Vdash_V G\alpha \Rightarrow y \Vdash_V \alpha)\ ]$;
    $(4)(\forall x, y \in W)[\ [x]_{\equiv_S} R_1 [y]_{\equiv_S} \Rightarrow (H\alpha \in S \& y \Vdash_V H\alpha \Rightarrow x \Vdash_V \alpha)\ ]$;
  if $\mathfrak{M}$ is an intuitionistic model then (1) and
    $(5)(\forall x, y \in W)[\ [x]_{\equiv_S} R_1 [y]_{\equiv_S} \Rightarrow (\alpha \in S \& x \Vdash_V \alpha \Rightarrow y \Vdash_V \alpha)\ ]$.

It is clear that V_S above is well-defined. The classes of filtrations are always non-empty. Indeed, it is easy to see that among all the filtrations on $\mathfrak{M}$ by S there is a *greatest filtration*: the filtration with the greatest accessibility relation R_1. This relation is obtained from (2) for the modal case, conjunction of (3) and (4) for the temporal case, and (5) for the intuitionistic case, by exchanging every the leftmost occurrence $\Rightarrow$ by $\Leftrightarrow$. The intersection of all R_1 in all possible filtrations on $\mathfrak{M}$ by S gives a smallest accessibility relation R_m which turns $W_{\equiv_S}$ into a filtration. We call this filtration the smallest filtration. It is not hard to see from (1) that R_m can be defined also as follows: $\langle X, Y \rangle \in R_m \Leftrightarrow (X = [a]_{\equiv_S} \& Y = [b]_{\equiv_S} \& (aRb))$. It is clear also that, for the case of canonical intuitionistic models, the smallest and the greatest filtrations coincide, that is, for the canonical intuitionistic models there is only a single filtration for any given S.

Lemma 2.6.8 *(Filtration Lemma) If $\mathfrak{M}_S = \langle W_{\equiv_S}, R_1, V_S \rangle$ is a filtration of a model $\mathfrak{M} = \langle W, R, V \rangle$ by S then, for every $a \in W$ and every $\alpha \in S$,*

$$a \Vdash_V \alpha \Leftrightarrow [a]_{\equiv_S} \Vdash_S \alpha. \tag{2.20}$$

Proof. This can be easily obtained by simple induction on the length the formula α in (2.20). Indeed the basis of the induction, the case when $\alpha = p$, follows directly from the definition of V_S. The inductive steps for the logical connectives $\vee$, $\wedge$ (and for the connective $\neg$ for modal and temporal models) are obvious in all (modal, temporal, and intuitionistic) cases.

Consider the inductive step for the modal case and the formula $\alpha = \Box\beta$. Since S is closed with respect to subformulas, we have $\beta \in S$. Suppose that $a \Vdash_V \Box\beta$. Then for all $b \in W$, $b \Vdash_V \beta$ provided aRb. Consider an arbitrary $[c]_{\equiv_S}$ such that $[a]_{\equiv_S} R_1 [c]_{\equiv_S}$. That $c \Vdash_V \beta$ follows directly by (2) from the definition of filtration, and using the inductive hypothesis we conclude $[c]_{\equiv_S} \Vdash_{V_S} \beta$. Hence $[a]_{\equiv_S} \Vdash_{V_S} \Box\beta$. Conversely, suppose that $[a]_{\equiv_S} \Vdash_{V_S} \Box\beta$ holds and aRb. Then by (1) from the definition of filtrations we have $[a]_{\equiv_S} R_1 [b]_{\equiv_S}$. Consequently

$[b]_{\equiv_S}$ ⊩$_{V_S}$ β and applying the inductive hypothesis we have b ⊩$_V$ β. Hence a ⊩$_V$ $\Box\beta$ and the modal case is complete. The inductive steps for the temporal case and formulas $G\beta$ and $H\beta$ have almost the same proof using (3) and (4) from the definition of filtrations instead of (2).

Consider the inductive step for the case of intuitionistic models and the formula $\alpha := \beta \to \gamma$. Note that then $\beta, \gamma \in S$ since S is closed with respect to subformulas. Let a ⊩$_V$ $\beta \to \gamma$. Suppose $[a]_{\equiv_S} R_1 [c]_{\equiv_S}$ and $[c]_{\equiv_S}$ ⊩$_{V_S}$ β. Using the inductive assumption we conclude c ⊩$_{V_S}$ β. According to (5) from the definition of filtrations it follows c ⊩$_V$ $\beta \to \gamma$. Consequently, c ⊩$_V$ γ. Applying the inductive hypothesis again the we derive $[c]_{\equiv_S}$ ⊩$_{V_S}$ γ. Hence we shown $[a]_{\equiv_S}$ ⊩$_{V_S}$ $\beta \to \gamma$. Conversely, suppose now that $[a]_{\equiv_S}$ ⊩$_{V_S}$ $\beta \to \gamma$. Consider any b that aRb and suppose that b ⊩$_V$ β. By the inductive hypothesis we get $[b]_{\equiv_S}$ ⊩$_{V_S}$ β. By (1) from the definition of filtrations we obtain $[a]_{\equiv_S} R_1 [b]_{\equiv_S}$. Therefore $[b]_{\equiv_S}$ ⊩$_{V_S}$ γ and inductive hypothesis yields b ⊩$_V$ γ. Thus a ⊩$_V$ $\beta \to \gamma$. The inductive step for an intuitionistic model and formulas of kind $\neg\beta$ follows immediately from the equivalence $\neg\beta \equiv_H \beta \to \bot$. ∎

Recall that for any model $\mathfrak{M}$ (frame $\mathcal{F}$), and any subset X of $\mathfrak{M}$ ($\mathcal{F}$), the open submodel $\mathfrak{A}_X$ (or the open subframe $\mathcal{F}_X$) generated by X is the smallest open submodel (subframe) of $\mathfrak{A}$ (of $\mathcal{F}$) containing X.

Definition 2.6.9 *Let λ be a modal, temporal or superintuitionistic logic. We say λ has the filtration property iff the following holds. For any finite set of formulas S closed with respect to subformulas, and for any element a of the canonical model $\mathcal{C}$ for λ, there is a filtration $\mathcal{C}_{a,S}$ by means of S of the open submodel $\mathcal{C}_a$ of $\mathcal{C}$ generated by the element a such that $\mathcal{C}_{a,S}$ is based on some λ-frame.*

Theorem 2.6.10 *If a logic λ has the filtration property then λ has the finite model property.*

Proof. In fact, if $\alpha \notin \lambda$ then α is falsified on the canonical Kripke model $\mathcal{C}$ for λ, because, as we know, the canonical model $\mathcal{C}$ characterizes λ. We define $Sub(\alpha)$ to be the set of all subformulas of α. Let a be an element of $\mathcal{C}$ which disproves α and $\mathcal{C}_a$ be the open submodel of $\mathcal{C}$ generated by a. Then, by the property of open submodels, $\mathcal{C}_a$ also disproves α. Since λ has the filtration property, there exists a filtration $\mathcal{C}_{a,Sub(\alpha)}$ of the $\mathcal{C}_a$ by $Sub(\alpha)$ such that $\mathcal{C}_{a,Sub(\alpha)}$ is based on a λ-frame. By definition the model $\mathcal{C}_{a,Sub(\alpha)}$ is finite. According to Lemma 2.6.8 the model $\mathcal{C}_{a,Sub(\alpha)}$ also disproves α. Therefore λ has the Kripke fmp and consequently by Theorem 2.6.5 λ has the fmp. ∎

Now we proceed to give some examples of logics with the finite model property on the basis of the filtration technique we have described above.

Lemma 2.6.11 *The modal logics K, T, $K4$, $S4$, $S5$, D, $K4.3$ and $S4.3$ have the filtration property.*

Proof. According to Theorems 2.3.27, 2.3.28, 2.3.48, and 2.3.51, the frames of the canonical models C_λ of these logics λ are λ-frames and

C_K is a model ;
C_T is a reflexive model;
C_{K4} is a transitive model;
C_{S4} is a reflexive and transitive model;
C_{S5} is a reflexive and transitive model and satisfies to the condition:
$\quad \forall x \forall y (x R y \Rightarrow y R x)$;
C_D satisfies to the condition $\forall x \exists y (x R y)$;
$C_{K4.3}$ is transitive model and its frame satisfies the condition
$\quad \forall x, y, z [(x R y) \& (x R z) \Rightarrow ((y R z) \vee (z R y))]$;
$C_{S4.3}$ is transitive, reflexive model and its frame satisfies
the condition
$\quad \forall x, y, z [(x R y) \& (x R z) \Rightarrow ((y R z) \vee (z R y))]$.

Given a finite set of formulas S closed with respect to subformulas. Considering the case of the minimal normal modal logic K we take the greatest filtration $C_{K,a,S}$ by S for the open submodel $C_{K,a}$ of the canonical model C_K generated by an arbitrary element a. It is clear that $C_{K,a,S}$ is a finite model and its frame is a K-frame. Hence K has the filtration property. Any filtration of any reflexive model must be a reflexive model. Therefore, using Theorem 2.3.28, the greatest filtration $C_{T,a,S}$ by S of any open submodel $C_{T,a}$ of the canonical model C_T is based on a certain T-frame, i.e. T has the filtration property. In the case when $\lambda := K4, S4$ we choose the following relation R_S between classes $[w]_{\equiv_S}$ of elements of the open submodel $C_{\lambda,a}$ of the canonical model C_λ for λ generated by an element a:

$$([w]_{\equiv_S} R_S [v]_{\equiv_S}) \Leftrightarrow \forall \Box \alpha \in S(w \Vdash_V \Box \alpha \Rightarrow v \Vdash_V \Box \alpha \wedge \alpha).$$

By Theorem 2.3.28 the canonical model C_λ is transitive. Therefore R_S satisfies (1) and (2) from the definition of filtrations and R_S gives a filtration $\mathfrak{M}$ of the model $C_{\lambda,a}$ by S. If $\lambda := K4$ then in order to show that $\mathfrak{M}$ is a transitive model assume $[b]_{\equiv_S} R_S [c]_{\equiv_S}$ and $[c]_{\equiv_S} R_S [d]_{\equiv_S}$. Let $\Box \alpha \in S$ and $a \Vdash_V \Box \alpha$. Then by definition of R_S it follows $c \Vdash_V \Box \alpha \wedge \alpha$, and using this and the definition of R_S again we infer $d \Vdash_V \Box \alpha \wedge \alpha$, in particular $[b]_{\equiv_S} R_S [d]_{\equiv_S}$. Hence the frame of $\mathfrak{M}$ is transitive and $K4$ has the filtration property. When $\lambda := S4$ we show that $\mathfrak{M}$ is transitive as above, and since C_{S4} is reflexive (see Theorem 2.3.28), and any filtration of reflexive model is reflexive, we conclude $S4$ has the filtration property.

For $S5$ we define the accessibility relation R_S between classes $[w]_{\equiv_S}$ of elements in $C_{S5,a}$ (which is the open submodel of the canonical model C_{S5} generated by an element a) as follows:

$$([w]_{\equiv_S} R_S [v]_{\equiv_S}) \Leftrightarrow [\forall \Box \alpha \in S(w \Vdash_V \Box \alpha \Leftrightarrow v \Vdash_V \Box \alpha)].$$

By Theorem 2.3.28 the canonical model C_{S5} has an equivalence relation as the accessibility relation, consequently R_S satisfies (1) and (2) from the definition of filtrations and R_S sets a filtration $\mathfrak{M}$ of $C_{S5,a}$ by S. Directly by definition of R_S, $\mathfrak{M}$ is based on a $S5$-frame. Hence $S5$ has the filtration property.

Now consider the modal logic D. The frame of C_D as we have seen above satisfies the condition $\forall x \exists y (x R y)$. Consider the greatest filtration $C_{D,a,S}$ by S of the open submodel $C_{D,a}$ of the canonical model C_D generated by an element a. According to (1) from the definition of filtrations, the frame of $C_{D,a,S}$ satisfies the condition $\forall x \exists y (x R y)$ also, which according to Theorem 2.3.51 entails the frame of $C_{D,a,S}$ is a D-frame. Thus D has the filtration property.

Now we turn to the case $\lambda := K4.3, S4.3$ and define the accessibility relation R_S between classes $[w]_{\equiv_S}$ of elements in $C_{\lambda,a}$ (which is the open submodel of the canonical model C_λ generated by an element a) as for $K4$ and $S4$:

$$([w]_{\equiv_S} R_S [v]_{\equiv_S}) \Leftrightarrow [\forall \Box \alpha \in S(w \Vdash_V \Box \alpha \Rightarrow v \Vdash_V \Box \alpha)].$$

We show as above for the case $\lambda := K4, S4$ that R_S defines a filtration $\mathfrak{M}$ by S on $C_{\lambda,a}$ and that $\mathfrak{M}$ is a transitive model for $\lambda := K4.3$ and $\mathfrak{M}$ is a transitive and reflexive model when $\lambda := S4.3$. We have seen above that the model C_λ satisfies the condition

$$\forall x, y, z[(x R y) \& (x R z) \Rightarrow ((y R z) \vee (z R y))].$$

Therefore the open subframe $C_{\lambda,a}$ of C_λ generated by a consists of elements which are comparable by the accessibility relation. Consequently, by (1) from the definition of filtrations, all elements of $\mathfrak{M}$ are also comparable by the accessibility relation. Thus in the frame of $\mathfrak{M}$ $\forall x, y, z[(x R y) \& (x R z) \Rightarrow ((y R z) \vee (z R y))]$ is valid and according to Theorem 2.3.51 the frame of $\mathfrak{M}$ is a certain λ-frame for $\lambda := K4.3, S4.3$. Hence $K4.3$ and $S4.3$ have the filtration property. ∎

Immediately from this lemma, Theorems 2.6.10 and 2.6.3 we derive

Corollary 2.6.12 *The modal logics K, T, $K4$, $S4$, $S5$, D, $K4.3$ and $S4.3$ have the finite model property and are decidable.*

Thus filtration property allows us very easily and simply to show that certain logics have the fmp and are decidable. At the same time, not all logics with the fmp have the filtration property. Very often we need some advanced modifications of filtrations in order to show possession of the fmp. We will present these tools below but first we briefly demonstrate the filtration property for superintuitionistic and temporal logics.

Lemma 2.6.13 *The superintuitionistic logics H, KC and LC have the filtration property.*

Proof. By Theorems 2.4.26 and 2.3.25 the frames of the canonical models C_λ of logics $\lambda = H, KC$ and LC are λ-frames and

C_H is an intuitionistic model ;
C_{KC} is an intuitionistic model and satisfies the condition:
$$\forall x \forall y \forall z[(xRy)\&(xRz) \Rightarrow \exists t((yRt)\&(zRt))];$$
C_{LC} is an intuitionistic model and satisfies the condition
$$\forall x \forall y \forall z[(xRy)\&(xRz) \Rightarrow ((yRz) \vee (zRy))];$$

Let S be a finite set of formulas closed with respect to subformulas. Let a be an element of the canonical model C_λ, $\lambda = H, KC, LC$. We take the greatest intuitionistic filtration $C_{\lambda,a,S}$ by S of the open submodel $C_{\lambda,a}$ of the canonical intuitionistic Kripke model C_λ for λ generated by an element a. Clearly that, for any $\lambda := H, KC, LC$, the model $C_{H,a,S}$ is a finite intuitionistic model. In particular, the frame of $C_{H,a,S}$ is a H-frame, so H has the filtration property. Consider the frame of $C_{KC,a,S}$. The frame of the model C_{KC} satisfies

$$\forall x \forall y \forall z[(xRy)\&(xRz) \Rightarrow \exists t((yRt)\&(zRt))].$$

Therefore, according to (1) from the definition of filtrations, for any pair of elements from $C_{KC,a,S}$ there is an element of $C_{KC,a,S}$ which is accessible from both elements of this pair. Therefore $C_{KC,a,S}$ also satisfies

$$\forall x \forall y \forall z[(xRy)\&(xRz) \Rightarrow \exists t((yRt)\&(zRt))]$$

and by Theorem 2.4.27 the frame of $C_{KC,a,S}$ is a KC-frame. Thus KC has the filtration property also. The model C_{LC} satisfies the condition

$$\forall x \forall y \forall z[(xRy)\&(xRz) \Rightarrow ((yRz) \vee (zRy))].$$

Therefore the frame of $C_{KC,a,S}$ according to (1) from the definition of filtrations satisfies the first order formula $\forall y \forall z((yRz) \vee (zRy))$. Hence by Theorem 2.4.27 $C_{KC,a,S}$ is an LC-frame and LC has the filtration property. ■

From this lemma, Theorems 2.6.10 and 2.6.3 we can immediately infer

Corollary 2.6.14 *The superintuitionistic logics H, KC and LC have the fmp and are decidable.*

Exercise 2.6.15 *Show the temporal logics T_0, $T_{0,W}$., $T_{0,K4}$ and $T_{0,S4}$ have the filtration property.*

Hint: By Theorem 2.3.46 the frames of the canonical models C_λ of the logics $\lambda = T_0, T_{0,W}, T_{0,K4}, T_{0,S4}$ are λ-frames and

C_{T_0} is a temporal model ;
$C_{T_{0,W}}$ is a reflexive temporal model ;
$C_{T_{0,K4}}$ is a transitive temporal model ;
$C_{T_{0,S4}}$ is a reflexive and transitive temporal model .

For any finite set S of temporal formulas closed with respect to subformulas and any element a from C_λ, for $\lambda = T_0, T_{0,W}$, follow directly the proof of Lemma 2.6.11. For $\lambda := T_{0,K4}, T_{0,S4}$, modify appropriately the definition of the relation R_S from Lemma 2.6.11 for the cases of modal logics $K4$ and $S4$, bearing in mind the necessity to satisfy direct and converse temporal conditions.

In the examples above it was sufficient, although not always, to deal only with the greatest filtration. Sometimes it is necessary to use other filtrations, as we have already seen, and we give a certain additional example of applications of filtrations concerning the smallest filtration.

Lemma 2.6.16 *The modal logic $B := T \oplus p \rightarrow \Box\Diamond p$ has the filtration property.*

Proof. According to Theorem 2.3.48 the frame of the canonical model C_B is a reflexive B-frame and satisfies the condition

$$\forall x \forall y (x R y \Rightarrow y R x).$$

Let S be a finite set of formulas closed with respect to subformulas. Suppose a is an element of C_B. We take the smallest filtration $C_{B,a,S}$ by S of the open submodel $C_{B,a}$ of the model C_B generated by a. It remains to show that the frame of $C_{B,a,S}$ is a B-frame. Since C_C is reflexive, according to (1) from the definition of filtrations, C_{B,a,S_1} is also reflexive. Suppose $[b]_{\equiv_S}, [c]_{\equiv_S} \in C_{B,a,S}$ and $\langle [b]_{\equiv_S}, [c]_{\equiv_S} \rangle \in R_S$. We must show that $\langle [c]_{\equiv_S}, [b]_{\equiv_S} \rangle \in R_S$. By definition of the smallest filtration there exist $b_1 \in [b]_{\equiv_S}$ and $c_1 \in [c]_{\equiv_S}$ such that $b_1 R c_1$. Since the formula $\forall x, \forall y (x R y \Rightarrow y R x)$ is valid on C_B we obtain $c_1 R b_1$ which yields $\langle [c]_{\equiv_S}, [b]_{\equiv_S} \rangle \in R_S$ by (1) from the definition of filtrations. Thus $C_{B,a,S}$ is a B-frame. ∎

Corollary 2.6.17 *The modal logic $B := T \oplus p \rightarrow \Box\Diamond p$ has the fmp and is decidable.*

Now we will present some more developed tools for proving possession of the fmp. We begin with a simple generalization of the filtration property.

Definition 2.6.18 *Suppose λ is a modal, temporal or superintuitionistic logic. We say λ has the weak filtration property iff the following holds. For any element a of the canonical model C of λ and any finite set of formulas S there is a finite set S_1 of formulas such that:*
   (a) S_1 extends S;
   (b) The number of elements of S_1 is effectively calculable by S ;
   (c) There is a filtration C_{a,S_1} by means of S_1 of the open submodel C_a of C which is generated by a such that the frame of C_{a,S_1} is a certain λ-frame.

Theorem 2.6.19 *Any logic λ with the weak filtration property has the finite model property.*

Proof. We simply replace each occurrence of S with S_1 in the proof of Theorem 2.6.10. ∎

The weak filtration property can be used to show possession of the fmp for a wider class of logics compared to just the filtration property. Note that sometimes in order to show the fmp we must combine our filtration technique with some additional construction on filtrations. A good example is the proof of the fmp for the Gödel-Löb provability logic GL, this normal modal logic has the following axiomatization

$$GL := S4 \oplus \Box(\Box p \rightarrow p) \rightarrow \Box p.$$

We will show the fmp for GL due to K.Segerberg [165].

Theorem 2.6.20 *Gödel-Löb modal logic $GL := K4 \oplus \Box(\Box p \rightarrow p) \rightarrow \Box p$ has the finite model property.*

Proof. Suppose $\alpha \notin GL$. We consider the canonical model C_{GL} for GL. Because C_{GL} is a characterizing model for GL (see Theorem 2.3.27) there exists some $a \in C_{GL}$ such that $a \nVdash_V \alpha$. Clearly, the formula α will be invalidated on the element a in the open submodel $C_{GL,a}$ of C_{GL} generated by a. Before making a filtration on $C_{GL,a}$ we prepare the set of formulas S. First we put $S_0 := \{\alpha, \Box\bot\}$, then we close S_0 with respect to subformulas and obtain the set S_1; and finally, $S := S_1 \cup \{\neg\beta \mid \beta \in S_1\}$. Now we take the filtration by S of the open submodel $C_{GL,a}$ of the model C_{GL} generated by the element a defining the accessibility relation R_S as follows:

$$([w]_{\equiv_S} R_S [v]_{\equiv_S}) \Leftrightarrow \forall \Box\alpha \in S(w \Vdash_V \Box\alpha \Rightarrow v \Vdash_V \Box\alpha \wedge \alpha).$$

Using the same arguments as in Lemma 2.6.11 we can show that the filtrated model $\mathfrak{M}$ is transitive. Therefore the notion of clusters is meaningful for $\mathfrak{M}$. By Lemma 2.6.8 this model rejects α.

We claim that every maximal cluster C of $\mathfrak{M}$ is single-element and consists of an irreflexive element, which we will denote $i(C)$. Indeed, otherwise there is some $b \in C$ which is reflexive. Then it follows $b = [b_1]_{\equiv_S}$, and $b \nVdash_{V_S} \Box\bot$. According to Lemma 2.6.8 $b_1 \nVdash_V \Box\bot$. Since $\Box(\Box\bot \rightarrow \bot) \rightarrow \Box\bot$ is a theorem of GL we get $b_1 \nVdash_V \Box(\Box\bot \rightarrow \bot)$, which yields there is some c_1 such that $b_1 R c_1$, and $c_1 \Vdash_V \Box\bot$. By Lemma 2.6.8 we conclude $[c_1]_{\equiv_V} \Vdash_{V_S} \Box\bot$. Therefore $[c_1]_{\equiv_V}$ is a maximal irreflexive element. By (1) from the definition of filtrations we have $[b_1]_{\equiv_S} R_S [c_1]_{\equiv_S}$ which contradicts C being a maximal cluster.

Suppose C is some non-maximal cluster from $C_{GL,a,S}$ which is not degenerated (i.e. is not a cluster consisting of an irreflexive element). Consider the set

$E_C := \{u \mid u \in |\mathfrak{M}|, [u]_{\equiv_S} \in C\}$. We will show that there is a maximal irreflexive element $i(C)$ in E_C. The set S is finite and $\mathfrak{M}$ is a filtration by S, therefore for every $v \in |\mathfrak{M}|$ and each $w \in |\mathfrak{M}|$ if $w \neq v$ then there is a formula $\beta_{u,w} \in S$, such that $v \Vdash_S \beta_{v,w}$ and $w \nVdash_S \beta$. Hence, for every $v \in |\mathfrak{M}|$, there is a conjunction γ_v of formulas from S such that γ_v is valid only v and false on other elements of $\mathfrak{M}$. Therefore the formula

$$\delta(C) := \bigvee_{v \in C_{GL,a},S, v \notin C} \gamma_v$$

is valid only on elements which are out of C. Using this, according to Lemma 2.6.8 we derive

$$\forall u \in C_{GL,a}, \ u \Vdash_V \delta(C) \Leftrightarrow [u]_{\equiv_S} \notin C. \tag{2.21}$$

(note that (2.21) holds in spite of $\delta(C)$ could be out of S since $\delta(C)$ is a boolean combination of formulas from S). If $u \in C_{GL,a}$ is a such that $[u]_{\equiv_S} \in C$ then by (2.21) $u \nVdash_V \delta(C)$. Therefore $a \nVdash_V \Box\delta(C)$. But we know that $\Box(\Box\delta(C) \rightarrow \delta(C)) \rightarrow \Box\delta(C)$ is a theorem of GL. Therefore $a \Vdash_V \Box(\Box\delta(C) \rightarrow \delta(C)) \rightarrow \Box\delta(C)$. This implies $a \nVdash_V \Box(\Box\delta(C) \rightarrow \delta(C))$, i.e. there is some b such that $b \Vdash_V \Box\delta(C)$ and $b \nVdash_V \delta(C)$. Then b must be irreflexive and according to (2.21) $[b]_{\equiv_S} \in C$. If bRd then $d \Vdash_V \delta(C)$ and by (2.21) $[d]_{\equiv_S} \notin C$. Thus $i(C) := b$ is some maximal irreflexive element in $E(C)$.

Now we define the accessibility relation R_1 on $|\mathfrak{M}|$ as follows: the relation R_1 on elements from distinct clusters coincides with R_S. For any non-degenerate cluster C, all elements of C are irreflexive by R_1 and form an ascending by R_1 chain with $[i(C)]_{\equiv_S}$ as maximal element. The model on $|\mathfrak{M}|$ with R_1 as the accessibility relation and the same valuation is denoted by $\mathfrak{M}_1$. It is clear that $\mathfrak{M}_1$ is a finite transitive irreflexive model (i.e. $\neg(xR_1x)$ for all elements x). From this it follows immediately that the model $\mathfrak{M}_1$ is based on a GL-frame (we can easily show directly that the frame of such model satisfies the additional axiom of GL comparing to $K4$)

The model $\mathfrak{M}_1$ disproves α also. In fact, for any formula $\beta \in S$ and any $b \in C_{GL,a}$,

$$[b]_{\equiv_S} \Vdash_S \beta \text{ in } \mathfrak{M} \Leftrightarrow [b]_{\equiv_S} \Vdash_S \beta \text{ in } \mathfrak{M}_1. \tag{2.22}$$

The proof of this equivalence can be given by induction on the length of β. The basis of induction and inductive steps for non-modal connectives are evident. Let $[b]_{\equiv_S} \Vdash_S \Box\gamma$ in $\mathfrak{M}$. Consider $[c]_{\equiv_S}$ such that $[b]_{\equiv_S} R_1 [c]_{\equiv_S}$. Then according to the definition of R_1 $[b]_{\equiv_S} R_S [c]_{\equiv_S}$ and consequently $[c]_{\equiv_S} \Vdash_S \gamma$ in $\mathfrak{M}$ which by inductive hypothesis yields $[c]_{\equiv_S} \Vdash_S \gamma$ in $\mathfrak{M}_1$. Thus $[b]_{\equiv_S} \Vdash_S \Box\gamma$ in $\mathfrak{M}_1$. Conversely, let $[b]_{\equiv_S} \Vdash_S \Box\gamma$ in $\mathfrak{M}_1$. Suppose that $[b]_{\equiv_S} \nVdash_S \Box\gamma$ in $\mathfrak{M}$. Then in particular $[i(C)]_{\equiv_S} \nVdash_S \Box\gamma$ in $\mathfrak{M}$. By Lemma 2.6.8 we have $i(C) \nVdash_S \Box\gamma$. Since $i(C)$ is irreflexive there is an element v such that $i(C)Rv, i(C) \neq v$ and

$v \nVdash_V \gamma$. Then $[i(C)]_{\equiv_S} R_S [v]_{\equiv_S}$ and $[v]_{\equiv_S} \nVdash_S \gamma$ in $\mathfrak{M}$. Hence $[i(C)]_{\equiv_S} R_1 [v]_{\equiv_S}$ and by inductive hypothesis $[v]_{\equiv_S} \nVdash_S \gamma$ in $\mathfrak{M}_1$ which contradicts to $[b]_{\equiv_S} \Vdash_S$ $\Box\gamma$ in $\mathfrak{M}_1$. Hence $[b]_{\equiv_S} \Vdash_S \Box\gamma$ in $\mathfrak{M}$. Thus, (2.22) holds and $\mathfrak{M}_1$ disproves α. ∎

Corollary 2.6.21 *The modal logic GL is decidable.*

We recall that

Definition 2.6.22 *A Kripke model $\mathcal{M} := \langle W, R, V \rangle$ (a frame $\mathcal{F} := \langle W, R \rangle$ is called* irreflexive *if $\forall a \in W \neg(aRa)$.*

A subset S of an irreflexive frame $\mathcal{F} := \langle F, R \rangle$ is called *an infinite ascending chain* if (a) S is a chain (i.e. $\forall a, b \in S$ aRb or bRa) and (b) for every $a \in S$, there is some $b \in S$ such that aRb. Using Theorem 2.6.3 and Corollary 2.6.21 we immediately infer

Corollary 2.6.23 *The modal logic GL is Kripke complete and the class Fr_{GL} of all transitive irreflexive frames which have no infinite ascending chains of elements characterizes GL.*

Proof. Bearing in mind all said above we need only to verify that on all frames form Fr_{GL} the axiom $\Box(\Box p \to p) \to \Box p$ is valid, which can be easily shown directly. ∎

Exercise 2.6.24 *Prove that the normal modal logic*

$$GL4.3 := GL \oplus \Box(\Box p \to q) \vee \Box(\Box q \to p)$$

has the fmp, is decidable and Kripke complete.

Hint: Use a technique from Theorem 2.6.20 and the filtration technique used to prove similar results for $K4.3$.

The next example of developed filtration methods is the technique of *virtually last element* which is also due to K.Segerberg [165] and is related to the method used to prove above that GL has the fmp. The relationship consists of necessity to redefine the accessibility relation on the clusters of the filtrated model and to convert them into partially ordered sets. Recall that the Grzegorczyk normal modal logic Grz has the following axiomatic system:

$$Grz := S4 \oplus \Box(\Box(p \to \Box p) \to p) \to p.$$

As we will see further this logic has an especial importance in view of its special role while interpreting intuitionistic logics in modal logics. Therefore we are interested to clarify its semantic properties, in particular, to prove the fmp.

Theorem 2.6.25 *The modal Grzegorczyk's logic*

$$Grz := S4 \oplus \Box(\Box(p \to \Box p) \to p) \to p$$

has the finite model property.

Proof. Let $\alpha \notin Grz$. The canonical model C_{Grz} for Grz rejects α because C_{Grz} is a characterizing model for Grz (cf. Theorem 2.3.27). Hence there exists some $a \in C_{Grz}$ such that $a \not\Vdash \alpha$. The formula α will also be disproved on the element a in the open submodel $C_{Grz,a}$ of C_{Grz} generated by a. Let S be the set of all subformulas of α and their negations. Now we define the filtration of $C_{Grz,a}$ by the set S by means of introducing the accessibility relation R_S as follows:

$$([w]_{\equiv_S} R_S [v]_{\equiv_S}) \Leftrightarrow \forall \Box \alpha \in S(w \Vdash_V \Box \alpha \Rightarrow v \Vdash_V \Box \alpha \wedge \alpha).$$

Using the same arguments as in Lemma 2.6.11 we can show that the filtrated model $\mathfrak{M}$ is reflexive and transitive. The model $\mathfrak{M}$ also rejects α by Lemma 2.6.8. But this model can have proper clusters, that is, its frame can be not a Grz-frame.

Let $\mathfrak{N}$ be a filtration of a generated rooted submodel $C_{\lambda,a}$ of a canonical model C_λ of some normal modal logic λ. Suppose $\mathfrak{N}$ is reflexive and transitive and C is some cluster of $\mathfrak{N}$. For $c \in C$, we say c is *virtually last* in C if there is an $u \in C_{\lambda,a}$ such that:

$(i) [u]_{\equiv_S} = c,$
$(ii) \forall v \in C_{\lambda,a}[(uRv) \& ([v]_{\equiv_S} \in C) \Rightarrow ([v]_{\equiv_S} = [u]_{\equiv_S})].$

In particular, it is clear that any single-element cluster has a unique virtually last element.

Lemma 2.6.26 *Let λ be a normal modal logic containing Grz, let $C_{\lambda,a}$ be the open submodel of the canonical model C_λ generated by some $a \in |C_\lambda|$. Let S be a finite set of formulas $\{\alpha_1, ..., \alpha_m\} \cup \{\neg\alpha_1, ..., \neg\alpha_m\}$ closed with respect to subformulas. Then, for any filtration $C_{\lambda,a,S}$ of $C_{\lambda,a}$ by S, any cluster from $C_{\lambda,a,S}$ contains some virtually last element.*

Proof. Note that using the same arguments as in the proof of Theorem 2.3.28 concerning the logic $S4$ we can show that the canonical model C_λ is reflexive and transitive. Suppose there are no virtually last elements in some proper cluster C from $C_{\lambda,a,S}$. This, in particular, means there is an infinite sequence $a_n, n \in N$ of elements from $C_{\lambda,a}$ and an element $u \in C$ such that

$(a) \forall n(a_n R a_{n+1}), \forall n([a_n]_{\equiv_S} \in C), \forall n([a_{2n+1}]_{\equiv_S} = u,$
$\quad \forall n([a_{2n}]_{\equiv_S} \neq u),$
$(b) \forall n \forall b[(a_n Rb) \& ([b]_{\equiv_S} \in C) \Rightarrow \exists b_1 \exists b_2 (bRb_1) \& (b_1 Rb_2) \&$
$\quad ([b_1]_{\equiv_S} = u) \& ([b_2]_{\equiv_S} \neq u) \& (([b_2]_{\equiv_S} \in C)].$

Since $C_{\lambda,a,S}$ is a filtration by finite S of special described above type, for every different $v, w \in C_{GL,a,S}$ there is a formula $\beta_{v,w} \in S$, such that $v \Vdash_S \beta_{v,w}$ and $w \nVdash_S \beta$. Hence, for any $v \in C_{\lambda,a,S}$, there is a conjunction γ_v of formulas from S such that γ_v is valid only on the element v and is false on all other elements. Hence the formula

$$\delta(C, u) := \bigvee_{(v \notin C) \vee (v = u)} \gamma_v$$

will be valid only on elements which are out of C or equal to u. Hence by Lemma 2.6.8 we conclude

$$\forall b \in C_{\lambda,a}, \ b \Vdash_V \delta(C, u) \Leftrightarrow (([b]_{\equiv_S} \notin C) \vee ([b]_{\equiv_S} = u)). \tag{2.23}$$

Let $\beta := \delta(C, u)$. Then $a_0 \nVdash_V \beta$. We claim that $a_0 \Vdash_V \Box(\Box(\beta \to \Box\beta) \to \beta)$. In fact, if $a_0 R b$ and $b \nVdash_V \beta$ then by (2.23) $[b]_\equiv \in C$. Therefore by (b) there are some b_1 and b_2 such that $b R b_1, b_1 R b_2$ and $([b_1]_{\equiv_S} = u), ([b_2]_{\equiv_S} \neq u), (([b_2]_{\equiv_S} \in C)$. So by (2.23) $b_1 \Vdash_V \beta, b_2 \nVdash_v \beta$ and $b_1 \nVdash_V \Box\beta$. Hence $b \nVdash_V \Box(\beta \to \Box\beta)$. Thus it follows $a_0 \nVdash_V \Box(\Box(\beta \to \Box\beta) \to \beta) \to \beta$, but the last formula is a theorem of Grz, a contradiction. ∎

We return to the proof of our theorem. By Lemma 2.6.26 any cluster of $\mathfrak{M}$ has a virtually last element. We fix one virtually last element in any cluster C and denote it by $vl(C)$. Now we define the new accessibility relation $\leq$ on $\mathfrak{M}$ as follows: the relation $\leq$ on elements from the different clusters coincides with R_S. For any cluster C, elements of C form an ascending chain by $\leq$, where the virtually last element $vl(C)$ of C becomes the maximal element of this chain. The model on $\mathfrak{M}$ with $\leq$ as the accessibility relation and the same valuation is denoted by $C^{\prec}_{Grz,a,S}$. It is clear that $C^{\prec}_{Grz,a,S}$ is a finite partially ordered set. It is easy to see that all axioms of Grz are valid on finite posets. Hence $C^{\prec}_{Grz,a,S}$ is based on a finite Grz-frame.

Our aim now is to show that $C^{\prec}_{Grz,a,S}$ also disproves α. For this it is sufficient to prove that for any formula $\beta \in S$ and any $b \in |\mathfrak{M}|$,

$$[b]_{\equiv_S} \Vdash_S \beta \text{ in } \mathfrak{M} \Leftrightarrow [b]_{\equiv_S} \Vdash_S \beta \text{ in } C^{\prec}_{Grz,a,S}. \tag{2.24}$$

We use induction on the length of β. The basis of induction and the inductive steps for non-modal connectives are evident. Suppose that $[b]_{\equiv_S} \Vdash_S \Box\gamma$ in $\mathfrak{M}$. Consider $[c]_{\equiv_S}$ such that $[b]_{\equiv_S} \leq [c]_{\equiv_S}$. By the definition of $\leq$ we immediately obtain $[b]_{\equiv_S} R_S [c]_{\equiv_S}$ and therefore $[c]_{\equiv_S} \Vdash_S \gamma$ in $\mathfrak{M}$. This by the inductive hypothesis directly implies $[c]_{\equiv_S} \Vdash_S \gamma$ in $C^{\prec}_{Grz,a,S}$. Thus we have proved $[b]_{\equiv_S} \Vdash_S \Box\gamma$ in $C^{\prec}_{Grz,a,S}$. Conversely, suppose now $[b]_{\equiv_S} \Vdash_S \Box\gamma$ in $C^{\prec}_{Grz,a,S}$. Assume that $[b]_{\equiv_S} \nVdash_S \Box\gamma$ in $\mathfrak{M}$. Suppose C is the cluster containing $[b]_{\equiv_S}$. Then $vl(C) = [r]_{\equiv_S}$ for some $r \in C_{Grz,a}$ and $[r]_{\equiv_S} \nVdash_S \Box\gamma$ in $\mathfrak{M}$. By Lemma 2.6.8 we have $r \nVdash_V \Box\gamma$. Hence there is some q such that $r R q$ and $q \nVdash_V \gamma$.

By Lemma 2.6.8 it follows $[q]_{\equiv_S} \not\Vdash_S \gamma$ in $\mathfrak{M}$. By inductive hypothesis we infer $[q]_{\equiv_S} \not\Vdash_S \gamma$ in $C^{\prec}_{Grz,a,S}$. Since $[r]_{\equiv_S}$ is a virtually last in the cluster C, either $[q]_{\equiv_S} = [r]_{\equiv_S}$ or $[q]_{\equiv_S} \notin C$. Thus in any case we obtain $[b]_{\equiv_S} \leq [q]_{\equiv_S}$. Hence $[b]_{\equiv_S} \not\Vdash_S \Box\gamma$ in $C^{\prec}_{Grz,a,S}$, a contradiction. Thus, (2.24) holds which completes the proof of our theorem. ∎

This theorem, Theorem 2.6.3 and Corollary 2.6.6 immediately yield

Corollary 2.6.27 *The modal logic Grz is decidable and Kripke complete, and the class of posets without infinite ascending chains characterizes Grz.*

Exercise 2.6.28 *Show that the normal modal logic $Grz.3 := Grz \oplus \Box(\Box p \to \Box q) \vee \Box(\Box q \to \Box p)$ has the fmp, is decidable and is Kripke complete.*

Hint: Use the technique of *virtually last elements* from Theorem 2.6.25 and the filtration technique used in the proof of similar results for $S4.3$.

2.7 Relation of Intuitionistic and Modal Logics

There is a deep relationship between modal and intuitionistic logics which is based both on the motivations for introducing corresponding axiomatic systems and on the similarity of the semantic tools. There is a well developed theory of the relations between the family of modal logics and the lattice of all superintuitionistic logics. In this section we present only a small number of the known results (which play however a basic role in this theory), which we will need for our investigation of inference rules. Some short historical comments are made in the final part of this section.

In a sense, the theory of relations between modal and intuitionistic logics is based on an idea of K.Gödel[57] concerning the possibility of expressing the intuitionistic truth of formulas by means of the modal logic, which was father developed by McKinsey and Tarski [101]. This approach uses the following translation T of propositional formulas in the language $\mathcal{L}_H$ into modal formulas. The Gödel-McKinsey-Tarski translation T can be inductively defined as follows.

Definition 2.7.1 *For any propositional letter p,*

$$T(p) := \Box p,$$
$$T(\alpha \wedge \beta) := T(\alpha) \wedge T(\beta),$$
$$T(\alpha \vee \beta) := T(\alpha) \vee T(\beta),$$
$$T(\alpha \to \beta) := \Box(T(\alpha) \to T(\beta)),$$
$$T(\neg\alpha) := \Box\neg T(\alpha).$$

Lemma 2.7.2 *For any formula α in the language $\mathcal{L}_H$,*

$$(T(\alpha) \equiv \Box T(\alpha)) \in S4.$$

Proof. This can be shown by means of simple induction on the length of the formula α. Basis of the induction and the inductive steps for connectives $\rightarrow$ and $\neg$ are evident. $\Box[T(\gamma) \wedge T(\delta)]$ is equivalent in $S4$ to $\Box T(\gamma) \wedge \Box T(\delta)$ by Proposition 1.3.16. The formula $\Box T(\gamma) \wedge \Box T(\delta)$ is equivalent in $S4$ to $T(\gamma) \wedge T(\delta)$ by the inductive hypothesis. Regarding $\vee$, $\Box(T(\alpha) \vee T(\delta)) \rightarrow T(\alpha) \vee T(\beta) \in S4$ since $S4$ has as axiom $\Box p \rightarrow p$. Conversely, using the inductive hypothesis we derive $T(\gamma) \vee T(\delta) \rightarrow \Box T(\gamma) \vee \Box T(\delta) \in S4$ and $\Box T(\gamma) \vee \Box T(\delta) \rightarrow \Box[T(\gamma) \vee T(\delta)] \in S4$ (using Kripke models for $S4$ to show this). Thus $T(\gamma) \vee T(\delta) \rightarrow \Box[T(\gamma) \vee T(\delta)] \in S4$. ∎

Note that it is possible to define other translations T_1 that are equivalent to T with respect to derivability in modal logical systems (for instance letting $T_1(p) = \Box p$, $T_1(\alpha \wedge \beta) := \Box T_1(\alpha) \wedge \Box T_1(\beta)$, $T_1(\alpha \vee \beta) := \Box T_1(\alpha) \vee \Box T_1(\beta)$, $T_1(\neg \alpha) := \Box \neg \Box T_1(\alpha)$ and $T_1(\alpha \rightarrow \beta) := \Box(\Box T_1(\alpha) \rightarrow \Box T_1(\beta))$). But we will consider here only the translation T (since this corresponds to our particular aims). The Gödel conjecture can be presented in the following form: for any formula α in the language $\mathcal{L}_H$, α is provable in the intuitionistic logic H iff the formula $T(\alpha)$ is a theorem of the Lewis modal logic $S4$. This conjecture was proved using semantic methods by McKinsey and Tarski [101], and was then generalized on all superintuitionistic logics by Dummett and Lemmon [34], and was developed then by Maksimova and Rybakov in [97]. To present this generalization we introduce the following mapping τ of the lattice $\mathcal{L}_{si}$ of all superintuitionistic logics into the lattice $\mathcal{L}_{ml}$ of all modal logics over $S4$:

Definition 2.7.3 *For any given superintuitionistic logic λ,*

$$\tau(\lambda) := S4 \oplus \{T(\alpha) \mid \alpha \in \lambda\}.$$

Theorem 2.7.4   *em Dummett-Lemmon [34] Translation Theorem.*

(i) For any superintuitionistic logic λ and any formula α in the language $\mathcal{L}_H$, $\alpha \in \lambda \Longleftrightarrow T(\alpha) \in \tau(\lambda)$. (ii) For any superintuitionistic logic $\lambda := H \oplus Ax$ and any formula α in the language $\mathcal{L}_H$,

$$\alpha \in H \oplus Ax \Longleftrightarrow T(\alpha) \in S4 \oplus \{T(\alpha) \mid \alpha \in Ax\}.$$

This theorem can be proved in a number of different ways. We prefer to prove it using certain semantic tools which we will also employ later for other purposes. The theorem will be obtained as a simple corollary of these mentioned semantic instruments. Recall that any modal algebra $\mathfrak{B}$ from $Var(S4)$ is called a *topoboolean algebra* and an *open element* of $\mathfrak{B}$ is an element $a \in |\mathfrak{B}|$ such that

$\Box a = a$. We use this names since in any algebra of $\mathfrak{B} \in Var(S4)$ operation $\Diamond$ satisfies all equations for the topological closure operator, and operation $\Box$ satisfies all equalities for topological operator of taking interior from the Kuratowski axiomatic system of topological spaces. First we show that:

Lemma 2.7.5 *For any modal algebra $\mathfrak{B}$ from the variety $Var(S4)$ the set $G(\mathfrak{B}) := \{\Box a \mid a \in |\mathfrak{B}|\}$ of all open elements from $\mathfrak{B}$ forms a pseudo-boolean algebra with respect to operations $\vee$, $\wedge$ and*

$$\Box a \to \Box b := \Box(\neg \Box a \vee \Box b), \quad \neg \Box a := \Box \neg \Box a.$$

Proof. By Lemma 2.1.13 it is sufficient to show that $G(\mathfrak{B})$ is a lattice with a greatest and a smallest element and that $\Box a \to \Box b$ is a pseudo-complement of $\Box a$ to $\Box b$. It is clear that $\bot = \Box \bot$ and $\top = \Box \top$ are the smallest and the greatest elements of $G(\mathfrak{B})$ and $G(\mathfrak{B})$ is a lattice with respect to $\vee$, $\wedge$ (since $\mathfrak{B}$ itself is a boolean algebra). We calculate directly:

$$\Box(\neg \Box a \vee \Box b) \wedge \Box a = \Box((\neg \Box a \vee \Box b) \wedge a) \leq \Box a.$$

Suppose that $\Box c \wedge \Box a \leq \Box b$. Then $\Box c \wedge \Box a \leq \Box c \wedge \Box b$. Using this we derive

$$\Box c \wedge \Box(\neg \Box a \vee \Box b) = \Box \Box c \wedge \Box(\neg \Box a \vee \Box b) =$$
$$\Box((\Box c \wedge \neg \Box a) \vee (\Box c \wedge \Box b)) \geq \Box((\Box c \wedge \neg \Box a) \vee (\Box c \wedge \Box a)) =$$
$$\Box(\Box c \wedge (\neg \Box a \vee \Box a)) = \Box \Box c = \Box c.$$

That is, $\Box a \supset \Box b := \Box(\neg \Box a \vee \Box b)$ is really a pseudo-complement in the lattice $G(\mathfrak{B})$ of $\Box a$ to $\Box b$. ∎

The connection between the truth of T-translated formulas in $\mathfrak{B}$ and the original formulas in the pseudo-boolean algebra $G(\mathfrak{B})$ is given by the following lemma.

Lemma 2.7.6 *For any modal algebra $\mathfrak{B} \in Var(S4)$ and any formula α in the language $\mathcal{L}_H$, the following holds:*

$$\mathfrak{A} \models T(\alpha) \Longleftrightarrow G(\mathfrak{B}) \models \alpha.$$

Proof. This follows immediately from the definition of operations on $G(\mathfrak{B})$ and the action of T on formulas from $\mathcal{L}_H$.

We recall also a simple connection between the validity of intuitionistic formulas in posets and T-translations of intuitionistic formulas in $S4$-frames. For any $S4$-frame $\mathcal{F} := \langle W, R \rangle$, we can introduce the frame $\mathcal{F}_c := \langle W_c, \preceq \rangle$ which is based upon the set W_c of all clusters from $\mathcal{F}$ and for all $C_1, C_2 \in W_c$, $C_1 \leq C_2$ $\Leftrightarrow \exists c_1 \in C_1, \exists c_2 \in C_2$ $(c_1 R c_2)$. That is, $\mathcal{F}_c$ is the poset of all clusters from $\mathcal{F}$ with partial order generated from $\mathcal{F}$. We call $\mathcal{F}_c$ by skeleton of $\mathcal{F}$.

Lemma 2.7.7 *For any S4-frame $\mathcal{F}$ and any formula $\alpha(p_1, ..., p_m)$ in the non-modal propositional language the following hold*

    (i) For any intuitionistic valuation V of letters $p_1, ..., p_n$ in $\mathcal{F}_c$ and any $C \in \mathcal{F}_c$, $C \Vdash_V \alpha$ in the intuitionistic model $\langle \mathcal{F}_c, V \rangle$ iff for every $a \in C$, $a \Vdash_S T(\alpha)$ in the modal Kripke model $\langle \mathcal{F}, S \rangle$, where $S(p_i) := \bigcup \{C_j \mid C_j \in |\mathcal{F}_c|, C_j \Vdash_V p_i\}$.

    (ii) For any valuation V of letters $p_1, ..., p_n$ in $\mathcal{F}$ and any $a \in \mathcal{F}_c$, $a \Vdash_V T(\alpha)$ in the modal Kripke model $\langle \mathcal{F}, V \rangle$ iff $C(a) \Vdash_S \alpha$ in the intuitionistic Kripke model $\langle \mathcal{F}_c, S \rangle$, where $C \Vdash_S p_i \Leftrightarrow \forall a \in C(a \Vdash_V \Box p_i)$.

Proof. The proof is by simple induction on the length of α. We leave the details as an exercise to the reader.

Now we define a converse construction to the construction of pseudo-boolean algebra $G(\beta)$, namely we introduce the wrapping modal algebras for pseudo-boolean algebras. Given a pseudo-boolean algebra $\mathfrak{A}$, we generate the boolean algebra $S(\mathfrak{A})$ as follows. Suppose $\mathcal{B}(\mathfrak{A})$ is the set of all boolean terms constructed out of elements of $\mathfrak{A}$ as constants. We define a binary relation $\sim$ on $\mathcal{B}(\mathfrak{A})$, supposing that $t_1 \sim t_2$ if we can transform t_1 into t_2 by a finite number of the following steps: (i) replacing any subterm t_3, which is a lattice's term constructed out of elements from $\mathfrak{A}$, by a certain lattice's term t_4 which is constructed out of elements of $\mathfrak{A}$ and has the same value in $\mathfrak{A}$ as t_3, (ii) replacing any subterm t_5 by any term t_6 whose value is equal to the value of t_5 in any boolean algebra when we consider the elements on which these terms are constructed as variables. It is clear that $\sim$ is a congruence relation on the algebra of boolean terms $\mathcal{B}(\mathfrak{A})$. Therefore it is possible to define the quotient algebra $\mathcal{B}(\mathfrak{A})/\sim$ which will be a boolean algebra. Let as agree for simplicity to denote elements $t(a_1, ..., a_n)$ from $\mathcal{B}(\mathfrak{A})$ and also the corresponding quotient classes $[t(a_1, ..., a_n)]_\sim$ from $\mathcal{B}(\mathfrak{A})/\sim$ by the same notation $t(a_1, ..., a_n)$, bearing in mind that any such element can have different but equal representations. Now we introduce the operation $\Box$ on $\mathcal{B}(\mathfrak{A})/\sim$ as follows. Any term $t(a_1, ..., a_n)$ from $\mathcal{B}(\mathfrak{A})/\sim$ can be represented in a conjunctive normal form $\bigwedge_{1 \leq i \leq n}(\neg a_i \vee b_i)$ were all a_i and b_i are elements of $\mathfrak{A}$ (which is possible since the values of lattice's terms on elements of $\mathfrak{A}$ again belong to $\mathfrak{A}$). We define

$$\Box \bigwedge_{1 \leq i \leq n} (\neg a_i \vee b_i) := \bigwedge_{1 \leq i \leq n} (a_i \rightarrow_{\mathfrak{A}} b_i). \qquad (2.25)$$

Lemma 2.7.8 *The operation $\Box$ is well-defined on $\mathcal{B}(\mathfrak{A})$ and does not depend on the representations of elements.*

Proof. Suppose $a \in \mathcal{B}(\mathfrak{A})/\sim$ and

$$a = \bigwedge_{1 \leq i \leq n} (\neg a_i \vee b_i) = \bigwedge_{1 \leq j \leq m} (\neg c_j \vee d_j).$$

Note that (i): for any elements $u, v \in |\mathfrak{A}|$:

$$[(u \to_{\mathfrak{A}} v) \leq \neg_{\mathcal{B}(\mathfrak{A})} u \vee v] \&$$

$$[\& \forall g \in |\mathfrak{A}|((g \leq \neg_{\mathcal{B}(\mathfrak{A})} u \vee v) \Rightarrow g \leq u \to_{\mathfrak{A}} v)] \tag{2.26}$$

In order to show this note that, for every element c from $\mathcal{B}(\mathfrak{A})$, $c \leq \neg_{\mathcal{B}(\mathfrak{A})} u \vee v \Leftrightarrow c \wedge u \leq v$. We will omit the subscription $\mathcal{B}(\mathfrak{A})$ in $\neg_{\mathcal{B}(\mathfrak{A})}$ for simplicity. We have if $c \leq \neg u \vee v$ then $u \wedge c \leq u \wedge (\neg u \vee v) = u \wedge v \leq v$. Conversely, if $u \wedge c \leq v$ then $c = (c \wedge (u \vee \neg u)) = (c \wedge u) \vee (u \wedge \neg u) \leq v \wedge (c \wedge \neg u) \leq \neg u \vee v$. Thus the necessary equivalence holds. This equivalence immediately implies (2.26).

From (2.26) we obtain, for every i (j), the element $a_i \to_{\mathfrak{A}} b_i$ $(c_j \to_{\mathfrak{A}} d_j)$ is the biggest element from $\mathfrak{A}$ contained in $\neg a_i \vee b_i$ (in $\neg c_j \vee d_j$ respectively). Using this observation we infer, for every i,

$$\bigwedge_{1 \leq j \leq m} (\neg c_j \vee d_j) \leq \neg a_i \vee b_i,$$
$$\bigwedge_{1 \leq j \leq m} (\neg c_j \to_{\mathfrak{A}} d_j) \leq \bigwedge_{1 \leq j \leq m} (\neg c_j \vee d_j)$$
$$\bigwedge_{1 \leq j \leq m} (\neg c_j \to_{\mathfrak{A}} d_j) \leq a_i \to_{\mathfrak{A}} b_i.$$

This yields

$$\bigwedge_{1 \leq j \leq m} (\neg c_j \to_{\mathfrak{A}} d_j) \leq [\bigwedge_{1 \leq i \leq n} (\neg a_i \to_{\mathfrak{A}} b_i)].$$

The converse inclusion can be shown in the same way. ∎

Definition 2.7.9 *For any pseudo-boolean algebra $\mathfrak{A}$, the wrapping modal algebra $S(\mathfrak{A})$ is the boolean algebra $\mathcal{B}(\mathfrak{A})/_\sim$ with the modal operation $\square$ defined by (2.25).*

Lemma 2.7.10 *The algebra $S(\mathfrak{A})$ is a modal algebra and $S(\mathfrak{A}) \in Var(S4)$.*

Proof. We already noted that this algebra is a boolean algebra. It follows by definition immediately that for any $b, c \in \mathfrak{A}$, $\square b \leq b$, $\square \square b = \square b$, $\square \top = \top$, and $b \leq c \Rightarrow \square b \leq \square c$. It remains only to check the inclusion $\square(a \to b) \leq (\square a \leq \square b)$. in order to show that $S(\mathfrak{A})$ is a modal algebra. Using above observations and (2.26), we derive:

$$\square[\bigwedge_{1 \leq i \leq n} (\neg a_i \vee b_i) \to \bigwedge_{1 \leq j \leq m} (\neg c_j \vee d_j)] \leq$$
$$\square([\bigwedge_{1 \leq i \leq n} (a_i \to_{\mathfrak{A}} b_i)] \to_{\mathcal{B}(\mathfrak{A})} \bigwedge_{1 \leq j \leq m} (c_j \to_{\mathcal{B}(\mathfrak{A})} d_j) \leq$$
$$\square(\bigwedge_{1 \leq j \leq m} (([\bigwedge_{1 \leq i \leq n} (a_i \to_{\mathfrak{A}} b_i)] \wedge c_j) \to d_j)) =$$
$$(\bigwedge_{1 \leq j \leq m} (([\bigwedge_{1 \leq i \leq n} (a_i \to_{\mathfrak{A}} b_i)] \wedge c_j) \to_{\mathfrak{A}} d_j)) \leq$$
$$\bigwedge_{1 \leq i \leq n} (a_i \to_{\mathfrak{A}} b_i) \to_{\mathfrak{A}} (\bigwedge_{1 \leq j \leq m} (c_j \to_{\mathfrak{A}} d_j) \leq$$
$$\bigwedge_{1 \leq i \leq n} (a_i \to_{\mathfrak{A}} b_i) \to_{\mathcal{B}(\mathfrak{A})} \bigwedge_{1 \leq j \leq m} (c_j \to_{\mathfrak{A}} d_j) =$$
$$\square(\bigwedge_{1 \leq i \leq n} (\neg a_i \vee b_i)) \to \square(\bigwedge_{1 \leq j \leq m} (\neg c_j \vee d_j))$$

and our lemma is proved. ∎

Note that $|G(S(\mathfrak{A}))| = |\mathfrak{A}|$ and the values of pseudo-boolean operations on $G(S(\mathfrak{A}))$ and $\mathfrak{A}$ itself coincide. Hence this observation, Lemma 2.7.10 and Lemma 2.7.6 immediately yields

Corollary 2.7.11 *The following hold:*

(i) Pseudo-boolean algebras $G(S(\mathfrak{A}))$ and $\mathfrak{A}$ coincide;

(ii) For any formula α, $\mathfrak{A} \models \alpha \Longleftrightarrow S(\mathfrak{A}) \models T(\alpha)$.

Using these initial simple observations we can easily proof Theorem 2.7.5.

Proof of Theorem 2.7.5: We begin with (ii). Suppose $\alpha \in H \oplus Ax$. Also suppose that $T(\alpha) \notin S4 \oplus \{T(\beta) \mid \beta \in Ax\}$. By the algebraic completeness theorem for modal logics (see Theorem 2.2.5) there exists a modal algebra $\mathfrak{B} \in Var(S4 \oplus \{T(\beta \mid \beta \in Ax\})$ such that $\mathfrak{B} \models S4 \oplus \{T(\beta \mid \beta \in Ax\}$ and $\mathfrak{B} \not\models T(\alpha)$. Then by Lemma 2.7.6 $G(\mathfrak{B}) \models Ax$, $G(\mathfrak{B}) \not\models \alpha$. Therefore $G(\mathfrak{B}) \in Var(\lambda)$ and $\alpha \notin \lambda$, a contradiction. Therefore we can conclude that $T(\alpha) \in S4 \oplus \{\beta \mid \beta \in Ax\}$.

Assume now that $\alpha \notin H \oplus Ax$. Then by the algebraic completeness theorem for superintuitionistic logics (cf. Theorem 2.1.10) there is a pseudo-boolean algebra $\mathfrak{A} \in Var(H \oplus Ax)$ such that $\mathfrak{A} \models Ax$ and $\mathfrak{A} \not\models \alpha$. Applying Corollary 2.7.11 we get $S(\mathfrak{A}) \models \{T(\beta) \mid \beta \in Ax\}$ and $S(\mathfrak{A}) \not\models T(\alpha)$. Thus $S(\mathfrak{A}) \in Var(S4 \oplus \{T(\beta) \mid \beta \in Ax\})$ and $T(\alpha) \notin S4 \oplus \{T(\beta) \mid \beta \in Ax\}$. Thus (ii) is proved. To show (i) note that if $\alpha \in \lambda$ then by (ii) $T(\alpha) \in S4 \oplus \{T(\beta) \mid \beta \in \lambda\} = \tau(\lambda)$. If $\alpha \notin \lambda$ then by (ii) $T(\alpha) \notin S4 \oplus \{T(\beta) \mid \beta \in \lambda\} = \tau(\lambda)$. ∎

Definition 2.7.12 *Let λ be a superintuitionistic logic. A modal logic λ_1 is said to be a modal counterpart for λ if for any formula α, $\alpha \in \lambda \Longleftrightarrow T(\alpha) \in \lambda_1$.*

By Theorem 2.7.5 the logic $\tau(\lambda)$ is the smallest modal counterpart for λ among modal logics extending $S4$. The family of modal counterparts for a logic λ can be very rich and complicated. We are going now to show that among modal counterparts for any superintuitionistic logic λ there is always a greatest modal counterpart. The introduction of the greatest modal counterpart is based on the definition of wrapping modal algebras of the kind $S(\mathfrak{A})$ which were introduced (in other technique) by L.L.Maksimova [97].

Definition 2.7.13 *Let λ be a superintuitionistic logic. The modal logic $\sigma(\lambda)$ has the following definition:*

$$\sigma(\lambda) := \lambda(\{S(\mathfrak{A}) \mid \mathfrak{A} \in Var(H), \mathfrak{A} \models \lambda\}.$$

To show that $\sigma(\lambda)$ is the greatest modal counterpart for λ among extensions $S4$ we need the following technical lemma.

Lemma 2.7.14 *If $\alpha(p_1, ..., p_u)$ is a modal formula in the modal language built up of propositional letters $p_1, ..., p_u$ then the formula $\Box\alpha(\Box p_1, ..., \Box p_u)$ is equivalent in S4 to the formula $T(\gamma(p_1, ..., p_u))$ for a certain effectively constructed propositional formula $\gamma(p_1, ..., p_u)$ in the propositional language of the intuitionistic logic.*

Proof. The proof proceeds by induction on the length of the given formula α. If $\alpha = p_j$ then $\Box p_j = T(p_j)$ and the basis of the induction holds. Let $\alpha(p_1, ..., p_u)$ be a modal formula. Suppose that for all formulas which have length strictly less than the length of $\alpha(p_1, ..., p_u)$ the conjecture of this lemma holds. If $\alpha = \Box\beta$ then by the inductive hypothesis $\Box\beta(\Box p_i) \equiv T(\gamma(p_i)) \in S4$ and $\Box\Box\beta(\Box p_i) \equiv \Box T(\gamma(p_i)) \in S4$. Since $\Box T(\delta) \equiv T(\delta) \in S4$ for any δ (see Lemma 2.7.2), it follows that $\alpha \equiv T(\gamma(p_i)) \in S4$. If our formula α is not a formula of the form $\Box\beta$, i.e. α has a main logical connective which differs from $\Box$, then α is constructed from subformulas of the form $\Box\delta$ and propositional letters p_i by means of only connectives $\wedge, \vee, \rightarrow$ and $\neg$. Thus

$$\alpha(\Box p_i) := \Box\varphi(\Box\delta_1(\Box p_i), ..., \Box\delta_k(\Box p_i), \Box p_l, ..., \Box p_t),$$

where φ is a formula in the language of the logic H. We transform the formula

$$\varphi(\Box\delta_1(\Box p_i), ..., \Box\delta_k(\Box p_i), \Box p_l, ..., \Box p_t)$$

into a conjunctive normal form and obtain a formula

$$\kappa := \bigwedge_m [\bigvee_{1 \le r \le n_m} \neg\Box\mu_r^m(\Box p_i) \vee \bigvee_{1 \le r \le k_m} \Box\nu_r^m(\Box p_i)],$$

which is equivalent to $\alpha(\Box p_i)$ in $S4$, where $\mu_r^m(\Box p_i))$ and ν_r^m are different formulas from the set $\Box\delta_1(\Box p_i), ..., \Box\delta_k(\Box p_i), \Box p_l, ..., \Box p_t$. At the same time the formula $\Box\kappa$ is equivalent in $S4$ to the formula

$$\kappa_1 := \bigwedge_m \Box(\bigwedge_{1 \le r \le n_m} \Box\mu_r^m(\Box p_i) \rightarrow \bigvee_{1 \le r \le k_m} \Box\nu_r^m(\Box p_i)).$$

By the inductive hypothesis, formulas $\Box\mu_r^m(\Box p_i)$ and $\Box\nu_r^m(\Box p_i)$ are equivalent in $S4$ to certain formulas of the form $T(\gamma_r^m(p_i))$ and $T(\epsilon_r^m(p_i))$, respectively for some effectively constructed formulas $\gamma_r^m(p_i)$ and $\epsilon_r^m(p_i)$ in the language of H. Therefore the formula κ_1 is equivalent in $S4$ to the formula

$$\kappa_2 := \bigwedge_m \Box(\bigwedge_{1 \le r \le n_m} T(\gamma_r^m(p_i)) \rightarrow \bigvee_{1 \le r \le k_m} T(\epsilon_r^m(p_i))).$$

It is clear that this formula coincides with the formula

$$T[\bigwedge_m (\bigwedge_{1 \le r \le n_m} \gamma_r^m(p_i) \rightarrow \bigvee_{1 \le r \le k_m} \epsilon_r^m(p_i))]$$

which establishes the conjecture of our lemma. ∎

Theorem 2.7.15 *(L.L.Maksimova [97]) For any superintuitionistic logic λ, the logic $\sigma(\lambda)$ is the greatest modal counterpart for λ over S4.*

Proof. Let $\alpha \in \lambda$. Consider an arbitrary pseudo-boolean algebra $\mathfrak{A}$ such that $\mathfrak{A} \models \lambda$. By Corollary 2.7.11 $S(\mathfrak{A}) \models T(\beta)$ for any $\beta \in \lambda$. Therefore $T(\alpha) \in \sigma(\lambda)$. If $\alpha \notin \lambda$ then by the algebraic completeness theorem for λ (see Theorem 2.1.10) there is a pseudo-boolean algebra $\mathfrak{A}$ such that $\mathfrak{A} \models \lambda$ and $\mathfrak{A} \not\models \alpha$. Again by Corollary 2.7.11 it follows that $S(\mathfrak{A}) \not\models T(\alpha)$. Hence $T(\alpha) \notin \sigma(\lambda)$ and $\sigma(\lambda)$ is a modal counterpart for λ. Suppose λ_1 is a certain modal counterpart for λ over S4. Consider any pseudo-boolean algebra $\mathfrak{A}$ such that $\mathfrak{A} \models \lambda$. Assume that

$$S(\mathfrak{A}) \not\models \alpha(p_1, ..., p_n)$$

for some modal formula $\alpha(p_1, ..., p_n) \in \lambda_1$. Since the algebra $S(\mathfrak{A})$ is generated by some open elements, there are formulas $\beta_j(\Box q_1, ..., \Box q_m), 1 \leq j \leq n$ such that

$$S(\mathfrak{A}) \not\models \Box\alpha(\beta_1(\Box q_1, ..., \Box q_m), ..., \beta_n(\Box q_1, ..., \Box q_m)).$$

According to Lemma 2.7.14, there is a formula $\gamma(q_1, ..., q_m)$ in the language $\mathcal{L}_H$ of intuitionistic logic H such that

$$\Box\alpha(\beta_1(\Box q_1, ..., \Box q_m), ..., \beta_n(\Box q_1, ..., \Box q_m)) \equiv T(\gamma(q_1, ..., q_m)) \in S4.$$

Thus $S(\mathfrak{A}) \not\models T(\gamma(q_1, ..., q_m))$. At the same time $T(\gamma(q_1, ..., q_m)) \in \lambda_1$. By Corollary 2.7.11 we derive $\mathfrak{A} \not\models \gamma(q_1, ..., q_m)$ and, in particular, the following holds: $\gamma(q_1, ..., q_m) \notin \lambda$. But $T(\gamma(q_1, ..., q_m)) \in \lambda_1$ and λ_1 is a modal counterpart for λ which implies $\gamma(q_1, ..., q_m) \in \lambda$, a contradiction. Thus all theorems of λ_1 are valid on all modal algebras from the definition of $\sigma(\lambda)$. Hence we obtain $\lambda_1 \subseteq \sigma(\lambda)$. ∎

Now we are in a position to further specify the semantic description of the greatest modal counterparts.

Lemma 2.7.16 *For any superintuitionistic logic λ and any family $\mathfrak{A}_i, i \in I$ of pseudo-boolean algebras, if $\lambda = \lambda(\{\mathfrak{A}_i \mid i \in I\})$ then $\sigma(\lambda) = \lambda(\{S(\mathfrak{A}_i) \mid i \in I\})$.*

Proof. It is clear by the definition of $\sigma(\lambda)$ that,

$$\sigma(\lambda) \subseteq \lambda(\{S(\mathfrak{A}_i) \mid i \in I\}).$$

At the same time $\lambda(\{S(\mathfrak{A}_i) \mid i \in I\})$ is a modal counterpart for λ. In fact, if a formula α is not a theorem of λ than there is some $\mathfrak{A}_i$ such that $\mathfrak{A}_i \not\models \alpha$. Then by (ii) from Corollary 2.7.11, $S(\mathfrak{A}_i) \not\models T(\alpha)$. Consequently $T(\alpha) \notin \lambda(\{S(\mathfrak{A}_i) \mid i \in I\})$. Thus the logic $\lambda(\{S(\mathfrak{A}_i) \mid i \in I\})$ is a modal counterpart for λ. Therefore, since by Theorem 2.7.15 $\sigma(\lambda)$ is the greatest modal counterpart for λ, we conclude $\lambda(\{S(\mathfrak{A}_i) \mid i \in I\}) \subseteq \sigma(\lambda)$ and $\lambda(\{S(\mathfrak{A}_i) \mid i \in I\}) = \sigma(\lambda)$ ∎

The following two observations may be also of interest and we will often use them in what follows.

Lemma 2.7.17 *For any modal algebra $\mathfrak{B} \in Var(S4)$, there is an isomorphism h of the algebra $S(G(\mathfrak{B}))$ into the algebra $\mathfrak{B}$ which is identical mapping on the set $G(\mathfrak{B})$. Hence $S(G(\mathfrak{B}))$ can be considered as a subalgebra of the algebra $\mathfrak{B}$ generated by the set $G(\mathfrak{B})$ of all open elements.*

Proof. Any element a of $S(G(\mathfrak{B}))$ has a certain representation of the form $a = \bigwedge_{1 \leq i \leq n}(\neg a_i \vee b_i)$, where $a_i, b_i \in G(\mathfrak{B})$. We define, for any $c \in G(\mathfrak{B})$,

$$h(c) := c \in |\mathfrak{B}| \text{ and } h(a) := \bigwedge_{1 \leq i \leq n} (\neg h(a_i) \vee h(b_i)),$$

where all the operations on the right hand side of the equality above are taken in algebra $\mathfrak{B}$. That h is well-defined follows from the fact that equal elements of $S(G(\mathfrak{A}))$ with different representations of the form displayed above can be obtained from each other by transformations of kind (i) and (ii) from the definition of the algebra $S(\mathfrak{A})$. And these transformations preserve the equality of the values of terms in $\mathfrak{B}$. It is easily seen from the definition of h, that h is a boolean homomorphism. As to the operation $\square$,

$$h(\square a) = h(\square \bigwedge_{1 \leq i \leq n}(\neg a_i \vee b_i)) = h(\bigwedge_{1 \leq i \leq n} \square(\neg a_i \vee b_i) =$$
$$\bigwedge_{1 \leq i \leq n} \square(\neg a_i \vee \overline{b_i}) = \bigwedge_{1 \leq i \leq n} \square(\neg h(a_i) \vee \overline{h}(b_i)) =$$
$$\square(\bigwedge_{1 \leq i \leq n}(\neg h(a_i) \vee h(b_i))) = \square h(a).$$

Hence h is a homomorphism of modal algebras. This isomorphism is a one-to-one mapping. Indeed, otherwise there are two distinct elements $a, b \in S(G(\mathfrak{B}))$ such that $a \neq b$ and $h(a) = h(b)$. Then $h((a \rightarrow b) \wedge (b \rightarrow a)) = \top$ and $h(\square((a \rightarrow b) \wedge (b \rightarrow a))) = \top$. At the same time $(a \rightarrow b) \wedge (b \rightarrow a) \neq \top$ and consequently $\square((a \rightarrow b) \wedge (b \rightarrow a)) \neq \top$. Since h is the identical mapping on $G(\mathfrak{A})$ we obtain $h(\square((a \rightarrow b) \wedge (b \rightarrow a))) \neq \top$, a contradiction. So, h is an isomorphic embedding. ∎

Lemma 2.7.18 *If a pseudo-boolean algebra $\mathfrak{A}_1$ is a subalgebra of a pseudo-boolean algebra $\mathfrak{A}_2$ then $S(\mathfrak{A}_1)$ is a subalgebra of $S(\mathfrak{A}_2)$.*

Proof. All elements of $S(\mathfrak{A}_1)$ are elements of $S(\mathfrak{A}_2)$. From the definition of the wrapping modal algebras it follows directly that the operations of modal algebras act identically on the elements from $S(\mathfrak{A}_1)$ in the algebra $S(\mathfrak{A}_1^*)$ and in the algebra $S(\mathfrak{A}_2)$. ∎

It is possible to define the converse mapping ρ of modal logics over $S4$ into superintuitionistic logics by using correspondence between such modal logics and their intuitionistic fragments.

Definition 2.7.19 *For any modal logic λ extending S4, its intuitionistic frag-ment $\rho(\lambda)$ is the set $\rho(\lambda) := \{\alpha \mid T(\alpha) \in \lambda\}$.*

According to Theorem 2.7.6 it is clear that $\rho(\lambda)$ is a superintuitionistic logic. It also follows immediately from our definitions that λ is a modal counterpart for $\rho(\lambda)$. Thus we have at our disposal three mappings τ, σ and ρ. Some basic properties of these mappings are summarized in the following theorem.

Theorem 2.7.20 *(Maksimova, Rybakov [97]) The following hold*

(i) *The mappings τ, σ, ρ are monotonic: $S4 \subseteq \lambda_1 \subseteq \lambda_2 \in \mathcal{L}_{ml}$ $\Rightarrow \tau(\lambda_1) \subseteq \tau(\lambda_2), \; \sigma(\lambda_1) \subseteq \sigma(\lambda_2); \; H \subseteq \lambda_1 \subseteq \lambda_2 \in \mathcal{L}_{si} \Rightarrow$ $\rho(\lambda_1) \subseteq \rho(\lambda_2);$*

(ii) *$\rho\tau(\lambda) = \lambda; \; \rho\sigma(\lambda) = \lambda; \; \tau\rho(\lambda) \subseteq \lambda; \; \lambda \subseteq \sigma\rho(\lambda);$*

(iii) *Mapping τ is a lattice's isomorphism which preserves infinite unions;*

(iv) *Mapping ρ is a lattice's homomorphism which preserves in-finite unions and intersections.*

(v) *Mapping σ is one-to-one and preserves infinite intersections.*

Proof. (i) follows directly from our definitions. (ii) follows directly from our definition and Theorem 2.7.15. (iii): by (i) the mapping τ preserves the order and by Theorem 2.7.4 is one-to-one. The inclusion $\tau(\lambda_1 \wedge \lambda_2) \subseteq \tau(\lambda_1) \wedge \tau(\lambda_2)$ is evident (by (i), say). By Lemma 2.2.26

$$\tau(\lambda_1) \wedge \tau(\lambda_2) = S4 \oplus \{\Box T(\alpha)^2 \vee \Box T(\beta)^3 \mid \alpha \in \lambda_1, \beta \in \lambda_2\}.$$

By Lemma 2.7.2 $T(\alpha) \equiv \Box T(\alpha) \in S4, T(\beta) \equiv \Box T(\beta) \in S4$. Therefore $T(\alpha)^2 \vee T(\beta)^3 \equiv \Box T(\alpha)^2 \vee T(\beta)^3 \in S4$. But $T(\alpha)^2 \vee T(\beta)^3 = T(\alpha^2 \vee \beta^3)$, and by Theorem 2.1.31 $\alpha^2 \vee \beta^3 \in \lambda_1 \wedge \lambda_2$. Thus

$$\Box T(\alpha)^2 \vee \Box T(\beta)^3 \in \tau(\lambda_1 \wedge \lambda_2),$$

i.e. $\tau(\lambda_1 \wedge \lambda_2) \subseteq \tau(\lambda_1) \wedge \tau(\lambda_2)$. Hence $\tau(\lambda_1 \wedge \lambda_2) = \tau(\lambda_1) \wedge \tau(\lambda_2)$. Let $\lambda_i, i \in I$ be a family of superintuitionistic logics. Then

$$\tau(\bigvee_{i \in I} \lambda_i) := S4 \oplus \bigcup_{i \in I} \{T(\alpha) \mid \alpha \in \lambda_i\} =$$
$$\bigvee_{i \in I} S4 \oplus \{T(\alpha) \mid \alpha \in \lambda_i\} = \bigvee_{i \in I} \tau(\lambda_i).$$

(iv): We simply calculate

$$\rho(\bigwedge_{i \in I} \lambda_i) := \{\alpha \mid T(\alpha) \in \bigwedge_{i \in I} \lambda_i\} = \{\alpha \mid T(\alpha) \in \bigcap_{i \in I} \lambda_i\} =$$
$$\bigcap_{i \in I} \{\alpha \mid T(\alpha) \in \lambda_i\} = \bigwedge_{i \in I} \rho(\lambda_i).$$

The case of $\vee$ is not so simple but also follows from the general results which we obtained before.

$$\rho(\bigvee_{i\in I}\lambda_i) \subseteq \rho(\bigvee_{i\in I}\sigma\rho(\lambda_i)) \text{ by (ii)}$$
$$\rho(\bigvee_{i\in I}\sigma\rho(\lambda_i)) \subseteq \rho\sigma(\bigvee_{i\in I}\rho(\lambda_i)) \text{ by (i)}$$
$$\rho\sigma(\bigvee_{i\in I}\rho(\lambda_i)) = (\bigvee_{i\in I}\rho(\lambda_i)) \text{ by (i)}$$
$$(\bigvee_{i\in I}\rho(\lambda_i)) \subseteq \rho(\bigvee_{i\in I}\lambda_i) \text{ by (i)} .$$

Thus $\rho(\bigvee_{i\in I}\lambda_i) \subseteq \bigvee_{i\in I}\rho(\lambda_i) \subseteq \rho(\bigvee_{i\in I}\lambda_i)$. Hence we have shown ρ is a homomorphism preserving all biggest lower and smallest upper bounds.

(v): According to Theorem 2.7.15 the mapping σ is one-to-one. σ preserves order by (i). To show σ preserves all infinite intersection we calculate:

$$\sigma(\bigwedge_{i\in I}\lambda_i) \subseteq \bigwedge_{i\in I}\sigma(\lambda_i) \text{ by (i)} ,$$
$$\bigwedge_{i\in I}\sigma(\lambda_i) \subseteq \sigma\rho(\bigwedge_{i\in I}\sigma(\lambda_i)) \text{ by (ii)} ,$$
$$\sigma\rho(\bigwedge_{i\in I}\sigma(\lambda_i)) \subseteq \sigma(\bigwedge_{i\in I}\rho\sigma(\lambda_i)) \text{ by (iv)}$$
$$\sigma(\bigwedge_{i\in I}\rho\sigma(\lambda_i)) \subseteq \sigma(\bigwedge_{i\in I}(\lambda_i)) \text{ by (ii)} .$$

Hence $\sigma(\bigwedge_{i\in I}\lambda_i) \subseteq \bigwedge_{i\in I}\sigma(\lambda_i) \subseteq \sigma(\bigwedge_{i\in I}\lambda_i)$ and σ indeed preserves all infinite intersections. ∎

From this theorem, Theorem 2.7.15 and Lemma 2.7.16 we immediately derive

Corollary 2.7.21 *(Maksimova, Rybakov [97]) The following hold*

> *(i) If $\lambda \in \mathcal{L}_{ml}$ is a tabular modal logic over S4 then its intuitionistic fragment $\rho(\lambda)$ is also tabular;*
>
> *(ii) If $\lambda \in \mathcal{L}_{ml}(S4)$ has the fmp then $\rho(\lambda)$ also has the fmp;*
>
> *(iii) If $\lambda \in \mathcal{L}_{ml}(S4)$ and $\rho(\lambda)$ is tabular then λ has the fmp;*
>
> *(iv) If $\lambda \in \mathcal{L}_H$ is tabular then $\sigma(\lambda)$ is tabular and $\tau(\lambda)$ has the fmp;*
>
> *(v) If $\lambda \in \mathcal{L}_H$ has the fmp then $\sigma(\lambda)$ has the fmp.*

Proof. (i) If $\lambda = \lambda(\mathfrak{B})$ where $\mathfrak{B}$ is a tabular modal algebra then we have $\rho(\lambda) := \{\alpha \mid \mathfrak{B} \models T(\alpha)\}$. By Lemma 2.7.6 $\{\alpha \mid \mathfrak{B} \models T(\alpha)\} = \{\alpha \mid G(\mathfrak{B}) \models \alpha\} = \lambda(G(\mathfrak{B}))$. (ii) follows from (i) and (iv) of Theorem 2.7.20.

(iii): Suppose that $\rho(\lambda)$ is a tabular logic. Then there is some n such that the formula $\alpha(n) := \bigvee_{1\le i<j\le n}((p_i \rightarrow p_j) \wedge (p_j \rightarrow p_i))$ is a theorem of $\rho(\lambda)$. Then $T(\alpha(n))$ is a theorem of λ. According to Theorem 1.2.57 the variety $Var(\lambda)$ is generated by a collection of subdirectly irreducible modal algebras $\{\mathfrak{B}_i \mid i \in I\}$. By Theorem 2.2.20 any $\mathfrak{B}_i$ has a greatest open element ω, i.e. $\square\omega = \omega$, among all open elements below the element $\top$. Therefore the fact that $\mathfrak{B}_i \models T(\alpha(n))$ implies that the number of open elements of $\mathfrak{B}_i$ is less than $n + 1$. Thus any subalgebra of $\mathfrak{B}_i$ generated by k elements is finite and has less than $2^{2^{k+n+1}}$

elements. Clearly λ is the logic of all finite generated subalgebras of algebras $\mathfrak{B}_i, i \in I$. Hence λ has the fmp.

(iv): the condition concerning $\sigma(\lambda)$ follows from Lemma 2.7.16, the condition concerning $\tau(\lambda)$ is a corollary of (iii) because $\rho\tau(\lambda) = \lambda$ by (ii) of Theorem 2.7.20. (v) follows directly from (v) of Theorem 2.7.20 and Lemma 2.7.16. ∎

To demonstrate briefly the developed technique we describe below axiomatic systems for the greatest modal counterparts $\sigma(\lambda)$ of certain superintuitionistic logics λ.

Lemma 2.7.22 *For any superintuitionistic logic λ, if $\tau(\lambda) \oplus \Box(\Box(p \to \Box p) \to \Box p) \to p$ has the fmp then $\sigma(\lambda) := \tau(\lambda) \oplus \Box(\Box(p \to \Box p) \to \Box p) \to p$.*

Proof. Indeed, according to Theorem 2.7.20 $\rho(\tau(\lambda) \oplus \Box(\Box(p \to \Box p) \to \Box p) \to p) = \rho(\tau(\lambda)) \vee \rho(Grz)$, where $Grz := S4 \oplus \Box(\Box(p \to \Box p) \to \Box p) \to p$ and $\rho(\tau(\lambda)) = \lambda$. By Theorem 2.6.5 and Theorem 2.6.25 the logic Grz has the Kripke finite model property. Since Grz is valid on a finite transitive reflexive frame $\mathcal{F}$ iff $\mathcal{F}$ is poset, it follows that $Grz := \lambda(\{\mathcal{F} \mid \mathcal{F}$ is a finite poset $\})$. Since H also has Kripke finite model property (see Corollary 2.6.14), we also have $H := \lambda(\{\mathcal{F} \mid \mathcal{F}$ is a finite poset $\})$ which by Lemma 2.5.67 yields

$$H := \lambda(\{\mathcal{F}_p^+ \mid \mathcal{F} \text{ is a finite poset }\}).$$

Using Lemma 2.7.16 it follows that $\sigma(H) := \lambda(\{S(\mathcal{F}_p^+) \mid \mathcal{F}$ is a finite poset $\})$. For any finite poset $\mathcal{F}$, using Lemma 2.7.17 it follows immediately that $S(\mathcal{F}_p^+) = \mathcal{F}^+$, therefore we have $\sigma(H) := \lambda(\{\mathcal{F}^+ \mid \mathcal{F}$ is a finite poset $\})$. Applying $Grz = \lambda(\{\mathcal{F} \mid \mathcal{F}$ is a finite poset $\})$ and Lemma 2.5.17 we conclude $\sigma(H) := Grz$. Therefore $\rho(\tau(\lambda)) \vee \rho(Grz) = \lambda$, and the logic $\rho(\tau(\lambda)) \vee \rho(Grz)$ is a modal counterpart for λ, and since $\sigma(\lambda)$ is the greatest modal counterpart for λ, we conclude $\rho(\tau(\lambda)) \vee \rho(Grz) \subseteq \sigma(\lambda)$. Suppose that $\alpha \in \sigma(\lambda)$ and $\alpha \notin \rho(\tau(\lambda)) \vee \rho(Grz)$. Since this logic has the fmp by our assumption, there is a finite frame $\mathcal{F}$, such that $\mathcal{F} \Vdash \rho(\tau(\lambda)) \vee \rho(Grz)$ but $\mathcal{F} \nVdash \alpha$. Then $\mathcal{F}$ is a poset and $\mathcal{F}$ is a λ-frame. Therefore by Lemma 2.7.16 $\mathcal{F}$ is a $\sigma(\lambda)$-frame, a contradiction. Thus $\rho(\tau(\lambda)) \vee \rho(Grz) = \sigma(\lambda)$. ∎.

Note that, since using Theorem 2.6.5 and Theorem 2.6.25 the logic Grz has Kripke finite model property, we can derive the already known us following result also from the just proved lemma.

Corollary 2.7.23 *The greatest modal counterpart $\sigma(H)$ of the intuitionistic logic H is the Grzegorczyk logic*

$$Grz := S4 \oplus \Box(\Box(p \to \Box p) \to \Box p) \to p.$$

Thus we proved $\sigma(\lambda) = \tau(H) \oplus \Box(\Box(p \to \Box p) \to \Box p) \to p$ for certain logics λ. This result is a partial case of the general theorem which says that, for any superintuitionistic logic λ,

$$\sigma(\lambda) = \tau(\lambda) \oplus \Box(\Box(p \to \Box p) \to \Box p) \to p.$$

It is the famous Esakia (1974) conjecture which has been first proved in the Ph.D. Dissertation of W.Blok [12].

Note that the lattice of all modal counterparts for a given superintuitionistic logic, has generally speaking complicated structure. For instance, the following proposition holds.

Proposition 2.7.24 *For any given superintuitionistic logic λ, the interval $[\tau(\lambda), \sigma(\lambda)]$, where*

$$[\tau(\lambda), \sigma(\lambda)] := \{\lambda \mid \lambda \text{ is a modal logic and } \tau(\lambda) \subseteq \lambda \subseteq \sigma(\lambda)\},$$

has infinitely many members, and, in particular, contains infinite descending chains.

Proof. Let logic λ be given. Consider the sequence $\mathcal{F}_n, n \in N$, where any $\mathcal{F}_n$ is the n-element reflexive transitive cluster. Also we define the following sequence of modal logics:

$$\sigma(\lambda) \wedge \lambda(\mathcal{F}_n), n \in N.$$

It is clear that all these logics are modal counterparts for λ, indeed,

$$\rho(\sigma(\lambda) \wedge \lambda(\mathcal{F}_n)) = \rho\sigma(\lambda) \wedge \rho(\lambda(\mathcal{F}_n))) =$$
$$\lambda \wedge \lambda(\langle\{1\}, \{\langle 1, 1\rangle\}\rangle) = \lambda$$

(see Theorem 2.7.20). Thus $\tau(\lambda) \subseteq \sigma(\lambda) \wedge \lambda(\mathcal{F}_n) \subseteq \sigma(\lambda)$. By definition we have $\sigma(\lambda) \wedge \lambda(\mathcal{F}_{n+1}) \subseteq \sigma(\lambda) \wedge \lambda(\mathcal{F}_n)$. It remains only to note that there is a formula from $\sigma(\lambda) \wedge \lambda(\mathcal{F}_n)$ which does not belong to $\sigma(\lambda) \wedge \lambda(\mathcal{F}_{n+1})$. Take the formula

$$\phi := \Box Grz^2 \vee \Box \bigvee_{1 \le i, j \le n+1} \Diamond((p_i \to p_j) \wedge (p_j \to p_i))^3,$$

see Theorem 2.2.26 to explain upper indexes $2, 3$. By Theorem 2.2.26 $\phi \in \sigma(\lambda) \wedge \lambda(\mathcal{F}_n)$ and simultaneously $\phi \notin \sigma(\lambda) \wedge \lambda(\mathcal{F}_{n+1})$. ∎

Historical Comments. The idea of interpretation intuitionistic logic in modal logic belongs to K.Gödel[57]. The development of ideas from this short note of K.Gödel is contained at detail article McKinsey and Tarski [101], which good embarrassed the topic and include all mathematical tools necessary to evolve this field further. At the same time, this paper actually had deal only with intuitionistic logic itself and modal logic $S4$ as its modal counterpart. However already in Dummett, Lemmon [34] this apparatus was extended to all superintuitionistic logics and the general translation theorem was obtained.

The further study of connections between modal and superintuitionistic logics on the base these ideas is contained in the paper Maksimova, Rybakov [97], where first the notion of modal counterparts for superintuitionistic logics, mappings τ, ρ and σ were introduced and studied. The major part of results proved at this section appeared first at that paper. Simultaneously the similar ideas appeared at the research of L.Esakia in 1970s, though it was no published in easy reachable literature.

The further study of the lattice of modal logic and their connections with superintuitionistic logics is the main topic of papers of L.Maksimova [92, 93, 94, 96] and L.Esakia and V.Meskhi [40]. About the same time W.Rautenberg published the book [122] devoted to the same field with a fundamental detailed exposition of the base of that theory. The Esakia conjecture concerning the axiomatization of the greatest modal counterparts of superintuitionistic logics has first been proved W.Blok [12].

It is appropriate to mention here the algorithmic questions concerning modal and superintuitionistic logics. The Harrop problem: whether any finitely axiomatizable superintuitionistic logic is decidable, was answered negatively by V.Shehtman [171]. This also gives, in particular, examples of finitely axiomatizable modal logics extending $S4$ which are undecidable. M.Zakharyaschev [213, 214] invented certain canonical formulas by means of which any modal logic over $K4$ can be axiomatized, and using this technique, he solved many problems concerning modal logics which were difficult to answer. Further research of modal pretabular logics which do not extend $S4$ and other questions concerning the lattice of modal logic at hole could be found at W.Blok [14, 15]. The research program concerning algorithmic recognizing of different properties of modal logics by their axiomatization is good presented in Chagrov, Zakharyaschev [27], were it is shown that the majority of such problems are undecidable.

2.8 Advanced Tools for the Finite Model Property

We will consider here only a small number of results concerning finite model property since this area is very extensively developed and there are a lot of different theorems within it's scope. We restrict our attention here to those results which will be applied in our research concerning inference rules. To begin we describe an approach based on the Glivenko theorem concerning the properties of negation in the intuitionistic logic. We extend the Glivenko Theorem to modal logics and develop here a technique which will give us general theorems concerning the preservation of the finite model property when a logic is extended by adding a finite set of axioms in a special form.

Glivenko's Theorem [56] has a number of different formulations. For example, it can be presented in the following way: for any formulas α and β, $\alpha \equiv \beta \in PC$ if and only if $\neg\alpha \equiv \neg\beta \in H$. In particular, for any formula α, $\alpha \in PC$ iff $\neg\neg\alpha \in H$. That is, the classical propositional logic can be embedded into intuitionistic logic. It turns out that this theorem can be proved very easily by employing semantic methods. We employ Kripke semantics in order to proof a generalization of this theorem to modal logics. First we describe this generalization and then extract as a corollary the original Glivenko Theorem.

Theorem 2.8.1 *[151, 132]* Modal Analog of the Glivenko Theorem. *Let α and β be any modal formulas. Then*

$$\Box\Diamond \to \Box\Diamond\beta \in K4 \text{ if and only if } \Diamond\alpha \to \Diamond\beta \in S5.$$

Proof. Suppose that $\Box\Diamond \to \Box\Diamond\beta \in K4$. Then since $\Diamond\gamma \equiv \Box\Diamond\gamma \in S5$ for any formula γ, we conclude $\Diamond\alpha \to \Diamond\beta \in S5$ because $K4 \subseteq S5$. Conversely, suppose that $\Diamond\alpha \to \Diamond\beta \in S5$. Since $K4$ has the fmp (cf. Corollary 2.6.12), in order to complete the proof of this theorem it is sufficient to show that, for any finite transitive Kripke model $\mathfrak{M} := \langle W, R, V\rangle$, the formula $\Box\Diamond \to \Box\Diamond\beta$ is valid in $\mathfrak{M}$. To show this suppose that $a \in |\mathfrak{M}|$ and $a \Vdash_V \Box\Diamond\alpha$. We have to show that $a \Vdash_V \Box\Diamond\beta$. If a is a maximal irreflexive element then obviously $a \Vdash_V \Box\Diamond\beta$. Otherwise no maximal clusters accessible from a by R can be degenerated clusters (i.e. be clusters consisting of reflexive elements). This directly follows from our assumption that $a \Vdash_V \Box\Diamond\alpha$.

Consider an arbitrary maximal by R cluster C accessible from a. The fact that $a \Vdash_V \Box\Diamond\alpha$ entails there is an element b from C such that $b \Vdash_V \alpha$ and $b \Vdash_V \Diamond\alpha$. Because C forms an $S5$-model (with the valuation generated by V) and $\Diamond\alpha \to \Diamond\beta \in S5$, it follows $b \Vdash_V \Diamond\beta$. Since this holds for any maximal cluster C of the kind mentioned and $\mathfrak{M}$ is a transitive model, we obtain $a \Vdash_V \Box\Diamond\beta$. Thus we can conclude that $\mathfrak{M} \Vdash (\Box\Diamond\alpha \to \Box\Diamond\beta)$. $\blacksquare$

Thus, for any formulas α and β, in order for the formulas $\Box\Diamond\alpha$ and $\Box\Diamond\beta$ to be equivalent in $K4$ it is necessary and sufficient that formulas $\Diamond\alpha$ and $\Diamond\beta$ are equivalent in $S5$. Also if $\Diamond\alpha \in S5$ then $\Diamond\top \to \Diamond\alpha \in S5$, and by Theorem 2.8.1 $\Box\Diamond\top \to \Box\Diamond\alpha \in K4$, i.e. $\Box\Diamond\alpha \vee \Diamond\Box\bot \in K4$. Conversely, if $\Box\Diamond\alpha \vee \Diamond\Box\bot \in K4$ then $\Box\Diamond\alpha \vee \Diamond\Box\bot \in S5$ and $\Box\Diamond\alpha \in S5$. And using this observations we immediately infer

Corollary 2.8.2 *For any formula α, $\Diamond\alpha \in S5 \iff \Box\Diamond\alpha \vee \Diamond\Box\bot \in K4$.*

It is also easy to extract the original Glivenko theorem for intuitionistic logic from Theorem 2.8.1.

Corollary 2.8.3 Glivenko's Theorem . *[56]. For any formulas α and β in the language $\mathcal{L}_{PC}$, $\neg\alpha \to \neg\beta \in H$ iff $\beta \to \alpha \in PC$.*

Proof. If $\neg\alpha \to \neg\beta \in H$ then by the contraposition law (see (16) from Lemma 2.1.9) we conclude $\neg\neg\beta \to \neg\neg\alpha \in H$. Hence $\neg\neg\beta \to \neg\neg\alpha \in PC$ and $\beta \to \alpha \in PC$. Conversely, suppose that $\beta \to \alpha \in PC$. Here we use the translation of superintuitionistic logics into modal logics from the previous section. Applying the mapping σ defined there and Lemma 2.7.16, it is easy to see that $\sigma(PC) = S5$. Also we have $T(\beta \to \alpha) = \Box(T(\beta) \to T(\alpha)) \in S5$. Then $\Diamond T(\beta) \to \Diamond T(\alpha) \in S5$ and by Theorem 2.8.1 it follows $\Box\Diamond T(\beta) \to \Box\Diamond T(\alpha) \in K4 \subseteq S4$. Hence $T(\neg\neg\beta \to \neg\neg\alpha) \in S4$. Since $S4$ is a modal counterpart for H (see Theorem 2.7.4) we have $\neg\neg\beta \to \neg\neg\alpha \in H$. ∎

Note that the semantic proof schema of the Theorem 2.8.1 is suitable for a straightforward proof of the Glivenko Theorem for intuitionistic logic without any using of modal logics.

Exercise 2.8.4 *Show that for any formulas α and β in the language $\mathcal{L}_{PC}$, $\beta \to \alpha \in PC$ iff $\neg\alpha \to \neg\beta \in H$ using finite intuitionistic Kripke models and the proof schema of Theorem 2.8.1.*

We now consider certain modal propositional formulas in a special form:

Definition 2.8.5 *A formula α is said to be a $\Box\Diamond$-formula in a modal logic λ if there is a formula $\beta(p_1, ..., p_n)$ built up of variables $p_1, ..., p_n$ and there is a tuple of formulas $\Box\Diamond\alpha_1, ..., \Box\Diamond\alpha_n$ such that the formula $\beta(\Box\Diamond\alpha_1, ..., \Box\Diamond\alpha_n)$ is equivalent to α in the logic λ. A formula α is just a $\Box\Diamond$-formula if α itself has the required form.*

It turns out that $\Box\Diamond$ formulas are very often used for the axiomatization of different modal logics. It turned out that these formulas possess certain well desirable properties. First note that the adjoining a finite number of $\Box\Diamond$-formulas to a modal logic as new axioms preserves decidability.

Theorem 2.8.6 *[132, 151] Suppose λ is a modal logic over $K4$ and $\delta_1, ..., \delta_m$ are $\Box\Diamond$-formulas. Then if λ is decidable then the logic $\lambda \oplus \{\delta_1, ... , \delta_m\}$ is also decidable.*

Proof. By assumption for every δ_i, $\delta_i := \gamma_i(\Box\Diamond\phi_1, ... , \Box\Diamond\phi_d)$ (here d is the maximal number of formulas of the form $\Box\Diamond\psi$ out of which all formulas δ_i are constructed; so it can be that some formulas $\Box\Diamond\phi_j$ will not really occur in a certain given formula δ_i). And all formulas δ_i are constructed from some variables q_j, where $1 \leq j \leq v$. By the deduction theorem for $K4$ (see Theorem 1.3.15) we have, for any formula $\alpha(p_1, ... , p_n)$,

$$\alpha(p_1, ... , p_n) \in \lambda \oplus \{\delta_1, ... , \delta_m\} \Longleftrightarrow \Box\Psi \wedge \Psi \to \alpha(p_1, ... , p_n) \in \lambda, (2.27)$$

where $\Psi := \bigwedge_{e_r \in R} e_r (\bigwedge_{1 \leq i \leq m} \delta_i)$ and R is a finite set of substitutions e_r of formulas built up only of variables $p_1, ..., p_n$ instead of all the variables q_j, $1 \leq j \leq v$ occurring in formulas $\delta_1, ..., \delta_m$.

There are only finitely many formulas built up out of the variables $p_1, ..., p_n$ up to equivalence in the modal logic $S5$ (exactly $2^{2^{2^n}}$). Indeed, the Kripke model $\mathfrak{M}$ consisting of all different clusters having no elements with the same valuation of all the letters $p_1, ..., p_n$ distinguishes all formulas non-equivalent in $S5$ that are built up out of the letters $p_1, ..., p_n$ (since $S5$ has the fmp, see Corollary 2.6.12). Thus there are not more than $2^{2^{2^n}}$ formulas non-equivalent in $S5$ and built up of n variables. Therefore we can pick one representative in any class of formulas, which are equivalent in $S5$ and built up out of n variables, in a such way that the length of any such representative formula is less than or equal to $n_1 := 2^{2^{2^n}} * n$. Hence, if we take the set U of all modal formulas constructed out of the letters $p_1, .., p_n$ with length of not more than n_1 then all the formulas constructed out of $p_1, ..., p_n$ are among formulas of U up to equivalence in $S5$. We set

$$\Psi_1 := \bigwedge_{e_g \in R_1} \bigwedge_{1 \leq i \leq m} \gamma_i (\Box \Diamond (\phi_1)(e_g(q_j)), ..., \Box \Diamond \phi_d (e_g(q_j))),$$

where R_1 is the set of all possible substitutions e_g of formulas from U in place of letters $q_1, ..., q_v$. Consider a member $e_r (\bigwedge_{1 \leq i \leq m} \delta_i)$ from the conjunct Ψ in (2.27). We have

$$e_r \Big(\bigwedge_{1 \leq i \leq m} \delta_i \Big) := \Big(\bigwedge_{1 \leq i \leq m} \gamma_i (\Box \Diamond \phi_1 (e_r(q_j)), ..., \Box \Diamond \phi_d (e_r(q_j))) \Big),$$

where all letters q_j are variables of the formula δ_i. By our choice of the set U, any formula $\Diamond \phi_t (e_r(q_j))$, $1 \leq t \leq d$ is equivalent in $S5$ to the formula $\Diamond \phi_t (e_g(q_j))$ for some $e_g \in R_1$. Therefore according to Theorem 2.8.1 the formula $\Box \Diamond \phi_t (e_r(q_j))$, where $1 \leq t \leq d$, is equivalent in $K4$ to the formula $\Box \Diamond \phi_t (e_g(q_j))$. Therefore by (2.27) if $\alpha(p_1, ..., p_n) \in \lambda \oplus \{\delta_1, ..., \delta_m\}$ then $\Box \Psi_1 \wedge \Psi_1 \to \alpha(p_1, ..., p_n) \in \lambda$. The converse conjecture is evident. So

$$\alpha(p_1, ..., p_n) \in \lambda \oplus \{\delta_1, ..., \delta_m\} \Longleftrightarrow \Box \Psi_1 \wedge \Psi_1 \to \alpha(p_1, ..., p_n) \in \lambda. \quad (2.28)$$

Since the modal logic λ itself is decidable, this equivalence gives the algorithm for recognizing the derivability of formulas in $\lambda \oplus \{\delta_1, ..., \delta_m\}$. ∎

In order to prove a similar result for superintuitionistic logics we must first specify a particular class of propositional formulas, which were first introduced by McKay [99].

Definition 2.8.7 *A formula α in the propositional language $\mathcal{L}_H$ is essentially negative in a superintuitionistic logic λ if there are a formula $\beta(p_1, ..., p_n)$*

in $\mathcal{L}_H$ built up out of the variables $p_1, \ldots, p_n$ and a tuple of certain formulas $\neg\alpha_1, \ldots, \neg\alpha_n$ such that the formula $\beta(\neg\alpha_1, \ldots, \neg\alpha_n)$ is equivalent to α in the logic λ. A formula α is simply essentially negative if α itself has the required form.

In a manner quite similar to Theorem 2.8.4, we can prove the following general result using Glivenko theorem.

Theorem 2.8.8 *[132, 151] Let λ be a certain decidable superintuitionistic logic. Then for every tuple of essentially negative formulas $\delta_1, \ldots, \delta_m$, the logic λ_1, where $\lambda_1 := \lambda \oplus \{\delta_1, \ldots, \delta_m\}$ is decidable.*

Proof. We have $\forall i, 1 \leq i \leq m, \delta_i := \gamma_i(\neg\phi_1, \ldots, \neg\phi_d)$. Suppose all letters occurring in all the formulas δ_i are q_j, where $1 \leq j \leq v$. By the deduction theorem for H (cf. Theorem 1.3.9), for any formula $\alpha(p_1, \ldots, p_n)$,

$$\alpha(p_1, \ldots, p_n) \in \lambda \oplus \{\delta_1, \ldots, \delta_m\} \Longleftrightarrow \Psi \to \alpha(p_1, \ldots, p_n) \in \lambda, \qquad (2.29)$$

where $\Psi := \bigwedge_{e_r \in R} e_r(\bigwedge_{1 \leq i \leq m} \delta_i)$ and R is a finite set of substitutions e_r of formulas built up only of variables $p_1, \ldots, p_n$ in place of all the variables $q_j, 1 \leq j \leq v$ occurring in formulas $\delta_1, \ldots, \delta_m$. It is well known that there are only finitely many formulas up to equivalence in PC (exactly 2^{2^n}) which are built up of variables $p_1, \ldots, p_n$. To show this it is sufficient to use the normal conjunctive form for formulas in PC. Therefore the set of all formulas in the language PC, which are built up out of letters $p_1, \ldots, p_n$ and have length less or equal to $n_1 := 2^{2^n} * n$, contains a formula equivalent in PC to γ for any given formula γ in the language of PC which is built up of letters $p_1, \ldots, p_n$. Let U be the set of all formulas constructed out of $p_1, \ldots, p_n$ with length of not more than n_1. Then all formulas constructed out of $p_1, \ldots, p_n$ are among the formulas of U up to equivalence in PC. Let

$$\Psi_1 := \bigwedge_{e_g \in R_1} \bigwedge_{1 \leq i \leq m} \gamma_i(\neg\phi_1(e_g(q_j)), \ldots, \neg\phi_d(e_g(q_j))),$$

where R_1 is the set of all possible substitutions e_g of formulas from U in place of letters $q_1, \ldots, q_v$. For any member $e_r(\bigwedge_{1 \leq i \leq m} \delta_i)$ from the conjunct Ψ in the relation (2.29),

$$e_r(\bigwedge_{1 \leq i \leq m} \delta_i) := (\bigwedge_{1 \leq i \leq m} \gamma_i(\neg\phi_1(e_r(q_j)), \ldots, \neg\phi_d(e_r(q_j)))),$$

where all the letters q_j are variables of the formula δ_i. By our choice of the set U any formula $\phi_t(e_r(q_j)), 1 \leq t \leq d$ is equivalent in PC to the formula $\phi_t(e_g(q_j))$ for some $e_g \in R_1$. Therefore according to the Glivenko Theorem (see Corollary 2.8.3) the formula $\neg\phi_t(e_r(q_j))$, where $1 \leq t \leq d$, is equivalent in H to the formula $\neg\phi_t(e_g(q_j))$. Therefore by (2.29) if $\alpha(p_1, \ldots, p_n) \in \lambda \oplus \{\delta_1, \ldots, \delta_m\}$ then $\Psi_1 \to \alpha(p_1, \ldots, p_n) \in \lambda$. It is evident that the converse also holds. Hence

$$\alpha(p_1, \ldots, p_n) \in \lambda \oplus \{\delta_1, \ldots, \delta_m\} \Longleftrightarrow [\Psi_1 \to \alpha(p_1, \ldots, p_n)] \in \lambda.$$

By the assumption that the logic λ is itself decidable, this equivalence yields an algorithm for recognizing formulas provable in $\lambda \oplus \{\delta_1, \ldots, \delta_m\}$. ∎

Before illustrating these results with examples, we consider the question of whether the adjoining of formulas in the form described above preserves the possession of the fmp.

Theorem 2.8.9 *[132, 151] Let λ be a modal logic over $K4$ with the fmp. Suppose some formulas $\delta_1, \ldots, \delta_m$ are $\Box\Diamond$-formulas in λ. Then the logic $\lambda \oplus \{\delta_1, \ldots, \delta_m\}$ also has the fmp.*

Proof. Since all the formulas $\delta_1, \ldots, \delta_m$ are $\Box\Diamond$-formulas in λ, we can assume without loss of generality that all formulas $\delta_1, \ldots, \delta_m$ are $\Box\Diamond$-formulas themselves. Suppose $\alpha(p_1, \ldots, p_n)$ is a modal formula which is not a theorem of $\lambda \oplus \{\delta_1, \ldots, \delta_m\}$. We will use now a part of the proof of Theorem 2.8.4. We showed there (cf. (2.28)) that

$$\alpha(p_1, \ldots, p_n) \in \lambda \oplus \{\delta_1, \ldots, \delta_m\} \Longleftrightarrow [\Box\Psi_1 \wedge \Psi_1 \to \alpha(p_1, \ldots, p_n) \in \lambda.$$

Therefore we conclude $\Box\Psi_1 \wedge \Psi_1 \to \alpha(p_1, \ldots, p_n) \notin \lambda$. Since λ has the fmp, there is a finite modal algebra $\mathfrak{B}$ such that $\mathfrak{B} \models \lambda$ and at the same time $\mathfrak{B} \not\models [\Box\Psi_1 \wedge \Psi_1 \to \alpha(p_1, \ldots, p_n)]$. Then there is a tuple $a_1, \ldots, a_n$ of elements from $|\mathfrak{B}|$ such that

$$\mathfrak{B} \not\models (\Box\Psi_1 \wedge \Psi_1)(a_1, \ldots, b_n) \leq \alpha(a_1, \ldots, a_n). \tag{2.30}$$

Let $\mathfrak{B}_1$ be the subalgebra of $\mathfrak{B}$ generated by $a_1, \ldots, a_n$. Then using (2.30) we infer

$$\mathfrak{B}_1 \not\models (\Box\Psi_1 \wedge \Psi_1)(a_1, \ldots, a_n) \leq \alpha(a_1, \ldots, a_n). \tag{2.31}$$

and $\mathfrak{B}_1 \models \lambda$. Consider the filter

$$\nabla := \{b \mid b \in |\mathfrak{B}_1| \text{ and } \Box\Psi_1 \wedge \Psi_1)(a_i) \leq b\}.$$

It is clear that $c \in \nabla$ implies $\Box c \in \nabla$, i.e. ∇ is a $\Box$-filter. Therefore we can take the quotient algebra $\mathfrak{B}_1/\nabla$ which is a homomorphic image of $\mathfrak{B}_1$ (see Theorem 2.2.23). Since homomorphic images preserve the truth of modal formulas we infer $\forall\mu \in \lambda, \mathfrak{B}_1/\nabla \models \mu$. At the same time $\alpha([a_i]\nabla) \neq [\top]\nabla$ by the relation (2.31).

Our aim now is to show that $\mathfrak{B}_1/\nabla \models \delta_1 \wedge \ldots \wedge \delta_m$. Note that by choice of ∇ we have

$$\Box\Psi_1 \wedge \Psi_1([a_1]_\nabla, \ldots, [a_n]_\nabla) = [\top]_\nabla. \tag{2.32}$$

Suppose that there are some elements $b_1, \ldots, b_v \in |\mathfrak{B}_1|$ such that

$$(\bigwedge_{1 \leq i \leq m} \delta_i)([b_1]_\nabla, \ldots, [b_v]_\nabla) \neq [\top]_\nabla.$$

Since the algebra $\mathfrak{B}_1$ is generated by elements $a_1, \ldots, a_n$, there are formulas $\mu_1, \ldots, \mu_v$ built up of letters $p_1, \ldots, p_n$ such that $b_j = \mu_j(a_1, \ldots, a_n)$ for any j, where $1 \leq j \leq v$. Hence, in particular we have

$$(\bigwedge_{1 \leq i \leq m} \delta_i)([\mu_1(a_1, \ldots, a_n)]_\nabla, \ldots, [\mu_v(a_1, \ldots, a_n)]_\nabla) \neq [\top]_\nabla.$$

Therefore $(\bigwedge_{1 \leq i \leq m} \delta_i)(\mu_1(a_1, \ldots, a_n), \ldots, \mu_v(a_1, \ldots, a_n)) \notin \nabla$. Also for any j, $1 \leq j \leq v$, there is some formula $\nu_j(p_1, \ldots, p_n)$ from the set U introduced in the proof of Theorem 2.8.4 such that $\mu_j(p_1, \ldots, p_n) \equiv \nu_j(p_1, \ldots, p_n) \in S5$. We replace all formulas $\mu_j(p_1, \ldots, p_n)$ by corresponding formulas $\nu_j(p_1, \ldots, p_n)$ in the formula $(\bigwedge_{1 \leq i \leq m} \delta_i)(\mu_1(p_1, \ldots, p_n), \ldots, \mu_v(p_1, \ldots, p_n))$. Since all formulas δ_i are $\Box\Diamond$-formulas, applying Theorem 2.8.1 it follows

$$(\bigwedge_{1 \leq i \leq m} \delta_i)(\mu_1(p_1, \ldots, p_n), \ldots, \mu_v(p_1, \ldots, p_n)) \equiv$$
$$(\bigwedge_{1 \leq i \leq m} \delta_i)(\nu_1(p_1, \ldots, p_n), \ldots, \nu_v(p_1, \ldots, p_n)) \in K4.$$

Hence, since $(\bigwedge_{1 \leq i \leq m} \delta_i)(\mu_1(a_1, \ldots, a_n), \ldots, \mu_v(a_1, \ldots, a_n)) \notin \nabla$, we conclude

$$(\bigwedge_{1 \leq i \leq m} \delta_i)(\nu_1(a_1, \ldots a_n), \ldots, \nu_v(a_1, \ldots a_n)) \notin \nabla.$$

This yields that $\Box\Psi_1 \wedge \Psi_1)(a_1, \ldots, a_n) \notin \nabla$ because

$$(\bigwedge_{1 \leq i \leq m} \delta_i)(\nu_1(p_1, \ldots, p_n), \ldots, \nu_v(p_1, \ldots, p_n))$$

is a conjunct member of $\Box\Psi_1 \wedge \Psi_1)(a_1, \ldots, a_n)$. This contradicts (2.32). Thus $\mathfrak{B}_1/\nabla \models \delta_1 \wedge \ldots \wedge \delta_m$ and consequently $\mathfrak{B}_1/\nabla \models \lambda \oplus \{\delta_1 \wedge \ldots \wedge \delta_m\}$, but $\mathfrak{B}_1/\nabla \not\models \alpha$ and $\mathfrak{B}_1/\nabla$ is finite. Hence the modal logic $\lambda \oplus \{\delta_1 \wedge \ldots \wedge \delta_m\}$ has the fmp. ∎

The following corollary is a general result for superintuitionistic logics concerning the preservation of the fmp. It is possible to give a straightforward proof of this result in a manner similar to the proof of Theorem 2.8.9, but using Theorem 2.8.8 in this case. However we prefer to extract this result as a corollary of the above result concerning the preservation of the fmp for modal logics using the technique of modal counterparts.

Corollary 2.8.10 *[132, 151] Let λ be a superintuitionistic logic which possesses the fmp. If some formulas $\delta_1, \ldots, \delta_m$ are essentially negative in λ then the logic $\lambda \oplus \{\delta_1, \ldots, \delta_m\}$ has the fmp as well.*

Proof. By our assumption, we have $\delta_i := \gamma_i(\neg\alpha_1, \ldots, \neg\alpha_n)$. Therefore using T-translation of propositional formulas into modal formulas and noting that for any formula γ, $T(\gamma) \equiv \Box T(\gamma) \in S4$ (Lemma 2.7.2) it follows

$$T(\delta_i) \equiv T(\gamma_i)(\Box\Diamond\neg T(\alpha_1), \ldots, \Box\Diamond\neg T(\alpha_n)) \in S4.$$

That is, any $T(\delta_i)$ is a $\Box\Diamond$-formula in $S4$. Suppose a formula α is not a theorem of $\lambda \oplus \{\delta_1, \ldots, \delta_m\}$. We now invoke the mapping σ of superintuitionistic logics in their greatest modal counterparts. We claim that

$$T(\alpha) \notin \sigma(\lambda) \oplus \{T(\delta_1), \ldots, T(\delta_m)\} = \sigma(\lambda) \vee (S4 \oplus \{T(\delta_1), \ldots, T(\delta_m)\}).$$

Indeed, otherwise, applying Theorem 2.7.20 and Theorem 2.7.4 we derive

$$\alpha \in \rho(\sigma(\lambda) \vee (S4 \oplus \{T(\delta_1), \ldots, T(\delta_m)\})) =$$
$$\rho\sigma(\lambda) \vee \rho(S4 \oplus \{T(\delta_1), \ldots, T(\delta_m)\}) = \lambda \vee (H \oplus \{\delta_1, \ldots, \delta_m\}) =$$
$$\lambda \oplus \{\delta_1, \ldots, \delta_m\},$$

a contradiction. By Corollary 2.7.21 the modal logic $\sigma(\lambda)$ also has fmp. We have showed that all formulas $T(\delta_i)$ are $\Box\Diamond$-formulae in $S4$. Therefore according to Theorem 2.8.9 the modal logic $\sigma(\lambda) \oplus \{T(\delta_1), \ldots, T(\delta_m)\}$ has fmp. Using this we infer that there is a finite modal algebra $\mathfrak{B}$ such that $\mathfrak{B} \models \sigma(\lambda) \oplus \{T(\delta_1), \ldots, T(\delta_m)\}$ but $\mathfrak{B} \not\models T(\alpha)$. Since $\sigma(\lambda)$ is a modal counterpart for λ we get $\mathfrak{B} \models T(\beta)$ for all $\beta \in \lambda$. By Lemma 2.7.6 we conclude $G(\mathfrak{B}) \models \beta$ for all $\beta \in \lambda$, $G(\mathfrak{B}) \models \delta_1 \wedge \ldots \wedge \delta_m$, but $G(\mathfrak{B}) \not\models \alpha$. Hence $G(\mathfrak{B})$ is a finite pseudo-boolean algebra that detaches α from $\lambda \oplus \{\delta_1, \ldots, \delta_m\}$. Thus the logic $\lambda \oplus \{\delta_1, \ldots, \delta_m\}$ has the fmp. ■

Now we briefly describe certain applications of the theorems we have obtained. For instance, because the modal systems $K4$, $S4$ and Grz have the fmp (see Corollary 2.6.12 and Theorem 2.6.25), using Theorem 2.8.9 we infer

Corollary 2.8.11 *Modal logics*

$$K4.1 := K4 \oplus \Box\Diamond p \to \Diamond\Box p,$$
$$K4.2 := K4 \oplus \Diamond\Box p \to \Box\Diamond, p$$
$$S4.1 := S4 \oplus \Box\Diamond p \to \Diamond\Box p,$$
$$S4.2 := S4 \oplus \Diamond\Box p \to \Box\Diamond p,$$
$$Grz.1 := Grz \oplus \Box\Diamond p \to \Diamond\Box p,$$
$$Grz.2 := Grz \oplus \Diamond\Box p \to \Box\Diamond p$$

have the fmp and are decidable.

Consider now the formulas

$$\alpha_n := \bigwedge_{0 \leq i \leq n} \Diamond\Box p_i \to \bigvee_{1 \leq i < j \leq n} \Diamond\Box(p_i \wedge p_j),$$

$$\beta_n := \Box\Diamond \bigvee_{0 \leq i \leq n} \Box(p_i \to \bigvee_{1 \leq i < j \leq n} p_j).$$

Since these formulas are $\Box\Diamond$-formulas in $K4$, we also infer from Theorem 2.8.9

Corollary 2.8.12 *For any n, the modal logics*

$$K4 \oplus \alpha_n, S4 \oplus \alpha_n, K4 \oplus \beta_n, S4 \oplus \beta_n$$

have the fmp and are decidable.

It is not difficult to see that the formula α_n expresses the following property of finite Kripke models: for any rooted reflexive transitive frame, the truth of α_n on this frame is equivalent to the number of maximal clusters is not more than n. The formula β_n expresses the following property: for any finite rooted reflexive and transitive frame $\mathcal{F}$, the truth of β_n in $\mathcal{F}$ is equivalent to the number of elements in any maximal cluster is not more than n.

Now we proceed to the next known general result concerning the fmp. We will describe the modal logics of finite depth which were introduced in articles of K.Segerberg [165] and L.L.Maksimova [93], and then also studied by W.Blok [15]. As a matter of fact, all the approaches to such logics take issue in the research of T.Hosoi [68] of superintuitionistic logics of finite depth. To define modal logics of finite depth we introduce the following sequence of modal formulas:

$$\sigma_o := \bot, \quad \sigma_{n+1} := p_{n+1} \vee \Box(\Box p_{n+1} \to \sigma_n).$$

Definition 2.8.13 *We say that a modal logic λ over $K4$ is a logic of depth n if $\sigma_n \in \lambda$ but $\sigma_{n-1} \notin \lambda$.*

We have to clarify why we mention *depth* in this definition. We recall the reader that a transitive frame $\mathcal{F}$ has the depth n if n is the upper bound of the number of clusters in ascending chains of clusters from $\mathcal{F}$. It is easy to show that the truth of formulas σ_n on transitive frames effects the depth of such frames:

Lemma 2.8.14 *For any transitive frame $\mathcal{F}$, $\mathcal{F} \Vdash \sigma_n$ for $1 \leq n$ if and only if the depth of $\mathcal{F}$ is not greater than n.*

Proof. Suppose σ_n is falsified on $\mathcal{F} := \langle W, R \rangle$ by a valuation V. Then there is an element c_n from $|\mathcal{F}|$ such that $c_n \nVdash_V \sigma_n$, i.e. $c_n \nVdash_V p_n$ and also $c_n \nVdash_V \Box(\Box p_n \to \sigma_{n-1})$. Therefore there is an c_{n_1} such that $c_n R c_{n_1}$, $\neg(c_{n_1} R c_n)$ and $c_{n_1} \nVdash_V \sigma_{n-1}$. Continuing this reasoning, we conclude that the frame $\mathcal{F}$ has depth $n + 1$ or more. Conversely, suppose that $\mathcal{F}$ has depth more than n and $c_{n+1}, \dots, c_1$ is a tuple of elements of $\mathcal{F}$ such that: $c_{i+1} R c_i$ and $\neg(c_i R c_{i+1})$ for all i, $1 \le i \le n$. We pick the following valuation V:

$$V(p_i) := c_i^{R \le}$$

for all i, $1 \le i \le n$. Then $c_1 \nVdash_V \sigma_0$. Suppose $c_k \nVdash_V \sigma_{k-1}$ has already been shown. Then $c_k \Vdash_V \Box p_k$ and $c_{k+1} \nVdash_V \Box(\Box p_k \to \sigma_{k-1})$. Consequently $c_{k+1} \nVdash_V \sigma_k$. Hence we infer $c_{n+1} \nVdash_V \sigma_n$. ∎

A similar result can be obtained for the depth of frames in the Stone representations of modal algebras regarding the truth of formulas σ_n on these algebras.

Lemma 2.8.15 *Suppose that $\mathfrak{B} \models \sigma_n$ for some modal algebra $\mathfrak{B} \in Var(K4)$. Then the frame of the Stone representation $\langle W, R, V \rangle^+$ of $\mathfrak{B}$ is transitive and has a depth of not more than n.*

Proof. First note that the frame of $\langle W, R, V \rangle$ is transitive. Indeed, suppose ∇_1, ∇_2 and ∇_3 are ultrafilters from W such that $\nabla_1 R \nabla_2$ and $\nabla_2 R \nabla_3$. Suppose $\Box a \in \nabla_1$. Then $\Box\Box a \in \nabla_1$ and $\Box a \in \nabla_2$. Consequently we obtain $a \in \nabla_3$. Hence $\nabla_1 R \nabla_3$. Suppose that the depth of the frame of $\langle W, R, V \rangle$ is strictly greater than n. Consider some sequence of elements $\nabla_1^t, \dots, \nabla_{n+1}^t$ from W such that $\nabla_{i+1}^t R \nabla_i^t$ but $\neg(\nabla_i^t R \nabla_{i+1}^t)$ for every i, $1 \le n$. By construction of the Stone representation $\langle W, R, V \rangle^+$ this, in particular, means for every i, $1 \le i \le n$, there exists an element $\Box a_i$ such that $\Box a_i \in \nabla_i^t$ and $a_i \notin \nabla_{i+1}^t$. At the same time, for all elements $\Box b$, if $\Box b \in \nabla_{i+1}$ then $b \in \nabla_i$.

Using elements $a_i, 1 \le i \le n+1$ we introduce the following elements of $\mathfrak{B}$: $\sigma_i^a := \sigma_i\binom{p_1, \dots, p_i}{a_1, \dots, a_i}$, $1 \le i \le n+1$. We claim that for all i, $0 \le i \le n$, $\sigma_i^b \notin \nabla_{i+1}$. We show this by induction on i. Indeed, $\sigma_0^a \notin \nabla_1$. Suppose we have already proved $\sigma_i^a \notin \nabla_{i+1}$, $i < n$. Then by definition of a_{i+1} we have $\Box a_{i+1} \in \nabla_{i+1}$, $a_{i+1} \notin \nabla_{i+2}$. Therefore $a_{i+1} \vee \Box(\Box a_{i+1} \to \sigma_i^a) \notin \nabla_{i+2}$, i.e. $\sigma_{i+1}^a \notin \nabla_{i+2}$. From our inductive procedure we obtain $\sigma_n^a \notin \nabla_{n+1}$ which contradicts $\mathfrak{B} \models \sigma_n$. ∎

Theorem 2.8.16 *[165, 93] For any modal algebra $\mathfrak{B}$ from $Var(K4)$ and any natural numbers n, k, if $\mathfrak{B} \models \sigma_n$ then the subalgebra $\mathfrak{B}_1$ of $\mathfrak{B}$ generated by arbitrary elements $a_1, \dots, a_k$ from $\mathfrak{B}$ is finite and the number of its elements can be effectively evaluated by k.*

Proof. Since $\mathfrak{B}_1$ is a subalgebra of $\mathfrak{B}$ it follows $\mathfrak{B}_1 \models \sigma_n$. By the Stone representation theorem (see Theorem 2.5.6), the modal algebra $\mathfrak{B}_1$ is isomorphic to the algebra $\langle W, R, V \rangle^+$, where V is the valuation of all letters p_a corresponding to all elements a of $\mathfrak{B}_1$, which maps each p_a into $\{ \nabla \mid a \in \nabla \} \subseteq W$. The isomorphism e of $\mathfrak{B}_1$ onto $\mathfrak{B}_1 = \langle W, R, V \rangle^+$ is the following: $\forall a \in \mathfrak{B}_1$, $e(a) := V(p_a) = \{ \nabla \mid a \in \nabla \}$. Any element a from $\mathfrak{B}_1$ has a representation $a = \alpha(a_1, \ldots, a_k)$. We claim that for any such element and any $\nabla \in W$,

$$\alpha(a_1, \ldots, a_k) \in \nabla \Leftrightarrow \nabla \Vdash_V \alpha(p_{a_1}, \ldots, p_{a_k}). \tag{2.33}$$

Indeed $\alpha(a_1, \ldots, a_k) \in \nabla \Leftrightarrow \nabla \in e(\alpha(a_1, \ldots, a_k)) \Leftrightarrow \nabla \in \alpha(e(a_1), \ldots, e(a_k))$ $\Leftrightarrow \nabla \in \alpha(V(p_{a_1}), \ldots, V(p_{a_k})) \Leftrightarrow \nabla \in V(\alpha(p_{a_1}, \ldots, p_{a_k}))$ (for this equivalence see Lemma 2.5.17) $\Leftrightarrow \nabla \Vdash_V \alpha(p_{a_1}, \ldots, p_{a_k})$.

We will show that the number of elements from W is finite and can be evaluated effectively by k. Applying Lemma 2.8.15 it follows that the depth of the frame $\langle W, R \rangle$ is not more than n. First we evaluate the number of elements in clusters from $\langle W, R \rangle$. Suppose ∇_1 and ∇_2 belong to the same cluster C and that in addition $\nabla_1 \Vdash_V p_{a_i} \Leftrightarrow \nabla_2 \Vdash_V p_{a_i}, \forall i, 1 \leq i \leq k$. Then for all formulas $\alpha(p_{a_1}, \ldots, p_{a_k})$ built up of $p_{a_1}, \ldots, p_{a_k}$,

$$\nabla_1 \Vdash_V \alpha(p_{a_1}, \ldots, p_{a_k}) \Leftrightarrow \nabla_2 \Vdash_V \alpha(p_{a_1}, \ldots, p_{a_k}). \tag{2.34}$$

This equivalence could be easily shown by induction on the length of the formula $\alpha(p_{a_1}, \ldots, p_{a_k})$. By (2.34) and (2.33) we infer $\nabla_1 = \nabla_2$. Thus any cluster C from $\langle W, R \rangle$ has not more than 2^k elements.

Now we evaluate the number of maximal clusters from $\langle W, R \rangle$. Suppose that some two maximal clusters C_γ and C_δ are isomorphic as models with valuation V of letters $p_{a_1}, \ldots, p_{a_n}$. Assume $\nabla_\gamma \in C_\gamma, \nabla_\delta \in C_\delta$ are identified by this isomorphism. Then by induction on the length of any given formula $\alpha(p_{a_1}, \ldots, p_{a_k})$ built up of $p_{a_1}, \ldots, p_{a_k}$ it is easy to show that

$$\nabla_\gamma \Vdash_V \alpha(p_{a_1}, \ldots, p_{a_k}) \Leftrightarrow \nabla_\delta \Vdash_V \alpha(p_{a_1}, \ldots, p_{a_k}).$$

This and (2.33) imply $\nabla_\gamma = \nabla_\delta$. Thus C_1 and C_2 coincide. Hence the number of maximal clusters from $\langle W, V \rangle$ is bounded by $(2^k) * (2^{2^k})$.

Suppose that we have already proved that the number of clusters from $\langle W, R \rangle$ of depth not more than m is finite and effectively evaluated by k. Consider two clusters C_1 and C_2 of depth $m+1$ which are isomorphic as models with valuation V of letters $p_{a_1}, \ldots, p_{a_n}$ by an isomorphism f and moreover $C_1^{R<} = C_2^{R<}$, where for $g := 1, 2$ $C_g^{R<} := \{ \nabla \mid \forall \nabla_1 \in C_g (\nabla_1 R \nabla \& \neg[\nabla R \nabla_1]) \}$. Suppose $\nabla_1 \in C_1$ and $f(\nabla_1) = \nabla_2$. Then again, for any formula $\alpha(p_{a_1}, \ldots, p_{a_k})$ built up of letters $p_{a_1}, \ldots, p_{a_k}, \nabla_1 \Vdash_V \alpha(p_{a_1}, \ldots, p_{a_k}) \Leftrightarrow \nabla_2 \Vdash_V \alpha(p_{a_1}, \ldots, p_{a_k})$, which could again be easily shown by induction on length of $\alpha(p_{a_1}, \ldots, p_{a_k})$. This equivalence and (2.33) yield $\nabla_1 = \nabla_2$. Thus C_1 and C_2 coincide. Because the number of clusters with a depth of not more than m is finite and effectively evaluated by k we conclude that the same result holds for the clusters of depth $m + 1$. Since the frame $\langle W, R \rangle$ has depth of not more than n this procedure terminates and we arrive at the required result. ∎

This theorem supplies us with a strong sufficient condition for a logic to have finite model property:

Corollary 2.8.17 *[165, 93, 15] Any modal logic over $K4$ with finite depth (i.e. containing σ_n for some n) has the finite model property.*

This result gives a stronger property of finiteness then just the finite model property. We specify this property in the following

Definition 2.8.18 *An algebraic logic is* locally finite *if for any algebra $\mathcal{A}$ from $Var(\lambda)$ and any finite tuple $a_1, \ldots, a_n$ of elements from $\mathcal{A}$, the subalgebra of $\mathcal{A}$ generated by $a_1, \ldots, a_n$ is finite*

From Theorem 2.8.16 it immediately follows that all modal logics with finite depth over $K4$ are locally finite. It is a remarkable fact that the possession of finite depth is not only a sufficient but also a necessary condition for a logic over $K4$ to be locally finite. We proceed to prove this conjecture but first we need to establish certain simple preliminary facts. We recall that the unary operation $\square_0$ has the following definition: $\square_0 x := x \wedge \square x$.

Lemma 2.8.19 *If $\mathfrak{B} \in Var(K4)$ then the algebra $\mathfrak{B}_{\square_0}$, which is the boolean algebra $\mathfrak{B}$ with the unary modal operation $\square_0$, forms an $S4$-modal algebra, that is $\mathfrak{B}_{\square_0} \in Var(S4)$.*

Proof. Indeed, $\square_0 \top = \top \wedge \square \top = \top \wedge \top = \top$. Also $\square_0 a = a \wedge \square a \leq a$. Furthermore, $\square_0 a = a \wedge \square a$, $\square_0 \square_0 a = \square_0 a \wedge \square \square_0 a = a \wedge \square a \wedge \square(a \wedge \square a) = a \wedge \square a \wedge \square a \wedge \square \square a = a \wedge \square a$. That is $\square_0 a = \square_0 \square_0 a$. At last,

$$\square_0(a \to b) = (a \to b) \wedge \square(a \to b) \leq$$
$$(a \to b) \wedge (\square a \to \square b) \leq a \wedge \square a \to b \wedge \square b = \square_0 a \to \square_0 b,$$

i.e. $\mathfrak{B}_{\square_0}$ is in fact a modal algebra from $Var(S4)$.

Lemma 2.8.20 *Suppose λ is a modal logic over $K4$ which is not a logic of finite depth, i.e. $\sigma_n \notin \lambda$ for any $n \in N$. Then there is an algebra $\mathfrak{B}$ form $Var(\lambda)$ which contains an infinite strictly ascending chain of elements which have the form $\square_0 a_n, n \in N$, that is for all n, $\square_0 a_n < \square_0 a_{n+1}$.*

Proof. Note first that the formula $\square_0 \sigma_n \to \square_0 \sigma_{n+1}$ is a theorem of $K4$ which can be easy verified using Kripke models. Therefore this formula is a theorem of the logic λ. Consider the free algebra $\mathfrak{F}_\lambda(\omega)$ of countable rank from $Var(\lambda)$. By our observation above, $\square_0 \sigma_n \leq \square_0 \sigma_{n+1}$ holds in $\mathfrak{F}_\lambda(\omega)$ when we consider all letters p_i from the above mentioned formulas as the free generators. Suppose that $\Box_0\sigma_{n+1} \leq \Box_0\sigma_n$ also holds in $\mathfrak{F}_\lambda(\omega)$. Then $\mathfrak{F}_\lambda(\omega) \models \Box_0\sigma_{n+1} \rightarrow \Box_0\sigma_n$. Making the substitution $p_{n+1} = \top$ in the formula of the right hand side of this relation we conclude $\mathfrak{F}_\lambda(\omega) \models \Box_0\sigma_n$ and $\sigma_n \in \lambda$ which contradicts the condition of this lemma. Thus $\Box_0\sigma_n < \Box_0\sigma_{n+1}$. ∎

Theorem 2.8.21 *If λ is a modal logic over $K4$ and $\sigma_n \notin \lambda$ for any n then the modal algebra $\mathfrak{F}_\lambda(1) \in Var(\lambda)$ is infinite.*

Proof. We begin our proof with a preliminary construction. Applying Lemma 2.8.20 we pick an algebra $\mathfrak{B}$ from $Var(\lambda)$ which has a strictly ascending sequence of elements $\Box_0 a_n, n \in N$ the first element of which is $\bot$. So we have $\Box_0 a_n < \Box_0 a_{n+1}$ for any n, $\Box_0 a_0 = \bot$. We add to this chain the element $\top$ and obtain a chain $\mathcal{L}$. $\mathcal{L}$ is a pseudo-boolean algebra consisting of certain open elements of the modal algebra $\mathfrak{B}_{\Box_0}$. That is, $\mathcal{L}$ is a subalgebra of the pseudo-boolean algebra $G(\mathfrak{B}_{\Box_0})$. Therefore by Lemma 2.7.18 the wrapping modal algebra $S(\mathcal{L})$ is a subalgebra of the modal algebra $S(G(\mathfrak{B}_{\Box_0}))$. By Lemma 2.7.17 the algebra $S(G(\mathfrak{B}_{\Box_0}))$ is a subalgebra of the modal algebra $\mathfrak{B}_{\Box_0}$ generated by the set $G(\mathfrak{B}_{\Box_0})$. Thus $S(\mathcal{L})$ is a subalgebra of the modal algebra $\mathfrak{B}_{\Box_0}$. Therefore, for any term $t(x_1, \dots, x_n)$ built up of variables $x_1, \dots, x_n$ by means of the boolean operations and $\Box_0$ only,

$$\forall c_1, \dots, c_n \in S(\mathcal{L})(val_{S(\mathcal{L})}t(c_1, \dots, c_n) = val_{\mathfrak{B}}t(c_1, \dots, c_n). \tag{2.35}$$

Now we consider the following sequence of formulas:

$$\beta_1 := \Box_0(\bot), \quad \alpha_k := \Box_0(p \lor \beta_k), \quad \beta_{k+1} := \Box_0(p \rightarrow \alpha_k), \quad k \in N.$$

We claim that all these formulas are pairwise not equivalent in λ. In order to show all formulas β_k, α_k are not equivalent in λ, for $1 \leq k \leq i+1$ and any given $i+1$, we interpret the letter p on $\mathfrak{B}$ as follows:

$$p := b = (\Box_0 a_1 \land \neg\Box_0 a_0) \lor \dots \lor (\Box_0 a_{2i+1} \land \neg\Box_0 a_{2i}).$$

We will show that the values of the introduced formulas on b are the following, for every k, $1 \leq k \leq i+1$

$$\alpha_k(b) = \Box_0 a_{2k-1}, \quad \beta_k(b) = \Box_0 a_{2k-2}. \tag{2.36}$$

Note that by (2.35) the values of formulas $\alpha_k(p), \beta_k(p)$ on $p := b$ belong to the algebra $S(\mathcal{L})$. The basis of induction: $\beta_1(b) = \bot = \Box_0 a_0$. Consider $\alpha_k(b)$.

$$\alpha_k(b) = \Box_0(b \lor \beta_k(b)) = \Box_0(b \lor \Box_0 a_{2k-2}) =$$
$$\Box_0[\Box_0 a_{2k-2} \lor (\Box_0 a_{2k-1} \land \neg\Box_0 a_{2k-2}) \lor \dots \lor (\Box_0 a_{2i+1} \land \neg\Box_0 a_{2i})] =$$
$$\Box_0[\Box_0 a_{2k-1} \lor (\Box_0 a_{2k+1} \land \neg\Box_0 a_{2k}) \lor \dots \lor (\Box_0 a_{2i+1} \land \neg\Box_0 a_{2i})]$$

Consequently $\Box_0 a_{2i+1} = \Box_0 \Box_0 a_{2i+1} \leq \alpha_k(b)$. Moreover

$$\alpha_k(b) \wedge \Box_0 a_{2k} = \Box_0[\Box_0 a_{2k-1} \vee (\Box_0 a_{2k+1} \wedge \neg \Box_0 a_{2k})$$
$$\vee ... \vee (\Box_0 a_{2i+1} \wedge \wedge \neg \Box_0 a_{2i})] \wedge \Box_0 a_{2k} \leq$$
$$\Box_0[(\Box_0 a_{2k-1} \vee (\Box_0 a_{2k+1} \wedge \neg \Box_0 a_{2k})$$
$$\vee ... \vee (\Box_0 a_{2i+1} \wedge \wedge \neg \Box_0 a_{2i})) \wedge \Box_0 a_{2k}] \leq$$
$$\Box_0[(\Box_0 a_{2k-1} \vee \neg \Box_0 a_{2k} \vee ... \vee \neg \Box_0 a_{2i}) \wedge \Box_0 a_{2k}] =$$
$$\Box_0[(\Box_0 a_{2k-1} \vee \neg \Box_0 a_{2k}) \wedge \Box_0 a_{2k}] =$$
$$\Box_0 \Box_0 a_{2k-1} = \Box_0 a_{2k-1}.$$

Thus $\Box_0 a_{2i+1} \leq \alpha_k(b) = \Box_0 \alpha_k(b)$ and $\Box_0 \alpha_k(b) \wedge \Box_0 a_{2k} \leq \Box_0 a_{2k-1}$. By (2.35) $\alpha_k(b)$ belongs to $S(\mathcal{L})$ and consequently $\Box_0 \alpha_k(b) \in \mathcal{L}$. This implies $\Box_0 \alpha_k(b) = \alpha_k(b) = \Box_0 a_{2i+1}$.

Now we consider $\beta_{k+1}(b)$ and merely calculate:

$$b_{k+1}(b) := \Box_0(b \to \alpha_k(b)) = \Box_0(b \to \Box_0 a_{2k-1}) =$$
$$\Box_0((\Box_0 a_1 \wedge \neg \Box_0 a_0 \to \Box_0 a_{2k-1}) \wedge \ ... \ \wedge (\Box_0 a_{2i+1} \wedge$$
$$\wedge \neg \Box_0 a_{2i} \to \Box_0 a_{2k-1})) =$$
$$\Box_0((\Box_0 a_1 \to \Box_0 a_0 \vee \Box_0 a_{2k-1}) \wedge \ ... \ \wedge (\Box_0 a_{2i+1} \to$$
$$\to \Box_0 a_{2i} \vee \Box_0 a_{2k-1})) =$$
$$\Box_0((\Box_0 a_{2k+1} \to \Box_0 a_{2k}) \wedge \ ... \ \wedge (\Box_0 a_{2i+1} \to \Box_0 a_{2i})) =$$
$$\Box_0(\Box_0 a_{2k+1} \to \Box_0 a_{2k}) \wedge \ ... \ \wedge \Box_0(\Box_0 a_{2i+1} \to \Box_0 a_{2i}) =$$
$$\Box_0 a_{2k} \wedge \ ... \ \wedge \Box_0 a_{2i} = \Box_0 a_{2k},$$

where the last but one equality holds because this equality holds in $S(\mathcal{L})$ (cf. the definition of the modal operator in the wrapping modal algebras), and therefore according to (2.35) it is also valid in $\mathfrak{B}$. So (2.36) is shown. Thus all formulas $\beta_k(p)$, $\beta_k(p)$, $1 \leq k \leq i+1$ take different values under the valuation $p := b$ in the algebra $\mathfrak{B}$. Therefore all these formulas are not equivalent in λ. Therefore the free modal algebra $\mathfrak{F}_\lambda(1)$ of rank 1 from $Var(\lambda)$ is infinite. $\blacksquare$

Using this theorem and Theorem 2.8.16 we obtain the final general description theorem:

Theorem 2.8.22 *[93, 15] A modal logic λ over $K4$ is locally finite iff λ is a logic of finite depth, i.e. iff $\sigma_n \in \lambda$ for some n.*

We have already mentioned that the definition of modal logics of finite depth historically arises from the definition of T.Hosoi [68] regarding superintuitionistic logics of finite depth. The modal logics of finite depth are, in a sense, more general objects. Therefore it would be relevant to extract now some appropriate corollaries for superintuitionistic logics. We begin with the definition of propositional non-modal formulas describing the depth of partially-ordered frames. We set:

$$\phi_0 := \bot, \quad \phi_{n+1} := p_{n+1} \vee (p_{n+1} \to \phi_n).$$

Definition 2.8.23 *A superintuitionistic logic λ is a logic of depth n if $\phi_n \in \lambda$ but $\phi_{n-1} \notin \lambda$.*

The formulas ϕ_n describe the depth of intuitionistic frames, i.e. posets, similar to the description of depth of transitive frames by modal formulas σ_n. We leave the readers to prove the following exercise which proof is quite similar to the proof of Lemma 2.8.14.

Exercise 2.8.24 *The formula ϕ_n is valid on some intuitionistic frame $\mathcal{F}$ iff the depth of $\mathcal{F}$ is not greater than n.*

Lemma 2.8.25 *If λ is a superintuitionistic logic and $\phi_n \in \lambda$ then $\sigma_n \in \sigma(\lambda)$, i.e. the greatest modal counterpart of λ is a modal logic of depth not greater than n.*

Proof. Indeed, since $\sigma(\lambda)$ is a modal counterpart for the logic λ, we have that $T(\phi_n) \in \sigma(\lambda)$. Note that $T(\phi_n) \to \sigma_n \in S4$ for every n. Indeed, using induction on n, we conclude $T(\phi_0) = \Box(\bot) = \bot = \sigma_0$.

$$T(\phi_n) := \Box p_n \vee \Box(\Box p_n \to T(\phi_{n-1})),$$
$$\Box p_n \vee \Box(\Box p_n \to T(\phi_{n-1})) \to p_n \vee \Box(\Box p_n \to T(\phi_{n-1})) \in S4,$$
$$T(\phi_{n-1}) \to \sigma_{n-1} \in S4, \text{ inductive hypothesis}$$
$$\Box p_n \vee \Box(\Box p_n \to T(\phi_{n_1})) \to p_n \vee \Box(\Box p_n \to \sigma_{n-1}) \in S4.$$

Thus we conclude $\sigma_n \in \sigma(\lambda)$. ∎

Theorem 2.8.26 *If λ is a superintuitionistic logic of depth n then λ is locally finite and λ has the finite model property.*

Proof. According to Lemma 2.8.25 $\sigma(\lambda)$ is a modal logic of depth not more than n over $S4$. Therefore by Theorem 2.8.16 $\sigma(\lambda)$ is locally finite. We pick any algebra $\mathfrak{A} \in Var(\lambda)$. Then $S(\mathfrak{A}) \in Var(\sigma(\lambda))$ and any finite generated modal subalgebra of $S(\mathfrak{A})$ is finite. We know that $G(S(\mathfrak{A})) = \mathfrak{A}$. Because the pseudo-complements for pairs of elements from $\mathfrak{A}$ are expressible by modal terms on $S(\mathfrak{A})$ we conclude that any finite generated subalgebra of $\mathfrak{A}$ is finite. Thus λ is locally finite. In particular, λ has the fmp. ∎

It is easy to see that, for certain superintuitionistic logics, to be of finite depth is not necessary condition for their being locally finite,

Exercise 2.8.27 *Show that the logic $LC := H \oplus (p \to q) \vee (q \to p)$ is a locally finite superintuitionistic logic but is not a logic of finite depth.*

Hint: Use (i) the fact that $LC = \lambda(\mathfrak{A})$ where $\mathfrak{A}$ is the linearly ordered pseudo-boolean algebra of all natural numbers with adjoined $\top$, (ii) the fact that any n-generated subalgebra of $\mathfrak{A}$ has not more than 2^{2^n} elements which are presented as conjunctive normal forms on generators, (iii) Birkhoff's theorem concerning the presentation of varieties generated by certain classes of algebras.

2.9   Kripke Incomplete Logics

We already mentioned that though the possession of Kripke semantics gives us very strong instruments to analyze logics not all logics are Kripke complete. The first examples of Kripke incomplete modal logics were discovered independently by K.Fine [45] and S.K.Thomason [179]. These examples were bases for many constructions used by other authors to study phenomena of Kripke incompleteness. However these examples have rather complicated axiomatic systems. Lately certain examples of Kripke incomplete logics with simple axiomatic systems where offered by J. van Benthem [190, 191]. We describe these examples below.

Let λ_1 be the normal modal logic based on the axiomatic system of the minimal normal modal logic K enriched by the following formulas

$$\Box p \to p,$$
$$\Box(\Box p \to \Box q) \vee \Box(\Box q \to \Box p),$$
$$\Diamond p \wedge \Box(p \to \Box p) \to p,$$
$$\Box \Diamond p \to \Diamond \Box p.$$

Theorem 2.9.1 *(van Benthem [190]) The modal logic λ_1 is a Kripke incomplete extension of the von Wright modal system T.*

Proof. We need some auxiliary results. First we introduce the notation of *recession frame* $F_r := \langle N, R \rangle$ due to D.Makinson [90] and S.K.Thomason [179]. This frame is based on the set of all natural numbers N and has the following accessibility relation: $\forall n, m \ (mRn \Leftrightarrow m - 1 \leq n)$. Hence, F_r is reflexive but not transitive. The modal algebra $\mathfrak{B}(F_r)$ is the subalgebra of the modal algebra F_r^+ associated with F_r, which has as its basis set $|\mathfrak{B}(F_r)|$ the set of all finite and cofinite subsets of N.

Lemma 2.9.2 *The following hold:*

(i) $F_r \Vdash \Box p \to p$,

(ii) $F_r \Vdash \Box(\Box p \to \Box q) \vee \Box(\Box q \to \Box p)$,

(iii) $F_r \Vdash \Diamond p \wedge \Box(p \to \Box p) \to p$,

(iv) $\mathfrak{B}(F_r) \models \Box \Diamond p \to \Diamond \Box p$.

(v) $\mathfrak{B}(F_r) \not\models p \to \Box p$.

Proof. (i) holds since F_r is transitive. (ii) is valid because, for any valuation V of p and q in F_r, $V(\Box p)$ and $V(\Box q)$ are either $\emptyset$ or certain intervals of the kind $[n, \infty) := \{m \mid n \leq m\}$, so either $V(\Box p) \subseteq V(\Box q)$ or $V(\Box p) \subseteq V(\Box q)$. Suppose $m \in N$ and $m \Vdash_V \Diamond p \wedge \Box(p \to \Box p)$ for some V. Then either $(m - 1) \Vdash_V p$ or for some $n, m \leq n$ and $n \Vdash_V p$. In the first case we conclude $m \Vdash_v \Box p$ and

$m \Vdash_V p$. In the second case either $m = n$ and $m \Vdash_V p$ or $n = m + k, k \geq 1$. In the last case it follows that $m + k - 1 \Vdash_V p$, etc., and finally $m \Vdash_V p$. Thus (iii) holds also. For any valuation V on F_r such that $V(p)$ is a finite or cofinite set, suppose that $m \Vdash_V \Box\Diamond p$. Then $V(p)$ must be cofinite, whence $m \Vdash_V \Diamond\Box p$. Hence (iv) holds as well. To show (v) it is sufficient to take the valuation V on F_r as follows: $V(p) := \{1\}$. ■

From this lemma we immediately infer that $(p \rightarrow \Box p) \notin \lambda_1$. Thus in order to show that λ_1 is Kripke incomplete it is sufficient to prove that for any λ_1-frame $\mathcal{F}$, $\mathcal{F}_1 \Vdash p \rightarrow \Box p$ holds. To show this we need the following definition. For any frame $\mathcal{F} := \langle F, R \rangle$ and $w, v \in F$,

$$wR^0 v \Leftrightarrow w = v, wR^{n+1}v \Leftrightarrow \exists u_{n+1} \in F(wR^n n_{n+1} \& u_{n+1}Rv).$$

Lemma 2.9.3 *Given a frame* $\mathcal{F} := \langle F, R \rangle$, *and*

$$\mathcal{F} \Vdash \Box p \rightarrow p, \mathcal{F} \Vdash \Box(\Box p \rightarrow \Box q) \vee \Box(\Box q \rightarrow \Box p).$$

Then for any valuation V and $w \in F$,

$$w \Vdash_V (\Diamond p \wedge \Box(p \rightarrow \Box p)) \rightarrow p$$

iff for any $v \in F$ such that wRv, there exists a natural number n such that $vR^n w$.

Proof. Suppose wRv but, for no natural number n, $vR^n w$. Then we define the valuation V: $V(p) := \{u \mid u \in F, \exists n(vR^n u)\}$. It is easy to see that $w \notin V(p)$ and $w \nVdash_V p$. At the same time $v \Vdash_V p$ and $w \Vdash_V \Diamond p$. Moreover if $u \Vdash_V p$ and uRz then $z \Vdash_V p$ consequently $\forall g \in F(g \Vdash_V p \rightarrow \Box p)$ and $w \Vdash_V \Box(p \rightarrow \Box p)$. Hence $\Diamond p \wedge \Box(p \rightarrow \Box p) \rightarrow p$ is falsified by V on w.

For the converse direction, suppose that for some valuation V and $w \in F$, $w \Vdash_V \Diamond p \wedge \Box(p \rightarrow \Box p)$ and $w \nVdash_V p$. By the condition of our lemma we have that (a) $\mathcal{F}$ is reflexive (since $\mathcal{F} \Vdash \Box p \rightarrow p$) and that (b) $\forall x, y, z, u, v \in F(xRy \& xRz \& \Rightarrow \forall u((yRu) \rightarrow (zRu)) \vee ((zRu) \rightarrow (yRu))$ (since $\mathcal{F} \Vdash \Box(\Box p \rightarrow \Box q) \vee \Box(\Box q \rightarrow \Box p)$, which can be easily verified directly). Let wRv hold with $v \Vdash_V p$. By induction on n, we will show that, for all n, if $vR^n u$, then wRu and $u \Vdash_V p$. For $n = 0$ it is obvious. Next, if $vRu^{n+1}u$ then for some z with $vR^n z$, zRu. By the inductive hypothesis, wRz and $z \Vdash_V p$. Therefore $z \Vdash_V \Box p$ since $w \Vdash_V \Box(p \rightarrow \Box p)$, and $u \Vdash_V p$. To show wRu we compare w and z. Since $\mathcal{F}$ is reflexive we conclude wRw and wRz, but not zRw (since $z \Vdash_V \Box p$ and $w \nVdash_V p$). Therefore by (b) it follows for all s with zRs, wRs holds, consequently wRu. Thus we proved that the necessity assertion holds for all n. Therefore for any n and u, if $vR^n u$ then $u \neq w$ since $u \Vdash_V p$ and $w \nVdash_V p$, which concludes the proof. ■

Lemma 2.9.4 *For any λ_1-frame $\mathcal{F}$, $\mathcal{F} \Vdash p \rightarrow \Box p$.*

Proof. Similarly to above we obtain (a) $\mathcal{F}$ is reflexive (since $\mathcal{F} \Vdash \Box p \to p$) and (b) $\forall x, y, z, u, v \in F(xRy \& xRz \& \Rightarrow \forall u((yRu) \to (zRu)) \vee ((zRu) \to (yRu))$ (since $\mathcal{F} \Vdash \Box(\Box p \to \Box q) \vee \Box(\Box q \to \Box p)$). Moreover by Lemma 2.9.3 for all $w, v \in |\mathcal{F}|$ such that wRv there exists a certain n with $vR^n w$. Consider any $w \in |\mathcal{F}|$. We will show that w has only single R-successor in $\mathcal{F}$, viz. w itself, whence $\mathcal{F} \Vdash p \to \Box p$. Define, for every n,

$$G_n(w) := \{v \mid v \in |\mathcal{F}|, vR^n w \& \neg(vR^k w), k < n\}.$$

Hence, $G_0(w) := \{w\}, G_1(w) := \{v \mid vRw, v \neq w\}$ etc., of course, $G_n(w)$ will turn out to be empty for $n > 0$. We define the valuation V on $\mathcal{F}$ as follows:

$$V(p) := \bigcup_{n := 2m, m \in N} G_n(w).$$

Then $w \Vdash_V \Box \Diamond p$. Indeed, if wRv then by shown above facts v belongs to some $G_n(w)$. It has to be shown that $v \Vdash_V \Diamond p$. If n is even, then we are ready since vRv. If n is odd, then there exists some $v_1 \in G_{n-1}(w)$ such that vRv_1 (just take any R-sequence with length n from v to w and consider the successor of v, it must be in $G_{n-1}(w)$). Then since $v_1 \Vdash_V p$ we obtain $v \Vdash_V \Diamond p$. Hence $w \Vdash_V \Box \Diamond p$, and since $\Box \Diamond p \to \Diamond \Box p \in \lambda_1$ it follows w
$V dash_V Diamond \Box p$. Thus there is a certain $v \in |\mathcal{F}|$ such that wRv and $v \Vdash_V \Box p$. Because vRv we infer $v \in V(p)$ and $v \in G_n(w)$ for some even number n. But if $n > 0$ then v has an R-successor v_1 in the preceding $G_{n-1}(w)$ with odd $n - 1$ and $v_1 \notin V(p)$, whence $v \nVdash_V \Box p$. Thus it follows $v \in G_0(w)$, i.e. $v = w$ and we arrive at $w \Vdash_V \Box p$. Suppose that there is $v \in W$ such that wRv and $w \neq v$. Then applying $w \Vdash_V \Box p$ we derive $v \in V(p)$, i.e. for some even $n > 0, v \in G_n(w)$. That is, $vRv_1 R \ldots Rv_{n-1} Rw$, where $v_1 \notin V(p)$, in particular, $\neg(wRv_1)$ holds. Thus, wRw, wRv, vRv_1, and $\neg(wRv_1)$. Therefore by (b) vRw, in contradiction with $w \Vdash_V \Box p$ and $v \in G_1(w)$. Hence there is no $v \in W$ such that wRv and $w \neq v$ and $\mathcal{F} \Vdash p \to \Box p$. ∎

Combining this lemma and $p \to \Box p \notin \lambda_1$ we immediately derive that λ is Kripke incomplete. ∎

The second example of a Kripke incomplete modal logic offered by J. van Benthem [190] is also very elegant and has a simple axiomatic system. Namely, let

$$\lambda_2 := K \oplus \{(\Box p \to p), \Box(\Box(p \to \Box p) \to \Box^3 p) \to p)\}.$$

Theorem 2.9.5 *(van Benthem [190]) The logic λ_2 is Kripke incomplete.*

Proof. Consider the recession frame F_r and the corresponding algebra $\mathfrak{B}(F_r)$ introduced at the beginning of this section. First what we need is

Lemma 2.9.6 $\mathfrak{B}(F_r) \models \lambda_2$.

Proof. The frame F_r is reflexive, hence $\mathfrak{B}(F_r) \models \Box p \to p$. Suppose that for some finite or cofinite subset $V(p)$ of N and an element $w \in N$, we have $w \Vdash_V \Box(\Box(p \to \Box p) \to \Box^3 p)$. In this case $V(p)$ can not be finite. Indeed, otherwise we take the greatest natural number v_1 from $V(p)$ and $v := max(w, v_1 + 2)$. Then $v \Vdash_V \Box \neg p$ and $v \Vdash_V \Box(p \to \Box p)$ and $v \Vdash_V \Box^3 p$, in contradiction with $v \nVdash_V p$. If $V(p)$ is cofinite , then let v be the greatest member of $(N - V(p))$. Suppose that $w \leq v$. Then $v + 3 \Vdash_V \Box(p \to \Box p)$ and $v \Vdash_V \Box^3 p$, consequently $v \Vdash_V p$, a contradiction. Thus only $v < m$ is possible and $w \Vdash_V p$. ∎

It is easy to see that $\mathfrak{B}(F_r) \nvDash \Box p \to \Box\Box p$, but at the same time

Lemma 2.9.7 *For any λ_2-frame $\mathcal{F}$, $\mathcal{F} \Vdash \Box p \to \Box\Box p$.*

Proof. Suppose that for some λ_2-frame $\mathcal{F}$, there are some $w, v, u \in |\mathcal{F}|$ such that wRv, vRu but not wRu. Since $\mathcal{F}$ is an λ_2-frame, $\mathcal{F}$ is reflexive. We define in a similar way to above: for every n,

$$G_n(u) := \{v \mid v \in |\mathcal{F}|, vR^n u \& \neg(vR^k u), k < n\},$$

i.e. $G_0(u) := \{u\}, G_1(u) := \{v \mid vRu, v \neq u\}$ etc., and put

$$V(p) :=:= (|\mathcal{F}| - \bigcup_{n:=2m, m \in N} G_n(u)).$$

In order to prove this lemma it is sufficient to show that $w \Vdash_V \Box(\Box(p \to \Box p) \to \Box^3 p)$. Indeed, by the definition of V, $w \notin V(p)$, since $w \in G_2(u)$, which falsifies λ_2 in $\mathcal{F}$. Assume w_1 be a certain R-successor of w for which $\Box^3 p$ does not hold under V. We have to show that $w_1 \nVdash_V \Box(p \to \Box p)$. Since $w_1 \nVdash_V \Box^3 p$, there is some $w_2 \in S_n(G_n(u))$ for some even n such that $w_1 R^3 w_2$. Consequently w_1 itself is placed in some $G_n(u)$ but not in $G_0(u)$ since not wRu. Hence it suffices to prove that for any $s \in G_n(u)$ with $n > 0$, $s \nVdash_V \Box(p \to \Box p)$, i.e. $s \Vdash \Diamond(p \wedge \Diamond\neg p)$. If $s \in G_n(u)$ with even n then for some $s_1 \in G_{n-1}(u)$, sRs_1 which yields $s_1 \Vdash p \wedge \Diamond\neg p$ and $s \Vdash \Diamond(p \wedge \Diamond\neg p)$. Suppose that $s \in G_n(u)$ and n is odd. Then $s \Vdash_V p$ by the definition of V, and for some $s_1 \in G_{n-1}(u)$, sRs_1. Since $n - 1$ is even we get $s_1 \nVdash_V p$, hence $s \Vdash_V p \wedge \Diamond\neg p$ and we have $s \Vdash_V (p \wedge \Diamond\neg p)$. ∎

This lemma immediately yields that λ_2 is Kripke incomplete.

2.10 Advanced Tools for Kripke Completeness

As we have seen the basic tool for showing Kripke completeness is the using of canonical models. In fact, canonical Kripke models are characterizing Kripke models for corresponding logics. Therefore, as soon as the axioms of considered logic are valid on the frame of the canonical Kripke model, the logic is Kripke complete. At the same time not all Kripke complete logics have this property.

We have already considered some ways to overcome this obstacle for individual logics. However there is a powerful technique for proving Kripke completeness developed using different approaches. In this section we present only one result of this kind which we will use in our further constructions and proofs. It is the Kripke completeness of all modal logics extending $K4$ with finite width, what has been discovered by K.Fine [46].

The definition of logics with finite width employs the following formulas: for $1 \leq n$,

$$\mathcal{W}_n := \bigwedge_{0 \leq i \leq n} p_i \to \bigvee_{0 \leq i \neq j \leq n} \Diamond(p_i \wedge (p_j \vee \Diamond p_j)).$$

Definition 2.10.1 *We say that a modal logic λ over $K4$ has width n if $\mathcal{W}_n \in \lambda$ but the formula $\mathcal{W}_{n-1}$ is not a theorem of λ.*

Recall that a transitive frame $\mathcal{F} = \langle F, R \rangle$ has width n if $\mathcal{F}$ does not contain $n+1$ mutually R-incomparable elements, i.e.

$$\forall x_1, ..., x_{n+1} \in |\mathcal{F}|(\bigvee_{1 \leq i \neq j \leq n+1} ((w_i R w_j) \vee (w_i = w_j))),$$

but $\mathcal{F}$ has elements $c_1, c_2, ..., c_n$ which belong to different clusters and are R-incomparable by R.

Exercise 2.10.2 *For any transitive rooted frame $\mathcal{F} := \langle F, R \rangle$, $\mathcal{F} \Vdash \mathcal{W}_n$ iff the width of $\mathcal{F}$ is not more than n.*

It turns out that formulas $\mathcal{W}_n$ have the same effect on the canonical models Cn_λ of logics λ which include these formulas. Before to show this, we introduce canonical models for restricted sets of propositional letters. Let P_1 be a subset of the set P of all propositional letters. For any modal logic λ, $\lambda(P_1)$ is the set of all formulas built up of variables from P_1 which are in λ. We call $\lambda(P_1)$ the restriction of the logic λ to the set P_1. It is easy to see that $\lambda(P_1)$ is merely a modal logic on restricted set variables. The canonical model $Cn_{\lambda(P_1)}$ has the same properties as the model Cn_λ. In particular, for any modal formula α built up out of letters from P_1, $\alpha \in \lambda(P_1)$ iff $Cn_\lambda \Vdash \alpha$. But note that all the formulas of $\lambda(P_1)$ are constructed out of a finite fixed set of letters when P_1 is finite, and this allows as to prove some additional properties of such models.

Lemma 2.10.3 *[46] If λ is a modal logic over $K4$ and $\mathcal{W}_n \in \lambda$ then any rooted open subframe of the model $Cn_{\lambda(P_1)}$ has width not more than n for any set P_1.*

Proof. Suppose otherwise. Then there are elements $T, T_0, T_1, ..., T_n$ in $Cn_{\lambda(P_1)}$ such that $T R T_i$, $0 \leq i \leq n$, but $\neg(T_i R T_j)$ and $T_i \neq T_j$ for $i \neq j$. Therefore there are formulas $\beta_{i,j}$ and $\gamma_{i,j}$ such that $\Box\beta_{ij} \in T_i$, $\neg\beta_{ij} \in T_j$, $\gamma_{ij} \in T_i$ and $\neg\gamma_{ij} \in T_j$. We set $\beta_i := \bigwedge_{j\neq i}\Box\beta_{ij} \wedge \neg\beta_{ji}$, $\gamma_i := \bigwedge_{j\neq i}\gamma_{ij} \wedge \neg\gamma_{ji}$, and $\alpha_i := \beta_i \wedge \gamma_i$, where $0 \leq i \leq n$. Then it follows that $\alpha_i \in T_i$ and $\Diamond\alpha_i \in T$ because $T R T_i$. If $i \neq j$ then $\Box\beta_{ij}$ is a conjunct of α_i and $\neg\beta_{ij}$ is a conjunct of α_j. Therefore $\neg\Diamond(\alpha_i \wedge \Diamond\alpha_j) \in T$. Similarly, for $i \neq j$, γ_{ij} is a conjunct of α_i and $\neg\gamma_{ij}$ is a conjunct of α_j, consequently $\neg\Diamond(\alpha_i \wedge \alpha_j) \in T$. Thus we conclude $\Diamond\alpha_i \in T$, $\neg\Diamond(\alpha_i \wedge \Diamond\alpha_j) \wedge \neg\Diamond(\alpha_i \wedge \alpha_j) \in T$. This contradicts the fact that $\mathcal{W}_n \in \lambda$ and T is a consistent maximal theory. ∎

We need several technical results concerning sufficient conditions for certain Kripke models not to have infinite R-ascending chains of clusters.

Lemma 2.10.4 *[46] Suppose $\langle W, R\rangle$ is a transitive frame of finite width and $w_0, w_1, ...$ is a non-descending sequence of distinct elements in W, i.e., $\neg(w_j R w_i)$ for $j > i$. Then this sequence contains an ascending subsequence $v_0, v_1, ...,$ i.e., $v_i R v_{i+1}$ for any i.*

Proof. We call an element w_i of this sequence *suitable* if the set of all numbers j such that $w_i R w_j$ is infinite. First we show that there are suitable w_i. Suppose otherwise. We define an infinite subsequence $w_{i_0}, w_{i_1}, ...$ of the sequence $w_0, w_1, ...$ as follows. We set $i_0 := 0$, $i_{n+1} := 1 + max(\{j \mid w_{i+n} R w_j\} \cup \{i_n\})$. Then for any $m < n$, $\neg(w_{i_m} R w_{i_n})$ by the choice of our subsequence, and $\neg(w_{i_n} R w_{i_m})$ because our original sequence is non-descending. Therefore the sequence $w_{i_0}, w_{i_1}, ...$ contains an infinite number of R-incomparable elements, which contradicts the fact that the frame $\langle W, R\rangle$ has a finite width.

Now we show that if w_i is suitable then there exists a suitable w_j such that $w_i R w_j$, $i < j$. In fact, if w_i is suitable then there is an infinite subsequence $v_1, v_2, ...$ of $w_0, w_1, ...,$ where $w_i R v_j$ for all j. Then we repeat our above reasoning and find a suitable element in $v_1, v_2, ...$ Continuing this procedure we construct an infinite ascending subsequence. ∎

Definition 2.10.5 *A Kripke model $\mathfrak{M} := \langle W, R, V\rangle$ is* differentiated *if for any $a, b \in W$, where $a \neq b$, there is a formula α such that $a \Vdash_V \alpha$ and $b \nVdash_V \alpha$.*

Theorem 2.10.6 *Let $\mathfrak{M} := \langle W, R, V\rangle$ be a transitive differentiated model of finite width, where $||Dom(V)|| < \omega$ (i.e. the valuation of which is defined only for finitely many variables). Then $\mathfrak{M}$ contains no infinite ascending chains of elements, i.e., no distinct elements $w_0, w_1, ...,$ such that $w_i R w_{i+1}$ for any i.*

Proof. We say an element w of W is *deep* if there is an infinite ascending chain of elements from W which are accessible from w. For any $w \in W$, we put $U_w :=$

$\{u \mid u \in W, wRu$, where u is not deep$\}$. An element $w \in W$ is *static* if $U_v = U_w$ for any deep v such that wRv. We begin the proof with the following observation

$$\text{If } w \text{ is deep then there is a deep and static } v \text{ such that } wRv. \qquad (2.37)$$

Suppose otherwise. Since w is deep, there exists a v such that wRv and v is deep. As v is not static, there is u such that vRu, u is deep and $U_u \neq U_w$. Since R is transitive we have $U_u \subset U_v \subseteq U_w$ and u is also not static. Continuing this reasoning there is an infinite sequence $w_0, w_1, \ldots$ such that every w_i is deep but not static, $w_i R w_{i+1}$ and $U_{w_{i+1}} \subset U_{w_i}$ for every i, $0 \leq i$. Thus for every i, there is a $u_i \in (U_{w_i} - U_{w_{i+1}})$. Clearly, all w_i are different. Also $\neg(u_j R u_i)$ for every $j > i$. Indeed, otherwise $u_i \in U_{w_j}$ since R is transitive, which contradicts $u_i \notin U_{w_{i+1}}$. Therefore by Lemma 2.10.4 there is an infinite ascending subsequence $v_0, v_1, \ldots$ of $u_0, u_1, \ldots$. Then $v_i R v_{i+1}$ for every i and v_0 is deep which contradicts the fact that all $u_0, u_1, \ldots$ are not deep. Thus (2.37) holds. Recall that for any element $w \in W$, $V(w)$ is the set of all propositional letters which are valid on w under V. For each $w \in W$, we let $X_w := \{V(u) \mid wRu, u \text{ is deep}\}$ We say w is *stationary* if $X_v = X_w$ for any deep v such that wRv.

$$\text{For any deep } w \text{ there is deep and stationary } v \text{ such that } wRv. \qquad (2.38)$$

Suppose (2.38) is false. Then in a similar way to the proof of (2.37), it can be shown that for any deep and nonstationary w, there is a deep and nonstationary v such that wRv and $X_v \subset X_w$. Hence there is a sequence $w_0, w_1, \ldots$ such that $w = w_0, X_{w_{i+1}} \subset X_{w_i}$ for any i. Since $Dom(V)$ is finite, every X_{w_i} also is finite, a contradiction. Thus (2.38) is shown.

Assume now that the conjecture of the theorem is false. Then there is a deep w in W. Applying (2.37) and (2.38) we conclude there is a deep w_0 in W which is both static and stationary. We put $H := \{v \mid w_0 R v, v \text{ is deep}\}$. Then

$$(\forall u, v \in H)(V(u) = V(v) \Rightarrow (u \Vdash_V \alpha \Leftrightarrow v \Vdash_V \alpha)) \qquad (2.39)$$

for any formula α. In order to prove this we introduce the relation S on elements of W as follows: $wSv \Leftrightarrow (w = v) \vee (w, v \in H \& V(w) = V(v))$. It is easy to see that S is an equivalence relation. We put $[w] := \{v \mid v \in W, wSv\}$. Also we define $X := \{[w] \mid w \in W\}$. Consider the model $\mathfrak{M}_1 := \langle X, R_1, \Psi \rangle$, where

$$a R_1 b \Leftrightarrow (\exists w \in a)(\exists v \in b)(wRb)$$

and $[w] \Vdash_\Psi p \Leftrightarrow w \Vdash_V p$. In order to show (2.39) it is sufficient to prove that the mapping $w \mapsto [w]$ is a p-morphism (since p-morphisms preserve the truth of formulas). All the requirements to be p-morphism evidently hold except the requirement $[w]R_1[v] \Rightarrow \exists v_1 \colon [v_1] = [v]$ and wRv_1. Suppose $[w]R_1[v]$. If $w \in (W-H)$ then $[w]$ is a singleton (i.e. is an one element set) and there exists v_1 such that wRv_1 and $[v] = [v_1]$, what we needed. Let $w \in H$. If $v \in (W-V)$ then wRv

since w is static. The other possibility is $v \in H$. Then $\exists u(wRu \& [u] = [v])$ since w is stationary. Thus the mapping $w \mapsto [w]$ is a p-morphism and consequently the relation (2.39) holds.

Since w_0 is deep and $Dom(V)$ is finite there are distinct elements a, b from H which have the same valuation of letters with respect to V. Then by (2.39) all formulas have the same truth value on a and b under V. This contradicts the claim that $\mathfrak{M}$ is differentiated. ∎

Definition 2.10.7 *Let* $\mathcal{F} := \langle W, R \rangle$ *be a transitive frame. Let* H *be a subset of* W. *A set* $U \subseteq H$ *is a* cover *for* H *if*

$$(\forall v \in H)(\exists u \in U)[(v = u) \vee (vRu)].$$

$\mathcal{F}$ *itself has the* finite cover property *(abbreviation - (fcp)) if each subset* H *of* W *has a finite cover. We say* $v \in H$ *is* maximal *in* H *if* $v \in H$ *and for any* $u \in H$, $vRu \Rightarrow uRv$.

Theorem 2.10.8 *Suppose* λ *is a modal logic of width* n *over* $K4$ *and* P_1 *is a finite set of propositional letters. Let* $\mathcal{F} := \langle W, R \rangle$ *be a rooted generated subframe of the canonical model* $Cn_{\lambda(P_1)}$ *of the logic* $\lambda(P_1)$. *Then* $\mathcal{F}$ *has the finite cover property.*

Proof. We pick an $H \subseteq W$. We know that the model $Ch_{\lambda(P_1)}$, as any canonical model, is differentiated and by Lemma 2.10.3 $\mathcal{F}$ has finite width. Also this model is transitive because λ extends $K4$. Therefore by Theorem 2.10.6, any cluster of this model is finite and, for any subset G of $Cn_{\lambda(P_1)}$, any element of G is a maximal in G or R-sees a maximal element of G. Therefore any element of H either is maximal in H or R-sees some maximal element of H. Consider the set X consisting of all maximal elements of H. By definition X is a cover for H. If we suppose X contains infinitely many R-incomparable elements, this contradicts Lemma 2.10.3. Thus X consists of certain subsets of a finite number of some clusters in $\mathcal{F}$, since we observed all clusters in $\mathcal{F}$ are finite, it yields X is also finite. ∎

Now we present a drop points technique which will allow us to omit certain elements from models while preserving the truth value of formulas. First we recall the following notation.

Definition 2.10.9 *If* $\mathcal{F} := \langle F, R \rangle$ *is a frame and* $X \subseteq F$ *then*

$$X^{R \leq} := \{a \mid a \in F \& (\exists b \in X)(bRa)\}$$
$$X^{R <} := \{a \mid a \in F \& (\exists b \in X)(bRa) \& (\forall b \in X) \neg (aRb) \& (b \notin X)\}.$$

Sometimes we write for short X^R *instead of* $X^{R \leq}$. *If* $X = \{a\}$ *we write merely* a *instead of* $\{a\}$ *in the above notation.*

In other words, $X^{R\leq}$ is the set of all elements R-seen from elements of X, and $X^{R<}$ is the set of all elements strictly R-seen from elements of X.

Definition 2.10.10 *Given a model $\mathfrak{M} := \langle W, R, V \rangle$, we say an element $w \in W$ is eliminable in the model $\mathfrak{M}$ if for any formula α if $w \Vdash_V \alpha$ then there is an element v such that $v \in w^{R<}$ and $v \Vdash_V \alpha$.*

Lemma 2.10.11 *Let λ be a logic over $K4$. Let P_1 be a subset of the set of all propositional variables P (it is possible that $P_1 = P$). Suppose T_w is eliminable in the canonical model $Cn_{\lambda(P_1)}$ and $\alpha \in T_w$. Then there exists $T_v \in Cn_{\lambda(P_1)}$ such that $T_w R T_v$, $\alpha \in T_v$ and T_v is noneliminable in $Cn_{\lambda(P_1)}$.*

Proof. Suppose otherwise. Consider the set

$$H := \{T_v \mid T_v \in |Cn_{\lambda(P_1)}|, T_w R T_v, \alpha \in T_v\}.$$

This set contains no maximal elements, otherwise these elements would be none-liminable. We claim that

$$H \text{ contains a maximal by including elements } R\text{-chain } U. \tag{2.40}$$

Indeed, consider the set $\mathcal{X}$ of all R-chains from the set H ordered by the set-theoretic inclusion $\subseteq$. Applying Zorn's Lemma to $\mathcal{X}$ we easily derive that $\mathcal{X}$ has a maximal element, so (2.40) holds. Consider the set

$$\Gamma := \{\beta \mid (\exists T_u \in U)(\forall T_t \in U)(T_u R T_t \Rightarrow \beta \in T_t\}.$$

We will show that

$$\Gamma \text{ is consistent with } \lambda(P_1). \tag{2.41}$$

Suppose otherwise. Then there exist some formulas $\beta_1, \ldots, \beta_n$ and elements $T_{u_1}, \ldots, T_{u_n}$ from U such that $\neg(\beta_1 \wedge \ldots \wedge \beta_n) \in \lambda(P_1)$ and for every $T_t \in U$, $(T_{u_i} R T_t \Rightarrow \beta_i \in T_t$, where $1 \leq i \leq n$. Consider a maximal element T_{u_i} of $\{T_{u_1}, \ldots, T_{u_n}\}$. T_{u_i} is not maximal in U. For otherwise, since U is the maximal chain, T_{u_i} would be a maximal element of H, but we proved that H has no maximal elements. Hence $\exists T_u \in U$ such that T_u is a proper successor of T_{u_i}. But then $\beta_1, \ldots, \beta_n \in T_u$ which contradicts the claim that T_u is inconsistent with $\lambda(P_1)$. Thus (2.41) holds.

Therefore (cf.Lemma 2.3.21), as any set consistent with $\lambda(P_1)$, Γ is a subset of a complete theory T_v of $Cn_{\lambda(P_1)}$. Furthermore, $(\forall T_t \in U)(T_w R T_t \Rightarrow \alpha \in T_t)$, therefore $\alpha \in \Gamma \subseteq T_v$. Hence $T_v \in H$. Also for every $T_u \in U$, $T_u R T_v$ holds. Indeed, suppose $\Box \delta \in T_u$ and $T_u \in U$. Then for every $T_t \in U$, $T_u R T_t \Rightarrow \delta \in T_t$. Therefore $\delta \in \Gamma \subseteq T_v$. So, T_v is maximal in U. Since U is a maximal R-chain in H, we conclude that T_v is maximal in H which contradicts the claim that H has no maximal elements. $\blacksquare$

For any $\lambda(P_1)$, we denote by W_{ne} the set of all noneliminable elements in $Cn_{\lambda(P_1)}$. We denote by $Cnr_{\lambda(P_1)}$ the submodel of the model $Cn_{\lambda(P_1)}$ which is based on the set W_{ne} of all noneliminable elements, i.e., which has the accessibility relation and valuation transferred from $Cn_{\lambda(P_1)}$.

Definition 2.10.12 *For any modal logic λ over $K4$ and any subset P_1 of the set of all propositional variables P, $Cnr_{\lambda(P_1)}$ is called the* reduced canonical model *for λ and P_1.*

Theorem 2.10.13 *Let λ be a modal logic extending $K4$ and let P_1 be a subset of the set P of all propositional letters. Suppose the model $Cnr_{\lambda(P_1)}$ is a submodel of a model $\mathfrak{M}$ which is a submodel of $Cn_{\lambda(P_1)}$. Then for any element T_w of $\mathfrak{M}$, and any formula α*

$$T_w \Vdash_V \alpha \text{ in } \mathfrak{M} \Leftrightarrow \alpha \in T_w \Leftrightarrow T_w \Vdash_V \alpha \text{ in } Cn_{\lambda(P_1)}.$$

Proof. We show the first equivalence. The proof is by induction on the length of α. The only non-trivial step is the case $\alpha = \Box\beta$. If $\Box\beta \in T_w$ then for every T_v such that $T_w R T_v$, $\beta \in T_v$ and by the inductive hypothesis we infer $T_v \Vdash_v \beta$. Hence $T_w \Vdash_V \Box\beta$ in $\mathfrak{M}$. Conversely, suppose $\Box\beta \notin T_w$. Using the fact that $\mathfrak{M}$ is a submodel of $Cn_{\lambda(P_1)}$ and the basic property of canonical models (cf. Lemma 2.3.25) there is a complete theory $T_v \in Cn_{\lambda(P_1)}$ such that $T_w R T_v$ and $\neg\beta \in T_v$. Applying Lemma 2.10.11 we infer there exists $T_u \in Cnr_{\lambda(P_1)}$ such that $T_v = T_u$ or $(T_v R T_u \& \neg\beta \in T_u)$. Since $\mathfrak{M}$ contains the model $Cnr_{\lambda(P_1)}$, we have $T_u \in \mathfrak{M}$. Hence, as $\mathfrak{M}$ is transitive (since $K4 \subseteq \lambda$), we obtain $T_w R T_u$. Also $T_u \nVdash_V \beta$ by the inductive hypothesis, consequently $T_w \nVdash_V \Box\beta$. The last equivalence is a well-known property of canonical models. ∎

Definition 2.10.14 *A Kripke model $\mathfrak{M}$ is* reduced *if $\mathfrak{M}$ contains no eliminable elements.*

By Theorem 2.10.13 the model $Cnr_{\lambda(P_1)}$ is characterizing for the logic $\lambda(P_1)$. The question is as to whether even the reduced canonical model $Cnr_{\lambda(P_1)}$ can be based on some non-$\lambda(P_1)$-frame. Hence we have to investigate the question further.

Definition 2.10.15 *A Kripke model $\mathfrak{M}$ is* tight *if for any pair of elements w and v, $\forall\alpha(w \Vdash_V \Box\alpha \Rightarrow v \Vdash_V \alpha) \Rightarrow wRv$. A model $\mathfrak{M}$ is* natural *if $\mathfrak{M}$ is both tight and differentiated.*

We recall that an element w of a model $\mathfrak{M} := \langle W, R, V \rangle$ is *definable* (or is *expressible*) if there is a formula α such that $\{w\} = V(\alpha)$, i.e., the set of elements on which α is valid under V consists of the single element w. We then say that α defines w. Similarly, a subset X of W is definable (or expressible) if there is a formula α such that $V(\alpha) = X$.

Theorem 2.10.16 *Let $\mathfrak{M} := \langle W, R, V \rangle$ be a natural and transitive model with a finite width and the finite cover property, where $Dom(V) < \omega$. Then any noneliminable element w from $\mathfrak{M}$ is definable in $\mathfrak{M}$.*

Proof. Since w is noneliminable, there exists a formula α such that $\forall v \in W (v \in w^{R<} \Rightarrow v \Vdash_V \neg\alpha)$. By Lemma 2.10.6 all clusters, in particular the cluster $C(w)$ containing w, are finite. Because $\mathfrak{M}$ is differentiated, there exists a formula β such that $w \Vdash \beta$ and β is false on all other elements of $C(w)$. Let $H := \{v \mid v \in W, w \neq v, \neg(wRv)\}$ and $U := \{v_0, v_1, ..., v_n\}$ is a finite cover for H which exists since $\mathfrak{M}$ has the fcp. Since v_i is not a proper successor of w and $\mathfrak{M}$ is tight, there is a formula $\Box\gamma_i$ such that $w \Vdash_V \Box\gamma_i$ and $v_i \Vdash_V \neg\gamma_i$. Furthermore, since $w \neq v_i$ and $\mathfrak{M}$ is differentiated, there are formulas δ_i such that $w \Vdash_V \delta_i$ and $v_i \Vdash_V \neg\delta_i$, $0 \leq i \leq n$. We set

$$\phi := \alpha \wedge \beta \wedge \bigwedge_{0 \leq i \leq n} \Box\gamma_i \wedge \bigwedge_{0 \leq i \leq n} \delta_i.$$

We claim that ϕ defines w in $\mathfrak{M}$. Indeed, it is clear that $w \Vdash_V \phi$. Suppose that $v \Vdash_v \phi$ and $w \neq v$. If v is a proper successor for w then $v \Vdash_V \neg\alpha$. If $v \in C(w)$ then $v \Vdash_V \neg\beta$. If v is not a proper successor of w then $v \in H$ and either $v = v_i$ for some i and $v \Vdash_V \neg\delta_i$, or vRv_i for some i and $v \Vdash_V \neg\Box\gamma_i$. Hence, in all possible cases, $v \Vdash_V \neg\phi$. ∎

Lemma 2.10.17 *Let $\mathfrak{M} := \langle W, R, V \rangle$ be a natural and transitive model with the finite cover property, the finite width, and $\|Dom(V)\| \leq \omega$. Let H be a finite set of noneliminable elements in $\mathfrak{M}$. Let U be a subset of H. Then the set $Z := \{w \mid w \in (W - H), w^{R\leq} \cap H = U\}$ is definable in $\mathfrak{M}$.*

Proof. Let $H := \{v_0, v_1, ..., v_n\}$ and $U := \{v_0, v_1, ..., v_m\}$, $m \leq n$. By Theorem 2.10.16 there exists formula α_i which defines v_i for each i. We set

$$\beta := \bigwedge_{0 \leq i \leq n} \neg\alpha_i \wedge \bigwedge_{0 \leq j \leq m} \Diamond\alpha_j \wedge \bigwedge_{m+1 \leq j \leq n} \neg\Diamond\alpha_j.$$

It is not hard to see that β defines Z in $\mathfrak{M}$. ∎

Theorem 2.10.18 *Let $\mathfrak{M}$ be a rooted open submodel of the reduced canonical model $Cnr_{\lambda(P_1)}$ for some modal logic λ with finite width, where $K4 \subseteq \lambda$, and P_1 is finite. Then the results of Theorems 2.10.16 and Lemma 2.10.17 hold for $\mathfrak{M}$.*

Proof. $\mathfrak{M}$ is transitive since $K4 \subseteq \lambda$. $\mathfrak{M}$ is natural since $Cnr_{\lambda(P_1)}$ by Theorem 2.10.13 is differentiated and tight since the model $Cn_{\lambda(P_1)}$ is tight itself. $\mathfrak{M}$ is of finite width by Lemma 2.10.3 and finally, $\mathfrak{M}$ has the fcp by Theorem 2.10.8 ∎

Before we turn to consider the problem of Kripke completeness for modal logics of finite width, we need some more information concerning the structure and properties of certain Kripke models. For this we need to define some special equivalence relations on Kripke models.

Given a pair of models $\mathfrak{M}_1 := \langle W_1, R_1, V_1 \rangle$ and $\mathfrak{M}_1 := \langle W_1, R_1, V_1 \rangle$ with $Dom(V_1) = Dom(V_2) = P$, i.e. having valuations defined on a same set propositional letters. We introduce the relations $\sim_0, \leq_n, \equiv_n$ and $\sim$ on the elements of these models as follows, for any $w \in W_1$ and $v \in W_2$

$$(\mathfrak{M}_1, w) \sim_0 (\mathfrak{M}_2, v) \Leftrightarrow \forall p \in P(w \Vdash_{V_1} p \Leftrightarrow v \Vdash_{V_2} p);$$
$$(\mathfrak{M}_1, w) \leq_n (\mathfrak{M}_2, v) \Leftrightarrow (\forall u \in W_1)(\exists t \in W_2)(w R_1 u \Rightarrow$$
$$(v R_2 t \,\&\, (\mathfrak{M}_1, u) \sim_{n-1} (\mathfrak{M}_2, t));$$
$$(\mathfrak{M}_1, w) \equiv_n (\mathfrak{M}_2, v) \Leftrightarrow [(\mathfrak{M}_1, w) \leq_n (\mathfrak{M}_2, v)) \,\&\, ((\mathfrak{M}_2, v) \leq_n (\mathfrak{M}_1, w))];$$
$$(\mathfrak{M}_1, w) \sim_n (\mathfrak{M}_2, v) \Leftrightarrow ((\mathfrak{M}_1, w) \sim_0 (\mathfrak{M}_2, v)) \,\&\, ((\mathfrak{M}_1, w) \equiv_n (\mathfrak{M}_2, v));$$
$$(\mathfrak{M}_1, w) \sim (\mathfrak{M}_2, v) \Leftrightarrow \forall n((\mathfrak{M}_1, w) \sim_n (\mathfrak{M}_2, v)).$$

For $(\mathfrak{M}, w) \sim_n (\mathfrak{M}, v)$ we write just $w \sim_n v$ (bearing in mind in $\mathfrak{M}$) or simply $w \sim_n v$ if $\mathfrak{M}$ is understood.

Definition 2.10.19 *Let α be a modal formula. The modal degree $deg(\alpha)$ of α is its maximal number of nested occurrences of the connective $\Box$. More precisely, $deg(p) = deg(\top) = deg(\bot) = 0$, $deg(\alpha \to \beta) = deg(\alpha \wedge \beta) = deg(\alpha \vee \beta) = max(deg(\alpha), deg(\beta))$, $deg(\neg\alpha) = deg(\alpha)$, $deg(\Box\alpha) = deg(\alpha)+1$.*

Lemma 2.10.20 *Suppose $\mathfrak{M}_1$ and $\mathfrak{M}_2$ are Kripke models with the same domain of valuations. Let α and β be a modal formulas and let $deg(\alpha) \leq n$, $deg(\beta) < n$. Then for any $w \in |\mathfrak{M}_1|$ and $v \in |\mathfrak{M}_2|$,*

(i) $((\mathfrak{M}_1, w) \sim_n (\mathfrak{M}_2, v)) \,\&\, (w \Vdash_{V_1} \alpha) \Rightarrow (v \Vdash_{V_2} \alpha)$;

(ii) $((\mathfrak{M}_1, w) \leq_n (\mathfrak{M}_2, v)) \,\&\, (v \Vdash_{V_2} \Box\beta) \Rightarrow w \Vdash_{V_1} \Box\beta$.

Proof. We show (i) and (ii) simultaneously by induction on length of α and $\mathfrak{B}$. The basis of induction, when $\alpha := p$ and $\beta := p$, follows directly by definition of $\sim_0$ and $\leq_n$. The inductive steps for (i) in the case for non-modal logical connectives again follow directly from definitions $\leq_n$ and $\sim_n$ and symmetry of $\sim_n$. Consider inductive step for (ii). Suppose $w \not\Vdash_{V_1} \Box\beta$. Then $w R_1 u$ where $u \not\Vdash_V \beta$. Since $(\mathfrak{M}_1, w) \leq_n (\mathfrak{M}_2, v)$, there exists t such that $v R_2 t$ and $(\mathfrak{M}_1, u) \sim_{n-1} (\mathfrak{M}_2, t)$. Now $deg(\beta) \leq n - 1$ and by the inductive hypothesis we conclude $t \not\Vdash_{V_2} \beta$ which yields $v \not\Vdash_{V_2} \Box\beta$. The inductive step for (i) in the case when the main logical connective of the formula α is $\Box$ follows from (ii) since $\sim_n$ implies $\leq_n$. ∎

Readers experienced in first-order model theory can easily discern the influence of technique of Fraisse and Ehrenfeucht in the above construction of n-equivalence $\sim_n$ and in this lemma. The converse of this theorem does not hold for arbitrary models (F. [46] p. 33). Anyway for some special models, it is possible to prove the converse conjecture.

Lemma 2.10.21 *Suppose* $\mathfrak{M}_1 := \langle W_1, R_1, V_1 \rangle$ *and* $\mathfrak{M}_2 := \langle W_2, R_2, V_2 \rangle$ *are Kripke models and* $Dom(V_1) = Dom(V_2) = \{q_0, q_1, ..., q_n\}$. *Suppose* $w \in W_1, v \in W_2$ *and* $\forall\alpha((deg(\alpha) = n)(w \Vdash_{V_1} \alpha \Rightarrow v \Vdash_{V_2} \alpha)$. *Then* $(\mathfrak{M}_1, w) \sim_n (\mathfrak{M}_2, v)$.

Proof. For any $n \geq 0$, we associate a formula to any element w of $\mathfrak{M}_1$ in such a way that for any n, the set of all associated formulas will be finite. We do this as follows. If $n = 0$ then α is the conjunction of all propositional letters from $Dom(V_1)$ which are valid in w. If $n > 0$ then we have a finite list $\beta_0, ..., \beta_m$ of formulas which are associated to all elements from $\mathfrak{M}_1$ for $n - 1$. The modal formula associated with w for n is the formula $\alpha := \beta \wedge \gamma_0 \wedge \gamma_1 \wedge ... \wedge \gamma_m$, where β is the formula associated with w for 0 in $\mathfrak{M}_1$; and $\gamma_i = \Diamond\beta_i$ if there exists v such that wR_1v and β_i is the formula associated to v for $n - 1$, otherwise $\gamma = \neg\Diamond\beta_i$. Now we can easily prove by induction on n that for any element w from $\mathfrak{M}_1$ and any model $\mathfrak{M}_1$, the formula α associated with w for n defines the set $\{v \mid v \in W_2, (\mathfrak{M}_2, v) \sim_n (\mathfrak{M}_1, w)\}$ in $\mathfrak{M}_2$. Therefore because $(\mathfrak{M}_1, w) \sim_n (\mathfrak{M}_1, w)$ Since $w \Vdash_{V_1} \alpha$ (using $\mathfrak{M}_1 = \mathfrak{M}_2$) and the formula $\mathfrak{A}$ associated to w has modal degree n we conclude $v \Vdash_{V_2} \alpha$ and $(\mathfrak{M}_1, w) \sim_n (\mathfrak{M}_2, v)$. ∎

Now we proceed to prove two crucial results, but to do this we need to recall one more definition. Recall that for any model $\mathfrak{M} := \langle W, R, V \rangle$ and $w \in W$, $[w]_n := \{v \mid v \in W, (\mathfrak{M}, w) \sim_n (\mathfrak{M}, v)\}$.

Definition 2.10.22 *A Kripke model* $\mathfrak{M} := \langle W, R, V \rangle$ *is n-simple if there is a finite* $H \subseteq W$ *such that*

    (i) $H \cap [w]_n$ *covers* $[w]_n := \{v \mid v \in W, (\mathfrak{M}, w) \sim_n (\mathfrak{M}, v)\}$ *for each* $w \in W$;

    (ii) $\forall u, t \in (W - H)(u^{R\leq} \cap H = t^{\leq} \cap H) \Rightarrow (\mathfrak{M}, u) \sim_0 (\mathfrak{M}, t)$.

$\mathfrak{M}$ *is just* simple *if for some* n, $\mathfrak{M}$ *is n-simple.*

Lemma 2.10.23 *Suppose* $\mathcal{F} := \langle W, R \rangle$ *is a transitive frame with the finite cover property. Then any modal formula* α *is valid in* $\mathcal{F}$ *if* α *is valid in all simple models* $\langle W, R, V \rangle$ *with finite* $Dom(V)$.

Proof. Suppose α is disproved in the frame $\mathcal{F}$, more precisely α is falsified in a model $\mathfrak{M}$ on $\mathcal{F}$ with a valuation Ψ. Of course we can suppose that the $Dom(\Psi)$ is the set of all letters having occurrences in α. Hence, for some $w_0 \in W$, $w_0 \nVdash_\Psi \alpha$. Clearly it is sufficient to find a simple model $\mathfrak{M}_1 := \langle W, R, \Phi \rangle$ such that this model also disproves α, and $Dom(\Phi)$ consists of only variables of α. Suppose $deg(\alpha) = n$ and $X := \{[w]_n \mid w \in |\mathfrak{M}|\}$. By Lemma 2.10.21 X is finite. Since $\mathcal{F}$ has the fcp, any $[w]_n \in X$ has a finite cover $H([w]_n)$. We set $H := \bigcup_{[w]_n \in X} H([w]_n)$. Suppose $v_0, v_1, ..., v_k$ is a list of all the distinct members of H. We introduce a function from W into W as follows:

  (*i*) for $w \in H$, f(w):=w,
  (*ii*) for $w \in (W - H)$, $f(w)$ is the first v in $\{v_0, ..., v_k\}$ such that
     wRv and $w \equiv_n v$.

The mapping f is well-defined. For suppose $w \in (W - H)$, then there exists v in $H([w]_n) \subseteq H$ such that wRv and $\neg(vRw)$. But $w \sim_n v$, consequently $w \equiv_n v$.

We now introduce the model $\mathfrak{M}_1 := \langle W, R, \Phi \rangle$, where for every $w \in W$ and any propositional letter p having an occurrence in α, $w \Vdash_\Phi p \Leftrightarrow f(w) \Vdash_\Psi p$.

$$\text{For any } k \leq n \text{ and } w \in W, ((\mathfrak{M}_1, w) \sim_k (\mathfrak{M}, f(w))) \tag{2.42}$$

We show this by induction on $k \leq n$. If $k = 0$ then it holds by definition of Φ. Let $k > 0$. Suppose that wRv. Then by the definition of f we have $v = f(v)$ or $vRf(v)$ and by transitivity of R, $wRf(v)$. Using the definition of f it follows that $(\mathfrak{M}, f(w)) \equiv_n (\mathfrak{M}, w)$. Therefore since $wRf(v)$, there exists u such that $f(w)Ru$ and $(\mathfrak{M}, u) \sim_{n-1} (\mathfrak{M}, f(v))$. By Lemmas 2.10.21 and 2.10.20, $\sim_{n-1}$ on $\mathfrak{M}$ contains $\sim_{k-1}$. Hence $(\mathfrak{M}, u) \sim_{k-1} (\mathfrak{M}, f(v))$. By inductive hypothesis we have $(\mathfrak{M}_1, v) \sim_{k-1} (\mathfrak{M}, f(v))$. Therefore $(\mathfrak{M}_1, v) \sim_{k-1} (\mathfrak{M}, u)$, as it was required.

Suppose now that $f(w)Rv$. Then there exists $u \in H$ such that $v = u$ or vRu and $(\mathfrak{M}, v) \sim_n (\mathfrak{M}, u)$. We know that $w = f(w)$ or $wRf(w)$, consequently wRu. Since $u \in H$, we have $f(u) = u$, and by inductive hypothesis we conclude $(\mathfrak{M}_1, u) \sim_{k-1} (\mathfrak{M}, u)$. Since $v = u$ or $(\mathfrak{M}, v) \sim_n (\mathfrak{M}, u)$ (i.e. in any case $(\mathfrak{M}, v) \sim_n (\mathfrak{M}, u))$, we infer $(\mathfrak{M}, v) \sim_{k-1} (\mathfrak{M}_1, u)$ as it was required. Hence (2.42) holds.

The model $\mathfrak{M}_1$ is simple. $\hspace{4cm}$ (2.43)

The property (i) of *n*-simplicity: by construction, we can conclude that $H \cap [w]_n$ covers $[w]_n$ in the model $\mathfrak{M}$ for any $w \in W$, where $[w]_n$ is generated in $\mathfrak{M}$. We consider again the set H to show the simplicity of $\mathfrak{M}_1$.

Consider any w. If $w \in H$ then we have the required property for covering immediately. If $w \notin H$ then we consider $f(w)$, by definition of f, we have $f(w) \in H$ and $wRf(w)$ and $(\mathfrak{M}, w) \equiv_n (\mathfrak{M}f(w))$. By (2.42) we conclude $(\mathfrak{M}_1, w) \sim_n (\mathfrak{M}, f(w))$ and $(\mathfrak{M}_1, f(w)) \sim_n (\mathfrak{M}, f(f(w)))$, i.e. since $f(f(w)) = f(w)$, $(\mathfrak{M}_1, f(w)) \sim_n (\mathfrak{M}, f(w))$. Consequently $(\mathfrak{M}_1, w) \sim_n (\mathfrak{M}_1, f(w))$ and $f(w)$ covers w in $\{a \mid (\mathfrak{M}_1, a) \sim_n (\mathfrak{M}_1, w)\}$. Thus H satisfies (i) in $\mathfrak{M}_1$ from definition of simple models in $\mathfrak{M}_1$.

To show property (ii) from the definition of *n*-simple models suppose that $u, t \in (W - H)$ and $u^{R \leq} \cap H = t^{R \leq} \cap H$. Then $u \equiv_n t$ in $\mathfrak{M}$ which follows from the choice of H. Consider the definition of f on u and t. To define $f(u)$ we take the first v_i in the list $v_0, v_1, ..., v_k$, where (i) uRv_i and $v_i \equiv_n u$. But any v_i with (i) satisfies the same property (i) concerning t instead of u. Thus $f(u) = f(t)$ and $(\mathfrak{M}_1, u) \sim_0 (\mathfrak{M}_1, t)$, hence (ii) for $\mathfrak{M}_1$ is shown, and $\mathfrak{M}_1$ is simple.

We were given that $w_0 \not\Vdash_\Psi \alpha$ in the model $\mathfrak{M}$. Since by choice the set $H([w_0]_n)$ is a cover of $[w_0]_n$, we have there is $v \in H([w_0]_n)$. Then by Lemma 2.10.20 $v \not\Vdash_P si\alpha$, and, since $f(v) = v$ by (2.42) we conclude $f(v) \not\Vdash_\Phi \alpha$ in the model $\mathfrak{M}_1$. ∎

Lemma 2.10.24 *Let* $\mathfrak{M} := \langle W, R, \Psi \rangle$ *be a transitive, reduced and natural model of finite width with the finite cover property and* $||Dom(\Psi)|| < \omega$. *Then any simple model* $\mathfrak{M}_1 := \langle W, R, \Phi \rangle$ *with finite* $Dom(\Psi)$ *is a definable variant of* $\mathfrak{M}$, *i.e., for any* $p \in Dom(\Psi)$, $\Psi(p)$ *is definable in* $\mathfrak{M}$.

Proof. Consider any $n \in N$. Suppose $\mathfrak{M}_1 := \langle W, R, \Phi \rangle$ is n-simple with respect to a finite subset H of W. For $w, v \in W$, we set

$$wSv \Leftrightarrow (u = t) \vee (u, v \in (W - H) \& (u^{R \leq} \cap H = v^{R \leq} \cap H)),$$

and $[u] := \{w \mid uSw\}$, $X := \{[w] \mid w \in W\}$. Every $[w]$ in X is definable in $\mathfrak{M}$. Indeed, if $[w] = \{w\}$ and $w \in H$ then $[w]$ is definable by Theorem 2.10.16. In other case there exists a $U \subseteq H$ such that $[w] = \{s \mid s \in (W - H), s^{R \leq} \cap H = U\}$. In this case $[w]$ is definable by Lemma 2.10.17. Since H is finite, X is also finite. Since $\mathfrak{M}_1$ is simple, wSv implies $w \sim_0 s$ in $\mathfrak{M}_1$. This yields $w \Vdash_\Phi p \Leftrightarrow v \Vdash_\Phi p$. Consequently, $\Phi(p) = [w_0] \cup [w_1] \cup ... \cup [w_k]$, where $\{[w_0], [w_1], ..., [w_k]\} := \{[w] \mid w \in W, w \Vdash_\Phi p\}$. Because every $[w_i]$ is definable and X is finite we conclude $\Phi(p)$ is definable. ∎

Theorem 2.10.25 *If* λ *is a modal logic of finite width extending* K *then the frame of* $Cnr_{\lambda(P_1)}$, *where* P_1 *is finite, is a* λ-*frame.*

Proof. Let α be a theorem of λ. Then any substitution instance of α is again a theorem of λ. By Theorem 2.10.13 it follows that α is valid in the model $Cnr_{\lambda(P_1)}$ while valuation of the propositional letters from α by the truth values in the model $Cnr_{\lambda(P_1)}$ of arbitrary formulas built up on letters from P_1. By Theorems 2.10.8 and 2.10.13 each rooted open submodel $\mathfrak{M}$ of the model $Cnr_{\lambda(P_1)}$ is transitive, reduced, natural and has the finite cover property and is of a finite width by Lemma 2.10.3. Therefore by Lemma 2.10.24 any simple model, which is based on the frame of $\mathfrak{M}$ and has a finite domain of the valuation, is a definable variant of the model $\mathfrak{M}$. Therefore, since we noted above that λ is valid in any definable variants of $\mathfrak{M}$, α is valid on all such simple models. Then by Lemma 2.10.23 α is valid in the frame of $\mathfrak{M}$. ∎

Now we are in a position to proof the main theorem of this section.

Theorem 2.10.26 *(K.Fine [46]) Any modal logic extending $K4$ and having finite width is Kripke complete and $\lambda := \lambda(\mathcal{K})$, where $\mathcal{K}$ is a class of transitive Kripke frames without ascending chains of clusters.*

Proof. We show that λ is Kripke complete proving that the logic of the class of Kripke frames of the models $Cnr(\lambda(P_1))$, where $||P_1|| < \omega$, coincides with λ. If for a formula α, $\alpha \notin \lambda$ then $\lambda \notin \lambda(P_1)$, where P_1 is the set of all propositional letters from α. By the main properties of canonical models α is false on the canonical model $Cn_{\lambda(P_1)}$. Using Theorem 2.10.13 it follows that α is false on the reduced canonical model $Cnr_{\lambda(P_1)}$. Conversely, by Theorem 2.10.25 all theorems of λ are valid on all frames of models $Cnr(\lambda(P_1))$, where $||P_1|| < \omega$. ∎

We conclude this section by transferring this fundamental result to superintuitionistic logics. Since posets are special transitive frames, therefore we have already the definition of posets of finite width. The formulas $W_{i,n}$, which play for superintuitionistic logics the role which formulas $\mathcal{W}_n$ play for modal logics in defining the width of logics, have the following definition: for $1 \le n$,

$$W_{i,n} := \bigvee_{0 \le i \le n} (p_i \to \bigvee_{0 \le j \le n, j \ne i} p_j).$$

Definition 2.10.27 *A superintuitionistic logic λ has width n if $W_{i,n} \in \lambda$ but $W_{i,n-1} \notin \lambda$.*

Exercise 2.10.28 *For any poset $\mathcal{F} := \langle F, R \rangle$, $\mathcal{F} \Vdash W_{i,n}$ iff the width of $\mathcal{F}$ is not more than n.*

Theorem 2.10.29 *Any superintuitionistic logic λ with a finite width is Kripke complete.*

Proof. Indeed if $W_{i,n} \in \lambda$, then $T(W_{i,n}) \in \sigma(\lambda)$, where $\sigma(\lambda)$ is the greatest modal counterpart for λ. That is

$$\bigvee_{0 \le i \le n} \Box(\Box p_i \to \bigvee_{0 \le j \le n, j \ne i} \Box p_j) \in \sigma(\lambda). \tag{2.44}$$

Suppose that $\mathcal{W}_n \notin \sigma(\lambda)$. Then the latter formula is invalid in the canonical Kripke model $Cn_{\sigma(\lambda)}$ for $\sigma(\lambda)$, which is reflexive and transitive. This means there is an element T of that model such that

$$T \Vdash_V \bigwedge_{0 \le i \le n} \Diamond p_i, \ T \Vdash_V \Box \neg(p_i \wedge \Diamond p_j), \ \forall i, j, \text{ where } 1 \le i \ne j \le n.$$

Therefore there are elements T_i from $Cn_{\sigma(\lambda)}$ such that

$$T_i \Vdash_V p_i \wedge \psi_i, \text{ where } 1 \le i \le n \text{ and}$$
$$\psi_i := \Box \bigwedge_{1 \le j \ne i \le n} \neg p_j.$$

Using (2.44) it follows that, for any $i \in \{0, 1, ..., n\}$, $T_i \Vdash_V \bigvee_{1 \leq j \neq i \leq n} \psi_j$. Therefore there is $k \neq i$ such that $T_i \Vdash_V \psi_k$. Then it follows that $T_i \Vdash_V \neg p_i$, a contradiction. Thus we have proved that $\mathcal{W}_n \in \sigma(\lambda)$. And by Theorem 2.10.26 $\sigma(\lambda)$ is Kripke complete. Applying Lemma 2.7.7 we obtain λ is also Kripke complete. ∎

Remarks Concluding the Chapter. To end this chapter, we again recall that all developed technique is a self-contained introduction in semantic tools of non-standard logic. Primary attention was paid to tools and results that will be used in our further research connected with inference rules. Readers interested to study the field we had dealt with in the first two chapters can use the cited literature. Except the reference already cited above we would recommend the following issues. To study general logical consequence relation, the summarizing books are Wojcicki [203] and Rasiowa [120], many interesting information cab be found in Rautenberg [123, 124, 125, 126], Herrmann and Rautenberg [66] Prucnal [117, 118, 119]. The paper Prucnal [118] contains a solution of two H.Friedman's problems which are very related to investigated in this book questions concerning admissibility inference rules. Readers interested in aspects connected with interaction of theory of quasi-varieties and non-standard logic will find many strong results in Pigozzi [112]. Concerning advanced technique of modal superintuitionistic and temporal logics, and deductive systems in whole, we would recommend Blok [14, 15], Block and Köhler [16], Bellissima [10, 11], Gabbay [53], Gabbay and de Jongh [54], Goldblatt [58, 59], and K.Fine [46, 47], Maksimova [91, 92, 93], Kracht [73, 74, 75, 76], de Jongh and Visser [52], de Jongh and Troelstra [50], Sahlqvist [160], see also Sambin and Vaccaro [163], Segerberg [166, 167, 168], Thomason [180, 181], Urquhart [183, 185], Vakarelov [187], van Benthem [189, 190, 193, 194], Wolter [207, 208, 209], Zakharyaschev [213, 214]. A good summarizing presentation of semantic technique for modal logics can be found in the book Chagrov and Zakharyaschev [28].

Readers interested in study relevant logics will find many strong results in Urquhart [184, 185, 186], in particular the paper of A.Urquhart [184] contains a negative solution of the long standing problem concerning decidability of relevant logics. Research connected with non-standard predicate logics has a good presentation in papers Ono [107, 108]. It is relevant to mention here that the study of the lattice of all superintuitionistic logics was very much initiated by paper Hosoi [68]. The readers interested to get acquaintance with propositional dynamic logic may find strong technical tools in Kozen and Parikh [72]. Many new information concerning substructural logics readers can meet in Dosen [33] and Ono and Komori [110]. An information on famous L.Maksimova's results concerning Kraig's interpolation theorem, amalgamation property and Beth's definability could be found, for instance, in Maksimova [94, 95].

An individual logics and particular classes of modal logics also attract many attention of researchers. For instance, the investigation concentrated around Gödel-Löb modal provability GL has generated whole research branch in non-standard logic which could be conventionally called *provability logic*. This field has dealt with modal interpretation of provability in formal Peano arithmetic and related first-order theories. A good line for study could go through the book Boolos [20] and papers Artemov [1, 2, 4], Beklemishev [7, 8], Sambin and Vaccaro [162], Shavrukov [164], Solovay [174], de Jongh and Visser [51], Visser [197, 198, 199]. Note also that a first example of Kripke incomplete superintuitionistic calculus was found by V.Shehtman [170], and the first example of undecidable finitely axiomatizable superintuitionistic logic was also constructed by V.Shehtman [171]. A deep research connected with impossibility to recognize effectively different properties of non-standard logics could be found in Chagrov and Zakharyaschev [27]. Probably the comment above even does not reflect precisely the advise of the author what to read. Many references and comments are disseminated though whole this book. Since the systematization of all research concerning non-standard logics and writing a survey are out of aims of this book, we restrict us in what is said above and refer interested reader to the literature.

Chapter 3

Criteria for Admissibility

3.1 Reduced Forms

In this section we investigate the possibility of reducing inference rules to more uniform and simple forms which are equivalent o original rules with respect to admissibility. As we know the admissibility of inference rules in an algebraic logic λ is equivalent to the validity of the corresponding quasi-identities in free algebras of variety $Var(\lambda)$. Therefore we begin with a description of reduced forms for quasi-identities.

Definition 3.1.1 *Suppose q_1 and q_2 are quasi-identities in a language $\mathcal{L}$. We say q_1 is equivalent (or semantically equivalent) to q_2 in a class of algebraic systems $\mathcal{K}$ if for every algebraic system $\mathcal{A} \in \mathcal{K}$, $\mathcal{A} \models q_1$ iff $\mathcal{A} \models q_2$.*

Definition 3.1.2 *Let $q := (\Phi \Rightarrow \Psi)$ be a quasi-identity in a language $\mathcal{L}$. The weak reduced form of the quasi-identity q is the quasi-identity $r_w(q)$ which is effectively constructed from q as follows. For any term t having occurrence in q (even a variable), we introduce the new variable x_t indexed by this term. If t is a term from q which is not a variable then t has the main functional symbol which is denoted below by f_t. Let $\mathcal{M}(\Phi)$ and $\mathcal{M}(\Psi)$ are lists of all final terms (which are not subterms of other terms) of formulas Φ and Ψ respectively. Let $Ter(q)$ be the set of all terms which are not variables from q. Further,*

> $r_w(\Phi)$*is the formula obtained from Φ by replacing all terms t from $\mathcal{M}(\Phi)$ by variables x_t.*
> $r_w(\Psi)$*is the formula obtained from Ψ by replacing all terms t from $\mathcal{M}(\Psi)$ by variables x_t.*
> $Dec(q) := \bigwedge \{x_t = f_t(x_{t_1}, ..., x_{t_n}) \mid t \in Ter(q), t = f_t(t_1, ..., t_n)\}$
> $r_w(q) = r_w(\Phi) \wedge Dec(q) \Rightarrow r_w(\Psi)$.

It is easily seen that weak reduced forms of quasi-identities have only simple terms, i.e. terms of the form: a variable or a term $f(x_1, ..., x_n)$, where $x_1, ..., x_n$ are variables and f is a functional symbol. Weak reduced forms have in the premise conjunctions of equalities of simple terms or a predicate symbol applied to simple terms, and the conclusions of weak reduced forms are equalities of simple terms or variables, or a predicate symbols applied to simple terms. It is clear that this form is the simplest, in the sense that the weak reduced form cannot be decomposed on more simple components.

Lemma 3.1.3 *Quasi-identities q and $r_w(q)$ are (semantically) equivalent in any class of algebraic systems $\mathcal{K}$. If $r_w(q)$ is falsified in an algebraic system $\mathcal{A}$ by a valuation $v : x_t \mapsto a_t \in |\mathcal{A}|$ then q is false in $\mathcal{A}$ under the valuation $u : z_i \mapsto v(x_{z_i}) = a_{x_{z_i}}$.*

Proof. Suppose $\mathcal{A}$ is an algebraic system of signature $\mathcal{L}$ and $\mathcal{A} \models q$. Let v be a valuation of all variables x_t of $r_w(q)$ on $\mathcal{A}$. Suppose the premise of $r_w(q)$ is valid in $\mathcal{A}$ with respect to v. That is

$$\mathcal{A} \models [r_w(\Phi) \wedge Dec(q)](^{x_t}_{v(x_t)}). \tag{3.1}$$

Consider the valuation u in $\mathcal{A}$ for variables z_i of q, where $u(z_i) := v(x_{z_i})$. We claim that for any term t of q

$$u(t) = v(x_t). \tag{3.2}$$

We show this by induction on the length of t. The basis of induction follows directly from the definition of u. Let t be a term constructed from a functional symbol f_t, as the main functional symbol of arity n, and the terms $t_1, ..., t_n$. Suppose that for terms $t_1, ..., t_n$, our proposition has been proved. Then

$$u(t_1) = v(x_{t_1}), ..., u(t_n) = v(x_{t_n}),$$

and by our hypothesis (3.1) $\mathcal{A} \models v(x_t) = f_t(v(x_{t_1}), ..., v(x_{t_n}))$. Combining these equalities we conclude

$$u(t) = f_t(u(t_1), ..., u(t_n)) = f_t(v(x_{t_1}), ..., v(x_{t_n})) = v(x_t).$$

Thus the equality (3.2) holds. It follows immediately from (3.2) and (3.1) that the premise Φ of q is also valid in $\mathcal{A}$ under the valuation u. Because $\mathcal{A} \models q$ it follows that the conclusion Ψ is also valid in $\mathcal{A}$ under u. Again applying (3.2) we arrive at $\mathcal{A} \models [r_w(\Psi)](^{x_t}_{v(x_t)})$. Thus $\mathcal{A} \models r_w(q)$.

Conversely, suppose that $\mathcal{A} \models r_w(q)$. Consider an arbitrary valuation u of variables z_i from q in $\mathcal{A}$. Suppose that

$$\mathcal{A} \models \Phi(^{z_i}_{u(z_i)}).$$

We introduce the valuation v of all variables x_t of the quasi-identity $r_w(q)$ as follows: $v(x_t) := u(t)$. It is clear that $\mathcal{A} \models [r_w(\Phi)](^{x_t}_{v(x_t)})$. Moreover for every $t \in Ter(q)$

$$v(x_t) := u(t) = f_t(u(t_1), ..., u(t_n)) = f_t(v(x_{t_1}), ..., v(x_{t_n})).$$

That is $\mathcal{A} \models [Dec(q)](^{x_t}_{v(x_t)})$. Hence the premise of $r_w(q)$ is valid in $\mathcal{A}$ under the valuation v. Since $\mathcal{A} \models r_w(q)$ we get $\mathcal{A} \models [r_w(\Psi)](^{x_t}_{v(x_t)})$. Since $v(x_t) := u(t)$, it follows $\mathcal{A} \models \Psi(^{z_i}_{u(z_i)})$. Hence $\mathcal{A} \models q$.

Now we turn to the second part of our lemma. Suppose $r_w(q)$ is falsified in $\mathcal{A}$ by a valuation v. The valuation u of variables from the quasi-identity q, where $u(z_i) := v(x_{z_i})$, will satisfy (3.2) as we have proved above. That is, for any term t from q, $u(t) = v(x_t)$. This implies that u falsifies q in $\mathcal{A}$. ∎

Now we will transfer the weak reduced forms of quasi-identities on inference rules. We recall that for any inference rule

$$r = \frac{\alpha_1(x_1, ...x_n), ..., \alpha_m(x_1, ...x_n)}{\beta(x_1, ...x_n)}$$

the corresponding quasi-identity $q(r)$ is the quasi-identity

$$q(r) := \bigwedge_{1 \le i \le m} \alpha_i(x_1, ...x_n) = \top \Rightarrow \beta(x_1, ...x_n) = \top.$$

When we consider the quasi-identities in the language $\mathcal{L}$ of the variety $Var(\lambda)$, where λ is an algebraic logic, it is possible properly to arrange corresponding tuple of inference rules $r(q)$ for any quasi-identity q. Namely, if

$$q = \bigwedge_{1 \le i \le m} f_i = g_i \wedge \bigwedge_{1 \le j \le k} (h_j = \top) \Rightarrow f = g,$$

where all f_i, g_i, h_j are different from $\top$ then

$$r(q) := \frac{\{f_i \equiv g_i \mid 1 \le i \le m\}, \{h_j \mid 1 \le j \le k\}}{\varphi},$$

where if f, g are different from $\top$ then $\varphi := (f \equiv g)$ (recall that, $f \equiv g$ can be a tuple of formulas, and in this case we consider the tuple of inference rules with conclusions from $f \equiv g$), otherwise φ is the term which is different from $\top$ part of $f = g$.

Definition 3.1.4 *Let r be an inference rule. The weak reduced form for r is the inference rule $wrf(r) := r(r_w(q(r)))$.*

It is now necessary to show that the weak reduced forms of inference rules are equivalent to the original rules. For this we need to clarify the notion of equivalence for inference rules because the meaning of equivalence in this case is not uniquely definable. We will consider here two forms of equivalence: semantic equivalence (in terms of truth) and equivalence by admissibility. Recall the definition of validness for inference rules. Let λ be an algebraic logic. Let $\mathcal{A} \in Var(\lambda)$ and

$$r = \frac{\alpha_1(x_1, ...x_n), ..., \alpha_m(x_1, ...x_n)}{\beta(x_1, ...x_n)}$$

be an inference rule in the language of λ. We say r is valid on $\mathcal{A}$ (denotation $\mathcal{A} \models r$) if the corresponding to r quasi-identity

$$q(r) := \bigwedge_{1 \le i \le m} \alpha_i(x_1, ...x_n) = \top \Rightarrow \beta(x_1, ...x_n) = \top$$

is valid in $\mathcal{A}$.

Let λ be an algebraic logic. We say a rule r is *valid* in λ (denotation $\lambda \models r$) if r is valid in any algebra $\mathcal{A}$ from the variety $Var(\lambda)$.

Lemma 3.1.5 *Any valid inference rule r of every algebraic logic λ is admissible for λ.*

Proof. If a rule r is valid in a logic λ then the quasi-identity $q(r)$, where $q(r) := \bigwedge_{1 \leq i < m} \alpha_i(x_1, ...x_n) = \top \Rightarrow \beta(x_1, ...x_n) = \top$ is valid in $Var(\lambda)$, and, in particular, it is valid in the free algebra $\mathfrak{F}_\lambda(\omega)$ of countable rank from $Var(\lambda)$. Then Theorem 1.4.5 yields that r is admissible for λ. ■

Definition 3.1.6 *We say a rule r_1 is semantically equivalent to a rule r_2 in an algebraic logic λ if for every algebra $\mathcal{A} \in Var(\lambda)$, $\mathcal{A} \models r_1$ iff $\mathcal{A} \models r_2$.*

Definition 3.1.7 *A rule r_1 is equivalent by admissibility to a rule r_2 in a logic λ if r_1 is admissible for λ iff r_2 is admissible in λ.*

Lemma 3.1.8 *Let λ be an algebraic logic. For any inference rule r, r is semantically equivalent to $wrf(r)$ in λ.*

Proof. Let $\mathcal{A}$ be an algebra from $Var(\lambda)$ and let r be an inference rule. Let $\mathcal{A} \models r$, i.e. $\mathcal{A} \models q(r)$. According to Lemma 3.1.3 $\mathcal{A} \models q(r) \Leftrightarrow \mathcal{A} \models r_w(q(r))$ and again immediately by definition

$$\mathcal{A} \models r_w(q(r) \Leftrightarrow \mathcal{A} \models r(w_r(q(r))).$$

Hence $\mathcal{A} \models r \Leftrightarrow \mathcal{A} \models wrf(r)$. ■

Corollary 3.1.9 *A rule r is valid in an algebraic logic λ iff the rule $wrf(r)$ is valid in λ.*

Corollary 3.1.10 *Let λ be an algebraic logic. For any rule r, rules r and $wrf(r)$ are equivalent by admissibility in λ.*

Proof. By Lemma 3.1.8 rules r and $wrf(r)$ are semantically equivalent in λ. Therefore $\mathfrak{F}_\lambda(\omega) \models r$ iff $\mathfrak{F}_\lambda(\omega) \models wrf(r)$. Therefore by Theorem 1.4.5 rules r and $wrf(r)$ are equivalent by admissibility in λ. ■

In general there is no way to simplify or to further specify the notion of reduced forms for inference rules of arbitrary algebraic logics. But in many particularly important cases it is possible to develop these approach further. We will begin with the case of modal logics. For any term (formula) α we fix the denotations $\alpha^0 := \alpha, \alpha^1 := \neg\alpha$.

Theorem 3.1.11 *There exists an algorithm which, for any quasi-identity q in a language of modal algebras, constructs a quasi-identity $rf(q)$ with the following properties: $rf(q)$ has the form*

$$[\bigvee_{1\leq j\leq n} \varphi_j] = \top \Rightarrow x_0 = \top,$$

where

$$\varphi_j := \bigwedge_{0\leq i\leq m} x_i^{k(j,i,1)} \wedge \bigwedge_{0\leq i\leq m} (\Diamond x_i)^{k(j,i,2)},$$

$k(j,i,1), k(j,i,2) \in \{0,1\}$, x_i are different variables, and all the variables of q are among these x_i. $rf(q)$ is semantically equivalent to r in the variety $Var(\lambda)$ for any normal modal logic λ. Moreover if $rf(q)$ is falsified in an algebra $\mathfrak{B} \in Var(\lambda)$ under a valuation $x_i \mapsto a_i \in |\mathfrak{B}|$, then q is also false in $\mathfrak{B}$ under this valuation.

Proof. Let λ be a normal modal logic and $Var(\lambda)$ be the variety of corresponding modal algebras. Let $q = \Phi \Rightarrow f = g$. It is easy to see that the quasi-identity $q_1 := \Phi \wedge x_0 = (f \equiv g) \Rightarrow x_0 = \top$ is semantically equivalent to q on $Var(\lambda)$. If we transform any quasi-identity q_i by replacing all terms by equivalent (as formulas in λ) other terms then we obtain certain quasi-identities semantically equivalent to q_i on $Var(\lambda)$

Using this observation we first transform q_1 by everywhere replacing $\Box$ with $\neg\Diamond\neg$ and obtain a quasi-identity q_2 semantically equivalent to q_1 in $Var(\lambda)$. According to Lemma 3.1.3 q_2 is semantically equivalent to $r_w(q_2)$. We alter $r_w(q_2)$ by transformation. the premise of $r_w(q_2)$ into disjunctive normal form and obtain a quasi-identity q_3 semantically equivalent to $r_w(q_2)$ (on $Var(\lambda)$). Now we replace all variables of the kind x_{z_i} (where z_i are variables of q_1) in q_3 by the variables z_i and obtain a quasi-identity q_4 equivalent to q_3 in $Var(\lambda)$. Thus $q_4 := \bigvee_{1\leq l\leq k} \theta_l \Rightarrow x_0 = \top$ and q_4 fails to conform to the required form of $rf(q)$ only by some terms of kind $x_i^{k_g}$ or $(\Diamond x_i)^{k_h}$ lacking in some θ_l, where $k_g, k_h \in \{0,1\}$, for some variables x_i. We replace every disjunct θ_l which has missing occurrences by the disjunction of disjuncts which are obtained from θ_l by adding conjunctions of the missing $x_i, \Diamond x_i$ with all the possible distributions of $\neg$. We then obtain a quasi-identity q_5 and it is not hard to see that q_5 is equivalent to q_4 on $Var(\lambda)$ and that q_5 has a form required for $rf(q)$. Hence we can set $rf(q) := q_5$.

If $\mathfrak{B}$ is an algebra from $Var(\lambda)$ and v is a valuation in $\mathfrak{B}$ which falsifies $rf(q)$ then v falsifies q_4. q_3 will be also false in $\mathfrak{B}$ under the valuation v extended to the variables x_{z_i} by $v(x_{z_i}) := v(z_i)$. Then $r_w(q_2)$ will be false in $\mathfrak{B}$ under v because q_2 is obtained from $r_w(q_2)$ by replacing terms by equivalent ones. Then by Lemma 3.1.3 q_2 is also false in $\mathfrak{B}$ under v which implies that q_1 and q are also falsified in $\mathfrak{B}$ by v. ∎

We will call $rf(q)$ the *reduced form* of the quasi-identity q. Now we will transfer the definition of reduced form from quasi-identities of modal algebras to inference rules of modal logics.

Definition 3.1.12 *We say an inference rule r in the language of modal logics has reduced form if*

$$r := \frac{\bigvee\{\varphi_j \mid 1 \leq j \leq n\}}{x_0}, \varphi_j := \bigwedge_{0 \leq i \leq m} x_i^{k(j,i,1)} \wedge \bigwedge_{0 \leq i \leq m} (\Diamond x_i)^{k(j,i,2)},$$

$k(j, i, 1), k(j, i, 2) \in \{0, 1\}$, x_i *are different variables and all variables of r are among these x_i.*

Corollary 3.1.13 *There is an algorithm which, for any given inference rule r in the language of modal propositional logic, constructs a suitable inference rule $rf(r)$ in reduced form which is semantically equivalent to the rule r in any normal modal logic λ.*

Proof. Given an inference rule r. We take $q(r)$ and $rf(q(r)$, by Theorem 3.1.11 these quasi-identities are semantically equivalent on $Var(\lambda)$. Therefore for any algebra $\mathfrak{B}$ from $Var(\lambda)$, $\mathfrak{B} \models q(r)$ iff $\mathfrak{B} \models rf(q(r)$. On the other hand, $\mathfrak{B} \models q(r) \Leftrightarrow \mathfrak{B} \models r$. Further, $\mathfrak{B} \models rf(q(r))$ iff $\mathfrak{B} \models r(rf(q(r)))$. Thus $\mathfrak{B} \models r \Longleftrightarrow \mathfrak{B} \models r(rf(q(r)))$. It is clear that the rule $r(rf(q(r)))$ has reduced form. Hence we can let $rf(r) := r(rf(q(r)))$. ∎

We call the inference rule $rf(r)$ the *reduced form* of the rule r. It is clear that the reduced forms of rules for modal logics have a very simple and clear structure: the premises are simply certain disjunctive normal forms and are built up out of variables, formulas of the kind $\Diamond x$, where x is a variable, and their negations. The conclusions are simply variables. It is also clear that reduced forms (as opposite to weak normal forms) are not uniquely defined: there are different but equivalent forms. This is connected with, there being, different possibilities to transform a formula into disjunctive normal form. Directly from Lemma 3.1.13 we infer

Corollary 3.1.14 *An inference rule r is valid in a normal modal logic λ iff its reduced form $rf(r)$ is valid in λ.*

Corollary 3.1.15 *Let λ be a normal modal logic. For any rule r, rule r and it's reduced form $rf(r)$ are equivalent in λ by admissibility.*

Proof. It follows directly from Lemma 3.1.13 and the fact that any rule r_1 is admissible in λ iff r_1 is valid in the free algebra of countable rank $\mathfrak{F}_\lambda(\omega)$ from $Var(\lambda)$ (see Theorem 1.4.5). ∎

It is easy to transfer this developed technique of normal forms to the case of temporal logics. We briefly describe below the reduced forms for inference rules of temporal logics.

Theorem 3.1.16 *There exists an algorithm which, for any quasi-identity q in the language of temporal algebras, constructs a quasi-identity $rf(q)$ with the following properties: $rf(q)$ has form $[\bigvee_{1 \leq j \leq n} \varphi_j] = \top \Rightarrow x_0 = \top$, where*

$$\varphi_j := \bigwedge_{0 \leq i \leq m} x_i^{k(j,i,1)} \wedge \bigwedge_{0 \leq i \leq m} (Fx_i)^{k(j,i,2)}, \bigwedge_{0 \leq i \leq m} (Px_i)^{k(j,i,3)},$$

$k(j,i,1), k(j,i,2), k(j,i,3) \in \{0,1\}$, x_i *are different variables, and all the variables of q are among these x_i. $rf(q)$ is semantically equivalent to r in the variety $Var(\lambda)$ for any temporal logic λ. Moreover if $rf(q)$ is falsified in an algebra $\mathfrak{B} \in Var(\lambda)$ under a valuation $x_i \mapsto a_i \in |\mathfrak{B}|$, then q is false in $\mathfrak{B}$ under this valuation as well.*

Proof. The proof is quite similar to the proof of Theorem 3.1.11 and is left to the readers as an exercise.

Definition 3.1.17 *An inference rule r in the language of temporal logic has reduced form if*

$$r := \frac{\bigvee_{1 \leq j \leq n} \varphi_j}{x_0}, \quad \text{where}$$

$$\varphi_j := \bigwedge_{0 \leq i \leq m} x_i^{k(j,i,1)} \wedge \bigwedge_{0 \leq i \leq m} (Fx_i)^{k(j,i,2)}, \bigwedge_{0 \leq i \leq m} (Px_i)^{k(j,i,3)},$$

$k(j,i,1), k(j,i,2), k(j,i,3) \in \{0,1\}$, x_i *are different variables and all the variables of r are among these x_i.*

Corollary 3.1.18 *There is an algorithm which, for any given inference rule r in the language of temporal logic, constructs an inference rule $rf(r)$ in reduced form which is semantically equivalent to the rule r in any temporal logic λ.*

Proof. The proof is similar to the proof of Corollary 3.1.13.

The inference rule $rf(r)$ is called the *temporal reduced form* of the rule r. And similar to above, we call $rfr(q)$ the reduced form of the quasi-identity q.

Corollary 3.1.19 *An inference rule r is valid in a temporal logic λ iff its temporal reduced form $rf(r)$ is valid in λ.*

Corollary 3.1.20 *Let λ be a temporal logic. For any given rule r, the rule r and it's temporal reduced form $rf(r)$ are equivalent in λ by admissibility.*

Proof. The proof is similar to the proof of Corollary 3.1.15. (cf. Theorem 1.4.5).

While constructing reduced forms we decomposed the terms completely until we were left with variables. Sometimes (in particular, while studying inference rules for superintuitionistic logics using modal inference rules) we will need other reduced forms in which terms are not completely decomposed. We now present such a kind of reduced form for the inference rules of normal modal logics. It is so-called semi-reduced forms. We begin with semi-reduced forms for quasi-identities. The our designating a term as

$$t(\Diamond x_1, ..., \Diamond x_m, x_{m+1}, ..., x_n)$$

means that this term is only built up from the terms $\Diamond x_1, ..., \Diamond x_m$ and only variables $x_{m+1}, ..., x_n$ by means of function symbols.

Theorem 3.1.21 *For any quasi-identity*

$$q := [\bigwedge_{1 \leq l \leq} f_l(\Diamond x_1, ..., \Diamond x_m, x_{m+1}, ..., x_n) =$$
$$g_l(\Diamond x_1, ..., \Diamond x_m, x_{m+1}, ..., x_n)] \Rightarrow$$
$$f(\Diamond x_1, ..., \Diamond x_m, x_{m+1}, ..., x_n) = g(\Diamond x_1, ..., \Diamond x_m, x_{m+1}, ..., x_n)$$

in the language of modal logics there is a quasi-identity $srf(q)$ semantically equivalent to q in $Var(K)$ of the form

$$[\bigvee_{1 \leq j \leq n} \varphi_j] = \top \Rightarrow x_0 = \top, \quad \text{where}$$

$$\varphi_j := \bigwedge_{m+1 \leq i \leq n+k} x_i^{k(j,i,1)} \wedge \bigwedge_{0 \leq i \leq n+k} (\Diamond x_i)^{k(j,i,2)},$$

$k(j,i,1), k(j,i,2) \in \{0,1\}$. *Moreover $srf(q)$ can be effectively constructed from q and if $srf(q)$ is falsified in an algebra $\mathfrak{B} \in Var(K)$ under a valuation $x_i \mapsto a_i \in |\mathfrak{B}|$, then q is also false in $\mathfrak{B}$ under this valuation.*

Proof. We just follow to the proof of Theorem 3.1.11, making only the following alterations. We do not take $r_w(q_2)$ but the similar $r_{sw}(q_2)$ which defers from $r_w(q_2)$ in that we treat $\Diamond x_1, ... , \Diamond x_m$ as variables and do not decompose they into simpler terms. That is, we omit in the premise of $r_w(q_2)$ the equalities $x_{\Diamond x_\psi} = \Diamond x_{x_\psi}, 1 \leq \psi \leq m$. Also $r_w(\Phi)$ and $r_w(\Psi)$, where Φ, Ψ are the premise and the conclusion of q_2, are obtained without replacing the terms $\Diamond x_1, ..., \Diamond x_m$ by something; they are present in these expressions themselves. The quasi-identity $r_{sw}(q_2)$ is evidently equivalent to $r_w(q_2)$ on $Var(K)$. We then reproduce the proof of Theorem 3.1.11 replacing $r_w(q_2)$ by $r_{sw}(q_2)$. The final quasi-identity $rf(q)$ will have the form required for $srf(q)$ and will be equivalent to q in $Var(K)$. The proof concerning the falsification of q will also be the same. ∎

We call the quasi-identity $srf(q)$ the *semi-reduced form* for the quasi-identity q, and define semi-reduced forms for inference rules in a similar way.

Definition 3.1.22 *An inference rule* r *in the language of modal logic has semi-reduced form if*

$$r := \frac{\bigvee_{1 \leq j \leq n} \varphi_j}{x_0}, \quad where$$

$$\varphi_j := \bigwedge_{m+1 \leq i \leq n+k} x_i^{k(j,i,1)} \wedge \bigwedge_{0 \leq i \leq n+k} (\Diamond x_i)^{k(j,i,2)},$$

$k(j,i,1), k(j,i,2) \in \{0,1\}.$

Corollary 3.1.23 *There is an algorithm which, for any given modal inference rule* r *in form*

$$r := \frac{\alpha_1(\Diamond x_1, ..., \Diamond x_m, x_{m+1}, ..., x_n), ..., \alpha_d(\Diamond x_1, ..., \Diamond x_m, x_{m+1}, ..., x_n)}{\beta(\Diamond x_1, ..., \Diamond x_m, x_{m+1}, ..., x_n)},$$

constructs a suitable inference rule $srf(r)$ *in semi-reduced form of the type*

$$\frac{\bigvee_{1 \leq j \leq n} \varphi_j}{x_0}, \quad where \; \varphi_j := \bigwedge_{m+1 \leq i \leq n+k} x_i^{k(j,i,1)} \wedge \bigwedge_{0 \leq i \leq n+k} (\Diamond x_i)^{k(j,i,2)},$$

which is semantically equivalent to the rule r *in any normal modal logic* λ.

Proof. The prof is parallel to the proof of Corollary 3.1.13.

Corollary 3.1.24 *The following hold*

(i) *For any rule* r, *the rule* r *and* $srf(r)$ *are equivalent with respect to validity in any normal modal logic* λ.

(ii) *For any rule* r, *the rule* r *and* $srf(r)$ *are also equivalent in any normal modal logic* λ *in terms of admissibility.*

The rule $srf(r)$ is the *semi-reduced form* of the rule r. Concerning the all reduced forms of rules, we have showed the semantic equivalence of inference rules and their distinct reduced forms and, as corollary, their equivalence with respect to admissibility. However we have not as yet consider the question how rules and their reduced forms behavior with respect to derivability. We know that the deduction theorem holds for normal modal logics (cf. Theorem 1.3.15). Therefore we can express the derivability of inference rules $r := \alpha_1, ..., \alpha_n/\beta$ in a modal logic λ which extends $K4$ in virtue of provability of corresponding formulas:

$$\alpha_1, ... \alpha_n \vdash_\lambda \beta \Longleftrightarrow \vdash_\lambda \left(\bigwedge_{1 \leq i \leq n} \Box \alpha_i \right) \wedge \left(\bigwedge_{1 \leq i \leq n} \alpha_i \right) \to \beta. \tag{3.3}$$

Using this observation we are in a position to prove the following property of reduced forms:

Theorem 3.1.25 *A rule r is derivable in a normal modal logic λ extending $K4$ iff its reduced form $rf(r)$ is derivable in λ. The same is true for semi-reduced forms.*

Proof. We need the following

Lemma 3.1.26 *Let λ be a normal modal logic extending $K4$. If the quasi-identities $\alpha_1 = \top \Rightarrow \beta_1 = \top$, $\alpha_2 = \top \Rightarrow \beta_2 = \top$ are semantically equivalent in $Var(\lambda)$ then the identities $(\Box\alpha_1 \wedge \alpha_1 \to \beta_1) = \top$, $(\Box\alpha_2 \wedge \alpha_2 \to \beta_2) = \top$ are also semantically equivalent in the variety $Var(\lambda)$.*

Proof. Suppose $\mathfrak{B} \in Var(\lambda)$ and the identity $(\Box\alpha_1 \wedge \alpha_1 \to \beta_1) = \top$ is falsified on $\mathfrak{B}$ under the valuation $v(x_i) = a_i \in |\mathfrak{B}|$, where x_i are variables of this identity. This means

$$\Box\alpha_1(a_i) \wedge \alpha_1(a_i) \not\leq \beta_1(a_i). \tag{3.4}$$

We take the prime filter

$$\nabla(\Box\alpha_1(a_i) \wedge \alpha_1(a_i)) := \{c \mid c \in |\mathfrak{B}|, \Box\alpha_1(a_i) \wedge \alpha_1(a_i) \leq c\}$$

generated by $\Box\alpha_1(a_i) \wedge \alpha_1(a_i)$. We claim that this filter is a $\Box$-filter. Indeed, if $c \in \nabla$ then $\Box\alpha_1(a_i) \wedge \alpha_1(a_i) \leq c$ and consequently $\Box(\Box\alpha_1(a_i) \wedge \alpha_1(a_i)) \leq \Box c$. By Proposition 1.3.16 $\Box(\Box\alpha_1(a_i) \wedge \alpha_1(a_i)) = \Box\Box\alpha_1(a_i) \wedge \Box\alpha_1(a_i)$. Also we have $\Box\alpha_1(a_i) \leq \Box\Box\alpha_1(a_i)$. Hence

$$\Box\alpha_1(a_i) \wedge \alpha_1(a_i) \leq \Box\alpha_1(a_i) \leq \Box\Box\alpha_1(a_i) \wedge \Box\alpha_1(a_i) \leq \Box c.$$

Thus $\nabla(\Box\alpha_1(a_i) \wedge \alpha_1(a_i))$ is a $\Box$-filter indeed. Therefore the relation

$$x \sim_\nabla y \Leftrightarrow (x \to y) \wedge (y \to y) \in \nabla(\Box\alpha_1(a_i) \wedge \alpha_1(a_i))$$

is a congruence relation (cf. Lemma 2.2.13). Consider the quotient algebra $\mathfrak{B}_{\sim_\nabla}$ of algebra $\mathfrak{B}$ by this congruence relation. Since the mapping $\varphi : a \mapsto [a]_{\sim_\nabla}$ is a homomorphism it follows from (3.4) that the quasi-identity $\alpha_1 = \top \Rightarrow \beta_1 = \top$ is false in $\mathfrak{B}_{\sim_\nabla} \in Var(\lambda)$. By the assumption of our lemma $\alpha_2 = \top \Rightarrow \beta_2 = \top$ is also false in $\mathfrak{B}_{\sim_\nabla} \in Var(\lambda)$. Then the identity $(\Box\alpha_2 \wedge \alpha_2 \to \beta_2) = \top$ must be also false in $\mathfrak{B}_{\sim_\nabla}$. Since $\mathfrak{B}_{\sim_\nabla}$ is a homomorphic image of $\mathfrak{B}$ (see Theorem 2.2.23) it follows that $(\Box\alpha_2 \wedge \alpha_2 \to \beta_2) = \top$ is false on $\mathfrak{B}$. ∎

We continue the proof of our theorem. According to Corollary 3.1.13 and Corollary 3.1.23 the rule $r := \alpha_1, ...\alpha_n/\beta$, its reduced form $rf(r) := \delta_1, ..., \delta_m/\rho$, and its semi-reduced form $srf(r) := \gamma_1, ..., \gamma_k/\theta$, are semantically equivalent in

$K4$. Therefore the quasi-identities $q(r)$, $q(rf(r)$ and $q(srf(r)$ are semantically equivalent in $Var(K4)$. By Lemma 3.1.26 identities

$$(\Box(\bigwedge_{1 \le i \le n} \alpha_i) \wedge (\bigwedge_{1 \le i \le n} \alpha_i)) \to \beta,$$
$$(\Box(\bigwedge_{1 \le i \le m} \delta_i) \wedge (\bigwedge_{1 \le i \le m} \delta_i)) \to \rho,$$
$$(\Box(\bigwedge_{1 \le i \le k} \gamma_i) \wedge (\bigwedge_{1 \le i \le k} \gamma_i)) \to \theta$$

are equivalent in $Var(\lambda)$. Taking into account (3.3) it follows that any of the rules r, $rf(r)$, $srf(r)$ is derivable in λ iff all others are also derivable in λ. ∎

As an application of this theorem we can show that any normal modal logic above $K4$ can be axiomatized by formulas having very simple (in a sense) canonical form. Indeed a formula α belongs to a modal logic λ iff the rule $x \to x/\alpha$ is derivable in λ. On the basis of this observation we infer the following description for axiomatization of modal logics. For every inference rule $r := \alpha_1, ..., \alpha_n/\beta$, the expressing modal formula $f(r)$ is defined as follows:

$$f(r) := \Box(\bigwedge_{1 \le i \le n} \alpha_i) \wedge (\bigwedge_{1 \le i \le n} \alpha_i) \to \beta.$$

By (3.3) r is derivable in a normal modal logic λ extending $K4$ iff the formula $f(r)$ is provable in λ.

Corollary 3.1.27 *For any normal modal logic λ above $K4$ and any modal formula α, $\lambda \oplus \alpha = \lambda \oplus f(rf(x \to x/\alpha)) = \lambda \oplus f(srf(x \to x/\alpha))$.*

Proof. The rule $r := p \to p/\alpha$ is derivable in $\lambda \oplus \alpha$. Therefore by Theorem 3.1.25 it follows that $rf(r)$ and $srf(r)$ are derivable in $\lambda \oplus \alpha$. By (3.3) formulas $f(rf(r))$ and $f(srf(r))$ are derivable in $\lambda \oplus \alpha$ which yields that $\lambda \oplus f(rf(x \to x/\alpha)) \subseteq \lambda \oplus \alpha$ and $\lambda \oplus f(srf(x \to x/\alpha)) \subseteq \lambda \oplus \alpha$. Conversely, the rule $rf(x \to x/\alpha)$ is derivable in $\lambda \oplus f(rf(x \to x/\alpha))$, and the rule $srf(x \to x/\alpha)$ is derivable in $\lambda \oplus f(srf(x \to x/\alpha))$ according to (3.3). Hence by Theorem 3.1.25 rule r is derivable in the logics $\lambda \oplus f(rf(x \to x/\alpha))$ and $\lambda \oplus f(srf(x \to x/\alpha))$ which means that the formula α is derivable in these logics. Thus $\lambda \oplus \alpha \subseteq \lambda \oplus f(rf(x \to x/\alpha))$ and $\lambda \oplus \alpha \subseteq \lambda \oplus f(srf(x \to x/\alpha))$. Hence $\lambda \oplus \alpha = \lambda \oplus f(rf(x \to x/\alpha)) = \lambda \oplus f(srf(x \to x/\alpha))$. ∎

Thus we have showed that every normal modal logic above $K4$ can be axiomatized by means of formulas of kind

$$\Box(\bigvee_{1 \le j \le n} \varphi_j) \wedge (\bigvee_{1 \le j \le n} \varphi_j) \to x_0, \text{ where}$$

$$\varphi_j := \bigwedge_{0 \le i \le m} x_i^{k(j,i,1)} \wedge \bigwedge_{0 \le i \le m} (\Diamond x_i)^{k(j,i,2)},$$

where $k(j, i, 1), k(j, i, 2) \in \{0, 1\}$, x_i are different variables (we use the corollary only for the case of reduced forms). We call such formulas the *canonical formulas*. These formulas have a rather simple and uniform structure: formulas have modal degree 2 and are implications which have as their closure merely a variable. The reader can compare this axiomatization of modal logics by means offered canonical formulas of modal degree 2 with canonical formulas in Zakharyaschev [214] which were placed in the basis of a new semantic investigation of modal logics.

3.2 T-translation of inference rules

As we know there is an embedding τ of superintuitionistic logics into modal propositional logics using the Gödel-McKinsey-Tarski translation T of formulas in the language of intuitionistic logic into modal propositional formulas. Recall that the translation T has the following definition:

$T(p) := \Box T$ for any propositional letter p;

$T(\alpha \wedge \beta) := T(\alpha) \wedge T(\beta)$;

$T(\alpha \vee \beta) := T(\alpha) \vee T(\beta)$;

$T(\alpha \to \beta) := \Box(T(\alpha) \to T(\beta))$;

$T(\neg \alpha) := \Box \neg T(\alpha)$.

According to Theorem 2.7.4, for any superintuitionistic logic λ, for any modal counterpart λ_1 for λ (i.e. $\tau(\lambda) \subseteq \lambda_1 \subseteq \sigma(\lambda)$), and for any formula α in the language $\mathcal{L}_H$, $\alpha \in \lambda \Leftrightarrow T(\alpha) \in \lambda_1$. Hence we can reduce the problem of recognizing the provability of formulas in a superintuitionistic logic λ to the problem of recognizing the provability of T-translations of these formulas in its modal counterparts. In this section we demonstrate that a similar approach in a restricted form is possible for the inference rules of superintuitionistic logics. First we simply transfer the definition of T-translation to inference rules.

Definition 3.2.1 *For any inference rule*

$$r := \frac{\alpha_1, ..., \alpha_n}{\beta}$$

in the language of $\mathcal{L}_H$, the Gödel-Mckinsey-Tarski translation $T(r)$ of r is the inference rule

$$r := \frac{T(\alpha_1), ..., T(\alpha_n)}{T(\beta)}.$$

Our aim now is to show that, for any superintuitionistic logic λ, a rule r is admissible for λ if and only if the rule $T(r)$ is admissible in the greatest modal counterpart $\sigma(\lambda)$ of λ. For this aim we will use Lemma 2.7.14. Recall that we will denote the fact that a formula (or a term) α is built up out of propositional letters or variables $p_1, ..., p_k$ by $\alpha(p_1, ..., p_k)$; if the number of variables above in not important for our reasoning then we write $\alpha(p_i)$, and if the variables are not important at all then we abbreviate the denotation by α. We proceed straight to the translation theorem.

Theorem 3.2.2 TRANSLATION THEOREM. *An inference rule r is admissible in a superintuitionistic logic λ if and only if the rule $T(r)$ is admissible in the greatest modal counterpart $\sigma(\lambda)$ of the logic λ.*

Proof. We pick out a superintuitionistic logic λ. Consider some inference rule $r := \alpha_1(p_i), ..., \alpha_n(p_i)/\beta(p_i)$ in the language of $\mathcal{L}_H$. We take its translation $T(r) := T(\alpha_1(p_i)), ..., T(\alpha_n(p_i))/T(\beta)(p_i)$. Suppose $T(r)$ is admissible in the modal logic $\sigma(\lambda)$. Take a substitution e of propositional formulas in place of the letters p_i from r. Suppose that $\forall j, 1 \leq j \leq n, \alpha_j(e(p_i)) \in \lambda$. By the translation theorem for formulas (cf. Theorem 2.7.15), it follows

$$\forall j, 1 \leq j \leq n, T(\alpha_j(e(p_i))) \in \sigma(\lambda).$$

Since $\Box T(\delta) \equiv T(\delta) \in S4$ for any formula δ (see Lemma 2.7.2), we conclude

$$\forall j, 1 \leq j \leq n, T(\alpha_j)(T(e(p_i))) \in \sigma(\lambda).$$

Since by assumption the rule $T(r)$ is admissible in the logic $\sigma(\lambda)$, it follows $T(\beta)(T(e(p_i))) \in \sigma(\lambda)$. Again because $\Box T(\delta) \equiv T(\delta) \in S4$ for any formula δ we have $T(\beta)(e(p_i)) \in \sigma(\lambda)$. By the translation theorem for formulas (Theorem 2.7.15), this yields $\beta(e(p_i)) \in \lambda$. Hence r is admissible for λ.

Conversely, suppose that r is admissible in λ and that $T(r)$ is not admissible for $\sigma(\lambda)$. Then there exists a substitution e of modal formulas in place of the propositional letters p_i from $T(r)$ such that $\forall j, 1 \leq j \leq n, T(\alpha_j)(e(p_i)) \in \sigma(\lambda)$ but $T(\beta)(e(p_i)) \notin \sigma(\lambda)$. Then, by the definition of $\sigma(\lambda)$, there exists a pseudo-boolean algebra $\mathfrak{A} \in Var(\lambda)$ such that $S(\mathfrak{A}) \not\models T(\beta)(e(p_i)) = \top$. We know that the algebra $S(\mathfrak{A})$ is generated by elements of the form $\Box a$, $\Box a \in \mathfrak{A}$. Therefore there are elements $\Box a_j \in \mathfrak{A}$ and a modal terms δ_k such that $T(\beta)(e(p_i)(\delta_k(\Box a_j))) \neq \top$ in $S(\mathfrak{A})$. According to Lemma 2.7.14 the formulas $\Box[e(p_i)(\delta_k(\Box q_j))]$ are equivalent in $S4$ to formulas $T(\gamma_i)(q_j)$. Therefore, for any algebra $\mathfrak{B} \in Var(S4)$,

$$\mathfrak{B} \models \Box[e(p_i)(\delta_k(\Box x_j))] = T(\gamma_i)(x_j) \text{ which yields}$$

$$S(\mathfrak{A}) \models \Box[e(p_i)(\delta_k(\Box a_j))] = T(\gamma_i)(a_j).$$

Therefore

$$S(\mathfrak{A}) \models T(\beta)(T(\gamma_i)(a_j)) \neq \top.$$

Since $\Box T(\phi) \equiv T(\phi) \in S4$ for any ϕ, it follows that

$$S(\mathfrak{A}) \models T(\beta(\gamma_i))(a_j)) \neq \top.$$

Therefore $T(\beta(\gamma_i)) \notin \sigma(\lambda)$. Because $\sigma(\lambda)$ is a modal counterpart for λ, we infer $\beta(\gamma_i) \notin \lambda$. Also we had $\forall j, 1 \leq j \leq n$, $(T(\alpha_j)(e(p_i)) \in \sigma(\lambda))$, therefore the fact that the formulas $\Box[e(p_i)(\delta_k(\Box q_j))]$ are equivalent in $S4$ to formulas $T(\gamma_i)(q_j)$ implies $\forall j, 1 \leq j \leq n$, $(T(\alpha_j(T(\gamma_i))) \in \sigma(\lambda))$ which yields $\forall j, 1 \leq j \leq n$, $T(\alpha_j(\gamma_i)) \in \sigma(\lambda)$. Hence $\forall j, 1 \leq j \leq n$, $\alpha_j(\gamma_i)) \in \lambda$. Thus r is not admissible in λ ∎

We have proved the translation theorem concerning the admissibility of inference rules only for the greatest modal counterparts of superintuitionistic logics, but not for all counterparts, as in the case concerning provability of formulas (it is the restriction which we mentioned above). Further we will see that for some particular superintuitionistic logics, it is sometimes possible to obtain the translation theorem about rules for other modal counterparts also, not only for the greatest ones. But, in general, the question of whether is it possible to extend the translation theorem for rules of superintuitionistic logics to all modal counterparts remains open.

Now we compare the Gödel-Mckinsey-Tarski translations of inference rules to these rules themselves, with respect to derivability. It is a simple task to show that this translation preserves derivability:

Proposition 3.2.3 *For any inference rule r in the language $\mathcal{L}_H$, r is derivable in a superintuitionistic logic λ iff the rule $T(r)$ is derivable in λ_1 for any modal counterpart λ_1 of the logic λ.*

Proof. Let $r := \alpha_1(p_i), ..., \alpha_n(p_i)/\beta(p_i)$. We take the translation $T(r) := T(\alpha_1(p_i)), ..., T(\alpha_n(p_i))/T((\beta)(p_i))$ for r. Assume r is derivable in λ. By the deduction theorem for λ (see Theorem 1.3.9) this means

$$\alpha_1(p_i) \wedge ... \wedge \alpha_n(p_i) \to \beta(p_i) \in \lambda.$$

If λ_1 is a modal counterpart for λ then

$$\Box(T(\alpha_1(p_i)) \wedge ... \wedge T(\alpha_n(p_i)) \to T(\beta(p_i))) \in \lambda_1.$$

Thus $T(r)$ is derivable in λ_1.

Suppose now that $T(r) := T(\alpha_1(p_i)), ..., T(\alpha_n(p_i))/T(\beta)(p_i)$ is derivable in a modal counterpart λ_1 for λ. Then according to the deduction theorem for modal logics (see Theorem 1.3.15) it follows that

$$\Box[T(\alpha_1(p_i)) \wedge ... \wedge T(\alpha_n(p_i))] \to T(\beta(p_i)) \in \lambda_1.$$

Since $\Box T(\delta) \equiv T(\delta)$ for any formula δ (see Lemma 2.7.2), from the inclusion above we infer

$$\Box[T(\alpha_1(p_i)) \wedge ... \wedge T(\alpha_n(p_i))) \rightarrow T(\beta(p_i))] \in \lambda_1$$

which means $T(\alpha_1(p_i)) \wedge ... \wedge T(\alpha_n(p_i)) \rightarrow T(\beta(p_i)) \in \lambda_1$. and, since λ_1 is a modal counterpart for λ, it follows $\alpha_1(p_i)) \wedge ... \wedge \alpha_n(p_i) \rightarrow \beta(p_i) \in \lambda$ which implies r is derivable in λ. $\blacksquare$

Theorem 2.7.14 can be transferred directly for the language of quasi-identities of pseudo-boolean algebras, corresponding Gödel-Mckinsey-Tarski translations of quasi-identities, and the truth of such quasi-identities in the corresponding free algebras.

Finally we clarify the behavior of quasi-identities in the language of pseudo-boolean algebras and their Gödel-Mckinsey-Tarski translations by T with respect to the semantic consequence. Here we provide the following lemma for use in the forthcoming sections, because the property it concerns is related to that studied above. We suppose T acts on quasi-identities in the language of pseudo-boolean algebras subsequently, on any term from the given quasi-identity.

Lemma 3.2.4 *Let λ be a superintuitionistic logic and let λ_1 be a modal counterpart for λ above S4. A quasi-identity*

$$T(q) := \bigwedge_{1 \leq m \leq n} T(\alpha_m) = T(\beta_m) \Rightarrow T(\alpha) = T(\beta)$$

is a corollary of some family of quasi-identities

$$T(q_i) := \bigwedge_{1 \leq j \leq n_i} T(\alpha_j^i) = T(\beta_j^i) \Rightarrow T(\alpha^i) = T(\beta^i), \ i \in I$$

in the variety $Var(\lambda_1)$ iff the quasi-identity $q := \bigwedge_{1 \leq m \leq n} \alpha_m = \beta_m \Rightarrow \alpha = \beta$ is a corollary of the family of quasi-identities $q_i := \bigwedge_{1 \leq j \leq n_i} \alpha_j^i = \beta_j^i \Rightarrow \alpha^i = \beta^i, \ i \in I$ in the variety $Var(\lambda)$.

Proof. Let $\mathfrak{B}$ be a modal algebra from $Var(S4)$. Consider the associated pseudo-boolean algebra $G(\mathfrak{B})$. Recall that $G(\mathfrak{B})$ consists of all open elements (elements of kind $\Box a$) from the algebra $\mathfrak{B}$. Let, for any algebra $\mathcal{A}$, $val_{\mathcal{A}}\mu(b_j)$ be the value of a term $\mu(x_j)$ on elements b_j in the algebra $\mathcal{A}$. Comparing the definition of the Gödel-Mckinsey-Tarski translation T for formulas and terms, and the definition of pseudo-boolean operations on the algebra $G(\mathfrak{B})$, it is easy to see that for any formula γ in the language $\mathcal{L}_H$,

$$val_{G(\mathfrak{B})}\gamma(\Box a_i) = val_{\mathfrak{B}}(T(\gamma)(a_i)). \tag{3.5}$$

Suppose now that $T(q)$ is not a semantic corollary of quasi-identities $T(q_i), i \in I$ in $Var(\lambda_1)$. Then there is a modal algebra $\mathfrak{B}$ from $Var(\lambda_1)$ such that $\mathfrak{B} \models$

$T(q_i)$ for all $i \in I$ and $\mathfrak{B} \not\models T(q)$. Consider the pseudo-boolean algebra $G(\mathfrak{B})$. According to Lemma 2.7.6 the algebra $G(\mathfrak{B})$ belongs to $Var(\lambda)$. From (3.5) it follows that all $q_i, i \in I$ are valid in $G(\mathfrak{B})$ but q is false in $G(\mathfrak{B})$. Thus q is not a semantic corollary of $q_i, i \in I$ on $Var(\lambda)$.

Conversely, suppose that q is not a semantic corollary of $q_i, i \in I$ in $Var(\lambda)$. Then there is an algebra $\mathfrak{A} \in Var(\lambda)$ such that q is false in $\mathfrak{A}$ but all $q_i, i \in I$ are valid in $\mathfrak{A}$. We take the wrapping topoboolean algebra $S(\mathfrak{A})$ for $\mathfrak{A}$. By the definition of $\sigma(\lambda)$, $S(\mathfrak{A}) \in Var(\sigma(\mathfrak{A}))$. Since $\sigma(\lambda)$ is the greatest counterpart for λ (Theorem 2.7.15), it follows $S(\mathfrak{A}) \in Var(\lambda_1)$. We know that $G(S(\mathfrak{A})) = \mathfrak{A}$ (see. Corollary 2.7.11). Therefore by (3.5) we conclude

$$\forall i \in I(S(\mathfrak{A}) \models T(q_i)), \ S(\mathfrak{A}) \not\models T(q).$$

Therefore $T(q)$ is not a corollary of $T(q_i), i \in I$ in $Var(\lambda_1)$. ∎

Now it is easy to derive a corollary concerning the validity of inference rules in logics. Recall that an inference rule $r := \alpha_1, ..., \alpha_n/\beta$ is *valid* in a logic λ if the quasi-identity $q(r) := \alpha_1 = \top \wedge ... \wedge \alpha_n = \top \Rightarrow \beta = \top$ is valid in the variety $Var(\lambda)$.

Corollary 3.2.5 *Suppose λ is a superintuitionistic logic and λ_1 is a modal counterpart for λ above S4. A rule $T(r) := \bigwedge_{1 \leq m \leq n} T(\alpha_m)/T(\alpha)$ is valid for λ_1 iff the rule $r := \bigwedge_{1 \leq m \leq n} \alpha_m/\alpha$ is valid for λ.*

Proof. It is sufficient to apply our Lemma 3.2.4 to the corresponding quasi-identities and empty family I.

3.3 Semantic Criteria for Admissibility

In this section we prepare a technique which will be used later in constructing algorithmic and semantic criteria for recognizing the admissibility of inference rules. This technique comes from the possibility of representing free algebras from varieties of modal algebras by means of special Kripke models. Some details on the origination of this technique are given in the historical comments in the end of this section.

The technique presented below is intended for normal modal logics with the finite model property and extending modal system $K4$, and also for superintuitionistic logics with the fmp. Therefore all frames and models considered here are transitive. First we recall some definitions and notation.

Let $\mathcal{F} := \langle F, R \rangle$ be a frame, and C be a non-empty subset of F. We say C is a *cluster* of $\mathcal{F}$ if $\forall x \in C, \forall y \in C \ (x \neq y \Rightarrow (xRy))$ and $\forall x \in C, \forall y \in F \ (xRy)\&(yRx) \Rightarrow y \in C$. A cluster C is *proper* if $||C|| \geq 2$. A cluster C is degenerated if C consists of a single irreflexive element. Suppose C_1 and C_2

are clusters of a frame $\mathcal{F} := \langle F, R \rangle$. We say a cluster C_2 is *accessible* or R-accessible or *seen* from a cluster C_1 if there are elements $a_1 \in C_1$ and $a_2 \in C_2$ such that $a_1 R a_2$. In this event we also say C_1 sees (or R-sees) cluster C_2. A chain of clusters from $\mathcal{F}$ is a family of clusters S such that, for any two different clusters in S, one of them is seen from the other. An ascending chain of clusters in a frame $\mathcal{F} := \langle F, R \rangle$ is a finite chain of clusters or an infinite chain S such that any cluster C from S sees some other cluster which differs from C.

We say a chain S begins with cluster C if C is a member of S which sees all other clusters in S. Any element x of a transitive frame $\mathcal{F}$ belongs to the uniquely definable cluster which we denote by $C(x)$ and call it the cluster originated by x. We let $Ch(x)$ be the family of all chains of clusters from $\mathcal{F}$ which begin with $C(x)$. Let x be an element of a transitive frame $\mathcal{F}$. We say x has depth n in $\mathcal{F}$ if n is the maximal number of clusters in chains from $Ch(x)$. We say x has infinite depth in $\mathcal{F}$ if there is an infinite ascending chain of clusters from $Ch(x)$.

Definition 3.3.1 *For any transitive frame* $\mathcal{M} := \langle M, R \rangle$ *(or transitive Kripke model* $\mathcal{M} := \langle M, R, V \rangle$*), the n-slice of* $\mathcal{M}$ *is the set of all elements of depth n from* $\mathcal{M}$*. We denote the n-slice of* $\mathcal{M}$ *by* $Sl_n(\mathcal{M})$*.* $S_n(\mathcal{M})$ *is the set of all elements from* $\mathcal{M}$ *with depth of not more than n.*

We will use again the following notation. Let $\mathcal{M}$ be a transitive frame $\langle M, R \rangle$ or a transitive model $\langle M, R, V \rangle$. If $c \in M$ then

$$c^{R<} := \{ x \mid x \in M, cRx, \neg (xRc) \},$$

$$c^{R\leq} := \{ x \mid x \in M, cRx \}.$$

If $Y \subseteq M$ then

$$Y^{R<} := \{ x \mid x \in M, \exists y \in Y ((yRx) \& \neg (xRy)) \},$$

$$Y^{R\leq} := \{ x \mid x \in M, \exists y \in Y (yRx) \}.$$

Sometimes we write for short Y^R and c^R instead of $Y^{R\leq}$ and $c^{R\leq}$. Now we introduce the main object of this section.

Definition 3.3.2 *A Kripke model* K_n *is called n-characterizing for a modal logic* λ *(any normal modal logic, not necessarily an extension of the system $K4$) if the domain of the valuation V from K_n is the set P which consists of n different propositional variables, and if the following holds. For any formula α which is build up of variables from P,*

$$\alpha \in \lambda \Longleftrightarrow K_n \Vdash \alpha.$$

Using n-characterizing models we can give a very simple criterion for the admissibility of inference rules which we will develop and specify in the following sections. First we recall a necessary notion.

Let $\mathcal{K} := \langle K, R, V \rangle$ be a Kripke model. A subset X of K is called definable (or expressible) iff there exists a formula α such that

$$X = \{x \mid x \in K, x \Vdash_V \alpha\}.$$

An element $x \in K$ is definable (or expressible) if the set $\{x\}$ is definable. Let S be a new valuation of certain propositional variables on the frame $\langle K, R \rangle$. The valuation S is called definable (or expressible) if and only if for any letter p_i from the domain of S, there exists a formula α_i such that $S(p_i) := V(\alpha_i)$.

Theorem 3.3.3 *Let $K_n, n \in N$ be a sequence of n-characterizing models for a modal logic λ. An inference rule $r := \alpha_1, ..., \alpha_m/\beta$ is admissible in λ iff for every $n \in N$ and each definable valuation S of variables from r in K_n the following holds. If $S(\alpha_1) = K_n, \ ... \ , S(\alpha_m) = K_n$ then $S(\beta) = K_n$ as well (that is, if r is valid in K_n with respect to all definable valuations).*

Proof. Consider an inference rule $r := \alpha_1, ..., \alpha_m/\beta$ with variables $x_1, ..., x_k$. Suppose r is not admissible in λ. Then there are formulas $\gamma_i, 1 \leq i \leq k$ such that for all $j, 1 \leq j \leq m$, $\alpha_j(\gamma_i) \in \lambda$ & $\beta(\gamma_i) \notin \lambda$. Suppose the number of propositional variables occurring in the above mentioned formulas is n. Model K_n is n-characterizing for λ, therefore we have

$$\forall j[(K_n \Vdash_V \alpha_j(\gamma_i)]\&K_n \nVdash_V \beta(\gamma_i).$$

Then the valuation $S(x_i) := V(\gamma_i)$ invalidates r in K_n: $\forall j[S(\alpha_j) = K_n]$, meanwhile $S(\beta) \neq K_n$. Moreover it follows directly from our definition that the valuation S is definable.

Conversely. Suppose there exist n and an definable valuation W of propositional letters $x_1, ..., x_k$ in the frame of K_n such that $W(\alpha_1) = K_n, ... , W(\alpha_m) = K_n$ but $W(\beta) \neq K_n$. Then according to the definability there exist formulas $\gamma_1(p_1, ..., p_n), ..., \gamma_k(p_1, ..., p_n)$ such that $W(x_i) = V(\gamma_i(p_1, ..., p_n))$. Using just the definition of the truth of formulas in Kripke models we obtain

$$\forall j[W(\alpha_j) = V(\alpha_j(\gamma_i(p_1, ..., p_n)))], \ W(\beta) = V(\beta(\gamma_i(p_1, ..., p_n))).$$

Therefore we arrive at

$$\forall j[V(\alpha_j(\gamma_i)) = K_n] \ \& \ V(\beta(\gamma_i)) \neq K_n.$$

This yields $\forall j[K_n \Vdash_V \alpha_j(\gamma_i)]$ and $K_n \nVdash_V \beta(\gamma_i)$. Since K_n is n-characterizing model for λ, this implies the rule r is not admissible in the logic λ. ■

Note that there is another equivalent definition of the definable valuation on Kripke models. We recall that, for any Kripke model $\mathcal{M} := \langle M, R, V \rangle$, where $\{p_1, ..., p_n\} \subseteq Dom(V)$, the modal algebra $M^+(V(p_1), ... , V(p_n))$ is the subalgebra of the associated modal algebra $\langle M, R \rangle^+$ generated by the elements $V(p_1)$, ... , $V(p_n)$. Let S be a definable valuation of the letters x_i in $\mathcal{M}$, that is there are formulas $\gamma_i(p_1, ...,_n)$ such that $S(x_i) = V(\gamma_i(p_1, ..., p_n))$. By Lemma 2.5.17 it follows

$$S(x_i) = \gamma_i(V(p_1), ..., V(p_n)) \in M^+(V(p_1), ..., V(p_n)).$$

That is, a definable valuation of x_i in $\mathcal{M}$ is a valuation of variables x_i by the elements of $M^+(V(p_1), ..., V(p_n))$. The converse proposition is also valid. Thus by Theorem 3.3.3 we can consider only such special valuations. That theorem has a direct application involving canonical models. We know (see Theorem 2.3.27) that, for any normal modal logic λ, the canonical model C_λ will be n-characterizing for any n (if we consider, for any given n, the valuation on C_λ of only the first n propositional variables $p_1, ..., p_n$). Therefore immediately from Theorem 3.3.3 we derive

Corollary 3.3.4 *For any normal modal logic λ extending the minimal normal modal logic K, a rule $r := \alpha_1, ..., \alpha_m / \beta$ is admissible for λ iff for every definable valuation S of variables from r in C_λ the following holds: if $S(\alpha_1) = C_\lambda, ..., S(\alpha_m) = C_\lambda$ then $S(\beta) = C_\lambda$ as well; i.e. if r is valid in C_λ with respect to definable valuations.*

Unfortunately this corollary does not give a convenient criterion for evaluating admissibility because, besides of complexity of the model C_λ, we consider only special expressible valuations which are not easy to handle. Therefore it looks a rather attractive idea to construct the possible simplest (in a sense) n-characterizing models for modal logics. We turn now to a solution that task along these lines for the case of modal logics with the finite model property. For any normal modal logic λ with the fmp and extending $K4$, we will effectively construct a certain n-characterizing Kripke model $Ch_\lambda(n)$.

CONSTRUCTION OF $Ch_\lambda(n)$

Recall that if $\mathcal{W}_1 := \langle W_1, R, V \rangle$ and $\mathcal{W}_2 := \langle W_2, R, V \rangle$ are Kripke models, $W_1 \subseteq W_2$, and R and V in both models on the same elements coincide then $\mathcal{W}_1$ is an open submodel of $\mathcal{W}_2$ iff the following holds

$$\forall a \in W_1, \forall b \in W_2 (aRb \Rightarrow b \in W_1).$$

The main property of an open submodel is that (see Exercise 2.5.12): for any formula α,

$$(\forall a \in W_1)[(a \Vdash_V \alpha) \text{ in } \mathcal{W}_1 \Leftrightarrow (a \Vdash_V \alpha) \text{ in } \mathcal{W}_2].$$

Let λ be a modal logic with the finite model property which extends $K4$. We denote by P_n the set $\{p_i \mid 1 \leq i \leq n\}$ consisting of first n propositional letters, and now we define $\mathcal{S}_1$ to be the set of all different (not isomorphic) models such that:

(a) any model is simply a cluster C with the valuation V of all letters from P_n,

(b) any cluster C has no elements with the same valuation of letters from P_n with respect to V,

(c) any model C is based on the cluster which is the λ-frame.

The model $\mathcal{M}_1$ is the model obtained from $\mathcal{S}_1$ as the disjoint union of all models contained in $\mathcal{S}_1$. Note that the frame of $\mathcal{M}_1$ is a λ-frame.

Suppose we have constructed already effectively our sequence of models $\mathcal{M}_1$, ... , $\mathcal{M}_k$ already and they all have the following properties:

(a) all models $\mathcal{M}_1$, ... , $\mathcal{M}_k$ are based on finite λ-frames,

(b) any $\mathcal{M}_i$ has a depth of not more than i,

(c) any $\mathcal{M}_i$ is an open submodel of $\mathcal{M}_{i+1}$.

Consider the model $\mathcal{M}_k$ which has the valuation V of letters from P_n. We take the set $\mathcal{A}(\mathcal{M}_k)$ of all possible antichains of clusters from $\mathcal{M}_k$ (an antichain of clusters is an arbitrary collection of clusters in which no one see any other, i.e. all the clusters in this collection are mutually incomparable by the relation R in $\mathcal{M}_k$) which have at least one cluster of depth k. The set $\mathcal{S}_1 \otimes \mathcal{A}(\mathcal{M}_k)$ is given by

$$\mathcal{S}_1 \otimes \mathcal{A}(\mathcal{M}_k) := \bigcup \{e(\langle C, A \rangle) \mid C \in \mathcal{S}_1, A \in \mathcal{A}(\mathcal{M}_k), \text{ if } \|A\| = 1$$
$$\text{then } C \text{ is not a submodel of } A,$$
$$e(\langle C, A \rangle) := \{\langle x, A \rangle \mid x \in C\}\}.$$

We define the model $\mathcal{G}_{k+1}$ as follows: the basic set of $\mathcal{G}_{k+1}$ is $|\mathcal{M}_k| \cup \mathcal{S}_1 \otimes \mathcal{A}(\mathcal{M}_k)$. The valuation V of any letter p_i from P_n is defined as follows: $V(p_i)$ is the set of all elements from $\mathcal{M}_k$ in which p_i is valid under V combined with the set of all new elements of the form $\langle x, A \rangle$, where $\langle x, A \rangle \in e(\langle C, A \rangle)$, $e(\langle C, A \rangle) \subseteq \mathcal{S}_1 \otimes \mathcal{A}(\mathcal{M}_k)$ and p_i is valid on x under V in C.

The accessibility relation R on $\mathcal{G}_{k+1}$ is the transitive closure of the accessibility relation in $\mathcal{M}_k$, the relation $\langle x, A \rangle R_1 \langle y, A \rangle$ for all $\langle x, A \rangle, \langle y, A \rangle$ belonging to $e(\langle C, A \rangle)$ for any $e(\langle C, A \rangle)$, and the relation Q, where

$$aQb \Leftrightarrow a = \langle x, A \rangle \in e(\langle C, A \rangle) \subseteq \mathcal{S}_1 \otimes \mathcal{A}(\mathcal{M}_k) \ \&$$
$$(x \in C) \ \& \ (b \in C_1 \in A).$$

The model $\mathcal{M}_{k+1}$ is obtained from $\mathcal{G}_{k+1}$ by deleting all clusters C of depth $k+1$ such that $C^{R \leq}$ (in $\mathcal{G}_{k+1}$) is not a λ-frame. It is clear that $\mathcal{M}_k$ is an open submodel of $\mathcal{M}_{k+1}$ and that the frame of $\mathcal{M}_{k+1}$ is a λ-frame.

In this way we can construct any model of the sequence $\mathcal{M}_k$, $k \in N$. To define the model $Ch_\lambda(n)$, we let

$$Ch_\lambda(n) := \bigcup_{k \in N} \mathcal{M}_k.$$

Since any $\mathcal{M}_k$ is an open submodel of the model $\mathcal{M}_{k+1}$, the model $Ch_\lambda(n)$ is well-defined. This completes the construction of $Ch_\lambda(n)$ ■

Note that a certain n-characterizing model $T(n)$ for the modal logic $S4$ was constructed in [135, 144] in a very similar way. Those construction was really quite similar to the general construction of $Ch_\lambda(n)$ given above. The difference, besides considering only reflexive transitive models, is merely that during the construction of $T(n)$ in [135, 144] it was not verified that initial clusters were λ-frames and whether rooted frames generated by adjoined clusters in any step were λ-frames. Therefore no clusters or elements were rejected during that construction.

We infer the following corollary directly from the constructions of the model $Ch_\lambda(n)$.

Corollary 3.3.5 *Suppose λ_1 and λ_2 are normal modal logics with the fmp and that both these logics extend $K4$. Suppose $\lambda_1 \subseteq \lambda_2$. Then for any n, $Ch_{\lambda_2}(n)$ is an open submodel of $Ch_{\lambda_1}(n)$.*

Now we show that the models $Ch_\lambda(n)$ are in fact n-characterizing for logics λ.

Theorem 3.3.6 *For any modal logic $\lambda \supseteq K4$, the Kripke model $Ch_\lambda(n)$ is n-characterizing for λ.*

Proof. All theorems of λ are valid in the model $Ch_\lambda(n)$ in virtue of the way we constructed this model, because we did not add clusters C such that $C^{R \leq}$ was not a λ-frame. If α is a formula with n propositional variables $p_1, ..., p_n$ and $\alpha \notin \lambda$ then by finite model property there is a finite Kripke model $\mathcal{M} := \langle M, R, V \rangle$ such that $\langle M, R \rangle$ is a frame for λ and α is not true in $\mathcal{M}$. Without loss of generality we can assume that the domain of the valuation V in $\mathcal{M}$ is exactly $p_1, ..., p_n$. We take a cluster of $\mathcal{M}$ that is maximal by R with an element a which disproves α. Then the generated submodel $a^{R \leq}$ of $\mathcal{M}$ falsifies α as well (see Exercise 2.5.12). Now we collapse the model $a^{R \leq}$ by the following procedure.

First in every cluster we identify all the elements that have the same valuation of variables. Then we contract the clusters of depth 1 which are isomorphic as models. This gives us a p-morphic image of the previous model. Therefore the resulting model F_1 is a p-morphic image of the original one. Consequently (cf. Lemma 2.5.15) the model F_1 disproves α as well. Also it is clear that $S_1(F_1)$ forms a generated (open) submodel of the model $Ch_\lambda(n)$.

Now suppose that we have already constructed the model F_i from the initial model on $a^{R\leq}$ such that: the frame of F_i is an λ-frame, the depth of F_i is not more than the depth of $\mathcal{M}$, F_i disproves α, and $S_i(F_i)$ forms an open submodel of $Ch_\lambda(n)$. We construct the model F_{i+1} from F_i in the following way.

(a) First we contract all clusters C from $Sl_{i+1}(F_i)$, each of which have only one immediate successor cluster C_1 such that C is a submodel of C_1 (more precisely, is isomorphic to a submodel), with the corresponding submodels of the clusters C_1.

(b) Then we contract clusters from $Sl_i(F_i)$ that are isomorphic as models and have the same sets of immediate successor clusters.

It is not hard to see that the resulting model will be a p-morphic image of F_i. We denote this model by F_{i+1}. Since the frame of F_{i+1} is a p-morphic image of the frame of F_i, the frame of F_{i+1} is a λ-frame (see Corollary Lemma 2.5.16). Also the model F_{i+1} disproves the formula α. It can be easily seen also that $Sl_{i+1}(F_{i+1})$ is a generated (i.e. open) submodel of $Ch_\lambda(k)$.

We continue this transformation and on some step m, which is less or equal to the depth of $\mathcal{M}$, obtain a model which rejects α and is also a generated (open) submodel of $Ch_\lambda(n)$. Thus (see Exercise 2.5.12) $Ch_\lambda(n)$ disproves α. ∎

Theorem 3.3.7 *Any element of the model $Ch_{K4}(n)$ is definable. In particular, for any normal modal logic λ extending $K4$ and having the fmp, every element of $Ch_\lambda(n)$ is definable (expressible).*

Proof. The proof is carried out by induction on the depth of elements in the model $Ch_{K4}(n)$. For any element $a \in |Ch_{K4}|$, we set $p(a) := \{i \mid a \Vdash_V p_i\}$ and

$$\alpha(a) := \bigwedge_{i \in p(a)} p_i \wedge \bigwedge_{i \notin p(a)} \neg p_i.$$

First we show that all elements of depth 1 are expressible. If a is an irreflexive element of depth 1 then the formula $\beta(a) := \alpha(a) \wedge \Box(p_1 \wedge \neg p_1)$ defines the element a, i.e. this formula is valid only on element a. For any non-degenerated cluster C from $Ch_{K4}(n)$, we introduce the formula $\alpha(C) := \bigwedge_{a \in C} \Diamond\alpha(a)$. Any element a is a member of the cluster $C(a)$ generated by a. It is not hard to see that for a reflexive element a of depth 1, a is definable by the formula

$$\beta(a) := \alpha(a) \wedge \alpha(C(a)) \wedge \Box(\alpha(C(a))).$$

Assume that we already shown that every element a of depth not greater than $i, 1 \leq i$, is expressible in $Ch_\lambda(n)$ and is definable by the formula $\beta(a)$. For any element a of depth more than 1, we introduce the formula

$$\delta(a) := \bigwedge_{c \in \{a\}^{R<}} \Diamond\beta(c) \wedge \bigwedge_{c \in S_i(Ch_{K4}(n)) \& c \notin \{a\}^{R<}} \neg\Diamond\beta(c).$$

For every irreflexive element a from $Sl_{i+1}(Ch_{K4}(n))$, we set

$$\beta(a) := \alpha(a) \wedge \delta(a) \wedge \Box(\bigvee_{c \in \{a\}^{R<}} \beta(c)) \wedge \bigwedge_{c \in S_i(Ch_{K4}(n))} \neg\beta(c).$$

The formula $\beta(a)$ defines the element a in $Ch_{K4}(n)$. Indeed, it is clear that $\beta(a)$ is valid on a. If $d \Vdash_V \beta(a)$ then a has a depth of more than i. Since the conjunct $\delta(a)$ is present in $\beta(a)$, d sees exactly those elements from the set $S_i(Ch(K4))$ that a sees. Any element accessible from d is an element of $S_i(Ch_{K4}(n))$ in view of the presence of the third conjunct in $\beta(a)$. Therefore d has depth $i+1$ and is an irreflexive element. As $\alpha(a)$ is a conjunct of $\beta(a)$, it follows $d = a$.

Consider a reflexive element a from $Sl_{i+1}(Ch_{K4}(n))$ and the cluster $C(a)$ containing a. For such elements a we fix the formulas:

$$\gamma(i) := \bigwedge_{c \in S_i(Ch_{K4}(n))} \neg\beta(c),$$

$$\theta(a) := \bigwedge\{\Diamond(\alpha(d) \wedge \delta(a) \wedge \gamma(i)) \mid d \in C(a)\}\wedge$$

$$\bigwedge\{\neg\Diamond(\alpha(u) \wedge \gamma(i)) \mid\mid u \in S_1(Ch_{K4}(n))\&$$

$$\& \ \forall d \in C(a)(p(u) \neq p(d))\},$$

$$\phi(a) := \Box(\bigvee_{c \in \{a\}^{R<}} \beta(c) \vee \bigvee_{d \in C(a)} (\alpha(d) \wedge \delta(a) \wedge \theta(a)),$$

$$\beta(a) := \alpha(a) \wedge \delta(a) \wedge \gamma(i) \wedge \theta(a) \wedge \phi(a).$$

We will show that the formula $\beta(a)$ defines a in $Ch_{K4}(n)$. It is easy to see that β is valid on a. If $d \Vdash_V \beta(a)$ then d has a depth of more than i by the presence of the conjunct $\gamma(i)$ in $\beta(a)$. d sees only those elements of depth not more than i which are seen from a as $\delta(a)$ is a conjunct of $\beta(a)$. If d is an element of depth $i+1$ then $C(d)$ is isomorphic to $C(a)$ because the conjunct $\theta(a)$ is present in $\beta(a)$ and by the property of the last conjunct member in $\beta(a)$. This means $C(d) = C(a)$ and since $\alpha(a)$ is a conjunct of $\beta(a)$, we infer $d = a$.

Suppose d is an element of depth $i + 2$ or more. Then d sees a certain element d_1 of depth $i + 1$. All such elements d_1 see only those elements of the set $S_i(Ch(K4)(n))$ which are seen from the element d (because $\phi(a)$ is a conjunct of $\beta(a)$). If d sees two different clusters C_{d_1}, C_{d_2} of depth $i+1$ then we have a contradiction the fact that $\phi(a)$ is a conjunct of $\beta(a)$. Thus d sees only one cluster $C(d_1)$ of depth $i+1$. Therefore there is a cluster $C(d_2)$ which is seen from d and which is the immediate predecessor of $C(d_1)$. Then d_2 has depth $i + 2$ and sees only cluster $C(d_1)$ among the clusters of depth $i+1$. As $\phi(a)$ is a conjunct of

$\beta(a)$, $C(d_2)$ has single immediate successor cluster, the cluster $C(d_1)$. Accord-ing to the construction of $Ch_{K4}(n)$ the cluster $C(d_2)$ has an element e such that $p(e) \neq p(c)$ for all $c \in C(d_1)$. This contradicts the fact that $\phi(a)$ is a conjunct of $\beta(a)$. Thus $\beta(a)$ is valid only on element a.

To end our theorem suppose that λ is a normal modal logic with the fmp and λ extends $K4$. According to Corollary 3.3.5 the model $Ch_\lambda(n)$ is an open submod-el of the model $Ch_{K4}(n)$. Therefore all the elements of $Ch_\lambda(n)$ are definable. ∎

For a modal logic λ, we can describe the free modal algebras of finite rank from $Var(\lambda)$ using n-characterizing models of λ.

Theorem 3.3.8 *Let λ be a normal modal logic and $\mathfrak{M}_n := \langle M, R, V \rangle$ be a n-characterizing model for λ, $Dom(V) := \{p_1, ..., p_n\}$. The subalgebra $\mathfrak{M}_n^+(V(p_1)..., V(p_n))$ of the associated wrapping algebra $\langle M, R \rangle^+$ generated by the elements $V(p_1), ..., V(p_n)$ is the free algebra $\mathfrak{F}_n(\lambda)$ of rank n from $Var(\lambda)$. Moreover, elements $V(p_1), ..., V(p_n)$ are free generators of $\mathfrak{F}_n(\lambda)$. In particu-lar, for any modal logic λ extending $K4$ with the fmp, $\mathfrak{F}_n(\lambda)$ is isomorphic to the modal algebra $Ch_\lambda(n)^+(V(p_1)..., V(p_n))$.*

Proof. The tuple of elements $V(p_1), ..., V(p_n)$ generate, as we know, the modal algebra $\mathfrak{M}_n^+(V(p_1), ... , V(p_n))$ immediately by it's definition. Consider an arbi-trary algebra $\mathfrak{B}$ from $Var(\lambda)$ and an arbitrary mapping s of elements $V(p_1), ... , V(p_i)$ from the algebra $\mathfrak{M}_n^+(V(p_1), ... , V(p_n))$ into $|\mathfrak{B}|$: $s(V(p_i)) := a_i \in |\mathfrak{B}|$. We must extend s to hole the algebra $\mathfrak{M}_n^+(V(p_1), ... , V(p_n))$ in order to produce a homomorphism. Any element c of $\mathfrak{M}_n^+(V(p_1)..., V(p_n))$ has a representation $c = \alpha(V(p_1), ..., V(p_n))$, where α is a term (formula). Let

$$s(c) = s(\alpha(V(p_1), ..., V(p_n))) := \alpha(a_1, ..., a_n).$$

We must show that this definition of s does not depend on our representation of the element c. Suppose that

$$c = \alpha_1(V(p_1), ..., V(p_n)) = \alpha_2(V(p_1), ..., V(p_n)).$$

Note that according to Lemma 2.5.17

$$\alpha_1(V(p_1), ..., V(p_n)) = V(\alpha_1(p_1, ..., p_n)),$$
$$\alpha_2(V(p_1), ..., V(p_n)) = V(\alpha_2(p_1, ..., p_n)).$$

Therefore

$$\alpha_1(p_1, ..., p_n) \to \alpha_2(p_1, ..., p_n) \in \lambda, \ \alpha_2(p_1, ..., p_n) \to \alpha_1(p_1, ..., p_n) \in \lambda$$

because $\mathfrak{M}_n$ is n-characterizing for λ. Thus $\alpha_1(p_i) \equiv \alpha_2(p_i) \in \lambda$. Therefore for any algebra $\mathfrak{B}_1 \in Var(\lambda)$,

$$\mathfrak{B}_1 \models \alpha_1(x_1, ..., x_n) = \alpha_2(x_1, ..., x_n).$$

In particular we have $\alpha_1(a_1, ..., a_n) = \alpha_2(a_1, ..., a_n)$ in the algebra $\mathfrak{B}$. Hence the mapping s from $\mathfrak{M}_n^+(V(p_1), ... , V(p_n))$ is well defined. Immediately from the definition of s it is easily seen that s is a homomorphism. Thus the algebra $\mathfrak{M}_n^+(V(p_1), ... , V(p_n))$ is the free algebra of rank n from $Var(\lambda)$ and $V(p_1), ..., V(p_n)$ are its free generators. The remaining part of this theorem holds because by Theorem 3.3.6 any $Ch_\lambda(n)$ is a n-characterizing model for the logic λ (placed above $K4$) which has the fmp. ∎

Now we turn to consider properties of n-characterizing Kripke models for superintuitionistic logics. A significant part of this material is like that in modal case and we will omit the common parts, concentrating on the distinctions. The definition of n-characterizing intuitionistic Kripke models for superintuitionistic logics is parallel to our definition for the modal case:

Definition 3.3.9 *An intuitionistic Kripke model $\mathcal{M}_n$ is n-characterizing for a superintuitionistic logic λ if the domain of the valuation V from $\mathcal{M}_n$ is p_1, ..., p_n and for any formula α built up out of these letters p_1, ..., p_n, $\alpha \in \lambda$ iff $\mathcal{M}_n \Vdash \alpha$.*

It is obvious that, like as in the case of modal logics, all superintuitionistic logics posses n-characterizing Kripke models for every n. It is sufficient to note that the canonical models C_λ are characterizing and consequently n-characterizing models for superintuitionistic logics λ. At the same time the structure of such models is complicated and not easy visible, therefore we will construct certain constructive n-characterizing models for superintuitionistic logics which possess the fmp. Recall that, for any intuitionistic Kripke model $\langle M, \leq, V \rangle$, a subset X (or an element a) of M is definable or expressible in $\langle M, \leq, V \rangle$ if there is a formula α such that $X = V(\alpha)$ ($\{a\} = V(\alpha)$ respectively). A new valuation S of a set P of propositional letters on the frame $\langle M, V \rangle$ is definable in the model $\langle M, \leq, V \rangle$ if for any letter p from the domain of S there exists a formula α such that $S(p) = V(\alpha)$. And as with modal logics we can give a semantic criterion for the admissibility of inference rules in superintuitionistic logics.

Theorem 3.3.10 *Suppose $\mathcal{M}_n, n \in N$ is a sequence of n-characterizing intuitionistic Kripke models for a superintuitionistic logic λ. An inference rule $r := \alpha_1, ..., \alpha_m/\beta$ is admissible for λ if and only if for every $n \in N$ and each definable valuation S of variables from r on $\mathcal{M}_n$ the following holds. If we have $S(\alpha_1) = \mathcal{M}_n, ..., S(\alpha_m) = \mathcal{M}_n$ then $S(\beta) = \mathcal{M}_n$.*

Proof. This is quite similar to the proof of Theorem 3.3.3 and is left to the readers as a simple exercise.

We know that intuitionistic Kripke models differ from usual Kripke models for modal logics in that (a) the accessibility relation for an intuitionistic model must be a partial order, and (b) the valuation on an intuitionistic model is *hereditary*: if in an element a a letter p is valid then p is valid on all elements that are greater than a. By this reason the construction of n-characterizing intuitionistic Kripke models for superintuitionistic logics will differ from the construction of models $Ch\lambda(n)$ for modal logics only in these respects. In a such way, for any given superintuitionistic logic λ with the finite model property, we can construct the model $Ch\lambda(n)$ in a quite similar way to the construction of the models $Ch_{\lambda_m}(n)$ for modal logics λ_m above $K4$ having fmp. We have only the following distinctions:

(a) All clusters are just reflexive elements (non-degenerated, not proper clusters);

(b) While adjoining to the model $\mathcal{M}_i$ elements of depth $i + 1$ in order to construct $\mathcal{M}_{i+1}$, we add only new elements c of depth $i + 1$ such that the open submodel generated by c has an intuitionistic valuation.

Theorem 3.3.11 *For any superintuitionistic logic λ with the fmp, the intuitionistic Kripke model $Ch_\lambda(n)$ is a certain n-characterizing model for λ.*

Proof. The proof is quite similar to the proof of Theorem 3.3.6. We use again the technique of corresponding contracting p-morphisms in analogous way. Therefore the proof is left to the readers as an exercise.

Immediately from the construction of the model Ch_λ for a superintuitionistic logic λ with the fmp we derive

Lemma 3.3.12 *Assume λ_1 and λ_2 are superintuitionistic logics possessing the fmp and $\lambda_1 \subseteq \lambda_2$. Then for every n the intuitionistic model $Ch_{\lambda_2}(n)$ is an open submodel of the intuitionistic model $Ch_{\lambda_1}(n)$.*

It is impossible to obtain a complete analog to the Theorem 3.3.7 because intuitionistic formulas define not elements but an open subsets of elements on which such formulas are valid. However a suitable analog can nevertheless be obtained by a similar reasoning to above. We will use the following notation for posets and intuitionistic Kripke models. If $\mathcal{M}$ is a poset or an intuitionistic Kripke model with partial order $\leq$ and $c \in |\mathcal{M}|, Y \subseteq |\mathcal{M}|$ then

$$c^< := \{x \mid x \in |\mathcal{M}|, c \leq x, c \neq x\},$$
$$c^\leq := \{x \mid x \in |\mathcal{M}|, c \leq x\},$$
$$Y^< := \{x \mid x \in |\mathcal{M}|, \exists y \in Y (y \leq x) \& (x \notin Y))\},$$
$$Y^\leq := \{x \mid x \in |\mathcal{M}|, \exists y \in Y (y \leq x)\}.$$

An subset X of $|\mathcal{M}|$ is a *sharp cone* (or is rooted) if there is an element a from $|\mathcal{M}|$ such that $X = a^\leq$.

Theorem 3.3.13 *Any sharp cone of $Ch_H(n)$ for any n is definable. In particular for any n, for every superintuitionistic logic λ with fmp, any sharp cone in $Ch_\lambda(n)$ is definable.*

Proof. We proceed by induction on the depth of the minimal element of the sharp cone in $Ch_H(n)$. For every $b \in Ch_H(n)$, let $p(b) := \{i \mid b \Vdash_V p_i\}$ and $\delta(b) := \bigwedge_{i \in p(b)} p_i$. Suppose that a is an element of the depth 1 from $Ch_H(n)$. We put $\alpha(a) := \delta(a) \wedge \bigwedge_{i \leq n, i \notin p(a)} \neg p_i$. It is clear that $a \Vdash_V \alpha(a)$ and that for every element c of depth 1 which differs from a, $c \nVdash_V \alpha(a)$.

Suppose b is an element of depth 2 or more from $Ch_H(n)$. If b sees some element c of depth 1 which differs with a then $\alpha(a)$ is false on c as we have seen. Since the truth of formulas in intuitionistic models is hereditary, we get $b \nVdash_V \alpha(a)$. So if $b \Vdash_V \alpha(a)$ then all elements which are seen from b are only a. Furthermore, b sees an element c of depth 2. Then c sees only elements c and a. Therefore the valuation of propositional letters on c has strongly fewer valid letters than on a. This contradicts $c \Vdash_V \alpha(a)$ and the presence of second conjunct member in $\alpha(a)$.

Suppose that we have already showed that all sharp cones with the depth of the generating elements not more than m are definable in $Ch_H(n)$. We fix the denotation $\alpha(a)$ for the formulas defining $a^\leq$, where a is an element of depth not more than m. For any element $x \in Ch_H(n)$, $d(x)$ denotes the depth of x. For every element x of depth $i + 1$, we introduce the formula

$$g(x) := \bigwedge_{x \nleq c, d(c) \leq m} (\alpha(c) \to \bigvee_{x < b} \alpha(b)).$$

It is clear that

$$\forall y(y \Vdash_V g(a)) \Leftrightarrow y^\leq \cap S_m(Ch_H(n)) \subseteq a^\leq \cap S_m(Ch_H(n)). \tag{3.6}$$

We also set for any a of depth $m + 1$,

$$\beta(a) := \bigwedge_{i \leq n, i \notin p(a)} (p_i \to \bigvee_{a \leq b} \alpha(b)),$$

$$\alpha(a) := \delta(a) \wedge g(a) \wedge \beta(a) \wedge \bigwedge\{(g(c) \to \bigvee_{r \in S_m(Ch_H(n))} \alpha(r) \mid c \in$$

$$Sl_{m+1}(Ch_H(n)), c^< \subseteq a^<\}).$$

It can be easily seen that $a \Vdash_V \alpha(a)$. Suppose that $d \in Sl_{m+1}(Ch_H(n))$ and $d^< \neq a^<$. Then either (a) $\exists c((d < c)\&(a \nleq c))$, or (b) $\exists c((a < c)\&(d \nleq c))$ and (a) is false.

In the case (a), we have $c \Vdash_V \alpha(c)$ and $c \nVdash_V \bigvee_{a \leq b} \alpha(b)$ (since when $c \Vdash_V \alpha(b)$, $b \leq c$ holds, and $a < c$, a contradiction). Therefore we conclude $d \nVdash_V \alpha(a)$ since

$g(a)$ presents as a conjunctive member in $\alpha(a)$. In the case (b) $d^<$ is a proper subset of $a^\leq$, i.e. it is a subset which is not equal to $a^<$. Therefore the fact that $d \Vdash_V g(d)$ and $d \nVdash_V \bigvee_{r \in S_m(Ch_H(n))} \alpha(r)$ imply $d \nVdash_V \alpha(a)$.

Suppose now that $d \in Sl_{m+1}(Ch_H(n))$ and $d^< = a^<$ but $d \neq a$. Then either (a) $\exists i \in p(a)$ that $i \notin p(d)$, $d \nVdash \delta(a)$ and $d \nVdash_V \alpha(a)$ or (b) $\exists i \in p(d)$ such that $i \notin p(a)$ and $p(a)$ is proper subset of $p(d)$. Therefore in the case (b), supposing $d \Vdash_V \alpha$, we get $d \Vdash_V \bigvee_{a<b} \alpha(b)$, which is evidently false. Hence $d \nVdash_V \alpha(a)$. Thus we have proved

$$\forall d \in Sl_{m+1}(Ch_H(n))(d \neq a \Rightarrow d \nVdash_V \alpha(a).$$

Assume $r \in Sl_{m+2}(Ch_H(n))$ and $r \Vdash_V \alpha(a)$. Then by (3.6) and what is shown above $r^\leq \cap Sl_{m+1}(Ch_H(n) = \{a\}$. Then there exists a $d \in Sl_{m+2}(Ch_H(n))$ such that $r \leq d$ and $d^< = a^\leq$. This fact and the construction of $Ch_H(n)$ entail $p(d)$ is a proper subset of $p(a)$. Therefore $d \nVdash_V \beta(a)$ and $r \nVdash_V \alpha(a)$. Suppose $c \in S_m(Ch_H(n))$ and $c \Vdash_V \beta(a)$. Then $c \Vdash_V \alpha(c)$ and if $a \not< c$ then $c \Vdash_V \alpha(b)$, where $(b \leq c)$ and $a < b$ i.e. $a \leq c$. Thus it is always the case that for any such c, $a \leq c$. Hence the formula α defines the sharp cone $a^\leq$. ∎

Finally, similar to the case of modal algebras we can describe free pseudo-boolean algebras by means of n-characterizing Kripke models as follows.

Theorem 3.3.14 *Let λ be a superintuitionistic logic and $\mathfrak{M}_n := \langle M, R, V \rangle$ be a certain n-characterizing model for λ, and $Dom(V) := \{p_1, ..., p_n\}$. The subalgebra $\mathfrak{M}_{np}^+(V(p_1)..., V(p_n))$ of the pseudo-boolean algebra $\langle M, R \rangle_{np}^+$ generated by the elements $V(p_1), ..., V(p_n)$ is the free algebra $\mathfrak{F}_n(\lambda)$ of rank n from $Var(\lambda)$. Elements $V(p_1), ..., V(p_n)$ are free generators of $\mathfrak{F}_n(\lambda)$. In particular, for any superintuitionistic logic λ having the fmp, $\mathfrak{F}_n(\lambda)$ is isomorphic to the pseudo-boolean algebra $Ch_\lambda(np)^+(V(p_1)..., V(p_n))$.*

Proof. The proof is parallel to the case of free modal algebras. Indeed, elements $V(p_1), ..., V(p_n)$ generate the algebra $\mathfrak{M}_{np}^+(V(p_1), ... , V(p_n))$. Let $\mathfrak{A}$ be an algebra from $Var(\lambda)$, and s be an arbitrary mapping of elements $V(p_1), ... , V(p_i)$ into $|\mathfrak{A}|$: $s(V(p_i)) := a_i \in |\mathfrak{A}|$. Every element c of $\mathfrak{M}_{np}^+(V(p_1)..., V(p_n))$ has a representation $c = \alpha(V(p_1), ..., V(p_n))$, where α is a term. We set $s(c) = s(\alpha(V(p_1), ..., V(p_n))) := \alpha(a_1, ..., a_n)$. This definition of s does not depend on the representation of the element c. Indeed, let $c = \alpha_1(V(p_1), ..., V(p_n)) = \alpha_2(V(p_1), ..., V(p_n))$. Using Lemma 2.5.67 we conclude

$$\alpha_1(V(p_1), ..., V(p_n)) = V(\alpha_1(p_1, ..., p_n)),$$

$$\alpha_2(V(p_1), ..., V(p_n)) = V(\alpha_2(p_1, ..., p_n)).$$

Therefore we the following holds

$$\alpha_1(p_1, ..., p_n) \rightarrow \alpha_2(p_1, ..., p_n) \in \lambda,$$

$$\alpha_2(p_1, ..., p_n) \to \alpha_1(p_1, ..., p_n) \in \lambda.$$

because $\mathfrak{M}_n$ is n-characterizing for λ. Thus $\alpha_1(p_i) \equiv \alpha_2(p_i) \in \lambda$ and, for any algebra $\mathfrak{B}_1 \in Var(\lambda)$, $\mathfrak{B}_1 \models \alpha_1(x_1, ..., x_n) = \alpha_2(x_1, ..., x_n)$. In particular we obtain $\alpha_1(a_1, ..., a_n) = \alpha_2(a_1, ..., a_n)$ in the algebra $\mathfrak{B}$. So the mapping s from $\mathfrak{M}_{np}^+(V(p_1), ... , V(p_n))$ is well defined, and the definition of s yields s is a homomorphism. The remaining assertion of our theorem holds since any $Ch_\lambda(n)$ is a n-characterizing model for the logic λ. ∎

Historical Comments The main content of this section are the construction of n-characterizing Kripke models, studying their properties and description the free algebras of finite rank by means of n-characterizing Kripke models. The description of free modal and pseudo-boolean algebras for varieties of modal and pseudo-boolean algebras corresponding to certain individual logics is, in a sense, a folklore of non-standard logic. Descriptions of modal free algebras from $Var(\lambda)$ when $\lambda = S4, S4.3$ can be found in Grigolia [62], Shehtman [172], Rybakov [133, 135]. The free pseudo-boolean algebras from $Var(H)$ are described in Bellissima [9] and Shehtman [172]. Elements of description for free temporal algebras can be found in Bellissima [11], and certain properties of free pseudo-boolean algebras connected with structure of their subalgebras are contained in Balbes and Horn [6]. The main distinction of research in this section from mentioned above ones consists of that we give description of free algebras not for individual varieties but for all varieties which correspond to logics with finite model property. However we merely modify the techniques used in previous research and this did not meet a serious technical problems.

3.4   Some Technical Lemmas

In this section we introduce two notions necessary for constructing criteria determining admissibility in modal and superintuitionistic logics: the branching below m property, and the effective m-drop points property. Studying these properties we prove several technical lemmas, two of which will play a central role when we will come to the criteria for admissibility. As a matter of fact for this we will need only Lemmas 3.4.9 and 3.4.10. The proof of these lemmas is rather long, therefore, on a first reading a reader might want to omit these proofs and concentrate only on the formulations of the lemmas.

We begin with some simple notes about the structure of p-morphisms of finite frames and other related frames which have only finite generated rooted subframes. It is well known that any p-morphism of a finite frame to another can be decomposed into a composition of the smallest (in a sense) permissible contractions (cf., for instance de Jongh and Troelstra [50]) Here we simply extend this notion of permissible contractions to special infinite models with the frames of the kind mentioned above.

Let $\mathfrak{M} := \langle M, R, V \rangle$ be a modal (or intuitionistic) Kripke model which has no infinite ascending chains of clusters. We define a special p-morphism from this model as the basis set. In the case of an intuitionistic model, all the below mentioned clusters are simply elements (single-element clusters). Pick an $m \in M$. Let $H_1 := M$. In step 1, we define a p-morphism from H_1 as follows:

1) first we contract some members in some clusters of depth $m + 1$ provided these members have the same valuation of the propositional letters in the model;

2) then we drop some clusters C of depth $m + 1$ provided each of them has exactly one covering cluster which contains C as submodel and C is not an irreflexive element;

3) then we identify some different clusters of depth $m + 1$ provided they have the same strong successors and are isomorphic as models.

We denote by f_1 the composition of these p-morphisms. And the corresponding p-morphic image of H_1 is denoted by H_2. We define f_2 and H_2 in a similar way working with clusters of depth $m + 2$ etc. This way we obtain a sequence $H_n, n \in N$ of models and a sequence $f_n, n \in N$ of corresponding p-morphisms. It is clear that f_n is isomorphic mapping on elements of depth of not more than n. Therefore the following composition $... \circ f_n \circ ... \circ f_1$ is meaningful and forms a certain p-morphism Cg from H_1, which we call a *condensing* of the model $\langle M, R, V \rangle$. Because Cg is a p-morphism, the following holds.

Lemma 3.4.1 Condensing Lemma. *For each $x \in M$ and every formula α with propositional letters from the domain of V, $x \Vdash_V \alpha$ iff $Cg(x) \Vdash_V \alpha$.*

In a similar way, we can define condensing mappings of any frame $\mathcal{F}$ all elements of which have finite depth. It is sufficient to introduce the valuation V of a single letter p_1 such that p_1 is valid under V on all elements of $\mathcal{F}$, and then to use the definition of condensing mappings for Kripke models.

We also need the following observation concerning invalidating of inference rules in the special n-characterizing models $Ch_\lambda(n)$ for modal and superintuitionistic logics λ with the fmp, which were introduced in the previous section.

Lemma 3.4.2 *Let λ be a modal logic extending $K4$ or a superintuitionistic logic which has the finite model property. If there is a valuation S of the variables of an inference rule $r = \alpha_1, ... \alpha_n / \beta$ in the frame of some n-characterizing model $Ch_\lambda(n)$ which invalidates r then there is a valuation S of these variables in $Ch_\lambda(k)$, where k is the number of variables in r, which also invalidates r.*

Proof. We give the proof for the more general case of modal logics λ, and the proof for superintuitionistic logics follows from this proof with the corresponding simplifications: any cluster is simply an element, etc. Consider the first case when $n < k$. It is not hard to see that there is a condensing mapping f of the frame of $Ch_\lambda(k)$ onto the frame of $Ch_\lambda(n)$. We know that this f is a p-morphism for frames. We transfer the valuation S from $Ch_\lambda(n)$ onto the frame of $Ch_\lambda(k)$ by $S(q_j) := f^{-1}(S(q_j))$. It is clear that f is a p-morphism of models with respect to the valuation S. Since any p-morphism preserves the truth of formulas, we have S invalidates r in $Ch_\lambda(k)$.

Assume now that $k < n$. We will construct a condensing p-morphism of the model $\langle Ch_\lambda(n), R, S \rangle$ onto a model with the frame $Ch_\lambda(k)$ and a corresponding valuation S. The p-morphism will be the composition of certain contracting p-morphisms h_i constructed as follows.

The definition of h_1 is the following. First we map some of the clusters from $S_1(Ch_\lambda(n))$ into some clusters from $S_1(Ch_\lambda(k))$. Namely, we fix one cluster C_Ψ in every class Ψ of clusters from $S_1(Ch_\lambda(n))$ which are isomorphic as models after contracting in any cluster all elements having the same valuation of letters under S. We consider a mapping h_1 which maps every C_Ψ onto the cluster $h_1(C_\Psi)$ obtained from C_Ψ by contracting all elements having the same valuation of propositional letters under S. We can assume h_1 is a p-morphism of models for these clusters with the valuation S, which is transferred onto the cluster $h_1(C_\Psi)$ by h_1, and that all clusters $h_1(C_\Psi)$ are different clusters from $S_1(Ch_\lambda(k))$.

It follows directly from our definition that, for any $d \leq 2^k$, the number of clusters with d elements in $S_1(Ch_\lambda(n))$ is greater than or equal to $2 * u$, where u is the number of such clusters in $S_1(Ch_\lambda(k))$. Therefore, for all clusters C from $Sl_1(Ch_\lambda(k))$ which do not coincide with clusters of the form $h_1(C_\Psi)$, we can choose different clusters C_1 from $Sl_1(Ch_\lambda(n))$ such that each C_1 differs from any C_Ψ and is isomorphic to the cluster C. We map every cluster C_1 onto C by h_1 and let the valuation S in C be the restriction of S to C_1. Given these definitions h_1 is a partial mapping from the model $Sl_1(Ch_\lambda(k))$ with the valuation S onto the model $Sl_1(Ch_\lambda(k))$ with the valuation S we introduced (transferred).

Finally extending the domain of h_1, we map by h_1 each cluster C_2 from the frame $Sl_1(Ch_\lambda(n))$, which is not yet included in the domain of h_1, onto the cluster $C_{2,\Psi}$ which is obtained from C_2 by contracting all elements which have the same valuation of letters under S (we map C_2 by means of this contraction). We know already that all such clusters $C_{2,\Psi}$ (as models with respect to S) belong to $Sl_1(Ch_\lambda(k))$. It is clear now that h_1 is a p-morphism from the model $Sl_1(Ch_\lambda(n))$ with the valuation S onto the model $Sl_1(Ch_\lambda(k))$ with the valuation S we introduced.

Suppose that we have already constructed a p-morphism h_i from the model $\langle S_i(Ch_\lambda(n)), R, S \rangle$ onto a model on the frame $S_i(Ch_\lambda(k))$ with the introduced (transferred) valuation S. We set h_{i+1} as identical to h_i on the set of elements with a depth of not more than i. Then we define h_{i+1} on elements of depth $i+1$ in the following way. Consider every antichain X of clusters from $S_i(Ch_\lambda(n))$ which has at least one cluster of depth i, and the set Y of all clusters C_2, where $C_2^{R<} = X^{R<} \cup X$. Then we define h_{i+1} on the clusters from Y in a similar way to the definition of h_1 on clusters of depth 1.

In this way we construct the p-morphisms h_i from the models $S_i(Ch_\lambda(n))$ with valuation S, onto the models $S_i(Ch_\lambda(k))$ with the valuation S for every $i \in N$. Finally, $\forall x \in Sl_j(Ch_\lambda(n))$, we put $h(x) := h_j(x)$. It is clear that the mapping h must be a p-morphism from $\langle Ch_\lambda(n), R, S \rangle$ onto $\langle Ch_\lambda(k), R, S \rangle$. Since p-morphisms preserve the truth of formulas, it follows that the valuation S in $Ch_\lambda(k)$ invalidates the rule r. ∎

Note that we can already now extract from this simple result an algorithm for determining the admissibility of inference rules for a number of representative infinite classes of modal and superintuitionistic logics.

Theorem 3.4.3 *For any modal logic λ extending $K4$ and any superintuitionistic logic λ such that λ has (i) the finite model property, (ii) a finite depth and (iii) a finite number of axioms, there is an algorithm which determines the admissibility of inference rules for λ. In particular, a rule r with k variables is admissible in λ iff r is valid in the finite frame of the k-characterizing model $Ch_\lambda(k)$.*

Proof. By Theorem 3.3.6 and Theorem 3.3.11 a rule r with k variables is admissible in λ iff r is valid in all models $Ch_\lambda(n)$ with respect to those valuations in $Ch_\lambda(n)$ which are expressible. All the models $Ch_\lambda(n)$ are finite because λ is a logic of finite depth and the number of their elements is effectively evaluated through n. For a modal logic λ of the mentioned above kind, all the elements from $Ch_\lambda(n)$ are definable (by Theorem 3.3.7). Similarly, for any superintuitionistic logic λ of the above mentioned kind, any sharp cone from $Ch_\lambda(n)$ is definable by Theorem 3.3.13. Using this, since the models $Ch_\lambda(n)$ are finite, we conclude that any valuation in $Ch_\lambda(n)$ is definable (i.e. expressible). Therefore we can specify our description of admissible rules as follows: a rule r with k variables is admissible in λ iff r is valid in every $Ch_\lambda(n)$ with respect to any valuation of variables from r. Applying Lemma 3.4.2 it follows that

$$r \text{ is admissible for } \lambda \Leftrightarrow Ch_\lambda(k) \models r,$$

which gives an algorithm for recognizing admissibility. Indeed, for finitely axiomatizable λ, all the models $Ch_\lambda(k)$ can be effectively constructed since in this case we can verify if a given finite frame is λ-frame. ∎

However in order to obtain a criteria for recognizing the admissibility of inference rules in logics of infinite depth we need to develop more powerful technique.

Definition 3.4.4 *Let λ be a modal or a superintuitionistic logic. Let m be a natural number. Suppose that there is a finite rooted λ-frame $\mathcal{F}$ of depth $k, m < k$, with an improper rooting cluster $C = \{a\}$ (i.e. the root cluster consists of a single element which is either reflexive or irreflexive) and with* *exactly d immediate successor clusters for C. Then we say that λ admits reflexive d-branching below m provided a is reflexive, otherwise we say λ admits irreflexive d-branching below m.*

If a frame (or a model) $\mathcal{F}$ has the form $\bigsqcup_{i \in I} \mathcal{F}_i$ we say $\mathcal{F}$ is decomposed in a disjoint union of frames (models) $\mathcal{F}_i$, and say that any $\mathcal{F}_i$ is a *component* of $\mathcal{F}$. If $\mathcal{F}$ is not decomposable into a disjoint union then the component of $\mathcal{F}$ is $\mathcal{F}$ itself. For example $Ch_{S4}(n)$ is not decomposable in a disjoint union and has a single component. For an example of the contrary, note that $Ch_{S4.2}(n)$ is decomposable into a disjoint union and has finitely many distinct components.

Definition 3.4.5 *We say that a logic λ with the fmp has the property of branching below m if the following holds. If λ admits reflexive (irreflexive) d-branching below m and $\mathcal{W}$ is a finite generated subframe of a component of $Ch_\lambda(k)$, where the depth of $\mathcal{W}$ is not less than m and $\mathcal{W}$ has n roots, where $n \leq d$, then the frame $\mathcal{W}_1$, which is obtained from $\mathcal{W}$ by adding a new single-element reflexive (irreflexive) root, is a λ-frame.*

Evident examples of logics with property of branching below 1 are $K4$, $S4$, GL, Grz, $S4.1$, $S4.2$, intuitionistic logic H, superintuitionistic logics KC and LC etc. Now we recall some necessary definitions connected with the dropping-points technique, which goes back to K.Fine [46]. The construction below is given for modal logics λ over $K4$, but it can be easy transferred onto superintuitionistic logics, in the case of superintuitionistic logics every cluster is merely a reflexive element etc. Let

(a) Y is a finite set of formulas closed with respect to subformulas.

(b) $\mathcal{W} := \langle W, R, V \rangle$ is a transitive model such that

    (b.1) $\mathcal{W}$ has no infinite ascending chains of clusters;

    (b.2) the domain of V includes all propositional letters having occurrences in formulas from Y;

    (b.3) every cluster from $\mathcal{W}$ has not more than 2^n elements, where n is the number of propositional letters in all formulas from Y.

For a subset A of W, we say an element $a \in W$ is *quasi-maximal* in A if $a \in A$ and $\forall x \in W, x \in a^{R<} \Rightarrow x \notin A$. We denote the set of all quasi-maximal elements of A by $qm(A)$. If $X \subseteq Y$ then

$$V_s(X) := \{a \mid a \in W, a \Vdash_V \bigwedge \{\alpha \mid \alpha \in X\}, \forall \beta \in (Y - X)(a \nVdash_V \beta)\}.$$

In other words, $V_s(X)$ is the set of all elements from $\mathcal{W}$ such that only members of X among formulas of Y are valid on these elements. For any subset X of Y, the set $X_{W,m}$ is defined as follows.

$$X_{W,m} := \{a \mid a \in (W - S_m(\mathcal{W})), a \in qm(V_s(X))\}.$$

That is, $X_{W,m}$ is the set of all elements of depth strongly more m which are quasi-maximal in $V_s(X)$.

We introduce now the submodel $max(Y, W, m)$ of the model W which consists of all clusters which include some elements of sets $X_{W,m}$ for all subsets X of Y (the accessibility relation and the valuation are transferred on $max(Y, W, m)$ from the original model W). Now we compound the model $S_m(W)$ and the model $max(Y, W, m)$ and obtain a submodel

$$max(Y, W, m) \bigcup S_m(W).$$

More precisely, the submodel $max(Y, W, m) \bigcup S_m(W)$ of the model W consists of all clusters from $max(Y, W, m) \bigcup S_m(W)$ with the accessibility relation and the valuation transferred from W. It is clear that $S_m(max(Y, W, m) \bigcup S_m(W)) = S_m(W)$. Also it is not hard to see that the model $max(Y, W, m) \bigcup S_m(W)$ has a finite depth which can be effectively evaluated (in fact, the depth is not more than $m + 2^v$, where v is the number of elements in Y).

The model $max(Y, W, m) \bigcup S_m(W)$ differs from the model W, in particular, in the point that we drop all enough deep elements from W. The main property of this model is the preservation of the truth for the formulas of Y. We show this in the following lemma which is due to K.Fine [46]).

Lemma 3.4.6 *Let $\mathfrak{M}$ be a submodel of the model W including the model $max(Y, W, m) \bigcup S_m(W)$. For any element $a \in |\mathfrak{M}|$ and any $\alpha \in Y$,*

$$a \Vdash_V \alpha \text{ in the model } \mathfrak{M} \Leftrightarrow a \Vdash_V \alpha \text{ in the model } W.$$

Proof. the proof is by induction on the length of the formula α. We begin with the case of modal formulas and modal Kripke models. For propositional letters, this lemma follows directly from the definitions. The inductive steps for non-modal connectives follow directly from the definition of the truth relation. If $a \Vdash_V \Box\gamma$ in the model W then by the inductive hypothesis we directly obtain $a \Vdash_V \Box\gamma$ in the model $\mathfrak{M}$. Suppose $a \nVdash_V \Box\gamma$ in W. Then there is an element $b \in W$ such that aRb and $b \nVdash_V \gamma$ in W. If b has depth m or less then $b \in |\mathfrak{M}|$ and using the inductive hypothesis we conclude $a \nVdash_V \Box\gamma$ in $\mathfrak{M}$. If b is a deeper element then we consider the set $X := \{\alpha \mid \alpha \in Y, b \Vdash_V \alpha\}$. According to our construction either (i) $C(b)$ contains a quasi-maximal element from $X_{W,m}$ or (ii) bRc, where $c \in X_{W,m}$. In case (i), if $C(b)$ is a degenerated cluster (i.e. $C(b)$ is simply an irreflexive element) we have $b \nVdash_V \gamma$, and $b \in |\mathfrak{M}|$. Then by the inductive hypothesis it follows $b \nVdash_V \gamma$ in $\mathfrak{M}$, and $a \nVdash_V \Box\gamma$ in $\mathfrak{M}$. If in case (i) $C(b)$ is not a degenerate cluster then bRc, where $c \in X_{W,m}$. In case (ii), also there exists a c such that bRc and $c \in X_{W,n}$. In both these cases we derive $c \in |\mathfrak{M}|$ and $c \nVdash_V \gamma$ in $\mathfrak{M}$ by the inductive hypothesis. Thus $a \nVdash_V \Box\gamma$ in $\mathfrak{M}$. This completes the proof for the modal case.

For the intuitionistic case, the only non-trivial steps are the cases concerning the connectives $\rightarrow$ and $\neg$. We will denote R in this case by $\leq$ because R is a partial order. If $a \Vdash_V (\gamma \rightarrow \delta)$ in $\mathcal{W}$ then by the inductive hypothesis we directly infer $a \Vdash_V (\gamma \rightarrow \delta)$ in $\mathfrak{M}$. Conversely, suppose that $a \nVdash_V (\gamma \rightarrow \delta)$ in $\mathcal{W}$. This implies there exists an element $b \in W$ such that $a \leq b$, $b \Vdash_V \gamma$ and $b \nVdash_V \delta$ in $\mathcal{W}$. If b has a depth of m or less then we obtain $b \in |\mathfrak{M}|$ and by the inductive hypothesis it follows $a \nVdash_V (\gamma \rightarrow \delta)$ in $\mathfrak{M}$. Suppose that b is a deeper element; consider the set $X := \{\alpha \mid \alpha \in Y, b \Vdash_V \alpha\}$. Then either (i) b is a quasi-maximal element from $X_{W,m}$ or (ii) $b \leq c$, where $c \in X_{W,m}$. In both these cases it follows that there is a $c \in |\mathfrak{M}|$ such that $c \Vdash_V \gamma$ and $c \nVdash_V \delta$ in $\mathcal{W}$. Then $c \Vdash_V \gamma$ and $c \nVdash_V \delta$ in $\mathfrak{M}$ by the inductive hypothesis. Since $a \leq c$, we conclude $a \nVdash_V \gamma \rightarrow \delta$ in $\mathfrak{M}$. The inductive step for $\neg$ can be established similarly. ∎

Thus if the model $\mathcal{W}$ invalidates certain formulas or inference rules then any submodel $\mathfrak{M}$ of the model $\mathcal{W}$ containing the submodel $max(Y, W, m) \bigcup S_m(\mathcal{W})$ will also invalidate these formulas and inference rules. However there is a problem in that sometimes, even if the model $\mathcal{M}$ is based upon a certain λ-frame for a logic λ, neither the submodel $max(Y, W, m) \bigcup S_m(\mathcal{W})$ nor all *good* submodels of $\mathcal{W}$ containing this submodel are based upon λ-frames. This leads us to refine our technique further.

Definition 3.4.7 *Let λ be a modal logic extending $K4$ or a superintuitionistic logic with the fmp. We say that λ has the* effective m-drop points property *if there is an effectively calculable (more precisely, a recursive) function $g(x, y)$ such that the following hold. For every finite set of formulas Y closed with respect to subformulas, and for each finite λ-model $\mathcal{W} := \langle Q, R, V \rangle$, such that*

> *(a) the domain of V includes the all propositional letters which occur in formulas from Y;*

> *(b) every cluster from $\mathcal{W}$ has not more than 2^n elements, where n is the number of propositional letters in all formulas from the set Y,*

the following hold. There exist a submodel $\mathfrak{M}$ of $\mathcal{W}$ containing the submodel $max(Y, W, m) \bigcup S_m(\mathcal{W})$ and a homomorphism f of the model $\mathfrak{M}$ onto a certain λ-model $\mathfrak{M}_1$ such that

> *(i) $\mathfrak{M}_1$ has not more than $g(v, u)$ elements, where v is the number of formulas in Y and u is the number of elements in $S_m(\mathcal{W})$.*

> *(ii) for any formula α from Y and for any $x \in |\mathfrak{M}|$, $x \Vdash_V \alpha$ in $\mathfrak{M}$ (or equivalently in $\mathcal{W}$, cf. Lemma 3.4.6) $\Leftrightarrow f(x) \Vdash_V \mathfrak{M}_1$.*

> *(iii) f is an isomorphism on $S_m(\mathcal{W})$ and f does not identify any element with a depth of strongly more than m with an element with a depth of not more than m.*

Thus the effective m-drop points property allows us, after dropping points from the original model W and obtaining the submodel $max(Y, W, m) \bigcup S_m(W)$, to extend this submodel to a submodel $\mathfrak{M}$ such that there is the required homomorphism (or p-morphism) from $\mathfrak{M}$ onto a λ-model $\mathfrak{M}_1$ which preserves the truth of the subformulas from Y. Moreover, the number of elements in $\mathfrak{M}_1$ is effectively evaluated by the upper part of W (the frame $S_m(W)$), and the number of subformulas from Y. This property, as well as the property of branching below m, will be actively employed in our later proofs and, in particular, in the proofs below of the two main lemmas of this section.

Note that, when the frame of the model $max(Y, W, m) \bigcup S_m(W)$ is a λ-frame, in order to show that λ has the effective m-drop points property, it is sufficient to find a p-morphism f of $max(Y, W, m) \bigcup S_m(W)$ on a λ-frame with an effectively bounded number of elements and such that f is an isomorphism on the set of all elements of depth not more than m. It is clear that this can be done for the modal logics $K4$, $S4$, Grz and GL by a condensing p-morphism. Therefore the logics $K4$, $S4$, Grz and GL have the effective m-drop points property. Some other examples will be given later.

Now we proceed to the proof of two crucial lemmas of this section which will give us some algorithms of a general kind for recognizing admissibility. In these following lemmas we consider only modal logics and modal Kripke models. In this part of the section it will be more convenient for us to consider modal logical connective $\Diamond$ (but not $\Box$) as the basic logical convective. This brings us no loss of generality since these modal connectives are mutually expressible. We also need the following notation and definition. For any frame (or model) $\mathcal{F}$, $d(\mathcal{F})$ denotes the depth of $\mathcal{F}$. Let $\mathcal{M}$ be a model and V be a valuation in this model. Let x be an element of $|\mathcal{M}|$ and Y be a set of formulas closed with respect to subformulas, and let all the propositional letters of formulas from Y be included in $Dom(V)$. We let

$$\Diamond(x, Y) := \{\alpha \mid \alpha \in Y, \Diamond\alpha \in Y, a \Vdash_V \Diamond\alpha\},$$

$$V(x, Y) := \{\alpha \mid \alpha \in Y, a \Vdash_V \alpha\}.$$

Let λ be a modal logic extending $K4$ and possessing the fmp. And also let $\mathcal{M} := \langle M, R, S \rangle = \bigsqcup M_j$ be a λ-model. Suppose that $Dom(S)$ is finite and Y is a finite set of formulas built up out of propositional letters from $Dom(S)$.

Definition 3.4.8 *We say that $\mathcal{M}$ has the* view-realizing *property for Y below m with respect to λ if the following hold.*

Let $\mathcal{F}$ be a finite generated subframe of a component M_j and $d(\mathcal{F}) \geq m$. Suppose that the set E of all minimal clusters from $\mathcal{F}$ consists of d clusters. Then if λ admits irreflexive d-branching below m then there exists an element x_E in the model M_j such that $m < d(x_E)$ and

$$\Diamond(x_E, Y) = \bigcup\{\Diamond(y, Y) \cup S(y, Y) \mid y \in E\}.$$

If λ admits reflexive d-branching below m then there exists an element x_E in the model M_j such that $m < d(x_E)$ and

$$\Diamond(x_E, Y) = \bigcup\{\Diamond(y, Y) \cup S(y, Y) \mid y \in E\} \cup S(x_E, Y).$$

Now we are in a position to prove our first main lemma. Let λ be a modal logic extending $K4$ such that λ has the fmp, the property of branching below m, and the effective m-drop points property. Suppose that g is an effectively calculable function for λ as given in the definition of the effective m-drop points property. And suppose that $r = \alpha_1, ..., \alpha_n/\beta$ is an inference rule with k variables, and $Sub(r)$ is the set of all subformulas of the rule r.

Lemma 3.4.9 *Suppose that there is a valuation S of k variables of the given inference rule r in the frame $\langle T, R \rangle$ of a certain special n-characterizing model $Ch_\lambda(n)$ for λ such that $\langle T, R, S \rangle$ invalidates r. Then there is a λ-model $\mathcal{M} := \langle M, R, S \rangle = \bigsqcup \mathcal{M}_j$ such that:*

(a) the frame $S_m(\mathcal{M})$ is isomorphic to the frame $S_m(Ch_\lambda(k))$;

(b) $\mathcal{M}$ has at most $g(e, q) * 2^{2^{k+1}}$ elements, where q is the number of subformulas of the rule r and e is the number of elements in $S_m(Ch_{K4}(k))$;

(c) the model $\mathcal{M}$ invalidates the rule r;

(d) the model $\mathcal{M}$ has the view-realizing property for $Sub(r)$ below m with respect to λ;

(e) each $\mathcal{M}_j$ is an open submodel of $Ch_\lambda(k)$.

Proof. By Lemma 3.4.2 there is a valuation S of k variables of the rule r in the model $Ch_\lambda(k)$ which invalidates the rule r. The model $Ch_\lambda(k)$ has a certain decomposition of the kind $Ch_\lambda(k) = \bigsqcup\{H_j \mid j \in J\}$. The number of elements in every component $S_m(H_j)$ is not more than the number of elements in $S_m(Ch_{K4}(k))$, and the number of components H_j in $Ch_\lambda(k)$ is not more than $2^{2^{k+1}}$. We choose in every H_j a certain finite submodel K_j as follows. For every $Z \subseteq Sub(r)$, we choose and fix in H_j some single element x_Z of depth more than m such that $S(x_Z, Sub(r)) = Z$ (if such elements exist for given Z). Then we take the open submodel K_j of H_j generated by $S_m(H_j)$ and all x_Z. It is clear that every K_j and $\bigsqcup K_j$ are finite, and the valuation S invalidates the rule r in the model $\bigsqcup K_j$.

Since λ has the effective m-drop points property, for each j, there exist a submodel G_j of the model K_j containing the set $max(Sub(r), K_j, m) \cup S_m(K_j)$ and a homomorphism f_j from G_j onto a λ-model D_j with the properties mentioned in the definition of the effective m-drop points property. We fix some G_j and D_j for each j and put

$$\mathcal{M}_j := D_j, \qquad \mathcal{W}_1 := \bigsqcup \{\mathcal{M}_j \mid j \in J\}.$$

By (ii) from the definition of the effective m-drop points property (abbreviation e-m-dpp) we have $S_m(D_j) = S_m(H_j)$, and part (a) of our lemma holds for $\mathcal{W}_1$. Part (b) holds for the model $\mathcal{W}_1$ by (i) from the definition of e-m-dpp. And part (c) holds for the model $\mathcal{W}_1$ by the choice of $\mathcal{W}_1$ and (ii) from the definition of e-m-dpp.

To prove part (d) suppose that $\mathcal{F}$ is a generated subframe of D_j, $m \le d(\mathcal{F})$ and $\mathcal{F}$ has d minimal clusters. We denote the antichain of minimal clusters of $\mathcal{F}$ by E. Suppose that λ admits irreflexive (reflexive) d-branching below m. For each cluster C from E, we pick a single representative z from C. We denote by U the set which consists of all these representatives z. For every $z \in U$, we pick a single cluster from G_j which contains an element from $f_j^{-1}(z)$. We denote the set of all such clusters by $U(f_j^{-1})$. It is clear that $U(f_j^{-1})$ is an antichain with d clusters (since f_j preserves the accessibility relation). Moreover, some such cluster has depth at least m by part (iii) of the definition of e-m-dpp.

Consider the generated subframe $U(f_j^{-1})^{R\le} \cup U(f_j^{-1})$ of the frame H_j. Since λ has the property of branching below m, and by our assumption λ admits irreflexive (reflexive) d-branching below m, there is a λ-frame $\mathcal{A}$ with a single-point irreflexive (reflexive) root cluster c such that the frame $c^{R<}$ is isomorphic to the frame $U(f_j^{-1})^{R\le} \cup U(f_j^{-1})$. Therefore by our construction of the models $Ch_\lambda(k)$, there exists an irreflexive (reflexive) element $y \in H_j$ such that $y^{R<}$ is exactly the set $U(f_j^{-1})^{R\le} \cup U(f_j^{-1})$. Then y has depth at least $m + 1$ (more than the depth of the clusters in $U(f_j^{-1})$). It is easy to see that the following equations hold: if y is irreflexive then with respect to the valuation S

$$\Diamond(y, Sub(r)) = \bigcup \{\Diamond(f_j^{-1}(z), Sub(r)) \mid z \in U\} \cup$$

$$\bigcup \{S(f_j^{-1}(z), Sub(r)) \mid z \in U, \} \tag{3.7}$$

if y is reflexive then with respect to the valuation S respectively

$$\Diamond(y, Sub(r)) = \bigcup \{\Diamond(f_j^{-1}(z), Sub(r)) \mid z \in U\} \cup$$

$$\bigcup \{S(f_j^{-1}(z), Sub(r)) \mid z \in U\} \cup S(y, Sub(r)). \tag{3.8}$$

Note that by choice of K_j there is an y_1 in K_j of depth more than m and such that $S(y_1, Sub(r)) = S(y, Sub(r))$. Therefore the model G_j has a certain element y_2 of depth more m such that $S(y_2, Sub(r)) = S(y, Sub(r))$. Besides we have that

$S(f_j(y_2), Sub(r)) = S(y, Sub(r))$ by (ii) from the definition of e-m-dpp and for each $z \in U$, $S(f_j^{-1}(z), Sub(r)) = S(z, Sub(r))$ (also by (ii) from the definition of e-m-dpp). We define $x := f_j(y_2)$. Then by (iii) from the definition of e-m-dpp, x has depth at least $m+1$, and using (3.7) and (3.8) we conclude that $x := f(y_1)$ satisfies the necessary property to be view realizing for $Sub(r)$ from (d). Thus part (d) of this lemma holds for $\mathcal{W}_1$. Now we make a corresponding condensing p-morphism of $\mathcal{W}_1$ starting with depth $m+1$ and obtain a model $Cg(\mathcal{W}_1)$ which satisfies (e). Let $\mathcal{M} := Cg(\mathcal{W}_1)$. Conditions (a) and (b) are evidently satisfied for $\mathcal{M}$. Conditions (c) and (d) are satisfied by Lemma 3.4.1. ∎

Now we proceed to our second crucial lemma. Because the proof has a constructive form and is long, we will divide it into several separate lemmas.

Lemma 3.4.10 *Let λ be a modal logic extending $K4$ with the fmp, the property of branching below m and the effective m-drop points property. Suppose that there exists a λ-model $\mathcal{M}$ meeting conditions (a) - (d) from Lemma 3.4.9. Then there is an effectively contractible definable valuation S of the variables from the rule r in the frame of the model $Ch_\lambda(k)$, where k is the number of variables of r, such that S invalidates r in $Ch_\lambda(k)$.*

Proof. Given an λ-model $\mathcal{M} = \langle M, R, S \rangle = \bigsqcup \mathcal{M}_j$ which satisfies properties (a) - (d) from Lemma 3.4.9. The model $Ch_\lambda(k)$ has a decomposition of the kind $Ch_\lambda(k) := \bigsqcup H_j$. Then, according to (a) and (e) from Lemma 3.4.9, we can assume that $S_m(H_j) = S_m(\mathcal{M}_j)$ (as frames) and that the frame of the model $\mathcal{M}_j$ is a generated subframe of the frame of H_j.

Without loss of generality we can consider all elements of $|\mathcal{M}|$ as some elements of $Ch_\lambda(k)$, and we will follow this agreement for the rest of our proof. According to Theorem 3.3.7 each element of $Ch_\lambda(k)$ is definable (i.e. expressible). Therefore we can effectively construct the definable valuation S of variables from r in the frame of $|\mathcal{M}|$, as a generated subframe of $Ch_\lambda(k)$, which coincides with the original valuation S in $\mathcal{M}$. Then by (c) from Lemma 3.4.9 the rule r is falsified by this S in the model $\mathcal{M}$. The main part of the proof of this lemma will consists of an extension of the valuation S onto the whole frame of $Ch_\lambda(k)$, under which the premise of r will be valid in $Ch_\lambda(k)$. Every component H_j has the corresponding component $\mathcal{M}_j$ in the model M, and the frame of $\mathcal{M}_j$ is an open subframe of H_j, as it is depicted in the picture Fig 3.1.

We fix some $Q = H_j$ and extend S from $\mathcal{M}_j$ onto whole the frame Q (recall that j is fixed) in the way described below. First we proceed to a construction of a sequence of subsets $\Sigma(x, t)$ of the frame Q, where x are all elements from $|\mathcal{M}_j|$, $0 \leq t \leq m_1$, where m_1 is the number of elements in $|\mathcal{M}_j|$.

Our sequence $\Sigma(x, t)$, $x \in |\mathcal{M}_j|$, $0 \leq t \leq m_1$ will have the following properties:

(a1) $\forall x_i \in |\mathcal{M}_j|(\Sigma(x, t) \subseteq \Sigma(x, t+1))$,
   $\forall x_i, x_j \, (x_i \neq x_j \Rightarrow \Sigma(x_i, t) \cap \Sigma(x_j, t) = \emptyset)$;

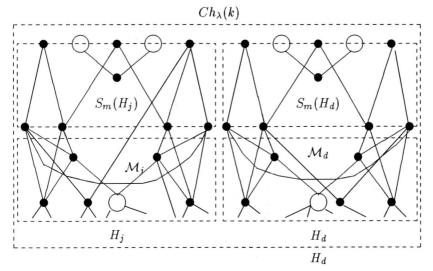

Figure 3.1: Fragment of $Ch_\lambda(k)$

(b1) $\forall \Sigma(x, t)$ there exists a modal formula $\alpha(x, t)$ such that under the original valuation V in the model $Q = H_j$
$$V(\alpha(x, t)) := \{y \mid y \in |Q|,\ y \Vdash_V \alpha(x, t)\} = \Sigma(x, t);$$

(c1) $\bigcup_{x_i \in M_j} \Sigma(x_i, t)$ forms a generated (i.e. open) subframe of the frame Q for any given t;

(d1) Let the valuation S of propositional letters from r in the frame $\bigcup\{\Sigma(x, t) \mid x \in |M_j|\}$ be the following: $\forall p \in Var(r),\ S(p) := \bigcup\{\Sigma(x, t) \mid x \in |M_j|, x \Vdash_S p\}$ in the model $M_j\}$. Then $\forall a \in \Sigma(x, t), \forall \alpha \in Sub(r)$ $(a \Vdash_S \alpha) \Leftrightarrow (x \Vdash_S \alpha$ in the model M_j);

(e1) $\forall t, \forall y \in |Q|$, if $(y \notin \bigcup\{\Sigma(x, t) \mid x \in M_j\}$
$\Rightarrow \exists\ t+1$-element subset W of $|M_j|$ such that $\forall x \in W$
$(y^{R<} \cap \Sigma(x, t) \neq \emptyset)$;

(f1) If $y \in (\Sigma(x, t) - \Sigma(x, 0))$ then the depth of x in M_j is at least m.

The first step of our construction of this sequence of sets is the following: we choose for all $x \in |M_j|$, $\Sigma(x, 0) := \{x\}$. It is easy to see that all mentioned above properties (a1) - (e1) are valid for $t = 0$.

Now suppose that the sets $\Sigma(x, \mu), x \in |M_j|$, where $1 \leq \mu \leq t$, are already constructed and that all properties (a1) - (e1) hold. Consider all $D \subseteq |M_j|$, where D consists of $t + 1$ elements and has at least one element with depth not less m. For each such D, we denote by $C(D)$ the set of all clusters from the frame of M_j which contain elements of D. It is not difficult to see that there are antichains E of elements (i.e. sets elements which are incomparable by the accessibility relation) of these clusters which have the following property (recall that the model $\mathcal{M}_j$ has the valuation S of letters from $Var(r)$). E has an element with depth at least m and

$$\bigcup\{\Diamond(y, Sub(r)) \cup S(y, Sub(r)) \mid y \in E\} = \tag{3.9}$$

$$\bigcup\{\Diamond(y, Sub(r)) \cup S(y, Sub(r)) \mid y \in D\}.$$

Indeed, the set E which consists of some representative elements of the antichain of all minimal clusters of $C(D)$ satisfies (3.9). For every mentioned above D, we choose and fix some antichain E which satisfies (3.9) and has minimal number of elements. Now we consider every chosen E. Assume that a given E has d elements. For any such E, by (d) from Lemma 3.4.9, if λ admits irreflexive (reflexive) d-branching below m then there is a $x_{E,i} \in |\mathcal{M}_j|$ (there is a $x_{E,r} \in \mathcal{M}_j$ correspondingly) with the properties described in Lemma 3.4.9 concerning the definition of view realizing property:

$$\Diamond(x_{E,i}, Sub(r)) = \bigcup\{\Diamond(y, Sub(r)) \cup S(y, Sub(r)) \mid y \in E\}, \tag{3.10}$$

$$\Diamond(x_{E,r}, Sub(r)) = \bigcup\{\Diamond(y, Sub(r)) \cup S(y, Sub(r)) \mid y \in E\}\cup \tag{3.11}$$

$$\cup S(x_E, Sub(r)).$$

We fix some single $x_{E,i}$ and some single $x_{E,r}$ (if they exist, of course). Now we introduce the formulas $\varphi(D)$ and $\chi(D)$ as follows. Let

$$q(D) := \bigwedge_{x \in D} \Diamond\alpha(x, t) \wedge \bigwedge_{x \notin D, x \in \mathcal{M}_j} \neg\Diamond\alpha(x, t) \wedge \neg(\bigvee_{x \in \mathcal{M}_j} \alpha(x, t)).$$

If $x_{E,i}$ and $x_{E,r}$ do not exist then $\varphi(D) := \bot$ and $\chi(D) := \bot$. If $x_{E,i}$ exists and

$$S(x_{E,i}, Sub(r)) \subseteq \bigcup\{\Diamond(x, Sub(r)) \cup S(x, Sub(r)) \mid x \in E\} \tag{3.12}$$

then

$$\varphi(D) := q(D) \wedge \Box(q(D) \vee (\bigvee_{x \in D} \alpha(x, t))), \chi(D) := \bot$$

and the definition is complete. If $x_{E,i}$ exists but (3.12) does not hold then

$$\varphi(D) := q(D) \wedge \Box(\bigvee_{x \in D} \alpha(x, t)).$$

If $x_{E,i}$ does not exist then $\varphi(D) := \perp$. Finally if the element $x_{E,r}$ exists (which is necessary in the latter case) then we put

$$\chi(D) := q(D) \wedge \Diamond q(D) \wedge \beta(D), \text{ where}$$

$$\beta(D) := \Box(\neg(\bigvee_{x \in \mathcal{M}_j} \alpha(x,t)) \rightarrow \Diamond q(D)) \wedge \Box(q(D) \vee (\bigvee_{x \in D} \alpha(x,t))).$$

Otherwise we put $\chi(D) := \perp$. This completes the definition (by the by, note that the formula $q(D)$ reflects the concept *being of span* for D and $\chi(D)$, $\varphi(D)$ reflect the concept *being of minimal span for D*).

Now we introduce the sets $\Sigma(x, t+1)$ for $x \in |\mathcal{M}_j|$ in the following way: if x is irreflexive then

$$\Sigma(x, t+1) := \Sigma(x,t) \cup \bigcup_{x_{E,i}=x} V(\varphi(D)).$$

If otherwise (x is reflexive) then

$$\Sigma(x, t+1) := \Sigma(x,t) \cup \bigcup_{x_{E,r}=x} V(\chi(D))$$

It is clear that $\Sigma(x, t+1) = V(\alpha(x,t+1))$, where

$$\alpha(x, t+1) := \alpha(x,t) \vee \bigvee_{x_{E,i}=x} \varphi(D) \vee \bigvee_{x_{E,r}=x} \chi(D).$$

Thus (b1) is true for the index $t+1$. It is clear also that $\Sigma(x,t) \subseteq \Sigma(x,t+1)$. The sets $V(\varphi(D))$ and $V(\chi(D))$ adjoined to the old sets are disjoint with the old sets by the presence of the last conjunct of the formula $q(D)$. When the sets D_1 and D_2 are distinct, the sets $V(\varphi(D_1))$ and $V(\varphi(D_2))$ are disjoint by the formulas $q(D_1)$ and $q(D_2)$. The same is true for the sets $V(\chi(D_1))$ and $V(\chi(D_2))$. It can be easily seen that the sets $V(\chi(D))$ and $V((\varphi(D))$ are disjoint as well. Thus the property (a1) holds for the sets with the index $t+1$.

We turn to show that (c1) holds for the sets of level $t+1$. If $x \in \bigcup \Sigma(x,t)$ and $x R y$ then y is a member of the set $\bigcup \Sigma(x,t)$ as well (by (c1) for t), and therefore y is included in the new sets. Now let $x R y$ and $x \in (\bigcup \Sigma(x, t+1) - \bigcup \Sigma(x,t))$. Then $x \Vdash_V \varphi(D) \vee \chi(D)$ for some D. If $y \notin \bigcup \Sigma(x,t)$ then it is easy to see that $y \Vdash_V \varphi(D) \vee \chi(D)$ and y is included in the constructed already sets of kind $\Sigma(x, t+1)$ as well. Thus (c1) holds for $t+1$. Now our aim is to show that (d1) holds for the sets of the kind $\Sigma(x, t+1)$. We do this in the following

Lemma 3.4.11 *Suppose the valuation S of propositional letters from r in* $\bigcup \{\Sigma(x, t+1) \mid x \in |\mathcal{M}_j|\}$ *is given by*

$$\forall p \in Var(r), S(p) := \bigcup \{\Sigma(x, t+1) \mid x \in |\mathcal{M}_j|, x \Vdash_S p\}$$

in the model $\mathcal{M}_j$ }.

Then the following holds

$$\forall a \in \Sigma(x, t+1), \forall \alpha \in Sub(r)[(a \Vdash_S \alpha) \Leftrightarrow (x \Vdash_S \alpha \text{ in the model } \mathcal{M}_j)].$$

Proof. For the case when $a \in \Sigma(x, t)$, this lemma follows from the inductive hypothesis for t. Now suppose that $a \in (\Sigma(x, t+1) - \bigcup \Sigma(x_1, t))$. Then we have $x = x_{E,\alpha}$, where $\alpha = i, r$ and

$$\exists D(a \Vdash_V \varphi(D) \vee \chi(D)) \tag{3.13}$$

We shall carry out the proof of our lemma for this element a by induction on the length of the subformulas α from $Sub(r)$. If a formula α is a variable then this lemma holds directly by the definition of the valuation S. The inductive steps for all nonmodal logical connectives are obvious.

We proceed to the case of modal logical connective $\Diamond$. Let $a \Vdash_S \Diamond \alpha$ in the model $\bigcup \{\Sigma(x, t+1) \mid x \in |\mathcal{M}_j|\}$. Then there is a b such that aRb and $b \Vdash_S \alpha$. If $b \notin \bigcup \Sigma(x_1, t)$ then by (3.13) $b \Vdash_V \varphi(D) \vee \chi(D)$ and b is included in the same set $\Sigma(x, t+1)$ with a. Then by the inductive hypothesis we have $x \Vdash_S \alpha$. If x is reflexive then $x \Vdash_S \Diamond \alpha$ which is what we need. Suppose that x is irreflexive. Then it follows $x = x_{E,i}$. If

$$S(x_{E,i}, Sub(r)) \subseteq \bigcup \{\Diamond(y, Sub(r)) \cup S(y, Sub(r)) \mid y \in E\} \tag{3.14}$$

holds then we have $\alpha \in \Diamond(y, Sub(r))$ or $\alpha \in S(y, Sub(r))$ for some $y \in E$. Then $x \Vdash_S \Diamond \alpha$, (again by (3.10)) what we needed. If (3.14) is false then, because $a \Vdash_V \varphi(D)$ and, in this case, $\varphi(D) = q(D) \wedge \Box(\bigvee_{x \in D} \alpha(x, t))$, b should be an element of $\Sigma(x_1, t)$, which contradicts our assumption that $b \notin \bigcup \Sigma(x_1, t)$.

Assume now that $b \in \bigcup \Sigma(x_1, t)$. By (3.13) $b \Vdash_V \alpha(x_l, t)$ for some $x_l \in D$. Then by the inductive hypothesis $x_l \Vdash_S \alpha$. It can be easily seen from (3.9) that in this case $\alpha \in \bigcup \{\Diamond(y, Sub(r)) \cup S(y, Sub(r)) \mid y \in E\}$. By the choice of $x_{E,\alpha}$ ($\alpha = i, r$) and (3.10) with (3.11), we have $\alpha \in \Diamond(x_{E,\alpha}, Sub(r))$, that is $x \Vdash_S \Diamond \alpha$, which is again what we needed. Thus we have proved that $a \Vdash_S \Diamond A \Rightarrow x \Vdash_S \Diamond A$ in the model $\mathcal{M}_j$.

Conversely, let $x \Vdash_S \Diamond \alpha$ and $x = x_{E,\alpha}$, where $\alpha = i, r$. First possible case is $x \Vdash_S \alpha$. Suppose that x is reflexive. Recall that by our assumption we have $a \in (\Sigma(x, t+1) - \bigcup \Sigma(x_1, t))$. That is, since x is reflexive (i.e., in particular, $x = x_{E,r}$), we get $a \Vdash \chi(D)$. Using the structure of the formula $\chi(D)$ it follows that there exists an element b such that aRb and $b \Vdash_V \chi(D)$, i.e. $b \in \Sigma(x, t)$. Then by the inductive hypothesis we infer $b \Vdash_S \alpha$ and consequently $a \Vdash_S \Diamond \alpha$. Other possible case is that $x \nVdash_S \alpha$ and xRy, where $y \Vdash_S \alpha$. Then by (3.10) and (3.11)

$$\alpha \in \bigcup \{\Diamond(y, Sub(r)) \cup S(y, Sub(r)) \mid y \in E\}.$$

Page 323 - The text on the left margin was inadvertently omitted during the printing process. For the missing text please consult this page.

That is, there is a certain $y \in E$ such that $y \Vdash_S \Diamond\alpha$ or $y \Vdash_S \alpha$. This yields there is an element $x_l \in D$ such that $x_l \Vdash_S \Diamond\alpha$ or $x_l \Vdash_S \alpha$. Then, by the choice of $q(D)$, aRb for some b such that $b \in \Sigma(x_l, t)$. By the inductive hypothesis we derive $b \Vdash_S \Diamond\alpha \vee \alpha$, which implies $a \Vdash_S \Diamond\alpha$. Hence the proof of our lemma is complete. ∎

We continue the proof of Lemma 3.4.10. By Lemma 3.4.11 the property (d1) holds for the sets $\Sigma(x, t+1)$ of level $t+1$. Now we prove that (e1) holds as well. Assume that

$$y \in (Q - \bigcup\{\Sigma(x, t+1) \mid x \in \mathcal{M}_j\}). \tag{3.15}$$

Then, in particular, y does not belong to the sets which we have constructed on step t. By (e1) for t, this means that there exists a $(t+1)$-element set D of elements from $|\mathcal{M}_j|$ such that

$$\forall x \in D[y^{R<} \cap \Sigma(x, t) \neq \emptyset].$$

Thus if there exists a $x_u \in (\mathcal{M}_j - D)$ such that $y^{R<} \cap \Sigma(x_u, t+1) \neq \emptyset$ then the conjecture (e1) holds for y. Suppose that there are no such elements x_u. Then it can be easily seen that $y \Vdash_V q(D)$. We consider all elements v from maximal clusters such that these v has property: yRv and $v \notin \bigcup \Sigma(x, t)$. If there are no such elements then we put $v = y$. Then for every v, $v \Vdash_V q(D)$ by (e1) for case t.

Lemma 3.4.12 *If there is a reflexive (irreflexive) v then there exists an $x_{E,r}$ ($x_{E,i}$, correspondingly).*

Proof. Let $G = \{z_1, ..., z_l\}$ be a set of representatives of all immediate successor clusters for $C(v)$. Note that v sees the same sets of the kind $\Sigma(g, t)$ as y. The frame $[\{v\}^{R<} - C(v)]$ has depth at least m. Therefore the logic λ admits reflexive (irreflexive) l-branching below m. Moreover, by choice of v, each $z \in G$ is an element of some $\Sigma(x_z, t)$, $x_z \in |\mathcal{M}_j|$. The picture illustrating the distribution of elements in the structure of the frame is depicted in Fig. 3.2. If all elements of G are included in $S_m(Q)$ then some x_z has depth m. Otherwise some $z \in G$ is included in $(\Sigma(x_z, t) - \Sigma(x_z, 0))$.

Then by (f1) for the index t, x_z has depth at least m. Thus, in any case, there is an x_z which has in $\mathcal{M}_j$ depth at least m. Let H be the set of all R-minimal x_z in $\mathcal{M}_j$. Then using property (d1) for the sets of level t we infer that H has property of the set E from (3.9) with respect to D. The set H has at most l elements, therefore the set E has at most l elements as well. By the property of l-branching below m and (d) from Lemma 3.4.9 we obtain that $x_{E,r}$ ($x_{E,i}$, correspondingly) exists. ∎.

We continue the proof of Lemma 3.4.10. Elements v are defined above. Assume that there are no irreflexive v. Then by Lemma 3.4.12 there is an $x_{E,r}$, and

'hat is, there is a certain $y \in E$ such that $y \Vdash_S \Diamond\alpha$ or $y \Vdash_S \alpha$. This yields lere is an element $x_l \in D$ such that $x_l \Vdash_S \Diamond\alpha$ or $x_l \Vdash_S \alpha$. Then, by the choice f $q(D)$, aRb for some b such that $b \in \Sigma(x_l, t)$. By the inductive hypothesis we erive $b \Vdash_S \Diamond\alpha \vee \alpha$, which implies $a \Vdash_S \Diamond\alpha$. Hence the proof of our lemma is ›mplete. ∎

We continue the proof of Lemma 3.4.10. By Lemma 3.4.11 the property (d1) olds for the sets $\Sigma(x, t+1)$ of level $t+1$. Now we prove that (e1) holds as well. .ssume that

$$y \in (Q - \bigcup\{\Sigma(x, t+1) \mid x \in \mathcal{M}_j\}). \tag{3.15}$$

'hen, in particular, y does not belong to the sets which we have constructed n step t. By (e1) for t, this means that there exists a $(t+1)$-element set D of .ements from $|\mathcal{M}_j|$ such that

$$\forall x \in D[y^{R<} \cap \Sigma(x, t) \neq \emptyset].$$

'hus if there exists a $x_u \in (\mathcal{M}_j - D)$ such that $y^{R<} \cap \Sigma(x_u, t+1) \neq \emptyset$ then the ›njecture (e1) holds for y. Suppose that there are no such elements x_u. Then can be easily seen that $y \Vdash_V q(D)$. We consider all elements v from maximal .usters such that these v has property: yRv and $v \notin \bigcup\Sigma(x, t)$. If there are no ıch elements then we put $v = y$. Then for every v, $v \Vdash_V q(D)$ by (e1) for case

emma 3.4.12 *If there is a reflexive (irreflexive) v then there exists an $x_{E,r}$ $v_{E,i}$, correspondingly).*

'roof. Let $G = \{z_1, ..., z_l\}$ be a set of representatives of all immediate successor .usters for $C(v)$. Note that v sees the same sets of the kind $\Sigma(g, t)$ as y. The ·ame $[\{v\}^{R<} - C(v)]$ has depth at least m. Therefore the logic λ admits reflexive rreflexive) l-branching below m. Moreover, by choice of v, each $z \in G$ is an .ement of some $\Sigma(x_z, t), x_z \in |\mathcal{M}_j|$. The picture illustrating the distribution f elements in the structure of the frame is depicted in Fig. 3.2. If all elements of '· are included in $S_m(Q)$ then some x_z has depth m. Otherwise some $z \in G$ is ıcluded in $(\Sigma(x_z, t) - \Sigma(x_z, 0))$.

Then by (f1) for the index t, x_z has depth at least m. Thus, in any case, there an x_z which has in $\mathcal{M}_j$ depth at least m. Let H be the set of all R-minimal $_z$ in $\mathcal{M}_j$. Then using property (d1) for the sets of level t we infer that H has roperty of the set E from (3.9) with respect to D. The set H has at most l lements, therefore the set E has at most l elements as well. By the property f l-branching below m and (d) from Lemma 3.4.9 we obtain that $x_{E,r}$ ($x_{E,i}$, ›rrespondingly) exists. ∎

We continue the proof of Lemma 3.4.10. Elements v are defined above. As-ıme that there are no irreflexive v. Then by Lemma 3.4.12 there is an $x_{E,r}$, and

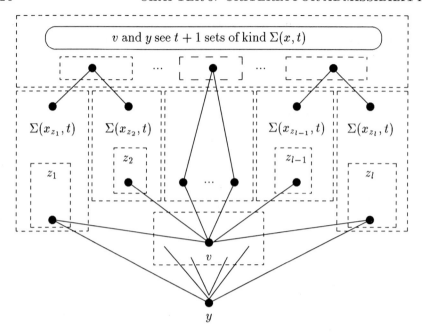

Figure 3.2: Placement of v

$y \Vdash_V \chi(D)$ that contradicts $y \in (Q - \bigcup\{\Sigma(x, t+1) \mid x \in |\mathcal{M}_j|\})$. Suppose that there is an irreflexive v. Then by Lemma 3.4.12 there is an $x_{E,i}$. Assume (3.12) holds. Then $y \Vdash_V \varphi(D)$, a contradiction. Assume (3.12) does not hold. Then $v \Vdash_V \varphi(D)$. If $v = y$ we have a contradiction. Another possible case is: yRv and $v \in \Sigma(x_v, t+1)$, where $S(x_v, Sub(r)) \not\subseteq \bigcup\{\Diamond(u, Sub(r)) \cup S(u, Sub(r)) \mid u \in E\} = \{\Diamond(u, Sub(r)) \cup S(u, Sub(r)) \mid u \in D\}$. The last means that $x_v \notin D$. Thus the property (e1) for the case of the sets of level $t+1$ holds.

If x is included in one of the sets $(\Sigma(x, t+1) - \Sigma(x, t))$ then $x := x_{E,\alpha}$, where the latter element $x_{E,\alpha}$ from $|\mathcal{M}_j|$ is chosen according to (d) from Lemma 3.4.9. This means, in particular, that the depth of this element is at least $m+1$. Thus (f1) holds for the sets of level $t+1$ as well.

In this way we construct the sequence of the sets $\Sigma(x, t), x \in |\mathcal{M}_j|$ and $1 \leq t \leq m_1$. Now note that if an element $v \in Q$ was not included in the constructed sets of the kind $\Sigma(x, t)$ until the step $m_1 - 1$ then v will be included in one of such sets constructed on step m_1. Indeed, if $y \notin \bigcup \Sigma(x, m_1 - 1)$ then $y \Vdash_V q(|\mathcal{M}_j|)$ by (e1) for the index $m_1 - 1$. Let z be an element from a R-maximal cluster C with the property:

$$(\exists c \in C)[(yRc)\&(c \notin \bigcup \Sigma(x, m_1 - 1)].$$

(if there exists such C). Then, in particular, yRz and $z \notin \bigcup \Sigma(x, m_1 - 1)$. If there is no such C we put $z = y$. Then $z \Vdash_V q(|\mathcal{M}_j|)$ by (e1) for the step $m_1 - 1$. Let $D := |\mathcal{M}_j|$. If z is reflexive then by Lemma 3.4.12 there is an $x_{E,r}$ and $y \Vdash_V \chi(D)$ as it required. Suppose z is irreflexive. Then by Lemma 3.4.12 there is an $x_{E,i}$ and of course (3.12) holds. Then $y \Vdash_V \varphi(D)$, which is what we needed. Hence we get $\bigcup \Sigma(x, m_1) = Q$.

Thus, in the way described, we have extended the valuation S onto the whole frame Q. We make in the way described above such extension for all H_j and $\mathcal{M}_j$. By (d1) for the sets of level m_1 and (c) from Lemma 3.4.9, it follows that the valuation S invalidates the inference rule r in a certain frame H_j, and that the premise of r is valid under S in every H_j. We let S_j be the expressible valuation S on H_j that we constructed. Then it follows $S_j(p) := V_j(\beta_{j,p})$, where $p \in Var(r)$, $\beta_{j,p}$ is the corresponding expressing formula. If $H_l \neq H_j$ then these models have no common maximal clusters. By Lemma 3.3.7 each element of $Ch_\lambda(k)$ is expressible. Therefore the set $S_1(H_l)$ of all elements of all maximal clusters from the model H_l is expressible by a formula δ_l for each index l. Finally, we choose the valuation S in $Ch_\lambda(k)$ as follows:

$$\forall p \in Var(r)(S(p) := V(\bigvee_j (\beta_{j,p} \wedge \bigwedge_{l \neq j} \neg(\delta_l \vee \Diamond \delta_l)))).$$

Then it can be easily seen that this expressible valuation S invalidates the inference rule r in $Ch_\lambda(k)$. ∎

3.5 Criteria for Determining Admissibility

Using results of the previous section, we are in a position now to construct algorithms for recognizing admissibility of inference rules in modal and superintuitionistic logics. We also give certain semantic criteria for admissibility which have no pure algorithmic nature but which are often useful when it comes really verifying the admissibility of given rules. We begin immediately with the following general theorems which follow from the technical lemmas which we have proved in the previous section.

Theorem 3.5.1 Semantic criterion for admissibility. *Let λ be a certain modal logic extending $K4$ and possessing (i) the fmp, (ii) the property of branching below m, and (iii) the effective m-drop points property. An inference rule r with k variables is admissible in λ if and only if r is valid in the frame of the model $Ch_\lambda(k)$ with respect to any valuation.*

Proof. First assume that r is not admissible in λ. Then Theorem 3.3.3 and Theorem 3.3.6 imply that there is a valuation of variables from r which invalidates

r in the frame of some model $Ch_\lambda(n)$. Then by Lemma 3.4.2 there is a valuation which invalidates r in $Ch_\lambda(k)$. Conversely, suppose there exists a valuation which invalidates r in the frame of $Ch_\lambda(k)$. Then by Lemma 3.4.9 there is a model $\mathcal{M} = \langle M, R, S \rangle$ which satisfies the conditions (a) - (e) of that lemma. Applying Lemma 3.4.10 there is a definable valuation of variables from r in the frame of $Ch_\lambda(k)$ which invalidates r. By Theorem 3.3.3 it follows that r is not admissible in λ. ∎

Theorem 3.5.2 Algorithmic Criterion for admissibility. *Let λ be a modal logic having the fmp, extending $K4$ and such that*

(i) λ has the property of branching below m;

(ii) λ has the effective m-drop points property;

(iii) there is an algorithm which determines whether a given finite frame is λ-frame (which holds if, in particular, λ is finitely axiomatizable).

Then there is an algorithm which determines the admissibility of inference rules in λ.

Proof. Assume we are given an inference rule r with k variables. We verify whether there is a certain λ-model $\langle M, R, S \rangle$ satisfying conditions (a) - (e) from Lemma 3.4.9 with at most $g(e, q) * 2^{2^{k+1}}$ elements (where g is a recursive function from the definition of the effective m-drop points property; e is the number of elements in $|S_m(Ch_{K4}(k))|$; and q is the number of subformulas in the rule r). If a certain such model exists then using Lemma 3.4.10 there exists a definable valuation S with effectively contrastable defining formulas which invalidates the rule r in the frame of $Ch_\lambda(k)$. Therefore by Theorem 3.3.3 the inference rule r is not admissible in λ.

Suppose that we discovered that such model with properties required above does not exist. We claim that then the inference rule r is admissible in λ. Indeed, assume that r is not admissible. By Theorem 3.3.3 and Lemma 3.3.6 there is a valuation of the variables from r in the frame of $Ch_\lambda(k)$ which invalidates r. Then Lemma 3.4.9 entails that the model with properties described above exists, a contradiction. ∎

We now intend to apply these two theorems to certain individual logics and whole classes of modal and superintuitionistic logics. The nature of the property of branching below m is intuitively clear. It is possibility of branching of the elements of a frame which are deep enough, with a depth of more than m. Moreover, for individual logics with the fmp, we usually know descriptions of all their finite frames, and we can easily determine whether a given logic has this property. The m-drop points property is also simple, but it is not always evident whether a given logic with the fmp has this property. Therefore we first describe a certain class of logics which have the effective m-drop points property and which are intuitively simpler. By the way, note that fortunately almost all the important individual logics which were considered in logical literature, and some of the most important infinite classes of modal logics with the fmp, have both the property of branching below m and the effective m-drop points property for some appropriate m.

Recall that in the definition of effective m-drop points property we consider the submodel $max(Y, Q, m) \cup S_m(Q)$ of a given finite model $Q = \langle Q, R, V \rangle$. It is easy to see that this submodel has a finite and effectively computable depth k. We consider the greatest condensing p-morphism ϕ of the model $max(Y, Q, m) \cup S_m(Q)$ with respect to the variables from Y starting with depth $m+1$ and without identifying elements of any depth strictly more than m with elements of a depth not more than m. ϕ condenses the model $max(Y, Q, m) \cup S_m(Q)$ into the model $mc(m, Y, Q)$.

Lemma 3.5.3 *If for given m and all Q and Y the frame of $mc(m, Y, Q)$ is a λ-frame then λ has the effective m-drop points property.*

Proof. The number of elements in the set

$$[mc(m, Y, Q) - S_m(mc(m, Y, Q))]$$

is bounded by the number of elements in $S_k(Ch_{K4}(v + w))$, where w is the number of elements in $S_m(Q)$ and v is the number of elements in Y. Moreover, $S_m(mc(m, Y, Q))$ is isomorphic to $S_m(Q)$ and the condensing p-morphism ϕ does not identify elements of depth m and less with deeper elements. Thus properties (i) and (iii) from the definition of the effective m-drop points property hold, and property (ii) holds by Lemma 3.4.6 and since p-morphisms preserve the truth of formulas. ■

Hence Lemma 3.5.3 shows that the model $mc(m, Y, Q)$ and the condensing p-morphism ϕ have all the properties of the model $\mathfrak{M}_1$ and the homomorphism f in our definition of the effective m-drop points property, except that $\mathfrak{M}_1$ is λ-model. This leads us to introduce the following definition.

Definition 3.5.4 *We say that a modal or a superintuitionistic logic λ has the strong effective m-drop-points property if it has the fmp and there is a recursive (i.e effectively calculable) function $h(x, y)$ such that the following hold. Suppose $\mathcal{F}$ is a finite λ-frame, $\mathcal{F}_1$ is a subframe of $\mathcal{F}$, and f is a p-morphism from $\mathcal{F}_1$ such that:*

(a) $S_m(\mathcal{F}) = S_m(\mathcal{F}_1)$;

(b) f is an isomorphism on the frame $S_m(\mathcal{F}_1)$ and f does not identify elements with different depths, that is, f is slice stable.

Then there are a finite subframe $\mathcal{F}_2$ of the frame $\mathcal{F}$ and a p-morphism f_1 from $\mathcal{F}_2$ such that:

(c) $\mathcal{F}_1 \subseteq \mathcal{F}_2$;

(d) $f_1(\mathcal{F}_2)$ *is a* λ-*frame and* $\|f_1(\mathcal{F}_2)\| \leq h(l, r)$, *where* l *is the number of elements in* $f(\mathcal{F}_1)$ *and* r *is the maximal number of elements in clusters from* $\mathcal{F}$;

(e) f_1 *does not identify elements with different depths, and* f_1 *coincides with* f *on* $\mathcal{F}_1$;

(g) f_1 *is a one-to-one mapping on the set* $(|\mathcal{F}_2| - |\mathcal{F}_1|)$ *and* f_1 *does not contract elements from* $\mathcal{F}_1$ *and* $(|\mathcal{F}_2| - |\mathcal{F}_1|)$.

Lemma 3.5.5 *If a modal logic* λ *extending* $K4$ *or a superintuitionistic logic* λ *with the fmp has the strong effective m-drop points property then* λ *has the effective m-drop points property.*

Proof. We take a finite λ-model $\mathcal{F} := \langle Q, R, V \rangle$. Consider the submodel

$$\mathcal{F}_1 := max(Y, Q, m) \bigcup S_m(Q)$$

of this model. Let f be the greatest condensing of the Kripke model $\mathcal{F}_1$ with respect to the valuation of the variables from Y and starting with the depth $m+1$ such that f does not identify elements of different depth. Then the number of elements in $f(\mathcal{F}_1)$ can be effectively evaluated through Y and $S_m(Q)$, and both properties (a) and (b) from the definition of the strong effective m-drop points property hold. Since λ has the strong effective m-drop points property (abbreviation se-m-dpp), there exists a subframe $\mathcal{F}_2$ of the frame of $\mathcal{F}$ and a p-morphism f_1 with the required properties. We will show that they satisfy the properties (i) - (ii) from the definition of the effective m-drop pints property. By (c) and (e) from the definition of se-m-dpp, the property (iii) holds. Moreover, by (e) and (g), f_1 can also be considered as a p-morphism of the model $\mathcal{F}_2$ with respect to the generated valuation. Therefore (ii) from the definition for the effective m-drop points property is valid. Finally, it is easy to see that (i) is valid for some recursive function g which can be simply constructed from h since (d) holds. ∎

On basis of this lemma we can transfer our criteria for the admissibility of inference rules from modal logics to superintuitionistic logics. To do this we need a simple technical lemma. Recall that for any superintuitionistic logic λ, $\sigma(\lambda)$ is the greatest modal counterpart for λ extending $S4$.

Lemma 3.5.6 *If a superintuitionistic logic* λ *has the property of branching below* m *and the strong effective m-drop-points property then* $\sigma(\lambda)$ *also has these properties.*

Proof. If λ has the fmp then $\sigma(\lambda)$ also has the fmp (see Corollary 2.7.21). Moreover, for any finite frame $\mathcal{F}$, $\mathcal{F}$ is a λ-frame iff $\mathcal{F}$ is a $\sigma(\lambda)$-frame. Therefore $\sigma(\lambda)$ has the property of branching below m because this property is simply a property of finite frames which holds since λ has the property of branching below m. Finally, the strong effective m-drop points property is again directly transferable from λ to $\sigma(\lambda)$ again for the same reason. ∎.

Now in order to apply our criteria for modal logics to superintuitionistic logics it is enough to use the analog of Gödel-McKinsey-Tarski translation theorem for inference rules. Recall that this analog (see Theorem 3.2.2) says that, for any inference rule $r := \alpha_1, ..., \alpha_n/\beta$, r is admissible in a superintuitionistic logic λ iff the rule $T(r) := T(\alpha_1), ..., T(\alpha_n)/T(\beta)$ is admissible in the greatest modal counterpart $\sigma(\lambda)$ of λ.

Theorem 3.5.7 *If a superintuitionistic logic λ has (i) the strong effective m-drop points property, (ii) the property of branching below m, and (iii) the finite λ-frames are effectively recognizable (in particular, if λ is finitely axiomatizable), then λ is decidable with respect to admissibility.*

Proof. As it is mentioned above, an inference rule r is admissible for our superintuitionistic logic λ if and only if the rule $T(r)$ is admissible in $\sigma(\lambda)$. Applying this observation, Lemma 3.5.6, Lemma 3.5.5 and Theorem 3.5.2 give us an algorithm for recognizing rules which are admissible in λ. ∎

Theorem 3.5.8 *Let λ be a certain superintuitionistic logic with the property of branching below m and also with the strong effective m-drop points property. Then an inference rule r with k variables is admissible in λ if and only if r is valid in the frame of the model $Ch_\lambda(k)$ with respect to arbitrary valuation.*

Proof. If r is not admissible in λ then the proof that r is invalid in the frame $Ch_\lambda(k)$ with respect to certain valuations is analogous to the proof of the similar part of Theorem 3.5.1. Suppose there is a valuation which invalidates r in the frame of $Ch_\lambda(k)$. Then the modal inference rule $T(r)$ will be also invalid in the frame of $Ch_\lambda(k)$. From our construction of n-characterizing models for logics with the fmp, it can be easily seen that the frame of $Ch_\lambda(k)$ is a p-morphic image of the frame of $Ch_{\sigma(\lambda)}(k)$. Therefore $T(r)$ is invalid in the frame of $Ch_{\sigma(\lambda)}(k)$ and Theorem 3.5.1 entails that $T(r)$ is not admissible in $\sigma(\lambda)$. This yields r is not admissible in λ. ∎.

Now we will simply apply the theorems obtained to important particular logical systems and some representative infinite series of modal and superintuitionistic logics. We begin with some individual logics. Recall that $Grz.2 := Grz \oplus \square\lozenge p \to \lozenge\square p$ and $Grz.3 := Grz \oplus \square(\square p \to \square q) \vee \square(\square q \to \square p)$.

Corollary 3.5.9 *The modal logics $K4$, $K4.1$, $K4.2$, $K4.3$, $S4$, $S4.1$, $S4.2$, $S4.3$, Grz, $Grz.2$, $Grz.3$ and GL have the property of branching below 1 and the strong effective 1-drop points property, and are decidable with respect to admissibility. Theorems 3.5.1 and 3.5.2 are applicable to all these logics.*

Proof. First we recall that all these logics have the fmp: $K4$, $S4$, $S4.3$, $K4.3$ - by Corollary 2.6.12, GL - by Theorem 2.6.20, Grz - by Theorem 2.6.25, $K4.1$, $K4.2$, $S4.1$, $S4.2$, and $Grz.2$ - by Theorem 2.8.9 and $Grz.3$ by Exercise 2.6.28. The fact that all these logics have the property of branching below 1 follows directly from the structure of the axiomatic systems of these logics. In order to show that they all have the effective 1-drop points property, note that for any logic λ given above, the corresponding frame $max(Y, Q, 1) \cup S_1(Q)$ is a λ-frame. Applying Lemma 3.5.3 it follows that all given modal logics have the effective 1-drop points property. Thus Theorems 3.5.1 and 3.5.2 are applicable to all these modal logics. ∎

Note that concerning superintuitionistic logics, it is not difficult to see that the superintuitionistic logics H, KC and LC, for example, have the fmp and the strong effective m-drop points property. Therefore Theorems 3.5.7 and 3.5.8 are applicable to these logics. More precisely, we have

Corollary 3.5.10 *The intuitionistic propositional logic H and the superintuitionistic logics $KC := H \oplus \neg p \vee \neg\neg p$ and $LC := H \oplus (p \to q) \vee (q \to p)$ are decidable with respect to admissibility and Theorems 3.5.7 and 3.5.8 are applicable to them.*

Proof. The proof is similar to the proof of the corollary above: the logics H, KC and LC have the fmp (see Corollary 2.6.14). They all have the property of branching below 1, it is easy to derive directly from axiomatic systems of these logics. Moreover it is also easy to see from our definitions immediately that all these logics have the strong effective 1-drop points property. Therefore Theorems 3.5.7 and 3.5.8 are applicable to all these superintuitionistic logics. ∎

Recall that a modal logic extending $K4$ is a logic of finite depth if $\sigma_n \in \lambda$, and respectively a superintuitionistic logic λ is a logic of finite depth if $\phi_n \in \lambda$).

Corollary 3.5.11 *All modal logics extending $K4$ and having a finite depth and all superintuitionistic logics of finite depth are logics to which Theorems 3.5.1 and 3.5.2, Theorems 3.5.7 and 3.5.8 respectively, are applicable. In particular all such finitely axiomatizable logics are decidable by admissibility.*

Proof. Consider any logic λ of depth m of mentioned above kind. By Corollary 2.8.17 or Theorem 2.8.26 λ has the fmp. The logic λ has the property of branching below m because λ has no λ-frames of depth more than m. Again λ has the strong effective m-drop points property because of the lack of λ-frames of depth more than m. Therefore if λ is a superintuitionistic logic then Theorems 3.5.7 and 3.4.10 are applicable to λ. If λ is a modal logic then by Lemma 3.5.3 λ has the effective m-drop points property and again we can use Theorems 3.5.1 and 3.5.2. ∎

An other interesting infinite class of superintuitionistic logics, which suits for application our technique, is the class of logics with bounded branching, which was first considered by Gabbay and de-Jongh [54]. They introduced a sequence of superintuitionistic logics which are generated by finite frames with uniformly bounded branching of elements.

Definition 3.5.12 *Let $\mathcal{F}$ be a frame and a be an element of $\mathcal{F}$. We say that a has the n-branching in $\mathcal{F}$ if the cluster containing a has exactly n immediate successor clusters.*

The sequence of logics $D_n, n \in N$ introduced by Gabbay and de-Jongh has the following semantic definition: $D_n = \lambda(\{\mathcal{F} \mid \mathcal{F}$ is a finite poset with a branching of elements of not more than $n + 1\}$. Thus simply dy definition any logic D_n has the fmp. Note that any D_n can also be defined in a syntactic way, by a single new axiom, i.e. these logics are finitely axiomatizable. Indeed, if

$$\alpha_n := \bigwedge_{0 \leq i \leq n+1} ((p_i \to \bigvee_{j \neq i} p_j) \to \bigwedge_{j \neq i} p_i) \to \bigwedge_{1 \leq i \leq n+1} p_i$$

than it is easy to see that for a poset $\mathcal{F}$, $\mathcal{F} \Vdash \alpha_n$ iff the branching of elements in $\mathcal{F}$ is not more than $n+1$. In particular, $\alpha_n \in D_n$ and $\alpha_n \notin D_{n+1}$. And it is possible to show (see [54]) that $D_n = H \oplus \alpha_n$. It is also very easy to show semantically that all logics D_n have the disjunction property. For any finite frame $\mathcal{F}$, the denotation $b(\mathcal{F}) \leq m$ means below that any element from $|\mathcal{F}|$ has a branching of not more than n.

In a similar way to that for the superintuitionistic logics D_n described we introduce the following modal logics:

$$
\begin{aligned}
K4_{b,n} \quad &:= \quad \lambda(\{\mathcal{F} \mid \|\mathcal{F}\| < \omega, b(\mathcal{F}) \leq n+1, \\
&\qquad \mathcal{F} \text{ is a transitive frame }\}); \\
S4_{b,n} \quad &:= \quad \lambda(\{\mathcal{F} \mid \|\mathcal{F}\| < \omega, b(\mathcal{F}) \leq n+1, \\
&\qquad \mathcal{F} \text{ is a reflexive and transitive frame }\}); \\
Grz_{b,n} \quad &:= \quad \lambda(\{\mathcal{F} \mid \|\mathcal{F}\| < \omega, b(\mathcal{F}) \leq n+1, \\
&\qquad \mathcal{F} \text{ is a poset }\}); \\
GL_{b,n} \quad &:= \quad \lambda(\{\mathcal{F} \mid \|\mathcal{F}\| < \omega, b(\mathcal{F}) \leq n+1, \\
&\qquad \mathcal{F} \text{ is a transitive and irreflexive frame }\});
\end{aligned}
$$

Exercise 3.5.13 *Show that all logics of the list D_n, $K4_{b,n}$, $S4_{b,n}$, $Grz_{b,n}$ and $GL_{b,n}$ have the disjunction property.*

Corollary 3.5.14 *For any n, the logics D_n, $K4_{b,n}$, $S4_{b,n}$, $Grz_{b,n}$ and $GL_{b,n}$ have the property of branching below 1 and the strong effective 1-drop point property. Therefore these logics are decidable with respect to admissibility and Theorems 3.5.7 and 3.5.8 are applicable to all the logics D_n, and Theorems 3.5.2 and 3.5.1 are applicable to all the logics $K4_{b,n}$, $S4_{b,n}$, $Grz_{b,n}$ and $GL_{b,n}$.*

Proof. By definition all the above mentioned logics have the fmp and have the property of branching below 1. We cannot directly use Lemma 3.5.3 to show that these logics have the strong effective 1-drop points property because the frames defined in that lemma can have a branching of more than $n + 1$. Therefore we need to establish possession of the strong effective 1-drop points property by means of more complicated reasoning.

Lemma 3.5.15 *The logics $K4_{b,n}$, $S4_{b,n}$, $Grz_{b,n}$, $GL_{b,n}$ and D_n have the strong effective m-drop points property for any $(m > 0)$.*

Proof. As a matter of fact, this proof has a relationship to the original proof given in Gabbay an de-Jongh [54] that logics $H \oplus \alpha_n = D_n$, i.e. that $H \oplus \alpha_n$ has fmp. Suppose λ is a logic from our list. Consider a finite λ-frame $\mathcal{Q}$, its subframe $\mathcal{D}$, and a p-morphism f from $\mathcal{D}$ with the properties (a) and (b) from the definition of the strong effective m-drop points property. We will add some additional elements from $\mathcal{Q}$ to $\mathcal{D}$ in the following way.

We consider any cluster C of the frame $f(\mathcal{D})$ of a depth more than m which has the covering clusters $C_1, ..., C_p$ where $p > n + 1$. Suppose B and B_i are clusters from $\mathcal{D}$ and $f(B) = C$ and $f(B_i) = C_i$. Let $\mathcal{F}$ be the subframe of $\mathcal{Q}$ which consists of B, $B_1, ..., B_p$ and all the clusters E such that:

$$\exists i[B_i \subseteq E^{R \le}] \& [E \subseteq B^{R \le}] \& \forall x \in (|\mathcal{D}| - B \cup B_1 \cup ... \cup B_p)$$
$$((x \in E^{R <}) \Rightarrow \exists i((x \in B_i^{R \le}) \vee (x \in B_i)).$$

Now we search for a maximal cluster C_1 from $\mathcal{F}$ such that at least two different clusters from the list $B_1, ..., B_p$ are accessible from C_1. Such a cluster exists because the frame $\mathcal{Q}$ is finite and has branching of at most $n + 1$. We fix some such C_1. By the maximality of C_1 and since the branching in $\mathcal{Q}$ is limited by $n + 1$, the number of clusters from the list $B_1, ..., B_p$ which are accessible from C_1 is more than 2 and less than $n + 1$. We add C_1 to the frame $\mathcal{D}$ and now consider the frame $\mathcal{F}_1$ which is constructed in similar way to $\mathcal{F}$ but by taking the cluster C_1 and clusters from $B_1, ..., B_p$ which are not accessible from C_1 instead of $B_1, ..., B_p$. Note that if r is the number of such clusters from $B_1, ..., B_p$ then $r + 1$ is strictly less than p. If $r + 1$ is greater than $n + 1$ we use similar reasoning again and obtain the frame $\mathcal{F}_2$, etc. Thus, in not more than $(p - n)$ steps, we add these clusters C_1 to $\mathcal{D}$ and obtain a subframe of $\mathcal{Q}$ which consists of B, $B_1, ..., B_p$

and all the added clusters of the kind C_1. Note that this resulting subframe has the branching of elements at most $n + 1$.

The subframe $\mathcal{D}_1$ of $\mathcal{Q}$ obtained by the described above procedure of adding clusters will have all the properties required in the definition of the strong effective m-drop points property. Indeed, we can extend f from $\mathcal{D}$ onto the whole frame $\mathcal{D}_1$ and obtain a p-morphism f_1 from $\mathcal{D}_1$ as follows: f_1 is the one-to-one mapping on the set of clusters of kind C_1 and f_1 coincides with f on $\mathcal{D}$. It is easy to see that (i) $f_1(\mathcal{D}_1)$ is a λ-frame for the given λ since this frame has branching limited by $n + 1$; that (ii) the number of elements in $f_1(\mathcal{D}_1)$ is effectively evaluated through the number of elements in $f(\mathcal{D})$, the maximal number of elements in clusters from $\mathcal{Q}$, and the number limiting the branching. Also it is not hard to see that (iii): all the other properties in the definition of the strong effective m-drop points property also hold. ∎.

Using this lemma, it follows that Theorems 3.5.7 and 3.5.8 are applicable to all the superintuitionistic logics D_n, and Theorems 3.5.1 and 3.5.1 can be applied to all the modal logics mentioned in the formulation of our corollary. ∎

Now we give a few more exotic examples. Let $K4_{b,n,m}$ be the modal logic of all finite transitive frames which at depths greater than m have a factor of branching of at most $n + 1$. The modal logics $S4_{b,n,m}$, $Grz_{b,n,m}$, $GL_{b,n,m}$ and the superintuitionistic logics $D_{b,n,m}$ are defined in a similarly way for frames with accessibility relations corresponding to the abbreviations for logics.

Exercise 3.5.16 *Show that for any n, all the logics of the list $K4_{b,n,m}$, $S4_{b,n,m}$, $GL_{b,n,m}$, $Grz_{b,n,m}$, $D_{b,n,m}$ have the property of branching below m and the strong effective m-drop points property.*

Hint: The property of branching below m for these logics follows from the fact that the property of having a maximum branching of $n + 1$ at a depth greater than m is expressible by a modal formula and by a pure propositional formula, like the formula α_n introduced above expresses the property to have a branching of not more than $n + 1$ everywhere. Possession of the strong effective m-drop points property can be established in a similar way to the proof of Lemma 3.5.15.

Thus all the criteria concerning the admissibility of inference rules obtained in this section are applicable to all the logics $K4_{b,n,m}$, $S4_{b,n,m}$, $GL_{b,n,m}$, $Grz_{b,n,m}$ and $D_{b,n,m}$ for any n and m.

We also present a few more examples of applications for our theorems. Consider the modal logics $K4I_n$, $S4I_n$, GLI_n, and the superintuitionistic logics I_n, where each of these logics is the logic of all finite rooted frames with an accessibility relation corresponding to the abbreviation for logics ($K4$, $S4$, GL, H-frames, respectively), and with at most n maximal clusters. It is easy to see that the following lemma holds.

Lemma 3.5.17 *For any* n, *the logics* $K4I_n$, $S4I_n$, GLI_n *and* I_n *have the property of branching below 1 and the strong effective 1-drop points property.*

Thus the theorems of this section concerning admissibility are applicable to these logics as well. Before we consider certain examples of recognizing admissibility for individual inference rules in some logics of the kind considered, we give a more general observation concerning admissibility. It is the following theorem which follows directly from Theorems 3.5.1 and 3.5.8.

Theorem 3.5.18 *Suppose* λ_1 *and* λ_2 *are both modal logics extending* $K4$ *or that both are superintuitionistic logics. Suppose they both have the fmp, the branching below m property, and the strong effective m-drop points property. Moreover suppose that there is a p-morphism of a disjoint union of frames of models* $Ch_{\lambda_1}(n)$ *onto the frame of the model* $Ch_{\lambda_2}(k)$, *for each* k. *Then every inference rule admissible in* λ_1 *is admissible in* λ_2 *as well.*

Applying this theorem we find that any inference rule admissible in $S4$ is admissible in Grz, $S4.1$ and $S4.2$, any inference rule admissible in $S4.2$ is admissible in $Grz.2$, any rule which is admissible in H is also admissible in KC, etc.

Now we will demonstrate how the methods developed for determining admissibility of inference rules work for certain particular interesting inference rules. Of course direct application of Theorems 3.5.2 and 3.5.7 is hampered because we would be forced to consider too many distinct finite frames, and to determine the validity of the inference rules in all of them, and also to verify that these frames possess various other necessary properties. One possible way of avoiding these difficulties is simply to anticipate the structure of such frames. But it is more convenient to apply Theorems 3.5.11 and 3.5.8, - our semantic criteria. Of course, these mentioned theorems have no pure algorithmic nature since the models $Ch_\lambda(n)$ can be infinite in general. But very often the validness of inference rules in frames of such models are easy recognizable. All what we need in this case is simply to have an intuition concerning the structure of the frames of $Ch_\lambda(k)$. The main principle here is simple: all possible co-covers for antichains of clusters must present.

We begin with well known and rather the first example of admissible but not derivable rule in the intuitionistic propositional logic H. It is so-called Harrop rule (cf. Harrop [64]):

$$r_H := \frac{\neg x \to y \vee z}{(\neg x \to y) \vee (\neg x \to z)}.$$

Example 3.5.19 *The rule* r_H *is admissible but not derivable in the intuitionistic propositional logic* H.

Assume that r_H is not admissible in H. Then by Theorem 3.5.8 there exists a valuation S of variables x, y, z in the frame of $Ch_H(3)$ which invalidates r_H. This means there exists an $a \in |Ch_H(3)|$ such that

$$a \nVdash_S (\neg x \to y) \vee (\neg x \to z)$$

Thus there are $a_1 \in |Ch_H(3)|$ and $a_2 \in |Ch_H(3)|$ such that $a_1 \nVdash_S y$ and $a_1 \Vdash_S \neg x$, and $a_2 \nVdash_S z$ and $a_2 \Vdash_S \neg x$. By the construction of $Ch_H(3)$ this model contains a co-cover b for the set $\{a_1, a_2\}$. Then $b \Vdash_S \neg x$ and $b \nVdash_S y \vee z$. Hence the premise of r_H is falsified by S. Thus we showed that r_H is admissible in H. In order to verify that r_H is not derivable in H we consider the model $\mathfrak{M}$ depicted in Fig. 3.3. It is easy to see that this model $\mathfrak{M}$ invalidates the formula

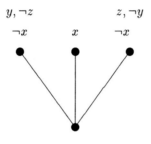

$$y, \neg z \qquad\qquad z, \neg y$$
$$\neg x \qquad x \qquad \neg x$$

Figure 3.3: Model $\mathfrak{M}$

$$(\neg x \to y \vee z) \to (\neg x \to y) \vee (\neg x \to z).$$

Therefore this formula is not provable in H and according to the deduction theorem for λ (cf. Theorem 1.3.9) the rule r_H is not derivable in H. ∎

Consider the Gödel-Mckinsey-Tarski translation T of the described above Harrop rule r:

$$T(r_H) := \frac{\Box(\Box\neg\Box x \to \Box y \vee \Box z)}{\Box(\Box\neg\Box x \to \Box y) \vee \Box(\Box\neg\Box x \to \Box z)}.$$

Example 3.5.20 *The rule $T(r_H)$ is admissible but not derivable in any modal logic from the list $K4$, $S4$, Grz, GL.*

Before to show these properties of $T(r_H)$, we note that by Example 3.5.19, Theorem 3.2.2, Proposition 3.2.3 and Corollary 2.7.23 the rule $T(r_H)$ is admissible but not derivable in Grz, since Grz is the greatest modal counterpart for H above $S4$. Nevertheless, for all the mentioned above modal logics, we can give a uniform proof of the admissibility of this rule. Let λ be any modal logics from our list. Assume that $T(r_H)$ is not admissible in λ. Then by Theorem 3.5.1 there is a valuation S of letters x, y, z in the frame of the model $Ch_\lambda(3)$ which invalidates $T(r_H)$. This means that there is an element $a \in |Ch_\lambda(3)|$ such that

$$a \not\Vdash_S \Box(\Box\neg\Box x \to \Box y) \vee \Box(\Box\neg\Box x \to \Box z).$$

This entails there are some elements a_1, a_2 in this model such that

$$a_1 \Vdash_S \Box\neg\Box x, a_1 \Vdash_V \neg\Box x, a_1 \not\Vdash_V y$$
$$a_2 \Vdash_S \Box\neg\Box x, a_1 \Vdash_V \neg\Box x, a_2 \not\Vdash_V z.$$

According to the construction of $Ch_\lambda(3)$, there is a reflexive or an irreflexive element b from $|Ch_\lambda(3)|$ which is co-cover for the set of clusters $C(a_1), C(a_2)$ containing our elements. Then it is easy to see that $b \Vdash_S \Box\neg\Box x$ and $b \not\Vdash_S \Box y$, $b \not\Vdash_V \Box z$. Hence it follows that the premise of $T(r_H)$ is falsified also, a contradiction. Thus the rule $T(r_H)$ is admissible in any logic λ from our list.

That the rule $T(\lambda)$ is not derivable in $S4$ and Grz we show using the model $\mathfrak{M}$ depicted in Fig. 3.3. It is easy to see that the formula

$$\Box\Box(\Box\neg\Box x \to \Box y \vee \Box z) \to \Box(\Box\neg\Box x \to \Box y) \vee \Box(\Box\neg\Box x \to \Box z)$$

is false in this model and consequently this formula is not a theorem nether $S4$ nor Grz. Therefore according to the deduction theorem for $S4$ and Grz (see Theorem 1.3.15) the rule $T(r_H)$ is not derivable in these logics. As to logics $K4$ and GL, we use the model which is similar to the model $\mathfrak{M}$ from Fig 3.3 but in which any reflexive element is replaced by two-element chain of irreflexive elements with the same valuation. It is easy to check that the formula

$$\Box\Box(\Box\neg\Box x \to \Box y \vee \Box z) \wedge \Box(\Box\neg\Box x \to \Box y \vee \Box z) \to$$
$$\Box(\Box\neg\Box x \to \Box y) \vee \Box(\Box\neg\Box x \to \Box z)$$

is false in this model and therefore this formula is not a theorem of both logics $K4$ and GL. Then by the deduction theorem for $K4$ (see Theorem 1.3.15) the rule $T(r_H)$ is not derivable in logics $K4$ and GL. ∎

Now we present Lemmon-Scott rule

$$r_{LS} := \frac{(\neg\neg p \to p) \to (p \vee \neg p)}{\neg\neg p \vee \neg p}.$$

Example 3.5.21 *The Lemmon-Scott rule r_{LS} is admissible but not derivable in the intuitionistic logic H.*

Indeed, suppose that r_{LS} is not admissible in H. Then by Theorem 3.5.8 there is a valuation S of the letter p in the frame of the model $Ch_H(1)$ which invalidates the rule r_{LS}. Then there is an $a \in |Ch_H(1)|$ such that $a \nVdash_S \neg\neg p \vee \neg p$. This entails there are elements a_1, a_2 from $|Ch_H(1)|$ such that

$$a_1 \Vdash_S p, \qquad a_2 \Vdash_S \neg p.$$

By construction of $Ch_H(1)$ there is an element b which is co-cover for the antichain $\{a_1, a_2\}$ (i.e. $b^{\leq <} = a_1^{\leq \leq} \cup a_2^{\leq \leq}$). Then, since the intuitionistic truth is stable (see Lemma 2.4.5), we conclude $b \nVdash_S \neg p \vee p$. At the same time $a_1 \Vdash_S \neg\neg p \to p$, and $a_2 \Vdash \neg p$ which entails $a_2 \Vdash_S \neg\neg p \to p$. Since $b \nVdash_S \neg\neg p$, it follows that $b \Vdash_S \neg\neg p \to p$ and

$$b \nVdash_s (\neg\neg p \to p) \to (p \vee \neg p),$$

that is the premise of r_{LS} is false, a contradiction. Therefore the rule r_{LS} is admissible in H. In order to show that this rule is not derivable in H we consider the model $\mathfrak{M}_{LS}$ depicted in Fig. 3.4. It is not difficult to see that this model $\mathfrak{M}_{LS}$

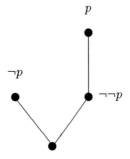

Figure 3.4: Model $\mathfrak{M}_{LS}$

invalidates the formula

$$((\neg\neg p \to p) \to (p \vee \neg p)) \to (\neg\neg p \vee \neg p).$$

Therefore this formula is not a theorem of H by deduction theorem for H (see Lemma 1.3.9), and the rule r_{LS} is not derivable in H. ∎

Now we will consider the Gödel-Mckinsey-Tarski translation $T(r_{LS})$ of the Lemmon-Scott rule:

$$T(r_{LS}) := \frac{\Box(\Box(\Box\Diamond\Box p \to \Box p) \to (\Box p \vee \Box\neg\Box p))}{\Box\Diamond\Box p \vee \Box\neg\Box p}.$$

Example 3.5.22 *The rule $T(r_{LS})$ is admissible but not derivable in the all modal logics S4, Grz and S4.1.*

We apply the same scheme of the proof as in several examples before. Let λ be a logic from our list. Suppose that $T(r_{LS})$ is not admissible in λ. According to Theorem 3.5.1 there is a valuation S of the letter p in the frame of the model $Ch_\lambda(1)$ which invalidates the rule $T(r_{LS}$. Then there is an element $a \in |Ch_\lambda(1)$ such that

$$a \not\Vdash_S \Box\Diamond\Box p \vee \Box\neg\Box pS.$$

This yields there are some elements a_1 and a_2 in this model such that

$$a_1 \Vdash_S \Box\neg\Box p, a_1 \Vdash_S \neg\Box p, a_1 \Vdash_S \neg p$$
$$a_2 \Vdash_S \Box p, a_2 \Vdash_S p.$$

According to our construction of $Ch_\lambda(1)$, there is a reflexive element b in this model which is co-cover for the set of clusters $\{C(a_1), C(a_2)\}$ containing the elements a_1, a_2. It is not hard to see that $b \not\Vdash_S \Box p \vee \Box\neg\Box p$. At the same time $b \Vdash_S \Box(\Box\Diamond\Box p \to \Box p)$. Hence the premise of the rule $T(r_{LS})$ is false on the element b under S, a contradiction. Thus the rule $T(r_{LS})$ is admissible in λ. In order to show that the rule $T(r_{LS})$ is not derivable in logics $S4$, Grz and $S4.1$ we use the model $\mathfrak{M}_{LS}$ pictured in Fig. 3.4. It can be easy verified directly that the formula

$$\Box\Box(\Box\Diamond\Box p \to \Box p) \to (\Box p \vee \Box\neg\Box p)) \to (\Box\Diamond\Box p \vee \Box\neg\Box p)$$

is false in $\mathfrak{M}_{LS}$ and, by this reason, is not a theorem of any logic from the list $S4$, $S4.1$, Grz. By the deduction theorem for $S4$ (see Theorem 1.3.15) the rule $T(r_{LS})$ is not derivable in these logics. ∎

Note that the rule $T(r_{LS})$ is not admissible in the modal logic GL. Indeed,

Lemma 3.5.23 *Let r be a rule*

$$r := \frac{\alpha_1, ..., \alpha_n}{\beta},$$

such that (i) β is a formula built up out of own subformulas of the form $\Box\Diamond\beta_i$ by means of only logical connectives $\vee, \wedge$, and (ii) there is a substitution which turns all premises of r into a theorems of GL. Then the rule r is not admissible in the Gödel-Löb modal system GL.

Proof. Suppose e is a substitution of variables $x_1, ..., x_n$ occurring in the rule r such that $e(\alpha_1) \in GL, ... , e(\alpha_n) \in GL$. Suppose that this substitution uses formulas which are constructed out of letters $p_1, ..., p_m$. Consider the model $Ch_{GL}(m)$ and the valuation S of the variables $x_1, ..., x_n$ of r, where $S(x_i) := V(e(x_i))$. Since our model $Ch_{GL}(m)$ is n-characterizing for GL and by the assumption that all $e(\alpha_j)$ are theorems of GL, it follows all premises of r are valid in $Ch_{GL}(m)$ under S. At the same time, there is an irreflexive element a in $Ch_{GL}(m)$ which has depth 2. It is clear that the formula β is false in a with respect to any valuation. Hence r is invalidated in $Ch_{GL}(m)$ by the valuation S. Applying Lemma 3.4.2 and Theorem 3.5.1 it follows r is not admissible in the logic GL. ∎

Turning to consider our rule $T(r_{LS})$, it is clear that this rule is equivalent to a rule satisfying the conditions of this lemma (it is sufficient to take the substitution $e(p) := \top$). Therefore $T(r_{LS})$ is not admissible in GL.

Now we pause briefly from these examples of admissible and non-admissible rules in order to use our method to prove a general result about derivable and admissible inference rules for superintuitionistic logics (which was proved by A.Citkin in 1976 for the intuitionistic logic H itself). We do this in order to demonstrate the convenience of the methods developed in this chapter.

Proposition 3.5.24 *Let λ be a superintuitionistic logic having (i) the fmp, (ii) the property of branching below m, and (iii) the strong effective m-drop points property. Any inference rule r of the form*

$$r := \frac{\neg\alpha_1, ..., \neg\alpha_m}{\beta}$$

is admissible in λ if and only if r is derivable in λ. (in particular, this holds for the logics H, KC, LC, etc.).

Proof. We know (see Proposition 1.4.3) that any derivable rule must be admissible. Suppose that r is not a rule derivable in λ. By the deduction theorem for H (see Lemma 1.3.9) it follows that the formula $\gamma := \neg\alpha_1 \wedge, ..., \wedge\neg\alpha_m \rightarrow \beta$ is not provable in λ. Suppose that r is built up out of the variables $p_1, ..., p_n$. Since $Ch_\lambda(n)$ is n-characterizing for λ (see Lemma 3.3.11), the formula γ is false in this model, i.e. there is an element $a \in |Ch_\lambda(n)|$ such that

$$a \nVdash_V \neg\alpha_1 \wedge, ..., \wedge\neg\alpha_m \rightarrow \beta.$$

Now we introduce a certain new intuitionistic valuation S of letters from r in the poset $Ch_\lambda(n)$ as follows. We fix some maximal element c in the set $a^{\leq\leq}$. We let S coincide in the set $a^{\leq\leq}$ with the original valuation V, and, for every maximal element b outside the set $a^{\leq\leq}$, we let S coincide on b with V on c. For all elements not considered yet, we let all letters be invalid by S. It is clear that β is false under S on a and that all the formulas $\neg\alpha_1, ..., \neg\alpha_m$ are valid on any element. Thus r is invalidated in $Ch_\lambda(n)$ by S, and by Theorem 3.5.8 r is not admissible in λ. ∎

It is not difficult to transfer this result to the corresponding modal logics. To do this we consider rules whose premises have a special form related to the form of the Gödel-Mckinsey-Tarski translations of the premises of rules mentioned in the lemma above.

Exercise 3.5.25 *Suppose λ is a modal logic extending Grz which possesses (i) the fmp, (ii) the property of branching below m and (iii) the effective m-drop points property. Then any inference rule r of the form*

$$r := \frac{\Box\neg\Box\alpha_1, ..., \Box\neg\Box\alpha_m}{\beta}$$

is admissible in λ if and only if r is derivable in λ. In particular, this holds for $\lambda := Grz, Grz.2$ and $Grz.3$.

Hint: the proof is similar to the proof of Proposition 3.5.24.

Now we give one more example of a rule which is admissible but not derivable in the intuitionistic logic H. We will also show that this rule has the Lemmon-Scott rule r_{LS} as a consequence. Let

$$r_{GLS} := \frac{((x \to y) \to (x \vee \neg y))}{\neg\neg x \vee \neg y}.$$

We will call this rule the *generalized Lemmon-Scott rule*. As before we can show that this rule is admissible in H directly using only technique of intuitionistic n-characterizing models $Ch_H(k)$. But here we prefer to establish admissibility of the Gödel-Mckinsey-Tarski translation $T(r_{GLS})$,

$$T(r_{GLS}) := \frac{\Box(\Box(\Box x \to \Box y) \to \Box(\Box x \vee \Box\neg\Box y))}{\Box\Diamond\Box x \vee \Box\neg\Box y},$$

of this rule in modal system Grz and to derive from this the admissibility of r_{GLS} in H.

Example 3.5.26 *The rule $T(r_{GLS})$ is admissible but not derivable in the modal system Grz.*

Proof. Suppose that $T(r_{GLS})$ is not admissible in Grz. By Theorem 3.5.1 there exists a valuation S of the letters x, y in the frame of the model $Ch_{Grz}(1)$ which invalidates the rule $T(r_{GLS}$. Then there is an element $a \in |Ch_{Grz}(1)|$ such that

$$a \nVdash_S \Box\Diamond\Box x \vee \Box\neg\Box y.$$

Then there are some elements a_1 and a_2 in this model satisfying

$$a_1 \Vdash_S \Box\neg\Box x, a_2 \Vdash_S \Box y.$$

In virtue of the construction of $Ch_{Grz}(1)$, in this model there is a reflexive element b which is co-cover for the set of clusters $\{C(a_1), C(a_2)\}$ containing the elements a_1 and a_2. Then it is easy to see that $b \nVdash_S \Box\Diamond\Box x \vee \Box\neg\Box y$ and that $b \nVdash_S (\Box(\Box x \vee \Box\neg\Box y))$. At the same time $b \Vdash_S \Box(\Box x \to \Box y)$. So the premise of the rule $T(r_{GLS})$ is false on b under S, a contradiction. Hence the rule $T(r_{GLS})$

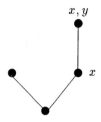

Figure 3.5: Model $\mathfrak{M}_{GLS}$

is admissible in Grz. The rule $T(r_{LS})$ is not derivable in the logic Grz. Indeed, consider the model $\mathfrak{M}_{GLS}$ which is displayed in Fig. 3.5. It can be easy verified directly that the formula

$$\Box\Box(\Box(\Box x \to \Box y) \to \Box(\Box x \vee \Box\neg\Box y)) \to \Box\Diamond\Box x \vee \Box\neg\Box y$$

is false in $\mathfrak{M}_{GLS}$. Thus this formula is not a theorem of the logic Grz. By the deduction theorem for $S4$ (see Theorem 1.3.15) the rule $T(r_{GLS})$ is not derivable in Grz. ∎

Immediately from this example, Theorem 3.2.2 and Proposition 3.2.3 we infer

Corollary 3.5.27 *The rule*

$$r_{GLS} := \frac{((x \to y) \to (x \vee \neg y))}{\neg\neg x \vee \neg y}$$

is admissible but not derivable in the intuitionistic logic H.

In the following proposition we compare the Lemmon Scott rule and its generalization

Proposition 3.5.28 *The rule r_{LS} is a corollary of the rule r_{GLS} in the intuitionistic logic H but not conversely.*

Proof. We can derive the conclusion of r_{LS} from its premise in H using the rule r_{GLS}. Indeed, let the premise of r_{LS}, i.e. $(\neg\neg p \to p) \to \neg p \vee p$, be given. Using the fact that $\neg p \vee p \to \neg\neg p \vee \neg p \in H$ we derive $(\neg\neg p \to p) \to \neg\neg p \vee \neg p$. We let $x := \neg\neg p, y := p$ and apply the rule r_{LS}, as result we obtain $\neg\neg\neg p \vee \neg\neg p$. Using $\neg\neg\neg p \to \neg p \in H$ (see Glivenko's Theorem) we can derive from this $\neg\neg p \vee \neg p$ which is the conclusion of r_{LS}. Hence r_{LS} is a corollary of r_{GLS}.

We recall that rules r_1 and r_2 are equivalent in a logic λ iff the corresponding quasi-identities $q(r_1)$ and $q(r_2)$ are semantically equivalent in the variety

$Var(\lambda)$. Therefore in order to show that r_{GLS} is not a corollary of r_{LS} it is sufficient to find a finite H-frame such that $q(r_{LS})$ is valid in $\mathcal{F}^+$ but $q(r_{GLS})$ is false
in $\mathcal{F}^+$. This is equivalent to the fact that the rule r_{LS} is valid in the frame $\mathcal{F}$ but
r_{GLS} is false in $\mathcal{F}$. We offer as a certain such $\mathcal{F}$ the frame pictured in Fig. 3.6.
We take the valuation $V(x) := \{c_2, c_3\}$, $V(y) := \{c_3\}$. It is easy to see that

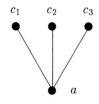

Figure 3.6: Frame $\mathcal{F}$

$a \Vdash_V \square\lozenge\square x \vee \square\neg\square y$.

At the same time the formula $\square x \vee \square\neg\square y$ is false under V only on a. But $a \Vdash_V$
$\square(\square x \rightarrow \square y)$. Therefore the premise of r_{GLS} is valid in this model. Thus r_{GLS}
is not valid in $\mathcal{F}$. Suppose that S is a valuation of the letter p in $\mathcal{F}$ and that the
premise $(\neg\neg p \rightarrow p) \rightarrow \neg p \vee p$ of r_{LS} is valid on all elements from $\mathcal{F}$ under S.
Suppose also that there is an element $b \in |\mathcal{F}|$ such that $b \Vdash_S \neg\neg p \vee \neg p$. Then
evidently $b = a$. This implies $a \Vdash_S \neg p \vee p$ and $a \Vdash_S \neg\neg p \rightarrow p$. Thus the premise
$(\neg\neg p \rightarrow p) \rightarrow \neg p \vee p$ of r_{LS} is false on a, a contradiction. Thus r_{LS} is valid in
the frame $\mathcal{F}$ and r_{GLS} is not a consequence of r_{LS} in H. ∎

We present one more example of an admissible but not derivable inference
rule for intuitionistic logic H. It is the rule which is discovered by G.Mints [104].
This rule will play an important role in our further research while studying tabular logics preserving admissible rules of H, and structurally complete fragments
of rules for H.

$$r_M := \frac{(x \rightarrow y) \rightarrow x \vee z}{((x \rightarrow y) \rightarrow x) \vee ((x \rightarrow y) \rightarrow z)}.$$

Example 3.5.29 *The rule r_M is admissible but not derivable in the intuitionistic logic H.*

Proof. We use the same scheme of the proof as before. Admit R_M is not admissible in H. Then be Theorem 3.5.8 there is a valuation S of letters x, y, z in the
frame of $Ch_H(3)$ which disproves r_M. Then,

$$\forall a \in |Ch_H(3)||[a \Vdash_S (x \rightarrow y) \rightarrow x \vee z] \text{ and}$$
$$\exists c_1 \in |Ch_(3)||[c \Vdash_S (x \rightarrow y) \rightarrow x],$$
$$\exists c_2 \in |Ch_(3)||[c \Vdash_S (x \rightarrow y) \rightarrow z].$$

Therefore, for some d_1, $c_1 \leq d_1$ and $d_1 \Vdash_S (x \to y)$ but $d_1 \nVdash_S x$, and respectively for some d_2, $c_2 \leq d_2$ and $d_2 \Vdash_S (x \to y)$ but $d_2 \nVdash_S x$. By contraction of $Ch_H(3)$ there is an element $b \in |Ch_H(3)|$ such that

$$b^{\leq <} = d_1^{\leq \leq} \cup d_2^{\leq \leq}.$$

It is easy to see that

$$b \Vdash_S (x \to y) \text{ and } b \nVdash_V x, \ b \nVdash_V z, b \nVdash_S x \lor z,$$

a contradiction. Thus r_M is admissible in H. To show that this rule is not derivable in H, we consider the intuitionistic Kripke model $\mathfrak{M}_M$ depicted below in Fig. 3.7. Direct calculation shows that the formula

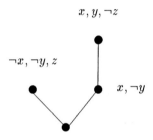

$$x, y, \neg z$$

$$\neg x, \neg y, z$$

$$x, \neg y$$

Figure 3.7: Model $\mathfrak{M}_M$

$$[(x \to y) \to x \lor z] \to [((x \to y) \to x) \lor ((x \to y) \to z)]$$

is falsified in $\mathfrak{M}_M$, and consequently the rule R_M is not derivable in H. ∎

Now we have already given many examples of useful applications of the criteria we have obtained for admissibility of inference rules in modal and superintuitionistic logics. These criteria will be actively used also in the following sections of this chapter and in the following chapters for studying the bases of admissible rules, logical equations and various other related questions.

We close this section with some results concerning modal counterparts with respect to admissibility of superintuitionistic logics. We know that an inference rule r is admissible for a superintuitionistic logic λ iff $T(r)$ is admissible in the greatest modal counterpart $\sigma(\lambda)$ of λ (see Theorem 3.2.2). In this sense, $\sigma(\lambda)$ is always a modal counterpart for λ with respect to admissibility. Are there such correspondences for arbitrary modal counterparts for λ? We cannot answer this question yet but we can give a restricted translation theorem for the smallest modal counterparts $\tau(\lambda)$ of superintuitionistic logics λ provided $\tau(\lambda)$ and λ are in the scope of the semantic criteria for admissibility obtained in this section. We begin with a simple technical lemma stated below.

Lemma 3.5.30 *If the smallest modal counterpart $\tau(\lambda)$ of a superintuitionistic logic λ has (i) the fmp, (ii) the property of branching below m, and (iii) the strong effective m-drop points property, then the logic λ also has these properties.*

Proof. Since $\tau(\lambda)$ has the fmp, the logic λ also have the fmp by Corollary 2.7.21. Any finite λ-frame is $\tau(\lambda)$-frame (see Lemma 2.7.7). Therefore the fact that $\tau(\lambda)$ has the property of branching below m entails that λ also possesses this property. Moreover from the fact that all λ-frames are $\tau(\lambda)$-frames we can directly infer that λ has the strong effective m-drop points property. ■

Theorem 3.5.31 Translation Theorem for the Smallest Modal Counterparts. *Suppose that λ is a superintuitionistic logic and that the modal logic $\tau(\lambda)$ has (i) the fmp, (ii) the property of branching below m, and (iii) the strong effective m-drop points property for some m. Then, for any inference rule r, r is admissible in λ if and only if $T(r)$ is admissible in $\tau(\lambda)$.*

Proof. Suppose that a rule r is admissible in λ and has k variables. By Lemma 3.5.30 λ has the fmp, the property of branching below m, and the strong effective m-drop points property. Then by Theorem 3.5.8 and Lemma 3.4.2 r is valid in the frame of n-characterizing model $Ch_\lambda(n)$ of λ for any n. For any finite $\tau(\lambda)$-frame $\mathcal{F}$, the frame $\mathcal{F}_c$ is a λ-frame (see Lemma 2.7.7). Therefore the frame of $Ch_{\tau(\lambda)}(k)_c$ is a p-morphic image of the frame of $Ch_\lambda(2^{2^n})$ under a certain condensing p-morphism (which is not hard to see from the definition of n-characterizing models $Ch_\lambda(k)$). Therefore by Lemma 2.7.7 the rule $T(r)$ is valid in $Ch_{\tau(\lambda)}(k)$. Then by Lemma 3.4.2 it follows $T(r)$ is admissible in $\tau(\lambda)$.

Conversely, suppose $T(r)$ is admissible in $\tau(\lambda)$. Then by Theorem 3.5.1 the rule $T(r)$ is valid in the frame of $Ch_{\tau(\lambda)}(k)$. Applying Lemma 2.7.7 it follows that r is valid in the frame $Ch_{\tau(\lambda)}(k)_c$. Again, since for any finite $\tau(\lambda)$-frame $\mathcal{F}$, the frame $\mathcal{F}_c$ is a λ-frame (see Lemma 2.7.7), we obtain there is a condensing p-morphism from $Ch_{\tau(\lambda)}(k)_c$ onto the frame of $Ch_\lambda(k)$. Therefore the rule r is valid in the frame of $Ch_\lambda(k)$ and according to Theorem 3.5.8 the rule r is admissible in λ. ■

Thus the smallest modal counterparts for superintuitionistic logics possessing special properties are also counterparts with respect to admissibility. We illustrate this fact in the following examples. We know that the smallest modal counterpart $\tau(H)$ for the intuitionistic logic H is the modal system $S4$ (see Theorem 2.7.4). The smallest modal counterpart for the superintuitionistic logic of the weak law of excluded middle $KC := H \oplus \neg p \vee \neg\neg p$ is the modal logic $S4.2 := S4 \oplus \Diamond\Box p \rightarrow \Box\Diamond p$ (it is easy to see that $S4 \oplus \Diamond\Box p \rightarrow \Box\Diamond p = S4\oplus\Diamond\Box p \rightarrow \Box\Diamond\Box p$). The superintuitionistic logic $LC := H\oplus(p \rightarrow q)\vee(q \rightarrow p)$ has as its smallest modal counterpart the logic $S4.3 := S4 \oplus \Box(\Box p \rightarrow \Box q) \vee \Box(\Box q \rightarrow \Box p)$. The logics $S4$, $S4.2$ and $S4.3$ have the fmp (see Corollary 2.6.12

and Theorem 2.8.9), the property of branching below 1 and the effective 1-drop points property (see Corollary 3.5.9). Therefore from our above theorem we immediately derive

Corollary 3.5.32 *The following hold*

(i) *An inference rule r is admissible in H $\Leftrightarrow T(r)$ is admissible in S4;*

(ii) *An inference rule is admissible in the logic $KC := H \oplus \neg p \vee \neg\neg p$ iff $T(r)$ is admissible in $S4.2 := S4 \oplus \Diamond\Box p \to \Box\Diamond p$.*

(ii) *An inference rule is admissible in the logic $LC := H \oplus (p \to q) \vee (q \to p)$ iff $T(r)$ is admissible in the modal logic $S4.3 := S4 \oplus \Box(\Box p \to \Box q) \vee \Box(\Box q \to \Box p)$.*

Final Remarks, Comments It is appropriate to recall now, that in a sense, the research of this section was initiated by Harvey Friedman problem concerning existence an algorithm recognizing admissibility inference rules in H. Also note that a time ago, rather, no algorithms for recognizing admissible rules in decidable non-trivial non-standard logics were known at all. First the solution of Harvey Friedman problem was obtained in Rybakov [135] and Rybakov [137], and then the solutions of the questions analogous to Friedman's one for different individual modal logics where obtained in Rybakov [133, 136, 140, 141, 144, 154]. However the technique of these solutions was, like that in this book, purely semantic. Therefore it is relevant to recall that the solution to H.Friedman question by pure semantic, proof theoretic methods, was given by P.Roziere in his Ph.D., this result was published in P.Roziere [128]. Note that this section not only provides us with general theorems which allow to determine admissibility in many individual logics and whole classes of modal and superintuitionistic logics, but also gives the line along which it is possible to investigate inference rules of many other related logical systems.

3.6 Elementary Theories of Free Algebras

As we have seen in Section 4 of Chapter 1 there is a closure correspondence between the admissible inference rules of algebraic logics λ and the quasi-identities in the signature of $Var(\lambda)$ which are valid in the corresponding free algebras $\mathfrak{F}_\lambda(\omega)$. More precisely, for any inference rule

$$r := \frac{\alpha_1, ..., \alpha_n}{\beta},$$

r is admissible in an algebraic logic λ iff the corresponding quasi-identity

$$r := [\alpha_1 = \top \wedge ... \wedge \alpha_n = \top \Rightarrow \beta = \top]$$

is valid in the free algebra $\mathfrak{F}_\lambda(\omega)$ from $Var(\lambda)$. Conversely, a quasi-identity

$$q := [\alpha_1 = \beta_1 \wedge ... \wedge \alpha_n = \beta_n \Rightarrow \alpha = \beta]$$

is valid in $\mathfrak{F}_\lambda(\omega)$ iff the corresponding inference rules

$$r(q) := \frac{\alpha_1 \equiv \beta_1, ..., \alpha_n \equiv \beta_n}{\alpha \equiv \beta}$$

are admissible in λ (see Lemma 1.4.6). Therefore the algorithms for determining admissibility of inference rules in modal and superintuitionistic logics, that we have found, bring the decidability of the quasi-equational theories of the corresponding free algebras (for any algebra $\mathcal{A}$ (or a class of algebras $\mathcal{K}$), the quasi-equational theory of $\mathcal{A}$ (of $\mathcal{K}$) is the set of all quasi-identities $Th_q(\mathcal{A})$ ($Th_k(\mathcal{K})$, correspondingly) which are valid in $\mathcal{A}$ (in $\mathcal{K}$, respectively)). Hence from Theorem 3.5.2 we can immediately infer

Theorem 3.6.1 *Let λ be a modal logic over $K4$ with the fmp, with the property of branching below m and the effective m-drop points property. Suppose also λ has an algorithm to determine whether a finite frame is λ-frame, which hods if λ is finitely axiomatizable. Then the quasi-equational theory $Th_q(\mathfrak{F}_\lambda(\omega))$ of the free modal algebra $\mathfrak{F}_\lambda(\omega)$ is decidable.*

Similarly from Theorem 3.5.7 we directly derive

Theorem 3.6.2 *Let λ be a superintuitionistic logic with the fmp, with the property of branching below m, and with the strong effective m-drop points property. Let also λ have an algorithm recognizing finite λ-frames. Then the quasi-equational theory $Th_q(\mathfrak{F}_\lambda(\omega))$ of the free pseudo-boolean algebra $\mathfrak{F}_\lambda(\omega)$ is decidable.*

Note that the quasi-equational theory of the free algebra of countable rank from a variety Var coincides with the quasi-equational theory of all the free algebras (of all possible ranks) from Var. Therefore the analogous of the theorems presented above hold for the quasi-equational theories of all free algebras from the corresponding varieties. By the results of the previous section theorems 3.6.1 and 3.6.2 are applicable to the free algebras of varieties corresponding to modal and superintuitionistic logics considered there. Using Corollaries 3.5.9, 3.5.10, 3.5.11, 3.5.14, Exercise 3.5.16 and Lemma 3.5.17 we can immediately infer

Corollary 3.6.3 *The quasi-equational theories $Th_q(\mathfrak{F}_\lambda(\omega))$ are decidable for all the modal logics $\lambda = K4$, $K4.1$, $K4.2$, $K4.3$, $S4$, $S4.1$, $S4.2$, $S4.3$, Grz, $Grz.2$, $Grz.3$, GL, $K4_{b,n}$, $S4_{b,n}$, $Grz_{b,n}$, $GL_{b,n}$, $K4I_n$, $S4I_n$, $GrzI_n$ and GLI_n, for all finitely axiomatizable modal and superintuitionistic logics of finite depth, and for all superintuitionistic logics $\lambda = H$, KC, LC, D_n, and I_n.*

These results can be generalized to logics which have the disjunction property. Recall that a superintuitionistic logic λ has the disjunction property if for any formulas α and β, $\alpha \vee \beta \in \lambda$ iff either $\alpha \in \lambda$ or $\beta \in \lambda$; similarly, a modal logic λ has the disjunction property if for any formulas α and β, $\Box\alpha \vee \Box\beta \in \lambda$ iff $\alpha \in \beta$ or $\beta \in \lambda$. Recall that the universal theory of an algebra $\mathcal{A}$ (or a class of algebras $\mathcal{K}$) is the set $Th_u(\mathcal{A})$ ($Th_u(\mathcal{K})$ respectively) of all universal formulas which are valid in $\mathcal{A}$ (in all algebras of $\mathcal{K}$).

Theorem 3.6.4 *Let λ be a modal logic over $K4$ with (i) the disjunction property, (ii) the finite model property, (iii) the property of branching below m and (iv) the effective m-drop points property (for some given m). Then the universal theory $Th_u(\mathfrak{F}_\lambda(\omega)$ of the free modal algebra $\mathfrak{F}_\lambda(\omega)$ is decidable.*

Proof. Given an universal formula Δ, we transform Δ into a conjunctive normal form

$$\Delta_1 := \bigwedge_{1 \le i \le n} \Omega_i, \text{ where}$$
$$\Omega_i := \bigvee_{1 \le j \le n_i} \alpha_j^i = \beta_j^i \vee \bigvee_{1 \le j \le m_i} \gamma_j^i \ne \delta_j^i.$$

It is easy to see that the formula Δ_1 is equivalent to the following formula $\Delta_2 := \bigwedge_{1 \le i \le n} \Psi_i$, where

$$\Psi_i := \bigwedge_{1 \le j \le m_i} \gamma_j^i = \delta_j^i \Rightarrow \bigvee_{1 \le j \le n_i} \alpha_j^i = \beta_j^i.$$

And it is clear that $\mathfrak{F}_\lambda(\omega) \models \Delta_2$ iff $\mathfrak{F}_\lambda(\omega) \models \Phi_i$ for all $i, 1 \le i \le n$. We claim that $\mathfrak{F}_\lambda(\omega) \models \Phi_i$ iff the quasi-identity

$$q_i := \bigwedge_{1 \le j \le m_i} \gamma_j^i = \delta_j^i \Rightarrow \bigvee_{1 \le j \le n_i} \Box((\alpha_j^i \to \beta_j^i) \wedge (\beta_j^i \to \alpha_j^i)) = \top$$

is valid in $\mathfrak{F}_\lambda(\omega)$. Indeed, if $\mathfrak{F}_\lambda(\omega) \models \Psi_i$ then $\mathfrak{F}_\lambda(\omega) \models q_i$ because $a = b$ yields $\Box((a \to b) \wedge (a \to b)) = \top$ in any modal algebra. Conversely, suppose that q_i is valid in $\mathfrak{F}_\lambda(\omega)$. Assume that e is an interpretation of all the variables x_v, which occur in the formula Φ_i, in the algebra $\mathfrak{F}_\lambda(\omega)$. Suppose that

$$\mathfrak{F}_\lambda(\omega) \models \bigwedge_{1 \le j \le m_i} \gamma_j^i([e(x_v)]_\sim) = \delta_j^i([e(x_v)]_\sim).$$

Since q_i is valid in $\mathfrak{F}_\lambda(\omega)$, we infer

$$\mathfrak{F}_\lambda(\omega) \models [\ \bigvee_{1\leq j\leq n_i} \Box((\alpha_j^i \to \beta_j^i) \wedge (\beta_j^i \to \alpha_j^i))]([e(x_v)]_\sim) = \top.$$

This implies $[\bigvee_{1\leq j\leq n_i} \Box((\alpha_j^i \to \beta_j^i) \wedge (\beta_j^i \to \alpha_j^i))](e(x_v)) \in \lambda$. Applying the fact that $\Box\mu \vee \Box\nu \vee \Box\phi \to \Box\mu \vee \Box(\Box\nu \vee \Box\psi) \in K4$ for any formulas μ, ν and ϕ, and the disjunction property of λ, it follows that there is some j such that $[(\alpha_j^i \to \beta_j^i) \wedge (\beta_j^i \to \alpha_j^i))](e(x_v)) \in \lambda$. Then $[\alpha_j^i \equiv \beta_j^i](e(x_v)) \in \lambda$ and consequently $\mathfrak{F}_\lambda(\omega) \models [\alpha_j^i = \beta_j^i]([e(x_v)]_\sim)$. Thus $\mathfrak{F}_\lambda(\omega) \models \Phi_i \Longleftrightarrow \mathfrak{F}_\lambda(\omega) \models q_i$. Therefore Theorem 3.6.1 entails the decidability of $Th_u(\mathfrak{F}_\lambda(\omega))$. ∎

Theorem 3.6.5 *Let λ be a superintuitionistic logic with (i) the disjunction property, (ii) the finite model property, (iii) the property of branching below m and (iv) the strong effective m-drop points property (for a certain m). Then the universal theory $Th_u(\mathfrak{F}_\lambda(\omega)$ of the free pseudo-boolean algebra $\mathfrak{F}_\lambda(\omega)$ is decidable.*

Proof. The proof is parallel to the proof of Theorem 3.6.4 with the following small alterations. We use (i) Theorem 3.6.2 instead of Theorem 3.6.1, (ii) the formula

$$q_i^1 := \bigwedge_{1\leq j\leq m_i} \gamma_j^i = \delta_j^i \Rightarrow \bigvee_{1\leq j\leq n_i} ((\alpha_j^i \to \beta_j^i) \wedge (\beta_j^i \to \alpha_j^i)) = \top$$

instead of the formula q_i, and (iii) the disjunction property of λ instead of the disjunction property for modal logics. The details are left to the reader as a simple exercise. ∎

Immediately from Theorems 3.6.4 and 3.6.5, and Corollaries 3.5.9, refc3s415 and Exercise 3.5.16 (the possession of disjunction property for logics below is well known and, follows easily, for instance, from the fmp of those logics and the structure of finite frames adequate for these logics) we derive the following

Corollary 3.6.6 *The universal theories $Th_u(\mathfrak{F}_\lambda(\omega)$ are decidable for all the modal logics $\lambda = K4, K4.1, S4, S4.1, Grz, GL, K4_{b,n}, S4_{b,n}, Grz_{b,n}, GL_{b,n},$ and for all superintuitionistic logics $H, D_n,\ n \geq 1$*

Hence we have established that the quasi-equational and the universal theories of the free algebras $\mathfrak{F}_\lambda(\omega)$ from many varieties of algebras corresponding to non-standard logics λ are decidable. The quasi-equational theories and the universal theories are simply subtheories of complete first order theories, i.e. elementary theories. The formulas of first-order language valid in $\mathfrak{F}_\lambda(\omega)$ express various meta-properties of the logic λ. Thus, for example, a superintuitionistic logic λ has the disjunction property iff the universal formula $x \vee y = \top \Rightarrow (x = \top) \vee (y =$

T) is valid in $\mathfrak{F}_\lambda(\omega)$. In fact, it is interesting to see to which extent we can effectively (i.e. by a certain algorithm) recognize the meta-properties of logics. Therefore it is also of interest to investigate the algorithmic complexity of the elementary theories of free algebras, and, in particular, to determine whether there are algorithms determining the truth of first-order formulas in the corresponding free algebras. Being motivated by this, in the remaining part of this section we study the elementary first-order theories of free algebras.

We recall certain notions from model theory which we will use. For a class of algebraic systems $\mathcal{K}$, $Th(\mathcal{K})$ denotes the elementary theory (which is often called simply *the theory*) of $\mathcal{K}$. More precisely, $Th(\mathcal{K})$ is the set of all first-order formulas which are valid in all algebraic systems from $\mathcal{K}$. The theory $Th(\mathcal{K})$ is undecidable if there is no algorithm which can determine for any given formula Ψ whether $\Psi \in Th(\mathcal{K})$.

Definition 3.6.7 *Let $\mathcal{K}$ be a class of algebraic systems. We say $Th(\mathcal{K})$ is hereditarily undecidable if for any class $\mathcal{K}_1$ containing $\mathcal{K}$ the theory $Th(\mathcal{K}_1)$ (which is a subtheory of $Th(\mathcal{K})$) is undecidable.*

We also need a definition of the elementary relative definability of one class of algebraic systems within another one. Here we will not give the strongest (and well known) definition but a simpler instance because it will be sufficient for our particular aim. Suppose $\mathcal{K}_1$ and $\mathcal{K}_2$ are certain classes of algebraic systems of the signatures Σ_1 and Σ_2 respectively. And suppose Σ_1 consists of only certain predicate symbols.

Definition 3.6.8 *The class $\mathcal{K}_1$ is elementarily relatively definable in $\mathcal{K}_2$ if the following holds. There is a formula $\Phi(x, y)$ in the signature of $\mathcal{K}_2$ and there are formulas $\Psi_P(x_1, ..., x_n, y, z)$ in the signature of $\mathcal{K}_2$ for every n-ary predicate symbol P from Σ_1 such that the following hold. For every model $\mathfrak{M}$ from $\mathcal{K}_1$, there is an algebraic system $\mathcal{A}$ from $\mathcal{K}_2$, and there are elements $a(\mathfrak{M}) \in |\mathcal{A}|$ and $a_P \in |\mathcal{A}|$ for all $P \in \Sigma_1$ such that the model*

$$\mathcal{A}_{a(\mathfrak{M}),a_P} := \langle \{a \mid a \in |\mathcal{A}|, \mathcal{A} \models \Phi(a(\mathfrak{M}), a)\}, \{\Psi(P) \mid P \in \Sigma_1\}\rangle$$

where $\Psi(P)(a_1, ..., a_n) = true \Longleftrightarrow \mathcal{A} \models \Psi_P(a_1, ..., a_n, a(\mathfrak{M}), a_P)$, is isomorphic to the model $\mathfrak{M}$.

Lemma 3.6.9 *If some class $\mathcal{K}_1$ is elementarily relatively definable in a class $\mathcal{K}_2$ and $Th(\mathcal{K}_1)$ is hereditarily undecidable then $Th(\mathcal{K}_2)$ is also hereditarily undecidable.*

Proof. We define a translation ε of all first-order formulas in the signature Σ_1 of the class $\mathcal{K}_1$ into formulas in the signature Σ_2 of the class $\mathcal{K}_2$ as follows. We reserve the variables x, y, z for our own use and consider only formulas in the signature Σ_1 which do not contain these variables. We introduce the translation ε of such formulas as follows:

(i) For all $P \in \Sigma_1$,
$$P(u_1, ..., u_n)^\varepsilon := \Psi_P(u_1, ..., u_n, y, z) \wedge \bigwedge_{1 \leq i \leq n} \Phi(u_i, y);$$
(ii) $(\Delta \vee \Omega)^\varepsilon := \Delta^\varepsilon \vee \Omega^\varepsilon$, $(\Delta \wedge \Omega)^\varepsilon := \Delta^\varepsilon \wedge \Omega^\varepsilon$,
$\qquad (\Delta \to \Omega)^\varepsilon := \Delta^\varepsilon \to \Omega^\varepsilon$, $(\neg \Delta)^\varepsilon := \neg \Delta^\varepsilon$;
(iii) $(\forall v \Delta)^\varepsilon := \forall v(\Phi(v, y) \to \Delta^\varepsilon)$;
(iv) $(\exists v \Delta)^\varepsilon := \exists v(\Phi(v, y) \wedge \Delta^\varepsilon)$;

Consider an arbitrary class $\mathcal{K}_3$ of algebras in the signature Σ_2 which contains $\mathcal{K}_2$. Using the formulas from the definition of elementary relative definability, we introduce the class $\mathcal{K}_4$ of models in the signature Σ_1 as follows. We set $\mathcal{K}_4 := \{\mathcal{A}_{b,c_P} \mid \mathcal{A} \in \mathcal{K}_3\}$, where $b \in |\mathcal{A}|$, for all $P \in \Sigma_1$, $c_P \in |\mathcal{A}|$,

$$\mathcal{A}_{b,c_P} := \langle \{a \mid a \in |\mathcal{A}|, \mathcal{A} \models \Phi(b, a)\}, \{\Psi(P) \mid P \in \Sigma_1\}\rangle$$

where (as in the definition of elementary relative definability) the predicates $\Psi(P)$ are defined by $\Psi(P)(d_1, ..., d_n) = \text{true} \Longleftrightarrow \mathcal{A} \models \Psi_P(d_1, ..., d_n, b, c_P)$. It is clear that $\mathcal{K}_4$ contains a subclass which consists of models isomorphic to the models of the class $\mathcal{K}_1$. Since $Th(\mathcal{K}_1)$ is hereditarily undecidable, the theory $Th(\mathcal{K}_4)$ also is undecidable. At the same time, it is not difficult to see that for any model $\mathcal{A}_{b,c_P} \in \mathcal{K}_4$ and any first-order formula Θ in the signature Σ_1 with free variables $u_1, ... u_k$ and without occurrences of variables x, y, z, and with predicate symbols $P_1, ..., P_m$,

$$\mathcal{A}_{b,c_P} \models \Theta(d_1, ..., d_m) \Longleftrightarrow \mathcal{A} \models \Theta^\varepsilon(d_1, ..., d_n, b, c_{P_1}, ..., c_{P_m}).$$

This equivalence can be simply verified by induction on the length of the formula Θ. Therefore, for any sentence Θ (formula without free variables) with predicate symbols $P_1, ..., P_m$, we conclude

$$\Theta \in Th(\mathcal{K}_4) \Longleftrightarrow \forall y \forall z_{P_1} ... \forall z_{P_m} \Theta^\varepsilon(y, z_{P_1}, ..., z_{P_m}) \in Th(\mathcal{K}_3).$$

Since $Th(\mathcal{K}_4)$ is undecidable, we immediately infer $Th(\mathcal{K}_3)$ is undecidable as well. Hence $Th(\mathcal{K}_2)$ is hereditarily undecidable. ∎

Now we introduce certain special classes of algebraic systems which will play a central role in our subsequent research of this section. Let $\mathcal{K}$ be a class of algebraic systems (or algebras, or models). Suppose that there is a first-order formula $\phi_\leq(x, y)$ in two free variables in the signature of $\mathcal{K}$ such that for any $\mathcal{A} \in \mathcal{K}$ the model $\mathcal{A}_{\phi_\leq} := \langle \{a \mid a \in |\mathcal{A}|\}, \phi_\leq(x, y)\rangle$ is a lattice which is partially ordered by the predicate $\phi_\leq(x, y)$, where by definition $\phi_\leq(a, b) \Leftrightarrow \mathcal{A} \models \phi_\leq(a, b)$. Then the smallest upper and the greatest lower bounds $\vee$ and $\wedge$ (for a pair of elements x, y) are expressible by certain first order formulas $\phi_\vee(x, y, z)$ and $\phi_\wedge(x, y, z)$ respectively. We fix the formulas $\phi_\leq, \phi_\vee(x, y, z)$ and $\phi_\wedge(x, y, z)$. In the following definition and what follows the predicate symbol $\leq$ and the function symbols $\vee$ and $\wedge$ have the meaning mentioned above, and are expressible by the corresponding first-order formulas.

Definition 3.6.10 *We say K is a CDL-class if the following hold.*

(a) There exists a term $f(x)$ in the signature of K which is built up out of a single variable x and has the following property. For every number n and for any $M \in K$, for every tuple $g_i, d \in |M|$, $1 \leq i \leq n$,

$$M \models [f(d) \leq \bigvee_{1 \leq i \leq n} f(g_i) \Rightarrow \tag{3.16}$$

$$\Rightarrow \bigvee_{1 \leq i \leq n} (f(d) \leq f(g_i))].$$

(b) For each $n \in N$ and every family of subsets $X_i \subseteq \{1, ..., n\}$, $1 \leq i \leq n$ such that

(i) $\forall i, j \in \{1, ..., n\}(i \in X_j \Leftrightarrow j \in X_i)$,
(ii) $\forall i \in \{1, ..., n\})(i \notin X_i)$,

there exists a sequence of terms $t_{X_i}(x_1, ..., x_n, y)$ (in the signature of K) corresponding to subsets X_i, and there exists an algebra $M \in K$ which has the following property: There is a tuple of elements $a_i, 1 \leq i \leq n$ from $|M|$ and there is an element $v \in |M|$ such that the elements $b_i, 1 \leq i \leq n$ from $|M|$, where $b_i := f(t_{X_i}(a_1, ..., a_n, v))$, satisfy (in M)

$$b_i \wedge b_j = v \iff (j \in X_i) \& (i \in X_j), \tag{3.17}$$

$$b_i \not\leq b_j, b_j \not\leq b_i. \tag{3.18}$$

Now we will prove the main technical theorem of this section which will allow us to transfer results concerning undecidability of elementary theories to the free algebras of varieties corresponding to certain algebraic logics.

Theorem 3.6.11 *The class K_0 of all finite models of the symmetric irreflexive binary predicate is elementarily relatively definable in each CDL-class of algebras K. In particular, $Th(K)$ is hereditarily undecidable for every CDL-class K.*

Proof. Let K be a CDL-class. We fix certain formulas $A(x, y)$ and $B(x, y, z)$ in the signature of K with the following property. $A(x, y)$ expresses the following property: y is a maximal element of the form $f(z)$ among all the elements of the form f(s) such that $f(s) \leq x$. And $B(x, y, z)$ is the formula which states that (a) z is the intersection of x and y, and that (b) x is not equal to y. Consider an arbitrary model $\langle \{i \mid 1 \le i \le n\}, P(x,y) \rangle$ from $\mathcal{K}_0$. We pick the subsets X_i, $1 \le i \le n$, of the set $\{1 \le m \le n\}$ as follows:

$$X_i := \{j \mid 1 \le j \le n \,\&\, P(i,j)\}.$$

By our assumption $P(x,y)$ is symmetric and irreflexive, and moreover we have $\forall i, j \in \{1, ..., n\}(i \in X_j \Leftrightarrow j \in X_i)$. Hence the sequence X_i, $1 \le i \le n$ satisfies the conditions (i) and (ii) from the definition of CDL-class. Since $\mathcal{K}$ is a CDL-class, there is a certain $\mathcal{M} \in K$ and there are some elements $b_i, v \in |\mathcal{M}|$, $1 \le i \le n$, where all $b_i := f(t_{X_i}(a_1, ..., a_n, v))$ have properties (3.17) and (3.18). Then by (3.17) it follows $b_i \wedge b_j = v \Longleftrightarrow ((j \in X_i) \& (i \in X_j))$. Therefore the following holds:

$$P(i,j) \Longleftrightarrow (j \in X_i) \& (i \in X_j) \Longleftrightarrow b_i \wedge b_j = v. \tag{3.19}$$

We consider the set

$$C := \{y \mid \mathcal{M} \models A(a,y)\}, \text{ where } a := \bigvee_{1 \le i \le n} b_i.$$

By (3.18) all elements b_i are $\le$-incomparable. Then applying (3.16) it follows that the set C is the set of all the elements b_i (in particular, any element of C is some element b_i).

Now we define a predicate $P_1(x_1, x_2)$ for elements of the set C as follows

$$P_1(x_1, x_2) \Leftrightarrow \mathcal{M} \models B(x_1, x_2, v).$$

From (3.19) it follows immediately that

$$\mathcal{M} \models B(b_i, b_j, v) \Leftrightarrow ((b_i \wedge b_j = v) \& (i \ne j)) \Leftrightarrow P(i,j)).$$

Thus $P_1(b_i, b_j) = \text{true} \Leftrightarrow P(i,j) = \text{true}$ and we have established that the models $\langle C, P_1 \rangle$ and $\langle \{i \mid 1 \le i \le n\}, P(x,y) \rangle$ are isomorphic. Hence the class $\mathcal{K}_0$ is elementarily relatively definable in the class $\mathcal{K}$ when

$$\Phi(x,y) := A(x,y),$$
$$\Psi_P(x_1, x_2, y, z) := B(x_1, x_2, z).$$

Since $\mathcal{K}_0$ is hereditarily undecidable (see, for instance, [38]), so is $\mathcal{K}$ (see Lemma 3.6.9). ∎

Now we will apply this theorem to classes of various free algebras having a lattice structure representation by first-order formulas. In particular, we will apply it to various free pseudo-boolean and modal algebras. It is not difficult to see that all following applications will be based on finding appropriate terms to imitate a lattice structure and on verifying the properties from the definition of CDL-classes. We begin with the free algebras from certain varieties of pseudo-boolean algebras. First we specify a class of superintuitionistic logics with which we will work.

Definition 3.6.12 *Let λ be a superintuitionistic logic. We say that λ is a certain $H(1,\infty)$-logic if all theorems of λ are valid in frames (posets) of the form $\bigsqcup_{1 \leq i \leq n} \mathcal{F}_i \odot 1$, where $\bigsqcup_{1 \leq i \leq n} \mathcal{F}_i$ is a certain disjoint union of single-element posets $\mathcal{F}_i$, and the frame $\bigsqcup_{1 \leq i \leq n} \mathcal{F}_i \odot 1$ is obtained from $\bigsqcup_{1 \leq i \leq n} \mathcal{F}_i$ by adjoining an additional smallest reflexive element.*

Theorem 3.6.13 *[137, 153] Let λ be a $H(1,\infty)$-logic and $\mathcal{K}$ be a class of free algebras from the variety $Var(\lambda)$, which contains among their subalgebras all free algebras of finite rank from $Var(\lambda)$. Then $Th(\mathcal{K})$ is hereditarily undecidable.*

Proof. We apply Theorem 3.6.11. To do this we need to show that $\mathcal{K}$ is a CDL-class. In other words, we must find special terms corresponding to the definition of CDL-class and prove necessary properties. We let $f(x) := \neg\neg x$ and show that (3.16) from the definition of CDL-class holds. Let $\mathfrak{F}_\lambda(\beta)$ be a free algebra from $\mathcal{K}$. Assume $d, g_i, 1 \leq i \leq n$ are certain elements of this algebra. Any element of every free algebra from $\mathcal{K}$ can be considered as a term, or formula of the logic λ, which is built up out of some free generators. Suppose that the premise of the formula from the relation (3.16) is valid in $\mathfrak{F}_\lambda(\beta)$ but the conclusion of this formula is false in $\mathfrak{F}_\lambda(\beta)$ for the chosen term $f(x)$ and for these d and g_i, i.e.

$$\mathfrak{F}_\lambda(\beta) \models \neg\neg d \leq \bigvee_{1 \leq i \leq n} \neg\neg g_i. \tag{3.20}$$

$$\mathfrak{F}_\lambda(\beta) \not\models \bigvee_{1 \leq i \leq n} \neg\neg d \leq \neg\neg g_i.$$

Then $\neg\neg d \to \neg\neg g_i$ is not a theorem of λ for any i. By the finite model property of the Heyting intuitionistic logic H (see Corollary 2.6.14), for each i, there is a rooted finite poset and a model $\mathfrak{M}_i$ which is based on this poset, such that the formula $\neg\neg d \to \neg\neg g_i$ is false in $\mathfrak{M}_i$. The presence of the prefix $\neg\neg$ in the premise and in the conclusion of this implication entails that the formula $\neg\neg d \to \neg\neg g_i$ is false in some single-element model $\mathcal{F}_i$ (more precisely in the open submodel of $\mathfrak{M}_i$ which consists of some maximal element of $\mathfrak{M}_i$). Using these models $\mathcal{F}_i$, we generate the model $\bigsqcup_{1 \leq i \leq n} \mathcal{F}_i \odot 1$, where $\bigsqcup_{1 \leq i \leq n} \mathcal{F}_i$ is the disjoint union of the models $\mathcal{F}_i$, and the model $\bigsqcup_{1 \leq i \leq n} \mathcal{F}_i \odot 1$ is obtained from $\bigsqcup_{1 \leq i \leq n} \mathcal{F}_i$ by adjoining a reflexive smallest element. The valuation of variables is directly transferred to the model $\bigsqcup_{1 \leq i \leq n} \mathcal{F}_i \odot 1$ from the models $\mathcal{F}_i$. It can be verified directly that the formula $\neg\neg d \to \bigvee_{1 \leq i \leq n} \neg\neg g_i$ is false in the model $\bigsqcup_{1 \leq i \leq n} \mathcal{F}_i \odot 1$ as well. Therefore this formula is not a theorem of λ because λ is a $H(1,\infty)$-logic. This contradicts (3.20). Thus the relation (3.16) holds in our case.

Now we prove that the relation (3.17) from the definition of CDL-class also holds. We need to specify the corresponding terms, algebras and elements. We set

$$t_{X_i}(x_1, ..., x_n, y) := (x_i \wedge \bigwedge_{j \in X_i} \neg x_j).$$

As $\mathcal{M}$ we take the algebra $\mathfrak{F}_\lambda(\beta)$ from $\mathcal{K}$, where $\mathfrak{F}_\lambda(n)$ is a subalgebra of $\mathfrak{F}_\lambda(\beta)$. For elements a_i, $1 \leq i \leq n$, we take the free generators of the free algebra $\mathfrak{F}_\lambda(n)$, which is a subalgebra of the chosen free algebra $\mathfrak{F}_\lambda(\beta)$ from $\mathcal{K}$. Finally we let $v := \perp$. According to our choice, the elements b_i have the following form:

$$b_i := \neg\neg(a_i \wedge \bigwedge_{j \in X_i} \neg a_j)$$

If $j \notin X_i$ then by the assumption in the definition of CDL-classes we directly derive $i \notin X_j$. Consider the single-element model $\mathcal{W}_1$ which is poset with the valuation of the propositional letters a_v, $1 \leq v \leq n$ such that only a_i and a_j are valid under this valuation on the single element of $\mathcal{W}_1$. Then it can be easily seen that the formula $b_i \wedge b_j \equiv \perp$ is false in $\mathcal{W}_1$ and, therefore, this formula is not a theorem of λ since λ is $H(1, \infty)$-logic. Then $b_i \wedge b_j \neq \perp$ in the free algebra $\mathfrak{F}_\lambda(\beta)$. Now let $j \in X_i$ and $i \in X_j$. Then $b_i \wedge b_j \leq \neg\neg a_i \wedge \neg\neg\neg a_i \leq \perp$. Thus (3.17) holds. In order to show (3.18) we take the model $\mathcal{Q}_1$ which is a single element poset, and the only variable a_i is valid in the model $\mathcal{Q}_1$. Then $\mathcal{Q}_1 \nVdash b_i \rightarrow b_j$. Therefore the formula $b_i \rightarrow b_j$ is not a theorem of λ since λ is a certain $H(1, \infty)$ logic. In a similar way we derive $b_j \rightarrow b_i \notin \lambda$. Thus therms b_i and b_i are incomparable in the algebra $\mathfrak{F}_\lambda(\beta)$. Hence (3.18) holds also. We have showed that $\mathcal{K}$ is a CDL-class. Therefore by Theorem 3.6.11 the theory $Th(\mathcal{K})$ is hereditarily undecidable. ∎

We will use the ideas and the proof scheme of this theorem in several following theorems. The experienced reader can easy transfer the proof of this theorem to the appropriate modal logics and free modal algebras by making simple adjustments in the technique used. However we prefer to give straightforward proofs of these theorems in order to demonstrate how the specific character of the modal logics works. First we consider the modal logics over $S4$ which are more closely related to superintuitionistic logics.

Definition 3.6.14 *A modal logic λ extending the normal modal logic $S4$ is said to be a $S4(1, \infty)$-logic if for any finite tuple of reflexive single-element frames $\mathcal{F}_i$, $1 \leq i \leq n$, all theorems of λ are valid in the partially ordered frame $\bigsqcup_{1 \leq i \leq n} \mathcal{F}_i \odot 1$ which is the frame obtained from the disjoint union $\bigsqcup_{1 \leq i \leq n} \mathcal{F}_i$ of frames $\mathcal{F}_i$ by adjoining a new reflexive smallest element.*

Theorem 3.6.15 *[137, 153] Let λ be a $S4(1, \infty)$-logic and $\mathcal{K}$ be a class of free algebras of variety $Var(\lambda)$ which contain among their subalgebras all the free algebras of finite rank of $Var(\lambda)$. Then the first-order theory of $\mathcal{K}$ is hereditarily undecidable.*

Proof. We again use Theorem 3.6.11 and show that $\mathcal{K}$ is a CDL-class. Let $f(x) := \Box\Diamond x$. We need to show (3.16). Suppose $\mathfrak{F}_\lambda(\beta)$ is a free modal algebra from $\mathcal{K}$ and $d, g_i, 1 \leq i \leq n$ are certain elements of $\mathfrak{F}_\lambda(\beta)$. We consider all these elements as terms, or formulas, which are built up out of free generators. Suppose (3.16) does not hold for these $f(x)$ and these elements d, g_i. This means

$$\mathfrak{F}_\lambda(\beta) \models \Box\Diamond d \leq \bigvee_{1 \leq i \leq n} \Box\Diamond g_i, \quad \mathfrak{F}_\lambda(\beta) \not\models \bigvee_{1 \leq i \leq n} (\Box\Diamond d \leq \Box\Diamond g_i). \qquad (3.21)$$

Then by the finite model property of the modal logic $S4$ itself (Corollary 2.6.12), for every i, there is a rooted finite transitive reflexive model $\mathfrak{N}_i$ such that the formula $\Box\Diamond d \rightarrow \Box\Diamond g_i$ is false in $\mathfrak{N}_i$. The structure of this formula allows us to conclude that the formula $\Box\Diamond d \rightarrow \Box\Diamond g_i$ is false in a single-element reflexive model $\mathcal{F}_i$. We take the model $\bigsqcup_{1 \leq i \leq n} \mathcal{F}_i \odot 1$ the frame of which is constructed as in the definition of $S4(1, \infty)$-logic and with the valuation transferred from the models $\mathcal{F}_i$. It can be verified directly that the formula $\Box\Diamond d \rightarrow \bigvee_{1 \leq i \leq n} \Box\Diamond g_i$ is false in the model $\bigsqcup_{1 \leq i \leq n} \mathcal{F}_i \odot 1$ on its smallest element. Hence this formula is not a theorem of λ because λ is a certain $S4(1, \infty)$-logic. This contradicts (3.21). Thus (3.16) holds.

We need to show (3.17). We choose:

$$t_{X_i}(x_1, ..., x_n, y) := (a_i \wedge \bigwedge_{j \in X_i} \neg\Diamond a_j)).$$

For $\mathcal{M}$ we pick the free modal algebra $\mathfrak{F}_\lambda(\beta)$ from the class $\mathcal{K}$ which has $\mathfrak{F}_\lambda(n)$ as an subalgebra. We let elements $a_i, 1 \leq i \leq n$ be the free generators of the algebra $\mathfrak{F}_\lambda(n)$, and $v := \bot$. We define elements b_i as follows:

$$b_i := \Box\Diamond(a_i \wedge \bigwedge_{j \in X_i} \neg\Diamond a_j).$$

Suppose $j \notin X_i$. Then by the definition of a CDL-classes it follows $i \notin X_j$. Consider a single-element reflexive model $\mathcal{U}_1$ with a valuation of the letters a_m, $1 \leq m \leq n$ such that only the variables a_i and a_j are valid on its single element. Then it can be easily seen that the formula $b_i \wedge b_j \equiv \bot$ is false in $\mathcal{U}_1$ and therefore this formula is not a theorem of λ since λ is a $S4(1, \infty)$-logic. Then $b_i \wedge b_j \neq \bot$ in the free algebra $\mathfrak{F}_\lambda(\beta)$.

Suppose now that $j \in X_i$ and $i \in X_j$. In this case $b_i \wedge b_j \leq \Box\Diamond a_i \wedge \Box\Diamond\neg\Diamond a_i \leq \bot$. Thus (3.17) holds. In order to check (3.18) consider a model $\mathcal{G}$ such that $\mathcal{G}$ has a single-element reflexive frame, and a valuation of letters a_m, $1 \leq m \leq n$, such that only the variable a_i is valid on the single basis element of $\mathcal{G}$. Then it can be easily seen that the formula $b_i \rightarrow b_j$ is false in $\mathcal{G}$. As λ is a $S4(1, \infty)$-logic we have $b_i \rightarrow b_j \notin \lambda$. We can similarly show that $b_j \rightarrow b_i \notin \lambda$. This entails the non-comparability of the elements b_i and b_j in the algebra $\mathfrak{F}_\lambda(\beta)$. Hence (3.18) holds. Using Theorem 3.6.11 we conclude our proof. ∎

The two previous theorems show that the elementary theories of free algebras from varieties corresponding to very strong and simple modal and superintuitionistic logics are already undecidable. Indeed, it is easy to see that the
minimal modal and superintuitionistic logics of depth 2 are among the logics to
which these theorems are applicable. These minimal logics of depth 2 are very
simple, locally finite and have finite axiomatization; but nevertheless even they
cannot have decidable elementary theories of the free algebras. It is reasonable
likely to expect that for weak logics the result must be the same, because weak
logics usually have more negative properties. It is really the case, and we intend
to show this now. But in this case we must specify more precisely the semantic
properties of logics that we consider.

Definition 3.6.16 *Let λ be a modal logic extending the minimal normal modal
logic K. We say λ is a $K(\infty)$-logic if the following hold.*

(1) all theorems of λ are valid in the reflexive single-point frame;

*(2) Suppose $\mathcal{F}_i$, $1 \le i \le n$ is a tuple of rooted finite λ-frames each of which is
not a single-point irreflexive frame. Then all theorems of λ are valid in
the frame $\biguplus_{1 \le i \le n} F_i \oplus 1$ which is constructed from frames $\mathcal{F}_i$ as follows.
Suppose the tuple $\mathcal{F}_i, 1 \le i \le n$ consist of a tuple $\mathcal{D}_{i,i \in I}$ of frames with
reflexive roots and a tuple $\mathcal{G}_j, j \in J$ of frames with irreflexive roots.
We construct frames $\mathcal{W}_j, j \in J$ from frames of $\mathcal{G}_j, j \in J$ by removing
from all frames $\mathcal{G}_j$ the irreflexive roots. The frame $\biguplus_{1 \le i \le n} \mathcal{F}_i \oplus 1$ is the
frame obtained from the disjoint union $\bigsqcup_{1 \le i \le n} \mathcal{E}_i$ of all $\mathcal{D}_i, i \in I$ and all
$\mathcal{W}_{j,j \in J}$ by adjoining the new irreflexive element a such that:*

> *(i) all elements of each $\mathcal{D}_i$ which are accessible in $\mathcal{D}_i$ from the
> root $\mathcal{D}_i$ are accessible now from a,*

> *(ii) all elements of every $\mathcal{W}_j$ which are accessible in $\mathcal{G}_j$ from
> the irreflexive root of $\mathcal{G}_j$ are accessible from a.*

Theorem 3.6.17 *Let λ be a modal $K(\infty)$-logic with the finite model property. Let $\mathcal{K}$ be a class of free modal algebras of variety $Var(\lambda)$ which contain
among their subalgebras all free algebras of finite rank from $Var(\lambda)$. Then the
elementary first-order theory of $\mathcal{K}$ is hereditarily undecidable.*

Proof. In order to use Theorem 3.6.11 we have to show that $\mathcal{K}$ is a CDL-class.
Let $f(x) := \Box x$. To show (3.16) we take a free algebra $\mathfrak{F}_\lambda(\beta)$ from $\mathcal{K}$ and some
elements d, g_i from this algebra. As before, all d, g_i can be understood as terms
or modal formulas built up out of free generators. Assume that (3.16) is false for
this $f(x)$ and the tuple of elements $d, g_i, 1 \le i \le n$, i.e.:

$$\mathfrak{F}_\lambda(\beta) \models \Box d \le \bigvee_{1 \le i \le n} \Box g_i, \quad \mathfrak{F}_\lambda(\beta) \not\models \bigvee_{1 \le i \le n} (\Box d \to \Box g_i). \tag{3.22}$$

Using the finite model property of λ we obtain, for each i, there is a rooted finite model $\mathcal{F}_i$ which disproves on the its root the formula $\Box d \to \Box g_i$, and which is based upon a finite λ-frame and is not a single-element irreflexive model. We take the model $\biguplus_{1 \leq i \leq n} \mathcal{F}_i \oplus 1$ with the frame which is constructed out of the frames of the models $\mathcal{F}_i$ as it was shown in the definition of a $K(\infty)$-logic, and with the valuation which is transferred from the models $\mathcal{F}_i$. It can be verified directly that the formula $\Box d \to \bigvee_{1 \leq i \leq n} \Box g_i$ is false in the model $\biguplus_{1 \leq i \leq n} \mathcal{F}_i \oplus 1$. Therefore this formula is not a theorem of λ because λ is a certain $K(\infty)$-logic, which contradicts (3.22). Thus (3.16) holds.

Now we will check (3.17). We fix the formulas, the algebra and the elements: we set as before

$$t_{X_i}(x_1, ..., x_n, y) := (x_i \wedge \bigwedge_{j \in X_i} \neg x_j)),$$

for $\mathcal{M}$ we choose the free modal algebra $\mathfrak{F}_\lambda(\beta)$ which contains $\mathfrak{F}_\lambda(n)$ as a subalgebra. As elements $a_i, 1 \leq i \leq n$ we take the free generators of the algebra $\mathfrak{F}_\lambda(n)$, and we let $v := \Box \perp$. According to this choice all elements b_i have the form

$$b_i := \Box(a_i \wedge \bigwedge_{j \in X_i} \neg a_j).$$

Assume that $j \notin X_i$, then by the definition of a CDL-class we conclude $i \notin X_j$. Consider the single-element reflexive transitive model $\mathcal{M}_1$ with a valuation of the letters a_m, $1 \leq v \leq n$, such that only the letters a_i and a_j are valid on the basis element of $\mathcal{M}_1$. Then it can be easily seen that the formula $b_i \wedge b_j \equiv \Box \perp$ is false in $\mathcal{M}_1$. Therefore, this formula is not theorem of λ since λ is a $K(\infty)$-logic. Then $b_i \wedge b_j \neq \Box \perp$ in the free algebra $\mathfrak{F}_\lambda(\beta)$.

Conversely, suppose $j \in X_i$ and $i \in X_j$. In this case, $\Box \perp \leq b_i \wedge b_j \leq \Box a_i \wedge \Box \neg a_i \leq \Box \perp$. Thus (3.6.2) holds. As for (3.6.4), we consider again the models of kind $\mathcal{G}$ from the final part of the proof of Theorem 3.6.15, and show in a similar way that the elements b_i and b_j are incomparable in $\mathfrak{F}_\lambda(\beta)$. Thus (3.18) holds, $\mathcal{K}$ is a CDL-class, and by Theorem 3.6.11 $Th(\mathcal{K})$ is hereditarily undecidable. ∎

Using this theorem we can extract a corollary for modal free algebras from varieties corresponding to (i) the smallest normal modal logic K, (ii) the modal logic $K4$ and (iii) the Von Wright modal logic T.

Corollary 3.6.18 *The free modal algebras* $\mathfrak{F}_K(\omega)$, $\mathfrak{F}_{K4}(\omega)$ *and* $\mathfrak{F}_T(\omega)$ *have hereditarily undecidable first-order theories.*

The theorems presented above concerning undecidability of elementary theories of free modal algebras do not work directly neither for the free modal algebras $\mathfrak{F}_\lambda(\omega)$ of varieties corresponding to modal logics λ over the Gödel-Löb provability logic GL nor for $\mathfrak{F}_{GL}(\omega)$ itself. But by appropriately choosing the therms concerning the definition of CDL-classes for such modal logics, it is possible to obtain a general description theorem using even simpler arguments. First we specify a class of provability modal logics.

Definition 3.6.19 *We say an extension λ of the modal provability logic GL is a $GL(2, \infty)$-logic if all the theorems of this logic are valid in each transitive irreflexive frame $\bigsqcup_{1 \leq i \leq n} \mathcal{F}_i \oplus 1$ which is obtained from the disjoint union $\bigsqcup_{1 \leq i \leq n} \mathcal{F}_i$ of the frames $\mathcal{F}_i$ each of which is an irreflexive element by adjoining to $\bigsqcup_{1 \leq i \leq n} \mathcal{F}_i$ an R-smallest irreflexive element.*

Theorem 3.6.20 *Let λ be a modal $GL(2, \infty)$-logic. Let $\mathcal{K}$ be a class of free modal algebras from $Var(\lambda)$ such that any free algebra of finite rank from $Var(\lambda)$ is embeddable in an appropriate algebra from $\mathcal{K}$. Then $Th(\mathcal{K})$ is hereditarily undecidable.*

Proof. Applying Theorem 3.6.11 we need to choose some terms corresponding to the definition of the CDL-class. We let

$$f(x) := \Box(\Box\bot \to x).$$

To prove relation (3.16) suppose d, g_i, $1 \leq i \leq n$ are elements of a free algebra $\mathfrak{F}_\lambda(\beta)$ from $\mathcal{K}$. We can consider all these elements as certain terms, or formulas of the logic λ, which are built up out of free generators. Suppose (3.16) is false for our choice of $f(x)$ and these elements. Then the premise in relation (3.16) is transformed in our case into

$$\mathfrak{F}_\lambda(\beta) \models \Box(\Box\bot \to d) \leq \bigvee_{1 \leq i \leq n} \Box(\Box\bot \to g_i), \tag{3.23}$$

$$\mathfrak{F}_\lambda(\beta) \not\models \bigvee_{1 \leq i \leq n} (\Box(\Box\bot \to d) \leq \Box(\Box\bot \to g_i)).$$

Then by the finite model property of the provability logic GL (Lemma 2.6.20), for each i, there is a finite transitive irreflexive model $\mathfrak{M}_i$ such that the formula $\Box(\Box\bot \to d) \to \Box(\Box\bot \to g_i)$ is false in $\mathfrak{M}_i$. This entails that there is a model $\mathcal{F}_i := \langle \{s_i\}, R, V_i \rangle$ which is based upon a single irreflexive element s_i and $s_i \Vdash_{V_i} d$, and $s_i \not\Vdash_{V_i} g_i$ (it is easy to see from the structure of our formula and $\mathfrak{M}_i$) is irreflexive.

We take the model $\bigsqcup_{1 \leq i \leq n} \mathcal{F}_i \oplus 1$, where the frame of this model is obtained from the disjoint union of the frames of the all given models $\mathcal{F}_i$ by adjoining a smallest irreflexive element, and which has the valuation transferred onto it from the models $\mathcal{F}_i$. It can be easily verified that the formula $\Box(\Box\bot \to d) \to \bigvee_{1 \leq i \leq n} \Box(\Box\bot \to g_i)$ is false in the model $\bigsqcup_{1 \leq i \leq n} \mathcal{F}_i \oplus 1$. Therefore this formula is not a theorem of λ because λ is a $GL(2, \infty)$-logic, which contradicts (3.23). Thus (3.16) holds.

In order to prove relation (3.17) we choose appropriate terms, algebras and elements. We put

$$t_{X_i}(x_1, ..., x_n, y) := (x_i \wedge \bigwedge_{j \in X_i} (\neg x_j)),$$

and take as the algebra $\mathcal{M}$ the algebra $\mathfrak{F}_\lambda(\beta) \in \mathcal{K}$, where $\mathfrak{F}_\lambda(n)$ is a subalgebra of $\mathfrak{F}_\lambda(\beta)$. We take all the elements $a_i, 1 \le i \le n$ to be free generators of the algebra $\mathfrak{F}_\lambda(n)$, and choose $v := \Box \perp$. For this case all b_i have the form

$$b_i := \Box(\Box\perp \to (a_i \wedge \bigwedge_{j \in X_i} \neg a_j)).$$

If $j \notin X_i$ then we have $i \notin X_j$. Consider the following model: $\mathcal{F}_2$ is based upon a two-element irreflexive transitive chain, and it has a valuation of the letters $a_m, 1 \le m \le n$ such that only the letters a_i and a_j are valid on the last (final) element of this chain. Then it can be easily seen that the formula $b_i \wedge b_j \equiv \Box\perp$ ($\equiv$ in the sense of GL) is false on $\mathcal{F}_2$. Consequently this formula is not a theorem of λ since $\mathcal{M}_2$ is based upon a λ-frame. Hence $b_i \wedge b_j \ne \Box \perp$ in the free algebra $\mathfrak{F}_\lambda(\beta)$.

Assume that $j \in X_i$ and $i \in X_j$. Then

$$\Box \perp \le b_i \wedge b_j \le \Box(\Box\perp \to a_i) \wedge \Box(\Box\perp \to \neg a_i) =$$
$$\Box(\Box\perp \to \perp) = \Box\neg\Box\perp = \Box\perp$$

The last equality holds in any modal algebra from $Var(GL)$. Thus the relation (3.17) holds for our choice. It remains only to check (3.18). We consider the following model $\mathcal{M}_2$: this model is based upon a two-element irreflexive transitive chain, and has a valuation of the letters $a_m, 1 \le m \le n$ such that only the variable a_i is valid on its last (final) element. Then it can be easily seen that the formula $b_i \to b_j$ is false in the model $\mathcal{M}_2$. The frame of $\mathcal{M}_2$ is a λ-frame. Therefore $b_i \to b_j$ is not a theorem of λ. In a similar way we show $b_j \to b_i \notin \lambda$. Thus elements b_i and b_j are incomparable in the algebra $\mathfrak{F}_\lambda(\beta)$. Hence the relation (3.18) also holds. We have proved that $\mathcal{K}$ is a CDL-class and applying Theorem 3.6.11 it follows that the first-order theory of $\mathcal{K}$ is hereditarily undecidable. ∎

Corollary 3.6.21 *The class of all free diagonalizable algebras $\mathfrak{F}_{GL}(n)$ of finite rank, as well as the free algebra $\mathfrak{F}_{GL}(\omega)$ of countable rank, have hereditarily undecidable elementary theories.*

It is relevant to note that Artemov and L.Beklemishev [3] studying propositional quantifiers in provability logic have showed that the elementary theory of every free diagonalizable algebra $\mathfrak{F}_{GL}(n)$ for $n > 1$ is undecidable. We shall close the section by considering one more type of non-standard logics which have undecidable theories of free algebras from the corresponding varieties. These logics, which are called *interpretability logics*, are related to modal provability logics and, in a sense, are based up on them. The language of interpretability logics contains the language of modal logic but is enriched by an additional single binary connective $\triangleright$. The informal meaning of $\alpha \triangleright \beta$ is the following (i.e. it can be read as asserting:): α is *interpretable* by β.

The Kripke semantics for interpretability logic consists of IL-frames. An IL-frame is a triple $\langle W, R, S \rangle$, where the frame $\langle W, R \rangle$ is a GL-frame and $S := \{ S_w \mid w \in W \}$ is a collection of binary relations on W with the following properties. Any S_w is a binary relation on elements accessible from w by R, S_w is reflexive and transitive, and for each two elements accessible from w by R elements a and b, $aRb \Rightarrow aS_w b$. An IL-model is an IL-frame with a given valuation of the propositional letters.

The truth relation of formulas in the language of interpretability logic in IL-Kripke models is defined in the same way as the truth of modal formulas, considering the truth for the additional connective $\triangleright$ as follows:

$$u \Vdash \alpha \triangleright \beta \iff \forall v[((uRv) \wedge v \Vdash \alpha) \Rightarrow \exists w((vS_u w) \,\&\, w \Vdash \beta)].$$

It is known [52] that the smallest interpretability logic IL can be defined as the set of all formulas which are valid in all finite IL-frames. An interpretability logic is a set of formulas which contains all the theorems of IL and which is closed with respect to modus ponens, the normalization rule $x/\Box x$, and substitution.

Definition 3.6.22 *We say that an interpretability logic λ is an $IL(1, \infty)$-logic if all the theorems of λ are valid in each IL-frame $\bigotimes_{1 \le i \le n} \mathcal{F}_i \oplus 1$ which has the following structure. Let $\mathcal{F}_i$ be a single-element IL-frame (i.e. the set of the additional binary relations S is empty for any frame $\mathcal{F}_i$) The frame $\mathcal{F}$ is the disjoint union of all these frames $\mathcal{F}_i$, $1 \le i \le n$; and the frame $\bigotimes_{1 \le i \le n} \mathcal{F}_i \oplus 1$ is obtained from $\mathcal{F}$ by adjoining a new R-smallest irreflexive element a and by setting $S_a := \emptyset$.*

Theorem 3.6.23 *Let λ be an $IL(1, \infty)$-logic. And let $\mathcal{K}$ be a class of free algebras of variety $Var(\lambda)$ which contain among their subalgebras all the free algebras of finite rank from $Var(\lambda)$. Then the first-order theory of $\mathcal{K}$ is hereditarily undecidable.*

Proof. We follow to the proof of Theorem 3.6.20. Hence, in particular, we set $f(x) := \Box(\Box\bot \to x)$. The relation (3.16) can be verified in a same way as in Theorem 3.6.20. Indeed, suppose d, g_i, $1 \le i \le n$ are elements of $\mathfrak{F}_\lambda(\beta) \in \mathcal{K}$. Assuming (3.16) is false for such choice we have

$$\mathfrak{F}_\lambda(\beta) \models \Box(\Box\bot \to d) \le \bigvee_{1 \le i \le n} \Box(\Box\bot \to g_i), \tag{3.24}$$

$$\mathfrak{F}_\lambda(\beta) \not\models \bigvee_{1\leq i\leq n} (\square(\square\bot \to d) \leq \square(\square\bot \to g_i)).$$

Then since IL is the set of all formulas which are valid in all finite IL-frames, for every i, there is a finite transitive irreflexive IL-model $\mathfrak{M}_i$ such that the formula $\square(\square\bot \to d) \leq \square(\square\bot \to g_i)$ is false in $\mathfrak{M}_i$. Then it is easy to see that there is an single element irreflexive IL-model $\mathcal{F}_i := \langle\{s_i\}, R, S, V_i\rangle$ such that $s_i \Vdash_{V_i} d$ and $s_i \nVdash_{V_i} g_i$. We take the model $\bigotimes_{1\leq i\leq n} K_i \oplus 1$ with the frame obtained from frames $\mathcal{F}_i$ as it is shown in the definition of a $IL(1,\infty)$-logic and with the valuation transferred from models $\mathcal{F}_i$. It can be directly verified that the formula $\square(\square\bot \to d) \leq \bigvee_{1\leq i\leq n} \square(\square\bot \to g_i)$ is false in the model $\bigotimes_{1\leq i\leq n} K_i \oplus 1$. Therefore this formula is not a theorem of λ since λ is a $IL(1,\infty)$-logic. This contradicts (3.24), thus (3.16) holds.

The modal logic $\lambda_\square$ which consists of all the formulas from λ which have no occurrences of $\rhd$ is a $GL(1,\infty)$-logic. This follows directly from our definitions. Moreover, for all formulas α and β in the language of $\lambda_\square$, $\alpha \equiv \beta \in \lambda_\square$ if and only if $\alpha \equiv \beta \in \lambda$. Therefore for any m, the algebra $\mathfrak{F}_{\lambda_\square}(m)$ is a subalgebra of $\mathfrak{F}_\lambda(m)$. From this observations it follows directly that the relations (3.17) and (3.18) can be proved in a similar way as the corresponding relations in Theorem 3.6.20. Applying Theorem 3.6.11 we establish the conjecture of this theorem. ∎

As we have seen above, all the applications of Theorem 3.6.11 to obtain results concerning undecidability of the first-order theories of free algebras from varieties of algebras corresponding to algebraic non-standard logics are given in a rather uniform manner. It looks likely that this theorem will also work for other algebraic logics with lattice structure representable by first-order formulas. Therefore the first-order meta-theories of non-standard logics would appeals to be often undecidable.

3.7 Scheme-Logics of First-Order Theories

In previous sections we have studied admissible rules of propositional algebraic logics and up to now we did not have precise information about admissible rules of first order logic. At the same time first order logic has powerful language by means of which we can precisely describe properties of models, and it looks quite natural to consider the admissibility for first order logic also. In this section we investigate admissibility of inference rules in first-order theories which are based upon classical predicate logic. We will work out a complete description of those first-order theories which have an algorithm for determining the admissibility of inference rules. For this purpose we will use formula schemes of first-order theories. The approach that we will use is based upon a simple observation that the set of all formula schemes for a given first-order theory forms a poly-modal propositional logic which we call a scheme-logic. Our interest to scheme-logics comes primary from the possibility of describing the notion of admissibility of inference rules within this language. It is also rather natural to clarify the recursive complexity of these logics, to investigate what semantics are adequate for them.

We begin with an opening definition of formula schemes. What is a formula scheme (or scheme of theorems) for a given first order theory T? As it is well known, a formula scheme (or a scheme of theorems) is an expression (or a formula) $A(z_1, ..., z_n)$, which is built up out of metavariables z_i by means of logical connectives, quantifiers, and equalities between variables, and which has the following property. For any tuple of formulas $\Psi_1, ..., \Psi_n$ in the first-order language of our theory T, $A(\Psi_1, ..., \Psi_n) \in T$ holds. Therefore we introduce the language SL of scheme-logics in the following way. This language has a countable set of metavariables z_i for formulas, the usual logical connectives, a countable set of unary logical connectives $\forall x_i$ (which simulate the universal quantifier) and the countable set of 0-place logical connectives (or logical constants) $(x_i = x_j)$ (which simulate equality between variables). The formulas in the language SL are built up out of the metavariables by using the above mentioned connectives in usual way. As often before, we will abbreviate in this section the formula $(x \to y) \land (y \to x)$ by $x \equiv y$, and the writing $A(z_i)$ means that the formula A is built up out of a certain set of meta-variables z_i. Hence

$$\forall x_1 (z_1 \& z_2) \equiv (\forall x_1 z_1) \& (\forall x_1 z_2), \quad \forall x_1 \exists x_1 z_1 \to \exists x_1 z_1$$

are examples of formulas in the language SL.

Definition 3.7.1 *The scheme-logic* $SL(T)$ *for a first-order theory* T *is the set*

$$\{A(z_i) \mid A(z_i) \text{ is a scheme formula}, A(\Psi_i) \in T$$
$$\text{for arbitrary formulas } \Psi_i \text{ in the language of } T \}.$$

We can consider each $SL(T)$ as a propositional logic with propositional variables z_i and the logical connectives mentioned above (including logical constants). The logical connectives $\forall x_i$ can be understood as modal *necessity* operators $\Box_i$. It is clear that every $\Box_i$ satisfies all the laws of the modal system $S4$. Moreover, for each theory T, we clearly have

$$(\exists x_i z_1 \to \forall x_i \exists x_i z_1) \in SL(T)$$

$$(\forall x_i \forall x_j z_1 \equiv \forall x_j \forall x_i z_1) \in SL(T).$$

Hence, if we consider connective $\forall x_i$ as the modal logical connective $\Box_i$,

$$\Diamond_i z \to \Box_i \Diamond_i z \in SL(T)$$

holds for any first-order theory T. This immediately yields

Proposition 3.7.2 *For any first-order theory T, $SL(T)$ is a polymodal propositional logic with constants which extends the poly-modal analog $S5_\infty$ (with constants) of the Lewis modal logic $S5$, and $SL(T)$ has the law of commutating modalities: $\Box_i\Box_j z \equiv \Box_j\Box_i z \in SL(T)$; i.e.*

$$S5_\infty \oplus \{(\Box_i\Box_j z \equiv \Box_j\Box_i z) \mid i, j \in N\} \subseteq SL(T).$$

An structural inference rule for a first-order theory (or logic) T is an expression of the form

$$\frac{A_1(z_i), \ldots, A_n(z_i)}{B(z_i)}$$

(or, more simply, of the form $A_1(z_i), \ldots, A_n(z_i)/B(z_i)$), where all $A_j(z_i)$ and $B(z_i)$ are formulas in the language SL. We will consider in this section only structural inference rules of first order theories, and will call they simply inference rules for short.

Definition 3.7.3 *An inference rule $A_1(z_i), \ldots, A_n(z_i)/B(z_i)$ is admissible in a first-order theory T if for any tuple of formulas Ψ_i in the language of T the following holds:*

$$A_1(\Psi_i) \in T, \ldots, A_n(\Psi_i) \in T \Rightarrow B(\Psi_i) \in T.$$

It is clear that the admissible inference rules for a theory T are exactly those which we can add to the set of postulated inference rules of T without increasing the set of provable theorems (though by adding such rules we will increase the power of the resulting deductive system for our first-order theory). Of course the rule $z \to z/A(z_i)$ is admissible in T iff $A(z_i) \in SL(T)$. Therefore if $SL(T)$ is undecidable then there exists no algorithm which could determine admissibility of inference rules in T. It is our primary motivation for studying the scheme-logics $SL(T)$ of first-order theories T: as soon as we show that $SL(T)$ for a given T is undecidable, we know that there can be no algorithm determining admissibility in T.

In order to investigate the logics $SL(T)$ enough deeply we will need to employ certain tools from the model theory.

Definition 3.7.4 *Suppose $\mathcal{M}_1$ and $\mathcal{M}_2$ are models (algebraic systems) in a signature Σ. Let g be a homomorphism from $\mathcal{M}_1$ onto $\mathcal{M}_2$. A homomorphism g is called strong if for any predicate $P_i(x_1, \ldots, x_{m_i})$ from Σ,*

$$\mathcal{M}_1 \models P_i(a_1, \ldots, a_{m_i}) \Leftrightarrow \mathcal{M}_2 \models P_i(g(a_1), \ldots, g(a_{m_i})).$$

The next lemma is well known (and can easily be verified by induction on the length of formulas, as a simple exercise).

Lemma 3.7.5 *Let g be a strong homomorphism of $\mathcal{M}_1$ onto $\mathcal{M}_2$. Suppose that $A(x_1, ..., x_n)$ is a formula with the free variables $x_1, ..., x_m$ which has no occurrences of equality, then, for any interpretation s of the free variables of A in $\mathcal{M}_1$, where $s(x_i) := a_i$,*

$$\mathcal{M}_1 \models A(a_1, ..., a_m) \Leftrightarrow \mathcal{M}_2 \models A(g(a_1), ..., g(a_m)).$$

Let K_1 and K_2 be certain classes of models in signatures Σ_1 and Σ_2, respectively, where Σ_1 has no constants or function symbols.

Definition 3.7.6 *The class K_1 is weakly reflected in K_2 if for any model $\mathcal{M}_1 \in K_1$ there exists a model $\mathcal{M}_2 \in K_2$ and there is a tuple $a_1, ..., a_m$ of elements from $\mathcal{M}_2$ such that the following holds. For all predicates $P_i(v_1, ..., v_{n_i})$ from Σ_1, there are certain first-order formulas $f_i(v_1, ..., v_{n_i}, y_1, ..., y_m)$ in the signature Σ_2 which have the following property. There is a strong homomorphism of the model $\langle \mathcal{M}_2, f_i(v_1, ..., v_{n_i}, a_1, ..., a_m) \rangle$ onto the model $\mathcal{M}_1$ (where $f_i(v_1, ..., v_{n_i}, a_1, ..., a_m)$ are predicates of arities n_i corresponding to predicates P_i, and which are generated in M_2 by the truth value of the formulas f_i under any value of the variables $v_1, ..., v_{n_i}$ and the value $a_1, ..., a_m$ for the variables $y_1, ..., y_m$).*

Readers experienced in model theory can easily recognize in this definition a partial case of relatively elementary definability. We recall also that certain two sets X, Y, where $X \subseteq Y$, are called *recursively inseparable* if there is no a decidable set Z such that $X \subseteq Z \subseteq Y$. And recall that, for a class of algebraic systems (or models) K, $Th(K)$ denotes the elementary theory of K. $Th_p(K)$ denotes the pure elementary theory of K (i.e. the set of all first-order formulas in the signature of K without equality which are true on K); PPC_Σ denotes the (classical) pure predicate calculus in the signature Σ (predicate calculus without equality and without function symbols). Let K_1 be a class of models in a signature Σ_1 which has no function and constant symbols.

Theorem 3.7.7 *If K_1 is weakly reflected in a class K_2 and $Th_p(K_1)$ and PPC_{Σ_1} are recursively inseparable then $SL(Th(K_2))$ is undecidable.*

Proof. Let A be a formula in the signature Σ_1 without equality, let $Var(A)$ be all the variables occurring in A, and let $P_i(x_{j1}, ..., x_{jn_i})$ be all the predicate letters from A with all the given occurrences of variables as they occur in the formula A. We define a translation of A into a certain formula A^t in the language of the scheme-logic, where

$$A^t := (P(A) \rightarrow V(A)),$$

$V(A)$ is obtained from A by replacing of all $P_i(x_{j1}, ..., x_{jn_i})$ by the corresponding meta-variables $z_{P_i(x_{j1}, ..., x_{jn_i})}$, and $P(A)$ is the conjunction of all the following formulas

$$\forall x_{j1} ... \forall x_{jn_i} [\forall x_1 ... \forall x_d z_{P_i(x_{j1}, ..., x_{jn_i})} \equiv \exists x_1 ... \exists x_d z_{P_i(x_{j1}, ..., x_{jn_i})}], \quad (3.25)$$

where $\{x_1, ..., x_d\} := [\, Var(A) - \{x_{j1}, ..., x_{jn_i}\} \,]$,

$$\forall x_{j1}...\forall x_{jn_i}\forall x_{k1}...\forall x_{kn_i}\forall y_1...\forall y_p[\,\bigwedge(x_{jr} = x_{kr}) \rightarrow \qquad (3.26)$$

$$\rightarrow (z_{P_i(x_{j1},...,x_{jn_i})} \equiv z_{P_i(x_{k1},...,x_{kn_i})})],$$

where $\{y_1, ..., y_p\} := [\, Var(A) - \{x_{j1}, ..., x_{jn_i}, x_{k1}, ..., x_{kn_i}\} \,]$,

with respect to all $z_{P_i(x_{j1},...,x_{jn_i})}$ and $z_{P_i(x_{k1},...,x_{kn_i})}$. We now need the following

Lemma 3.7.8 *If A is a theorem of PPC_{Σ_1} then, for each first-order theory T, A^t is included in $SL(T)$.*

Proof. Suppose that $A^t \notin SL(T)$. This means there is a tuple of first-order formulas $B_{P_i(x_{j1},...,x_{j_in})}$ in the signature of the theory T which have the following property. The formula

$$(A^t)(B_{P_i(x_{j1},...,x_{jn_i})} \hookleftarrow z_{P_i(x_{j1},...,x_{jn_i})})$$

which is obtained from A^t by substitution of the formulas $B_{P_i(x_{j1},...,x_{jn_i})}$ in place of the meta-variables $z_{P_i(x_{j1},...,x_{jn_i})}$ is not a theorem of T. Therefore there is a model M of the theory T and there is an interpretation $s : y_l \mapsto b_l$ of all the free variables y_l of the formula $(A^t)\,(B_{P_i(x_{j1},...,x_{jn_i})} \hookleftarrow z_{P_i(x_{j1},...,x_{jn_i})})$ in the model M such that

$$M \not\models (A^t)(B_{P_i(x_{j1},...,x_{jn_i})} \hookleftarrow z_{P_i(x_{j1},...,x_{jn_i})})\binom{y_l}{b_l}.$$

We now introduce new predicate relations $Q_i(x_1, ..., x_{n_i})$ in M as follows

$$\forall c_{j1} \in M, ..., \forall c_{jn_i} \in M \; (M \models Q_i(c_{j1}, ..., c_{jn_i})) \Longleftrightarrow$$

$$M \models B_{P_i(x_{j1},...,x_{jn_i})}\binom{x_{j1},...,x_{jn_i}}{c_{j1},...,c_{jn_i}}\binom{y}{c}\binom{y_l}{b_l},$$

where

any y is a variable from $(Var(A) - \{x_{j1}, ..., x_{jn_i}\})$,

any y_l is a free variable of the formula $B_{P_i(x_{j1},...,x_{jn_i})}$ which differs from variables of $Var(A)$,

c is a fixed element of M,

any b_l is the above given interpretation under s of the variable y_l.

It is easy to see that our definition of Q_i above depends formally on the choice of the tuple of variables $x_{j1}, ..., x_{jn_i}$ and the corresponding formula $B_{P_i(x_{j1},...,x_{jn_i})}$. Now we show that, in fact, the choice of the tuple $x_{j1}, ..., x_{jn_i}$ is not essential, and that, for any formula $B_{P_i(x_{k1},...,x_{kn_i})}$ corresponding to any metavariable $z_{P_i(x_{k1},...,x_{kn_i})}$, the definition of Q_i is the same as for $B_{P_i(x_{j1},...,x_{jn_i})}$. Indeed, the fact that $M \not\models (A^t)(B_{P_i(x_{j1},...,x_{jn_i})} \leftrightarrow z_{P_i(x_{j1},...,x_{jn_i})})\binom{y_l}{b_l}$ implies

$$M \models P(A)_t\binom{y_l}{b_l}, \tag{3.27}$$

$$M \not\models V(A)_t\binom{y_l}{b_l}, \tag{3.28}$$

where formulas $P(A)_t$ and $V(A)^t$ are the premise and the conclusion of the formula $(A^t)(B_{P_i(x_{j1},...,x_{jn_i})} \leftrightarrow z_{P_i(x_{j1},...,x_{jn_i})})$ respectively. Using (3.27) it is easy to see that, for any $c_{j1}, ..., c_{jn_i} \in M$ and any $c \in M, b_l \in M, b_g \in M$,

$$M \models B_{P_i(x_{j1},...,x_{jn_i})}\binom{x_{j1},...,x_{jn_i}}{c_{j1},...,c_{jn_i}}\binom{y}{c}\binom{y_l}{b_l} \Leftrightarrow$$

$$\Leftrightarrow M \models B_{P_i(x_{k1},...,x_{kn_i})}\binom{x_{k1},...,x_{kn_i}}{c_{j1},...,c_{jn_i}}\binom{w}{c}\binom{w_g}{b_g}, \tag{3.29}$$

where w are variables from $(Var(A) - \{x_{k1}, ..., x_{kn_i}\})$, w_g are all the free variables from $B_{P_i(x_{k1},...,x_{kn_i})}$ which differ from the variables of $Var(A)$, and b_g are the elements of M which our interpretation s assigns to the variables w_g. Thus we can take either $B_{P_i(x_{j1},...,x_{jn_i})}$ or $B_{P_i(x_{k1},...,x_{kn_i})}$ for our definition of Q_i and we will still obtain the same predicate (in this sense, the truth of Q_i in M is expressible by means of formulas of the form $B_{P_i(x_{j1},...,x_{jn_i})}$). This observation, together with (3.28) and (3.29), implies that the formula A is false in the model of signature Σ_1 which is obtained by introducing predicates $Q_i(x_1, ..., x_n)$ in M, while interpretation of any predicate P_i as Q_i. This contradicts the fact that the formula A is a theorem of PPC_{Σ_1}. ∎

To continue the proof of our theorem, we assume $SL(Th(K_2))$ is decidable. The translation t is a one-to-one mapping which is effectively defined. Therefore the set $t^{-1}(SL(Th(K_2))$ is also decidable. We will prove that

$$t^{-1}(SL(Th(K_2)) \subseteq Th_p(K_1) \tag{3.30}$$

Indeed, let A be a member of $t^{-1}(SL(Th(K_2))$. Then $A^t \in SL(Th(K_2))$. Suppose that A does not belong to $Th_p(K_1)$. Then there exists a model M_1 of K_1 which falsifies A. By our assumption, K_1 is weakly reflected in K_2. Therefore for our given model M_1, there is a model M_2 from K_2 (related to M_1) with those properties which are required by our definition of weak reflection. In particular, there is a strong homomorphism g of the model $\langle M_2, f_i(v_1, ..., v_{n_i}, a_1, ..., a_m)\rangle$ onto the model M_1 (where $f_i(v_1, ...v_{n_i}, a_1, ..., a_m)$ are predicates generated by the truth value of formulas f_i under any values of the variables $v_1, ...v_{n_i}$ and fixed elements $a_1, ..., a_m$, according to the definition of the weak reflection). Since A is false in M_1, according to Lemma 3.7.5,

$$\langle M_2, f_i(v_1, ..., v_{n_i}, a_1, ..., a_m)\rangle \not\models A\binom{y_r}{b_r}),$$

where b_r are the elements of M_2 which are interpretation of the free variables y_r of the formula A. Thus if we denote by $A^{P_i(x_{j1},...,x_{jn_i})}_{f_i(x_{j1},...,x_{jn_i},u_1,...,u_m)}$ the formula obtained from A by substituting the formulas $f_i(x_{j1}, ..., x_{jn_i}, u_1, ..., u_m)$ (where $u_1,...,u_m$ are new variables which have no occurrences in all the above mentioned formulas) in place of the subformulas $P_i(x_{j1}, ..., x_{jn_i})$ then

$$\langle M_2, f_i(v_1, ..., v_n, a_1, ..., a_m)\rangle \not\models \nabla_1 \text{ where} \tag{3.31}$$

$$\nabla_1 := A^{P_i(x_{j1},...,x_{jn_i})}_{f_i(x_{j1},...,x_{jn_i},u_1,...,u_m)}\binom{u_1,...,u_m}{a_1,...,a_m}\binom{y_r}{b_r}.$$

We take an interpretation of all the meta-variables from A^t as follows:

$$z_{P_i(x_{j1},...,x_{jn_i})} \mapsto f_i(x_{j1}, ..., x_{jn_i}, w_1, ..., w_m),$$

where $w_1, ..., w_m$ are new variables which have no occurrences in the formulas $f_j(v_1, ..., v_{n_j}, y_1, ..., y_m)$ from the definition of weak reflection and which differ from the variables $x_{j1}, ..., x_{jn_i}$ and all the variables of formulas A. For any formula α in the language of the scheme-logic built up out of the metavariables $z_{P_i(x_{j1},...,x_{jn_i})}$,

$$\alpha^{z_{P_i(x_{j1},...,x_{jn_i})}}_{f_i(x_{j1},...,x_{jn_i},w_1,...,w_m)}$$

denotes the formula which is obtained from the formula α by substituting the formulas $f_i(x_{j1}, ..., x_{jn_i}, w_1, ..., w_m)$ in place of the metavariables $z_{P_i(x_{j1},...,x_{jn_i})}$. We choose the following interpretation s of the new variables $w_1, ..., w_m$ in M_2: $s(w_1) := a_1, ..., s(w_m) := a_m$. It can be easily verified that

$$\langle M_2, f_i(v_1, ..., v_n, a_1, ..., a_m)\rangle \models \nabla_2 \text{ where} \tag{3.32}$$

$$\nabla_2 := P(A)^{z_{P_i(x_{j1},...,x_{jn_i})}}_{f_i(x_{j1},...,x_{jn_i},w_1,...,w_m)}\binom{w_1,...,w_m}{a_1,...,a_m}\binom{w}{b}),$$

and b are the elements of the model M_2 which are the (arbitrary) interpretation of all those free variables w of the formula

$$P(A)^{z_{P_i(x_{j1},...,x_{jn_i})}}_{f_i(x_{j1},...,x_{jn_i},w_1,...,w_m)}$$

which differ from $w_1, ..., w_m$. However, from (3.31) together with (3.32) we can derive immediately that

$$\langle M_2, f_i(v_1, ..., v_n, a_1, ..., a_m)\rangle \not\models$$

$$V(A)^{^{z}P_i(x_{j1},...,x_{jn_i})}_{f_i(x_{j1},...,x_{jn_i},w_1,...,w_m)}\binom{w_1,...,w_m}{a_1,...,a_m}\binom{y_r}{b_r}.$$

and from this and (3.32) we infer

$$\langle M_2, f_i(v_1,...,v_n,a_1,...,a_m)\rangle \not\models A^t \binom{^{z}P_i(x_{j1},...,x_{jn_i})}{f_i(x_{j1},...,x_{jn_i},w_1,...,w_m)}\binom{w_1,...,w_m}{a_1,...,a_m}\binom{y_r}{b_r},$$

which contradicts $A^t \in SL(Th(K_2))$. Hence the inclusion (3.30) holds. From this and Lemma 3.7.8 we obtain that

$$\text{PPC}_{\Sigma_1} \subseteq t^{-1}(SL(Th(K_2))) \subseteq Th_p(K_1), \qquad (3.33)$$

which contradicts the recursive inseparability of the first and the last set in the inclusions (3.33). This completes the proof of our theorem. ■

Now we extract some corollaries from the theorem proved above. First, it is rather interesting to compare the complexity of a first-order theory T and its scheme-logic $SL(T)$. At first glance, $SL(T)$ is much simpler than T itself because (i) $SL(T)$ is simply a poly-modal propositional logic which extends the poly-modal analog of the modal logic $S5$ (which is a very simple modal logic); (ii) the scheme-logic cannot express any concrete, precise properties of the models because all predicates letters have been removed from its language. However surprisingly, all this does not really matter. To be more precise, we have the following theorems.

Theorem 3.7.9 *[146, 149] Let K be a class of models. Assume that, for all $m \in N$, there is a model M in K such that K has at least m elements. Then $SL(Th(K))$ is undecidable.*

Proof. Consider the class K_0 of all finite models of the binary irreflexive symmetric predicate. It is well known that the pure first-order elementary theory $Th_p(K_0)$ of this class and the pure predicate calculus PPC of a binary predicate letter are recursively inseparable (see, for instance, [38]). Now we will use Theorem 3.7.7. To apply this theorem it is sufficient to show that the class K_0 is weakly reflected in the class K.

Let $M_0 = \langle\{1,...,m\}, P(x,y)\rangle$ be a model from K_0. By assumption, there is a model M from K which has m distinct elements, say $a_1,...,a_m$. We introduce the set R and the formula F as follows

$$R := \{j \mid P(1,j)\&(1 \leq i \leq m)\},$$

$$F(x,y,a_1,...,a_m) := \bigvee\{(x = a_i)\&(y = a_j) \mid P(i,j)\}\vee$$

$$(\bigwedge_{a_i} \neg(x = a_i) \wedge \bigvee_{j \in R}(y = a_j)) \vee (\bigwedge_{a_i} \neg(y = a_i) \wedge \bigvee_{j \in R}(x = a_j)).$$

The mapping g of M onto the given model M_0 from K_0 is defined as follows: $g(a_i) := i$, and for all $b \neq a_i$, $1 \leq i \leq m$, we put $g(b) := 1$. It is easy to see that, for all $c, d \in M$,

$$M \models F(c, d) \iff M_0 \models P(g(c), g(d)).$$

The latter means that g is a strong homomorphism onto. Thus by Theorem 3.7.7 we have that the scheme-logic $SL(Th(T))$ is undecidable. ∎

Corollary 3.7.10 *If a first-order theory T is undecidable then $SL(T)$ is undecidable as well.*

The proof is evident because every theory of a finite number of finite models is decidable; hence every undecidable theory satisfies the condition of Theorem 3.7.9.

Theorem 3.7.11 *[146, 149] Let T be a decidable first-order theory and for all $m \in N$ let there be a model of the theory T which has at least m elements (or, equivalently, let T have infinite models). Then $SL(T)$ is not recursively enumerable. In particular, $SL(T)$ has no recursive axiomatization (and, in particular, no finite axiomatization).*

Proof. The fact that T is decidable implies that the complement of the corresponding scheme-logic $SL(T)$ is recursively enumerable. By theorem 3.7.9 $SL(T)$ is not decidable. Therefore $SL(T)$ is not recursively enumerable. ∎

Note that the same result for first-order theories in only monadic signature follows from Theorem 2.1(iv)(b) of Nemeti [105].

Examples 3.7.12 *It is well known that the following first-order theories are decidable and have infinite models (cf. for instance, [37, 38])*

   a) the theory BA of boolean algebras;

   b) the theory DLO of dense unbounded linear order;

   c) the theory AG of abelian groups.

Hence by Theorem 3.7.11 all these theories do not have recursively enumerable scheme-logic.

The same is true for theories of all finite boolean algebras, or all finite cyclic groups, for instance. Thus, contrary to our previous intuitions, we see that scheme-logics for first-order theories are much more complicated than the theories themselves.

Theorem 3.7.13 *If M is a finite model then the scheme-logic $SL(Th(M))$ is decidable and there is an algorithm for determining admissibility of inference rules in $Th(M)$.*

Proof. Assume the basis set of M has k elements $\{1, ..., k\}$. Let $P(x_1, ..., x_n)$ be a predicate relation in $\{1, ..., k\}$ (possibly new, which is not a predicate relation of the model M). Any such relation P can be expressed in M by a first-order formulas using only equality. More precisely, we put

$$F_P(x_1, ..., x_n, y_1, ..., y_k) := \bigwedge\{\neg(y_i = y_j) \mid 1 \le i < j \le k\} \wedge$$

$$\wedge(\bigvee\{(x_1 = y_{i_1}) \wedge ... \wedge (x_n = y_{i_n}) \mid i_1, ..., i_n \in \{1, ..., k\},$$

$$M \models P(i_1, ..., i_n)\}).$$

Under the interpretation $y_i \mapsto i$ the truth of F_P in M coincides with the truth of P in M :

$$\forall a_i \in M(M \models F_P(a_1, ..., a_n, 1, ..., k) \Leftrightarrow M \models P(a_1, ..., a_n)) \qquad (3.34)$$

Taking into account (3.34), we have the following: for any formula $A(z_i)$ (in the language of the scheme-logics)

$$A(z_i) \in SL(Th(M)) \Longleftrightarrow \forall Q_i(A(Q_i) \in Th(M)), \qquad (3.35)$$

where Q_i are arbitrary atomic formulas of the form $P_i(x_{i1}, ..., x_{in})$, where P_i are any predicate letters (not only from the signature of M). Assume that a given formula $A(z_i)$ has the quantifiers (that is, connectives simulating quantifier) $\forall x_1, ... \forall x_m$. It can be easily seen that it is sufficient to consider in (3.35) only those atomic formulas Q_i which depend only on variables from the list $x_1, ..., x_n$. This gives an algorithm for recognizing whether $A(z_i) \in SL(Th(M))$ holds. Hence $SL(Th(M))$ is decidable.

Any given inference rule $A_1, ..., A_n/B$ is admissible in $Th(T)$ iff the following holds: if all formulas A_i are true in the model M under some interpretation of the metavariables by means of unary predicates, which are expressible by formulas, then B is true in M under this interpretation as well. As we have seen above, any predicate in M can be expressed by a first-order formula which employ only equality. Therefore we can above exchange the phrase *predicates expressible by formulas* in the condition above describing admissibility with simply the phrase *predicates*. Reasoning similarly, we conclude that it is sufficient to consider only predicates with effectively bounded dimension, which conclude our proof. ∎

Thus we have a complete description of first-order theories which are decidable with respect to the admissibility of inference rules.

Theorem 3.7.14 *[159] A first-order theory T has an algorithm recognizing the admissibility of inference rules in T if and only if $T = \bigcap_{1 \le i \le n} Th(M_i)$ where all M_i are finite.*

Proof. This follows immediately from Theorems 3.7.13 and 3.7.9.

In particular, as a corollary, we have a description of first-order theories with decidable scheme-logics.

Corollary 3.7.15 *[159] A first-order theory T has a decidable scheme-logic if and only if $T = \bigcap_{1 \leq i \leq n} Th(M_i)$ where all M_i are finite.*

Note that this corollary also can be extracted from a result concerning the properties of algebraic logics in I.Nemeti [105] (see p. 248). (The decidability of scheme-logics of first-order theories can be expressed in the algebraic language: the decidability of a scheme-logic is equivalent to the solvability of the equational theory of a variety generated by cylindric-like algebras.)

The results obtained above show that the scheme-logics are very complicated from an algorithmic point of view. We can make the language of scheme-logics weaker by removing the logical constants ($x_i = x_j$) simulating equalities. It is rather natural to consider such language because equality can be simply understood as a certain binary predicate relation. Hence we are not indebted specially distinguish the equality predicate from other predicates. Another reason for considering such a language is the existence of a number of well-known and well studied first-order theories in the language of pure predicate calculus. A final reason is that it allows us to study the question of to what extent the expressive power of the language of the scheme-logics is influenced by the presence of logical constants simulating equalities.

Let PSL be the language of the pure scheme-logics, i.e., the language obtained from SL by removing of all logical constants $x_i = x_j$. The pure scheme-logic $PSL(T)$ of any first order theory T (with an arbitrary first-order language, with equality or without it, and in arbitrary signature) is defined like $SL(T)$ but in its own language. The question arises: which results concerning scheme-logics in the complete language are true for PSL. Some answers are obvious. For instance,

Proposition 3.7.16 *If M_1 and M_2 are certain models in the pure predicate signature and there is a strong homomorphism of M_1 onto M_2 then $PSL(Th_p(M_1)) = PSL(Th_p(M_2))$.*

Proof. By Lemma 3.7.5 the truth of formulas without equality in the models M_1 and M_2 coincides. Hence the pure scheme-logics of the pure first-order theories of M_1 and M_2 coincide as well. ∎

Because there are strong homomorphisms from certain infinite models to finite models the last proposition and Theorem 3.7.13 mean that there exist infinite models with decidable pure scheme-logics. Thus there are counterexamples to the Theorem 3.7.9 and Corollary 3.7.15, for the case of pure scheme-logics. Hence the language of pure scheme-logics is really weaker than the language of scheme-logics. Do analogies exist for pure scheme-logics of the results which we obtained for scheme-logics? We will proceed to answer this question for the pure predicate calculus PPC itself (from the viewpoint of algebraic logic we will now investigate equational theories of diagonal-free cylindric algebras (see [65])).

Let $A(z_i)$ be a formula in the language of the pure scheme-logic which has only the connectives $\forall x_1, .., \forall x_n$ among its connectives simulating quantifiers. Then we say this formula is an *n-meta-formula*. Let $t(A(z_i))$ be the first-order formula obtained from $A(z_i)$ by replacing all metavariables z_i by corresponding predicate letters $P_i(x_1, .., x_n)$. We call $t(A(z_i))$ the *first-order track* of $A(z_i)$.

Lemma 3.7.17 $A(z_i) \in PSL(PPC)$ *if and only if* $t(A) \in PPC$.

Proof. The entailment from left to right follows immediately. Suppose that the inclusion in left-hand side of the proposition above does not hold. Then there is a tuple of formulas $g_i(x_1, .., x_n, y_i)$ in the language of PPC such that $A(g_i(x_1, .., x_n, y_i)) \notin$ PPC. Hence there is a model M of PPC and there is a valuation of the free variables of this formula such that

$$M \not\models A(g_i(a_1, ..., a_n, b_i)).$$

We introduce a new model M_1 on the set $|M|$ by taking new predicates P_i, where

$$M_1 \models P_i(c_1, ..., c_n) \Longleftrightarrow M \models g_i(c_1, ..., c_n, b_i)).$$

Now it can easily be established by induction on the length of the subformulas $B(z_i)$ of the formula $A(z_i)$ that

$$M_1 \models B(P_i(c_1, ..., c_n)) \Longleftrightarrow M \models B(g_i(c_1, ..., c_n, b_i))).$$

Thus $M_1 \not\models A(P_i(c_1, .., c_n))$ and $t(A) \notin$ PSL(PPC), which conclude the proof of our lemma. ∎

Thus PSL(PPC) can be embedded (in a sense) into PPC itself. This allows us to use known results about PPC for describing the pure scheme-logic of PPC.

Theorem 3.7.18 *The subclass* $PSL_3(PPC)$ *of* $PSL(PPC)$ *which consists of all formula schemes, having as connectives simulating quantifiers only connectives* $\forall x_i$ *from the set* $\{\forall x_1, \forall x_2, \forall x_3\}$, *is undecidable. But* $PSL(PPC)$ *is recursively enumerable and undecidable.*

Proof. By Lemma 3.7.17 the question about the decidability of PSL(PPC) is reduced to the question about the decidability of a subclass of PPC which consists of first-order tracks of formula schemes. The first-order tracks are certain first-order formulas, all the predicates of which have *frozen* occurrences of all variables. These formulas are called homogeneous in [111]. In [111] it is proved to be undecidable whether a given homogenous formula of the degree 3 (i.e., having only ternary predicate) is valid in PPC. This together with Lemma 3.7.17 gives us the undecidability of $PSL_3(PPC)$. By Lemma 3.7.17 the pure scheme-logic $PSL(PPC)$ is recursively enumerable. ∎

In a private conversation Dr. Max Urchs drew my attention to the possibility using this result of Pieczkowski [111] together with Lemma 3.7.17. Note that Theorem 3.7.18 can also be extracted from Theorem 5.1.66 (concerning equational theories of diagonal-free cylindric algebras) of Henkin, Monk and Tarski [65] (Part 2 (1985), p. 202). Theorem 3.7.18 has some consequences for polymodal propositional logic. As we have seen above, $PSL(PPC)$ is a poly-modal propositional logic without any logical constants and with commuting modalities. Therefore Theorem 3.7.18 entails

Corollary 3.7.19 *For any $n > 2$, there are undecidable n-poly-modal propositional logics extending $S5_n \oplus \{\Box_i \Box_j p \equiv \Box_j \Box_i p) \mid i, j \in N\}$.*

It is interesting to look at Kripke semantics for $PSL(PPC)$. Let M be a model of PPC with n-place predicate letters $P_1, ..., P_m$. We call any such M the nm-model. The associated n-poly-modal Kripke model $K(M)$ for M (or n-cube, cf. Venema [195]) is the set $\{(a_1, ..., a_n) \mid a_i \in M\}$ with n accessibility relations $R_1, ..., R_n$, where

$$(a_1, ..., a_n) R_i (b_1, ..., b_n) \iff \forall j((i \neq j) \Rightarrow (a_j = b_j)),$$

and with the valuation V of variables $z_1, ..., z_m$ in this model given by

$$(a_1, ..., a_n) \Vdash_V z_i \iff M \models P_i(a_1, ..., a_n).$$

Now, if we suppose the connectives $\forall x_i$ to be modal logical operators $\Box_i$, we can define the truth in $K(M)$ of every formula scheme $A(z_i)$ of the pure scheme-logic, which is built up out of variables $z_1, ..., z_m$.

Let M be a nm-model, let $K(M)$ be its associated n-poly-modal Kripke model, and let $A(z_i)$ be an n-meta-formula in the variables given.

Lemma 3.7.20 $K(M) \Vdash A(z_i) \iff M \models t(A).$

Proof. Let $B(z_i)$ be a subformula of $A(z_i)$. Then

$$(a_1, ..., a_n) \Vdash_V B(z_i) \iff M \models t(B)_{x_1, ..., x_n}^{a_1, ..., a_n}.$$

The proof can be directly given by induction on the length of B. Indeed, the inductive basis and the cases of the non-modal logical connectives are obvious. Suppose that $(a_1, ..., a_n) \Vdash_V \Box_r B(z_i)$. Then

$$\forall b_r (a_1, ..., a_{r-1}, b_r, a_{r+1}, ..., a_n) \Vdash_V B(z_i)$$

By inductive hypotheses this entails

$$\forall b_r \, M \models t(B)^{x_1,\ldots,x_r,\ldots,x_n}_{a_1,\ldots,b_r,\ldots,a_n}$$

This relation yields $M \models t(\forall x_r B)^{x_1,\ldots,x_n}_{a_1,\ldots,a_n}$. The converse implication can be shown step by step in a way similar to above, but in the converse direction. From the inductive process we finally obtain the formula $A(z_i)$. ∎

From Lemmas 3.7.17 and 3.7.20 we obtain a semantic description of the rule scheme logic PSL(PPC):

Theorem 3.7.21 *For an arbitrary n-meta-formula $A(z_i)$ in the language of pure scheme-logic and with m metavariables, $A(z_i) \in PSL(PCC)$ if and only if $K(M) \Vdash A(z_i)$ holds for every nm-model M.*

So, in particular, the class of S5-n-poly-modal Kripke frames mentioned in Theorem 3.7.21 has an undecidable modal theory for every $n > 2$ (using Theorem 3.7.18). Note that the same description is possible for scheme-logic of PC. We only briefly sketched here the research of poly-modal Kripke frames (simply in order to give some examples of applications of results concerning scheme-logics). A more detailed investigations of such frames (n-cubes) can be found in Venema [195]. Thus, as we have seen above, some of the results concerning scheme-logics are not preserved for pure scheme-logics, but certain results do have analogies (see for instance Theorem 3.7.18).

Historical Comments. The study of formula schemes in algebraic language goes back to the famous monograph Henkin, Monk and Tarski [65], which contains a deep investigation of this field from an algebraic point of view. The study of substitutionless predicate logic, see Monk [103], and the study of the algebraic counterparts of logical systems (see Block, Pigozzi [18]) are also closely related to this field. The valid formula schemes, generally speaking, correspond to valid identities of cylindric algebras. A significant number of the results concerning formula schemes in algebraic form known currently are given in Nemeti [105]. Also, in [195] Venema studies cylindric modal logics in detail; the completeness results of this work give, in particular, an axiomatization for some scheme-logics. The question of the axiomatization of classes consisting of frames in multi-modal logic by derivation rules is considered in Venema [196] where a meta-theorem concerning completeness is given. Some of the results of this section, as we noted above, can also be extracted from Henkin, Monk, Tarski [65] (Part 2, 1985) and Nemeti [105]. However our proofs are based on different ideas from those of [65, 105] and are shorter and more straightforward. A part of the results of this section was announced in [146, 149].

3.8 Some Counterexamples

In this section we will employ some counterexamples to demonstrate illustrate difference between the decidability of a given logic λ and the decidability of λ with respect to the admissibility of inference rules. It is clear that the decidability of a logic λ with respect to admissibility (i.e. the existence of an algorithm which determines the admissible inference rules) implies the decidability of this logic. Indeed, for any given formula α, the rule x/α, where x is a variable which does not occur in α, is admissible in λ if and only if the formula α is a theorem of λ. The converse is not true, there are logics which are decidable but are not decidable with respect to admissibility. In order to show this we will present some examples discovered by A.Chagrov [26].

First we present a simple counterexample which is based upon an idea which will be developed later to produce a stronger and more complicated counterexample for the moment. We consider a propositional non-algebraic logic L_1 with a language $\mathcal{L}$ including variables, brackets, unary logical connectives f, g, h, k, the logical constant symbol 0 and a binary logical connective which we denote by $=$. For any unary logical connective t, t^n stands for t written n-times. In order to introduce this logic L_1, we need to employ a notion from the theory of recursive functions. It is well known (cf. [129] for instance) that there is a recursive set of pairs of natural numbers M with a non-recursive projection of this set onto the second components. This means that there is an algorithm which decides whether $\langle n, m \rangle \in M$ holds; but there is no algorithm which decides whether or not n belongs to $Pr_2 M := \{ n \mid \exists m \langle m, n \rangle \in M \}$. Using this set we can introduce the following axioms and inference rules for L_1. The axioms schemes for L_1 are:

(1) $x = x$, reflexivity;

(2) $fg^{n+1}h^{m+2}0 = fk^{n+1}h^{m+2}0$ for $\langle m, n \rangle \in M$.

And the inference rules for L_1 are

(i) $x = y/y = x$, symmetricity;

(ii) $x = y, y = x/x = z$, transitivity ;

(iii $\alpha = \beta, \gamma = \delta/\alpha(\frac{x}{\gamma}) = \beta(\frac{x}{\delta})$, replacement of equals.

Certainly L_1 forms a deductive axiomatic system, and the corresponding logic (the set of theorems for L_1) is closed with respect to substitution. The logic of L_1 as like often before we denote also by L_1.

Theorem 3.8.1 *(A.Chagrov, [26]) The logic L_1 is decidable but the problem of admissibility of inference rules in L_1 is undecidable. The problem: for any given formula $\alpha(x_1, ..., x_n)$ to determine whether there are formulas $\beta_1, ..., \beta_n$ such that $\alpha(\beta_1, ..., \beta_n) \in L_1$ also is undecidable.*

Proof. For any two formulas α and β, $\alpha \doteq \beta$ means that α and β coincide graphically. Suppose t_1 and t_2 are formulas constructed out of variables and 0 by means of f, g, h and k. It is not difficult to check by induction on the length of the proof of formulas of the form $t_1 = t_2$ in the deductive system of L_1 that

$$t_1 = t_2 \in L_1 \Leftrightarrow (t_1 \doteq t_2) \vee \tag{3.36}$$

$$\vee [(t_1 = t_2 \doteq tfg^{n+1}h^{m+2}0 = tfk^{n+1}h^{m+2}0) \vee$$

$$(t_1 = t_2 \doteq tfk^{n+1}h^{m+2}0 = tfg^{n+1}h^{m+2}0)],$$

where t is a term constructed by using connectives from $\{f, g, h, k\}$, and $\langle n, n \rangle \in M$. Since M is recursive, it follows L_1 itself is decidable. Consider an inference rules of the kind:

$$R_n := \frac{fg^{n+1}hx = fk^{n+1}hx}{h0 = 0}.$$

We will show that R_n is admissible in L_1 iff $n \notin Pr_2(M)$. Indeed, let $n \in Pr_2(M)$, then there is an m such that $\langle m, n \rangle \in M$. Then it follows that

$$fg^{n+1}h^{m+2}0 = fk^{n+1}h^{m+2}0 \in L_1,$$

but $h0 = 0 \notin L_1$ as we have seen above. Thus R_n is not admissible in L_1. Suppose that $n \notin Pr_2(M)$. As we have proved above, in this case,

$$fg^{n+1}ht = fk^{n+1}ht \notin L_1$$

for any formula t. Thus, since the premise of R_n is never fulfilled in L_1, we conclude R_n is admissible for λ_1. Hence R_n is admissible in L_1 if and only if $n \in Pr_2(M)$. And since $Pr_2(M)$ is non-recursive, L_1 is undecidable with respect to admissibility of rules. Again, using (3.36) we infer that $fg^{n+1}ht = fk^{n+1}ht \in L_1$ iff $n \in Pr_2(M)$, which implies the last assertion of our theorem. ∎

Now we intend to extend the idea which was used in the above counterexample to modal logics, and to give an example of a modal logic which itself is decidable but is undecidable with respect to admissibility of inference rules. For this purpose we employ the frames $\mathcal{F}(m, n)$ of the structure displayed in the picture in Fig. 3.8. There $\mathcal{F}(m, n)$ is a transitive frame with the single reflexive element b, and all other elements are irreflexive. We will not write out the precise formal definition of these frames, everything is clear from the picture.

Definition 3.8.2 *The modal logic L_2 is the set of all modal formulas which are valid in all frames $\mathcal{F}(m, n)$ such that $\langle \frac{m-1}{2}, \frac{n-1}{2} \rangle \notin M$ (where M is the fixed before recursive set of pairs of natural numbers with the non-recursive projection $Pr_2(M)$ of M onto the second components).*

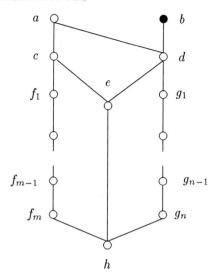

Figure 3.8: Frame $\mathcal{F}(m, n)$.

Note that the set of frames $\mathcal{F}(m, n)$ from the definition above is the following: $\mathcal{F}(m, n)$ belongs to this set if and only if at least one of the numbers n and m is even, or both of them are odd but $\langle \frac{m-1}{2}, \frac{n-1}{2} \rangle \notin M$. The logic L_2 is an example which we need: it is a decidable logic which is not decidable with respect to admissibility of inference rules.

Theorem 3.8.3 *(A.Chagrov, [26]) The logic L_2 is decidable but there is no an algorithm which determines admissibility of inference rules in L_2.*

Proof. Since L_2 is a logic which is determined by a recursive set of finite frames, the set of formulas which are not theorems of L_2 is recursively enumerable (i.e. we can effectively enumerate all non-theorems of L_2). Therefore in order to show that L_2 is decidable it is sufficient to find a recursive axiomatization for L_2 (an effectively enumerable set of axioms). Indeed, then the both sets of theorems and of non-theorems L_2 will be recursively enumerable which gives an algorithm recognizing theorems of L_2. Therefore we are going to find a recursive axiomatization for L_2. First we define certain formulas connected with some elements of the frame $\mathcal{F}(m, n)$. Namely, any formula denoted below by a capital letter is valid exactly at one element of $\mathcal{F}(m, n)$ denoted by the same, but lowercase, latter:

$$A := \Box\top, \quad B := \Diamond\top \wedge \Box\Diamond\top, \quad C := \Box\Box\bot \wedge \Diamond\top,$$
$$D := \Diamond A \wedge \Diamond B \wedge \neg\Diamond\Diamond A, \quad E := \Diamond C \wedge \Diamond D \wedge \neg\Diamond\Diamond D, \quad H := \Diamond E.$$

Now we consider formulas which describe the structure of the frames of the form $\mathcal{F}(m, n)$. All frames $\mathcal{F}(n, m)$ are transitive, therefore

$$\Box p \to \Box\Box p \in L_2. \tag{3.37}$$

All frames $\mathcal{F}(m, n)$ have width 3, therefore (see Exercise 2.10.2)

$$\mathcal{W}_3 := \bigwedge_{0 \leq i \leq 3} \Diamond p_i \to \bigvee_{0 \leq i < j \leq 3} \Diamond(p_i \wedge (p_j \vee \Diamond p_j)) \in L_2. \tag{3.38}$$

Recall that $\Box_0 x := x \wedge \Box x$, and correspondingly $\Diamond_0 x := \neg\Box_0\neg x$. Then it is clear that $\Diamond_0 x \equiv x \vee \Diamond x \in K4$. We let $K := A \vee B \vee \Diamond_0 C \vee \Diamond_0 D \vee \Diamond_0 E$. It can be easily seen that this formula is valid in all frames $\mathcal{F}(m, n)$. Hence

$$K \in L_2. \tag{3.39}$$

Let Ψ be a formula built up out of $\bot$ and having no variables. Then the formula $U(\Psi) := \Box_0(p \to \neg\Psi) \vee \Box_0(\neg p \to \neg\Psi)$ is valid in a given transitive rooted frame $\mathcal{F}$ iff $\mathcal{F}$ contains at most one element a such that Ψ is valid on a. From this observation we infer

$$U(A) \wedge U(B) \wedge U(C) \wedge U(D) \wedge U(E) \wedge U(H) \in L_2. \tag{3.40}$$

In order to fix *rigidly* the structure of the top of $\mathcal{F}(m, n)$ we use the formulas of the kind

$$Alt_n(\Psi) := \Psi \to \neg(\Diamond_0(\neg p_1 \wedge p_2 \wedge ... \wedge p_{n+1}) \wedge \Diamond_0(p_1 \wedge \neg p_2 \wedge p_3 \wedge ...$$
$$\wedge p_{n+1}) \wedge ... \wedge \Diamond_0(p_1 \wedge ... \wedge p_n \wedge \neg p_{n+1})),$$

where Ψ is a constant formula (formula built up out of $\bot$ and $\top$, and having no variables). It can be easily verified directly, that $Alt_n(\Psi)$ is valid in a transitive rooted frame $\mathcal{F}$ iff, for every element a of $\mathcal{F}$ on which Ψ is valid, the subframe generated by a contains not more than n elements. Therefore we conclude

$$Alt_1(B) \wedge Alt_2(C) \wedge Alt_3(D) \wedge Alt_5(E) \in L_2. \tag{3.41}$$

It is not hard to see also that

$$\Diamond C \wedge \Diamond D \to E \vee H \in L_2. \tag{3.42}$$

The sets $\{x \mid c \in \mathcal{F}(m, n), c \in x^{R<}, d \notin x^{R<}\}$, $\{x \mid c \in \mathcal{F}(m, n), d \in x^{R<}, c \notin x^{R<}\}$ are linearly ordered in $\mathcal{F}(m, n)$ (but consists of irreflexive elements). It is easy to verify directly that the formula

$$Lin(\Psi) := \neg(\Diamond_0(\Psi \wedge \Box_0 p \wedge \neg q) \wedge (\Diamond_0(\Psi \wedge \Box_0 q \wedge \neg p)),$$

where Ψ is a constant formula, is valid in a transitive rooted frame $\mathcal{F}$ iff there is no a pair elements in $\mathcal{F}$ which are incomparable but Ψ is valid on both them. Therefore it follows that

$$Lin(\Diamond C \wedge \neg \Diamond D) \wedge Lin(\Diamond D \wedge \neg \Diamond C) \in L_2. \tag{3.43}$$

Furthermore, all elements of $\mathcal{F}(M, n)$, except b, are irreflexive. Being irreflexive is not expressible by modal formulas. Anyway, it is not hard to see that, for transitive frames $\mathcal{F}$, formula $\Box(\Box p \rightarrow p) \rightarrow \Box p$ is valid in $\mathcal{F}$ iff $\mathcal{F}$ does not contain an infinite ascending chain of elements: $a_1 R a_2$, $a_2 R a_3$, ... , in particular, any such $\mathcal{F}$ does not contain reflexive elements. We fix denotation: for any formulas α and β, $\Box_\alpha \beta := \Box(\alpha \rightarrow \beta)$. Using mentioned above observation it is not difficult to see that the formula $LF_{\neg B} := \Box_{\neg B}(\Box_{\neg B} p \rightarrow p) \rightarrow \Box_{\neg B} p$ is valid in $\mathcal{F}(m, n)$, i.e.

$$LF_{\neg B} \in L_2. \tag{3.44}$$

Now, when we described in general the structure of frames $\mathcal{F}(m, n)$ by means of formulas, we intend to *prohibit* the presence of frames $\mathcal{F}(2m + 1, 2n + 1)$ while $\langle m, n \rangle \in M$. For this we introduce formulas

$$Ban(m, n) := \neg(\Diamond E \wedge \Diamond(\neg \Diamond D \wedge \Diamond^{2m+1} C) \wedge \neg \Diamond(\neg \Diamond D \wedge \Diamond^{2m+2} C) \wedge$$
$$\Diamond(\neg \Diamond C \wedge \Diamond^{2n+1} D) \wedge \neg \Diamond(\neg \Diamond C \wedge \Diamond^{2n+2} C)).$$

It is easy to verify directly that

$$\mathcal{F}(s, t) \nVdash Ban(m, n) \Leftrightarrow s = 2m + 1 \,\&\, i = 2n + 1, \tag{3.45}$$

$$Ban(m, n) \in L_2 \Leftrightarrow \langle m, n \rangle \in M.$$

We introduce a recursively axiomatizable normal modal logic L_3 with axioms:

$$\Box p \rightarrow \Box\Box p, \; W_3, \; K, \; U(A), \; U(B), \; U(C), \; U(D), \; U(E), \; U(H),$$
$$\Diamond C \wedge C \rightarrow E \vee H, \; Alt_1(B), \; Alt_2(C), \; Alt_3(D), \; Alt_5(E),$$
$$Lin(\Diamond C \wedge \neg \Diamond D), \; Lin(\Diamond D \wedge \neg \Diamond C), \; LF_{\neg B},$$
$$Ban(m, n) \text{ for all } \langle m, n \rangle \in M.$$

By definition L_3 is recursively axiomatizable. Therefore in order to complete our proof that L_2 is decidable it is sufficient to prove the following

Lemma 3.8.4 *Logics L_2 and L_3 coincide.*

Proof. From (3.37) - (3.45) it follows $L_3 \subseteq L_2$. Suppose Φ is a formula such that $\Phi \notin L_3$. By definition L_3 is a modal logic extending $K4$ with width of not more than 3. Therefore by Theorem 2.10.26 there is a rooted frame $\mathcal{F}$ such that $\mathcal{F} \Vdash L_3$ but $\mathcal{F} \nVdash_V \Phi$. Then $\mathcal{F}$ is transitive (since $\Box p \rightarrow \Box\Box p \in L_3$), $\mathcal{F}$ has width of not more than 3 (since $W_3 \in L_3$). Consider transitive frames $\mathcal{F}(\infty, n)$, $\mathcal{F}(m, \infty)$, $\mathcal{F}(\infty, \infty)$ of the special kind depicted in Fig. 3.9. The presence of ∞ there means the presence of an infinite chain of elements which contains no infinite ascending subchains; the presence of m or n means the presence of an irreflexive chain consisting of m or n elements respectively. The only reflexive element of these frames is b. By simple exhaustive search of alternatives, it can

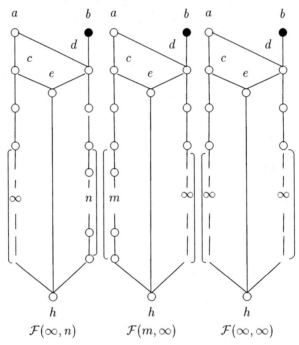

Figure 3.9: Structure of frames $\mathcal{F}(\infty, n)$, $\mathcal{F}(m, \infty)$, $\mathcal{F}(\infty, \infty)$

be shown that the presence of the axioms $K, U(A), U(B), ..., LF_{\neg B}$ imply that $\mathcal{F}$ can only be of the form $\mathcal{F}(m, n)$ or $\mathcal{F}(\infty, n)$ or $\mathcal{F}(m, \infty)$ or $\mathcal{F}(\infty, \infty)$, where three last mentioned frames are designated in Fig. 3.9, or to be a rooted open subframe of these frames.

Consider all these possibilities, starting with the latter. Let $\mathcal{F}$ be a rooted generated subframe of one of the frames $\mathcal{F}(m, n)$, $\mathcal{F}(\infty, n)$, $\mathcal{F}(m, \infty)$ or $\mathcal{F}(\infty, \infty)$. If $\mathcal{F}$ is finite, then $\mathcal{F}$ is a proper rooted generated subframe of a frame $\mathcal{F}(2k, 2k)$ for sufficiently large k. By definition of L_2, $\mathcal{F}(2k, 2k) \Vdash L_2$, hence $\Phi \notin L_2$. Consider the case when $\mathcal{F}$ is an infinite frame. Then $\mathcal{F}$ is a frame of the form (i) or (ii) depicted in Fig. 3.10. For each of these types (i), (ii) we proceed as follows. Assume first that Φ is false under a valuation V on the root of $\mathcal{F}$. Suppose $\Box\Phi_1, ..., \Box\Phi_k$ are exactly all subformulas of Φ with the main (final) connective $\Box$. For every i, $1 \leq i \leq k$, we choose and fix the R-maximal element out of those,

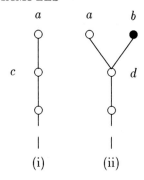

Figure 3.10: Frames of kind (i) and (ii).

if any, at which Φ_i is false under V. Then we cross out of $\mathcal{F}$ the elements to leave only those selected, the root, and also a, b and d in the case of type (ii). Denote the resulting frame with $\mathcal{F}$-induced accessibility relation by $\mathcal{F}_1$. The valuation V we also transfer directly on the frame $\mathcal{F}_1$. It is easy to verify by induction on length of subformulas Φ_s of the formula Φ that the truth value of any Φ_s under V in $\mathcal{F}$ and in $\mathcal{F}_1$ on the same elements coincides. Hence we obtain $\mathcal{F}_1 \nVdash \Phi$. Since $\mathcal{F}_1$ is finite, $\mathcal{F}$ is a rooted open subframe of some $\mathcal{F}(2k, 2k)$ for some sufficiently large k. Thus $\Phi \notin L_2$.

Consider the next possible alternative: $\mathcal{F}$ is $\mathcal{F}(\infty, \infty)$ and Φ is falsified in h by the valuation V. We denote by Γ_1 the set of all elements from $\mathcal{F}(\infty, \infty)$ which are placed between h and c, and which differ from e. And Γ_2 is the set of all elements from $\mathcal{F}(\infty, \infty)$ which are placed between h and d, and are different from e.

Again consider all subformulas $\Box\Phi_1, ..., \Box\Phi_k$ of Φ which has the final connective $\Box$ (which begins with $\Box$). For every i, $1 \leq i \leq k$, we choose and fix in $\mathcal{F}(\infty, \infty)$ the all R-maximal elements out of those, if any, at which Φ_i is false under V. Now we define the frame $\mathcal{F}_1$ as follows. We take (a) all selected above elements; (b) elements a, b, c, d, e, h; (c) a new arbitrarily element from Γ_1 if at the previous stage an odd number of elements from Γ_1 were selected; (d) a new arbitrary element from Γ_2 if at the previous stage an odd number of elements from Γ_2 were selected. Again we transfer the valuation V from $\mathcal{F}(\infty, \infty)$ in the frame $\mathcal{F}_1$. It can be easily verified by induction on the length of any subformula Φ_s of Φ that the truth values of Φ_s under V at the same elements of $\mathcal{F}(\infty, \infty)$ and $\mathcal{F}_1$ coincide. Therefore we conclude $\mathcal{F}_1 \nVdash \Phi$. But by the choice, $\mathcal{F}_1 = \mathcal{F}(2r, 2t)$, hence $\mathcal{F}(2r, 2t) \nVdash \Phi$, and since $\mathcal{F}(2r, 2t) \Vdash L_2$, it follows that $\Phi \notin L_2$.

The remaining possibilities $\mathcal{F} = \mathcal{F}(\infty, n)$ and $\mathcal{F} = \mathcal{F}(m, \infty)$ can be considered in a same way, except that at the end we have $\mathcal{F}_1 = \mathcal{F}(2r, n)$ and $\mathcal{F}_1 = \mathcal{F}(m, 2t)$ respectively. By definition, both these frames are L_2-frames which gives us required result. The last possibility is $\mathcal{F} = \mathcal{F}(m, n)$. If at least one of numbers m or n is even, then $\mathcal{F}(m, n) \Vdash L_2$ by definition of L_2. If both m and n are odd, then, since $\mathcal{F}(m, n) \Vdash L_3$ and (3.45), it follows $\langle (m-1)/2, (n-1)/2 \rangle \notin M$. Then according to definition of L_2 we obtain $\mathcal{F}(m, n) \Vdash L_2$. Hence we again arrive at $\Phi \notin L_2$. We have exhausted all the possibilities for $\mathcal{F}$, and each of which ends with $\Phi \notin L_2$. Hence $L_2 = L_3$. ∎

This lemma completes the proof of decidability of L_2 Now we turn to prove that L_2 is undecidable with respect to admissibility. We introduce the following sequence of inference rules: $P_n(p)/Q(p)$, where

$$P_n(p) := \neg(\lozenge E \wedge \lozenge(\neg \lozenge D \wedge (\lozenge C \wedge p)) \wedge \neg\lozenge(\neg\lozenge D \wedge \lozenge(\lozenge C \wedge p)) \wedge$$
$$\lozenge(\neg\lozenge C \wedge \lozenge^{2n+1} D) \wedge \neg\lozenge(\neg\lozenge C \wedge \lozenge^{2n+2} D);$$
$$Q(p) := \neg\lozenge(\neg\lozenge D \wedge \lozenge C \wedge p).$$

Lemma 3.8.5 *The rule R_n is admissible in L_2 if and only if $n \notin Pr_2(M)$.*

Proof. Suppose that $n \in Pr_2(M)$. This means there is a certain m such that $\langle m, n \rangle \in M$. By (3.45) it follows that $Ban(m, n) \in L_2$. It is not hard to see also that

$$(Ban(m, n) \to P_n(\lozenge^{2m+1}C)) \wedge (P_n(\lozenge^{2m+1}C) \to Ban(m, n)) \in L_2$$

using axiom $\Box p \to \Box\Box p$ of L_2. Therefore we obtain $P_n(\lozenge^{2m+1}C) \in L_2$. Now our aim is to show that $Q(\lozenge^{2m+1}C) \notin L_2$ which would mean that R_n is not admissible in L_2. In order to invalidate $Q(\lozenge^{2m+1})$ it is sufficient to take any finite frame of length not less than $2m+4$, in which the accessibility relation is a strong linear order. There are such frames among L_2-frames, for instance, some rooted open subframe of $\mathcal{F}(2k, 2k)$ with $k > m+2$ is a such one. Thus R_n is not admissible in L_2.

Suppose now that $n \notin Pr_2(M)$. We will show that R_n is admissible in L_2. Let $Q(\Phi) \notin L_2$ for some formula Φ. This means there is a frame $\mathcal{F}(s, t)$ such that $\mathcal{F}(s, t) \nVdash Q(\Phi)$, i.e. there is a valuation V in $\mathcal{F}(s, t)$ such that, for some element x,

$$x \Vdash_V \neg\lozenge D, \quad x \Vdash_V \lozenge C, \quad x \Vdash_V \Phi.$$

The first two truth relations above imply $\neg(xRd)$ and xRc, i.e. $x = f_i, 0 \le i \le s$. We fix the smallest such index i, i.e., if $0 \le j < i$, then $f_j \nVdash_V \Phi$. By our assumption we have $n \notin Pr_2(M)$. This implies that $\mathcal{F}(m, 2n+1) \Vdash L_2$ for any m. In particular, we conclude $\mathcal{F}(i, 2n+1) \Vdash L_2$.

We introduce a new valuation V_1 in $\mathcal{F}(i, 2n+1)$ as follows. We set V_1 so that it coincides with V on the elements $a, c, f_0, ..., f_i$ and we let V_1 take arbitrary values on all the other elements. Then

$$\{x \mid x \Vdash_{V_1} \neg\lozenge D \wedge \lozenge C \wedge \Phi\} = \{f_i\}, \text{ which yields}$$

$h \nVdash_{V_1} P_n(\Phi)$.

Hence $\mathcal{F}(i, 2n + 1) \nVdash_V P_n(\Phi)$ and $P_n(\Phi) \notin L_2$. ∎

Using Lemma 3.8.5 we can conclude that there is no algorithm which could recognize the admissibility of inference rules in L_2. This completes the proof of our theorem. ∎

There are many interesting open questions related to the topic of this sections. For instance, the following questions are non-answered yet:

(a) Does there exist a modal logic λ extending $S4$ which is itself decidable but which is undecidable with respect to admissibility?

(b) Does there exist a finitely axiomatizable modal logic over $K4$ which itself is decidable but has no algorithm which determines admissibility of inference rules?

(c) Does there exist a decidable superintuitionistic logic λ which is undecidable with respect to admissibility?

3.9 Admissibility through Reduced Forms

For certain classes of modal and superintuitionistic logics we have found algorithmic and semantic criteria for recognizing the admissibility of inference rules. These criteria have, generally speaking, a pure semantic form and are connected with evaluating the validity of inference rules in special classes of frames. As we know, inference rules in modal logics have an equivalent reduced form (see Section 1 of the current chapter), and it turns out that, for some modal logics, we can find an algorithm for determining admissibility which operates only on these reduced forms. Historically, the first algorithms for recognizing admissibility of inference rules in modal logics $S4$, Grz, GL in [135], [140], [143] and [147] where found exactly by means of using reduced forms for inference rules. Here we do not present those original algorithms but we transfer the criteria for admissibility already obtained in Section 5 of this chapter to inference rules in reduced form.

Algorithms for determining admissibility, which will be offered in this section, depend on given modal system. We begin with the modal system $K4$. Let

$$r := \frac{\bigvee_{1 \le j \le n} \varphi_j}{x_0}, \text{ where } \varphi_j := \bigwedge_{0 \le n \le m} x_i^{k(j,i,1)} \wedge \bigwedge_{0 \le n \le m} (\Diamond x_i)^{k(j,i,2)},$$

and $k(j, i, 1), k(j, i, 2) \in \{0, 1\}$, $t^0 := t$, $t^1 := \neg t$, be an inference rule in the reduced form. We put $Dpr(r) := \{\varphi_1, ..., \varphi_n\}$, i.e. $Dpr(r)$ is the set of all disjuncts in the premise of r. For every φ_j we fix the following denotation:

$$\theta_1(\varphi_j) = \{x_i \mid 0 \le i \le m, k(j, i, 1) = 0\},$$
$$\theta_2(\varphi_j) = \{x_i \mid 0 \le i \le m, k(j, i, 2) = 0\}.$$

The Kripke model $\mathfrak{M}(K4, r, X)$ is defined as follows. Let X be a subset of the set $Dpr(r) := \{\varphi_1, ..., \varphi_d\}$. The model $\mathfrak{M}(K4, r, X)$ is based upon the set X and has the following accessibility relation R and valuation V:

$$\forall \varphi_j, \varphi_k \in X(\varphi_i R \varphi_k \Leftrightarrow \theta_2(\varphi_k) \cup \theta_1(\varphi_k) \subseteq \theta_2(\varphi_j)),$$
$$\forall \varphi_j \in X, \forall i \in \{0, ..., m\}(\varphi_j \in V(x_i) \Leftrightarrow x_i \in \theta_1(\varphi_j)).$$

Lemma 3.9.1 *If a rule r having a reduced form is not admissible in $K4$ then there is a subset X of the set $Dpr(r)$ such that the model $\mathfrak{M}(K4, r, X)$ has the following properties:*

(i) *$\exists \varphi_j \in X$ such that $x_0 \notin \theta_1(\varphi)$,*

(ii) *$\forall \varphi_j \in X$; $\varphi_j \Vdash_V \varphi_j$ in the model $\mathfrak{M}(K4, r, X)$;*

(iii) *For any subset D of the model $\mathfrak{M}(K4, r, X)$, there is a reflexive element $e(r, D)$ and there is an irreflexive element $e(i, D)$ (in the model $\mathfrak{M}(K4, r, X)$) such that*

$$\theta_2(e(r, D)) = \theta_1(e(r, D)) \cup \bigcup\{\theta_1(\varphi_j) \mid \varphi_j \in D\} \cup \bigcup\{\theta_2(\varphi_j) \mid \varphi_j \in D\},$$
$$\theta_2(e(i, D)) = \bigcup\{\theta_1(\varphi_j) \mid \varphi_j \in D\} \cup \bigcup\{\theta_2(\varphi_j) \mid \varphi_j \in D\}.$$

Proof. Suppose the rule r is not admissible in $K4$. Then by Theorem 3.5.1 there is a valuation S of the variables from r in the frame of $Ch_{K4}(m_1)$ for some finite m_1, which disproves the rule r. Consider the set X consisting of all φ_j such that $S(\varphi_j) \neq \emptyset$. We claim that $\mathfrak{M}(K4, r, X)$ possesses all required properties. Since S disproves r, there is an element a from Ch_{K4} such that $a \notin S(x_0)$. At the same time, $a \in S(\varphi_j)$ for some j because the premise of r is valid under S on all elements. Then $x_0 \notin \theta_1(\varphi_j)$ and (i) holds.

To show (ii) note that the nonmodal part of the conjuncts from φ_j is valid on φ_j under V by definition. Consider the modal part. Suppose that $\Diamond x_i$ is a conjunctive member of φ_j. By our definition, there is an element $a \in S(\varphi_j)$. Therefore $a \Vdash_S \Diamond x_i$. This yields there is an b such that aRb and $b \Vdash_S x_i$. b is a member of $S(\varphi_k)$ for some k. Certainly then $x_i \in \theta_1(\varphi_k)$, therefore we can conclude that $\varphi_k \Vdash_V x_i$. Furthermore, $\varphi_j R \varphi_k$. Indeed, if $x_t \in \theta_2(\varphi_j) \cup \theta_1(\varphi_j)$ then $b \Vdash_S \Diamond x_t$ or $b \Vdash_V x_t$. Consequently $a \Vdash_S \Diamond x_t$. Therefore $x_t \in \theta_2(\varphi_j)$. Thus $\varphi_j R \varphi_k$ and $\varphi_j \Vdash_V \Diamond x_i$ in the model $\mathfrak{M}(K4, r, X)$. Conversely, suppose that $\varphi_j \Vdash_V \Diamond x_i$. This means there is some $\varphi_k \in X$ such that $\varphi_j R \varphi_k$ and $\varphi_k \Vdash_V x_i$. Then $x_i \in \theta_1(\varphi_k)$ and $x_i \in \theta_2(\varphi_j)$ i.e., $\Diamond x_i$ is a conjunctive member of φ_j and (ii) is established.

In order to show (iii) we pick a subset D of X. For every $\varphi_j \in D$, we pick and fix an element $e_j \in S(\varphi_j)$. Since the set $e(D) := \{e_j \mid e_j \in S(\varphi_j), \varphi_j \in D\}$ is finite, according to the construction of $Ch_{K4}(m_1)$, there is a reflexive element

$a(r, D) \in Ch_{K4}(m_1)$ and there is an irreflexive element $a(i, D) \in Ch_{K4}(m_1)$ such that

$$a(r, D)^{R \leq} := \{b \mid b \in Ch_{K4}(m_1), a(r, D)Rb\} =$$
$$\{a(r, D)\} \cup \bigcup\{e_j^{R \leq} \mid e_j \in e(D)\},$$
$$a(i, D)^{R <} := \{b \mid b \in Ch_{K4}(m_1), a(i, D)Rb, \neg(bRa(i, D))\} =$$
$$\bigcup\{e_j^{R \leq} \mid e_j \in e(D)\} \cup \bigcup\{\{e_j\} \mid e_j \in e(D)\}.$$

Then $a(r, D) \in S(\varphi_l)$ for some $\varphi_l \in X$. It is easy to verify immediately that φ_l has all properties required for $e(r, D)$. Similarly, $a(i, D)$ is an element of $S(\varphi_h)$ for some $\varphi_h \in X$, and it is not hard to see that φ_h has all the properties required for $e(i, D)$. ∎

Lemma 3.9.2 *If r is an inference rule in reduced form and there is a subset X of the set $Dpr(r)$ such that the model $\mathfrak{M}(K4, r, X)$ has properties (i) -(iii) from Lemma 3.9.1 then r is not admissible for $K4$.*

Proof. It follows directly from (ii) and (i) that r is invalid in the frame of the model $\mathfrak{M}(K4, r, X)$. Now we apply Lemmas 3.4.9 and 3.4.10. By definition the model $\mathfrak{M}(K4, r, X)$ is transitive. The modal logic $K4$ has the property of branching below 1 and the effective 1-drop points property (see Corollary 3.5.9). It is easy to see that the model $\mathfrak{M}(K4, r, X)$ has properties (a) - (d) from Lemma 3.4.9. Therefore by Lemma 3.4.10 there is an expressible valuation of the rule r which invalidates r in a model $Cn_{K4}(k)$. Then by Lemma 3.3.3 the rule r is not admissible in $K4$. ∎

Immediately from Lemmas 3.9.1 and 3.9.2 we immediately derive

Theorem 3.9.3 *A rule r in reduced form is admissible in $K4$ iff, for any set X of disjuncts of the premise of r, the model $\mathfrak{M}(K4, r, X)$ does not satisfy at least one from the properties (i) - (iii) of Lemma 3.9.1.*

Since, according to Corollary 3.1.13, there is an algorithm transforming rules into reduced forms, and any rule and its reduced form are equivalent in terms of their admissibility in $K4$ (see Corollary 3.1.15), Theorem 3.9.3 gives us another algorithm for determining admissibility in $K4$.

Below we present several similar results for other basic (in a sense) modal logics. The next example is an algorithm for recognizing admissibility in the Lewis modal logic $S4$. We preserve the denotation accepted above for the case of the modal system $K4$, but we define appropriate Kripke models which are based upon disjuncts of premises of reduced forms of inference rules in slightly other way. Let(r) be an inference rule in the reduced form in our modal language. Let X be a subset of the set $Dpr(r)$ of all disjuncts of the premise of r such that

$\forall \varphi_j \in X, \theta_1(\varphi_j) \subseteq \theta_2(\varphi_j)$. The model $\mathfrak{M}(S4, r, X)$ is based upon the set X and has the accessibility relation R and the valuation V defined as follows:

$$\forall \varphi_j, \varphi_k \in X(\varphi_i R \varphi_k \Leftrightarrow \theta_2(\varphi_k) \subseteq \theta_2(\varphi_j)),$$
$$\forall \varphi_j \in X, \forall i \in \{0, ..., m\}(\varphi_j \in V(x_i) \Leftrightarrow x_i \in \theta_1(\varphi_j)).$$

It is clear that by our definition above $\mathfrak{M}(S4, r, X)$ is reflexive and transitive.

Lemma 3.9.4 *Let r be a rule in reduced form. If r is not admissible in the modal logic S4 then there is a subset X of the set $Dpr(r)$ such that $\forall \varphi_j \in X, \theta_1(\varphi_j) \subseteq \theta_2(\varphi_j)$ and the model $\mathfrak{M}(S4, r, X)$ has the following properties:*

(i) $\exists \varphi_j \in X$ such that $x_0 \notin \theta_1(\varphi)$;

(ii) $\forall \varphi_j \in X$, $\varphi_j \Vdash_V \varphi_j$ in the model $\mathfrak{M}(S4, r, X)$;

(iii) For any subset D of the model $\mathfrak{M}(S4, r, X)$, there is a reflexive element $e(r, D)$ in the model $\mathfrak{M}(S4, r, X)$ such that

$$\theta_2(e(r, D)) = \theta_1(e(r, D)) \cup \bigcup \{\theta_2(\varphi_j) \mid \varphi_j \in D\}.$$

Proof. Our proof is merely a simplified variant of the proof of Lemma 3.9.1. We offer to prove it as an exercise. For any case, nevertheless, we allocate below the proof. Since the rule r is not admissible in $S4$, by Theorem 3.5.1 there is a valuation S of the variables from r in the frame of $Ch_{S4}(m_1)$ for some finite m_1 such that S disproves the rule r. We let X be the set of all φ_j such that $S(\varphi_j) \neq \emptyset$. It is clear that $\forall \varphi_j \in X, \theta_1(\varphi_j) \subseteq \theta_2(\varphi_j)$ since $Ch_{S4}(m_1)$ is reflexive. The model $\mathfrak{M}(S4, r, X)$ has all necessary properties. Indeed, since S disproves r, there is an element a from $Ch_{S4}(m_1)$ such that $a \notin S(x_0)$. But $a \in S(\varphi_j)$ for some j since the premise of r is valid under S on all elements. So $x_0 \notin \theta_1(\varphi_j)$ and (i) holds.

To show (ii) note that by definition the nonmodal part of the conjuncts from φ_j is valid on φ_j under V. Assume $\Diamond x_i$ is a conjunctive member of φ_j. By the definition of X, there is an element $a \in S(\varphi_j)$. Consequently $a \Vdash_S \Diamond x_i$ and there exists some b such that aRb and $b \Vdash_S x_i$. b is a member of $S(\varphi_k)$ for some k. Then $x_i \in \theta_1(\varphi_k)$ and we have $\varphi_k \Vdash_V x_i$. We will show $\varphi_j R \varphi_k$. If $x_t \in \theta_2(\varphi_j)$ then $b \Vdash_S \Diamond x_t$. Therefore $a \Vdash_S \Diamond x_t$ and $x_t \in \theta_2(\varphi_j)$. Hence $\varphi_j R \varphi_k$ and $\varphi_j \Vdash_V \Diamond x_i$ in the model $\mathfrak{M}(S4, r, X)$. Conversely, let $\varphi_j \Vdash_V \Diamond x_i$. This means there is a $\varphi_k \in X$ such that $\varphi_j R \varphi_k$ and $\varphi_k \Vdash_V x_i$. Then $x_i \in \theta_1(\varphi_k)$ and $x_i \in \theta_2(\varphi_j)$, i.e. $\Diamond x_i$ is a conjunctive member of φ_j and (ii) is shown.

To show (iii) we choose a subset D of X. For any $\varphi_j \in D$, we pick and fix a representative e_j from $S(\varphi_j)$. We set $e(D) := \{e_j \mid e_j \in S(\varphi_j), \varphi_j \in D\}$. $e(D)$ is finite and by our construction of $Ch_{S4}(m_1)$ there is a reflexive element $a(r, D) \in Ch_{S4}(m_1)$ such that

$$a(r, D)^{R\leq} := \{b \mid b \in Ch_{S4}(m_1), a(r, D)Rb\} =$$
$$\{a(r, D)\} \cup \bigcup \{e_j^{R\leq} \mid e_j \in e(D)\}.$$

Then $a(r, D) \in S(\varphi_l)$ for some $\varphi_l \in X$, and it is a routine matter to check that φ_l has all necessary properties required to $e(r, D)$. ∎

The next lemma is similar to Lemma 3.9.2.

Lemma 3.9.5 *Let r be an inference rule in reduced form. If there is a subset X of the set $Dpr(r)$ such that $\forall \varphi_j \in X, \theta_1(\varphi_j) \subseteq \theta_2(\varphi_j)$ and the model $\mathfrak{M}(S4, r, X)$ has properties (i)-(iii) from Lemma 3.9.4 then r is not admissible for $S4$.*

Proof. From (ii) and (i) it follows r is disprovable in the frame of the model $\mathfrak{M}(S4, r, X)$. By definition the model $\mathfrak{M}(S4, r, X)$ is transitive and reflexive. And the logic $S4$ has the branching property below 1 and the also the effective 1-drop points property (see Corollary 3.5.9). It is easy to check that the model $\mathfrak{M}(S4, r, X)$ has properties (a) - (d) from Lemma 3.4.9. Hence by Lemma 3.4.10 there is an expressible valuation of the rule r which disproves r in a model $Cn_{S4}(k)$. Then by Lemma 3.3.3 the rule r is not admissible in $S4$. ∎

Using Lemmas 3.9.4 and 3.9.5 we can immediately derive

Theorem 3.9.6 *A rule r having a reduced form is admissible in $S4$ iff, for any set X of disjuncts of the premise of r such that $\forall \varphi_j \in X, \theta_1(\varphi_j) \subseteq \theta_2(\varphi_j)$, the model $\mathfrak{M}(S4, r, X)$ fails to have at least one of the properties (i) - (iii) of Lemma 3.9.4.*

Now we demonstrate this approach for Grzegorczyk modal logic Grz which is the modal counterpart of the intuitionistic logic H. In this case we will again slightly vary the definition of the appropriate Kripke models which are based upon sets of disjuncts of premises of rules in reduced form. Suppose r is a rule in reduced form and X is a subset of the set $Dpr(r)$ consisting of all disjuncts of the premise of the rule r such that $\forall \varphi_j \in X, \theta_1(\varphi_j) \subseteq \theta_2(\varphi_j)$. We construct the Kripke model $\mathfrak{M}(Grz, r, X)$ as follows. $\mathfrak{M}(Grz, r, X)$ is based upon the set X and has as the accessibility relation an arbitrary partial order $\leq$ with properties:

$$(\forall \varphi_j, \varphi_k \in X)(\theta_2(\varphi_k) \subset \theta_2(\varphi_j) \Rightarrow \varphi_j \leq \varphi_k)),$$
For any A, the set $\{\varphi_j \mid \varphi_j \in X, \theta_2(\varphi_j) = A\}$
is linearly ordered by $\leq$.

The valuation of this model is defined as follows:

$$\forall \varphi_j \in X, \forall i \in \{0, ..., m\}(\varphi_j \in V(x_i) \Leftrightarrow x_i \in \theta_1(\varphi_j)).$$

Lemma 3.9.7 *Suppose r is a rule in reduced form and r is not admissible in Grz. Then there is a subset X of the set $Dpr(r)$ such that $\forall \varphi_j \in X, \theta_1(\varphi_j) \subseteq \theta_2(\varphi_j)$ and there is a certain model $\mathfrak{M}(Grz, r, X)$ which is based upon X and which has the following properties:*

(i) $\exists \varphi_j \in X$ such that $x_0 \notin \theta_1(\varphi)$;

(ii) $\forall \varphi_j \in X$, $\varphi_j \Vdash_V \varphi_j$ in the model $\mathfrak{M}(Grz, r, X)$;

(iii) For any subset D of the model $\mathfrak{M}(Grz, r, X)$, there is an element $e(r, D)$ in the model $\mathfrak{M}(Grz, r, X)$ such that

$$\theta_2(e(r, D)) = \theta_1(e(r, D)) \cup \bigcup \{\theta_2(\varphi_j) \mid \varphi_j \in D\}.$$

Proof. Our proof has the same form as the proofs of Lemmas 3.9.1 and 3.9.4 but there are more differences between it and the proof of Lemma 3.9.4. Suppose the rule r is not admissible in Grz, then by Theorem 3.5.1 there is a valuation S of the variables of r in the frame of $Ch_{Grz}(m_1)$, for some finite m_1, such that S disproves the rule r. We let X be the set of all φ_j such that $S(\varphi_j) \neq \emptyset$.

Clearly $\forall \varphi_j \in X, \theta_1(\varphi_j) \subseteq \theta_2(\varphi_j)$ because $Ch_{Grz}(m_1)$ is reflexive. We take the Kripke model $\mathfrak{M}(S4, r, X)$ and redefine the accessibility relation. Consider any $\varphi_j \in X$ and the cluster $C(\varphi_j)$ in the model $\mathfrak{M}(S4, r, X)$ containing φ_j. We replace the relation R from $\mathfrak{M}(S4, r, X)$ in any cluster $C(\varphi_j)$ by a certain linear order. Consider the set

$$e(C(\varphi_j)) := \{a \mid a \in Ch_{Grz}(m_1), (\exists \varphi_k \in C(\varphi_j))(a \in S(\varphi_k)).\}$$

The set $e(C(\varphi_j))$ has some maximal elements with respect to the partial order of $Ch_{Grz}(m_1)$ since by our construction $Ch_{Grz}(m_1)$ has no infinite ascending chains. We pick a maximal element in $e(C(\varphi_j))$ and denote it by $m(C(\varphi_j))$. There is only a single φ_m from $C(\varphi_j)$ such that $m(C(\varphi_j)) \in S(\varphi_m)$. We fix this φ_m and denote it by $m(\varphi_j)$. Now we take the arbitrary linear order in $C(\varphi_j)$ under which $m(\varphi_j)$ is the maximal element in $C(\varphi_j)$. In this way we impose the linear order $\leq$ in all $C(\varphi_j)$. And we let, for all φ_j, φ_k from distinct clusters $C(\varphi_j), C(\varphi_k), \varphi_j \leq \varphi_k \Leftrightarrow \varphi_i R \varphi_k \Leftrightarrow \theta_2(\varphi_k) \subset \theta_2(\varphi_j)$. This completes our definition of the model $\mathfrak{M}(Grz, r, X)$. We will show that $\mathfrak{M}(Grz, r, X)$ has properties (i)-(iii). Since S disproves r, $\exists a \in Ch_{Grz}(m_1)$ such that $a \notin S(x_0)$. Since the premise of r is valid under S on all elements, it follows that $a \in S(\varphi_j)$ for some j. Hence $x_0 \notin \theta_1(\varphi_j)$ and (i) holds.

(ii): It is evident that by definition the nonmodal part of φ_j is valid on φ_j under V. Suppose $\Diamond x_i$ is a conjunctive member of φ_j. By the definition of X, there is an element $a \in S(\varphi_j)$. Therefore $a \in e(C(\varphi_j))$. Moreover $a \leq m(C(\varphi_j))$ and $\Diamond x_i$ is a conjunct of $m(\varphi_j)$. Therefore there is an element c such that $m(C(\varphi_j)) \leq c, c \Vdash_S x_i$. Since the premise of r is valid under S on all elements, there is a $\varphi_v \in X$ such that $c \in S(\varphi_v)$. Then $x_i \in \theta_1(\varphi_v)$ and $\varphi_v \Vdash_V x_i$. And by the choice of $m(C(\varphi_j))$ we get either $m(C(\varphi_j)) = c$ or $m(\varphi_j) < \varphi_v$. If $m(\varphi_j) < \varphi_v$ then we directly obtain form $\varphi_j \leq m(\varphi_j)$ that $\varphi_j < \varphi_v$ and $\varphi_j \Vdash_V \Diamond x_i$. If $m(C(\varphi_j)) = c$ then $c \in S(m(\varphi_j))$ and $x_i \in \theta_1(m(\varphi_j))$, and consequently $m(\varphi_j) \Vdash_V x_i$. Since $\varphi_j \leq m(\varphi_j)$ we again conclude $\varphi_j \Vdash_V \Diamond x_i$.

Conversely, let $\varphi_j \Vdash_V \Diamond x_i$. This means there is a $\varphi_k \in X$ such that $\varphi_j \leq \varphi_k$ and $\varphi_k \Vdash_V x_i$. Then $x_i \in \theta_1(\varphi_k) \subseteq \theta_2(\varphi_k)$ and consequently $x_i \in \theta_2(\varphi_j)$, i.e. $\Diamond x_i$ is a conjunctive member of φ_j and (ii) is shown.

To show (iii) we choose a subset D of X. For any $\varphi_j \in D$, we pick and fix a representative e_j from $S(\varphi_j)$. We set $e(D) := \{e_j \mid e_j \in S(\varphi_j), \varphi_j \in D\}$. $e(D)$ is finite and by the construction of $Ch_{Grz}(m_1)$ there is an element $a(r, D) \in Ch_{Grz}(m_1)$ such that

$$a(r, D)^{R \leq} := \{b \mid b \in Ch_{Grz}(m_1), a(r, D)Rb\} =$$
$$\{a(r, D)\} \cup \bigcup \{e_j^{R \leq} \mid e_j \in e(D)\},$$

Then $a(r, D) \in S(\varphi_l)$ for some $\varphi_l \in X$ and it can be verified directly that φ_l has all necessary properties required to $e(r, D)$. ∎

Lemma 3.9.8 *Let r be an inference rule in reduced form. If there is a subset X of the set $Dpr(r)$ such that $\forall \varphi_j \in X, \theta_1(\varphi_j) \subseteq \theta_2(\varphi_j)$, and a model $\mathfrak{M}(Grz, r, X)$ has properties (i) -(iii) from Lemma 3.9.7 then r is not admissible in Grz.*

Proof. From (ii) and (i) we infer that r is disprovable in the frame of the model $\mathfrak{M}(Grz, r, X)$. The model $\mathfrak{M}(Grz, r, X)$ is based upon a certain finite poset, i.e., a certain Grz-frame. Logic Grz has the property of branching below 1 and the effective 1-drop points property (see Corollary 3.5.9). It is easy to check that the model $\mathfrak{M}(Grz, r, X)$ has properties (a) -(d) from Lemma 3.4.9. Therefore by Lemma 3.4.10 there is an expressible valuation of the rule r which disproves r in some model $Ch_{Grz}(k)$. Hence by Lemma 3.3.3 the rule r is not admissible in Grz. ∎

Lemmas 3.9.7 and 3.9.8 immediately imply

Theorem 3.9.9 *A rule r in reduced form is admissible in Grz iff, for any set X of disjuncts of the premise of r such that $\forall \varphi_j \in X, \theta_1(\varphi_j) \subseteq \theta_2(\varphi_j)$ and any model $\mathfrak{M}(Grz, r, X)$ on X, the model $\mathfrak{M}(Grz, r, X)$ does not satisfy at least one from the properties (i) - (iii) of Lemma 3.9.7.*

As a final example of application the technique of finite Kripke models based upon the disjuncts of premises for reduced forms of derivation rules, we present a result for Gögel-Löb provability logic GL. Suppose r is a rule in reduced form and X is a subset of the set $Dpr(r)$ of all disjuncts of the premise of r. We introduce the Kripke model $\mathfrak{M}(GL, r, X)$ which is based upon X as follows. $\mathfrak{M}(GL, r, X)$ has the set X as the basis set and $\mathfrak{M}(GL, r, X)$ has the accessibility relation $<$ and the valuation V defined below:

$$\forall \varphi_j, \varphi_k \in X (\varphi_j < \varphi_k \Leftrightarrow \theta_2(\varphi_k) \subset \theta_2(\varphi_j) \& \theta_1(\varphi_k) \subseteq \theta_2(\varphi_j)),$$
$$\forall \varphi_j \in X, \forall i \in \{0, ..., m\}(\varphi_j \in V(x_i) \Leftrightarrow x_i \in \theta(\varphi_j)).$$

It is clear that the introduced $<$ is irreflexive and transitive relation.

Lemma 3.9.10 *Let r be a rule in reduced form. Suppose that r is not admissible in GL. Then there exists a subset X of the set $Dpr(r)$ such that the model $\mathfrak{M}(GL, r, X)$ has the following properties:*

(i) *$\exists \varphi_j \in X$ such that $x_0 \notin \theta_1(\varphi)$;*

(ii) *$\forall \varphi_j \in X$, $\varphi_j \Vdash_V \varphi_j$ in the model $\mathfrak{M}(Grz, r, X)$;*

(iii) *For any subset D of the model $\mathfrak{M}(Grz, r, X)$, there is an element $e(i, D)$ in the model $\mathfrak{M}(GL, r, X)$ such that*

$$\theta_2(e(i, D)) = \bigcup \{\theta_2(\varphi_j) \mid \varphi_j \in D\}$$
$$\cup \bigcup \{\theta_1(\varphi_j) \mid \varphi_j \in D\}.$$

Proof. Since the rule r is not admissible in GL, by Theorem 3.5.1 there exists a valuation S of the variables of r in the frame of $Ch_{GL}(m_1)$, for some finite m_1, such that S disproves the rule r. We let X be the set of all φ_j such that $S(\varphi_j) \neq \emptyset$. We show below that the model $\mathfrak{M}(GL, r, X)$ has properties (i) - (iii). Since S disproves r, it follows $\exists a \in Ch_{GL}(m_1)$ such that $a \notin S(x_0)$. Since the premise of r is valid under S on all elements, $a \in S(\varphi_j)$ for some j. Hence $x_0 \notin \theta_1(\varphi_j)$ and (i) holds.

To show (ii) note that, by the definition of V it immediately follows that on any $\varphi_j \in X$, regarded as an element of the model $\mathfrak{M}(GL, r, X)$, the nonmodal part of the conjunction φ_j is valid under V. We consider the modal part. Assume that $\varphi_j \in X$ and $\varphi_j \Vdash_V \Diamond x_i$. Then there exists a $\varphi_k \in X$ such that $\varphi_i < \varphi_k$ and $\varphi_k \Vdash_V x_i$. Then $x_i \in \theta_1(\varphi_k)$. From $\varphi_j < \varphi_k$ it follows $x_i \in \theta_2(\varphi_j)$, i.e. $\Diamond x_i$ is a conjunct of φ_j.

Conversely, suppose that $\Diamond x_i$ is a conjunct of φ_j, i.e. $x_i \in \theta_2(\varphi_j)$. By definition of X there is an element $a \in S(\varphi_j)$. By our construction of $Ch_{GL}(m_1)$ any element b of the set

$$e(C(\varphi_j)) := \{a \mid a \in Ch_{GL}(m_1),$$
$$(\exists \varphi_k \in X)((\theta_2(\varphi_k) = \theta_2(\varphi_j) \& (a \in S(\varphi_k))\}$$

is a maximal element of $e(C(\varphi_j))$ or $b < c$, where c is a maximal element of this set. In particular, either a is a maximal element of $e(C(\varphi_j))$ or $a < m(C(\varphi_j))$, where $m(C(\varphi_j))$ is a maximal element of $e(C(\varphi_j))$. Depending on which is the case, let $u := a$ or $u := m(C(\varphi_j))$. Then $u \in S(\varphi_g)$, where $\theta_2(\varphi_g) = \theta_2(\varphi_j)$. In particular, it follows $x_i \in \theta_2(\varphi_g)$ and $u \Vdash_S \Diamond x_i$. Hence there is an element v such that $v \Vdash_S x_i$ and $u < v$. Since the premise of r is valid under S on all elements, there is some $\varphi_h \in X$ such that $v \in S(\varphi_h)$. Then $v \Vdash_S x_i$ yields $x_i \in \theta_1(\varphi_h)$ and $\varphi_h \Vdash_V x_i$. Since $u < v$, we conclude $\theta_2(\varphi_h) \subseteq \theta_2(\varphi_g)$ and $\theta_1(\varphi_h) \subseteq \theta_2(\varphi_g)$. As the element u was chosen to be maximal in $e(C(\varphi_j))$ and $u < v$, we obtain $\theta_2(\varphi_h) \neq \theta_2(\varphi_g) = \theta_2(\varphi_j)$. Hence $\varphi_j < \varphi_h$ and $\varphi_j \Vdash_V \Diamond x_i$, and (ii) is proved.

In order to show (iii) we choose a subset D of X. For any $\varphi_j \in D$, we pick and fix a representative e_j from the set $S(\varphi_j)$. We let $e(D) := \{e_j \mid e_j \in S(\varphi_j), \varphi_j \in$

$D\}$. $e(D)$ is finite and by the construction of $Ch_{CL}(m_1)$ there is an irreflexive element $a(i, D) \in Ch_{GL}(m_1)$ such that

$$a(i, D)^< := \{b \mid b \in Ch_{GL}(m_1), a(i, D) < b\} =$$
$$\bigcup\{e_j^< \mid e_j \in e(D)\} \cup \bigcup\{\{e_j\} \mid e_j \in e(D)\}.$$

Then $a(i, D) \in S(\varphi_l)$ for some $\varphi_l \in X$ and it can be easily verified directly that φ_l has all necessary properties which were required to $e(i, D)$. ■

Lemma 3.9.11 *Suppose r is an inference rule in reduced form. If there is a subset X of the set $Dpr(r)$ such that the model $\mathfrak{M}(GL, r, X)$ has the properties (i) - (iii) from Lemma 3.9.10 then r is not admissible in GL.*

Proof. Properties (ii) and (i) imply r is disprovable in the frame of the model $\mathfrak{M}(GL, r, X)$ which is based upon a finite GL-frame. Logic GL has the property of branching below 1 and the effective 1-drop points property (see Corollary 3.5.9). It is easy to check that the model $\mathfrak{M}(GL, r, X)$ has properties (a) - (d) from Lemma 3.4.9. Therefore by Lemma 3.4.10 there is an expressible valuation of the variables of the rule r which disproves r in a certain model $Ch_{GL}(k)$. Hence by Lemma 3.3.3 the rule r is not admissible in GL. ■

Applying Lemmas 3.9.10 and 3.9.11 we immediately derive

Theorem 3.9.12 *A rule r in reduced form is admissible in Gödel-Löb provability logic GL iff, for any set X of disjuncts of the premise of r, the model $\mathfrak{M}(GL, r, X)$ fails to have at least one of the properties (i) - (iii) from Lemma 3.9.10.*

Note that in this chapter we only slightly touched the temporal logics although it seems the methods developed could be adopted to the temporal logics. Anyway up to now no non-trivial examples of decidable by admissibility temporal logics are known.

Chapter 4

Bases for Inference Rules

4.1   Initial Auxiliary Results

In Section 4 Chapter I we described simple general results concerning structure of bases for admissible rules and connection of such bases with bases of quasi-identities of free algebras. This chapter is devoted to more deeper studying of bases for admissible and valid inference rules and is intended primary to problem of existence of finite bases. In this section we will consider simple general questions concerning the bases for quasi-identities and we will develop some simple tools for further research. It is obvious that certain inference rules can be the consequences of other inference rules in a given logic λ and we have already considered the relation: a rule r *is a consequence* of a collection of rules $\mathcal{R}$ in a logic λ. Recall that an inference rule $r := \alpha_1, ..., \alpha_n/\beta$ is derivable in a logic λ from a collection of rules $\mathcal{R}$ (or equivalently, r is a consequence of $\mathcal{R}$ in λ, abbreviation: $\mathcal{R} \vdash_\lambda r$), if there is a derivation of the conclusion β from the premises $\alpha_1, ..., \alpha_n$, as a set of hypothesis, in λ using rules from $\mathcal{R}$ and postulated rules of λ.

Recall also that a collection of inference rules $\mathcal{G}$ in the language of a logic λ is a *basis (or base)* in λ for a set $\mathcal{X}$ of rules iff $\mathcal{G} \subseteq \mathcal{X}$ and every rule $r \in \mathcal{X}$ is a consequence of $\mathcal{G}$ in λ. The central question for our investigation is whether there exists a finite basis for admissible inference rules of a given logic, simultaneously we will consider various related questions.

For any algebraic logic λ (in particular, any modal or superintuitionistic logic), it follows by Theorem 1.3.21 that a collection of inference rules $\mathcal{R}$ is a basis for all admissible in λ inference rules if and only if

$$\mathcal{R}^* := \{q(r) \mid (r \in \mathcal{R}) \vee (r \in R_p(\lambda))\} \cup Th_e(\mathrm{Var}(\lambda))$$

is a basis of the quasi-identities which are valid in $Th_q(\mathfrak{F}_\lambda(\omega))$. As a rule, in order to find a finite basis for the admissible rules of a logic λ, we have to use some particular properties of $\mathrm{Var}(\lambda)$ and its free algebras. In order to show that a logic λ has no finite bases for its admissible rules, and, equivalently, that $\mathfrak{F}_\lambda(\omega)$ has no finite bases for its valid quasi-identities, we can employ general methods which originate in universal algebra. In this section we present these general methods and adopt them to Kripke models.

Lemma 4.1.1 *A class of algebras K does not have bases in finitely many variables for its valid quasi-identities iff there exists an infinite sequence of algebras $\mathfrak{B}_n$, $n \in N$, $n \geq d$ for some $d \in N$ with the following properties:*

> *(i) Any algebra $\mathfrak{B}_n$ is $n + m + 1$-generated for some m and is not a member of K^Q;*

> *(ii) Any n-generated subalgebra of $\mathfrak{B}_n$ belongs to the quasi-variety K^Q.*

Proof. Suppose that there is a sequence $\mathfrak{B}_n, n \in N$ with properties (i) and (ii). Assume that the quasi-variety K^Q has a basis B_q in the variables $x_1, ..., x_n$ and consider the algebra $\mathfrak{B}_n$. By (ii) all quasi-identities from B_q are valid in $\mathfrak{B}_n$. Consequently $\mathfrak{B}_n \in K^Q$, which contradicts (i). Hence K does not have bases in finitely many variables for its quasi-identities.

Conversely, suppose that K has no bases in finitely many variables for its quasi-identities. We define Q_n to be the set of all the quasi-identities in n variables which are valid in K. By our assumption Q_n is not a basis for the quasi-identities of K. Therefore there is an algebra $\mathfrak{B}$ such that $\mathfrak{B}$ is not a member of K^Q but all quasi-identities from Q_n are valid in $\mathfrak{B}$. Then all n-generated subalgebras of $\mathfrak{B}$ are contained within K^Q. Since $\mathfrak{B} \notin K^Q$ there is a quasi-identity q in $n + k$ variables, $1 < k$, which is valid in K but is false in $\mathfrak{B}$. This implies that there is a subalgebra $\mathfrak{B}_n$ of $\mathfrak{B}$ such that: (a) $\mathfrak{B}_n$ does not belong to K^Q, (b) $\mathfrak{B}_n$ is $n + m + 1$-generated, and (c) all n-generated subalgebras of $\mathfrak{B}_n$ belong to K^Q, where $0 \leq m$. The constructed sequence $\mathfrak{B}_n, n \in N$ evidently has properties (i) and (ii). ∎

For finite algebras $\mathfrak{A}$ which have a finite basis for identities, this lemma can be made more precise. First we note a well known general fact that we will need later on. Recall that a variety of algebras V is said to be *uniformly locally finite* if there is an effectively calculable function f such that any n generated subalgebra of any algebra from V is finite and has not more than $f(n)$ elements.

Lemma 4.1.2 *Let Var be a variety generated by a finite algebra $\mathfrak{A}$ which has m elements. Then Var is uniformly locally finite: for any n, every n-generated algebra from Var has not more than m^{m^n} elements.*

Proof. By Theorem 1.2.36 the free algebra $\mathfrak{F}_{Var}(n)$ is a certain subalgebra of $\mathfrak{A}^{|\mathfrak{A}|^n}$. Therefore this free algebra has not more than m^{m^n} elements. Any n-generated algebra $\mathfrak{B}$ from Var is a homomorphic image of $\mathfrak{F}_{Var}(n)$, therefore $\mathfrak{B}$ has not more than m^{m^n} elements. ∎

Lemma 4.1.3 *A finite algebra $\mathfrak{A}$, which has a finite basis for its identities, does not have bases in finitely many variables for quasi-identities iff there is an infinite sequence of algebras $\mathfrak{B}_n, n \in N$, $n \geq d$ for some $d \in N$ with the properties:*

    (i) Any algebra $\mathfrak{B}_n$ is $n + m + 1$-generated for some m and is not a member of K^Q;

    (ii) Any n-generated subalgebra of $\mathfrak{B}_n$ belongs to the quasi-variety K^Q;

    (iii) Any proper subalgebra of $\mathfrak{B}_n$ belongs to the quasivariety $\mathfrak{A}^Q$ generated by $\mathfrak{A}$.

Proof. The sufficiency of the condition follows directly from Lemma 4.1.1. For its necessity, note that if $\mathfrak{A}$ has no bases in finitely many variables for quasi-identities then by Lemma 4.1.1 there exists a sequence of algebras $\mathfrak{A}_n, n \in N$

with properties (i) and (ii). Since $\mathfrak{A}$ has a finite basis for its identities, we can assume without loss of generality that all algebras $\mathfrak{A}_n$ are members of the variety $\mathrm{Var}(\mathfrak{A})$ generated by $\mathfrak{A}$. If some $\mathfrak{A}_n$ does not satisfy (iii) then taking the subsequent decreasing chain of subalgebras of $\mathfrak{A}_n$ which satisfy (i) and (ii) but do not satisfy (iii) we obtain an algebra $\mathfrak{B}_n$ which has all properties (i), (ii) and (iii) (the chain terminates because $\mathfrak{A}_n$, being a finitely generated algebra from $\mathrm{Var}(\mathfrak{A})$, is finite by Lemma 4.1.2). Now we simply replace $\mathfrak{A}_n$ by $\mathfrak{B}_n$. This transformation gives us the sequence $\mathfrak{B}_n, n \in N$ with the necessary properties. ∎

This result can be naturally and directly transferred to Kripke models and inference rules. To do this we use the arbitrary frames for a given logic λ, and we will give straightforward proofs below rather than to extract the results as a corollaries of Lemma 4.1.1.

Lemma 4.1.4 *Suppose that λ is a modal or a superintuitionistic logic and $\mathcal{R}$ is a family of inference rules in the language of λ. Suppose there is a sequence of λ-frames $\mathcal{F}_n, n \in N$, $n \geq d$ for some $d \in N$, with the properties:*

> *(i) There is an inference rule from $\mathcal{R}$ in $n+m+1$ variables which is invalid in $\mathcal{F}_n$;*
>
> *(ii) All inference rules of $\mathcal{R}$ in n variables are valid in $\mathcal{F}_n$.*

Then the family $\mathcal{R}$ has no bases in finitely many variables in the logic λ.

Proof. Suppose there is a basis B_n for $\mathcal{R}$ in λ containing only rules with formulas that are built up out of n variables. We take the frame $\mathcal{F}_n$. By (ii) all the rules from B_n are valid in $\mathcal{F}_n$. By (i) there is an inference rule $r \in \mathcal{R}$ which is built up out of $n + m + 1$ variables and which is invalid in $\mathcal{F}_n$. Since by assumption $B_n \vdash_\lambda r$, it follows there is a derivation $\mathcal{S}$ in λ of the conclusion of r from the premises of r using inference rules from B_n. Suppose V is a valuation of all the variables from r in $\mathcal{F}_n$ under which all the premises of r are valid on any element of $\mathcal{F}_n$. We extend V to all variables occurring in S letting $V(x) = |\mathcal{F}_n|$ for any variable x which does not occur in r. Then any premise of r is valid in $\mathcal{F}_n$ under V on any element by our assumption, and any theorem from λ is valid on any element of $\mathcal{F}_n$ under V since $\mathcal{F}_n$ is a certain λ-frame. The postulated inference rules of λ preserve the validity, and any rule from B_n preserves validity at its application in $\mathcal{S}$ since all the rules from B_n are valid in $\mathcal{F}_n$. Therefore it follows that the conclusion of r, i.e. the terminating formula of $\mathcal{S}$, is valid under V on all elements of $\mathcal{F}_n$. Hence we conclude that r is valid in $\mathcal{F}_n$, a contradiction. ∎

We will use this lemma further to prove that certain modal and certain superintuitionistic logics have no bases in finitely many variables for admissible inference rules. It is also of interest to investigate whether there exist certain independent bases for the admissible inference rules. Later we will show that even very simple tabular logics sometimes do not have a finite and even an independent basis for admissible inference rules. In this section we develop some common but simple tools for this research. We recall the following definition.

Definition 4.1.5 *Let λ be a modal or a superintuitionistic logic. Suppose $\mathcal{R}$ is a family of inference rules in the language of λ. A subset $\mathcal{B}$ of $\mathcal{R}$ is an* independent basis for inference rules *of $\mathcal{R}$ if*

(i) $\mathcal{B}$ *is a basis for $\mathcal{R}$ in λ;*

(ii) *for any rule $r \in \mathcal{B}$, r is not derivable from $\mathcal{B} - \{r\}$ in λ.*

Thus, if a set $\mathcal{R}$ has a finite basis in λ then, by removing from this basis all the members which are not independent, we obtain an independent basis of $\mathcal{R}$ in λ. Conversely, if a set $\mathcal{R}$ has no finite bases in λ the procedure of deleting members which are not independent can be never terminated. Therefore the study of existence independent bases is meaningful only for logics which do not possess finite bases for admissible rules. Note also that in a similar way we can define the notion of independent basis for families of quasi-identities.

Definition 4.1.6 *Suppose Var is a variety of algebras. Suppose Q is a family of quasi-identities in the language of Var. A subset Q_1 of Q is an* independent basis (or base) *for Q in Var if*

(i) Q_1 *is a basis for Q in Var, i.e., for any algebra $\mathcal{A} \in Var$, if all quasi-identities from Q_1 are valid in $\mathcal{A}$ then $\mathcal{A} \models q$, $\forall q \in Q$;*

(ii) *for any $q \in Q_1$, $Q_1 - \{q\}$ in not a basis for Q_1 in Var.*

The connection between the independence of inference rules and the independence of quasi-identities corresponding to them can be described as follows:

Lemma 4.1.7 *A family of inference rules $\mathcal{R}$ is an independent basis for the admissible inference rules of an algebraic logic λ iff the set of quasi-identities*

$$\mathcal{R}_q := \{q(r) := \bigwedge_{1 \leq i \leq n} \alpha_i = \top \Rightarrow \beta = \top \mid r := \frac{\alpha_1, ..., \alpha_n}{\beta} \in \mathcal{R}\}$$

is an independent basis of the set of all quasi-identities which are valid in $\mathfrak{F}_\lambda(\omega)$, i.e., for $Th_q(\mathfrak{F}_\lambda(\omega))$, in $Var(\lambda)$.

Proof. By Theorem 1.4.15 $\mathcal{R}$ is a basis for the set of admissible rules of λ iff $\mathcal{R}_q$ is a basis for the quasi-identities of $\mathfrak{F}_\lambda(\omega)$ in $Var(\lambda)$. Using this observation it follows by Theorem 1.4.11 that $\mathcal{R}$ is an independent basis of rules admissible in λ iff $\mathcal{R}_q$ is an independent basis for the quasi-identities of $\mathfrak{F}_\lambda(\omega)$ in $Var(\lambda)$. ∎

Thus the existence of an independent basis for admissible rules of λ depends on existence an independent basis for quasi-identities of the free algebras from $Var(\lambda)$. In universal algebra there is common way to show that a quasi-variety has no independent bases for quasi-identities.

Definition 4.1.8 *Let Var be a variety of algebras and let Q be a quasi-variety contained in Var. A quasi-variety Q_1 is a cover for Q in Var if $Q \subset Q_1 \subseteq Var$, and there is no quasi-variety Q_2 such that $Q \subset Q_2 \subset Q_1$.*

Lemma 4.1.9 *Suppose that Var is a variety of algebras and Q is a quasi-variety contained in Var. Suppose Q has no finite basis for the quasi-identities in Var. Then if Q has only finitely many covering quasi-varieties in Var then Q has no independent bases for the quasi-identities in Var.*

Proof. We denote the basis for $Th_e(Var)$ by E. Suppose Q has an independent basis $\mathcal{B} := \{q_i \mid i \in I\}$ for the quasi-identities in Var. By our assumption I must be infinite. For any $j \in I$ we let $T_j := \mathcal{B} - \{q_j\}$. Then any quasi-variety $Q_j := Mod(T_j \cup E)$ is a subclass of Var and this subclass contains Q. Since $\mathcal{B}$ is an independent basis for Q in Var we have $Q \subset Q_j \subseteq Var$ for any j. For any $j \in I$, consider the set S_j of all quasi-varieties Q_s such that $Q \subset Q_s \subseteq Q_j$. Consider any $Q_s \in S_j$. Since $\mathcal{B}$ is a basis for Q in Var, there is a $q_k, k \in I$ which is false in Q_s, and since $Q_s \subseteq Q_j$, we conclude $k = j$. Hence, for any $Q_s \in S_j$, q_j is false in Q_s. We introduce the poset $\mathcal{A} := \langle \{Th_q(Q_s) \mid Q_s \in S_j\}, \subseteq \rangle$. As we have noted above, $q_j \notin Th_q(Q_s)$ for any $Q_s \in S_j$. Therefore applying Zorn's Lemma to $\mathcal{A}$ it follows that $\mathcal{A}$ has certain maximal elements. Let $Th_q(Q_{sj})$ be a maximal element from $\mathcal{A}$. Then Q_{sj} is a cover for Q in Var. It is not difficult to see that, for distinct $j_1, j_2 \in I$, the covers Q_{sj_1}, Q_{sj_2} are different since q_{j1} is false in Q_{sj_1} but q_{j_1} is valid in Q_{sj_2}. Therefore we conclude that Q has infinitely many covering quasi-varieties in Var which contradicts the condition of our lemma. ∎

We closure this section with the following lemma which will be helpful in studying of bases for admissible rules in tabular logics.

Lemma 4.1.10 *For any finite algebra $\mathfrak{A}$ with m elements, the quasivariety $(\mathfrak{F}_\omega(\mathfrak{A}^Q))^Q$ generated by the free algebra $\mathfrak{F}_\omega(\mathfrak{A}^Q)$ coincides with $(\mathfrak{F}_m(\mathfrak{A}^Q))^Q$.*

Proof. The free algebra $\mathfrak{F}_m(\mathfrak{A}^Q)$ is a subalgebra of $\mathfrak{F}_\omega(\mathfrak{A}^Q)$. Therefore we need only to show that, for every quasi-identity q,

$$\mathfrak{F}_m(\mathfrak{A}^Q) \models q \implies \mathfrak{F}_\omega(\mathfrak{A}^Q) \models q.$$

Suppose that

$$q(x_1, ..., x_k) := \bigwedge_i f_i(\overline{x}) = g_i(\overline{x}) \implies f(\overline{x}) = g(\overline{x})$$

is a quasi-identity which is false in $\mathfrak{F}_\omega(\mathfrak{A}^Q)$. This means there exists a tuple $t_1, ..., t_k$ of terms (which are built upon free generators) such that the premise of q is true in $\mathfrak{F}_\omega(\mathfrak{A}^Q)$ but the conclusion of q is false in $\mathfrak{F}_\omega(\mathfrak{A}^Q)$ after replacing all x_j by corresponding t_j. This means that the identity $f(t_1, ..., t_k) = g(t_1, ..., t_k)$ is false in the algebra $\mathfrak{F}_\omega(\mathfrak{A}^Q)$. Then this identity will be false in the algebra

$\mathfrak{A}$ under a valuation $y_l \mapsto a_l$ of all the free generators y_l having occurrences in the terms $t_1, ..., t_k$. We let any term v_j be the term obtained from t_j by means of replacing each y_l by y_s, where s is the smallest index such that $a_s = a_l$. Then the identity $f(v_1, ..., v_k) = g(v_1, ..., v_k)$ will be false in $\mathfrak{A}$ under the same valuation $y_j \mapsto a_j$ of its variables. Then this identity is also false in the algebra $\mathfrak{F}_m(\mathfrak{A}^Q)$ under the valuation $y_s \mapsto y_s$. The premise of q was true in $\mathfrak{F}_\omega(\mathfrak{A}^Q))$ with respect to the valuation $x_i \mapsto t_i$, therefore the formula

$$\bigwedge_i f_i(v_1, ..., v_k) = g_i(v_1, ..., v_k)$$

is true in any algebra from $\mathrm{Var}(\mathfrak{A})$ under any valuation. Thus the quasi-identity $q(v_1, ..., v_k)$ is false in the algebra $\mathfrak{F}_m(\mathfrak{A}^Q)$, therefore q itself is false in $\mathfrak{F}_m(\mathfrak{A}^Q)$ as well. ∎.

4.2 The Absence of Finite Bases

Using criteria for determining the admissibility of inference rules in logics which where developed in Chapter 3 and the simple technical tools from the previous section, we are now in a position to solve the A.Kuznetsov-H.Friedman problem: whether the intuitionistic logic H has a finite basis for its admissible rules, and to develop a strong technique for studying of bases for admissible rules. In particular, here we present a negative solution to A.Kuznetsov-H.Friedman problem and to analogous questions for a wide classes of superintuitionistic and related modal logics. We do this by proving general theorems which says that all modal and superintuitionistic logics satisfying certain specific conditions cannot have bases for admissible rules in finitely many variables. As a corollary, we obtain that the free algebras from the varieties of algebras corresponding to these logics have no finite bases for quasi-identities.

In our proof a special role will be played by a certain sequence of frames $E_n, 1 \leq n$ the structure of which is described below. We prefer in the beginning to describe the structure of these frames E_n informally, before then giving their complete formal definition. Any poset E_n is obtained as the union of its finite open subframes $E_n^j, j \in N$, where E_n^j is the set of all elements of E_n with a depth of not more than j. The posets E_n^j are constructed by induction on j as follows.

(i) E_n^1 is the single element poset;

(ii) E_n^2 is obtained from E_n^1 by adding $2^n + 2$ new elements which form an antichain, and each of which is strictly less than the element of E_n^1. Clearly, E_n^2 has depth 2.

(iii) In order to construct E_n^3 we add to E_n^2 new elements a_i^3, where i is a non-trivial antichain of elements of depth 2 from E_n^2 such that $1 < ||i|| \neq 2^n + 1$. We do this so that: (a) the partial order in E_n^3 extends the partial order in E_n^2; (b) all the new elements added have depth 3 in E_n^3 and form an antichain; (c) any new element added a_i^3 is less than an element $b \in E_n^2$ iff b is an element of the antichain i, or there is an element c of the antichain i which is less than b.

(iv) To construct E_n^{j+1} from E_n^j where $3 \leq j$ we add new elements a_i^{j+1} to E_n^j, where i is an arbitrary non-trivial antichain of E_n^j containing at least one element of depth j from E_j. To define the partial order in E_n^{j+1} we ensure that (a) the partial order in E_n^{j+1} extends the partial order in E_n^j, (b) all the new elements added have depth $j + 1$ in E_n^{j+1} and form an antichain, (c) any new element a_i^{j+1} is less than an element $b \in E_n^j$ iff b is an element of the set i or there is an element $c \in i$ which is less than b.

(v) We define E_n as the union of all posets E_n^m, $m \in N$ with the partial order transferred from all E_n^m, $m \in N$.

It is clear that all the posets E_n^m $m \in N$ have uniform definitions, only the poset E_n^3 more precisely the set of elements of E_n of depth 3, is special: we do not include the elements of depth 3 which see $2^n + 1$ elements of depth 2. In a sense, we are spoiling our uniform definition by omitting such elements of depth 3. But it is an important trick in our construction, and it will allow us to prove the absence of finite bases for rules admissible in H.

In any case, now we give a precise formal description of posets $E_n, n \in N$:

(i) $E_n^1 := \langle \{a_1^1\}, \leq_1 \rangle$, where $a_1^1 \leq a_1^1$;

(ii) $E_n^2 := \langle |E_n^2|, \leq_2 \rangle$, where $|E_n^2| := \{a_1^1\} \cup \{a_1^2, ..., a_{2^n+1}^2, a_{2^n+2}^2\}$, $\leq_2$ is the partial order in $|E_n^2|$, where $a_i^2 \leq a_1^1$ for any i from $\{1, ..., 2^n+2\}$, and all distinct a_i^2 form an antichain with respect to $\leq_2$;

(iii) $E_n^3 := \langle |E_n^2| \cup \{a_i^3 \mid i \subseteq |Sl_2(E_n^2)| \& 1 < ||i|| \neq 2^n + 1\}, \leq_3 \rangle$, where $\leq_3$ is the partial order in E_n^3, $\leq_2 \subseteq \leq_3$, and $\forall a_i^3, \forall a_k^2$ $(a_i^3 \leq_3 a_k^2 \Leftrightarrow a_k^2 \in i)$, and all distinct a_i^3 form an antichain with respect to $\leq_3$;

(iv) Suppose E_n^j is already constructed and $3 \leq j$.
$E_n^{j+1} := \langle |E_n^j| \cup \{a_i^{j+1} \mid i$ is antichain in E_n^j , $(\exists a_h^j \in Sl_j(E_n^j))(a_h^j \in i) \& 1 < ||i||\}, \leq_{j+1} \rangle$, where $\leq_{j+1}$ is the partial order on E_n^{j+1}, $\leq_j \subseteq \leq_{j+1}$, and $\forall a_i^{j+1}, \forall b \in E_n^j$ $(a_i^3 \leq_3 b \Leftrightarrow (b \in i) \vee (\exists c \in i(c \leq_j b)))$; all distinct a_i^{j+1} form an antichain with respect to $\leq_{j+1}$;

(v) Given the description of all finite posets E_n^m, $m \in N$ which is specified in (i) - (iv) above we let

$$E_n := \langle \bigcup_{m \in N} |E_n^m|, \leq \rangle, \text{ where } \leq := \bigcup_{m \in N} \leq_m.$$

It is clear that any E_n^j is the set of all elements of E_n^{j+1} with a depth of not more than j, that is $S_j(E_n^{j+1}) = E_n^j$ and $(E_n^{j+1} - E_n^j) = Sl_{j+1}(E - n^{j+1})$, and, in particular, E_n^j is an open subframe of E_n^{j+1} and E_n.

Definition 4.2.1 *We say that a modal logic λ extending $S4$ or a superintuitionistic logic λ is an $(\omega, 1, 3)$-logic if any frame $E_{3,n}$ obtained in the same way as E_n except that in (iii) we take as i any set of elements of depth 2, is a λ-frame. A modal logic λ extending $K4$ or a superintuitionistic logic λ is an $(m, 1, 3)$-logic if λ is a logic of depth m, $3 \leq m$, and for any n, the frame $S_m(E_{3,n})$ is a λ-frame.*

In the first part of this section we focus our attention only on modal logics over $S4$ and superintuitionistic logics. Lately on we will show also that the results obtained for modal logics over $S4$ have complete analogy for modal logics extending $K4$. The methods of proofs will remain the same as a matter of fact, the proofs only will have more technical details because of the presence in this case of irreflexive elements in the frames, which we have to handle in special way. Therefore first we will develop the central idea for a simpler case of logics extending $S4$ and now we turn to prove the first technical lemma.

Lemma 4.2.2 *Let λ be a modal logic extending $S4$ or a superintuitionistic logic such that*

(i) λ is an $(\omega, 1, 3)$-logic or an $(m, 1, 3)$-logic;

(ii) λ has the fmp and the property of branching below 1;

(iii) for some k, λ has the effective k-drop points property if λ is a modal logic, and the strong effective k-drop points property if λ is a superintuitionistic logic.

Then the following holds. If λ is an $(\omega, 1, 3)$-logic then any inference rule r admissible for λ with n variables is valid in the frame E_n. If λ is an $(m, 1, 3)$-logic then any inference rule r in n variables and admissible for λ is valid in the frame $S_m(E_n)$.

Proof. First we consider the case when λ is an $(\omega, 1, 3)$-logic. Suppose a rule $r := \alpha_1, ..., \alpha_m/\beta$ is admissible in λ and has variables $x_1, ..., x_n$. Assume that V is a valuation of the variables $x_1, ..., x_n$ in E_n such that

$$V(\alpha_1) = |E_n|, V(\alpha_2) = |E_n|, V(\alpha_m) = |E_n|. \tag{4.1}$$

Since E_n has exactly $2^n + 2$ distinct elements of depth 2, there are two distinct elements a_k^2, a_j^2 of depth 2 from E_n such that

$$\forall x_i, 1 \leq i \leq n(a_k^2 \Vdash_V x_i \Leftrightarrow a_j^2 \Vdash_V x_i). \tag{4.2}$$

We modify the model $\langle E_n, V \rangle$ as follows. We contract the elements a_k^2 and a_j^2 into a single element b_2 preserving the partial order and valuation of E_n. Since (4.2) holds, this contraction is well-defined with respect to V. We denote the resulting frame by $\mathcal{F}_n$ and the resulting model by $\mathfrak{M}_n$. It is not hard to show by induction on the length of any formula γ built up out of the variables $x_1, ..., x_n$ that

$$\forall a \in |E_n| \, (a \neq a_k^2, a_j^2 \Rightarrow a \Vdash_V \gamma \text{ in } \langle E_n, V \rangle \Leftrightarrow a \Vdash_V \gamma \text{ in } \langle \mathcal{F}_n, V \rangle), \tag{4.3}$$

$$b_2 \Vdash_V \gamma \text{ in } \langle \mathcal{F}_n, V \rangle \Leftrightarrow \forall c = a_k^2, a_j^2 (c \Vdash_V \gamma \text{ in } \langle E_n, V \rangle).$$

Therefore using this coincidence of the validity and (4.1) it follows

$$V(\alpha_1) = |\mathcal{F}_n|, \ V(\alpha_2) = |\mathcal{F}_n|, \ V(\alpha_m) = |\mathcal{F}_n|. \tag{4.4}$$

We will show that the frame $\mathcal{F}_n$ is a p-morphic image of the disjoint union of three copies of the frames $E_{3,n}$. In fact, it is sufficient to make a condensing p-morphism f preserving the depth of elements $\mathcal{F} := E_{3,n} \sqcup E_{3,n} \sqcup E_{3,n}$. To define f we let f maps $S_1(\mathcal{F})$ in the maximal element of $\mathcal{F}_n$ and f acts on elements of depth 2 as follows. f maps the corresponding chosen above elements a_k^2, a_j^2 from any component $E_{3,n}$ of $\mathcal{F}$ into the element $b_2 \in Sl_2(\mathcal{F}_n)$. For other elements of depth 2 from any component $E_{3,n}$ of $\mathcal{F}$, f is the identical mapping in the set $Sl_2(\mathcal{F}_n)$. Clearly, f is a p-morphism of $S_2(\mathcal{F})$ onto the frame $S_2(\mathcal{F}_n)$. Now we define f on the antichain $Sl_3(\mathcal{F})$ as follows.

For any element $a \in Sl_3(\mathcal{F}_n)$, either there are two distinct from a elements c, d of depth 3 such that $(a^{\leq \leq} - \{a\}) = (c^{\leq \leq} - \{c\}) = (d^{\leq \leq} - \{d\})$, or there are no such elements at all. If for given a, elements c, d with above mentioned properties exist then we map the element a_1 from the first component $E(3, n)$ of $\mathcal{F}$ such that $f(a_1^{\leq \leq} - \{a_1\}) = (a^{\leq \leq} - \{a\})$ onto a; we map the element c_1 from the second component $E(3, n)$ of $\mathcal{F}$ such that $f(c_1^{\leq \leq} - \{c_1\}) = (c^{\leq \leq} - \{c\})$ onto c, and we map the element d_1 from the third component $E(3, n)$ of $\mathcal{F}$ such that $f(d_1^{\leq \leq} - \{d_1\}) = (d^{\leq \leq} - \{d\})$ onto d. These elements a_1, c_1 and d_1 exist since our choice the contracting p-morphism f from $S_2(\mathcal{F})$ onto $S_2(\mathcal{F}_n)$ and since, for any antichain $\mathcal{A}$ of elements form $Sl_2(E_{3,n})$, there is an element u of depth 3 from $E_{3,n}$ such that $(u^{\leq \leq} - \{u\}) = \bigcup_{v \in \mathcal{A}} v^{\leq \leq}$ by our construction of $E_{3,n}$. Finally, if there are no elements c and d distinct from a and with above described properties then we merely map by f the element a_1 from the first component $E(3, n)$ of $\mathcal{F}$ such that $f(a_1^{\leq \leq} - \{a_1\}) = (a^{\leq \leq} - \{a\})$ onto a;

After this extending of the domain of f, we obtain that f is a partial mapping, which maps $S_3(\mathcal{F})$ onto $S_3(\mathcal{F}_n)$. Now we extend f on whole set $|S_3(\mathcal{F})|$ as follows. Consider any element e of depth 3 out of the domain f. The set $f(e^{\leq\leq} - \{f(e)\})$ is a subset of $S_2(\mathcal{F}_n)$ containing elements of depth 2. Since we contracted the elements a_k^2 and a_j^2 into b_2 in E_n while constructing $\mathcal{F}_n$, there is an element e_1 of depth 3 in $\mathcal{F}_n$ such that $(e_1^{\leq\leq} - \{e_1\}) = f(e^{\leq\leq} - \{f(e)\})$. We map e into e_1. Now f maps the whole set $S_3(\mathcal{F})$ onto $S_3(\mathcal{F}_n)$, and by its definition f is a p-morphism. The definition f in other $Sl_j(\mathcal{F})$, where $3 < j$, is similar to the case $Sl_3(\mathcal{F})$. Hence there exists a p-morphism of $\mathcal{F}$ onto $\mathcal{F}_n$.

Since λ is an $(\omega, 3, 1)$-logic, $E_{3,n}$ is a λ-frame. λ has the fmp and therefore the n-characterizing model $Ch_\lambda(n)$ is defined. It is easy to see that there is a p-morphism from $Ch_\lambda(\omega)$ onto $E_{3,n}$. Therefore there is a p-morphism from the disjoint union of three copies of $Ch_\lambda(n)$ onto $\mathcal{F}_n$. Since λ has the fmp and λ has the property of branching below 1, λ has the property of branching below d for any d. Also λ has the effective k-drop points property if λ is a modal logic, and the strong effective k-drop points property if λ is a superintuitionistic logic. Therefore the following holds. By Theorem 3.5.1 or Theorem 3.5.8, being admissible in λ, the rule r is valid in $Ch_\lambda(n)$. Therefore r is valid in $\mathcal{F}_n$ as well. And (4.4) entails $V(\beta) = |\mathcal{F}_n|$. Then by (4.3) it follows $V(\beta) = |E_n|$. Thus r is valid in the frame E_n. For the case when λ is an $(m, 1, 3)$-logic the proof is the same but we consider only elements of depth not more than m. ∎

Lemma 4.2.3 *Let λ be a modal logic extending $S4$ or a superintuitionistic logic such that*

(i) λ is an $(\omega, 1, 3)$-logic or is an $(m, 1, 3)$-logic;

(ii) λ has the fmp and the property of branching below 1;

(iii) for some k, λ has the effective k-drop points property if λ is a modal logic, and the strong effective k-drop points property if λ is a superintuitionistic logic.

Then the following holds. If λ is an $(\omega, 1, 3)$-logic then there is an inference rule r admissible in λ and having $2^n + 4$ variables which is false in the frame E_n. If λ is an $(m, 1, 3)$-logic then there exists a certain rule r admissible for λ in $2^n + 4$ variables which is invalid in the frame $S_m(E_n)$.

Proof. We define the rule r with $2^n + 4$ variables as follows.

$$\psi := \bigwedge_{1 \leq i \leq 2^n + 2} \Box x_i \wedge \Box y_0 \wedge \Box x_0; \quad \forall i \in \{1, ..., 2^n + 2\}$$
$$\xi_i := \Box x_i \wedge \bigwedge_{1 \leq j \leq 2^n + 2, j \neq i} \neg\Box x_j \wedge \neg\Box y_0 \wedge \Box x_0;$$

$$\forall \rho, \rho \subseteq \{1, ..., 2^n + 2\} \& (1 < ||\rho|| \leq 2^n),$$

$$\mu_\rho \;\; := \;\; \Box x_0 \wedge \neg \Box y_0 \wedge \bigwedge_{j \in \rho} \neg \Box \neg \xi_i \wedge \bigwedge_{j \in \{1,...,2^n+2\}, j \notin \rho} \Box \neg \xi_j \wedge$$
$$\bigwedge_{1 \leq i \leq 2^n + 2} \neg \Box x_i;$$

$$\forall \rho, \rho \subseteq \{1, ..., 2^n + 2\} \& (||\rho|| = 2^n + 1),$$

$$\theta_\rho \;\; := \;\; (\bigvee \{\neg \Box \neg \mu_{\rho_1} \wedge \neg \Box \neg \mu_{\rho_2} \mid \rho_1 \subset \rho, \rho_2 \subset \rho, 1 < ||\rho_1||,$$
$$1 < ||\rho_2||, \rho_1 \neq \rho_2 \}) \wedge \Box x_0 \wedge \neg \Box y_0,$$

$$\pi \;\; := \;\; \bigwedge_{1 \leq i \leq 2^n + 2} \neg \Box \neg \xi_i \wedge \neg \Box y_0 \wedge \neg \Box x_0;$$

$$r := \frac{\Box (\bigvee_i \xi_i \vee \bigvee_\rho \mu_\rho \bigvee_\rho \theta_\rho \vee \pi \vee \psi)}{\Box x_0}.$$

We will show that r is false in E_n and $S_m(E_n)$. For this purpose we define the following valuation V in E_n and $S_m(E_n)$.

$$V(x_i) := (a_i^2)^{\leq \leq},$$
$$V(y_0) := S_1(E_n),$$
$$V(x_0) := \{v \mid v \in E_n, v^{\leq \leq} \cap Sl_2(E_n) \neq Sl_2(E_n).\}$$

It can be easily verified directly that

$$a_1^1 \Vdash_V \psi,$$
$$\forall i \in \{1, ..., 2^n + 2\} \; a_i^2 \Vdash_V \xi_i,$$
$$\forall a_j^k \in |E_n|(\{i \mid a_i^2 \in (a_j^k)^{\leq \leq} \cap Sl_2(E_n)\} = \rho \& (1 \leq ||\rho|| \leq 2^n) \Rightarrow$$
$$a_j^k \Vdash_V \mu_\rho),$$
$$\forall a_j^k \in |E_n|(\{i \mid a_i^2 \in (a_j^k)^{\leq \leq} \cap Sl_2(E_n)\} = \rho \& ||\rho|| = 2^n + 1 \Rightarrow$$
$$a_j^k \Vdash_V \theta_\rho),$$
$$\forall a_j^k \in |E_n|(\{i \mid a_i^2 \in (a_j^k)^{\leq \leq} \cap Sl_2(E_n)\} = \rho \& ||\rho|| = 2^n + 2 \Rightarrow$$
$$a_j^k \Vdash_V \pi).$$

In particular, it follows $V(\Box(\bigvee_i \xi_i \vee \bigvee_\rho \mu_\rho \bigvee_\rho \theta_\rho \vee \pi \vee \psi)) = E_n$. At the same time we obtain $V(x_0) \neq E_n$. Hence r is invalid in E_n. Using similar arguments, it is easy to see that r is also invalid in the frame $S_m(E_n)$, $3 \leq m$ by V.

Suppose λ is a modal logics with the properties from the assertion of our lemma. We will show that r is admissible in λ. Since λ has the above mentioned properties, by Theorem 3.5.1 it is sufficient to show that the rule r is valid in the frame of the model $Ch_\lambda(2^n + 4)$. Suppose S is a valuation of all variables which occur in the rule r in $Ch_\lambda(2^n + 4)$, and

$$S(\Box(\bigvee_i \xi_i \vee \bigvee_\rho \mu_\rho \bigvee_\rho \theta_\rho \vee \pi \vee \psi)) = |Ch_\lambda(2^n + 4)|, \text{ but} \qquad (4.5)$$

$$S(\Box x_0) \neq |Ch_\lambda(2^n + 4)|.$$

Using (4.5) we get

$$(\exists a \in |Ch_\lambda(2^n + 4)|) \; a \Vdash_S \pi.$$

Then there exist elements $a_i \in |Ch_\lambda(2^n + 4)|, i \in \{1, 2, ..., 2^n + 2\}$ such that aRa_i for any i, and $a_i \Vdash_S \xi_i$ where R is the accessibility relation from the model $Ch_\lambda(2^n + 4)$. Then the tuple of all a_i has the following property: $i_1 \neq i_2 \Rightarrow \neg(a_{i_1} Ra_{i_2}) \& \neg(a_{i_2} Ra_{i_1})$. Since λ has the fmp and the property of branching below 1, there is an element $b \in |Ch_\lambda(2^n + 4)|$ such that

$$(b^{R\leq\leq} - \{b\}) = \bigcup_{1 \leq i \leq 2^n + 1} a_i^{R\leq\leq}. \tag{4.6}$$

By (4.5) there is a formula α, where $\alpha \in \{\xi_i, \mu_\rho, \theta_\rho, \pi, \psi\}$, such that $b \Vdash_S \alpha$. The case $\alpha = \psi$ is impossible since for distinct $a_{i_1}, a_{i_2}, 1 \leq i_1 < i_2 \leq 2^n + 1$, $a_{i_1} \Vdash \neg\Box x_{i_2}$ and $a_{i_2} \Vdash \neg\Box x_{i_1}$. The case $\alpha = \pi$ is impossible also. Indeed, suppose $b \Vdash_S \pi$. Then for some c, bRc and $c \Vdash_S \xi_{2^n+2}$. This yields $\neg(cRa_i)$ for $i \in \{1, ..., 2^n + 1\}$ because $a_i \Vdash_S \neg\Box x_{2^n+2}$. Therefore from (4.6) it follows that there is an $i \in \{1, 2, ..., 2^n + 1\}$ such that $a_i Rc$. But $c \Vdash_S \neg\Box x_i$ and $a_i \Vdash_S \Box x_i$, a contradiction.

Assume that $\alpha = \xi_h$ for some suitable h. Then it follows that $a_i \Vdash_S \Box x_h$ for all $i \in \{1, ..., 2^n + 1\}$ which contradicts $a_i \Vdash_S \neg\Box x_h$ when $i \neq h$. Suppose now that $\alpha = \mu_\rho$ for some $\rho \subseteq \{1, 2, ..., \}, 1 < ||\rho|| \leq 2^n$. Since (4.6) holds it follows that there is some $j \in \{1, 2, ..., 2^n + 1\}, j \notin \rho$ such that bRa_j, $a_j \Vdash_S \xi_j$. But since by our assumption $b \Vdash_S \mu_\rho$ we conclude $b \Vdash_S \Box\neg\xi_j$, a contradiction.

Only the following possibility remains: $\alpha = \theta_\rho$ for some suitable ρ, where $||\rho|| = 2^n + 1$. Then there are some $\rho_1 \subset \rho$ and $\rho_2 \subset \rho$, where $1 < ||\rho_1||$ and $1 < ||\rho_2||$, such that

$$bRa_{\rho_1}, \quad bRa_{\rho_2}, \quad a_{\rho_1} \Vdash_S \mu_{\rho_1}, \quad a_{\rho_2} \Vdash_S \mu_{\rho_2}.$$

The case where $a_{\rho_u} Ra_i$, $u = 1, 2, i \in \{1, 2, ..., 2^n + 1\}$ is impossible. Indeed, if $a_{\rho_u} Ra_i, u = 1, 2, i \in \{1, 2, ..., 2^n+1\}$ it follows by (4.6) that $a_{\rho_u} = b$ or $a_{\rho_u} = a_i$. In the first case we obtain a contradiction from the fact that $||\rho_u|| \leq 2^n$ and $a_v \Vdash_V \xi_v$ for all $v \in \{1, 2, ..., 2^n + 1\}$. In second case, when $a_{\rho_u} = a_i$, we obtain a contradiction from the fact that $a_i \Vdash_S \Box x_i$ and $a_{\rho_u} \Vdash_S \mu_{\rho_u}$. Thus according to (4.6) it follows that there is some $a_i, i \in \{1, 2, ..., 2^n + 1\}$ such that $a_i Ra_{\rho_u}$ for $u = 1, 2$. But $a_i \Vdash_S \xi_i$ and, consequently, $a_i \Vdash_S \Box x_i$. At the same time $a_{\rho_u} \Vdash_S \bigwedge_{1 \leq j \leq 2^n+2} \neg\Box x_j$, a contradiction. Thus the case where $\alpha = \theta_\rho$ also is impossible. Thus we have exhausted all the possibilities and proved that none of the disjuncts of the premise of r can be valid on b under S. Therefore r is valid in the frame of $Ch_\lambda(2^n + 4)$ and, as we have noted before, this entails that r is admissible in λ.

Now consider the case when λ is a superintuitionistic $(\omega, 1, 3)$-logic with the properties mentioned in the formulation of this theorem. Consider the rule r. According to Lemma 2.7.14 there is a formula α in the language of intuitionistic logic H such that $T(\alpha) \equiv \beta \in S4$, where

$$\beta = \Box(\bigvee_i \xi_i \vee \bigvee_\rho \mu_\rho \bigvee_\rho \theta_\rho \vee \pi).$$

Therefore $T(\alpha)/T(x_0)$ is invalid in the frame E_n and in the frame $S_m(E_n)$. By Lemma 2.7.7, which connects the truth of intuitionistic formulas in Kripke models and their T-translations in the related modal Kripke models, it follows that the rule α/x_0 is invalid in the intuitionistic frame E_n. Further, by Theorem 3.2.2 α/x_0 is admissible in λ iff $T(\alpha)/T(x_0)$ is admissible in the greatest modal counterpart $\sigma(\lambda)$ of λ. By Theorem 2.7.20, Corollary 2.7.21 and Lemma 3.5.6 the greatest modal counterpart $\sigma(\lambda)$ has all the properties which are required for modal logics in the formulation of this lemma. Therefore applying the part of the proof which we have done already for modal logics, we arrive at the fact that r is admissible in $\sigma(\lambda)$ and consequently $T(\alpha)/T(x_0)$ is admissible in $\sigma(\lambda)$. As we have noted above, this entails that α/x_0 is admissible in λ. ■

Theorem 4.2.4 *Let λ be a modal logic extending $S4$ or a superintuitionistic logic. Suppose that*

(i) λ is an $(\omega, 1, 3)$-logic or an $(m, 1, 3)$-logic;

(ii) λ has the fmp and the property of branching below 1;

(iii) for some k, λ has the effective k-drop points property if λ is a modal logic, and the strong effective k-drop points property if λ is a superintuitionistic logic.

Then λ has no bases for admissible rules in finitely many variables, and, in particular, λ has no finite bases for admissible rules.

Proof. The proof follows immediately from Lemmas 4.2.2 and 4.2.3 and Lemma 4.1.4.

The following algebraic analog of Theorem 4.2.4 follows immediately from this theorem and Theorem 1.4.15.

Corollary 4.2.5 *If λ is a modal logic or a superintuitionistic logic with the proprieties stated in Theorem 4.2.4 then the free algebra $\mathfrak{F}_\lambda(\omega)$ from the variety $Var(\lambda)$ has no bases for quasi-identities in finitely many variables, and in particular, no finite bases.*

Now we will apply these general results to certain particular logics. This way we will obtain a negative solution to the A.Kuznetsov-H.Friedman problem concerning whether there is a finite basis for rules admissible in H, and to versions of this problem for other related logics.

Corollary 4.2.6 *The following hold*

    (i) The superintuitionistic logics H, KC, any logic I_n, where $n \in N$, have no bases in finitely many variables for admissible inference rules.

    (ii) The modal logics $S4$, $S4.1$, $S4.2$, any logic $S4I_n$, for $n \in N$, have no bases in finitely many variables for admissible rules.

Proof. To prove (i) it is sufficient to apply Theorem 4.2.4 for the logics mentioned and to recall that these logics have the fmp, the property of branching below 1 and the strong effective 1-drop points property. Similarly, for the modal logics mentioned above, all these logics have the fmp, the property of branching below 1 and the effective 1-drop points property. ∎

Before modifying this technique to establish a similar result for the logic $K4$ and related non-reflexive logics, we will slightly modify our tools in order to prove non-existence of finite bases for admissible rules in certain other reflexive modal and superintuitionistic logics. In the approach we used we rigidly fixed the *top* of frames E_n and the *top* of frames in the definition of $(\omega, 1, 3)$-logics and $(m, 1, 3)$-logics. However this poses a problem to our establishing the lack of finite bases for admissible rules of certain very simple and well known logics. But it is sufficient to make only a simple modification in order to remove this difficulty.

We define U_n to be the frame obtained from E_n by removing the greatest element of E_n; respectively, $U_{3,n}$ is the frame obtained from $E_{3,n}$ by removing the greatest element of $E_{n,3}$.

Definition 4.2.7 *A modal logic λ extending $S4$ or a superintuitionistic logic λ is said to be an $(\omega, 2)$-logic if any frame $U_{3,n}$ is a λ-frame. A modal logic λ extending $S4$ or a superintuitionistic logic λ is an $(m, \omega, 2)$-logic if λ is a logic of depth m, $3 \le m$, and any frame $S_m(U_{3,n})$ is a λ-frame.*

Lemma 4.2.8 *Let λ be a modal logic extending $S4$ or a superintuitionistic logic such that the following hold:*

(i) λ is an $(\omega, 2)$-logic or an $(m, \omega, 2) - logic$;

(ii) λ has the fmp and the property of branching below 1;

(iii) for some k, λ has the effective k-drop points property if λ is a modal logic, and the strong effective k-drop points property if λ is a superintuitionistic logic.

Then if λ is an $(\omega, 2)$-logic then any inference rule r in n variables and admissible in λ is valid in the frame U_n. If λ is an $(m, \omega, 2)$-logic then any inference rule r admissible in λ and having n variables is valid in the frame $S_m(U_n)$.

Proof. The proof of this lemma is a simplified variant of the proof of Lemma 4.2.2. We merely omit any mention of the single maximal element of E_n, since generally speaking U_n has more than one maximal element. We proceed as in earlier proof beginning with depth 2 in that our frames. ∎

Lemma 4.2.9 *Let λ be a modal logic extending $S4$ or a superintuitionistic logic such that the following hold:*

(i) λ is an $(\omega, 2)$-logic or an $(m, \omega, 2)$-logic;

(ii) λ has the fmp and the property of branching below 1;

(iii) for some k, λ has the effective k-drop points property if λ is a modal logic, and the strong effective k-drop points property if λ is a superintuitionistic logic.

Then the following hold. If λ is an $(\omega, 2)$-logic then there is an inference rule r admissible in λ and having $2^n + 4$ variables which is invalid in the frame U_n. If λ is an $(m, \omega, 3)$-logic then the same holds for the frame $S_m(U_n)$.

Proof. We follow the proof of Lemma 4.2.3 very closely. We define a rule r with $2^n + 4$ variables as in the proof of that lemma but we omit the disjunct ψ in the premise. In order to invalidate this rule in U_n and $S_m(U_n)$ we define a valuation V exactly as in Lemma 4.2.3 in the family of the elements of U_n which are present in E_n. We show that V invalidates r as in Lemma 4.2.3. In order to prove that r is admissible in λ we repeat the corresponding part of the proof of Lemma 4.2.3, once again ignoring the disjunct ψ. ∎

From Lemmas 4.2.9 and 4.2.8 we immediately derive the following theorem.

Theorem 4.2.10 *Let λ be a modal logic extending $S4$ or a superintuitionistic logic such that:*

(i) λ is an $(\omega, 2)$-logic or an $(m, \omega, 2)$-logic;

(ii) λ has the fmp and the property of branching below 1;

(iii) for some k, λ has the effective k-drop points property if λ is a modal logic, and the strong effective k-drop points property if λ is a superintuitionistic logic.

Then (iv): λ has no bases for admissible rules in finitely many variables, and, in particular, no finite bases. (v): the free algebra $\mathfrak{F}_\lambda(\omega)$ from the variety $Var(\lambda)$ has no bases for valid quasi-identities in finitely many variables, and, in particular, no finite bases.

Now we extract results concerning certain particular logics from this general theorem. Clearly we can present many series of logics to which this theorem is applicable, varying requirements on the number elements in clusters for frames generating such logics, etc. We restrict us with series presented in the following

Corollary 4.2.11 *The following hold*

(i) *The smallest superintuitionistic logics* $H \oplus \phi_n$) *of depth* n *for any* $n > 1$ *have no bases in finitely many variables for admissible inference rules.*

(ii) *The smallest modal logics* $S4 \oplus \sigma_n$ *of depth* n *extending* $S4$ *for any* $n > 1$ *have no bases in finitely many variables for admissible rules.*

Proof. The superintuitionistic and modal logics mentioned above have the fmp, the property of branching below 1 and the strong effective 1-drop points property or simply the effective 1-drop points property for the case of modal logics (see Lemma 3.5.3). Therefore to establish the absence of finite basis it is sufficient to apply Theorem 4.2.10. ∎

Note that the classes of modal and superintuitionistic logics to which the results concerning the lack of finite bases described above are applicable, nevertheless, seem to be rather narrow. This is a case in fact, but fortunately, as we have seen above, many important logics belong to these classes. The diversity of modal logics in such classes is rather representative (There are infinitely many different such modal logics. For instance, this is a corollary of the fact that it is possible to bound the number of elements in clusters of distinct depth (going from the top to the bottom) by certain modal formulas, and other related facts).

Now we intend to extend the results obtained to the minimal transitive modal logic $K4$ and various similar modal logics. This construction is more detailed since we have deal also with irreflexive elements, and we must treat these elements in a different way. First we introduce certain frames similar to frames E_n. Their construction is similar to one described before for E_n and the difference consists in that now we add some new irreflexive elements. More precisely, the construction of the frames $\mathcal{K}_n$, $1 \le n$ is as follows. For any given n, we introduce a sequence of finite transitive frames $\mathcal{K}_n^m := \langle K_n^m, R_m \rangle$:

(i) K_n^1 consists of two elements a_1^1 and b_1^1, where a_1^1 is reflexive and b_1^1 is irreflexive with respect to R_1, and $\{a_1^1, b_1^1\}$ is an antichain with respect to R_1.

(ii) K_n^2 includes $\mathcal{K}_n^1$, as its first slice, and the antichain of elements of depth 2, which consists of elements

$$\{a_0^2, a_1^2, ..., a_{2^n+1}^2, a_{2^n+2}^2\} \cup \{b_0^2, b_1^2, ..., a_{b^n+1}^2, a_{2^n+2}^2\}.$$

All a_i^2 are reflexive and all b_j^2 are irreflexive by R_2 and $\forall i$, where $1 \le i \le 2^n + 2$, $(a_i^2)^{R_2\le} = \{a_i^2\} \cup \{a_1^1, b_1^1\}$; $\forall i, 1 \le i \le 2^n + 2$, $(b_i^2)^{R_2\le} = \{a_1^1, b_1^1\}$, and $(a_0^2)^{R_2<} = \{b_1^1\}$, $(b_0^2)^{R_2<} = \{b_1^1\}$.

(iii) $\mathcal{K}_n^3 := \langle K_n^3, \le_3 \rangle$, where

$$
\begin{aligned}
K_n^3 := K_n^2 &\cup \{a_i^3 \mid i \text{ is an antichain of } Sl_2(\mathcal{K}_n^2) \& 1 \le ||i|| \\
&\& ||i|| \ne 2^{n+1} + 5\} \cup \{b_i^3 \mid i \text{ is an antichain of } Sl_2(\mathcal{K}_n^2) \& 1 \le ||i|| \\
&\& ||i|| \ne 2^{n+1} + 5\} \cup \{c_i^3 \mid i \text{ is an antichain of } \mathcal{K}_n^2 \text{ containing } a_1^1 \\
&\text{but not } b_1^1 \& 1 < ||i||\}\{d_i^3 \mid i \text{ is an antichain of } \mathcal{K}_n^2 \text{ containing } a_1^1 \\
&\text{but not } b_1^1 \& 1 < ||i||\};
\end{aligned}
$$

R_3 extends R_2, and all a_i^3, c_i^3 are reflexive with respect to R_3; all b_i^3, d_i^3 are irreflexive with respect to R_3; and they all form an antichain of all the elements of depth 3 in the frame $\mathcal{K}_n^3$ (with respect to R_3). Moreover, $\forall y_i = a_i^3, b_i^3, c_i^3, d_i^3, \forall x \in K_n^2, (y_i R_3 x) \Leftrightarrow (x \in i) \vee \exists z \in i(z R_2 x)$

(iv) Suppose $\mathcal{K}_n^j$ is already constructed and $3 \le j$. We define

$$
\begin{aligned}
K_n^{j+1} := K_n^j &\cup \{a_i^{j+1} \mid i \text{ is an antichain in } \mathcal{K}_n^j, 1 \le ||i||, \\
&(\exists a_h^j \in Sl_j(\mathcal{K}_n^j))(a_h^j \in i)\} \cup \{b_i^{j+1} \mid i \text{ is an antichain in } \mathcal{K}_n^j, \\
&(\exists x \in Sl_j(\mathcal{K}_n^j))(x \in i).\}
\end{aligned}
$$

We let R_{j+1} extend R_j, all a_i^{j+1} be reflexive with respect to R_3, all b_i^{j+1}, be irreflexive with respect to R_3, and all of them form the antichain of depth $j+1$ in the frame $\mathcal{K}_n^{j+1}$ (with respect to R_{j+1}). Moreover, $\forall y_i = a_i^{j+1}, b_i^{j+1}$, $\forall x \in K_n^j, (y_i R_{j+1} x) \Leftrightarrow (x \in i) \vee \exists z \in i(z R_j x)$.

(v) Given the description of all finite frames $\mathcal{K}_n^m$, $m \in N$ in (i) - (iv) we let

$$
\mathcal{K}_n := \langle \bigcup_{m \in N} K_n^m, R \rangle, \text{ where } R := \bigcup_{m \in N} R_m.
$$

As before it follows that any $\mathcal{K}_n^j$ is the set of all elements of $\mathcal{K}_n^{j+1}$ with a depth of not more than j, that is $S_j(\mathcal{K}_n^{j+1}) = \mathcal{K}_n^j$ and $(\mathcal{K}_n^{j+1} - \mathcal{K}_n^j) = Sl_{j+1}(\mathcal{K}_n^{j+1})$, and, in particular, $\mathcal{K}_n^j$ is an open subframe of $\mathcal{K}_n^{j+1}$ and of $\mathcal{K}_n$.

The frame $\mathcal{G}_n$ is obtained from the frame $\mathcal{K}_n$ by the following procedure. First we contract elements a_1^1 and b_1^1 into a single element b_1^1 supposing it to be irreflexive (of course this contraction is not a p-morphism), we denote the resulting frame by $\mathcal{F}_n$. Then we take the open subframe of $\mathcal{F}_n$ consisting of only irreflexive elements which do not see (a) elements a_2^0 and b_2^0, (b) any reflexive element of the frame $\mathcal{F}_n$, and (c) any irreflexive element of $\mathcal{F}_n$ which is a co-cover for an $2^n + 1$-elements antichain of irreflexive elements each of which has depth 2.

Definition 4.2.12 *We say that a modal logic λ extending $K4$ is an $(\omega, 2, 3)$-logic if any frame $\mathcal{K}_{3,n}$ obtained in the same way as the frame $\mathcal{K}_n$ except that in step (iii) we omit the limitations $||i|| \neq 2^{n+1} + 5$ is an λ-frame. A modal logic λ over $K4$ is an $(m, 2, 3)$-logic if λ is a logic of depth m, $3 \leq m$, and, for any n, the frame $S_m(\mathcal{K}_{3,n})$ is a λ-frame. Similarly, a modal logic λ extending $K4$ is an $(\omega, i, 3)$-logic if the frame $\mathcal{G}_{3,n}$ obtained from $\mathcal{G}_n$ by removing the limitation concerning $2^n + 1$-elements antichains of depth 2 in the definition of $\mathcal{G}_n$ is a λ-frame. Finally, λ, where $K4 \subseteq \lambda$, is an $(m, i, 3)$-logic if λ is a logic of depth m, $3 \leq m$, and, for any n, the frame $S_m(\mathcal{G}_{3,n})$ is a λ-frame.*

Lemma 4.2.13 *Let λ be a modal logic extending $K4$ such that*

(i) λ is an $(\omega, 2, 3)$-logic or an $(m, 2, 3)$-logic;

(ii) λ has the fmp and the property of branching below 1;

(iii) for some k, λ has the effective k-drop points property

If λ is an $(\omega, 2, 3)$-logic then any inference rule r in n variables and admissible in λ is valid in the frame of $\mathcal{K}_n$. If λ is an $(m, 2, 3)$-logic then any inference rule r admissible in λ with n variables is valid in the frame $S_m(\mathcal{K}_n)$.

Proof. We follow the proof-scheme of Lemma 4.2.2 closely, but our proof now is more complicated in view of presence irreflexive elements. Let λ be an $(\omega, 2, 3)$-logic. Suppose a rule $r := \alpha_1, ..., \alpha_m/\beta$ is admissible in λ and r has variables $x_1, ..., x_n$. Let V be a valuation of variables $x_1, ..., x_n$ in $\mathcal{K}_n$ and

$$V(\alpha_1) = |\mathcal{K}_n|, \ V(\alpha_2) = |\mathcal{K}_n|, ..., V(\alpha_m) = |\mathcal{K}_n|. \tag{4.7}$$

Since $\mathcal{K}_n$ has exactly $2^n + 2$ distinct reflexive elements of depth 2 which see both elements of depth 1 from $\mathcal{K}_n$, there are two distinct elements a_k^2, a_j^2 of depth 2 from K_n such that

$$\forall x_i, 1 \leq i \leq n(a_k^2 \Vdash_V x_i \Leftrightarrow a_j^2 \Vdash_V x_i). \tag{4.8}$$

We modify the model $\langle \mathcal{K}_n, V \rangle$ as follows. We contract the elements a_k^2 and a_j^2 into a single reflexive element w_2 preserving the relation R and the valuation from $\mathcal{K}_n$. Since (4.8) holds, this contracting is well-defined with respect to V. We will denote the resulting frame by $\mathcal{F}_n$ and the resulting model by $\mathfrak{M}_n$. Since this contracting is a p-morphism, it follows that for any formula γ built up out of variables $x_1, ..., x_n$,

$$\forall a \in K_n \ (a \neq a_k^2, a_j^2 \Rightarrow a \Vdash_V \gamma \text{ in } \langle \mathcal{K}_n, V \rangle \Leftrightarrow a \Vdash_V \gamma \text{ in } \langle \mathcal{F}_n, V \rangle), \tag{4.9}$$

$$w_2 \Vdash_V \gamma \text{ in } \langle \mathcal{F}_n, V \rangle \Leftrightarrow \forall c = a_k^2, a_j^2(c \Vdash_V \gamma \text{ in } \langle \mathcal{K}_n, V \rangle).$$

Using this coincidence of validity and (4.7) we conclude

$$V(\alpha_1) = |\mathcal{F}_n|, \ V(\alpha_2) = |\mathcal{F}_n|, \ ..., V(\alpha_m) = |\mathcal{F}_n|. \qquad (4.10)$$

Now we prove that the frame $\mathcal{F}_n$ is a p-morphic image of the disjoint union of three copies of the frames $\mathcal{K}_{3,n}$. In fact, it is sufficient to make the condensing p-morphism f preserving the depth of elements from $\mathcal{F} := \mathcal{K}_{3,n} \sqcup \mathcal{K}_{3,n} \sqcup \mathcal{K}_{3,n}$ which is described below. We let f map all reflexive elements from $S_1(\mathcal{F})$ into the maximal reflexive element of $\mathcal{F}_n$. We map by f all the irreflexive elements of $S_1(\mathcal{F})$ into the irreflexive maximal element of $\mathcal{F}_n$. Then we define f on elements of depth 2 from $\mathcal{F}$: f maps the corresponding chosen above elements a_k^2 and a_j^2 from any component $\mathcal{K}_{3,n}$ of $\mathcal{F}$ into the element $w_2 \in Sl_2(\mathcal{F}_n)$. For other elements of depth 2 from any component $\mathcal{K}_{3,n}$ of $\mathcal{F}$, f is the identical (one-to-one) mapping on $Sl_2(\mathcal{F}_n)$. It is clear that f is a p-morphism from $S_2(\mathcal{F})$ onto the frame $S_2(\mathcal{F}_n)$.

Then we extend the domain of f to all elements of depth 3 as follows. For any reflexive element $a_r \in Sl_3(\mathcal{F}_n)$, either there are two reflexive elements c_r, d_r of depth 3 distinct from a_r such that

$$(a_r^{R\le} - \{a_r\}) = (c_r^{R\le} - \{c_r\}) = (d_r^{R\le} - \{d_r\}),$$

or there are no such elements distinct from a_r at all. Similarly, for any irreflexive element $a_i \in Sl_3(\mathcal{F}_n)$, either there are two irreflexive elements c_i, d_i of depth 3 distinct from a_i such that

$$(a_i^{R\le} - \{a_i\}) = (c_i^{R\le} - \{c_i\}) = (d_i^{R\le} - \{d_i\}),$$

or there are no such elements distinct from a_i at all. If, for given a_r (a_i), elements c_r, d_r (c_i, d_i) with properties mentioned above exist then we map

(1) a reflexive element a_{r1} (an irreflexive element a_{i1}) from the first component $\mathcal{K}_{3,n}$ of $\mathcal{F}$ such that $f(a_{r1}^{R\le} - \{a_{r1}\}) = (a_r^{R\le} - \{a_r\})$ $(f(a_{i1}^{R\le} - \{a_{i1}\}) = (a_i^{R\le} - \{a_i\})$) onto a_r (onto a_i, respectively);

(2)) a reflexive element c_{r1} (an irreflexive element c_{i1}) from the second component $\mathcal{K}_{3,n}$ of $\mathcal{F}$ such that $f(c_{r1}^{R\le} - \{c_{r1}\}) = (a_r^{R\le} - \{a_r\})$ (such that $f(c_{i1}^{R\le} - \{c_{i1}\}) = (a_i^{R\le} - \{a_i\})$) onto c_r (onto c_i, respectively);

(3)) a reflexive element d_{r1} (an irreflexive element d_{i1}) from the third component $\mathcal{K}_{3,n}$ of $\mathcal{F}$ such that $f(d_{r1}^{R\le} - \{d_{r1}\}) = (a_r^{R\le} - \{a_r\})$ (such that $f(d_{i1}^{R\le} - \{d_{i1}\}) = (a_i^{R\le} - \{a_i\})$) onto d_r (onto d_i, respectively).

The mentioned elements a_{r1}, c_{r1}, d_{r1} a_{i1}, c_{i1} and d_{i1} exist since our choice of the contracting p-morphism f from $S_2(\mathcal{F})$ onto $S_2(\mathcal{F}_n)$ and since, for any antichain $\mathcal{A}$ of elements form $S_2(\mathcal{K}_{3,n})$ having elements of depth 2, there is an element u of depth 3 from $\mathcal{K}_{3,n}$ such that $(u^{R\leq} - \{u\}) = \bigcup_{v \in \mathcal{A}} v^{R\leq}$ by our construction of $\mathcal{K}_{3,n}$.

If there are no elements c_r and d_r (c_i and d_i) distinct from a_r (a_i) with property described above then we map a reflexive element a_{r1} (an irreflexive element a_{i1}) from the first component $\mathcal{K}_{3,n}$ of $\mathcal{F}$ such that $f(a_{r1}^{R\leq} - \{a_{r1}\}) = (a_r^{R\leq} - \{a_r\})$ (respectively, such that $f(a_{i1}^{R\leq} - \{a_{i1}\}) = (a_i^{R\leq} - \{a_i\})$) onto a_r (onto a_i, respectively); the existence of elements a_{r1} and a_{i1} can be shown as above.

Doing this extending of the domain of f, we obtain that f now is a partial mapping, which maps $S_3(\mathcal{F})$ onto $S_3(\mathcal{F}_n)$. Now we extend the domain of f to the whole set $S_3(\mathcal{F})$ by letting f on any element e of depth 3 which is out the domain of f yet as follows. The set $f(e^{R\leq} - \{e\})$ is a subset of $S_2(\mathcal{F}_n)$ containing elements of depth 2. Since we contracted the elements a_k^2 and a_j^2 into w_2 in $\mathcal{K}_n$ constructing $\mathcal{F}_n$, there is an element e_1 of depth 3 in $\mathcal{F}_n$ such that $(e_1^{R\leq} - \{e_1\}) = f(e^{R\leq} - \{e\})$. Moreover, if e is reflexive, there is a reflexive e_1 with property pointed above, and if e is irreflexive then there is an irreflexive element e_1 with the similar property. We map e by f into e_1. Now f maps $S_3(\mathcal{F})$ onto $S_3(\mathcal{F}_n)$ and, according to our definitions, f is a p-morphism. The definition of f on other $Sl_j(\mathcal{F})$, where $3 < j$, is similar to the case of $Sl_3(\mathcal{F})$. After defining f on elements of any finite depth we obtain that f is required p-morphism from $\mathcal{F}$ onto $\mathcal{F}_n$.

Since λ is an $(\omega, 2, 3)$-logic, $\mathcal{K}_{3,n}$ is a λ-frame. λ has the fmp and therefore the n-characterizing model $Ch_\lambda(n)$ is defined. It is easy to see that there is a p-morphism from the frame of $Ch_\lambda(n)$ onto the frame $\mathcal{K}_{3,n}$. Therefore there is a p-morphism from the disjoint union of three copies of $Ch_\lambda(n)$ onto $\mathcal{F}_n$. We know that λ has the fmp, the property of branching below k for any k, (since it has for 1) and, for some k, λ has the effective k-drop points property. Therefore according to Theorem 3.5.1, being admissible in λ, the rule r is valid in $Ch_\lambda(n)$. Hence r is valid in $\mathcal{F}_n$ and (4.4) entails $V(\beta) = |\mathcal{F}_n|$. Then by (4.9) it follows that $V(\beta) = \mathcal{K}_n$. Thus r is valid in the frame $\mathcal{K}_n$. For the case when λ is an $(m, 2, 3)$-logic, the proof is the same but we consider only elements of depth not more than m. $\blacksquare$

Lemma 4.2.14 *Let λ be a modal logic extending $K4$ and*

> *(i) λ is an $(\omega, i, 3)$-logic or an $(m, i, 3)$-logic;*
>
> *(ii) λ has the fmp and the branching below 1 property;*
>
> *(iii) for some k, λ has the effective k-drop points property.*

If λ is an $(\omega, i, 3)$-logic then any inference rule r admissible in λ and having n variables is valid in the frame of $\mathcal{G}_n$. If λ is an $(m, i, 3)$-logic then any inference rule r admissible in λ having n variables is valid in the frame $S_m(\mathcal{G}_n)$.

Proof. The proof is a simplified version of the proof of Lemma 4.2.13, we only omit all parts concerning reflexive elements. ∎

Lemma 4.2.15 *Let λ be a modal logic extending $K4$ such that*

(i) *λ is an $(\omega, 2, 3)$-logic or an $(m, 2, 3)$-logic;*

(ii) *λ has the fmp and the property of branching below 1;*

(iii) *for some k, λ has the effective k-drop points property*

If λ is an $(\omega, 2, 3)$-logic then there is an inference rule r admissible in λ and having $2^{n+1} + 10$ variables which is invalid in the frame $\mathcal{K}_n$. If λ is an $(m, 2, 3)$-logic then there exists a rule r in $2^{n+1} + 10$ variables admissible in λ which is invalid in the frame $S_m(\mathcal{K}_n)$.

Proof. We define the rule r with $2^{n+1} + 10$ variables as follows.

$$\psi_1 := v_1 \wedge \Box v_1 \wedge \neg v_2 \wedge \neg \Diamond v_2 \wedge \neg \Box \bot \wedge \neg \Diamond \Box \bot \wedge$$
$$\bigwedge_{1 \leq i \leq 2^n+2}(\neg x_i \wedge \neg \Diamond x_i) \wedge \bigwedge_{1 \leq i \leq 2^n+2}(\neg y_i \wedge \neg \Diamond y_i) \wedge$$
$$\wedge \neg z_1 \wedge \neg \Diamond z_1 \wedge \neg z_2 \wedge \neg \Diamond z_2 \wedge \neg z_3 \wedge \neg \Diamond z_3 \wedge x_0;$$

$$\psi_2 := \Box \bot \wedge v_2 \wedge \neg v_1 \wedge \bigwedge_{1 \leq i \leq 2^n+2} \neg x_i \wedge \bigwedge_{1 \leq i \leq 2^n+2} \neg y_i \wedge$$
$$\wedge \neg z_1 \wedge \neg z_2 \wedge \neg z_3 \wedge \neg \Diamond z_3 \wedge x_0;$$

$$\psi_3 := \neg \Box \bot \wedge \Box \Box \bot \wedge \Diamond v_2 \wedge \neg \Diamond v_1 \wedge \neg v_1 \wedge \neg v_2 \wedge \Diamond \psi_2 \wedge \neg \Diamond \psi_1$$
$$\wedge z_1 \wedge \neg \Diamond z_1 \wedge \neg z_2 \wedge \neg \Diamond z_2 \wedge \neg z_3 \wedge \neg \Diamond z_3 \wedge \bigwedge_{1 \leq i \leq 2^n+2}(\neg x_i \wedge$$
$$\neg \Diamond x_i) \wedge \bigwedge_{1 \leq i \leq 2^n+2}(\neg y_i \wedge \neg \Diamond y_i) \wedge \Box \psi_2 \wedge x_0;$$

$$\psi := \neg \Box \bot \wedge \neg \Box \Box \bot \wedge \Diamond v_2 \wedge \neg \Diamond v_1 \wedge \neg v_1 \wedge \neg v_2 \wedge \Diamond \psi_2$$
$$\wedge \neg \Diamond \psi_1 \wedge z_2 \wedge \Diamond z_2 \wedge \neg z_1 \wedge \neg \Diamond z_1 \wedge \neg z_3 \wedge \neg \Diamond z_3 \wedge$$
$$\bigwedge_{1 \leq i \leq 2^n+2}(\neg x_i \wedge \neg \Diamond x_i) \wedge \bigwedge_{1 \leq i \leq 2^n+2}(\neg y_i \wedge \neg \Diamond y_i) \wedge$$
$$\Box \psi_2 \wedge x_0; \psi_4 := \psi_4 \wedge \Diamond \psi_4 \wedge \Box(\psi_4 \vee \psi_2);$$

$$\forall i \in \{1, ..., 2^n + 2\}$$
$$\xi_{i0} := \Diamond \psi_1 \wedge \neg \psi_1 \wedge \Diamond \psi_2 \wedge \neg \psi_2 \wedge \neg \Diamond \psi_3 \wedge \neg \psi_3 \wedge \neg \Diamond \psi_4 \wedge \neg \psi_4 \wedge$$
$$x_i \wedge \Diamond x_i \wedge \bigwedge_{1 \leq j \leq 2^n+2, j \neq i}(\neg \Diamond x_j \wedge \neg x_j) \wedge$$
$$\bigwedge_{1 \leq j \leq 2^n+2}(\neg \Diamond y_j \wedge \neg y_j) \wedge \wedge \neg z_1 \wedge$$
$$\neg \Diamond z_1 \wedge \neg z_2 \wedge \neg \Diamond z_2 \wedge \neg z_3 \wedge \neg \Diamond z_3 \wedge \wedge \neg v_1 \wedge \neg v_2 \wedge x_0;$$

$$\xi_i := \xi_{i0} \wedge \Diamond \xi_{i0}$$

$$\forall i \in \{1, ..., 2^n + 2\}$$
$$\delta_i := \Diamond \psi_1 \wedge \neg \psi_1 \wedge \Diamond \psi_2 \wedge \neg \psi_2 \wedge \neg \Diamond \psi_3 \wedge \neg \psi_3 \wedge \neg \Diamond \psi_4 \wedge \neg \psi_4 \wedge$$
$$y_i \wedge \neg \Diamond y_i \wedge \bigwedge_{1 \leq j \leq 2^n+2, j \neq i}(\neg \Diamond y_j \wedge \neg y_j) \wedge$$
$$\bigwedge_{1 \leq j \leq 2^n+2}(\neg \Diamond x_j \wedge \neg x_j) \wedge \Box(\psi_1 \vee \psi_2) \wedge$$
$$\neg z_1 \wedge \neg \Diamond z_1 \wedge \neg z_2 \wedge \neg \Diamond z_2 \wedge \neg z_3 \wedge \neg \Diamond z_3 \wedge \neg v_1 \wedge \neg v_2 \wedge x_0;$$

$$\forall \rho_1, \rho_2, \rho_i \subseteq \{1, ..., 2^n + 2\}, \text{ for } i = 1, 2, \forall \rho_3 \subseteq \{3, 4\}$$
$$(1 \leq ||\rho_1 \cup \rho_2||) \& (||\rho_1 \cup \rho_2 \cup \rho_3|| \leq 2^{n+1} + 4) \Rightarrow$$

$$\mu^1_{\rho_1, \rho_2, \rho_3} \ := \ \bigwedge_{i \in \rho_1} \Diamond \xi_i \wedge \bigwedge_{i \notin \rho_1} \neg \Diamond \xi_i \wedge \bigwedge_{i \in \rho_2} \Diamond \delta_i \wedge \bigwedge_{i \notin \rho_2} \neg \Diamond \delta_i \wedge$$
$$\bigwedge_{i \in \rho_3} \Diamond \psi_i \wedge \bigwedge_{i \in (\{3,4\} - \rho_3)} \neg \Diamond \psi_i \wedge \bigwedge_{1 \leq i \leq 2^n + 2} \neg x_i \wedge$$
$$\bigwedge_{1 \leq i \leq 2^n + 2} \neg y_i \wedge \neg z_1 \wedge \neg z_2 \wedge z_3 \wedge \neg v_1 \wedge \neg v_2 \wedge x_0;$$

$$\mu_{\rho_1, \rho_2, \rho_3} \ := \ \mu^1_{\rho_1, \rho_2, \rho_3} \wedge \Diamond \mu^1_{\rho_1, \rho_2, \rho_3} \wedge x_0;$$
$$\nu_{\rho_1, \rho_2, \rho_3} \ := \ \mu^1_{\rho_1, \rho_2, \rho_3} \wedge \neg \Diamond \mu^1_{\rho_1, \rho_2, \rho_3} \wedge x_0;$$

$$\forall \rho \subseteq \{3, 4\} \& (1 \leq ||\rho||)$$

$$\mu_{\emptyset, \emptyset, \rho} \ := \ \bigwedge_{i \in \rho} \Diamond \psi_i \wedge \bigwedge_{i \in (\{3,4\} - \rho_3)} \neg \Diamond \psi_i \wedge \Diamond \psi_2 \wedge \neg \Diamond \psi_1 \wedge$$
$$\bigwedge_{1 \leq i \leq 2^n + 2} \neg \Diamond \xi_i \wedge$$
$$\bigwedge_{1 \leq i \leq 2^n + 2} \neg \Diamond \delta_i \wedge \bigwedge_{1 \leq i \leq 2^n + 2} \neg x_i \wedge \bigwedge_{1 \leq i \leq 2^n + 2} \neg y_i \wedge$$
$$\neg z_1 \wedge \neg z_2 \wedge z_3 \wedge \neg v_1 \wedge \neg v_2 \wedge x_0;$$

$$\forall \rho_1, \rho_2, \text{ where } \rho_i \subseteq \{1, ..., 2^n + 2\}, \text{ for } i = 1, 2, \forall \rho_3 \subseteq \{3, 4\}$$
$$(1 \leq ||\rho_1 \cup \rho_2 \cup \rho_3||) \& (||\rho_1 \cup \rho_2 \cup \rho_3|| = 2^{n+1} + 5) \Rightarrow$$

$$\theta_{\rho_1, \rho_2, \rho_3} := [\bigvee \{\Diamond \beta_{\eta_1, \eta_2, \eta_3} \mid \beta \in \{\mu, \nu\}, \eta_i \subseteq \rho_i, \eta_v \neq \rho_v \text{ for some } v \}]$$
$$\wedge \bigwedge_{1 \leq i \leq 2^n + 2} \neg x_i \wedge \bigwedge_{1 \leq i \leq 2^n + 2} \neg y_i \wedge \neg z_1 \wedge \neg z_2$$
$$\wedge \neg v_1 \wedge \neg v_2 \wedge x_0;$$

$$\pi := \bigwedge_{1 \leq i \leq 2^n + 2} \Diamond \xi_i \wedge \bigwedge_{1 \leq i \leq 2^n + 2} \Diamond \delta_i \wedge \bigwedge_{1 \leq i \leq 4} \Diamond \psi_i \wedge$$
$$\bigwedge_{1 \leq i \leq 2^n + 2} \neg x_i \wedge \bigwedge_{1 \leq i \leq 2^n + 2} \neg y_i \wedge$$
$$\neg z_1 \wedge \neg z_2 \wedge \neg v_1 \wedge \neg v_2 \wedge \neg x_0;$$

$$\alpha := \bigvee_i \xi_i \vee \bigvee_i \delta_i \vee \bigvee_i \psi_i \vee \bigvee_{(\rho_1, \rho_2, \rho_3)} \mu_{\rho_1, \rho_2, \rho_3} \vee$$
$$\bigvee_{(\rho_1, \rho_2, \rho_3)} \nu_{\rho_1, \rho_2, \rho_3} \vee \bigvee_{(\rho_1, \rho_2, \rho_3)} \theta_{\rho_1, \rho_2, \rho_3} \vee \pi,$$

$$r := \frac{\alpha}{x_0}.$$

In order to falsify the rule r in the frame $\mathcal{K}_n$ and in the frame $S_m(\mathcal{K}_n)$ we define the following valuation V in $\mathcal{K}_n$ and in $S_m(\mathcal{K}_n)$:

$$V(v_1) := \{a_1^1\}, \ V(v_2) := \{b_1^1\}, \ V(z_1) := \{b_0^2\}, \ V(z_2) := \{a_0^2\},$$
$$V(z_3) := (\mathcal{K}_n - S_2(\mathcal{K}_n)), \ V(x_i) := \{a_i^2\}, \ V(y_i) := \{b_i^2\},$$
$$V(x_0) := \{v \mid v \in \mathcal{K}_n, v^{R \leq} \cap Sl_2(\mathcal{K}_n) \neq Sl_2(\mathcal{K}_n).\}$$

It is not difficult to verify directly that

$$\forall c(c \Vdash_V \psi_1 \Leftrightarrow c = a_1^1), \ \forall c(c \Vdash_V \psi_2 \Leftrightarrow c = b_1^1),$$
$$\forall c(c \Vdash_V \psi_3 \Leftrightarrow c = b_0^2), \ \forall c(c \Vdash_V \psi_4 \Leftrightarrow c = a_0^2),$$
$$\forall c(c \Vdash_V \xi_i \Leftrightarrow c = a_2^i), \ \forall c(c \Vdash_V \delta_i \Leftrightarrow c = b_2^i),$$

$\forall c \in K_n, ((c^{R \leq} - \{c\}) \cap S_2(K_\backslash) \subseteq \{a_1^1\} \cup \{a_0^2, b_0^2\}$
$\& a_1^1 \in (c^{R \leq} - \{c\}) \cap S_2(K_\backslash) \& g_0^2 \in (c^{R \leq} - \{c\}) \cap S_2(K_\backslash)$
for some $g = a, b$
$\& \rho := \{3, 4\}$ for $a_0^2, b_0^2 \in (c^{R \leq} - \{c\}) \& \rho := \{3\}$ for only
$a_0^2 \in (c^{R \leq} - \{c\}) \& \rho := \{4\}$ for only $b_0^2 \in (c^{R \leq} - \{c\}) \Rightarrow$
$c \Vdash_V \mu_{\emptyset, \emptyset, \rho},$

$\forall c \in K_n, ((c^{R \leq} - \{c\}) \cap Sl_2(K_n) = \{a_i^2 \mid i \in \rho_1, 1 \leq i \leq 2^n + 2\} \cup$
$\{b_i^2 \mid i \in \rho_2, 1 \leq i \leq 2^n + 2\} \cup \{g_i^2 \mid i \in \rho_3 \subseteq \{3, 4\}, g_3^2 = a_0^2, g_4^2 = b_0^2\}$
$\& 1 \leq ||\rho_1 \cup \rho_2|| \& ||\rho_1|| + ||\rho_2|| + ||\rho_3|| \leq 2^n + 4 \Rightarrow$
$c \Vdash_V \mu_{\rho_1, \rho_2, \rho_3} \vee \nu_{\rho_1, \rho_2, \rho_3},$

$\forall c \in K_n, ((c^{R \leq} - \{c\}) \cap Sl_2(K_n) = \{a_i^2 \mid i \in \rho_1, 1 \leq i \leq 2^n + 2\} \cup$
$\{b_i^2 \mid i \in \rho_2, 1 \leq i \leq 2^n + 2\} \cup \{g_i^2 \mid i \in \rho_3 \subseteq \{3, 4\}, g_3^2 = a_0^2, g_4^2 = b_0^2\}$
$\& ||\rho_1|| + ||\rho_2|| + ||\rho_3|| = 2^n + 5 \Rightarrow c \Vdash_V \theta_{\rho_1, \rho_2, \rho_3}$
$\forall c \in K_n, ((c^{R \leq} - \{c\}) \cap Sl_2(K_n) = Sl_2(K_n) \Rightarrow c \Vdash_V \pi.$

In particular, it follows that $V(\alpha) = K_n$ and $V(x_0) \neq K_n$. Hence r is invalid in K_n. Since the model $S_m(K_n)$ with the valuation V is an open submodel of K_n with the valuation V, we have r is also invalid in $S_m(K_n)$ by V.

Now we will prove that r is admissible in λ. Since λ has the properties mentioned above, by Theorem 3.5.1 it is sufficient to show that the rule r is valid in the frame of the model $Ch_\lambda(2^{n+1} + 10)$. Suppose S is a valuation of all variables occurring in r in $Ch_\lambda(2^{n+1} + 10)$ and

$$S(\alpha) = |Ch_\lambda(2^{n+1} + 10)|, \quad \text{but} \quad S(x_0) \neq |Ch_\lambda(2^{n+1} + 10)|. \tag{4.11}$$

Using (4.11) we conclude $(\exists a \in |Ch_\lambda(2^{n+1} + 10)|) \, a \Vdash_S \pi$. This entails that there exist elements $a_i \in |Ch_\lambda(2^{n+1} + 10)|, i \in \{1, 2, ..., 2^n + 2\}$ and elements $b_i \in |Ch_\lambda(2^{n+1} + 10)|, i \in \{1, 2, ..., 2^n + 2\}$ such that aRa_i, aRb_i for any $i, 1 \leq i \leq 2^n + 2$ (where R is the accessibility relation from $Ch_\lambda(2^{n+1} + 10)$), and

$$a_i \Vdash_S \xi_i, \quad b_i \Vdash_S \delta_i.$$

Then the tuples of all such a_i and all such b_i has properties:

$$i_1 \neq i_2 \Rightarrow \neg(a_{i_1} R a_{i_2}) \& \neg(a_{i_2} R a_{i_1}) \& a_{i_1} \neq a_{i_2},$$

$$i_1 \neq i_2 \Rightarrow \neg(b_{i_1} R b_{i_2}) \& \neg(b_{i_2} R b_{i_1}) \& b_{i_1} \neq b_{i_2},$$

$$\forall i_1, i_2 (\neg(a_{i_1} R b_{i_2}) \& \neg(b_{i_2} R a_{i_1}) \& a_{i_1} \neq b_{i_2}).$$

Thus all a_i and all b_j form an antichain of distinct R-incomparable elements (which has $2^{n+1} + 4$ elements). Also there are elements $a_0, b_0 \in Ch_\lambda(2^{n+1} + 10)$ such that aRa_0, aRb_0 and $a_0 \Vdash_S \psi_4, b_0 \Vdash_S \psi_3$. Then a_0 and b_0 are distinct and R-see no each other. Moreover the elements a_0 and b_0 are distinct and R-incomparable from elements a_i, b_i for $1 \leq i$ which is easily seen by the structure of formulas δ_i, ξ_i and ψ_3, ψ_4. Thus the tuple of elements $a_0, a_1, ..., a_{2^n+2}$, $b_0, b_1, ..., b_{2^n+2}$ is an $2^{n+1} + 6$-elements antichain by R which is R-seen from the element a.

Since λ has the fmp and the property of branching below 1 there is either an reflexive or an irreflexive element $b \in |Ch_\lambda(2^{n+1} + 10)|$ such that

$$(b^{R\leq} - \{b\}) = \bigcup_{0 \leq i \leq 2^n+1} a_i^{R\leq} \cup \bigcup_{0 \leq i \leq 2^n+2} b_i^{R\leq}. \tag{4.12}$$

By (4.11) there is a formula γ, where

$$\gamma \in \{\xi_i, \delta_i, \psi_i, \mu_{\rho_1,\rho_2,\rho_3}, \nu_{\rho_1,\rho_2,\rho_3}, \theta_{\rho_1,\rho_2,\rho_3}, \pi\}, \quad \text{and } b \Vdash_S \gamma.$$

The case $\gamma = \psi_1, \psi_2$ is impossible since bRa_1, $a_1 \Vdash_S x_1 \wedge \Diamond x_1$. The case $\alpha = \psi_3, \psi_4$ is impossible again by the same reason: since bRa_1 and $a_1 \Vdash_S x_1 \wedge \Diamond x_1$.

We will show that the case $\gamma = \pi$ also is impossible. Indeed, suppose that $b \Vdash_S \pi$. Then, for some c, bRc and $c \Vdash_S \xi_{2^n+2}$. This yields $\neg(cRa_i)$, for $i \in \{1, ..., 2^n + 1\}$, because otherwise, we have $c \Vdash_S \neg \Diamond x_i$ (since $i \neq 2^n + 2$) and $a_i \Vdash_S x_i$. Also we similarly obtain $\neg(cRb_i)$ for $i \in \{1, ..., 2^n+2\}$ since $c \Vdash_S \neg \Diamond y_i$ and $b_i \Vdash_S y_i$. Moreover, $\neg(cRb_0)$ holds since $b_0 \Vdash_S z_1$ and $c \Vdash_S \neg \Diamond z_1$. Finally, it is also impossible that cRa_0 because $a_0 \Vdash_S z_2$ and $c \Vdash_S \neg \Diamond z_2$. Therefore from (4.12) we infer that there is an $i \in \{0, 1, 2, ..., 2^n + 1\}$ such that a_iRc or there exists some $i \in \{0, 1, 2, ..., 2^n + 2\}$ such that b_iRc. But $c \Vdash_S x_{2^n+2}$ and, for any a_i, $i \in \{0, 1, ..., 2^n + 1\}$, $a_i \Vdash_V \neg \Diamond x_{2^n+2}$, and also, for any b_i, $i \in \{0, 1, 2, ..., 2^n + 2\}$ and $b_i \Vdash_S \neg \Diamond x_{2^n+2}$, a contradiction. Hence $\gamma =\neq \pi$.

Assume that $\gamma = \xi_h$ for some suitable h. Then because $a_i \Vdash_S \Diamond x_i \wedge x_i$, and bRa_i for all $i \in \{1, ..., 2^n+1\}$, we obtain $b \Vdash_S \Diamond x_i$, which contradicts $b \Vdash_S \neg \Diamond x_i$ for $i \neq h$. Suppose that $\gamma = \delta_h$ for some suitable h. Then, since we have $b_i \Vdash_S y_i$ for all $i \in \{1, ..., 2^n + 2\}$, we derive $b \Vdash_S \Diamond y_i$ for all such i, which contradicts $b \Vdash_S \neg \Diamond y_i$ for all i.

Assume now that $\gamma = \mu_{\rho_1,\rho_2,\rho_3}$ for some suitable ρ_k, $k = 1, 2, 3$. By our choice of ρ_1, ρ_2 and ρ_3, it follows either (a): there is a $j \in \{1, 2, ..., 2^n + 1\}, j \notin \rho_1$ such that bRa_j, or (b) there is some $j \in \{1, 2, ..., 2^n+2\}, j \notin \rho_2$ such that bRb_j, or (c) $3 \notin \rho_3$, or (d): $4 \notin \rho_3$. If (a) holds, we have bRa_j, $1 \leq j, j \notin \rho_1$, and $a_j \Vdash_S \xi_j$. By the assumption that $b \Vdash_S \mu_{\rho_1,\rho_2\rho_3}$ it follows that $b \Vdash_S \neg \Diamond \xi_j$ for such j, a contradiction. If (b) holds then $j \in \{1, 2, ..., 2^n + 2\}, j \notin \rho_2$, and bRb_j and $b_j \Vdash_S \delta_j$ and, since $b \Vdash_S \mu_{\rho_1,\rho_2\rho_3}$, we obtain $b \Vdash_S \neg \Diamond \delta_j$, again a contradiction. If (c) holds, we have $b \Vdash_S \neg \Diamond \psi_3$, which contradicts $b_0 \Vdash_S \psi_3$. If (d) holds, we obtain $b \Vdash_S \neg \Diamond \psi_4$, which contradicts $a_0 \Vdash_S \psi_4$. Thus $\gamma \neq \mu_{\rho_1,\rho_2,\rho_3}$. The assumption that $\gamma = \nu_{\rho_1,\rho_2,\rho_3}$ for some ρ_k, $k = 1, 2, 3$ can be considered and rejected by the same arguments as with the case when $\gamma = \mu_{\rho_1,\rho_2,\rho_3}$.

The only single possible remaining case is that $\gamma = \theta_{\rho_1,\rho_2,\rho_3}$ for some ρ_k, $k = 1, 2, 3$. Then, under this assumption, there is some disjunct $\Diamond\beta_{\eta_1,\eta_2,\eta_3}$ in the formula $\theta_{\rho_1,\rho_2,\rho_3}$ which is valid on b and $\eta_g \subset \rho_g$ for some g. Therefore there is an element a_{η_1,η_2,η_3} such that

$$b R a_{\eta_1,\eta_2,\eta_3}, a_{\eta_1,\eta_2,\eta_3} \Vdash_S \beta_{\eta_1,\eta_2,\eta_3}.$$

The case when $a_i R a_{\eta_1,\eta_2,\eta_3}$, or $a_i = a_{\eta_1,\eta_2,\eta_3}$ for some $i \in \{0, 1, 2, ..., 2^n + 1\}$, is impossible. In fact, if $a_i R a_{\eta_1,\eta_2,\eta_3}$ or $a_i = a_{\eta_1,\eta_2,\eta_3}$ for $i \in \{0, 1, 2, ..., 2^n + 1\}$ then the fact that $a_{\eta_1,\eta_2,\eta_3} \Vdash_S z_3$ contradicts the fact that $a_i \Vdash_S \xi_i$ or $i = 0$ and $a_0 \Vdash_S \psi_4$ and $a_i \Vdash_S \neg z_3 \wedge \neg\Diamond z_3$. The case when $b_i R a_{\eta_1,\eta_2,\eta_3}$, or $b_i = a_{\eta_1,\eta_2,\eta_3}$ for some $i \in \{0, 1, 2, ..., 2^n + 2$ is again impossible. Indeed, if $b_i R a_{\eta_1,\eta_2,\eta_3}$ or $b_i = a_{\eta_1,\eta_2,\eta_3}$ for $i \in \{0, 1, 2, ..., 2^n + 2\}$ then we know that $a_{\eta_1,\eta_2,\eta_3} \Vdash_S z_3$. This contradicts the fact that, in this case, $b_i \Vdash_S \delta_i$ or $i = 0$ and $b_0 \Vdash_S \psi_3$ and $b_i \Vdash_S \neg z_3 \wedge \neg\Diamond z_3$.

Thus, using (4.12), the only possible case is b is reflexive and $b = a_{\eta_1,\eta_2,\eta_3}$. Then we have $b \Vdash_S \mu_{\eta_1,\eta_2,\eta_3}$ or $b \Vdash_S \nu_{\rho_1,\rho_2,\rho_3}$, what, as we have showed above, is impossible.

We exhausted all possibilities and proved that non of the disjuncts of the premise of r can be valid on b under S. Therefore r is valid in the frame of $Ch_\lambda(2^{n+1} + 10)$ and, as we have noted before, this entails that r is admissible in λ. ∎

Theorem 4.2.16 *Let λ be a modal logic extending $K4$ such that*

    (i) λ is an $(\omega, 2, 3)$-logic or an $(m, 2, 3)$-logic;

    (ii) λ has the fmp and the property of branching below 1;

    (iii) for some k, λ has the effective k-drop points property

Then λ has no bases for admissible rules in finitely many variables, and in particular, λ has no finite bases. The free algebra $\mathfrak{F}_\lambda(\omega)$ from the variety $Var(\lambda)$ has no bases in finitely many variables for quasi-identities.

Proof. The proof follows immediately from Lemmas 4.2.13 and 4.2.15, Lemma 4.1.4 and Theorem 1.4.15. ∎

Corollary 4.2.17 *Modal Logics $K4$, $K4.1$, $K4.2$ have no bases in finitely many variables for admissible rules.*

Proof. These logics have the fmp (Corollary 2.6.12 and Theorem 2.8.9), the property of branching below 1 and the effective 1-drop points property (cf. Lemma 3.5.3). ∎

Lemma 4.2.18 *Let λ be a modal logic extending $K4$ such that* *(i)* λ *is an* $(\omega, i, 3)$*-logic or an* $(m, i, 3)$*-logic;*

(ii) λ *has the fmp and the property of branching below* 1*;*

(iii) for some k, λ *has the effective* k*-drop points property*

If λ *is an* $(\omega, i, 3)$*-logic then there is an inference rule* r *admissible in* λ *and having* $2^n + 5$ *variables which is invalid in the frame* $\mathcal{G}_n$. *If* λ *is an* $(m, i, 3)$*-logic then there exists a rule* r *admissible in* λ *and having* $2^n + 5$ *variables which is invalid in the frame* $S_m(\mathcal{G}_n)$.

Proof. We follow the proof of Lemma 4.2.15 with simplifications connected with omitting all reflexive elements and all parts of the proof concerning such elements. To begin our proof we introduce the following rule r with $2^{n+1} + 5$ variables.

$$\psi := \Box\bot \wedge v_1 \wedge \bigwedge_{1 \le i \le 2^n+2} \neg y_i \wedge \neg z_1 \wedge \neg \Diamond z_1 \wedge x_0;$$

$$\forall i \in \{1, ..., 2^n + 2\}$$
$$\delta_i := \Diamond\psi \wedge \neg\psi \wedge y_i \wedge \neg\Diamond y_i \wedge \bigwedge_{1 \le j \le 2^n+2, j \ne i}(\neg\Diamond y_j \wedge \neg y_j) \wedge$$
$$\Box\psi \wedge \neg v_1 \wedge \neg z_1 \wedge \neg\Diamond z_1 \wedge x_0;$$

$$\forall\rho \subseteq \{1, ..., 2^n + 2\}, 1 \le ||\rho|| \le 2^n \Rightarrow$$
$$\mu_\rho := \bigwedge_{i \in \rho} \Diamond\delta_i \wedge \bigwedge_{i \notin \rho} \neg\Diamond\delta_i \wedge \bigwedge_{1 \le i \le 2^n+2} \neg y_i$$
$$\wedge \neg v_1 \wedge z_1 \wedge x_0;$$

$$\forall\rho \subseteq \{1, ..., 2^n + 2\}, ||\rho|| = 2^n + 1 \Rightarrow,$$
$$\theta_\rho := [\bigvee\{\Diamond\mu_\eta \mid \eta \subset \rho, \}] \wedge \bigwedge_{1 \le i \le 2^n+2} \neg y_i \wedge$$
$$\neg v_1 \wedge x_0;$$

$$\pi := \bigwedge_{1 \le i \le 2^n+2} \Diamond\delta_i \wedge \bigwedge_{1 \le i \le 2^n+2} \neg\delta_i \wedge \neg x_0$$
$$\alpha := \psi \wedge \bigwedge_{1 \le i \le 2^n+2} \delta_i \wedge \bigwedge_\rho \mu_\rho \wedge \bigwedge_\rho \theta_\rho \wedge \pi.$$

$$r := \frac{\alpha}{x_0}$$

We will show that the rule r is invalidated in the frame $\mathcal{G}_n$ by the following valuation:

$$V(v_1) := \{b_1^1\}, \quad V(y_i) := \{b_i\}, \quad V(z_1) := (\mathcal{G}_n - S_2(\mathcal{G}_n))$$
$$V(x_0) := \{u \mid u \in \mathcal{G}_n, u^{R\le} \cup Sl_2(\mathcal{G}_n) \ne Sl_2(\mathcal{G}_n)\}.$$

It is not hard to check that according to the definition of the disjuncts in the premise of r above, and to the structure of $\mathcal{G}_n$, the following relations in the frame $\mathcal{G}_n$ hold under V:

$$\forall c(c \Vdash_V \psi \Leftrightarrow c = b_i^1),$$
$$\forall c(c \Vdash_V \delta_i \Leftrightarrow c = b_i^2),$$
$$\forall c \forall \rho \subseteq \{1, 2, ..., 2^n + 2\} \& (1 \leq ||\rho|| \leq 2^n)(c \Vdash_V \mu_\rho \Leftrightarrow$$
$$c \in (\mathcal{G}_n - S_2(\mathcal{G}_n)) \& (\rho = \{i \mid b_i^2 \in c^{R \leq} \cap Sl_2(\mathcal{G}_n)\}),$$
$$\forall c \forall \rho \subseteq \{1, 2, ..., 2^n + 2\}((||\rho|| = 2^n + 1) \&$$
$$(\rho = \{i \mid b_i^2 \in^{R \leq} \cap Sl_2(\mathcal{G}_n)\}) \Rightarrow c \Vdash_V \theta_\rho),$$
$$\forall c(\rho = \{i \mid b_i^2 \in c^{R \leq} \cap Sl_2(\mathcal{G}_n)\} \& ||\rho|| = 2^n + 2 \Rightarrow c \Vdash_V \pi).$$

So we obtain $V(\psi \wedge \bigwedge_{1 \leq i \leq 2^n + 2} \delta_i \wedge \bigwedge_\rho \mu_\rho \wedge \bigwedge_\rho \theta_\rho \wedge \pi) = |\mathcal{G}_n|$ and $V(x_0) \neq |\mathcal{G}_n|$, i.e. V disproves r in $\mathcal{G}_n$ and in the frame $S_m(\mathcal{G}_n)$, when $3 \leq m$.

Now our aim is to show that r is admissible in the logic λ. Given all properties of λ from the formulation of our lemma and Theorem 3.5.1, it is sufficient to prove that r is valid in the frame of the model $Ch_\lambda(2^n + 5)$. Let S be a valuation of all variables from r in the frame of $Ch_\lambda(2^n + 5)$, and

$$S(\alpha) = |Ch_\lambda(2^n + 5)|, \quad S(x_0) \neq |Ch_\lambda(2^n + 5)|. \tag{4.13}$$

From (4.13) it follows that there is an $a \in |Ch_\lambda(2^n + 5)|$ such that $a \Vdash_S \pi$. This implies that there are elements $b_i \in |Ch_\lambda(2^n + 5)|$, $i \in \{1, ..., 2^n + 2\}$ such that aRb_i and $b_i \Vdash_V \delta_i$. The elements b_i are distinct and R-incomparable because formulas δ_i do prohibit the coinciding and comparability by R. Hence all b_i form an antichain. Since λ has the property of branching below 1 there is an element b in $|Ch_\lambda(2^n + 5)|$ such that

$$(b^{R \leq} - \{b\}) = \bigcup_{1 \leq i \leq 2^n + 1} (b_i^{R \leq} \cup \{b_i\}). \tag{4.14}$$

By (4.13) there is a formula γ, where $b \Vdash_S \gamma$ and

$$\gamma \in \{\psi, \delta_i, \mu_\rho, \theta_\rho, \pi\} \text{ for some } i \text{ and } \rho.$$

The case $\gamma = \psi$ is impossible since ϕ has the conjunct $\Box\bot$. The case $\gamma = \delta_i$ is impossible also. In fact, there is an b_j such that aRb_j and $i \neq j$. Moreover, since $b_j \Vdash_S \delta_j$, it follows that $b_j \Vdash_S y_j$, and $\neg\Diamond y_j$ is a conjunct of δ_i. Suppose that $\gamma = \mu_\rho$. Then ρ must include $\{1, 2, ..., 2^n + 1\}$ which is impossible.

Assume that $\gamma = \theta_\rho$. Then there is an element a_η, where $\eta \subset \rho$, $a_\eta \Vdash_S \mu_\eta$ and bRa_η. The case when $b_i Ra_\eta$ or $b_i = a_\eta$, where $i \in \{1, 2, ..., 2^n + 1\}$, is impossible since $a_\eta \Vdash_S z_1$ and $b_i \Vdash_S \delta_i$, and consequently $b_i \Vdash_S \neg z_1 \wedge \Diamond z_1$. Therefore by (4.14), only the case $b = a_\eta$ remains. But then we have $b \Vdash_V \mu_e ta$, which is also impossible as we showed above.

Suppose finally that $\gamma = \pi$. Then there is a c such that bRc and $c \Vdash_S \delta_{2^n + 2}$. The case when $c = b$ is impossible since bRb_1 and $b_1 \Vdash_S \delta_1$ and $\neg\Diamond y_1$ is a conjunct of $\delta_{2^n + 2}$. Therefore according to (4.14) for some i, $b_i Rc$ or $b_i = c$. This contradicts the fact that $b_i \Vdash_S \delta_i$ and $i < 2^n + 2$. Thus we exhausted all the possibilities, and non of the disjuncts from the premise of r could be valid under S on b. Hence r is valid in $Ch_\lambda(2^n + 5)$ which, as we noted before, entails that r is admissible in λ. ∎

From Lemmas 4.2.14, 4.2.18 and 4.1.4 and Theorem 1.4.15 we immediately obtain

Theorem 4.2.19 *Let λ be a modal logic extending $K4$ such that*

(i) λ is an $(\omega, i, 3)$-logic or an $(m, i, 3)$-logic;

(ii) λ has the fmp and the property of branching below 1;

(iii) for some k, λ has the effective k-drop points property

Then λ has no bases in finitely many variables, and, in particular, finite bases, for admissible inference rules. The free algebra $\mathfrak{F}_\lambda(\omega)$ from the variety $Var(\lambda)$ has no bases in finitely many variables for quasi-identities.

We can now apply Theorem 4.2.19 to the provability modal logic GL and other logics related to GL. Recall that the logics GLI_n are the modal logic generated by all finite irreflexive transitive rooted frames the number of maximal elements of which is not more than n. Using Lemma 3.5.17, we immediately derive

Corollary 4.2.20 *The Gödel-Löb provability logic GL, and logics GLI_n, for any $n \in N$, have no bases in finitely many variables for admissible inference rules.*

Final Comments. As a matter of fact the research of this section adjoins to the well-developed field of universal algebra connected with studying existence or non-existence of finite bases for quasi-identities of specific or general classes of algebraic systems. We give no any survey on this subject since the field is very large, and there are no a chance to describe it briefly. Our interest to this field wis motivated by A.Kuznetsov-H.Friedman question on whether there is a finite bases for admissible rules of intuitionistic logic H, i.e. whether the absolutely free pseudo-boolean algebra has a finite bases for quasi-identities. Note by, the way, that this question first has been answered in Rybakov [139], and also in Rybakov [141]. This section gives many theorems which allow us to determine the lack of bases for admissible rules. However, the main point here is that the methods we used here are flexible and it is possible to apply them to many other related logical calculi.

4.3   Bases for Some Strong Logics

In the previous section we proved various general theorems which show that if a logic λ has certain common properties then λ has no finite bases for admissible rules. Therefore logics with finite bases for admissible rules are more specific and are stronger (i.e. placed in the upper part of the lattice of logics). In this section we prove several general theorems concerning existence of finite bases for admissible rules. The first class of logics we will investigate is the class consisting of all modal logics extending the modal logic $S4.3$, where $S4.3 := S4 \oplus \Box(\Box p \rightarrow \Box q) \vee \Box(\Box q \rightarrow \Box p)$. This class of logics attracted the attention of researches for a long time since the additional axiom of $S4.3$, which be seen as expressing a linear current of events, is very natural, and $S4.3$-frames must have a linear accessibility relation.

The first important step have been made by R.Bull [21] who demonstrated that all normal modal logics extending the modal system $S4.3$ have the finite model property. It turns out that the situation with these logics is even better. K.Fine [43] has proved that all these logics are finitely axiomatizable although there are infinitely many such logics (for this he used his method of proof the possession of the fmp for logics extending $S4.3$ by Kripke models). We can also continue this trend and show that all modal logics extending $S4.3$ have certain simple finite bases for admissible inference rules. For this we need the results of R.Bull and K.Fine concerning the fmp and finite axiomatization, and we begin by showing that all modal logics extending $S4.3$ have the fmp. To do this we will follow the proof of K.Fine.

Theorem 4.3.1 *Bull [21], Fine [43]. For any modal logic λ extending $S4.3$, λ has the finite model property.*

Proof. Suppose a formula α is not a theorem of λ and has n propositional letters. By the completeness theorem for Kripke models (see Theorem 2.3.27), α is false in the canonical Kripke model for λ. Let w_0 be an element of this model such that $w_0 \nVdash_V \alpha$. Suppose $\mathfrak{M} := \langle W, R, V \rangle$ is the open submodel of the canonical model generated by w_0, i.e. the cluster containing w_0 is the root of $\mathfrak{M}$. The validity of formulas in open submodels corresponds to their validity in the model itself, therefore $\mathfrak{M}$ invalidates the formula α, and all the formulas from λ are valid in $\mathfrak{M}$. The model $\mathfrak{M}$ also is *differentiated*, i.e., for any two distinct elements a, b, there is a formula which is valid on a and is false on b (this follows from Theorem 2.3.25). Furthermore, the model $\mathfrak{M}$ is reflexive, transitive and connected, the later means, for all elements t, c and d, if tRc and tRd then either cRd or dRc (this follows from Theorem 2.3.51). We need the following

Definition 4.3.2 *The modal degree $d(\delta)$ of a formula δ is defined as follows. If $\delta = p$, where p is a propositional letter then $d(\delta) = 0$. $d(\beta \wedge \gamma) = d(\beta \vee \gamma) = d(\beta \rightarrow \gamma) = max(d(\beta), d(\gamma))$, $d(\neg\beta) = d(\beta)$, $d(\Box\beta) = d(\Diamond\beta) = d(\beta) + 1$.*

Suppose α has a modal degree n_0 and Γ is the set of all modal formulas built up out of the variables occurring in α which have a modal degree of not more than $n_0 + 1$. We take the greatest filtration of $\mathfrak{M}$ by the set of formulas Γ. That is, we suppose $a \sim b$ iff, for any formula $\beta \in \Gamma$, $a \Vdash_V \alpha \Leftrightarrow \Vdash_V \beta$; and we assume that

$$[a]_\sim R[b]_\sim \text{ iff } \forall \Box \beta \in \Gamma(a \Vdash_V \Box\beta \Rightarrow b \Vdash_V \beta),$$

and $[a]_\sim \Vdash_V p \Leftrightarrow a \Vdash_V p$ for any propositional letter $p \in \Gamma$. The result of this filtration is the model $\mathfrak{N} := \langle \{[a]_\sim \mid a \in W\}, R, V\rangle$. The model $\mathfrak{N}$ is reflexive, transitive and connected since filtrations preserve the accessibility relation. Moreover $\mathfrak{N}$ is finite since there are only finitely many non-equivalent in $S4$ formulas of modal degree not more than $n_0 + 1$ constructed out of fixed n propositional letters. Therefore any element of $\mathfrak{N}$ is definable (i.e. expressible) according to the definition of filtration and finitude of $\mathfrak{N}$. Now we will eliminate some elements form $\mathfrak{N}$. We say an element $[a]_\sim$ is a *subordinate* of $[b]_\sim$ (denotation $[a]_\sim sub[b]_\sim$) if there is an $c \in [b]_\sim$ such that, for all $d \in [a]_\sim$, $\neg(cRd)$. An element $[a]_\sim$ is *eliminable* if for certain $[b]_\sim$, $[b]_\sim R[a]_\sim$ and $[a]_\sim sub[b]_\sim$.

Lemma 4.3.3 *For any $w \in W$ and any $\Diamond\beta \in w \cap \Gamma$, there is a $[v]_\sim \in |\mathfrak{N}|$ such that $[w]_\sim R[v]_\sim$, $\beta \in v$ and $[v]_\sim$ is not eliminable.*

Proof. We define the following (possibly terminating) sequence $S := [a_1]_\sim$, $[a_1]_\sim$, ... of elements from the model $\mathfrak{N}$:

(i) : $[a_0]_\sim := [w]_\sim$,
(ii) : if i is odd and $i > 0$ then $[a_i]_\sim$ is such that
$\quad \neg([a_i]_\sim sub[a_{i-1}]_\sim)$ and $\beta \in a_i$,
(iii) : if i is even and $i > 0$ then $[a_i]_\sim$ is such that
$\quad [a_{i-1}]_\sim sub[a_i]_\sim$ and $[a_i]_\sim R[a_{i-1}]_\sim$.

For $i \geq 0$, $\Diamond\beta \in a_i$ if i is even and $\beta \in a_i$ if i is odd. $\quad$ (4.15)

Indeed, for $i = 0$, $\Diamond\beta \in a_0$ since $[a_0]_\sim = [w]_\sim$. If $i > 0$ and i is odd then $\beta \in a_i$ by the definition of the sequence. If $i > 0$ and i is even then $\beta \in a_{i-1}$ and consequently $[a_i]_\sim \Vdash_V \Diamond\beta$ and $\Diamond\beta \in a_i$.

The sequence does not terminate on even i $\quad$ (4.16)

Suppose otherwise. Assume $\{[b_1]_\sim, ..., [b_m]_\sim\} = \{[a]_\sim \mid [a]_\sim sub[a_i]_\sim\}$. Then for every $k = 1, 2, ..., m$, there is a $v_k \in [a_i]_\sim$ such that $\neg(v_k Ru)$ for all $u \in [b_k]_\sim$. Let v be a v_k such that $v_l Rv_k$ for $l = 1, 2, ..., m$, v is well-defined since $\mathfrak{N}$ is connected. Then by (4.15) $\Diamond\beta \in v$. Since the model $\mathfrak{M}$ is the canonical model, this yields, for some u, $\beta \in u$ and vRu. Then $[u]_\sim$ differs from any $[b_k]_\sim$, i.e. $\neg([u]_\sim sub[a_i]_\sim)$. Therefore we can assume $[a_{i+1}]_\sim := [u]_\sim$ and continue the sequence.

The sequence terminates. $\quad$ (4.17)

Suppose i is an even number. By the definition of the sequence, it immediately follows that $\neg([a_{i+1}]_\sim sub[a_i]_\sim)$ and $[a_{i+1}]_\sim sub[a_{i+2}]_\sim$. Since the relation R is connected, the fact that $\neg([a_{i+1}]_\sim sub[a_i]_\sim)$ entails, for every $v \in [a_i]_\sim$, there is a $u \in [a_{i+1}]_\sim$ such that vRu. Therefore, since R is transitive, it follows that $[a_i]_\sim sub[a_{i+2}]_\sim$. It is not hard to see that the relation sub is transitive and antisymmetric. Therefore the all $[a_i]_\sim$, for i even, are distinct. Since $|\mathfrak{N}|$ is finite, the sequence terminates.

By (4.16) and (4.17) the sequence terminates by $[a_i]_\sim$ for some odd i. According to (4.15), $\beta \in a_i$. Since $[a_{i+1}]$ does not exist, $[a_i]_\sim$ is not eliminable. Because $[a_0]_\sim R[a_1]_\sim R...R[a_i]_\sim$, it follows that $[w]_\sim R[a_i]_\sim$ since R is transitive. ∎

Now we pick the model $\mathfrak{N}_1$ which is the submodel of the model $\mathfrak{N}$ consisting of all not eliminable elements. Our aim is to show that the model $\mathfrak{N}_1$ also invalidates α and that $\mathfrak{N}_1$ is based up on a λ-frame.

Lemma 4.3.4 *For any $\beta \in \Gamma$ and any $[w]_\sim \in |\mathfrak{N}_1|$,*

$$[w]_\sim \Vdash_V \beta \text{ in } \mathfrak{N}_1 \Leftrightarrow \beta \in w.$$

Proof. We show this by induction on the length of α. The only non-trivial step is the case when $\beta = \Box\gamma$. Let $\Box\gamma \notin w$. Since w is a maximal consistent for λ theory, we obtain $\Diamond\neg\gamma \in w$. According to Lemma 4.3.3, for some $[u]_\sim$, $[w]_\sim R[u]_\sim$, $\neg\gamma \in u$ and $[u]_\sim$ is not eliminable. Using the inductive hypothesis it follows $[u]_\sim \nVdash_V \gamma$ in the model $\mathfrak{N}_1$. Therefore we conclude $[w]_\sim \nVdash_v \Box\gamma$ in $\mathfrak{N}_1$.

Conversely, suppose that $\Box\gamma \in w$. Let $[u]_\sim$ be an element of the model $\mathfrak{N}_1$ and $[w]_\sim R[u]_\sim$. Since $\Box\gamma \in w$, we get $\gamma \in u$. By inductive hypothesis we obtain $[u]_\sim \Vdash_V \gamma$ in $\mathfrak{N}_1$ and consequently $[w]_\sim \Vdash_V \Box\gamma$. ∎

Since the formula α was invalid on the root element w_0 in $\mathfrak{M}$, it follows that $\neg\alpha \in w_0$. Consequently it follows that $\Diamond\neg\alpha \in w_0$. Since $\Diamond\neg\alpha \in \Gamma$ by the definition of Γ, applying Lemma 4.3.3 we infer that there is a not eliminable element $[w]_\sim$ such that $\neg\alpha \in w$. Therefore by Lemma 4.3.4 $[w]_\sim \nVdash_V \alpha$ in the model $\mathfrak{N}_1$. The crucial point is to show that the frame of $\mathfrak{N}_1$ is a λ-frame. Suppose that $[a_1]_\sim, ..., [a_m]_\sim$ is an arbitrary ordering of the elements of $\mathfrak{N}_1$. Define a mapping f from W onto $|\mathfrak{N}_1|$:

> (i) : if $[w]_\sim \in |\mathfrak{N}_1|$ then $f(w) = [w]_\sim$,
> (ii) : if $[w]_\sim \notin |\mathfrak{N}_1|$ then $f(w) = [a_i]_\sim$,
> where $[a_i]_\sim$ is the first element in the above ordering
> which is an R-first element of $\{[a_j]_\sim \mid [w]_\sim R[a_j]_\sim\}$.

According to Lemma 4.3.3, f is well-defined. Now we redefine the valuation V in the model $\mathfrak{M}$. Namely, we let $\mathfrak{M}_1 := \langle W, R, S \rangle$, where, for all $p \in \Gamma$,

$$S(p) := \{w \mid w \in W, f(w) \Vdash_V p\}.$$

Lemma 4.3.5 *The following hold*

(i) *The mapping f is a p-morphism from the model $\mathfrak{M}_1$ onto the model $\mathfrak{N}_1$.*

(ii) *The valuation S of the model $\mathfrak{M}_1$ is definable: for any $p \in \Gamma$ the set $S(p)$ is definable by a formula, i.e. $\mathfrak{M}_1$ is a definable variant of $\mathfrak{M}$.*

Proof. Clearly, f is a mapping onto, and f is consistent with the valuation. If wRv holds in $\mathfrak{M}_1$ then $[w]_\sim R[v]_\sim$ in the model $\mathfrak{N}$. Moreover, $[v]_\sim Rf(v)$ in $\mathfrak{N}$. Therefore $[w]_\sim Rf(v)$ in $\mathfrak{N}$. But then $f(w)Rf(v)$ since otherwise $f(w)$ is not a R-first element of $\{[a_j]_\sim \mid [w]_\sim R[a_j]_\sim\}$. Suppose now that $f(w)Rf(v)$. We will verify the second property from the definition of p-morphisms. First consider the case when $\neg(f(v)Rf(w))$. Then $\neg(vRw)$, since we already showed that f preserves the accessibility relation. Since R is connected it follows that wRv, which is what we need. Suppose $f(v)Rf(w)$. Assume that, for every $u \in f(v)$, $\neg(wRu)$ holds. Then $f(v)subf(w)$ and $f(v)$ is eliminable, which contradicts the definition of f and $\mathfrak{N}_1$. Hence, for some $u \in f(v)$, we have wRu and $f(u) = f(v)$. Thus f is in fact a p-morphism from the model $\mathfrak{M}_1$ onto the model $\mathfrak{N}_1$.

In order to show that $\mathfrak{M}_1$ is a definable variant of $\mathfrak{M}$, consider the set $S(p) := \{w \mid w \in W, f(w) \Vdash_V p\}$, where $p \in \Gamma$. Since our mapping f preserves the equivalence relation $\sim$ in $\mathfrak{M}$, i.e. if $a \sim b$ then $f(a) = f(b)$, $S(p)$ is an union of a family of sets $[w]_\sim$ which are some elements of the model $\mathfrak{N}$. This union is finite since the model $\mathfrak{N}$ is finite. Every element $[w]_\sim$ from $\mathfrak{N}$ was definable in $\mathfrak{N}$ by a formula from Γ. Since $\mathfrak{N}$ is a filtration of $\mathfrak{M}$ by Γ, the validity of formulas from Γ is preserved under the filtration procedure. Therefore any set $[w]_\sim$ is definable in $\mathfrak{M}$ by a formula from Γ. In particular, the set $S(p)$ is definable in $\mathfrak{M}$ by a suitable finite disjunction of formulas from Γ. ∎

Suppose that a certain formula β is a theorem of λ. Let V_1 be a valuation of letters which occur in β in the frame of $\mathfrak{N}_1$. Suppose that V_1 disproves β in $\mathfrak{N}_1$. By Lemma 4.3.4 any element of $\mathfrak{N}_1$ is definable in $\mathfrak{N}_1$. Therefore there is a substitution e of some formulas which are built up out of letters which occur in formulas from Γ, in place of letters from β, such that $e(\beta)$ is false in the model $\mathfrak{N}_1$ with respect to the original valuation V of $\mathfrak{N}_1$. Applying (i) of Lemma 4.3.5 it follows $e(\beta)$ is false in the model $\mathfrak{M}_1$. By (ii) of Lemma 4.3.5 the valuation S of $\mathfrak{M}_1$ is definable in $\mathfrak{M}$. Therefore there is a substitution e_1 of certain formulas in place of the variables of $e(\beta)$ such that the formula $e_1(e(\beta))$ is false in the model $\mathfrak{M}$ with respect to the original valuation V of $\mathfrak{M}$. Since the model $\mathfrak{M}$ is an open submodel of the canonical Kripke model of λ, we conclude that $e_1(e(\beta))$ is not a theorem of λ, which contradicts the claim that $e_1(e(\beta))$ is a substitution-example of the theorem β of λ. Thus the frame of $\mathfrak{N}_1$ is a finite λ-frame and we have proved that α is invalid in $\mathfrak{N}_1$. Hence λ has the finite model property, which conclude the proof of Theorem 4.3.1. ∎

Now we present a proof that all extensions of $S4.3$ are finitely axiomatizable originally due to K.Fine [43]. First we prove certain properties concerning finite

sequences of positive natural numbers. It is possible to obtain these properties
as a corollary of Kruskal's Three Theorem ([79], pp. 210, 212) which is well
known to readers experienced in discrete mathematics. But we prefer to follow
a direct proof due to K.Fine [43] in order to give the reader an experience of
such combinatorial reasoning. If the reader is not interested in our proof of these
properties and knows Kruskal's Three Theorem well, then he can skip the proofs
of Lemma 4.3.7 and Corollary 4.3.9 (but not our definitions appearing) and go
directly to Lemma 4.3.10 and Theorem 4.3.11.

Definition 4.3.6 *Suppose $t = a_1, .., a_n$, $s = b_1, ..., b_m$ are finite sequences of
natural numbers. We say t contains s if $m \leq n$ and there is a subsequence
$a_{j1}, ..., a_{jm}$ of t such that $\forall i, 1 \leq i \leq m$, $b_i \leq a_{ji}$.*

Lemma 4.3.7 *K.Fine [43] Suppose $s = t_1, t_2, ...$ is an infinite sequence of
finite sequences of positive natural numbers. Then there is an infinite subsequence $r = t_{j1}, t_{j2}, ...$ of s such that t_{ji} contains t_{jh} for $i > h$.*

Proof. If for any sequences t_i from s there is a sequence $t_j \in s$ placed after t_i
which contains t_i then the assertion of our lemma evidently holds for s. Otherwise s contains some t_i which is not contained in all $t_j \in s$ placed after t_j.
Without loss of generality we can assume $t_j = t_1$ and further we will consider
only such sequences s. Let m be the maximal number from t_1 and l be the length
of t_1. We will show that the assertion of our lemma holds by induction on (m, l)
which we call the factor of s. We will conduct induction at first increasing l and
then increasing m. Clearly, that the case $m = l = 1$ is impossible.

It is easy to see that, for sequences with $m = 1$ and any $l > 1$, the assertion of our lemma also holds. For any sequence s with factor $(2, 1)$, there is an
infinite subsequence s_1 of s such that any $t \in s_1$ is a finite sequence consisting
of 1. It is evident that s_1 contains an infinite subsequence which satisfies the
required properties. Thus the basis of the induction is verified and we consider
the inductive steps.

Case 1. Suppose that, for all sequences s with factors (m_1, l_1), where $m_1 \leq
m$ and $l_1 \leq l$ and $1 \leq l$, or $m_1 < m$, the assertion of our lemma holds. Consider
a sequence s with factor $(m, l + 1)$.

Suppose $t_1 = r_1, x_1$, where x_1 is the last number of t_1 and r_1 is the prefix of
t_1, i.e. r_1 is the sequence obtained from t_1 by removing x_1. We distinguish two
cases.

(a) For some i, t_j does not contain r_1 for all $j \geq i$. Then we may apply the
inductive hypothesis to the sequence $r_1, t_i, t_{i+1}, ...$.

(b) For each i, there is a $j > i$ such that t_j contains r_1. By thinning we may
suppose that t_j contains r_1 for each $j \geq 1$. For $i > 1$, let $t_i = r_i x_i s_i$, where $r_i x_i$
is the shortest initial subsequence of t_i which contains r_1, and x_i is a number,
and s_i is the remaining part of t_i. Suppose that there is an infinite sequence
of non-empty sequences r_i. Then we may suppose that all r_i are non-empty.

By inductive hypothesis applied to the sequence $r_1, r_2, \ldots$, there is an infinite ascending subsequence (each member of this subsequence is contained in latter ones) of this sequence. By thinning we can assume that r_j contains r_i for $j > i > 1$. Also by thinning we may suppose that $x_j \geq x_i$ for $j > i > 1$. Note that since $t_1 := r_1, x_1$ is not contained in any r_i, x_i, s_i, x_1 is not contained in any s_i for $i > 1$. Again by inductive hypothesis applied to the sequence $x_1, s_2, s_3, \ldots$ and thinning, we may suppose that s_j contains s_i for $j > i > 1$. But now it follows that $t_j = r_j, x_j, s_j$ contains r_i, x_i, s_i for $j > i > 1$ and our lemma holds for s.

If there is no an infinite sequence of non-empty sequences r_i then we may assume that all t_i have the structure $x_i s_i$, i.e. r_1 is contained in any x_i. Suppose that there is an infinite sequence of non-empty sequences s_i. Then we may assume that all s_i are non-empty. Since $t_1 := r_1 x_1$ is not contained in any sequence $x_i s_i$, x_1 is not contained in any s_i. Applying inductive hypothesis to the sequence $x_1, s_2, s_3, ..$ and the thinning, we may suppose that s_j contains s_i for $j > i > 1$. Again by the thinning we may assume that $x_j \geq x_i$ for $j > i > 1$, and we conclude that $t_j = x_j, s_j$ contains $t_i = x_i, s_i$ for $j > i > 1$. Finally, if there is no infinite sequence of non-empty sequences s_i then we may assume that all t_i have form x_i and then we clearly can suppose by thinning that x_j contains x_i for $j > i > 1$.

Case 2. Now the assertion of our lemma is proved by induction for all sequences s with factors (m_1, l_1), where $m_1 \leq m$. Consider a sequence s with factor $(m + 1, 1)$.

Since $t_1 := m + 1$ is not contained in all t_j for $j > 1$, for any $k \in t_j$, $k < m + 1$ holds for $j > 1$. Then either the sequence $t_2, t_3, \ldots$ satisfies the requirements of our lemma, or contains a subsequence s_1 with factor (m_1, l_1), where $m_1 \leq m$, and we can apply to s_2 the proposition proved above. ∎

Definition 4.3.8 *Given two finite sequences t and s of positive natural numbers, we say the sequence t covers s if t contains s and the last number of t is bigger or equal to the last number of s.*

Corollary 4.3.9 *Let $s = t_1, t_2, \ldots$ be an infinite sequence of finite sequences of positive natural numbers. Then there is an infinite subsequence $r = t_{j1}, t_{j2}, \ldots$ of s such that t_{ji} covers t_{jh} for $i > h$.*

Proof. Let $t_i = s_i x_i$, where x_i is the last number in t_i. By thinning we may suppose that $x_j \geq x_i$ for $j > i$. If there is no an infinite sequence of nonempty s_i we may suppose that all t_i have form x_i and then by thinning we can assume that any x_j covers x_i while $j > i$. Otherwise we may suppose by thinning that all s_i are non-empty. By Lemma 4.3.7 and thinning we may assume that any s_j contains s_i for $j > i$. But then t_j covers t_i while $j > i$. ∎

We will associate any finite $S4.3$-frame $\mathcal{F}$ and the finite sequence of numbers $t(\mathcal{F}_i)$ such that (i) $t(\mathcal{F}_i)$ consists of numbers which are numbers of elements in clusters of $\mathcal{F}$, and (ii) the numbers are placed in $t(\mathcal{F}_i)$ according to the partial order on clusters of $\mathcal{F}$ generated by the accessibility relation of $\mathcal{F}_i$. More precisely, if $\mathcal{F} := \langle \bigcup_{1 \leq i \leq n} C_i, R \rangle$, C_i are distinct clusters of $\mathcal{F}$, R is reflexive, transitive relation and $\forall a \in C_i, \forall b \in C_{i+1}\ aRb$ then $t(\mathcal{F}) := \{\|C_1\|, \|C_2\|, ..., \|C_n\|\}$.

Lemma 4.3.10 *Suppose* $\mathcal{F}_1 := \langle W_1, R_1 \rangle$, $\mathcal{F}_2 := \langle W_2, R_2 \rangle$ *are certain finite* $S4.3$*-frames and* $t(\mathcal{F}_1)$ *covers* $t(\mathcal{F}_2)$. *Then there is a p-morphism from* $\mathcal{F}_1$ *onto* $\mathcal{F}_2$.

Proof. Suppose $t(\mathcal{F}_1) := t_1, t_2, ..., t_n$, $t(\mathcal{F}_2) := s_1, s_2, ..., s_m$. Since $t(\mathcal{F}_1)$ covers $t(\mathcal{F}_2)$, there are $j_1, j_2, ..., j_m$ such that $1 \leq j_1 < j_2 < ... < j_m \leq n$ and $t_{j_i} \geq s_i$ for $i = 1, 2, ..., m$. Let $t_j = \|C_j\|$, for $j = 1, 2, ..., n$, and $s_i = \|Q_i\|$ for $i = 1, 2, ..., m$, where C_j and Q_i are corresponding clusters from frames $\mathcal{F}_1$ and $\mathcal{F}_2$, respectively. Clearly, we may suppose $Q_i \subseteq C_{j_i}$ for $i = 1, 2, ..., m$ and that R_2 is R_1 restricted to $W_2 \subseteq W_1$. We define the mapping f as follows: if $w \in W_2$ then $f(w) := w$; if $w \in (W_1 - W_2)$ then $f(w) :=$ an R_1-first element of $\{v \mid v \in W_2, wR_1v\}$. Clearly f is a mapping onto. If wRg then $f(w)R_2f(g)$ since otherwise $f(w)$ would be not a R_1-first element of $\{v \mid v \in W_2, wR_1v\}$. Suppose that $f(w)R_2f(g)$. Then $wR_1f(w)R_1f(g)$. But $f(g) = f(f(g))$ which completes the proof. ∎

Theorem 4.3.11 *K.Fine [43]. There is no infinite sequence of modal logics* $\lambda_1, \lambda_2, ...$ *extending* $S4.3$ *such that* λ_i *is properly included in* λ_{i+1}.

Proof. Suppose that such sequence exists. Then, for every i, there is a formula α_i such that $\alpha_i \in \lambda_{i+1}$ but $\alpha_i \notin \lambda_i$. Since by Theorem 4.3.1 any logic λ_i has the fmp, there is a finite $S4.3$-frame $\mathcal{F}_i$ such that $\mathcal{F}_i \Vdash \lambda_i$ and $\mathcal{F}_i$ falsifies α_i. Suppose $t(\mathcal{F}_i)$ is the finite sequence of numbers associated to $\mathcal{F}_i$. Then by Corollary 4.3.9, or by Kruskal's Three Theorem ([79], p. 212), for some i, j, such that $i < j, t(\mathcal{F}_j)$ covers $t(\mathcal{F}_i)$. Then Lemma 4.3.10 entails that there is a p-morphism f from $\mathcal{F}_j$ onto $\mathcal{F}_i$. We have $\alpha_i \in \lambda_{i+1} \subseteq \lambda_j$, i.e. $\alpha_i \in \lambda_j$. Then $\mathcal{F}_j \Vdash \alpha_i$, and, since $\mathcal{F}_i$ is a p-morphic image of $\mathcal{F}_j$, it follows that $\mathcal{F}_i \Vdash \alpha_i$, a contradiction. ∎

Corollary 4.3.12 *K.Fine [43]. The following hold:*

(i) *All* $S4.3$*-logics are finitely axiomatizable.*

(ii) *All* $S4.3$*-logics are decidable.*

(iii) *There are at most* ω $S4.3$*-logics.*

Proof. (i) follows directly from Theorem 4.3.11; (ii) follows from (i) and Theorem 4.3.1, (iii) is a corollary of (i). ∎

Now, having at our disposal that all modal logic extending $S4.3$ have the fmp and finite axiomatization, we can decide the question concerning bases for admissible rules of such logics. First we precisely describe the structure of n-characterizing models $Ch_\lambda(n)$ for logics λ extending $S4.3$ (the structure of these models is constructive since any logic λ in question has the fmp and a finite number of axioms). We pick a modal logic λ extending $S4.3$. According to the procedure of our constructing models $Ch_\lambda(n)$, the model $Ch_\lambda(n)$ has the following structure:

$$Ch_\lambda(n) := \mathfrak{M}_1 \bigsqcup \mathfrak{M}_2 \bigsqcup .... \bigsqcup \mathfrak{M}_{2^{2^n}},$$

where

(i) any cluster of $\mathfrak{M}_i$ does not contain distinct elements with the same valuation of the letters;

(ii) for any element $a \in |\mathfrak{M}_i|$, the frame $a^{R\leq}$ generated in $\mathfrak{M}_i$ by a is some reflexive, transitive, connected and finite λ-frame, and the submodel of $\mathfrak{M}_i$ which is based upon the cluster $C(a)$ containing a is not a submodel of the model which is based upon the immediate successor cluster $C(b)$ for $C(a)$;

(iii) If C_1 and C_2 are distinct clusters of $\mathfrak{M}_i$ of depth n then one of them has an element a which has the valuation of letters such that all elements of other cluster have valuations of letters which are distinct from the valuation on a.

(iv) Any two models which are based upon the maximal clusters of distinct models $\mathfrak{M}_i$ and $\mathfrak{M}_j$ are not isomorphic.

We can illustrate the structure of such models in Fig. 4.1, where • represents a reflexive element, and big circles are proper clusters. Hence these models

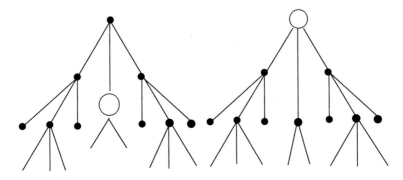

Figure 4.1: The draft of the structure of the frame of $Ch_\lambda(n)$

looks like turned over trees with their roots placed at the top. For this reason the similar constructions were called *Christmas Trees* in Meskhi, Esakia [39]. Note that since $Ch_\lambda(n)$ is a certain n-characterizing model for λ, the free algebra $\mathfrak{F}_\lambda(m)$ of rank n from the variety $Var(\lambda)$ is isomorphic to the subalgebra $Ch_\lambda(n)^+(V(p_1), ..., V(p_n))$ of the algebra $Ch_\lambda(n)^+$ which is n-generated by the elements $V(p_1), ..., V(p_n)$ as free generators (see Theorem 3.3.8).

Let $\mathcal{F}$ be a frame, by $\mathcal{F} \circ 1$ we will denote the frame $\mathcal{F} \bigsqcup \mathcal{F}_1$, where $\mathcal{F}_1$ is a reflexive single-element frame. For our investigation of bases for admissible rules we need the following lemma.

Lemma 4.3.13 *Suppose $\mathcal{F}_i$, $1 \le i \le m$ are distinct open subframes with no common elements of the model $Ch_\lambda(n)$, each of which is rooted, i.e. generated by some element from $Ch_\lambda(n)$. The algebra $\bigsqcup_{1 \le i \le m} \mathcal{F}_i \circ 1^+$ is isomorphic to a subalgebra of the free algebra $Ch_\lambda(n+1)^+(V(p_1), ..., V(p_{n+1}))$ of the variety $Var(\lambda)$ for some m.*

Proof. It is clear that $\bigsqcup_{1 \le i \le m} \mathcal{F}_i \circ 1$ is an open subframe of the frame of the model $Ch_\lambda(n+1)$, so we can consider elements of the frame $\bigsqcup_{1 \le i \le m} \mathcal{F}_i \circ 1$ as elements of $Ch_\lambda(n+1)$. Suppose that e is an element of $Ch_\lambda(n+1)$ corresponding to the single-element frame (i.e. 1) in the decomposition of $\bigsqcup_{1 \le i \le m} \mathcal{F}_i \circ 1$. According to Lemma 3.3.7 any element a of $\bigsqcup_{1 \le i \le m} \mathcal{F}_i \circ 1$ is definable in the model $Ch_\lambda(n+1)$ by a formula. For all elements distinct from e, by $\alpha(a)$ we denote a formula defining a. Since all clusters of the frame $\bigsqcup_{1 \le i \le m} \mathcal{F}_i \circ 1$ are finite, each cluster C is definable by a formula $\beta(C)$. For any cluster C from $\bigsqcup_{1 \le i \le m} \mathcal{F}_i$, we fix an element $e(C) \in C$ and introduce the following formulas

$$\gamma(C) := \gamma_1(C) \wedge \gamma_2(C) \wedge \gamma_3(C), \text{ where,}$$
$$\gamma_1(C) := \bigwedge\{\Diamond\alpha(x) \mid x \in C\},$$
$$\gamma_2(C) := \bigwedge\{\neg\alpha(x) \mid x \in C \& x \ne e(C)\},$$
$$\gamma_3(C) := \bigwedge\{\neg\Diamond\alpha(x) \mid x \in |\bigsqcup_{1 \le i \le m} \mathcal{F}_i|\&(xRe(C))\}\&,$$
$$\neg(e(C)Rx)\&\forall y \in |\bigsqcup_{1 \le i \le m} \mathcal{F}_i \circ 1|(yRe(C))\&$$
$$\&\neg(e(C)Ry)\&(xRy) \Rightarrow yRx)\}.$$

We pick the subalgebra $\mathfrak{B}$ of algebra $Ch_\lambda(n+1)^+$ generated by elements

(i) $\gamma(C)$, where C is any cluster of $\bigsqcup_{1 \le i \le m} \mathcal{F}_i$;

(ii) $\alpha(x)$, where $x \in |\bigsqcup_{1 \le i \le m} \mathcal{F}_i|$ and $x \ne e(C)$ for all $e(C)$;

We will show that $\mathfrak{B}$ is isomorphic to $\bigsqcup_{1 \le i \le m} \mathcal{F}_i \circ 1^+$. We define the mapping f of the algebra $Ch_\lambda(n+1)^+$ onto the algebra $\bigsqcup_{1 \le i \le m} \mathcal{F}_i \circ 1^+$ as follows

$$\forall X \subseteq |Ch_\lambda(n+1)|, f(X) := X \cap |\bigsqcup_{1 \le i \le m} \mathcal{F}_i \circ 1|.$$

Since $\bigsqcup_{1 \le i \le m} \mathcal{F}_i \circ 1$ is an open subframe of the frame of $Ch_\lambda(n+1)$, f is a homomorphism onto (see Lemma 2.5.9). Consider the restriction of f onto the algebra $\mathfrak{B}$. Clearly, f is a homomorphism from $\mathfrak{B}$ into $\bigsqcup_{1 \leq i \leq m} \mathcal{F}_i \circ 1^+$. Our aim now is to show that f maps $\mathfrak{B}$ onto $\bigsqcup_{1 \leq i \leq m} \mathcal{F}_i \circ 1^+$.

Let x be an element of $|\bigsqcup_{1 \leq i \leq m} \mathcal{F}_i \circ 1|$. Suppose that $x \neq e(C)$ for all elements $e(C)$. Since $x \Vdash_V \alpha(x)$ and $\alpha(x)$ defines x, it follows that $V(\alpha(x)) = \{x\}$, and $f(V(\alpha(x))) = \{x\}$. Consider the element e among those that we reserved above, which corresponds to the frame 1 in the decomposition $\bigsqcup_{1 \leq i \leq m} \mathcal{F}_i \circ 1$. By the definition, $e \nVdash_V \alpha(x)$ for all formulas $\alpha(x)$, therefore $e \in \neg(V(\alpha(x)))$ for any formula $\alpha(x)$. Also, for any cluster C of $\bigsqcup_{1 \leq i \leq m} \mathcal{F}_i$, it follows that $e \nVdash_V \gamma(C)$ since e does not see any element of C. Hence $e \in \neg(\gamma(C))$ for any such C. Since $x \in V(\alpha(x))$ for any $x \in |\bigsqcup_{1 \leq i \leq m} \mathcal{F}_i|$, it follows that

$$\{e\} = V\left(\bigwedge_{\alpha(x), \forall e(C)(x \neq e(C))} \neg\alpha(x) \wedge \bigwedge_C \neg\gamma(C)\right) \cap |\bigsqcup_{1 \leq i \leq m} \mathcal{F}_i \circ 1|,$$

and consequently $f(V(\bigwedge_{\alpha(x), \forall e(C)(x \neq e(C))} \neg\alpha(x) \wedge \bigwedge_C \neg\gamma(C))) = \{e\}$.

Consider now an element of kind $e(C)$. By the definition of $\gamma(C)$ we obtain $e(C) \Vdash_V \gamma(C)$. Assume that $y \in |\bigsqcup_{1 \leq i \leq m} \mathcal{F}_i \circ 1|$ and $y \neq e(C)$. If $\neg(yRe(C))$ then $y \Vdash_V \neg\gamma(C)$ by the definition of $\gamma_1(C)$. If $y \in C$ and $y \neq e(C)$ then $y \nVdash_V \gamma(C)$ by the definition of $\gamma_2(C)$. Finally, if $yRe(C)$ and $y \notin C$ then y R-sees some element of the cluster C_1 from $\bigsqcup_{1 \leq i \leq m} \mathcal{F}_i \circ 1$ which is immediate predecessor of C. Then $y \nVdash_V \gamma(C)$ by the definition of $\gamma_3(C)$. At the same time we know that $e(C) \Vdash_V \gamma(C)$, therefore

$$\gamma(C) \cap |\bigsqcup_{1 \leq i \leq m} \mathcal{F}_i \circ 1| = \{e(C)\},$$

i.e. $f(\gamma(C)) = \{e(C)\}$. Thus we showed that any element of $|\bigsqcup_{1 \leq i \leq m} \mathcal{F}_i \circ 1|$ has a pro-image in $\mathfrak{B}$. Because $|\bigsqcup_{1 \leq i \leq m} \mathcal{F}_i \circ 1|$ is finite, it follows that f maps $\mathfrak{B}$ onto the algebra $\bigsqcup_{1 \leq i \leq m} \mathcal{F}_i \circ 1^+$. It order to complete the proof of our lemma we need the following

Lemma 4.3.14 *The mapping f in one-to-one on the carrier of $\mathfrak{B}$.*

Proof. Since f is a homomorphism, it is sufficient to show that f maps any non-empty set (as an element of $\mathfrak{B}$) into certain non-empty set. Let Y be an element of $\mathfrak{B}$ and $Y \neq \emptyset$. Then $Y = V(t(\gamma(C_i), \alpha(x)))$, where $t(z, v)$ is a modal term constructed out of a tuple of variables z and a tuple of variables v. Hence, there is an element y such that

$$y \Vdash_V t(\gamma(C_i), \alpha(x)). \tag{4.18}$$

If $y \in |\bigsqcup_{1 \leq i \leq m} \mathcal{F}_i \circ 1|$ then $y \in f(Y)$ and $f(Y) \neq \emptyset$. Conversely, suppose now that the element y does not belong to $|\bigsqcup_{1 \leq i \leq m} \mathcal{F}_i \circ 1|$. Consider the case when

y sees no elements from $|\bigsqcup_{1 \leq i \leq m} \mathcal{F}_i|$. By the definition of the formulas $\gamma_1(C)$ it follows that

$$\forall z (yRz) \Rightarrow z \not\Vdash_V \gamma(C),$$

for every formula $\gamma(C)$. For every formula $\alpha(x)$, $y \not\Vdash_V \alpha(x)$ holds, and we also have $\forall z$, $(yRz) \Rightarrow z \not\Vdash_V \alpha(x)$. It is not hard to infer from these relations that, for any modal formula $\delta(z, v)$ built up out of any tuple of variables z and any tuple of variables v

$$e \Vdash_V \delta(\gamma(C), \alpha(x)) \Leftrightarrow y \Vdash_V \delta(\gamma(C), \alpha(x)).$$

This and (4.18) imply

$$e \Vdash_V t(\gamma(C_i), \alpha(x)),$$

consequently, $e \in V(t(\gamma(C_i), \alpha(x)))$ and $e \in f(Y)$, so $f(Y)$ is non-empty.

Suppose now that y sees some elements from . Let C_1 be a first with respect to R cluster from $|\bigsqcup_{1 \leq i \leq m} \mathcal{F}_i|$ which is seen from y. Then, for every element $z \in |Ch_\lambda(n + 1)|$, if yRz then either $z \notin |\bigsqcup_{1 \leq i \leq m} \mathcal{F}_i|$ and z R-sees C_1 or $z \in |\bigsqcup_{1 \leq i \leq m} \mathcal{F}_i|$ and z is R-seen from elements of $\overline{C_1}$. Therefore

$$\forall z (yRz \Rightarrow (e(C_1)Rz) \vee (z \Vdash_V \gamma(C_1) \wedge$$
$$\bigwedge \{\neg \alpha(x) \mid x \in |\bigsqcup_{1 \leq i \leq m} \mathcal{F}_i| \& x \neq e(C), \forall e(C)\}).$$

Besides, for every z, if yRz and $z \notin |\bigsqcup_{1 \leq i \leq m} \mathcal{F}_i|$ then $z \not\Vdash_V \gamma(C)$ for all $C \neq C_1$. Moreover,

$$e(C_1) \Vdash_V \neg\gamma(C) \wedge \bigwedge_{\alpha(x)} \neg\alpha(x), e(C_1) \Vdash_V \gamma(C_1).$$

Furthermore, for all such z and all C, either z and $e(C_1)$ do not R-see C (but do see C_1), or z and $e(C_1)$ both R-see C. From these observations it is easy to show by induction on the length of any formula $\delta(u, v)$ built up out of any tuple of variables u and any tuple of variables v that

$$e(C_1) \Vdash_V \delta(\gamma(C)), \alpha(x)) \Leftrightarrow y \Vdash_V \delta(\gamma(C)), \alpha(x)).$$

Therefore from (4.18) it follows $e(C_1) \Vdash_V t(\gamma(C)), \alpha(x))$, consequently we obtain $e(C_1) \in V(t(\gamma(C)), \alpha(x)))$ and $f(Y) \neq \emptyset$. This exhausts all the possibilities and f maps non-empty sets from $|\mathfrak{B}|$ into non-empty sets. Therefore f is one-to-one mapping defined on $\mathfrak{B}$. ∎

Now we are in a position to prove the following lemma.

Lemma 4.3.15 *The universal theory of the free algebra $\mathfrak{F}_\lambda(\omega)$ for a certain λ, where $S4.3 \subseteq \lambda$, and the universal theory of the class of all modal algebras of the kind $\bigsqcup_{1 \leq i \leq m} \mathcal{F}_i \circ 1^+$, $m \in N$, defined in Lemma 4.3.13, coincide.*

Proof. We have showed in Lemma 4.3.13 that all algebras $\bigsqcup_{1 \leq i \leq m} \mathcal{F}_i \circ 1^+$ are subalgebras of certain free algebras $\mathfrak{F}_\lambda(g)$ of a certain finite rank g. Therefore the universal theory of $\mathfrak{F}_\lambda(\omega)$ is a subtheory of the universal theory of the above mentioned class of modal algebras. Conversely, let ϕ be a universal formula which is false in $\mathfrak{F}_\lambda(\omega)$. Using the standard equivalences which hold for modal $S4$-algebras, it is easy to see that ϕ (as an arbitrary universal formula) has a prenex normal form, which is equivalent in $Var(S4)$ to a formula Ψ of the kind

$$\psi := \forall x_1...\forall x_l (\bigwedge_k (f^k = \top) \rightarrow \bigvee_{j=1}^{m_k} (\Box g_j^k = \top)).$$

Since ϕ is false in a $\mathfrak{F}_\lambda(d)$ for a finite d, and $Ch_\lambda(d)^+(V(p_1), ..., V(p_d))$ is isomorphic to $\mathfrak{F}_\lambda(d)$ (see Lemma 3.3.8) it follows that, for a valuation S of variables of ψ in the frame of $Ch_\lambda(d)$,

$$\exists k, S(f^k) = |Ch_\lambda(d)|, \text{ and } \bigwedge_{j=1}^{m_k} (\Box g_j^k) \neq |Ch_\lambda(d))|.$$

This means that there are clusters C_j in $Ch_\lambda(n)$ such that for all a_j from C_j, $(a_j \not\Vdash_S \Box g_j^k)$. It is easy to see from this observation that the formula ψ is false in the algebra $\bigsqcup_{j=1}^{m_k} C_j^{R \leq} \circ 1^+$. Then the formula ϕ, being equivalent to ψ, is also false in this algebra. ∎

Now we have developed all the tools which are necessary to describe bases of admissible rules in modal logics extending $S4.3$. Moreover these tools also allow us to solve the problem concerning decidability of the universal theories of free algebras from varieties of modal algebras corresponding to $S4.3$-logics. Of course it is possible to show the decidability of quasi-equational theories of these free algebras easily using algorithms for recognizing admissible rules which are found in Section 5 of Chapter 3 since they are applicable to all $S4.3$-logics. But since these logics do not have the disjunction property, this would not give us the decidability of universal theories of these free algebras. Therefore we now have a possibility to obtain a stronger result concerning those free algebras.

Theorem 4.3.16 *For any modal logic λ extending $S4.3$ the universal theory of the free algebra $\mathfrak{F}_\lambda(\omega)$ is decidable.*

Proof. Consider a universal formula ϕ. Transforming this formula into the prenex normal form and making the usual equivalent transformations in the quantifier-free part of the formula we obtain a formula

$$\psi := \forall x_1 ... \forall x_l (\bigwedge_k (f^k = \top) \to \bigvee_{j=1}^{m_k} (\Box g_j^k = \top)),$$

equivalent to ϕ in $Var(S4)$. By Lemma 4.3.15 we need only to verify whether ψ is valid in every modal algebra of the kind $\bigsqcup_{1 \leq i \leq r} \mathcal{F}_i \circ 1^+$. Suppose that $\bigsqcup_{1 \leq i \leq r} \mathcal{F}_i \circ 1^+ \not\models \psi$. Then, for some valuation V of variables from ψ in the frame $\bigsqcup_{1 \leq i \leq r} \mathcal{F}_i \circ 1$, there is some k such that

$$V(f_i^k) = |\bigsqcup_{1 \leq i \leq r} \mathcal{F}_i \circ 1| \text{ and } \bigwedge_{j=1}^{m_k} (\Box g_j^k \neq |\bigsqcup_{1 \leq i \leq r} \mathcal{F}_i \circ 1|).$$

We may assume that $r = m_k$, and using that ψ has n variables, we can suppose that any cluster of $\bigsqcup_{1 \leq i \leq m} \mathcal{F}_i \circ 1$ has no distinct elements with the same valuation of the variables of ψ with respect to V, i.e. any cluster of this frame has not more than 2^n elements.

For any $x \in |\bigsqcup_{1 \leq i \leq m} \mathcal{F}_i \circ 1|$, Ter denotes the set of all subterms of all formulas from ψ and

$$S(x) := \{\neg\alpha \mid \alpha \in Ter, x \not\Vdash_V \alpha\} \cup \{\alpha \mid \alpha \in Ter, x \Vdash_V \alpha\}.$$

For all $x \in |\bigsqcup_{1 \leq i \leq m} \mathcal{F}_i \circ 1|$, we put $D(x) := \bigcup_{xRy} S(y)$; for any cluster C from $\bigsqcup_{1 \leq i \leq m} \mathcal{F}_i \circ 1$, we set by definition $D(C) := D(x)$ for certain fixed $x \in X$. Now we will transform the frame $\bigsqcup_{1 \leq i \leq r} \mathcal{F}_i \circ 1$ by removing all elements x, except the elements of R-minimal cluster, such that $\exists y(xRy) \& (x \neq y) \& (D(x) = D(y))$ when we move from the bottom of the frame to the top. The resulting frame $\mathcal{F}$ with the valuation V transferred from the original frame has a form similar to the original frame, and it is also easy to see that $\mathcal{F} \not\models \phi$. Note that if C_1 is a cluster of $\mathcal{F}$ which is an immediate predecessor in $\mathcal{F}$ for a cluster C from $\mathcal{F}$, then

$$\bigcup_{x \in C_1} D(x) \supset \bigcup_{x \in C} D(x).$$

Indeed, since $\bigcup_{x \in C_1} D(x) = D(y)$ for some $y \in C_1$ and $\bigcup_{x \in C} D(x) = D(z)$ for some $z \in C$, if $\bigcup_{x \in C_1} D(x) = \bigcup_{x \in C} D(x)$ then $D(y) = D(z)$, and we must remove y when we construct $\mathcal{F}$, a contradiction. Thus the inclusion above is proper in fact.

Hence the frame $\mathcal{F}$ has the form $\bigsqcup_{1 \leq i \leq r} \mathcal{F}_i \circ 1$, where any $\mathfrak{F}_i$ is a linearly quasi-ordered finite frame, which is a chain of clusters $C_1, C_2, ..., C_{r_i}$, that is $C_1 R C_2 R ... R C_{r_i}$, it follows

$$D(C_1) \supset D(C_2) \supset ... \supset D(C_{r_i}).$$

This, in particular, entails that the number of clusters in any frame $\mathcal{F}_i$ is not more than $2 * g$, where g is the number of subterms of subformulas of ψ. Thus we conclude (recall we may assume $r = m_k$)

$$\bigsqcup_{1 \le i \le m_k} \mathcal{F}_i \circ 1^+ \not\models \phi,$$

where $\mathcal{F}_i$ are certain linearly-quasi-ordered finite λ-frames with not more than $2 * g$ clusters, every of which has not more than 2^n elements. Since λ by Corollary 4.3.12 has a finite number of axioms, this gives an algorithm for recognizing whether ψ is valid in $\mathfrak{F}_\lambda(\omega)$. ∎

From this theorem and Theorem 1.4.5 we obtain another proof of the fact that all modal logics extending $S4.3$ are decidable with respect to admissibility:

Corollary 4.3.17 *[136] For any modal logic λ extending $S4.3$, there is an algorithm which determines admissibility of inference rules in λ.*

In order to describe bases of admissible rules in logics extending the modal logic $S43$ we need a result concerning structure of modal algebras satisfying a condition which is expressible by a quasi-identity (which also will play an especial role in our considerations further).

Lemma 4.3.18 *Let $\mathfrak{B}$ be a finitely-generated modal algebra from $Var(S4)$ and $||\mathfrak{B}|| > 1$. Suppose $\mathfrak{B} \models \Diamond x \wedge \Diamond\neg x = \top \Rightarrow y = \top$. Then the frame $\mathfrak{B}^+$ has a single-element R-maximal cluster.*

Proof. Recall that the frame of $\mathfrak{B}^+$ has the structure $\langle T_\mathfrak{B}, R \rangle$, where $T_\mathfrak{B}$ is the set of all ultrafilters on $\mathfrak{B}$ and

$$\forall \nabla_1, \nabla_2 \in T_\mathfrak{B}, [\nabla_1 R \nabla_2 \Leftrightarrow \forall \Box x (\Box x \in \nabla_1 \Rightarrow x \in \nabla_2)].$$

Moreover by Theorem 2.5.6 the mapping $i : \mathfrak{B} \mapsto (\mathfrak{B}^+)^+$, where

$$i(a) := \{\nabla \mid a \in \nabla\}$$

is an isomorphic embedding. Suppose that a finite set consisting of elements a_i is a set of generators of $\mathfrak{B}$. We introduce a valuation V for propositional letters p_i in $\mathfrak{B}^+$ by $V(p_i) := \{\nabla \mid a_i \in \nabla\}$. Using Lemma 2.5.17 and that i is an isomorphism, it is easy to show that, for any formula $\varphi(p_i)$ constructed out of letters p_i,

$$\nabla \Vdash_V \varphi(p_i) \Leftrightarrow \varphi(a_i) \in \nabla. \tag{4.19}$$

Since $S4 \subseteq \lambda$ it follows that

$$\forall \nabla_1, \nabla_2 \in T_\mathfrak{B}[\nabla_1 R \nabla_2 \Leftrightarrow \forall \Box x (\Box x \in \nabla_1 \Rightarrow \Box x \in \nabla_2)].$$

Therefore applying Zorn's Lemma, it is easy to show that any cluster of $\mathfrak{B}^+$ R-sees some maximal cluster of $\mathfrak{B}^+$. And (4.19) entails that all maximal clusters of $\mathfrak{B}^+$ have not more than 2^n elements and the number of maximal clusters in $\mathfrak{B}^+$ is finite. Suppose that $C_1, C_2, ..., C_k$ are exactly all the maximal clusters of $\mathfrak{B}^+$ and that there is no single-element cluster among them. We pick an element ∇_i in any maximal cluster C_i, and denote by $\psi(\nabla_i)$ an element which is a member of ∇_i but does not occur in all other elements of the cluster C_i. Suppose that $\Box\phi(i, j)$ is an element which occurs in ∇_i but does not occur in elements of all maximal clusters C_j which are distinct from C_i. The existence of all mentioned above elements obviously follows from the definition of $\mathfrak{B}^+$ and finititude of the set of all maximal clusters, and finiteness of the clusters. Consider the element

$$a := \bigvee_{1 \leq i \leq k} [\psi(\nabla_i) \wedge (\bigvee_{j \neq i} \Box\phi(i, j))].$$

Suppose that $\Box a \neq \bot$. Then $\Box a$ is a member of some ultrafilter ∇ on the algebra $\mathfrak{B}$ (see Lemma 2.5.7). ∇ R-sees some maximal cluster C_m of $\mathfrak{B}^+$. Then $\Box a \in \Delta$ for any $\Delta \in C_m$. Let $\Delta \in C_m$ and $\Delta \neq \nabla_m$. Then $\Box a \in \Delta$ and for some i we have

$$\psi(\nabla_i) \wedge (\bigwedge_{j \neq i} \Box\phi(i, j)) \in \Delta.$$

Using the definition of $\phi(i, j)$ and this observation we infer that $\Delta \in C_i$, i.e. $m = i$. Then since $\Delta \neq \nabla_m$, we get $\psi(\nabla_i) \notin \Delta$, a contradiction. Hence we conclude $\Box a = \bot$.

Suppose that $\Box\neg a \neq \bot$. Then, as before, there is a maximal cluster C_m such that $\forall \Delta \in C_m$ ($\Box\neg a \in \Delta$). Let $\Delta = \nabla_m$. Then $\psi(\nabla_m) \in \nabla_m$ and $\Box\phi(m, j) \in \nabla_m$ when $j \neq m$. Therefore we obtain $a \in \nabla_m$ which contradicts $\Box\neg a \in \nabla_m$. Thus $\Box\neg a = \bot$. Consequently we have showed that $\Box a = \bot$ and $\Box\neg a = \bot$ and consequently $\Diamond a \wedge \Diamond\neg a = \top$ which contradicts the fact that $\mathfrak{B} \models \Diamond x \wedge \Diamond\neg x = \top \Rightarrow y = \top$. ∎

Theorem 4.3.19 *[136] For any modal logic λ extending S4.3, the free algebra $\mathfrak{F}_\lambda(\omega)$ has a finite basis for quasi-identities consisting of a finite basis for identities of $\mathfrak{F}_\lambda(\omega)$ and the quasi-identity $\Diamond x \wedge \Diamond\neg x = \top \Rightarrow y = \top$.*

Proof. By Corollary 4.3.12 there is a finite basis of identities for $Var(\lambda)$ and, consequently, there is a finite basis of identities for $\mathfrak{F}_\lambda(\omega)$. Further, using Lemma 4.3.15 it follows that $\mathfrak{F}_\lambda(\omega) \models \Diamond x \wedge \Diamond\neg x = \top \Rightarrow y = \top$ since it is easy to see that this quasi-identity is valid in all algebras of kind $\bigsqcup_{1 \leq i \leq m} \mathcal{F}_i \circ 1^+$ from Lemma 4.3.15. Suppose $\mathfrak{B}$ is a finitely-generated algebra $\mathfrak{B} \in \bar{V}ar(\lambda)$ and $\mathfrak{B} \models \Diamond x \wedge \Diamond\neg x = \top \Rightarrow y = \top$. Suppose that

$$\mathfrak{B} \not\models \bigwedge_{1 \leq i \leq m} f_i(x_1, ..., x_n) = \top \Rightarrow g(x_1, ..., x_n) = \top.$$

(It is clear that any quasi-identity in the modal language is equivalent in $Var(\lambda)$ to a quasi-identity of the form displayed on the right hand side of the relation above). We may assume that $\mathfrak{B}$ is generated by elements a_i, $i \leq n$, where $\bigwedge_{1 \leq i \leq m} f_i(a_1, ..., a_n) = \top$ and $g(a_1, ..., a_n) \neq \top$. Also it is clear that

$$\Box \bigwedge_{1 \leq i \leq m} f_i(x_1, ..., x_n) \to g(x_1, ..., x_n) \notin \lambda.$$

Now, since λ has the fmp (Theorem 4.3.1), there is a finite linearly quasi-ordered frame $\mathcal{F}_1$ such that $\mathcal{F}_1 \nVdash \Box \bigwedge_{1 \leq i \leq m} f_i \to g$. Since $\mathcal{F}_1$ is linearly quasi-ordered, this implies that there is an rooted open subframe $\mathcal{F}$ of $\mathcal{F}_1$ such that

$$\mathcal{F}^+ \nvDash (\bigwedge_{1 \leq i \leq m} f_i) = \top \Rightarrow g = \top. \tag{4.20}$$

Using the Stone Representation Theorem (see Theorem 2.5.6), we consider the isomorphic embedding i of the algebra $\mathfrak{B}$ into the modal algebra $(\mathfrak{B}^+)^+$. By (4.19) from Lemma 4.3.18 it follows that

$$\nabla \Vdash_V \bigwedge_{1 \leq i \leq m} f_i(p_1, ..., p_n), \text{ where}$$

$$V(p_i) := \{\nabla_1 \mid a_i \in \nabla_1\}$$

for all $\nabla \in |\mathfrak{B}^+|$. By Lemma 4.3.18 there is a single-element R-maximal cluster C in $\mathfrak{B}^+$. Therefore $\bigwedge_{1 \leq i \leq m} f_i(p_1, ..., p_n)$ is valid in the single-element reflexive frame for certain valuation of variables. This and (4.20) yield

$$(\mathcal{F} \circ 1)^+ \nvDash (\bigwedge_{1 \leq i \leq m} f_i) = \top \Rightarrow g = \top.$$

Using Lemma 4.3.15 we conclude that $\mathfrak{F}_\lambda(\omega) \nvDash (\bigwedge_{1 \leq i \leq m} f_i) = \top \Rightarrow g = \top$. Hence the finite basis of identities for the variety $Var(\lambda)$ and the quasi-identity $\Diamond x \wedge \Diamond \neg x = \top \Rightarrow y = \top$ form a finite basis for quasi-identities of the free algebra $\mathfrak{F}_\lambda(\omega)$. $\blacksquare$

Immediately from this theorem and Theorem 1.4.15 we derive

Corollary 4.3.20 *[136] Any modal logic λ above S4.3 has the basis for admissible inference rules which consists of the single inference rule $\Diamond x \wedge \Diamond \neg x / y$.*

The structure of this rule, which forms a basis for admissible rules, shows that its premise is satisfied in any consistent S4.3-logic and in any consistent modal logic at all (which means there is no a substitution which turns the premise of this rule into a theorem of any given consistent modal logic). This observation leads us to clarify the structure of the rules admissible in $S4.3$-logics. The rule $\Diamond x \wedge \Diamond \neg x / y$ is obviously admissible but not derivable rule in any consistent $S4.3$-logic. Therefore it looks likely that all the rules admissible but not derivable in $S4.3$-logics do not have premises which can be satisfied, and are consequently useless in derivations. And it turns out that this is really the case and we show this in the following proposition.

Proposition 4.3.21 *Let λ be a modal logic extending $S4.3$. Suppose that a rule $\alpha_1(x_i), ..., \alpha_n(x_i)/\beta(x_i)$ is admissible but not derivable in λ. Then, for every tuple of formulas γ_i, $\neg(\alpha_1(\gamma_i) \wedge ... \wedge \alpha_n(\gamma_i)) \in \lambda(\mathcal{F})$, where $\mathcal{F}$ is the single-element reflexive frame, and in particular, $(\alpha_1(\gamma_i) \wedge ... \wedge \alpha_n(\gamma_i)) \notin \lambda$ if λ is a consistent logic.*

Proof. Since the rule in question is not derivable in λ, we conclude

$$\Box(\alpha_1(x_i), ..., \alpha_n(x_i)) \rightarrow \beta(x_i) \notin \lambda,$$

(see Theorem 1.3.15) and, since λ has the fmp, (Theorem 4.3.1), there is a finite linearly quasi-ordered frame $\mathcal{F}_z$ such that $\mathcal{F}_z \nVdash \Box \bigwedge_{1 \leq j \leq n} \alpha_j \rightarrow \beta$. Since $\mathcal{F}_z$ is linearly quasi-ordered, this implies that there is a rooted open subframe $\mathcal{F}$ of $\mathcal{F}_z$ such that

$$\mathcal{F}^+ \nvDash (\bigwedge_{1 \leq j \leq n} \alpha_j) = \top \Rightarrow \beta = \top,$$

and $\mathcal{F}$ also is a finite linearly quasi-ordered λ-frame.

Suppose that the formula $\bigwedge_{1 \leq j \leq n} \alpha_j(\gamma_i)$ is satisfied in the single-element reflexive frame $\mathcal{F}_1$ under a certain valuation. Then the facts shown above and this assumption together give us

$$(\mathcal{F} \circ 1)^+ \nvDash (\bigwedge_{1 \leq j \leq n} \alpha_j = \top \Rightarrow \beta = \top),$$

which by Lemma 4.3.15 implies $\mathfrak{F}_\lambda(\omega) \nvDash \bigwedge_{1 \leq j \leq n} \alpha_j = \top \Rightarrow \beta = \top$. Then by Theorem 1.4.5 the rule $\alpha_1, ..., \alpha_n/\beta$ is not admissible in λ, a contradiction. ∎

Concerning this result, a number of relevant observations can be maid. The fact that the premises for rules admissible but not derivable in $S4.3$-logics can not be satisfied in these logics. Hence such rules are not applicable in derivations, but this does not demonstrate deficiency of such logics or rules but rather something quite opposite. In fact it means that such logics are in a sense self contained, they are almost structurally complete, i.e. all admissible rules which are really applicable are derivable. Hence these logics have certain additional positive properties, in the tradition originated from Bull [21] and Fine [43].

The question: whether the rules which are admissible but not derivable in Lewis's modal system $S5$ (which extends $S4.3$) have premises which cannot be satisfied in $S5$ was set by J.Port in [115]. Second relative question posed in [115] is the following one: whether the modal system $S5$ with the added inference rule $\neg(\Diamond x \to \Box x)/y$ is structurally complete. It is clear that this rule is equivalent to the rule $\Diamond x \wedge \Diamond\neg x/y$. Therefore, having proved already results concerning this rule, it looks very likely that the answer to the J.Port's questions is affirmative, even in the more general situation concerning all $S4.3$-logics. It does really a case.

Theorem 4.3.22 *For any modal logic* $\lambda \supseteq S4.3$, *the logic* λ *with the added inference rule* $\Diamond x \wedge \Diamond\neg x/y$ *is structurally complete and has the same theorems as* λ.

Proof. Suppose $r := \alpha_1(x_i), ..., \alpha_n(x_i)/\beta(x_i)$ is an admissible rule in the logic which is obtained from a logic $\lambda \supseteq S4.3$ by adding the rule $\Diamond x \wedge \Diamond\neg x/y$, i.e. in the logic $\lambda + \Diamond x \wedge \Diamond\neg x/y$. By Corollary 4.3.20 the rule r is admissible in λ. Then according to Proposition 4.3.21 either r is derivable in λ or, for arbitrary formulas γ_i substituted in place of variables x_i in the formulas $\alpha_1, ..., \alpha_n$, $\neg(\bigwedge_{1 \leq j \leq n} \alpha_j(\gamma_i)) \in \lambda(\mathcal{F})$, where $\mathcal{F}$ is the single-element reflexive frame. If r is derivable in λ then r is also derivable in $\lambda + \Diamond x \wedge \Diamond\neg x/y$. If r is not derivable in λ then we can derive from the observation above that

$$\Box \bigwedge_{1 \leq j \leq n} \alpha_j \to \Box \bigvee_{1 \leq l \leq n} (\Diamond p_l \wedge \Diamond\neg p_l) \in \lambda, \qquad (4.21)$$

where p_i, $1 \leq i \leq m$, are all the propositional variables from $\alpha_1, ..., \alpha_n$. In fact, otherwise, since λ has the fmp (Theorem 4.3.1), there is a finite linearly quasi-ordered frame $\mathcal{F}_c$ which disproves the formula $\Box \bigwedge_{1 \leq j \leq n} \alpha_j \to \Box \bigvee_{1 \leq l \leq n} (\Diamond p_l \wedge \Diamond\neg p_l)$. From this it is easily seen that this formula is invalid in the single-element reflexive frame $\mathcal{F}$. In particular, it follows $\neg(\bigwedge_{1 \leq j \leq n} \alpha_j) \notin \lambda(\mathcal{F})$, a contradiction. Thus (4.21) is proved.

Consider the free algebra $\mathfrak{F}_\lambda(m)$ of rank m from $Var(\lambda)$. We know that $\mathfrak{F}_\lambda(m)$ is isomorphic to the algebra $Ch_\lambda(m)^+(V(p_1), ..., V(p_m))$ (see Theorem 3.3.8). Besides, according to Theorem 3.3.7, any element of the model $Ch_\lambda(m)$ is definable (expressible) by certain formula. We pick out some element a_i, $1 \leq i \leq k$, in any maximal cluster C_i of $Ch_\lambda(m)$. Let φ_i be a formula defining a_i in $Ch_\lambda(m)$. Then

$$Ch_\lambda(m) \Vdash \Box(\bigvee_{1 \leq l \leq n} (\Diamond p_l \wedge \Diamond\neg p_l)) \to$$
$$\to \Diamond(\bigvee_{1 \leq i \leq k} \varphi_k) \wedge \Diamond\neg(\bigvee_{1 \leq i \leq k} \varphi_k).$$

Since $Ch_\lambda(m)$ is m-characterizing for λ, the formula from the right hand side of the truth relation above is a theorem of λ. Using this and (4.21) we immediately infer $\Box \bigwedge_{1 \leq j \leq n} \alpha_j \to \Diamond\psi \wedge \Diamond\neg\psi \in \lambda$, where $\psi := \bigvee_{1 \leq i \leq k} \varphi_k$. Hence using the postulated inference rules of λ we can derive the formula $\Diamond\psi \wedge \Diamond\neg\psi$ from

$\alpha_1, ..., \alpha_n$ in λ. Then applying the rule $\Diamond x \wedge \Diamond\neg x/y$ we derive β. Thus the rule $\alpha_1, ..., \alpha_n/\beta$ is derivable in $\lambda + \Diamond x \wedge \Diamond\neg x/y$. ∎

Thus the conjecture of J.Port holds not only for $S5$ itself but also for all modal logics above $S4.3$. Note however that if we consider not only the usual finite structural inference rules but also inference rules with infinite countable sets of formulas as premises then this reverses the situation: answers to all the questions similar to of J.Port's ones are negative. We illustrate this by an example below. Note that the notions of admissibility and derivability are directly transferable to inference rules with infinite premises.

Examples 4.3.23 *The rule*

$$r := \frac{(\bigwedge_{1 \leq i \leq n} \Diamond(p_i \wedge (\bigwedge_{j \leq n, j \neq i} \neg p_j))) \vee \Box x, n \in N}{\Box x}$$

is admissible but not derivable in the axiomatic system $S5 + \Diamond x \wedge \Diamond\neg x/y$. Hence this system is not structurally complete with respect to rules with infinite premises. Moreover, the premise of r can be satisfied in $S5$.

Proof. Suppose that the rule r is derivable in $S5 + \Diamond x \wedge \Diamond\neg x/y$. Then the conclusion of r is derivable in $S5 + \Diamond x \wedge \Diamond\neg x/y$ from a finite number of premises of r. Therefore, for some k, the rule

$$r_1 := \frac{\bigwedge_{1 \leq n \leq k}((\bigwedge_{1 \leq i \leq n} \Diamond(p_i \wedge (\bigwedge_{j \leq n, j \neq i} \neg p_j))) \vee \Box x)}{\Box x}$$

is also derivable in $S5 + \Diamond x \wedge \Diamond\neg x/y$. Then by Corollary 4.3.20 the rule r_1 is admissible in $S5$. But at the same time, using Lemma 4.3.15, it is easy to see that r is not admissible in $S5$, a contradiction. Hence r is not derivable in $S5 + \Diamond x \wedge \Diamond\neg x/y$.

In order to show that r is admissible in this system assume that $\Box\alpha \notin S5$. Since $S5$ has the fmp (see, for instance, Corollary 2.6.12) this means that $\Box\alpha$ is false in a frame $\mathcal{F}$ consisting of single finite cluster having, say, k elements. Under any valuation V of the variables of the rule r in $\mathcal{F}$ the formulas of the premise of r which have the numbers of variables n, for $n > k$, are false on any element of $\mathcal{F}$ with respect to V. Thus, for any substitution of certain formulas γ_i in place of the variables p_i in formulas of the premise of r and of the formula α in place of x, the resulting formulas which initially had numbers of variables greater than k are false in $\mathcal{F}$. Thus if a substitution turns the conclusion of the rule r in a formula which is not a theorem of $S5$ then this substitution also turns some formulas of the premise of r into formulas which are not theorems of $S5$. Thus r is admissible in $S5 + \Diamond x \wedge \Diamond\neg x/y$. Clearly that while the substitution $x := p \to p, p_i := \top$, all formulas in the premise of r turn to theorems of $S5$. ∎

We also note that although $S4.3$ extends $S4$ there are some inference rules which are admissible in $S4$ but not admissible in almost all extensions of $S4.3$. More precisely, we have the following example.

Examples 4.3.24 *The rule r with the premises*

$$\bigwedge_{0 \leq i \leq 2} \Diamond x_i \wedge \neg x_1 \wedge x_2 \wedge \neg \Diamond x_3 \wedge \neg x_3,$$
$$\bigwedge_{0 \leq i \leq 2} \Diamond x_i \wedge \neg x_2 \wedge x_1 \wedge \neg \Diamond x_3 \wedge \neg x_3,$$
$$\bigwedge_{1 \leq i \leq 2} \Diamond x_i \wedge \neg \Diamond x_0 \wedge x_1 \wedge x_2 \wedge x_3 \wedge \Diamond x_3$$

and the conclusion $\neg \Diamond x_0$ is admissible in $S4$ but is not admissible in any modal logic λ over $S4.3$ which has a two-element cluster as a λ-frame.

Proof. It is not hard to show that this rule is admissible in $S4$ using Theorem 2.4.1. That this rule is not admissible in any modal logic λ over $S4.3$ with pointed property follows easily from Lemma 4.3.15. ∎

Modal $S4.3$-logics form an representative but considerably narrow class of strong $S4$-logics. The strongest modal logics extending $S4$, as we know, are tabular logics. Recall that a modal logic λ (or a superintuitionistic logic λ) is tabular if there is a finite modal (pseudo-boolean) algebra $\mathfrak{B}$ such that $\lambda = \lambda(\mathfrak{B})$ or, equivalently, there is a finite frame $\mathcal{F}$ such that $\lambda = \lambda(\mathcal{F})$ (see the Stone Representation Theorem). Non-tabular logics are weaker than tabular logics in the sense that any non-tabular logic is a sublogic of a certain tabular logic. Moreover, there are some maximal non-tabular logics between tabular and non-tabular logics.

Definition 4.3.25 *A modal logic λ (a superintuitionistic logic λ) is pretabular if λ is non-tabular but any logic which is proper extension of λ is tabular.*

Proposition 4.3.26 *If λ is a non-tabular modal logic over $K4$ or a nontabular superintuitionistic logic then there is a pretabular logic λ_1 which extends λ.*

Proof. It is not hard to see that λ is a tabular modal logic over $K4$ iff, for some n, the formula

$$\phi_n := \bigvee_{1 \leq i,j \leq n, i \neq j} [(p_i \rightarrow p_j) \wedge (p_j \rightarrow p_i) \wedge \qquad (4.22)$$

$$\Box((p_i \rightarrow p_j) \wedge (p_j \rightarrow p_i))]$$

is a theorem of λ (see Theorem 2.2.15 which describes subdirectly irreducible modal algebras from $Var(K4)$ and Theorem 1.2.57 which says that any $Var(\lambda)$

is generated by own subdirectly irreducible algebras). A superintuitionistic logic λ is tabular iff, for some n, the formula

$$\psi_n := \bigvee_{1 \leq i,j \leq n, i \neq j} [(p_i \to p_j) \wedge (p_j \to p_i)] \tag{4.23}$$

is a theorem of λ (by Lemma 2.1.23 which describes all subdirectly irreducible pseudo-boolean algebras, and by the fact that $Var(\lambda)$ is generated by own subdirectly irreducible algebras). Therefore, applying Zorn's Lemma to the family of non-tabular modal (superintuitionistic) logics containing logic λ (partially ordered by set-theoretic inclusion), we obtain that this family has maximal elements, which are certain pretabular logics. ∎

The first description of pretabular superintuitionistic logics was given in the article L.L.Maksimova [91], it turned out that there are exactly three such logics. For modal logics over $S4$, it turned out that there are exactly five pretabular logics, a fact which was independently established by L.Maksimova in [92] and L.Esakia and V.Meskhi in [40]. In weaker logics the situation is the opposite: it has been shown by W.Blok [14] that among modal logics over $K4$ there are already 2^{ω} many pretabular logics. Below we first give a short proof of Maksimova-Esakia-Meskhi theorem and then study the bases of admissible rules for pretabular logics using their described semantics.

We define the modal logics which will be shown to be pretabular.

$PT_1 := \lambda(\{\mathcal{F} \mid \mathcal{F}$ is a rooted finite linearly ordered frame$\})$;

$PT_2 := \lambda(\{\mathcal{F} \mid \mathcal{F}$ is a rooted finite poset of depth 2 $\})$;

$PT_3 := \lambda(\{\mathcal{F} \mid \mathcal{F}$ is a rooted finite poset of depth 3 which has single greatest element $\})$;

$PT_4 := \lambda(\{\mathcal{F} \mid \mathcal{F}$ is a rooted finite $S4$-frame of depth 2 with a single greatest single-element cluster $\})$;

$PT_5 := \lambda(\{\mathcal{F} \mid \mathcal{F}$ is a finite cluster $\})$, i.e. $PT_5 = S5$.

Exercise 4.3.27 *Show that the logics $PT_1, ..., PT_5$ are incomparable and that any logic PT_i is not tabular.*

Hint: To verify that none of the logics $PT_1, ..., PT_5$ are tabular see propositions placed near relations (4.22) and (4.23). To show that these logics are not comparable, use finite frames from the semantic description of logics PT_i above and the appropriate detaching formulas.

Lemma 4.3.28 *All the logics $PT_1, ..., PT_5$ are pretabular.*

Proof. Let a logic λ be a proper extension of some PT_i, $i = 1, ..., 5$. If λ is not a logic of finite depth then by Lemma 2.8.20 λ is a sublogic of some logic $\lambda(\mathfrak{B}))$, where $G(\mathfrak{B})$ has an infinite linearly ordered subalgebra. Then (see Lemma 2.7.18 and Lemma 2.7.17) any topo-boolean algebra $S(L_n)$ (where L_n is the linearly-ordered pseudo-boolean algebra consisting of n elements) is a subalgebra of $S(G(\mathfrak{B}))$ and $\mathfrak{B}$. Therefore $\lambda \subseteq \lambda(\{S(L_n) \mid n \in N\}) = PT_1$, a contradiction.

If λ is a logic of finite depth then by Corollary 2.8.17 λ has the fmp and consequently is generated by the set of all finite λ-frames. All finite λ-frames must be PT_i-frames.

We claim that any finite rooted PT_i-frame $\mathcal{F}_1$ is an open subframe of a frame from the set of finite frames from the definition of PT_i. Indeed, we use Birkhoff's Representation Theorem for varieties (Theorem 1.2.24), and have that

$$\mathcal{F}_1^+ \in HS \prod_{i \in I} \mathcal{F}_i^+,$$

i.e. $\mathcal{F}_1^+$ is a homomorphic image of a subalgebra of $\prod_{i \in I} \mathcal{F}_i^+$ (where $\mathcal{F}_i$ are certain frames from the definition of PT_i). Since $\mathcal{F}_1^+$ is finite we can assume that the direct product above has only finitely many components, i.e. I is finite. Therefore using Lemmas 2.5.21, 2.5.23 and 2.5.7 we infer that $\mathcal{F}_1$ is a rooted open subframe of a p-morphic image of some frame $\mathcal{F}_i$. Since all p-morphic images of frames $\mathcal{F}_i$ have the structure similar to $\mathcal{F}_i$ or are open subframes of such frames, it follows that $\mathcal{F}_1$ is a rooted open subframe of a frame of kind $\mathcal{F}_i$.

If a rooted finite frame $\mathcal{F}$ from the semantic definition of PT_i is not a λ-frame then there are only finitely many rooted finite λ-frames. This follows from the fact that $\mathcal{F}$ is a p-morphic image of all finite rooted PT_i-frames except a finite set of such frames (to show this we again use the fact that any rooted finite PT_i-frame is on open subframe of a finite frame $\mathcal{F}_i$ from the set of frames generating PT_i). Therefore λ is tabular. For otherwise, all the finite rooted PT_i-frames are λ-frames and we obtain $\lambda = PT_i$. Hence any logic PT_i is pretabular. ∎

Definition 4.3.29 *Let $\lambda := \langle W, R \rangle$ be a certain S4-frame. We say that the external branching of $\mathcal{F}$ is equal to n if n is the maximal number of R-maximal R-incomparable clusters in $\mathcal{F}$ which are immediate R-successors for some cluster of $\mathcal{F}$.*

Definition 4.3.30 *Let $\lambda := \langle W, R \rangle$ be a S4-frame. The internal branching of $\mathcal{F}$ is equal to n if n is the maximal number of non R-maximal R-incomparable clusters in $\mathcal{F}$ which are immediate R-successors for some cluster of $\mathcal{F}$.*

Now we are in a position to prove our final theorem describing the pretabular modal logics over $S4$.

Theorem 4.3.31 *Maksimova [92], Esakia and Meskhi [40]. The logics of the list $PT_1,..., PT_5$ are exactly all the pretabular logics among the modal logics extending Lewis's system S4.*

Proof. Suppose λ is a pretabular logic. Assume that λ is not a logic of finite depth. Then by Lemma 2.8.20 λ is a sublogic of a certain logic $\lambda(\mathfrak{B}))$, where the pseudo-boolean algebra $G(\mathfrak{B})$ has an infinite linearly ordered subalgebra. Then (see Lemma 2.7.18 and Lemma 2.7.17) any modal algebra $S(L_n)$ (where L_n is the linearly ordered poset with n elements) is a subalgebra of the algebra $S(G(\mathfrak{B}))$ and the algebra $\mathfrak{B}$. This yields $\lambda \subseteq \lambda(\{S(L_n) \mid n \in N\})=PT_1$, hence $PT_1 = \lambda$ since PT_1 is not tabular.

Otherwise λ is a logic of some finite depth n and by Theorem 2.8.22 λ is locally finite, and, consequently, has the fmp. First, suppose that the number of elements in maximal clusters of finite λ-frames is unbounded. Then $\lambda \subseteq PT_5 = S5$ and $\lambda = S5$. Now assume that the number of elements in non-maximal clusters of rooted finite λ-frames is unbounded. Then taking the corresponding open rooted subframes of such frames and contracting all elements above the root-cluster of the resulting frames into a single reflexive element by the corresponding p-morphisms, we obtain $\lambda \subseteq PT_4$ and $\lambda = PT_4$.

Suppose now that the number of elements in maximal clusters of finite λ-frames is bounded by n and the number of elements in non-maximal clusters of finite λ-frames is bounded by k. Assume that all finite λ-frames have unbounded external branching. Then taking the open rooted subframes of such frames and the p-morphisms contracting clusters into simply reflexive elements we obtain $\lambda \subseteq PT_2$ and $\lambda = PT_2$. Now suppose that all finite λ-frames have unbounded internal branching. Then taking the corresponding rooted open subframes of a depth of strictly more 2 and generating p-morphic images of such frames by

> (i) contracting all clusters into simply reflexive elements;
>
> (ii) contracting all elements above the immediate successors of the root-element of the frames obtained into a single reflexive element;
>
> (iii) contracting all the immediate successors of the root which have no immediate successor with the maximal element in the frames obtained

we have $\lambda \subseteq PT_3$ and $\lambda = PT_3$.

Finally, suppose that all rooted finite λ-frames have the external branching of not more than n and the internal branching of not more than n for some n. Then we claim that any rooted finite λ-frame $\mathcal{F}$ has not more than k elements for some fixed k. Indeed, the depth of $\mathcal{F}$ is bounded since λ is a logic of finite depth. Since the internal branching of $\mathcal{F}$ is bonded, the number of non-maximal clusters of $\mathcal{F}$ is effectively bounded. And, since the external branching of $\mathcal{F}$ is bounded, the number of maximal clusters of $\mathcal{F}$ is bounded. Thus we conclude that the number of clusters of $\mathcal{F}$ is bounded. We have also by assumption above that the number of elements in clusters of $\mathcal{F}$ is bounded as well. Hence the number of elements of any rooted finite λ-frame $\mathcal{F}$ is bounded by some fixed number k. This entails λ is tabular, a contradiction. Thus $PT_1, ..., PT_5$ are only pretabular logics extending $S4$. ∎

Using this result it is easy to describe the pretabular superintuitionistic logics.

Theorem 4.3.32 *Maksimova [91]. The only superintuitionistic pretabular logics are*

$\mathcal{L}_\infty := LC = \lambda(\{\mathcal{F} \mid \mathcal{F}$ *is a rooted finite linearly ordered frame*$\})$;

$\mathcal{L}_2 := \lambda(\{\mathcal{F} \mid \mathcal{F}$ *is a rooted finite poset of depth 2* $\})$;

$\mathcal{L}_3 := \lambda(\{\mathcal{F} \mid \mathcal{F}$ *is a rooted finite poset of depth 3 with a single greatest element* $\})$

Proof. It is easy to see that these logics are non-tabular (see proposition near (4.23)) and incomparable. Since PT_1, PT_2 and PT_3 have the fmp (see their definition above), using Corollary 2.7.21 it follows that $\sigma(\mathcal{L}_1) = PT_1$, $\sigma(\mathcal{L}_2) = PT_2$, and $\sigma(\mathcal{L}_3)) = PT_3$ (therefore, in particular, above mentioned superintuitionistic logics are incomparable). If λ is a proper extension of λ_1, were $\lambda_1 = \mathcal{L}_1, \mathcal{L}_2, \mathcal{L}_3$ then, since σ in one-to-one mapping which preserves the order (Theorem 2.7.20), we obtain $\sigma(\lambda) \supset PT_i$, $i = 1, 2, 3$. Then using Theorem 4.3.31 we conclude that $\sigma(\lambda)$ is tabular. Because $\rho(\sigma(\lambda)) = \lambda$ (cf. Theorem 2.7.20) and ρ preserves the property to be tabular (see Corollary 2.7.21), we have λ is tabular. Hence logics $\mathcal{L}_1, \mathcal{L}_2$ and $\mathcal{L}_3$ are pretabular.

Suppose λ is a pretabular superintuitionistic logic. Then $\sigma(\lambda)$ is also non-tabular (otherwise $\rho\sigma(\lambda) = \lambda$ would be tabular as well). By Theorem 4.3.31 it follows $\sigma(\lambda) \subseteq PT_i$ for some i. If $\sigma(\lambda) \subseteq PT_i$ and $i = 1, 2, 3$ then, since ρ preserves the order (cf. Theorem 2.7.20), we have

$$\rho\sigma(\lambda) = \lambda \subseteq \rho(PT_i) = \rho\sigma(calL_i) = \mathcal{L}_i,$$

what we needed. Suppose that $\sigma(\lambda) \not\subseteq PT_i$ for $i = 1, 2, 3$. Then, as we have showed in the proof of Theorem 4.3.31, $\sigma(\lambda)$ has a finite depth and has the fmp. Then (see Corollary 2.7.21) λ also has the fmp and also has a finite depth. Since $\sigma(\lambda) \not\subseteq PT_i$ for $i = 1, 2, 3$ we conclude $\lambda \not\subseteq L$ for $L = \mathcal{L}_1, \mathcal{L}_2, \mathcal{L}_3$. Therefore all rooted finite λ-frames have bounded depth, have bounded external branching, and bounded internal branching. This implies that there are only finitely many non-isomorphic rooted finite λ-frames, and consequently λ is tabular, a contradiction. ∎

Now we investigate the question concerning bases of admissible rules in the pretabular logics extending $S4$. Since we now know the semantic description of these logics, and we have proved necessary tool's theorems, this can be done relatively simply.

Theorem 4.3.33 *[133] The following hold*

(i) *The logics PT_2 and PT_3 have no finite bases for admissible inference rules (even bases in finitely many variables);*

(ii) *The logics PT_1, PT_4 and PT_5 have finite bases for admissible inference rules;*

(iii) *The logics PT_1 and PT_4 are structurally complete, but the logic PT_5 and logics PT_2 and PT_3 are structurally incomplete.*

Proof. The logic PT_2 is a $(\omega, 2)$-logic (see definition in the previous section), has the fmp and the effective 1-drop points property, therefore by Theorem 4.2.10 PT_2 has no finite bases for admissible rules and has no bases in finitely many variables. The logic PT_3 is an $(3, \omega, 1, 3)$-logic, has the fmp and the effective 1-drop points property, therefore by Theorem 4.2.4 PT_3 has no finite bases for admissible rules and no bases in finitely many variables. Hence we have established (i). Any logic from the list PT_1, PT_4, PT_5 extends $S4.3$, therefore by Corollary 4.3.20 any logic of this list has a finite basis for admissible rules which consists of the single inference rule $\Diamond x \wedge \Diamond \neg x / y$, i.e. (ii) also holds.

The logic PT_5 is structurally incomplete. Indeed, the rule $\Diamond x \wedge \Diamond \neg x / y$ is admissible for PT_5 but the formula $\Box(\Diamond x \wedge \Diamond \neg x) \rightarrow y$ is not a theorem of $S5$ since it false in the two-elements cluster. Hence the rule $\Diamond x \wedge \Diamond \neg x / y$ is not derivable in $PT_5 = S5$. Since PT_2 and PT_3 have no finite bases for admissible rules they cannot be structurally complete. In order to show that PT_1 is structurally complete suppose that a rule $\alpha_1, ..., \alpha_n / \beta$ is not derivable in PT_1. Then $\Box(\bigwedge_{1 < i \leq n} \alpha_i) \rightarrow \beta \notin PT_1$. Then according to the definition of PT_1 there is a rooted linearly ordered finite frame $\mathcal{F}$ such that

$$\mathcal{F} \nVdash \Box(\bigwedge_{1 \leq i \leq n} \alpha_i) \rightarrow \beta.$$

Since $\mathcal{F}$ is linearly ordered, this entails that r is invalid in a finite open subframe $\mathcal{F}_1$ of $\mathcal{F}$. Since $\mathcal{F}_1$ is a poset, this implies that the premise of r is can be satisfied in the reflexive single-element frame $\mathcal{F}_e$. Therefore the frame $\mathcal{F}_1 \circ 1$ also invalidates the rule r. Applying Lemma 4.3.15 it follows that r is not admissible in PT_1. Hence PT_1 is structurally complete.

It remains to show that PT_4 is structurally complete. Again let $\alpha_1, ..., \alpha_n / \beta$ be a rule which is non-derivable in PT_4. Then (see Theorem 1.3.15)

$$\Box(\bigwedge_{1 \leq i \leq n} \alpha_i) \rightarrow \beta \notin PT_4.$$

By definition of PT_4 this entails that there is a rooted linearly quasi-ordered finite frame $\mathcal{F}$ which has the depth 2 and such that (a) its maximal cluster is a reflexive element, and (b)

$$\mathcal{F} \not\Vdash \Box(\bigwedge_{1 \leq i \leq n} \alpha_i) \to \beta.$$

Sine $\mathcal{F}$ is linearly quasi-ordered, this entails that r is invalid in a finite open subframe $\mathcal{F}_1$ of $\mathcal{F}$. Since $\mathcal{F}_1$ has a single-element maximal (by the accessibility relation R) cluster, this implies that the premise of r is can be satisfied in the reflexive single-element frame $\mathcal{F}_e$. Therefore the frame $\mathcal{F}_1 \circ 1$ also invalidates r. Using Lemma 4.3.15 and this fact we obtain that the rule r is not admissible in PT_4. Hence PT_4 is structurally complete. ∎

Now it is already quite simple, to answer questions concerning bases for admissible rules of pretabular superintuitionistic logics, as a corollary of the results presented.

Theorem 4.3.34 *[133]. The following hold*

(i) The pretabular superintuitionistic logics $\mathcal{L}_2$ and $\mathcal{L}_3$ have no finite bases for admissible inference rules and have no bases in finitely many variables;

(ii) The pretabular superintuitionistic logic $\mathcal{L}_1 := LC$ has a finite basis for admissible inference rules and is structurally complete.

Proof. The logic $\mathcal{L}_2$ is a superintuitionistic $(\omega, 2)$-logic (see definition in the previous section), has the fmp and the strong effective 1-drop points property, therefore by Theorem 4.2.10 $\mathcal{L}_2$ has no bases for admissible rules in finitely many variables. The logic $\mathcal{L}_3$ is a superintuitionistic $(\omega, 1, 3)$-logic, has the fmp and the strong effective 1-drop points property, therefore by Theorem 4.2.4 $\mathcal{L}_3$ also has no bases for admissible rules in finitely many variables, and we have established (i).

We know that the superintuitionistic logic LC has the fmp and is finitely axiomatizable (see Corollary 2.6.14). To show that LC is structurally complete suppose a rule $\alpha_1, ..., \alpha_n/\beta$ is not derivable in LC. Then by deduction theorem for H we have $\bigwedge_{1 \leq i \leq n} \alpha_i \to \beta \notin LC$. By the fmp for LC this entails that there is a rooted linearly ordered finite poset $\mathcal{F}$ such that

$$\mathcal{F} \not\Vdash \bigwedge_{1 \leq i \leq n} \alpha_i \to \beta.$$

This implies r is invalid in a rooted linearly ordered finite poset $\mathcal{F}_1$ (which is an open subframe of $\mathcal{F}$). Clearly that $\mathcal{F}_1$ is a p-morphic image of the frame of $Ch_{LC}(k)$ for some k. Therefore r is invalid in the frame of $Ch_{LC}(k)$. Applying

Theorem 3.5.8 and Lemma 3.4.2 we obtain that r is not admissible in LC. Hence LC is structurally complete and consequently has a finite basis for its admissible rules. ∎

It is noticeable that we have paid no attention here to those modal and super-intuitionistic logics which are stronger than pretabular logics, namely to tabular logics. However this is because we intend to consider bases of inference rules for tabular logics in the following two sections.

4.4   Bases for Valid Rules of Tabular Logics

In the previous sections we have already investigated various questions concerning structure of the bases for admissible inference rules in certain strong non-tabular logics. Tabular logics are the strongest logics in the sense that any non-tabular logic is a sublogic of an appropriate pretabular logic, and all the extensions of this pretabular logic are tabular logics.

For any tabular logic λ, there is always an algorithm for deciding the admissibility of inference rules in λ (see, for instance, Lemma 4.1.10). At the same time it is not so simple to determine whether there is a finite basis for admissible inference rules of a given tabular logic. On the one hand, any tabular algebraic logic which is generated by an algebra with the distributive lattice of congruences has a finite axiomatization by Baker's theorem [5]. At the same time, the quasi-varieties which are generated by such finite algebras do not always have finite bases for their quasi-identities. We will illustrate this below for certain finite algebras which are actively employed as semantic tools for modal and superintuitionistic logics.

As with question of whether there are finite bases for the admissible rules and for valid rules of tabular logics, this question is considerably more difficult to answer. Really, in this case we must consider the bases for quasi-identities not of a finite algebra itself, but the bases of quasi-identities for free algebras of countable rank from the varieties, which are generated by finite tuples of finite algebras. It turns out that, in a sense, in these different situations the answers are opposite. More precisely, there exist finite modal algebras and finite pseudo-boolean algebra which have no finite bases for their valid inference rules. At the same time, any tabular modal logic extending T and any tabular superintuitionistic logic have finite bases for their valid inference rules, and anyway there are tabular modal and superintuitionistic logics which do not have finite bases for admissible inference rules. In this section we will consider valid inference rules for tabular logics and then in the next section we will give examples of tabular logics without finite bases for admissible inference rules. Recall that an inference rule

$$r := \frac{\alpha_1, ..., \alpha_n}{\beta}$$

is said to be valid in an algebra $\mathfrak{B} \in Var(\lambda)$, where λ is an algebraic logic, iff the quasi-identity

$$q(r) := \alpha_1 = \top, ..., \alpha_n = \top \Rightarrow \beta = \top$$

is valid in $\mathfrak{B}$. This notion can be naturally extended to logics themselves.

Definition 4.4.1 *A rule r is valid in a logic λ if for any $\mathfrak{B} \in Var(\lambda)$, the rule r is valid in $\mathfrak{B}$.*

In other words, a rule r is valid in λ if the quasi-identity $q(r)$ is valid in the variety $Var(\lambda)$. Therefore it follows (see Theorem 1.4.5) that any valid inference rule is also admissible in λ for any algebraic logic λ. The converse does not hold, for instance the rule $\Diamond x \wedge \Diamond \neg x / y$ is admissible in $S5$ (see Corollary 4.3.20) but is not valid in $S5$. If a rule r is derivable in λ and all inference rules postulated for λ are valid in $Var(\lambda)$ then r is also valid in λ (since, in this case, derivability preserves the truth of formulas in algebras from $Var(\lambda)$). In particular, for all modal and superintuitionistic logics, all derivable rules are valid in these logics. Conversely, if a rule r is not derivable in a modal logic λ over $K4$, for example, then

$$\Box(\alpha_1 \wedge ... \wedge \alpha_n) \wedge \alpha_1 \wedge ... \wedge \alpha_n \rightarrow \beta$$

is not a theorem of λ. Then there is an algebra $\mathfrak{B} \in Var(\lambda)$ such that

$$\mathfrak{B} \not\models \Box(\alpha_1 \wedge ... \wedge \alpha_n) \wedge \alpha_1 \wedge ... \wedge \alpha_n \leq \beta.$$

Therefore for some interpretation V of the variables of the formula above,

$$\Box \bigwedge_{1 \leq i \leq n} V(\alpha_i) \wedge \bigwedge_{1 \leq i \leq n} V(\alpha_i) \not\leq V(\beta).$$

We take the filter ∇ generated by $\Box \bigwedge_{1 \leq i \leq n} V(\alpha_i) \wedge \bigwedge_{1 \leq i \leq n} V(\alpha_i)$. It is easy to see that ∇ is a $\Box$-filter. Therefore the quotient-algebra $\mathfrak{B}/\nabla$ will be an algebra from $Var(\lambda)$ and will invalidate the rule r. Hence any valid rule for a modal logic λ over $K4$ is derivable in λ. By means of similar argument it is easy to show that any rule r which is valid in a superintuitionistic logic λ is derivable in λ. Hence

Proposition 4.4.2 *For any modal logic λ extending $K4$ and for any superintuitionistic logic λ, a rule r is valid in λ iff r is derivable in λ.*

By Theorem 1.4.11, for any algebraic logic λ, a set $\mathcal{R}$ is a basis for the set of all inference rules valid in λ iff $\{q(r) \mid r \in \mathcal{R}\}$ and the set of all identities corresponding to the axioms of λ form a basis for the quasi-identities which are valid in $Var(\lambda)$. Therefore any tabular modal, or superintuitionistic logic λ has a finite basis for its valid inference rules. Namely, a finite set of axioms for λ provides a required finite basis.

The situation is quite opposite when it comes to the bases of valid inference rules of finite modal and pseudo-boolean algebras, but not tabular logics. The first examples of finite modal and pseudo-boolean algebras without finite bases for their valid inference rules were independently discovered in Dziobiak [35] and Rybakov [134]. Here we present examples from [134].

Let $\mathcal{A}$ be a finite algebra from $Var(\lambda)$, where λ is a modal or superintuitionistic logic. By Theorem 1.4.11 a set $\mathcal{R}$ is a basis in $\lambda(\mathcal{A})$ for the set of inference rules which are valid in $\mathcal{A}$ iff $\{q(r) \mid r \in \mathcal{R}\}$ and the set of all the identities corresponding to the axioms of $\lambda(\mathcal{A})$ form a basis for the quasi-identities which are valid in $\mathcal{A}$. Therefore in order to show that there is a finite modal or pseudo-boolean algebra without a finite bases for its valid inference rules it is sufficient to find a finite algebra which has no finite bases for its quasi-identities. And we now are going to construct some such examples.

We introduce the following poset $\mathcal{F}_{10}$ with 10 elements:

$$\mathcal{F}_{10} := \langle F_{10}, \leq \rangle, F_{10} := \{a_i \mid 1 \leq i \leq 5\} \cup \{b_i \mid 1 \leq i \leq 5\},$$
$$b_1 < a_1, b_1 < a_2, b_1 < a_3, b_2 < a_2, b_2 < a_3, b_3 < a_2, b_3 < a_3, b_3 < a_4,$$
$$b_4 < a_2, b_4 < a_3, b_4 < a_5, b_5 < a_1, b_5 < a_4, b_5 < a_5.$$

The structure of the frame $\mathcal{F}_{10}$ is depicted in Fig. 4.2. Thus $\mathcal{F}_{10}$ is a poset of depth 2.

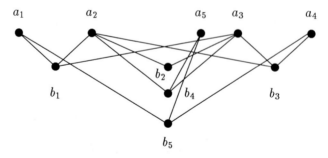

Figure 4.2: The structure of the frame $\mathcal{F}_{10}$.

Theorem 4.4.3 *[134] The quasi-variety generated by the finite modal algebra $\mathcal{F}_{10}^+$ does not have bases for quasi-identities in finitely many variables, in particular, it has no finite basis.*

Proof. In order to prove this theorem it is sufficient (by Lemma 4.1.1) to construct a sequence of modal algebras $\mathfrak{B}_n$, $n \in N$, such that any $\mathfrak{B}_n$ is not a member of $(\mathcal{F}_{10}^+)^Q$ but all n-generated subalgebras of $\mathfrak{B}_n$ are included in the quasi-variety $(\mathcal{F}_{10}^+)^Q$. Using this approach we define the posets $\mathcal{G}_n$ as follows:

$$\mathcal{G}_n := \langle A_n \cup Z, \leq \rangle, A_n := \{c_i \mid 1 \leq i \leq 2^n + 1\},$$
$$Z := \{X \mid X \subseteq A_n, \|X\| = 3\}, a \leq b \Leftrightarrow (a = b) \vee (a \in Z \& b \in a).$$

We will show that the sequence $\mathcal{G}_n^+$, $2 \leq n < \omega$ has the necessary properties. First we need the following lemma.

Lemma 4.4.4 $\mathcal{G}_n^+ \notin (\mathcal{F}_{10}^+)^Q$.

Proof. Suppose otherwise. Then by the theorem concerning the structure of quasi-variety generated by an algebra (see Theorem 1.2.24)

$$\mathcal{G}_n^+ \in S \prod_{i \in I} (\mathcal{F}_{10}^+)_i, \tag{4.24}$$

i.e. $\mathcal{G}_n^+$ is a subalgebra of a direct product of algebras $(\mathcal{F}_{10}^+)_i$ each of which is isomorphic to $\mathcal{F}_{10}^+$. Since $\mathcal{G}_n^+$ is finite we may suppose that I is finite. Therefore we can apply the duality between subalgebras and p-morphic images, and direct products and disjoint unions (see Lemmas 2.5.23 and 2.5.7). As the result we obtain

$$\mathcal{G}_n = f(\bigsqcup_{1 \leq i \leq n} \mathcal{A}_i, \ \mathcal{A}_i := \mathcal{F}_{10}).$$

where f is a p-morphism. Consider the f-image of i-th component $\mathcal{A}_i$ form $\bigsqcup_{1 \leq i \leq n} \mathcal{A}_i$. Since f preserves the order, it follows that $f(b_2) \leq f(a_2)$ and $f(b_2) \leq f(a_3)$. Suppose first that $f(a_2) \neq f(a_3)$. Then $f(b_2) \not\leq z$ when $z \neq f(a_2), f(a_3)$. But by the definition of $\mathcal{G}_n$, there is no element of depth 2 in $\mathcal{G}_n$ which has exactly two immediate successors. Therefore we conclude $f(a_2) = f(a_3)$.

Again since f preserves the order we obtain that $f(b_1) \leq f(a_1)$, $f(b_1) \leq f(a_2)$ and $f(b_1) \leq f(a_3)$. Suppose that $f(a_1) \neq f(a_2)$, then since $f(a_2) = f(a_3)$, it follows that $f(b_1) \not\leq z$ for all $z \neq f(a_1), f(a_2)$. Again, since by the definition of $\mathcal{G}_n$, there exists no element of depth 2 in $\mathcal{G}_n$ which has exactly two immediate successors, we obtain a contradiction. Hence $f(a_1) = f(a_2)$. By similar reasoning we show that $f(a_3) = f(a_4)$. Besides we have $f(b_4) \leq f(a_2)$, $f(b_4) \leq f(a_5)$ and $f(b_1) \leq f(a_3)$. Supposing $f(a_2) \neq f(a_5)$ and reasoning as above we again get a contradiction. Hence $f(a_2) = f(a_5)$. We have established that

$$\forall i, j, 1 \leq i \neq j \leq 5, f(a_j) = f(a_j).$$

Assume $x, y, z \in |\mathcal{G}_n|$ and $x \leq y, x \leq z$, and $y \neq z$. Then there is a certain $\mathcal{A}_i$ such that there exists an $a \in |\mathcal{A}_i|$ such that $f(a) = x$. Then according to the definition of p-morphisms, there are some elements $b, c \in |\mathcal{A}_i|$ such that $a \leq b$, $a \leq c$, and simultaneously $f(b) = y$ and $f(c) = z$. But $y \neq z$, consequently $b = a_i$ and $c = a_j$ for some $i \neq j$. At the same time we have showed above that $f(a_i) = f(a_j)$, a contradiction. ∎

We also need the following lemma.

Lemma 4.4.5 *Any n-generated subalgebra of $\mathcal{G}_n^+$ belongs to $(\mathcal{F}_{10}^+)^Q$.*

Proof. Suppose $\mathfrak{B}$ is a n-generated subalgebra of $\mathcal{G}_n^+$. By Lemma 2.5.23 $\mathfrak{B}^+$ is a p-morphic image of $\mathcal{G}_n^{++}$ with respect to a p-morphism, and $\mathcal{G}_n^{++}$ is isomorphic to $\mathcal{G}_n$ (see Theorem 2.5.6). Thus $f(\mathcal{G}_n) = \mathfrak{B}^+$, where f is a p-morphism. Since $\mathcal{G}_n$ is a poset, $\mathfrak{B}_n$ also must be a poset. Since $\mathfrak{B}$ is n-generated, the frame $\mathfrak{B}^+$ has not more than 2^n maximal elements. Indeed, maximal elements of $\mathfrak{B}^+$ are certain ultrafilters in $\mathfrak{B}$. Suppose that two such R-maximal ultrafilters ∇_1 and ∇_2 have the same valuation V of propositional variables p_{a_i}, $1 \leq i \leq n$, where a_i, $1 \leq i \leq n$ are the generators of $\mathfrak{B}$, and $\nabla_j \Vdash_V p_{a_i} \Leftrightarrow a_i \in \nabla_j$, $j = 1, 2$. By Lemma 2.5.17, for any formula $\varphi(p_{a_1}, ..., p_{a_n})$,

$$\nabla_j \Vdash_V \varphi(p_{a_1}, ..., p_{a_n}) \Leftrightarrow \varphi(a_1, ..., a_n) \in \nabla_j, \quad j = 1, 2.$$

Therefore since ∇_1 and ∇_2 are R-maximal ultrafilters with the same valuation by V of all letters p_{a_i}, $1 \leq i \leq n$, we obtain $\nabla_1 = \nabla_2$ since $a_1, ..., a_n$ generate the algebra $\mathfrak{B}$.

Therefore $||S_1(\mathfrak{B})|| \leq 2^n$, and there are elements c_i, c_j of depth 1 in $|\mathcal{G}_n|$ such that $f(c_i) = f(c_j)$ and $i \neq j$. Denote the frame obtained from $\mathcal{G}_n$ by contracting of elements c_i and c_j in a single element c_{ij} by $\mathcal{G}$. Clearly there is a p-morphism f_1 from $\mathcal{G}$ onto $\mathfrak{B}^+$. Namely, we let $f_1(c_{ij}) := f(c_i)$ and, for $c \neq c_{ij}$, we let $f_1(c) := f(c)$. By Lemma 2.5.20 $\mathfrak{B}$ is a subalgebra of the algebra $\mathcal{G}^+$. Therefore in order to complete the proof of our lemma it is sufficient to prove that $\mathcal{G}^+ \in (\mathcal{F}_{10}^+)^Q$.

According to Lemmas 2.5.20 and 2.5.27 it is sufficient to show that there is a p-morphism from $\bigsqcup_{1 \leq i \leq m} \mathcal{A}_i$, where $\mathcal{A}_i = \mathcal{F}_{10}$, onto $\mathcal{G}$. Denote by $\mathcal{M}$ the frame which is the p-morphic image of $\mathcal{F}_{10}$ with respect to the following p-morphism $g_1 \colon g_1(a_2) = g_1(b_2) = g_1(a_3)$, $g_1(x) \neq g_1(y)$ while $x \neq y$, $x, y \neq a_2, a_3, b_3$. It is not hard to see that there is a p-morphism g_2 of a frame of the form $\bigsqcup_{1 \leq i \leq k} \mathcal{N}_i$, where $\mathcal{N}_i = \mathcal{M}$, for some k, onto the frame $\mathcal{G}$. We know that $\mathcal{N}_i$ are p-morphic images of $\mathcal{F}_{10}$. Therefore the composition of p-morphisms g_1 and g_2 maps the frame $\bigsqcup_{1 \leq i \leq k} \mathcal{A}_i$ onto $\mathcal{G}$. The composition of p-morphisms is a p-morphism. Thus we conclude $\mathfrak{B} \in (\mathcal{F}_{10}^+)^Q$. ∎

Directly from Lemmas 4.4.4, 4.4.5 and 4.1.1 it follows that $(\mathcal{F}_{10}^+)^Q$ has no bases in finitely many variables for quasi-identities, in particular, no finite bases. The proof of Theorem 4.4.3 is complete. ∎

In particular, as we have seen above, this entails that the finite modal algebra $\mathcal{F}_{10}^+$ cannot have finite bases for valid inference rules. The frame $\mathcal{F}_{10}$ also gives an example of finite pseudo-boolean algebra which has no finite bases for its quasi-identities.

Theorem 4.4.6 *[134]. The quasi-variety generated by the pseudo-boolean algebra $G(\mathcal{F}_{10}^+)$ does not have bases for quasi-identities in finitely many variables, in particular, it has no finite bases.*

Proof. The proof will follow easily from the proof of Theorem 4.4.3 and relations connecting modal and pseudo-boolean algebras. We consider the sequence of pseudo-boolean algebras $G(\mathcal{G}_n^+)$, $n \in N$. We need to show that the sequence $G(\mathcal{G}_n^+)$, $2 \leq n < \omega$ has the properties required in Lemma 4.1.1. We claim that

$$G(\mathcal{G}_n^+) \notin G(\mathcal{F}_{10}^+)^Q.$$

Indeed, otherwise by the theorem describing the structure of the quasi-variety generated by an finite algebra (see Theorem 1.2.30)

$$G(\mathcal{G}_n^+) \in S \prod_{i \in I} G(\mathcal{F}_{10}^+)_i,$$

i.e. $G(\mathcal{G}_n^+)$ is a subalgebra of a direct product of certain pseudo-boolean algebras $G(\mathcal{F}_{10}^+)_i$ each of which is isomorphic to the algebra $G(\mathcal{F}_{10}^+)$. Since $G(\mathcal{G}_n^+)$ is finite we can assume that I is finite. So $G(\mathcal{G}_n^+)$ is a subalgebra of $\prod_{1 \leq i \leq m} G(\mathcal{F}_{10}^+)_i$. Then by Lemma 2.7.18 it follows that the modal algebra $S(G(\mathcal{G}_n^+))$ is a subalgebra of the modal algebra $S(\prod_{1 \leq i \leq m} G(\mathcal{F}_{10}^+)_i)$. It is clear that

$$\prod_{1 \leq i \leq m} G(\mathcal{F}_{10}^+)_i = G(\prod_{1 \leq i \leq m} (\mathcal{F}_{10}^+)_i).$$

Therefore again by Lemma 2.7.18 $S(\prod_{1 \leq i \leq m} G(\mathcal{F}_{10}^+)_i)$ is a subalgebra of the algebra $S(G(\prod_{1 \leq i \leq m} (\mathcal{F}_{10}^+)_i)$. Using Lemma 2.7.17 we conclude that the algebra $S(G(\prod_{1 \leq i \leq m} (\mathcal{F}_{10}^+)_i)$ is a subalgebra of $\prod_{1 \leq i \leq m} (\mathcal{F}_{10}^+)_i$. Thus $S(G(\mathcal{G}_n^+))$ is a subalgebra of $\prod_{1 \leq i \leq m} (\mathcal{F}_{10}^+)_i$ according to Lemma 2.7.17. Also using this lemma we conclude $\bar{S}(\bar{G}(\mathcal{G}_n^+))$ is a subalgebra of $\mathcal{G}_n^+$ generated by all its elements of the kind $\square a$. Since $\mathcal{G}_n$ is a finite poset, it is easy to see that such subalgebra coincides with $\mathcal{G}_n^+$. Thus $\mathcal{G}_n^+$ is a subalgebra of the algebra $\prod_{1 \leq i \leq m} (\mathcal{F}_{10}^+)_i$ which contradicts Lemma 4.4.4. Now we show that

Any n-generated subalgebra of $G(\mathcal{G}_n^+)$ belongs to $G(\mathcal{F}_{10}^+)^Q$.

Consider a certain n-generated subalgebra $\mathfrak{A}$ of $G(\mathcal{G}_n^+)$. Then the algebra $S(\mathfrak{A})$ is a n-generated subalgebra of $S(G(\mathcal{G}_n^+))$ by Lemmas 2.7.17 and 2.7.18. We have showed above that $S(G(\mathcal{G}_n^+))$ is isomorphic to $\mathcal{G}_n^+$. Therefore by Lemma 4.4.5 $S(\mathfrak{A})$ belongs to the quasi-variety $(\mathcal{F}_{10}^+)^Q$. Then by Theorem 1.2.30 $S(\mathfrak{A})$ is a subalgebra of a direct product of algebras $\mathcal{F}_{10}^+$. Since $S(\mathfrak{A})$ is finite we can assume that the product has finitely many components. Hence $S(\mathfrak{A})$ is a subalgebra of $\prod_{1 \leq i \leq m}(\mathcal{F}_{10}^+)_i$, $(\mathcal{F}_{10}^+)_i = \mathcal{F}_{10}^+$. Therefore $G(S(\mathfrak{A}))$ is a subalgebra of the algebra $\prod_{1 \leq i \leq m} G((\mathcal{F}_{10}^+)_i)$. Since $G(S(\mathfrak{A})) = \mathfrak{A}$ by Corollary 2.7.11, it follows that $\mathfrak{A}$ is a subalgebra of $\prod_{1 \leq i \leq m} G((\mathcal{F}_{10}^+)_i)$.

Now applying Lemma 4.1.1 we conclude that the quasi-variety generated by the algebra $G(\mathcal{F}_{10}^+)$ does not have bases for quasi-identities in finitely many variables. ∎

So finite pseudo-boolean algebras also do not always have finite bases for valid inference rules. Nevertheless, there are certain sufficient conditions for a finite modal or pseudo-boolean algebra to have a finite basis for valid inference rules, or equivalently for a finite topoboolean or pseudo-boolean algebra to have a finite basis for quasi-identities. Note that in algebraic logic there are several such conditions. For instance, in Block and Pigozzi [17] such a condition is given for algebras with the equationally definable principal congruence meet property. And an especially strong sufficient condition is given in Pigozzi [112]. Here we will demonstrate a weaker special case of such conditions. We choose to consider this case because it has a shorter and more straightforward proof, and concerns modal and pseudo-boolean algebras, the algebras which we actively use throughout all this book. Note also that this result is a consequence of a more general theorem from Blok and Pigozzi [17].

Theorem 4.4.7 *[134] For any finite tuple $\mathfrak{B}_1, ..., \mathfrak{B}_n$ of finite subdirectly irreducible modal algebras from $Var(T)$, the quasi-variety Q generated by the tuple of algebras $\mathfrak{B}_1, ... , \mathfrak{B}_n$ has a finite basis for quasi-identities.*

Proof. First, note that $\lambda(\{\mathfrak{B}_1, ..., \mathfrak{B}_n\})$ is a tabular modal logic, and therefore $Var(\lambda(\{\mathfrak{B}_1, ..., \mathfrak{B}_n\}))$ has a finite basis for identities (see Theorem 1.3.23). We also know (Theorem 2.2.20) that the finite subdirectly irreducible algebras from $Var(T)$ are exactly those that have the greatest element ω among their stable elements which differs from $\top$ (recall, a stable element a is an element a such that $\Box a = a$). Also we need the following lemma.

Lemma 4.4.8 *If $\mathfrak{B}$ is a finite subdirectly irreducible algebra from the variety $Var(\lambda(\{\mathfrak{B}_1, ..., \mathfrak{B}_n\})) \subseteq Var(T)$ then $||\mathfrak{B}|| \leq max(||\mathfrak{B}_1||, ..., ||\mathfrak{B}_n||)$.*

Proof. Since $\mathfrak{B} \in Var(\lambda(\{\mathfrak{B}_1, ..., \mathfrak{B}_n\})) \subseteq Var(T)$, it follows

$$\mathfrak{B} \models \bigvee_{1 < i \neq j \leq m+1} \Box^m (x_i \equiv x_j) = \top,$$
$$\mathfrak{B} \models \Box^{m+1} x = \Box^m x,$$

where $m := max(||\mathfrak{B}_1||, ..., ||\mathfrak{B}_n||)$. Suppose that $||\mathfrak{B}|| > m$. Then there are elements $a_i \in ||\mathfrak{B}||$, $1 \leq i \neq j \leq m + 1$, such that $a_i \equiv a_j \neq \top$. Then $\square^m(a_j \equiv a_j)$ are stable elements of $\mathfrak{B}$ and they all are distinct from $\top$. Since $\mathfrak{B}$ has a greatest element among those elements which are stable and distinct from $\top$, it follows that

$$\mathfrak{B} \not\models \bigvee_{1 \leq i \neq j \leq m+1} \square^m(x_i \equiv x_j) = \top,$$

a contradiction. ∎

Lemma 4.4.9 *The quasi-variety* $Q := \{\mathfrak{B}_1, ..., \mathfrak{B}_n\}^Q$ *has a basis for quasi-identities in finitely many variables.*

Proof. Let $m := max(||\mathfrak{B}_1||, ..., ||\mathfrak{B}_n||, k) + 1$, where k is the number of variables in a finite equational base for $Var(\lambda(\{\mathfrak{B}_1, ..., \mathfrak{B}_n\}))$ (which exists as we noted above). Suppose Q^m is the quasi-variety of all modal algebras which satisfy all quasi-identities in m variables which are valid in $\mathfrak{B}_1, ..., \mathfrak{B}_n$. In order to prove this lemma it is sufficient to show that $Q = Q^m$. It is clear that $Q \subseteq Q^m$.

Suppose that $Q^m \not\subseteq Q$. Then there is an algebra $\mathfrak{B} \in Q^m$ such that $\mathfrak{B} \notin Q$. Then there is a quasi-identity q which is built up out of n_1 variables, $n_1 > m$, such that $\mathfrak{B} \not\models q$ and $\mathfrak{B}_i \models q$, $1 \leq i \leq n$. Without loss of generality we may assume that $\mathfrak{B}$ is a n_1-generated algebra. Since $m > k$, it follows $\mathfrak{B}$ is a modal algebra from $Var(\lambda(\{\mathfrak{B}_1, ..., \mathfrak{B}_n\}))$. Therefore by Lemma 4.4.8 $\mathfrak{B}$ is finite. We will now decompose $\mathfrak{B}$ in a subdirect product of subdirectly irreducible algebras. The fact of the pure existence of such decomposition by Birkhoff's theorem gives no necessary information. We need to know the structure of subdirectly irreducible components more precisely.

Note that the set of all the stable elements of $\mathfrak{B}$ form a sublattice of the algebra $\mathfrak{B}$. Indeed, if x, y are stable elements then $\square(x \wedge y) = \square x \wedge \square y = x \wedge y$. Furthermore $\square(x \vee y) \leq (x \vee y)$, besides $\square(x \vee y) \geq \square x = x$, $\square(x \vee y) \geq \square y = y$. Consequently, $\square(x \vee y) = x \vee y$. Suppose that $\mathcal{L}$ is the lattice of all stable elements of $\mathfrak{B}$ and $\mathcal{L}_{ca}$ is the set of all co-atoms of $\mathcal{L}$ (recall, a is a co-atom if for all b, if $a < b \leq \top$ then $b = \top$). For any $a_i \in \mathcal{L}_{ca}$, $\nabla_{a_i} := \{x \mid x \in \mathcal{L}, x \vee a_i = \top\}$ is a proper filter in $\mathcal{L}$ since $a_i \notin \nabla_{a_i}$. The filter ∇_{a_i} is a prime filter in $\mathcal{L}$. Indeed, suppose that $x \vee y \in \nabla_{a_i}$. Then $a_i \vee (x \vee y) = (a_i \vee x) \vee y = \top$. Since a_i is a co-atom, either $a_i \vee x = \top$ and $x \in \nabla_{a_i}$, or $a_i \vee x = a_i$ and $y \in \nabla_{a_i}$. Denote by Δ_i the filter in $\mathfrak{B}$ generated by ∇_{a_i}. Then Δ_i is a proper $\square$-filter in $\mathfrak{B}$. In fact, if $a_i \in \Delta_i$ then $a_i \geq y \in \nabla_{a_i}$. Then $\square y = y \leq \square a_i = a_i$, consequently, $a_i \in \nabla_{a_i}$, a contradiction.

We will show that $\bigcap_{a_i \in \mathcal{L}_{ca}} \Delta_i = \{\top\}$. Suppose that x is an element of the mentioned intersection. Then $\square^m x \in \bigcap_{a_i \in \mathcal{L}_{ca}} \Delta_i$. Besides $\mathfrak{B} \models \square^m y = \square^{m+1} y$, therefore $\square^m x$ is a stable element of $\mathfrak{B}$, i.e. $\square^m x \in \mathcal{L}$. Suppose $x \neq \top$. Then $\square^m x \neq \top$. Since $\square^m x \in \bigcap_{a_i \in \mathcal{L}_{ca}} \Delta_i$, we get $\square^m x \in \nabla_{a_i}$ for all ∇_{a_i}. Thus

$a_i \vee \Box x^m = \top$ for all co-atoms a_i. Therefore $\Box^m x = \top$, a contradiction. Hence $\bigcap_{a_i \in \mathcal{L}_{ca}} \Delta_i = \{\top\}$ holds.

Consequently, the family of all the congruences $\sim_{\Delta_i}$ corresponding to all $\Box$-filters Δ_i is a detaching family of congruences. Then by Birkhoff's theorem (Theorem 1.2.55) the algebra $\mathfrak{B}$ is a subdirect product of the quotient-algebras $\mathfrak{B}/_{\sim_{\Delta_i}}$. We claim that any algebra $\mathfrak{B}/_{\sim_{\Delta_i}}$ is subdirectly irreducible. Indeed, $a_i \notin \Delta_i$, therefore $a_i/_{\sim_{\Delta_i}} \neq \top$. Moreover, $a_i/_{\sim_{\Delta_i}}$ being a homomorphic image of a stable element also is stable. Consider an arbitrary stable element $x/_{\sim_{\Delta_i}}$ of algebra $\mathfrak{B}/_{\sim_{\Delta_i}}$. The element $\Box^m x$ is stable in $\mathfrak{B}$ and

$$\Box^m x \vee a_i \in \Delta_i \Leftrightarrow \Box x^m \vee a_i/_{\sim_{\Delta_i}} = \top \Leftrightarrow \Box x^m/_{\sim_{\Delta_i}} \vee a_i/_{\sim_{\Delta_i}} = \top.$$

Besides, $\Box^m x \vee a_i \in \Delta_i \Leftrightarrow \Box^m x \vee a_i \in \nabla_{a_i}$. Since ∇_{a_i} is a proper prime filter, we conclude

$$\Box^m x \vee a_i \in \nabla_{a_i} \Leftrightarrow (\Box^m x \in \nabla_{a_i}) \vee (a_i \in \nabla_{a_i}).$$

But we know that $a_i \notin \nabla_{a_i}$, therefore

$$\Box x^m/_{\sim_{\Delta_i}} \vee a_i/_{\sim_{\Delta_i}} = \top \Leftrightarrow \Box^m x \in \nabla_{a_i} \Leftrightarrow \Box^m x/_{\sim_{\Delta_i}} = \top.$$

Since $x/_{\sim_{\Delta_i}}$ is a stable element in $\mathfrak{B}/_{\sim_{\Delta_i}}$, we obtain $\Box^m x/_{\sim_{\Delta_i}} = x/_{\sim_{\Delta_i}}$. Hence

$$x/_{\sim_{\Delta_i}} \vee a_i/_{\sim_{\Delta_i}} = \top \Leftrightarrow x/_{\sim_{\Delta_i}} = \top.$$

Consequently $a_i/_{\sim_{\Delta_i}}$ is the greatest stable element of $\mathfrak{B}/_{\sim_{\Delta_i}}$ among stable elements distinct from $\top$. Then by Theorem 2.2.20 the algebra $\mathfrak{B}/_{\sim_{\Delta_i}}$ is subdirectly irreducible.

Applying Lemma 4.4.8 it follows $\|\mathfrak{B}/_{\sim_{\Delta_i}}\| \leq m - 1$. Consider the quasi-identity q. We may assume that $q := \alpha(x_1, ..., x_{n_1}) = \top \Rightarrow \beta(x_1, ..., x_{n_1}) = \top$. Since $\mathfrak{B} \not\models q$ and $\mathfrak{B}$, being a subdirect product of algebras $\mathfrak{B}/_{\sim_{\Delta_i}}$, is a subalgebra of the direct product of these algebras, there is an algebra $\mathfrak{B}/_{\sim_{\Delta_i}}$ such that $\mathfrak{B}/_{\sim_{\Delta_i}} \not\models q$. Then

$$\mathfrak{B}/_{\sim_{\Delta_i}} \not\models \Box^m \alpha(x_1, ..., x_{n_1}) = \top \Rightarrow \Box^m \beta(x_1, ..., x_{n_1}) = \top.$$

Since $\|\mathfrak{B}/_{\sim_{\Delta_i}}\| \leq m - 1$,

$$\mathfrak{B}/_{\sim_{\Delta_i}} \not\models \Box^m \alpha_1(y_1, ..., y_h) = \top \Rightarrow \Box^m \beta_1(y_1, ..., y_h) = \top,$$

where $h \leq m - 1$ and formulas α_1 and β_1 are obtained from α and β by some substitution of variables $y_1, ..., y_h$ in place of variables $x_1, ..., x_{n_1}$. Hence there are elements $b_i \in \beta$, $1 \leq i \leq h$ such that

$$\Box^m \alpha_1(b_1/_{\sim_{\Delta_i}}, ..., b_h/_{\sim_{\Delta_i}}) = \top,$$
$$\Box^m \beta_1(b_1/_{\sim_{\Delta_i}}, ..., b_h/_{\sim_{\Delta_i}}) \neq \top.$$

Hence $\square^m \alpha_1(b_1, ..., b_h) \in \Delta_i$ and $\square^m \beta_1(b_1, ..., b_h) \notin \Delta_i$. Because we know that elements $\square^m \alpha_1(b_1, ..., b_h)$ and $\square^m \beta_1(b_1, ..., b_h)$ are stable elements of $\mathfrak{B}$, we obtain $\square^m \alpha_1(b_1, ..., b_h) \in \nabla_{a_i}$ and $\square^m \beta_1(b_1, ..., b_h) \notin \nabla_{a_i}$. Then according to the definition of ∇_{a_i}, it follows that $\square^m \alpha_1(b_1, ..., b_h) \vee a_i = \top$ and $\square^m \beta_1(b_1, ..., b_h) \vee a_i \neq \top$. Hence we have proved

$$\mathfrak{B} \not\models \square^m \alpha_1(y_1, ..., y_h) \vee \square^m y = \top \Rightarrow \square^m \beta_1(y_1, ..., y_h) \vee \square^m y = \top.$$

Algebras $\mathfrak{B}_1, ..., \mathfrak{B}_n$ were supposed to be subdirectly irreducible, and therefore any this algebra has a greatest element ω among stable elements which differ from $\top$. Therefore, since q is valid in algebras $\mathfrak{B}_1, ..., \mathfrak{B}_n$, we obtain

$$\forall i, 1 \leq i \leq n, \mathfrak{B}_i \models \square^m \alpha_1(y_1, ..., y_h) \vee \square^m y = \top \Rightarrow$$
$$\square^m \beta_1(y_1, ..., y_h) \vee \square^m y = \top.$$

The quasi-identity above has not more m than variables. Therefore we obtain a contradiction the fact that $\mathfrak{B} \in Q^m$. ∎

Now we are in a position to complete the proof of our theorem. By Lemma 4.4.9 Q coincides with a quasi-variety Q^m with a quasi-equational basis B in m variables. Consider any quasi-identity q in m variables. Since the free algebra $\mathfrak{F}(m)$ of rank m from $Var(\lambda(\{\mathfrak{B}_1, ..., \mathfrak{B}_n\}))$ is finite (see Theorem 1.2.36), there are only finitely many modal formulas in m variables which are non-equivalent in the logic $\lambda(\{\mathfrak{B}_1, ..., \mathfrak{B}_n\})$. Therefore among the quasi-identities of B there are only finitely many non-equivalent in $Var(\lambda(\{\mathfrak{B}_1, ..., \mathfrak{B}_n\}))$. Therefore picking a representative in any class of quasi-identities from B which are equivalent in $Var(\lambda(\{\mathfrak{B}_1, ..., \mathfrak{B}_n\}))$, and adding to them a finite basis of identities for $\{\mathfrak{B}_1, ..., \mathfrak{B}_n\}$ we obtain a finite basis for the quasi-variety Q. ∎

Note that it is not difficult to obtain a similar result for pseudo-boolean algebras. Indeed, it is possible to transform the proof of the previous theorem into the proof of the similar result for subdirectly irreducible pseudo-boolean algebras making word by word a few appropriate modifications. However we prefer to give a proof in a slightly different form which is a little bit shorter (but which uses the same ideas and tools). Note that a proof of this result in a style which is similar to used below was first offered by A.Citkin in 1977.

Theorem 4.4.10 *(A.Citkin). For a finite set $\mathfrak{A}_1, ..., \mathfrak{A}_n$ of any finite subdirectly irreducible pseudo-boolean algebras, the quasi-variety Q generated by $\mathfrak{B}_1, ..., \mathfrak{B}_n$ has a finite basis for quasi-identities.*

Proof. Let $\mathcal{U}$ be the set of all quasi-identities of the kind

$$\alpha_i(x_1, ..., x_m) \vee y = \top \Rightarrow \beta_i(x_1, ..., x_m) \vee y = \top$$

which are valid in the algebras $\mathfrak{A}_1, ..., \mathfrak{A}_n$, where $m := max(||\mathfrak{A}_1||, ..., ||\mathfrak{A}_n||)$. Since the free algebra of rank $m + 1$ from $Var(\lambda(\mathfrak{A}_1, ..., \mathfrak{A}_n))$ is finite (see Theorem 1.2.36), there are only finitely many distinct quasi-identities from $\mathcal{U}$ by modulo of semantic equivalence in $Var(\lambda(\mathfrak{A}_1, ..., \mathfrak{A}_n))$. We denote by $\mathcal{W}$ a maximal set of non-equivalent quasi-identities from $\mathcal{U}$. We will show that the set $\mathcal{B} := \mathcal{W} \cup \mathcal{E}$, where $\mathcal{E}$ is a finite equational basis for the algebras $\mathfrak{A}_1, ..., \mathfrak{A}_n$ (which exists by Theorem 1.3.23), is a finite basis for quasi-identities of $\mathfrak{A}_1, ..., \mathfrak{A}_n$. Clearly, all the formulas from $\mathcal{B}$ are valid in $\mathfrak{A}_1, ..., \mathfrak{A}_n$.

Suppose q is a quasi-identity which is valid in $\mathfrak{A}_1, ..., \mathfrak{A}_n$ but is not a consequence of $\mathcal{B}$. We may assume that q has the form

$$\alpha(x_1, ..., x_u) = \top \Rightarrow \beta(x_1, ..., x_u) = \top.$$

Then there is a finitely generated pseudo-boolean algebra $\mathfrak{A}$ from the variety $Var(\lambda(\mathfrak{A}_1, ..., \mathfrak{A}_n))$ which detaches q from $\mathcal{B}$, i.e. $\mathfrak{A} \not\models q$, and for all $\phi \in \mathcal{B}$, $\mathfrak{A} \models \phi$. We can assume that $\mathfrak{A}$ is finitely generated, and then it follows by Theorem 1.2.36 that $\mathfrak{A}$ is finite. We will decompose $\mathfrak{A}$ into a subdirect product in a special way. First we take the antichain $\omega_1, ..., \omega_k$ of all maximal union-irreducible elements of $\mathfrak{A}$. It is not hard to see that $\omega_1 \vee ... \vee \omega_k = \top$. Moreover, since all ω_i are union-irreducible, for any j, $\bigvee_{i \neq j} \omega_i \neq \top$.

Any element ω_i generates a filter ∇_i and the corresponding congruence relation $\sim_i$ (see Lemma 2.1.19). Since there is an isomorphism between the lattice of all congruences in $\mathfrak{A}$ and the lattice of all filters in $\mathfrak{A}$, we obtain that $\bigcap_i \sim_i$ is the smallest (trivial) congruence relation. Therefore by Theorem 1.2.55 $\mathfrak{A}$ is a subdirect product of quotient-algebras $\mathfrak{A}/_{\sim_i}$. Moreover, any algebra $\mathfrak{A}/_{\sim_i}$ is subdirectly irreducible since ω_i is union-irreducible and Theorem 2.1.23. In particular, it follows that q is false in some $\mathfrak{A}/_{\sim_j}$. We also need the following lemma.

Lemma 4.4.11 *If $\mathfrak{A}$ is a finite subdirectly irreducible pseudo-boolean algebra from $Var(\lambda(\mathfrak{A}_1, ..., \mathfrak{A}_n))$ then $||\mathfrak{A}|| \leq max(||\mathfrak{A}_1||, ..., ||\mathfrak{A}_n||)$.*

Proof. From $\mathfrak{A} \in Var(\lambda(\{\mathfrak{A}_1, ..., \mathfrak{A}_n\})) \subseteq Var(H)$, we infer

$$\mathfrak{A} \models \bigvee_{1 \leq i \neq j \leq m+1} (x_i \equiv x_j) = \top,$$

where $m := max(||\mathfrak{B}_1||, ..., ||\mathfrak{B}_n||)$. Suppose that $||\mathfrak{A}|| > m$. Then there are elements $a_i \in ||\mathfrak{A}||$, $1 \leq i \neq j \leq m + 1$, such that $a_i \equiv a_j \neq \top$. Then any $(a_j \equiv a_j)$ is distinct from $\top$. Since $\mathfrak{A}$ is subdirectly irreducible, by Theorem 2.1.23 there is a greatest element in $\mathfrak{A}$ among distinct from $\top$. Hence we it follows that

$$\mathfrak{A} \not\models \bigvee_{1 \leq i \neq j \leq m+1} (x_i \equiv x_j) = \top,$$

a contradiction. ∎

Thus, being a subdirectly irreducible algebra from $Var(\lambda(\mathfrak{A}_1, ..., \mathfrak{A}_n))$, $\mathfrak{A}/{\sim_j}$ has not more than m elements. Therefore the quasi-identity

$$q_1 := \alpha(s(x_1), ..., s(x_u)) = \top \Rightarrow \beta(s(x_1), ..., s(x_u)) = \top,$$

where s is a mapping of $x_1, ..., x_u$ into the new variables $y_1, ..., y_m$, is false in $\mathfrak{A}_{\sim_j}$ under the interpretation $s(x_i) \mapsto b_i$. Then it is not hard to see that the quasi-identity

$$q_2 := \alpha(s(x_1), ..., s(x_u)) \vee y = \top \Rightarrow \beta(s(x_1), ..., s(x_u)) \vee y = \top$$

is false in $\mathfrak{A}$. In fact, we take the valuation ψ of all letters $s(x_i)$ in $\mathfrak{A}$ as some elements a_i which have π_j-projection on $\mathfrak{B}_{\sim_j}$ equal to b_i (all such elements exist since $\mathfrak{A}$ is a subdirect product of the algebras $\mathfrak{A}_j$), and $\psi(y) := \bigvee_{i \neq j} \omega_i$. Then the π_j-projection on $\mathfrak{A}/{\sim_j}$ of the element $\beta(\psi s(x_1), ..., \psi s(x_u)) \vee \psi(y)$ take the value distinct from $\top$ since $\mathfrak{B}/{\sim_j}$ is subdirectly irreducible and has a greatest element among elements distinct from $\top$. Thus the conclusion of q_2 is false in $\mathfrak{A}$ under ψ. At the same time, π_j-projection of $\alpha(\psi s(x_1), ..., \psi s(x_u)))$ is equal to $\top$, consequently

$$\alpha(\psi s(x_1), ..., \psi s(x_u))) \geq \omega_j.$$

Therefore

$$\alpha(\psi s(x_1), ..., \psi s(x_u))) \vee \psi(y) \geq \bigvee_i \omega_i = \top.$$

Hence q_2 is false in $\mathfrak{A}$. Clearly q_2 is valid in all algebras $\mathfrak{A}_1, ..., \mathfrak{A}_n$ since q and q_1 where valid in all $\mathfrak{A}_i$ and all $\mathfrak{A}_i$, being subdirectly irreducible, have certain greatest elements among those elements which are distinct from $\top$. Furthermore, q_2 has not more than $m + 1$ variables, which contradicts our choice of $\mathcal{B}$. Hence $\mathcal{B}$ is a required finite base. ∎

Note that any tabular logic is generated by a finite tuple of finite subdirectly irreducible algebras. We have showed that any finite tuple of finite subdirectly irreducible modal T-algebras and any similar tuple of pseudo-boolean algebras have finite bases for valid inference rules. At the same time we have presented examples of finite algebras without finite bases for their valid rules. Moreover, we know, for any tabular modal $K4$-logic λ and any tabular superintuitionistic logic λ, there is a finite basis for the inference rules valid in λ. However this is not contradictory since the set of valid inference rules of a tabular logic λ is the set of all rules which are valid in all algebras from $Var(\lambda)$, i.e. is an invariant. And the set of rules which are valid in a finite algebra $\mathfrak{A}$ from $Var(\lambda)$ generating $Var(\lambda)$ is not an invariant set and depends on the choice of $\mathfrak{A}$. In particular, we see that, for any superintuitionistic tabular logic λ, there is always a finite number of finite pseudo-boolean algebras generating λ for which there is a finite basis for valid inference rules. Nevertheless sometimes it is possible to find a finite pseudo-boolean algebra $\mathfrak{A}$ which has no bases in finitely many variables for valid inference rules. All of this is a consequence of the fact that for quasi-identities which are valid in an algebra, their truth is not always preserved under homomorphic images. Therefore the rules which are valid in the algebras generating logic λ can be false in other algebras from $Var(\lambda)$.

4.5 Tabular Logics without Independent Bases

In this section we answer the question of whether there exist finite bases for admissible rules in tabular logics. In the previous sections we studied the bases for various weak modal and superintuitionistic logics, and the bases for certain strong logics, in particular, for pretabular logics. But we did not consider yet the question about admissible rules for tabular logics. We do this here by giving counterexamples, that is we present examples of a tabular modal logic and a tabular superintuitionistic logic which have no finite bases for their admissible rules (more precisely, which have no bases in finitely many variables). We will also show that they do not have even independent bases for their admissible rules. We already have provided in Section 1 of this chapter all general technical lemmas which we will need.

According to Theorem 1.4.15 and Lemma 4.1.10 in order to find tabular logics without finite bases for their admissible rules it is sufficient to find some finite modal (pseudo-boolean) algebra $\mathfrak{A}$ with m elements such that $\mathfrak{F}_m(\mathfrak{A}^Q)$ has no finite bases for quasi-identities. We will use this approach to find the counterexamples mentioned above.

In fact, both modal and superintuitionistic logics without finite bases for admissible inference rules will be generated by the same rooted finite partially ordered frame, namely by the frame depicted below in Fig. 4.3. The formal definition of this frame is as follows: the frame $\mathcal{K}_6$ is the rooted poset of depth 3 such that:

$$\mathcal{K}_6 := \langle K_6, \leq \rangle, K_6 := \{e_1, u_1, u_2, w_1, w_2, w_3 \,|\},$$
$$e_1 \leq u_1, e_1 \leq u_2, e_1 \leq w_3,$$
$$u_1 \leq w_1, u_1 \leq w_2, u_2 \leq w_1, u_2 \leq w_2,$$

We will apply Lemmas 4.1.1, 4.1.2, and Lemma 4.1.3 in order to show that the modal logic and the superintuitionistic logic generated by $\mathcal{K}_6$ have no basses in finitely many variables for admissible rules, which by Lemma 4.1.10 is equivalent to the free algebra $\mathfrak{F}_{\lambda(\mathcal{K}_6)}(6)$ from the variety of modal or, respectively, pseudo-boolean, algebras has no bases in finitely many variables for quasi-identities. We will use Lemma 4.1.3, therefore we need a sequence of finite algebras with the

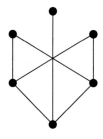

$$\mathcal{K}_6$$

Figure 4.3: The structure of frame $\mathcal{K}_6$

properties stated in Lemma 4.1.3. We introduce this sequence by writing out a sequence of finite frames each of which is a poset. The frame $\mathcal{F}_n$ is a poset of the following structure: it is the poset of depth 3 such that

$$S_1(\mathcal{F}_n) := \{a_1, a_2, a_3\},$$
$$Sl_2(\mathcal{F}_n) := \{b_{i,j} \mid i \in \{1, 2, ..., 2^n + 1\}, \ j \in \{1, 2, 3\}\},$$
$$Sl_3(\mathcal{F}_n) := \{c_{x,y} \mid x = b_{i,j}, \ y = b_{m,j}, \ i \neq m, \ j \in \{1, 2, 3\}\} \cup$$
$$\cup \{d_{x,y,z} \mid x = b_{i,j}, \ y = b_{m,j}, \ i \neq m, \ j \in \{1, 2, 3\}, \ z = a_j\}.$$

The accessibility relation between the elements of distinct slices in $\mathcal{F}_n$ is as follows:

$$b_{i,j} < a_r, \text{where } r \in (\{1, 2, 3\} - \{j\}),$$

$$c_{x,y} < x, \ c_{x,y} < y, \ d_{x,y,z} < x, \ d_{x,y,z} < y, \ d_{x,y,z} < z.$$

(of course the partial order $\leq$ in this frame is the transitive reflexive closure of the relations mentioned). The structure of the frame $\mathcal{F}_n$ is sketched in Fig. 4.4 below. For simplicity of notation, for any algebra $\mathfrak{A}$, we will denote the variety generated by the algebra $\mathfrak{A}$ by $\mathfrak{A}^V$. The modal algebra generated by $\mathcal{K}_6$ and the pseudo-boolean algebra generated by $\mathcal{K}_6$ will be denoted by $\mathcal{K}_6^+$, we will make it clear which object we are dealing with whenever we need to.

Lemma 4.5.1 *For any n, the modal algebra $\mathcal{F}_n^+$ (pseudo-boolean algebra $\mathcal{F}_n^+$) does not belong to the quasi-variety of modal algebras (pseudo-boolean algebras, respectively) $(\mathfrak{F}_{(\mathcal{K}_6^+)^V}(\omega))^Q$.*

Proof. We carry out the proof for modal and pseudo-boolean algebras simultaneously. From the context it will be clear what we mean by $\mathcal{F}_n^+$ and $(\mathcal{K}_6^+)^V$, i.e. whether we mean it to be the modal or pseudo-boolean algebra. In cases where we do not explicitly say whether we are dealing with a modal or pseudo-boolean algebra this is because there is no difference in the proof for both these cases. According to Lemma 4.1.10 it is sufficient to show that $\mathcal{F}_n^+$ is not a member of

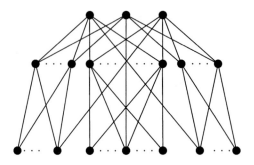

Figure 4.4: Structure of the frame $\mathcal{F}_n$ (draft)

the quasi-variety $Q_6 := (\mathfrak{F}_{(\mathcal{K}_6^+)^V}(6))^Q$. Suppose that this is not the case and $\mathcal{F}_n^+$ belongs to this quasi-variety. By Theorem 1.2.30 the quasi-variety Q_6 has the following description:

$$Q_6 := \mathbf{S\Pi}\mathrm{Mod}(\mathrm{Th}_U(\mathfrak{F}_{(\mathcal{K}_6^+)^V}(6))_e.$$

Therefore $\mathcal{F}_n^+$ is a subalgebra of a direct product of some algebras from the class $\mathrm{Mod}(\mathrm{Th}_U(\mathfrak{F}_{(\mathcal{K}_6^+)^V}(6))_e$. Because $\mathcal{F}_n^+$ is finite we can assume that all these algebras are finitely generated and this direct product has only finitely many components. Hence we can assume

$$\mathcal{F}_n^+ = \mathbf{S}(\mathfrak{A}_1 \times \ldots \times \mathfrak{A}_m),$$

where any $\mathfrak{A}_i$ is a finitely generated algebra from $\mathrm{Mod}(\mathrm{Th}_U(\mathfrak{F}_{(\mathcal{K}_6^+)^V})(6))_e$. Any algebra from the last class belongs to the variety $(\mathcal{K}_6^+)^V$. This variety is locally finite by Lemma 4.1.2. Therefore every algebra $\mathfrak{A}_i$ is finite. Then by Stone's Representation Theorem we have $\mathfrak{A}_i = \mathcal{G}_i^+$, where $\mathcal{G}_i$ is a finite frame (poset for the case of pseudo-boolean algebras). Thus $\mathcal{F}_n^+ = \mathbf{S}(\mathcal{G}_1^+ \times \ldots \times \mathcal{G}_m^+)$. From this observation, using Corollary 2.5.24 and Lemma 2.5.27, or Corollary 2.5.73 and Lemma 2.5.75 we infer that there exists a p-morphism f from $\mathcal{G}_1 \sqcup \ldots \sqcup \mathcal{G}_m$ onto the frame $\mathcal{F}_n$.

It can be easily seen that any algebra $\mathcal{G}_i^+$ must be a subalgebra of the algebra $\mathfrak{F}_{(\mathcal{K}_6^+)^V}(6)$ (or simply the single-element algebra) since (1) $\mathcal{G}_i^+$ is a member of the class $\mathrm{Mod}(\mathrm{Th}_U(\mathfrak{F}_{(\mathcal{K}_6^+)^V}(6))_e$ and (2) the property of being a finite subalgebra is definable by a universal formula (see Proposition 1.2.26). The algebra $\mathfrak{F}_{(\mathcal{K}_6^+)^V}(6)$ is isomorphic to the algebra $Ch_{\lambda(\mathcal{K}_6)}(6)^+$ according to Theorem 3.3.8 (or Theorem 3.3.14). Thus by Corollary 2.5.24 or Corollary 2.5.73) it follows that there exists a p-morphism f_i from $Ch_{\lambda(\mathcal{K}_6)}(6)$ onto the frame $\mathcal{G}_i$ for every i.

Assume that $\alpha \in \mathcal{G}_i$ and $f(\alpha) = d_{x,y,z}$, where $x = b_{1,1}$, $y = b_{2,1}$, $z = a_1$. Suppose that $\beta \in Ch_{\lambda(\mathcal{K}_6)}(6)$ and $f_i(\beta) = \alpha$. Then the element β has depth 3 and there are $\beta_{1,1}, \beta_{2,1}, \alpha_1, \alpha_2, \alpha_3 \in Ch_{\lambda(\mathcal{K}_6)}(6)$ such that:

$$f(f_i(\beta_{k,1})) = b_{k,1}, \ k = 1, 2; \quad f(f_i(\alpha_k)) = a_k, k = 1, 2, 3$$

$$\beta < \beta_{k,1}, \ k = 1, 2; \quad \beta < \alpha_k, \ k = 1, 2, 3 \qquad (4.25)$$

$$\alpha_k \not\leq \alpha_n, \ k \neq n; \quad \forall n (\alpha_n \not\leq \beta_{k,1}, \ k = 1, 3)$$

$$\beta_{k,1} < \alpha_m, \ k = 1, 2; \quad m = 2, 3, \quad \forall k (\beta_{k,1} \not\leq \alpha_1).$$

According to the construction of $Ch_{\lambda(\mathcal{K}_6)}(6)$ we have the following: (i) the elements mentioned above which are placed over β and β itself form in $Ch_{\lambda(\mathcal{K}_6)}(6)$ a cone, i.e. rooted open subframe); (ii) there are elements $\delta, \gamma \in |Ch_{\lambda(\mathcal{K}_6)}(6)|$ such that $\delta^{\leq <} = \{\alpha_2, \alpha_3\}$ and $\gamma^{\leq <} = \{\delta \cup \delta^{\leq, <} \cup \{\alpha_1\}\}$. To describe visually the location of these elements the accessibility relation between them is displayed in Fig. 4.5. Then f_i-images of elements δ, γ must be a members of $\mathcal{G}_i$. The f_i-

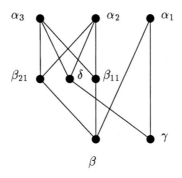

Figure 4.5:

images of all the elements from $\gamma^{\leq \leq}$ (see their placement depicted in Fig. 4.5) must be different and the accessibility relation between them in $\mathcal{G}_i$ is exactly those which is between their pro-images under f_i in Fig. 4.5 (this follows from $f(f_i(\beta)) = d_{x,y,z}$ and relations (4.25)). By the same reason, it is easy to see that the f-images of f_i-images of all elements from $\gamma^{\leq \leq}$ will be different and will have accessibility relations between them in $\mathcal{F}_n$ the same as their pro-images under $f \circ f_i$ have in Fig. 4.5. This contradicts the structure of $\mathcal{F}_n$ because in $\mathcal{F}_n$ there are no elements u such that $u^{\leq \leq}$ is isomorphic to $\gamma^{\leq \leq <}$. Therefore $\mathcal{F}_n^+$ is not a member of $(\mathfrak{F}_{(\mathcal{K}_6^+)^V}(\omega))^Q$. ∎

Lemma 4.5.2 *Every n-generated subalgebra of the modal algebra $\mathcal{F}_n^+$ (of the pseudo-boolean algebra $\mathcal{F}_n^+$) is included in the quasi-variety of modal algebras $(\mathfrak{F}_{(K_6^+)^V}(\omega))^Q$ (in the quasi-variety of pseudo-boolean algebras $(\mathfrak{F}_{(K_6^+)^V}(\omega))^Q$, respectively).*

Proof. We assume below that the algebra $\mathcal{F}_n^+$ is either modal or pseudo-boolean. Note that any algebra $\mathcal{F}_n^+$ is not n-generated because any two distinct elements from $b_{1,1}, ..., b_{1,2^n+1}$ cannot be distinguished by formulas in n variables. Let $\mathfrak{A}$ be a n-generated subalgebra of $\mathcal{F}_n^+$ which is generated by elements $X_1, ..., X_n$, each of which is a subset of $|\mathcal{F}_n|$. This, in particular, implies that any $Y \in |\mathfrak{A}|$ is the value of a term $t(y_1, ..., y_n)$ on the elements $X_1, ..., X_n$, that is $Y = t(X_1, ..., X_n)$. In other words, if we introduce the valuation V in $\mathcal{F}_n$ as follows

$$V(p_1) = X_1, ..., V(p_n) := X_n$$

then for any $Y \in |\mathfrak{A}|, Y = t(X_1, ..., X_n) = \{x \mid x \in \mathcal{F}_n, x \Vdash_V t(p_1, ..., p_n)\}$.

Now we transform $\mathcal{F}_n^+$ as follows. We contract all elements which cannot be distinguished by sets $X_1, ..., X_n$ (i.e., which have the same valuation of the letters $p_1, ..., p_n$ with respect to V) in any set $\{b_{j,i} \mid j \in \{1, ..., 2^n+1\}\}$. It is easy to see that this contraction gives us a p-morphic image $\mathcal{E}_n$ of the Kripke model which is based upon $\mathcal{F}_n$ and has the valuation $V(p_1) := X_1, ..., V(p_n) := X_n$. We denote contracting p-morphism by f.

It is not hard to see that $g(Y) := \{f(x) \mid x \in Y\}$, $Y \in |\mathfrak{A}|$ is an isomorphic embedding of the algebra $\mathfrak{A}$ into the algebra $\mathcal{E}_n^+$. Indeed, for every $Y \in |\mathfrak{A}|$, $Y = t(X_1, ..., X_n)$ and $g(t(X_1, ..., X_n)) = f(t(X_1, ..., X_n))$. Furthermore, we have $f(t(X_1, ..., X_n) = f(\{x \mid x \Vdash_V t(p_1, ..., p_n)\}$, and, being a p-morphism, f preserves the truth of formulas. That is,

$$f(\{x \mid x \Vdash_V t(p_1, ..., p_n)\}) = \{y \mid y \in \mathcal{E}_n, y \Vdash_V t(p_1, ..., p_n)\}.$$

Thus these observations show that g is a homomorphism into. If

$$t(X_1, ..., X_n) \neq q(X_1, ..., X_n)$$

then there is a certain $x \in \mathcal{F}_n$ such that

$$x \Vdash_V t(p_1, ..., p_n), \quad x \nVdash_V q(p_1, ..., p_n)$$

(or conversely). Since f preserves the truth of formulas, we have

$$g(t(X_1, ..., X_n)) \neq g(q(X_1, ..., X_n)).$$

Thus, in fact, g is an isomorphic embedding. Thus $\mathfrak{A}$ is a subalgebra of $\mathcal{E}_n^+$. Therefore, to complete the proof of our lemma, it is sufficient to show that $\mathcal{E}_n^+$ is a member of $(\mathfrak{F}_{(K_6^+)^V}(\omega))^Q$.

In order to prove this it is sufficient (by Theorem 3.3.8 or Theorem 3.3.14 and Lemmas 2.5.20, 2.5.27 for the case of modal algebras, and Lemmas 2.5.69, 2.5.75 for the case of pseudo-boolean algebras) to show that there is a p-morphism from a disjoint union of a sufficiently large finite number of the copies of the frame $Ch_{\lambda(\mathcal{K}_6)}(6)$ onto the frame $\mathcal{E}_n$. Suppose that α_2 and α_3 are arbitrary elements of depth 1 from $S_1(\mathcal{E}_n)$ and α_1 is the other element of depth 1 from this frame. Then there are elements

$$\delta, \beta_{11}, \beta_{12}, \beta, \gamma \in \mathcal{E}_n$$

which have the accessibility relation between them and α_i, where $i = 1, 2, 3$, exactly those which are depicted in Fig. 4.5. This follows immediately from the definition of $\mathcal{F}_n$ and the contraction at least two elements from any set $\{b_{j,i} \mid j \in \{1, ..., 2^n+1\}\}$ while defining of $\mathcal{E}_n$. The above mentioned contraction took place since at least two elements from any mentioned set had the same valuations of the letters $p_1, ..., p_n$ with respect to V. Using existence elements $\delta, \beta_{11}, \beta_{12}, \beta, \gamma$ in $\mathcal{E}_n$ with mentioned above properties for any $\alpha_i \in S(\mathcal{E}_n, i = 1, 2, 3$, it is not hard to see that there is a p-morphism from a disjoint union of a sufficiently large finite number of copies of the frame of $Ch_{\lambda(\mathcal{K}_6)}(6)$ onto the frame $\mathcal{E}_n$. ∎

Theorem 4.5.3 *The tabular modal logic $\lambda(\mathcal{K}_6)$ (the superintuitionistic tabular logic $\lambda_s(\mathcal{K}_6)$) has no bases for admissible rules in finitely many variables, in particular, it has no finite bases.*

Proof. It immediately follows by using Theorem 1.4.15 and Lemma 4.1.1, and Lemmas 4.5.1 and 4.5.2.

We are now in a position to prove stronger result concerning these tabular logics: we will prove that the tabular modal and superintuitionistic logics generated by $\mathcal{K}_6$ do not have even independent bases for admissible inference rules. We will apply Theorem 4.5.3 and Lemmas 4.1.7, 4.1.9. Therefore we need the following preliminary lemma.

Lemma 4.5.4 *The quasi-variety of modal algebras (pseudo-boolean algebras) generated by $\mathfrak{F}_{(\mathcal{K}_6^+)^V}(6)$ has not more than 5 covering quasi-varieties in the variety of modal (pseudo-boolean) algebras $(\mathcal{K}_6^+)^V$ generated by $\mathcal{K}_6^+$.*

Proof. We carry out the proof for modal and pseudo-boolean algebras simultaneously. Any quasi-variety Q which is a cover of the quasi-variety $(\mathfrak{F}_{(\mathcal{K}_6^+)^V}(6))^Q$ in $(\mathcal{K}_6^+)^V$ must be generated by certain single algebra $\mathfrak{A}$ together with the algebra $\mathfrak{F}_{(\mathcal{K}_6^+)^V}(6)$. Moreover, we can assume that this $\mathfrak{A}$ is finitely generated. Because $\mathfrak{A} \in (\mathcal{K}_6^+)^V$ and the variety $(\mathcal{K}_6^+)^V$ is locally finite (Lemma 4.1.2) we can assume that $\mathfrak{A}$ is finite. Furthermore, since Q is a cover, any subalgebra $\mathfrak{B}$ of

$\mathfrak{A}$ which is not a member of $(\mathfrak{F}_{(\mathcal{K}_6^+)^V}(6))^Q$ and the algebra $\mathfrak{F}_{(\mathcal{K}_6^+)^V}(6)$ must generate the same quasi-variety Q. Because $\mathfrak{A}$ is finite we can put $\mathfrak{A} = T^+$ where T is a finite frame (according to Theorem 2.5.6 or Theorem 2.5.60).

Now we will investigate the structure of T. First, by Theorem 3.3.8 or Theorem 3.3.14 $\mathfrak{F}_{(\mathcal{K}_6^+)6V}(6)^+$ is isomorphic to $Ch_{\lambda(\mathcal{K}_6^+)}(6)$. Because $\mathfrak{A} = T^+ \notin \mathfrak{F}^Q_{(\mathcal{K}_6^+)^V}$ there are no p-morphisms from any finite disjoint union of isomorphic copies of the frame of $Ch_{(\lambda(\mathcal{K}_6)_+)}(6)$ onto the frame T (to show this use Theorem 2.5.6 and 2.5.60, Lemmas 2.5.20 and 2.5.27, Lemmas 2.5.69 or 2.5.75). It is easy to see that any rooted finite frame $\mathcal{A}$ such that $\mathcal{A}^+ \in (\mathcal{K}_6^+)^V$ and such that there are no p-morphisms from the frame of $Ch_{(\alpha(\mathcal{K}_6^+)}(6)$ onto $\mathcal{A}$ is a certain frame pictured in Fig. 4.6. This observation and the fact that there are no

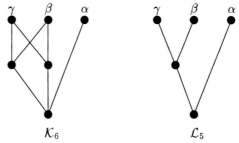

Figure 4.6:

p-morphisms from any finite disjoint union of isomorphic copies of the frame of $Ch_{(\lambda(\mathcal{K}_6)_+)}(6)$ onto the frame T have the following consequences: (a) there exist a rooted open subframe $\mathcal{A}$ of the frame T of the kind $\mathcal{K}_6$ or $\mathcal{L}_5$, (b) there exist a rooted open subframe $\mathcal{A}$ of the frame T of the kind $\mathcal{K}_6$ or $\mathcal{L}_5$ such that there is no a p-morphism f from $Ch_{\lambda(\mathcal{K}_6^+)}(6)$ into T such that $\mathcal{A} \subseteq f(Ch_{\lambda(\mathcal{K}_6^+)}(6))$.

We fixe some such $\mathcal{A}$ satisfying (b). Then exactly one from the following assertions holds:

(i) *There are no co-covers in T for some two elements α, β of depth 1 from the frame $\mathcal{A}$*

(ii) *Any two elements of depth 1 from $\mathcal{A}$ have in T some co-cover. But there are two elements $\alpha, \beta \in S_1(\mathcal{A})$ such that in T there is no co-cover for any co-cover δ of the pair α, β in T and the element $(S_1(\mathcal{A}) - \{\alpha, \beta\})$.*

Suppose (i) holds. Then we carry out consequently the following contractions of elements in T. For simplicity of notation, we below denote in any step the results of contracting in T also by T.

(1) we contract all elements of depth 1 from $\mathcal{T}$ which are outside of $\mathcal{A}$ with the element $\gamma \in S_1(\mathcal{A})$, where $\gamma \neq \alpha, \beta$;

(2) we contract all elements x of depth 2 from $\mathcal{T}$ which see only single element y of depth 1 with this y (for all possible y)

(3) we contract all co-covers for two-elements antichains of depth 1 in $\mathcal{T}$ that see the same elements from $S_1(\mathcal{A})$.

(4) we contract all elements x of depth 3 in $\mathcal{T}$, which see only a single element y of depth 2 and elements which are seen from y, with this y, and then we contract all elements of depth 3 which see the same elements from $S_2(\mathcal{T})$.

It is not hard to see that these contractions produce a p-morphism from $\mathcal{T}$ on either the frame $\mathcal{L}_5$ (see Fig. 4.6) or some frame of the kind $\mathcal{L}_6$, $\mathcal{L}_7$ (which are pictured in Fig. 4.7). It is easily seen that it is impossible to arrange a p-morphism

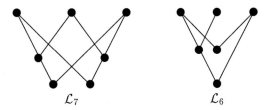

$$\mathcal{L}_7 \qquad\qquad \mathcal{L}_6$$

Figure 4.7:

from any finite disjoint union of frames of $Ch_{\lambda(\kappa_6^+)}(6)$ onto any frame from the list $\mathcal{L}_5$, $\mathcal{L}_6$, $\mathcal{L}_7$ (because there are no co-covers for some two element antichain of depth 1 in these frames). Therefore applying Theorem 2.5.6 or Theorem 2.5.60, and Corollary 2.5.24 and Lemma 2.5.7 in the case of modal algebras, and Corollary 2.5.73 and Lemma 2.5.75 in the case of pseudo-boolean algebras, we obtain that algebras $\mathcal{L}_5^+$, $\mathcal{L}_6^+$ and $\mathcal{L}_7^+$ do not belong to the quasi-variety $(\mathfrak{F}_{(\kappa_6^+)^\vee}(6))^Q$. Since we have shown that some frame $\mathcal{L}_i, i = 5, 6, 7$ is a p-morphic image of $\mathcal{T}$, the corresponding algebra $\mathcal{L}_i^+$ is a subalgebra of $\mathcal{T}^+$ (Lemma 2.5.20 or Lemma 2.5.69). Since we assumed that Q is the co-cover, Q must be the quasi-variety generated by $\mathfrak{F}_{\lambda(\kappa_6^+)}(6)$) and this algebra $\mathcal{L}_i^+$, where $i \in \{5, 6, 7\}$. Thus when (i) holds Q can be only one from the quasivarieties mentioned above.

Now suppose that (ii) holds. In this case we make the following contractions in the frame $\mathcal{T}$ (the result of contracting of $\mathcal{T}$ in any step below is again denoted by $\mathcal{T}$ for the sake of simplicity):

(5) we contract all elements of depth 1 from $\mathcal{T}$ that are outside of $\mathcal{A}$ with the element $\gamma \in \mathcal{A}$ that $\gamma \neq \alpha, \beta$;

(6) we contract all elements x of depth 2 from $\mathcal{T}$, which see only a single
     element y of depth 1 with this y (for all possible y);

(7) for any two-element antichain X consisting of elements of depth 1
     such that X is not $\{\alpha, \beta\}$, we contract all co-covers for X in $\mathcal{T}$;

(8) we contract all elements x of depth 3 in $\mathcal{T}$, which see only a single
     element y of depth 2 and elements which are seen from y, with this
     element y;

(9) we contract all elements x of depth 2, which see only α, β and are seen
     only from elements of depth 3 which see no γ, and then contract the
     elements of depth 3, which R-saw these elements x, with elements x;

(10) we contract all elements of depth 2, which see only elements α, β and
     are not seen from elements of depth 3 of $\mathcal{T}$;

(11) we contract all elements of depth 3, which see the same elements from
     $S_2(\mathcal{T})$.

It is not hard to see that all these contractions give us a p-morphism from the
frame $\mathcal{T}$ onto some frame of the kind $\mathcal{U}_i$, $i = 1, ..., 10$, where the frames $\mathcal{U}_1$, $\mathcal{U}_3$
are depicted in Fig. 4.8 and the other frames are obtained from $\mathcal{U}_1$ or $\mathcal{U}_3$ in the
way described below (note immediately that the p-morphisms on any $\mathcal{U}_i$ except
the case when $i = 1, 2$ are impossible, which we will show below). To clarify the
structure of $\mathcal{U}_3$ note that in $\mathcal{U}_3$ obligatory $n > 1$ holds and the elements $a_1, ..., a_n$
have the following property: for every two distinct a_i and a_j there exists an ele-
ment of depth 3 which is a co-cover for $\{a_i, a_j\}$ and γ. The frame $\mathcal{U}_2$, is a frame
obtained from $\mathcal{U}_1$ by deleting of the element φ. The frames $\mathcal{U}_i$, $i = 4, ..., 10$ are
frames obtained from $\mathcal{U}_3$ by all possible deleting of elements φ, ψ and c_1.

It is easy to see that there is no p-morphism from any finite disjoint union
of the frame $Ch_{\lambda(\mathcal{K}_6^+)}(6)$ onto any frame from the list $\mathcal{U}_i$, $i = 1, ..., 10$ (since
there are no co-covers for any a_i and γ in the case of frames $\mathcal{U}_i$, $i = 3, ..., 10$,
and there are no co-covers for c_1 and γ for the frames $\mathcal{U}_i$, $i = 1, 2$). Again using
Theorems 2.5.9 or 2.5.60, and Lemmas 2.5.20 and 2.5.27 in the case of modal
algebras and Lemmas 2.5.69 and 2.5.75 in the case of pseudo-boolean algebras,
it follows that the algebras of the list $\mathcal{U}_i^+$, $i = 1, ..., 10$ are not members of the
quasi-variety $(\mathfrak{F}_{\lambda(\mathcal{K}_6^+)}(6))^Q$.

Since $\mathcal{U}_i$, for some i where $i \in \{1, ..., 6\}$, is a p-morphic image of $\mathcal{T}$, it follows
that the corresponding algebra $\mathcal{U}_i^+$ is a subalgebra of $\mathcal{T}^+$ (see Lemma 2.5.20
or 2.5.69). Thus Q must be the quasi-variety generated by the algebra $\mathfrak{F}_{\lambda(\mathcal{K}_6^+)}(6)$
and the algebra $\mathcal{U}_i^+$ for some i:

$$Q = \{\mathfrak{F}_{\lambda(\mathcal{K}_6^+)}(6), \mathcal{U}_i^+\}^Q.$$

Clearly that it could be infinitely many different frames of kind $\mathcal{U}_i$, where $i \in
\{3, ..., 10\}$. But we will prove that in the description of Q above the cases $i =$

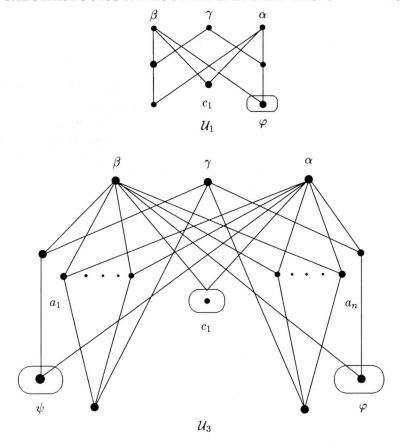

Figure 4.8:

$3, ..., 10$ are impossible. Indeed, let $i \in \{3, ..., 10\}$ and let $\mathcal{M}_{k,i}$, where $k > 1$, be the frame obtained from the disjoint union of k copies of $\mathcal{U}_i$ by the p-morphism which contracts all corresponding elements of distinct copies of $\mathcal{U}_i$ except all elements $a_1, ..., a_n$ and elements of depth 3 which $\leq$-see some elements from the list $a_1, ..., a_n$. Then the algebra $\mathcal{M}_{k,i}^+$ is a member of Q. The fact that $\mathcal{M}_{k,i}^+$ does not belong to the quasi-variety $(\mathfrak{F}_{\lambda(\mathcal{K}_6^+)}(6))^Q$ we can prove in the same way how we have showed that $\mathcal{U}_i^+$, $i = 3, ..., 10$ are not members of the quasi-variety $(\mathfrak{F}_{\lambda(\mathcal{K}_6^+)}(6))^Q$. Thus

$$(\mathfrak{F}_{\lambda(\mathcal{K}_6^+)}(6))^Q \subsetneqq \{\mathcal{M}_{k,i}^+, \mathfrak{F}_{\lambda(\mathcal{K}_6^+)}(6)\}^Q \subseteq Q.$$

We claim that the latter inclusion also is proper. Indeed, otherwise applying Theorem 2.5.6 or Theorem 2.5.60, and Lemmas 2.5.23 and 2.5.27 in the case of modal algebras, and Lemmas 2.5.73 and 2.5.75 in the case of pseudo-boolean algebras we would obtain that there exists a p-morphism f from a finite disjoint union of the copies of the frames $Ch_{\lambda(\mathcal{K}_6^+)}(6)$ and the frame $\mathcal{M}_{k,i}$ onto the frame $\mathcal{U}_i$ (i is fixed and $i \in \{3, ..., 10\}$). Then f does not map elements from copies of $Ch_{\lambda(\mathcal{K}_6^+)}(6)$ into elements $\delta_{t,j}$ of $\mathcal{U}_i$ which are co-covers for distinct a_t, a_j and γ. In fact, otherwise $\mathcal{U}_i$ must have an element which is co-cover for some co-cover of the pair α, β and the element γ which contradicts the structure of $\mathcal{U}_i$. Therefore $f(u) = \delta_{t,j}$ for some element u of depth 3 from a certain copy of $\mathcal{M}_{k,i}$. Then f is an identity mapping on $S_1(\mathcal{M}_{k,i})$. Note that some two distinct elements of kind a_v from $\mathcal{M}_{k,i}$ must be contracted by f. This holds because there are strongly more elements of this kind in $\mathcal{M}_{k,i}$ than in $\mathcal{U}_i$. This implies existence a co-cover in $\mathcal{U}_i$ for a co-cover of $\{\alpha, \beta\}$ and the element γ, a contradiction. Thus

$$\{\mathcal{M}_{n,i}^+, \mathfrak{F}_6(\lambda(\mathcal{K}_6))\}^Q \subsetneq Q,$$

which contradicts the fact that Q is a cover for $(\mathfrak{F}_{\lambda(\mathcal{K}_6^+)}(6))^Q$. Thus the only case $i = 1, 2$ is possible. Therefore Q has at most 5 covers. ∎

Theorem 4.5.5 *(Rybakov [158]) The tabular modal logic $\lambda(\mathcal{K}_6)$ (the tabular superintuitionistic logic $\lambda(\mathcal{K}_6)$) has no independent bases for admissible rules. The quasi-variety of modal (pseudo-boolean) algebras $(\mathfrak{F}_{\lambda(\mathcal{K}_6^+)}(\omega))^Q$ has no independent bases for quasi-identities.*

Proof. The second part of the theorem follows from Lemmas 4.1.9 and 4.1.10 and Theorems 4.5.3 and Lemma 4.5.4. The first part of this theorem follows from the second part and Lemma 4.1.7. ∎

The presented examples of tabular modal and superintuitionistic logics without finite and independent bases for admissible rules are logics of depth 3. It is possible to show that 3 is the explicit bound of the depth for tabular modal logics over $S4$ and tabular superintuitionistic logics with such properties. That is, all the logics of depth not more than 2 have finite bases for admissible rules. This has been shown recently by graduate student V.Remazky under authors supervision.

Theorem 4.5.6 *(V.Remazky) For any tabular modal logic $\lambda \supseteq S4$ and any tabular superintuitionistic logic λ, if λ is of depth not more than 2 then λ has a finite basis for admissible inference rules.*

Proof. Since $Var(\lambda)$ is locally finite (Lemma 4.1.2) and λ is finitely axiomatizable (Theorem 1.3.23), it is sufficient to show that λ has a basis for admissible

rules in finitely many variables. By Theorem 1.4.15 and Lemma 4.1.10 it is sufficient to prove that $\mathfrak{F}_\lambda(m)$, where m is equal to the number of elements of a finite
algebra generating λ, has a basis for quasi-identities in finitely many variables.
Suppose that it does not a case. Then by Lemma 4.1.3 there is a sequence $\mathfrak{A}_n$,
$n \in N$, $n \geq m_0$, $m_0 \in N$. of algebras such that any $\mathfrak{A}_n$ is finitely generated and
does not belong to $\mathfrak{F}_\lambda(m)^Q$; and any n-generated subalgebra of $\mathfrak{A}_n$ as well as
any proper subalgebra of $\mathfrak{A}_n$ already belong to $\mathfrak{F}_\lambda(m)^Q$. Since all $\mathfrak{A}_n$ are finitely
generated and $Var(\lambda)$ has a finite basis for identities and is locally finite, we can
assume that all $\mathfrak{A}_n$, while $n > k$, for some k, are finite and belong to $Var(\lambda)$. According to Theorem 1.2.30 and Theorem 3.3.8 or Theorem 3.4.6 together with
Lemma 4.1.10, it follows that

$$\mathfrak{F}_\lambda(m)^Q = \mathbf{S\Pi}\mathrm{Mod}(\mathrm{Th}_U(Ch_\lambda(m)^+)_e.$$

Hence $\mathfrak{A}_n$ is not a subalgebra of any direct product of any algebras from the class
$\mathrm{Mod}(\mathrm{Th}_U(Ch_\lambda(m)^+))_e$. Therefore by Corollary 2.5.24 and Lemma 2.5.27 or
Corollary 2.5.73 and Lemma 2.5.75) we obtain that there exists no p-morphism
f from any frame of kind $Ch_\lambda(m) \sqcup \ldots \sqcup Ch_\lambda(m)$ onto the finite frame $\mathfrak{A}_n^+$ for
$n > k$. Using similar way we show that there exists a p-morphism f from some
$Ch_\lambda(m) \sqcup \ldots \sqcup Ch_\lambda(m)$ onto any proper p-morphic image of the finite frame $\mathfrak{A}_n^+$
for $n > k$.

Let λ be a logic of depth 1. Then all $\mathfrak{A}_n^+$ are frames of depth 1 and the fact of
non-existence of p-morphisms onto $\mathfrak{A}_n^+$, $n > k$, from $Ch_\lambda(m) \sqcup \ldots \sqcup Ch_\lambda(m)$
means that $\mathfrak{A}_n^+$ has no a single-element cluster. The existence of a p-morphism
onto any proper p-morphic image of $\mathfrak{A}_n^+$, $n > k$, from $Ch_\lambda(m) \sqcup \ldots \sqcup Ch_\lambda(m)$
implies that $\mathfrak{A}_n^+$ has some two-elements cluster. If $\mathfrak{A}_n^+$, $n > k$, has two different clusters we get a contradiction since the proper p-morphism contracting all
clusters of $\mathfrak{A}_n^+$ into two-element cluster gives a frame which is not a p-morphic
image of the frames of kind $Ch_\lambda(m) \sqcup \ldots \sqcup Ch_\lambda(m)$. Thus any $\mathfrak{A}_n^+$ consists of
only one 2-element cluster for $n > k$. This contradicts the fact that $\mathfrak{A}_n$ is not
n-generated algebra while $n > k, n > 2$.

Consider now the case when λ is a logic of depth 2. The fact of non-existence of
p-morphisms onto given $\mathfrak{A}_n^+$, $n > k$, from the frames of the structure $Ch_\lambda(m) \sqcup$
$\ldots \sqcup Ch_\lambda(m)$ means that one from the following assertions holds.

(i) $\mathfrak{A}_n^+$ has no single-element cluster of depth 1.

(ii) there is a cluster C in $\mathfrak{A}_n^+$ of depth 2 such that either

    (a) there is a subset X of $C^{R\leq} \cap S_1(\mathfrak{A}_n^+)$ such that $\|X\| >$
    1 and there is no single-element cluster $\{a\}$ in $\mathfrak{A}_n^+$ such
    that $\{a\}^{R\leq} \cap S_1(\mathfrak{A}_n^+) = X$, or

    (b) $C^{R\leq} \cap S_1(\mathfrak{A}_n^+)$ consists of a tuple of proper clusters and
    the following holds. For any single-element cluster $\{u\} \in$

$S_1(\mathfrak{A}_n^+)$, where $\{u\} \not\subseteq C^{R\leq} \cap S_1(\mathfrak{A}_n^+))$, there is a subset X of clusters from $C^{R\leq} \cap S_1(\mathfrak{A}_n^+)$ such that: (b_1): there is no single-element cluster $\{a\}$ in $\mathfrak{A}_n^+$ such that $\{a\}^{R\leq} \cap S_1(\mathfrak{A}_n^+) = X \cup \{u\}$, and (b_2): the frame $\mathcal{F}$ of depth 2, where $S_1(\mathcal{F}) = X \cup \{u\}$ and the root of $\mathcal{F}$ is single-element and is co-cover for $S_1(\mathcal{F}) = X \cup \{b\}$, is a λ-frame.

Indeed, it is not hard to see that provided both (i) and (ii) are false it is possible to arrange a p-morphism onto $\mathfrak{A}_n^+$, $n > k$, from a certain tuple of frames of the kind $Ch_\lambda(m) \sqcup \ldots \sqcup Ch_\lambda(m)$. Thus we have that either (i) or (ii) holds for $\mathfrak{A}_n^+$. The existence of a p-morphism onto any proper p-morphic image of $\mathfrak{A}_n^+$, $n > k$, from $Ch_\lambda(m) \sqcup \ldots \sqcup Ch_\lambda(m)$ implies that both the assertions analogous to (i) and (ii) are false for any proper p-morphic image of $\mathfrak{A}_n^+$, $n > k$.

Suppose that (i) holds for some infinite subsequence $\mathcal{S}$ of the sequence $\mathfrak{A}_n^+$, $n > k$. If there is a member of $\mathcal{S}$ which has at least two maximal clusters we obtain a contradiction by contracting two maximal clusters into a single two-element cluster. Thus any frame from $\mathcal{F}$ has single maximal cluster and this cluster in proper. Since proper p-morphic images of frames from $\mathcal{S}$ do not satisfy (i), the maximal cluster C in any $\mathfrak{A}_n^+ \in \mathcal{S}$ is two-element. Contracting any cluster of depth 2 from any $\mathfrak{A}_n^+ \in \mathcal{S}$ which has not less that 3 elements with an element of maximal cluster C we again derive a contradiction the fact that (i) must be false for the result of this contracting. Therefore there is an infinite subsequence $\mathcal{S}$ of $\mathfrak{A}_n^+$, $n > k$ such that (i) is false for its frames.

Suppose that there is an infinite subsequence $\mathcal{S}_1$ of $\mathcal{S}$ such that (a) from (ii) holds for its frames. We fix a cluster C such that C satisfies (a) from (ii) in any frame from $\mathcal{S}_1$. We may suppose that frames $C^{R\leq}$ taken in all frames of $\mathcal{S}_1$ are isomorphic (using that λ is tabular). We also may suppose that there is only a single (if any) cluster C_2 of depth 1 outside of $C^{R\leq} \cap S_1(\mathfrak{A}_n^+)$ in frames of $\mathcal{S}_1$ and that this cluster is improper. Now we obtain a contradiction by contracting in enough big members of $\mathcal{S}_1$ (1) all proper clusters of depth 2 which see the same elements of depth 2 into a single two-element cluster, and then (2) all single-element clusters of depth 2 which see the same elements of depth 2 into a single reflexive element. Hence there is an infinite subsequence $\mathcal{S}_1$ of $\mathcal{S}$ such that (i) and (a) from (ii) are false in its frames.

In this case the only possibility is (b) from (ii) holds for all members of $\mathcal{S}$. Again we fix a cluster C in any frame from $\mathcal{S}_1$ such that C satisfies (b) from (ii). Now, using that λ is tabular, we may assume that all the subframes $C^{R\leq}$ in frames of $\mathcal{S}$ are isomorphic. Since only (b) of (ii) holds for C in frames of $\mathcal{S}_1$ all the clusters of $C^{R\leq} \cap S_1(\mathfrak{A}_n^+)$ are proper. In each frame from $\mathcal{S}$ consider all elements u corresponding to the same X in (b) of (ii). All u must be outside of $C^{R\leq} \cap S_1(\mathfrak{A}_n^+)$ since all cluster of this set are proper. Suppose that the contracting of some two such elements u in a single reflexive element has the effect of appearing a certain single-element co-cover for u and the set X. Then we obtain a contradiction the fact that (a) from (ii) is false in $\mathcal{S}_1$. Therefore we may assume that, for any X from (b) of (ii), the corresponding element u is outside of $C^{R\le} \cap S_1(\mathfrak{A}_n^+)$ and is unique. Consequently in every frame from $\mathcal{S}$ the number of elements u from (b) of (ii) is bounded. Furthermore, we may assume that the number of elements of depth 1 in frames of $\mathcal{S}$ is bounded also (by contracting all proper clusters of depth 1 outside of $C^{R\le} \cap S_1(\mathfrak{A}_n^+)$ into a two-element cluster). Now we derive a contradiction by possibility of contracting (3) all proper clusters of depth 2 which see the same elements of depth 2 into a single two-element cluster, and then (4) all single-element clusters of depth 2 which see the same elements of depth 2 into a single reflexive element. Thus all possibilities are exhausted, and our assumption about absence of the basis in finitely-many variables is false. ∎

Note that this theorem contrasts with Theorems 4.4.3 and 4.4.6 in which we showed that certain modal and pseudo-boolean algebras generated by frames of depth 2 have no finite bases for their valid inference rules.

Chapter 5

Structural Completeness

In the previous part of this book we have already met with the notation of structural completeness. This notion connects the derivability and admissibility of rules in a logic. As we know, an inference rule r is derivable in a logic λ if the conclusion of r can be derived from the premises of r by using theorems of λ and the postulated inference rules of λ. If a logic λ admits a variant of deduction theorem then the determining of derivability of rules in λ can be reduced to determining derivability in λ of certain special formulas. A logic λ is structurally complete iff every rule admissible in λ is derivable in λ, i.e. if the families of derivable rules and admissible rules coincide. Hence the structurally complete logics are exactly those that cannot be enlarged by adding new inference rules while preserving the set of provable theorems.

How we can evaluate the possession of a logic λ structural completeness? Is being structurally complete an advantage or a disadvantage? Such structurally complete logics λ are very stable, maximally consistent since we cannot enlarge they by adding new non-derivable rules that preserve the set of their theorems (note that many basic logics like H, $K4$, $S4$, $S5$ etc. are not structurally complete). Structurally complete logics are, in a sense, self contained and entirely complete, because they contain inside themselves all the rules which could be needed. Therefore structurally complete logics have very strong deductive systems, and are unique in this sense. Admittedly it is clear that the notion of structural completeness is not an invariant of a given logic, and depends on the choice of the axiomatic system, indeed it is extremely sensitive in this respect. However, for many popular classes of logics, the inference rules are usually fixed and the choice of axiomatic system depends on varying the set of axioms. Therefore the notion of structural completeness is meaningful for such classes of logics and indeed is a very desirable property in the light of what was said above. Some initial simple results concerning structural completeness for individual logics have already been presented in the previous chapters. Here, developing general theory, we are going to investigate structural completeness more deeply. In particular, we study in details the structural completeness of modal and superintuitionistic logics. Our main result consists of establishing the necessary and sufficient conditions for modal logics over $K4$ to be hereditarily structurally complete: a modal logic λ is hereditarily structurally complete iff λ is not included in any logic from a list of twenty special tabular logics. Hence there are exactly twenty maximal structurally incomplete modal logics above $K4$ and they are all tabular. Similar results are also obtained for superintuitionistic logics.

5.1 General Properties, Descriptions

In this section we present several general but simple results concerning the description of structurally complete logics which will often be used in later sections.

Regarding the derivability of rules, we recall that an inference rule

$$r = \frac{\alpha_1, ..., \alpha_m}{\beta}$$

is called *derivable* in a logic λ if β is derivable from $\alpha_1, ..., \alpha_m$ by using the theorems and postulated inference rules of λ, i.e.

$$\alpha_1, ..., \alpha_m \vdash_\lambda \beta.$$

Hence, in fact, the notion of derivability is quite sensitive to a given axiomatization of the logic in question, namely it depends on the postulated inference rules. We will consider here only classical Hilbert-style axiomatizations. For instance, for a modal logic *postulated rules* to be modus ponens and the normalization rule $x/\Box x$, and for superintuitionistic logics only modus ponens is a postulated inference rule. In the context of admissibility and derivability, if a logic λ has the property $\alpha \wedge \beta \in \lambda \Leftrightarrow (\alpha \in \lambda)\&(\beta \in \lambda)$, and $\{\alpha, \beta\}$ and $\alpha \wedge \beta$ are mutually derivable, then we can consider rules with only a single formula in the premises of rules. For this reason we can deal below with only one-premise rules considering modal and superintuitionistic logics. For a rule r, let $Var(r)$ denote the set of all the variables of r, and let $Sub(r)$ be the set of all subformulas of r.

Recall that a logic λ is *structurally complete* (in finitary sense) if each inference rule admissible in λ (and finite) is derivable in λ. Thus our definitions of derivability and structural completeness are pure syntactic. Clearly, it is desirable to find a semantic characterization for such logics. Such characterization are possible provided that an analog of the deduction theorem holds for the given logic λ.

Definition 5.1.1 *We say that an algebraic logic λ possess* general deduction theorem *if for any n there is a formula $\phi(p_1, ..., p_n, q)$ such that for all formulas $\alpha_1, ..., \alpha_n, \beta$,*

$$\alpha_1, ..., \alpha_n \Vdash_\lambda \beta \Leftrightarrow \phi(\alpha_1, ..., \alpha_n, \beta) \in \lambda.$$

We will call $\phi(p_1, ..., p_n, q)$ the n-deduction formula.

Note that the formulas $\phi(p_1, ..., p_n, q)$ in this definition (for distinct n) can have a non-homogeneous, arbitrary, structure, although usually, for individual logics and classes of logics, the structure of such formulas is homogeneous. For instance, all modal logics λ extending $K4$ possess the general deduction theorem (see Theorem 1.3.15) and

$$\alpha_1, ..., \alpha_n \Vdash_\lambda \beta \Leftrightarrow (\alpha_1 \wedge ... \wedge \alpha_n \wedge \Box\alpha_1 \wedge ... \wedge \Box\alpha_n \rightarrow \beta) \in \lambda.$$

All superintuitionistic logics λ also possess the general deduction theorem (see Theorem 1.3.9) and have homogenous n-deduction formulas, more precisely

$$\alpha_1, ..., \alpha_n \Vdash_\lambda \beta \Leftrightarrow (\alpha_1 \wedge ... \wedge \alpha_n \wedge \rightarrow \beta) \in \lambda.$$

Thus, in all these cases the structure of the n-deduction formulas $\phi(p_1, ..., p_n, q)$ is homogeneous.

Definition 5.1.2 *If λ is an algebraic logic which satisfies the general deduction theorem then, for any inference rule $r := \alpha_1, ..., \alpha_n/\beta$, the formula expressing r is the formula*

$$D(r) := \beta(\alpha_1, ..., \alpha_n, \beta),$$

where $\beta(p_1, ..., p_n, q)$ is the corresponding n-deduction formula.

Recall that, for algebraic logics, $p \equiv q$ represents a tuple of formulas expressing the equivalence of the propositions p and q. For instance, for superintuitionistic logics, $p \equiv q$ is the formula $(p \rightarrow q) \wedge (q \rightarrow p)$; for modal logics over $K4$, $p \equiv q$ is the formula

$$(p \rightarrow q) \wedge \Box(p \rightarrow q) \wedge (q \rightarrow p) \wedge \Box(q \rightarrow p).$$

An algebraic term (or a tuple of terms) $f \equiv g$ in the corresponding signature has the same meaning. If $r := \alpha_1, ..., \alpha_n/\beta$ is a rule in the language of an algebraic logic λ then the quasi-identity corresponding to r is the quasi-identity

$$q(r) := \alpha_1 \equiv \top \wedge ... \wedge \alpha_n = \top \Rightarrow \beta = \top,$$

Conversely, if

$$q := ((\bigwedge_{i \in I}(f_i = g_i) \rightarrow (f = g))$$

is a quasi-identity in the language of $Var(\lambda)$ for an algebraic logic λ then the tuple of inference rules corresponding to q is the tuple of rules

$$r(q) := \{f_i \equiv g_i \mid i \in I\}/(f \equiv g)),$$

where, for any formula δ from $f \equiv g$, the rule with the closure δ and the premise described above is presented. We know that q is valid in any free algebra $\mathfrak{F}_\lambda(\kappa)$ from $Var(\lambda)$ iff all the rules of $r(q)$ are valid in $\mathfrak{F}_\lambda(\kappa)$. Conversely, the rule r is valid in some $\mathfrak{F}_\lambda(\kappa)$ iff $q(r)$ is valid in $\mathfrak{F}_\lambda(\kappa)$. In particular, r is admissible in λ iff $q(r)$ is valid in $\mathfrak{F}_\lambda(\omega)$, and q is valid in $\mathfrak{F}_\lambda(\omega)$ iff all rules from $q(r)$ are admissible in λ. We must consider some additional property of algebraic logics λ possessing general deduction theorem in order to give a semantic description of structural completeness.

Definition 5.1.3 *We say that an algebraic logic λ is proper if λ possesses the general deduction theorem and*

(i) for any algebra $\mathfrak{A} \in Var(\lambda)$ and any $a, b \in |\mathfrak{A}|$, $\mathfrak{A} \models a = b \Leftrightarrow$
$\forall c \in (a \equiv b)\mathfrak{A} \models c = \top$;

(ii) for any rule r and any algebra $\mathfrak{A} \in Var(\lambda)$, if $\mathfrak{A} \models D(r) = \top$
then r is valid in $\mathfrak{A}$;

(iii) for any algebra $\mathfrak{A}$ form $Var(\lambda)$ and any rule r, if $\mathfrak{A} \not\models D(r) =$
$\top$ then there is a homomorphic image $\mathfrak{A}_1$ of the algebra $\mathfrak{A}$ such
that the rule r is false in $\mathfrak{A}_1$.

Theorem 5.1.4 *A proper algebraic logic λ is structurally complete if and only if any subdirectly irreducible algebra $\mathfrak{A}$ from $Var(\lambda)$ is a member of the quasivariety $(\mathfrak{F}_\lambda(\omega))^Q$ generated by the free algebra $\mathfrak{F}_\lambda(\omega)$, or, equivalently, if $Var(\lambda) = (\mathfrak{F}_\lambda(\omega))^Q$.*

Proof. Suppose that λ is structurally complete. Assume $\mathfrak{A}$ is any, in particular subdirectly irreducible, algebra from $Var(\lambda)$. Suppose $\mathfrak{A}$ is not a member of $(\mathfrak{F}_\lambda(\omega))^Q$. This implies that there is a quasi-identity q which is true in $\mathfrak{F}_\lambda(\omega)$ and is false in $\mathfrak{A}$. Then all inference rules from $r(q)$ are admissible in λ (see Theorem 1.4.5). Since λ is structurally complete,

$$(\forall \gamma \in \{D(r_1) \mid r_1 \in r(q)\})(\gamma \in \lambda).$$

In particular, all formulas γ from $\{D(r_1) \mid r_1 \in r(q)\}$ are valid in $\mathfrak{A}$. By (ii) from the definition of proper algebraic logics it follows that all rules from $\{r_1 \mid r_1 \in r(q)\}$ are valid in $\mathfrak{A}$. This and (i) from the definition of proper algebraic logics imply q is valid in $\mathfrak{A}$, a contradiction.

Conversely, suppose that any subdirectly irreducible algebra $\mathfrak{A}$ from $Var(\lambda)$ belongs to the quasivariety $(\mathfrak{F}_\lambda(\omega))^Q$. Assume that r is a not derivable in λ rule. Sine λ possesses the general deduction theorem, it follows that $D(r)$ is not a theorem of λ. Then by algebraic completeness theorem (see Theorem 1.3.20) there is an algebra $\mathfrak{B}$ from $Var(\lambda)$ such that the formula $D(r)$ is false in $\mathfrak{B}$. By (iii) from the definition of proper algebraic logics there is a homomorphic image $\mathfrak{A}_1$ of $\mathfrak{B}$ such that the rule r is false in $\mathfrak{A}_1$. By Birkhoff's representation theorem (see Theorem 1.2.57)

$$\mathfrak{A}_1 \in S \prod_{i \in I} \mathfrak{B}_i, \text{ any } \mathfrak{B}_i \text{ is subdirectly irreducible, } \mathfrak{B}_i \in Var(\lambda).$$

More precisely the algebra $\mathfrak{A}_1$ is a subdirect product of certain subdirectly irreducible algebras $\mathfrak{B}_i$ from $Var(\lambda)$. Using this observation we conclude that there is a subdirectly irreducible algebra $\mathfrak{A}$ from $Var(\lambda)$ which is a component of the decomposition of $\mathfrak{A}_1$ into the subdirect product of subdirectly irreducible algebras mentioned above, such that r is false in $\mathfrak{A}$. Then the quasi-identity $q(r)$ is also false in $\mathfrak{A}$. But, by our assumption, $\mathfrak{A}$ is a member of $(\mathfrak{F}_\lambda(\omega))^Q$. Hence $q(r)$ is false in $\mathfrak{F}_\lambda(\omega)$, and consequently r is not admissible in λ (see Theorem 1.4.5).

Thus λ is structurally complete. That the condition: all subdirectly irreducible algebras from $Var(\lambda)$ belong to $(\mathfrak{F}_\lambda(\omega))^Q$ is equivalent to $Var(\lambda) = (\mathfrak{F}_\lambda(\omega))^Q$ follows directly from Birkhoff's representation theorem (see Theorem 1.2.57) concerning decomposition of algebras into subdirect products of subdirectly irreducible algebras. ∎

Lemma 5.1.5 *All superintuitionistic logics and all modal logics over $K4$ are proper algebraic logics.*

Proof. We know that all these logics are algebraic and possess general deduction theorem (see Theorems 1.3.9 and 1.3.15). That (i) and (ii) hold for these logics follows directly from the structure of the algebraic representation formulas $p \equiv q$ for such logics. Let λ be a superintuitionistic logic and $\mathfrak{A} \in Var(\lambda)$. Suppose that given a rule $r := \alpha_1, ..., \alpha_n/\beta$ and $D(r)$ is false in $\mathfrak{A}$. Then there is an interpretation e of the variables from r in $\mathfrak{A}$ such that

$$\alpha_1(e(x_i)) \wedge ... \wedge \alpha_n(e(x_i) \not\leq \beta(e(x_i)),$$

where x_i are all the variables from r. Let $a := \alpha_1(e(x_i)) \wedge ... \wedge \alpha_n(e(x_i))$. We take the principle filter $\nabla(a)$ generated by a and the quotient-algebra $\mathfrak{A}/\nabla(a)$ which is a homomorphic image of $\mathfrak{A}$ (see Lemma 2.1.19). It is easy to see that under the interpretation h in $\mathfrak{A}/\nabla(a)$ of all the variables x_i, given by $h(x_i) := [e(x_i)]_{\nabla(a)}$ the rule r is false.

Let λ be a modal logic over $K4$ and $\mathfrak{B} \in Var(\lambda)$. Assume, for some rule $r := \alpha_1, ..., \alpha_n/\beta$, that $D(r)$ is false in $\mathfrak{B}$. Again this means that there exists an interpretation e of variables from r in $\mathfrak{B}$ such that

$$\bigwedge_{1 \leq i \leq n} \alpha_i(e(x_i)) \wedge \bigwedge_{1 \leq i \leq n} \Box\alpha_i(e(x_i)) \not\leq \beta(e(x_i)),$$

where x_i are all the variables from r. Let

$$a := \bigwedge_{1 \leq i \leq n} \alpha_i(e(x_i)) \wedge \bigwedge_{1 \leq i \leq n} \Box\alpha_i(e(x_i)).$$

We take the principle filter $\nabla(a)$ generated by a. It is easy to check that this filter is a $\Box$-filter. Therefore we can take the quotient-algebra $\mathfrak{B}/\nabla(a)$ which is a homomorphic image of $\mathfrak{B}$ (see Lemma 2.2.13). By straightforward calculation we show that under the interpretation h in $\mathfrak{B}/\nabla(a)$ of all the variables x_i, where $h(x_i) := [e(x_i)]_{\nabla(a)}$, the rule r is false. ∎

Thus we can apply Theorem 5.1.4 to describe structurally complete modal logics extending $K4$ and superintuitionistic logics. Taking into account certain well known theorems concerning the description of varieties and quasi-varieties generated by a class of algebraic systems (see Theorems 1.2.24 and 1.2.30) we can derive from Theorem 5.1.4 and Lemma 5.1.5 the following corollary.

Corollary 5.1.6 *A modal logic* λ *over* $K4$ *or a superintuitionistic logic* λ *is structurally complete if and only if, for any subdirectly irreducible algebra* $\mathfrak{A}$ *from* $Var(\lambda)$,

$$\mathfrak{A} \in S\Pi Mod(Th(\mathfrak{F}_\lambda(\omega)))_e$$

(this is equivalent to $HSP\mathfrak{F}_\lambda(\omega) = S\Pi Mod(Th(\mathfrak{F}_\lambda(\omega)))_e$ *).*

The above results concerning structural completeness are general and simple. Unfortunately the using of these results to determine structural completeness is confined. It is not easy to use they for straightforward examination of structural completeness for given logics since the structure of quasi-varieties generated by free algebras of countable rank is usually very complicated. Fortunately, it turns out that it is possible to clarify the given above descriptions for the case when logics have the finite model property. To prove this results we need the following definition.

Definition 5.1.7 *An algebraic logic* λ *has the* si-join-property *if the language of* λ *contains* $\vee$ *and* $\wedge$ *and, for any finite subdirectly irreducible algebra* $\mathfrak{A}$ *from* $Var(\lambda)$ *and for any tuple of elements* $a_j, b_j, 1 \leq j \leq m$, *from* $\mathfrak{A}$

$$\mathfrak{A} \models \bigvee_{1 \leq j \leq m} \bigwedge (a_j \equiv b_j) = \top \Rightarrow \mathfrak{A} \models \bigvee_{1 \leq j \leq m} (a_j = b_j).$$

Theorem 5.1.8 *Let* λ *be a proper algebraic logic with the si-join-property and the finite model property. The logic* λ *is structurally complete iff, for any subdirectly irreducible finite algebra* $\mathfrak{A}$ *from* $Var(\lambda)$, $\mathfrak{A}$ *is isomorphically embeddable in the free algebra* $\mathfrak{F}_\lambda(q)$ *of finite rank* q *from* $Var(\lambda)$ *for some rank* q.

Proof. Suppose that λ is structurally incomplete. Let r be an admissible but not derivable in λ inference rule. Then since λ is a proper algebraic logic, $D(r)$ is not a theorem of λ. Since λ has the finite model property, there exists a finite algebra $\mathfrak{A} \in Var(\lambda)$ such that $D(r)$ is false in $\mathfrak{A}$. By (iii) of the definition of proper algebraic logics, there is a homomorphic image $\mathfrak{A}_1$ of $\mathfrak{A}$ such that the rule r is false in $\mathfrak{A}_1$. By Birkhoff's representation theorem (see Theorem 1.2.57), $\mathfrak{A}_1$ is a subdirect product of some subdirectly irreducible algebras $\mathfrak{B}_i$ from $Var(\lambda)$. Every $\mathfrak{B}_i$ is a homomorphic image of $\mathfrak{A}_1$ by means of the i-th projection homomorphism π_i, in particular, all $\mathfrak{B}_i$ are finite. Then r is false in some $\mathfrak{B}_i$. Suppose that any finite subdirectly irreducible algebra form $Var(\lambda)$ is embeddable into a free algebra of finite rank form $Var(\lambda)$. Then r is false in a certain $\mathfrak{F}_\lambda(q)$, $||q|| < \omega$. Hence by Theorem 1.4.5 r is not admissible in λ, a contradiction.

Assume that λ is structurally complete and has the fmp. Let $\mathfrak{A}$ be a finite subdirectly irreducible algebra from $Var(\lambda)$. Suppose that $\mathfrak{A}$ is not a subalgebra of any free algebra of finite rank from $Var(\lambda)$. For any finite algebra $\mathfrak{B}$, the property *not being a subalgebra* is expressible by a universal formula $u(\mathfrak{B})$: $\mathfrak{B}$ is not a subalgebra of an algebra $\mathfrak{B}_1$ iff $u(\mathfrak{B})$ is valid in $\mathfrak{B}_1$ (see Proposition 1.2.26). In particular, $u(\mathfrak{A})$ is valid in $\mathfrak{F}_\lambda(\omega)$. We can assume that

$$u(\mathfrak{A}) = \bigwedge_k (\bigwedge_i (f_i^k = g_i^k) \Rightarrow \bigvee_j (h_j^k = s_j^k)),$$

where $f_i^k, g_i^k, h_j^k, s_j^k$ are certain terms. Because $\mathfrak{A}$ is a subalgebra of $\mathfrak{A}$, $\mathfrak{A} \not\models u(\mathfrak{A})$ holds. Moreover, since the algebra $\mathfrak{A}$ is finite and subdirectly irreducible, and λ has *si-join-property*, we infer that

$$\mathfrak{A} \not\models \bigwedge_k (\bigwedge_i (f_i^k = g_i^k) \Rightarrow [\bigvee_j (\bigwedge (h_j^k \equiv s_j^k)) = \top]).$$

Hence the quasi-identity $q_v := \bigwedge_i (f_i^v = g_i^v) \Rightarrow \bigvee_j (\bigwedge (h_j^v \equiv s_j))^v = \top$ is false in $\mathfrak{A}$ for some v. On the other hand, $u(\mathfrak{A})$ is valid in $\mathfrak{F}_\lambda(\omega)$, therefore any quasi-identity

$$q_k := \bigwedge_{i \in I} (f_i^k = g_i^k) \Rightarrow \bigvee_j (\bigwedge (h_j^k \equiv s_j^k)) = \top$$

also is valid in $\mathfrak{F}_\lambda(\omega)$ (which follows directly from (i) of the definition for proper algebraic logics). Consequently, all rules from $r(q_k)$ for any k are admissible in λ and, since λ is structurally complete, all formulas $D(\gamma)$, for $\gamma \in r((q_k))$ are theorems of λ for any k. In particular, all $D(\gamma)$ for $\gamma \in r(q_v)$, are valid in $\mathfrak{A}$. By (ii) from the definition of proper algebraic logics, we obtain that all rules from $r(q_v)$ are valid in $\mathfrak{A}$. These rules have the structure

$$\frac{(f_i^v \equiv g_i^v), i \in I}{\delta}$$

where δ is a single formula from $\bigvee_j (\bigwedge (h_j^v \equiv s_j^v)) \equiv \top$. Therefore applying (i) from the definition of proper algebraic logics we infer that q_v also is valid in $\mathfrak{A}$, which contradicts $\mathfrak{A} \not\models q_v$. Thus any finite subdirectly irreducible algebra from $Var(\lambda)$ is a subalgebra of a certain free algebra of finite rank from $Var(\lambda)$. ∎

Lemma 5.1.9 *Any superintuitionistic logic and any modal logic over $K4$ has si-join-property.*

Proof. Indeed, any finite subdirectly irreducible pseudo-boolean algebra has a greatest element ω among elements not equal to $\top$ according to Theorem 2.1.23. This evidently implies the si-join-property for superintuitionistic logics. Consider the modal case, suppose that $\mathfrak{B}$ is a finite subdirectly irreducible modal algebra from $Var(K4)$ and

$$\mathfrak{B} \models \bigvee_{1 \le j \le m} [(a_j \to b_j) \wedge (b_j \to a_i) \wedge \square(a_j \to b_j) \wedge \square(b_j \to a_j)] = \top.$$

The principal filters ∇_{c_j} generated by elements

$$c_j := (a_j \rightarrow b_j) \wedge (b_j \rightarrow a_i) \wedge \Box(a_j \rightarrow b_j) \wedge \Box(b_j \rightarrow a_j)$$

are $\Box$-filters (see Lemma 2.2.19). Suppose that all c_i differ from $\top$. By Theorem 2.2.15 $\mathfrak{B}$ has the smallest $\Box$-filter ∇ among $\Box$-filters which differ from $\{\top\}$. Since $\mathfrak{B}$ is finite, ∇ is a principal filter generated by c, $c \neq \top$ and $c_i \leq c$ for all c_i, a contradiction. ∎

From this lemma, Lemma 5.1.5 and Theorem 5.1.8 we immediately obtain

Corollary 5.1.10 *Let λ be a modal logic extending $K4$ or a superintuitionistic logic and let λ have the fmp. Then λ is structurally complete iff, for any subdirectly irreducible finite algebra $\mathfrak{A}$ from $Var(\lambda)$, $\mathfrak{A}$ is isomorphically embeddable in the free algebra $\mathfrak{F}_\lambda(q)$ of finite rank q from $Var(\lambda)$ for some q.*

Using this result it is simple to describe several examples of structurally complete logics.

Exercise 5.1.11 *Show that all the logics from the list below are structurally complete.*

(i) Classic propositional calculus PC;

(ii) Superintuitionistic logic LC and any of its extension;

(iii) Modal logic $S4.3 \oplus Grz$;

(iv) Minimal modal logic over $S4$ with width 2 and depth 2.

Show that the following logics are structurally incomplete:

(i) Any tabular modal logic extending $S5$ except the logic all formulas and the logic $\sigma(PC)$.

(iii) Any tabular superintuitionistic logic λ for which the rooted poset of depth 2 with three maximal elements is λ-frame.

Hint: use Corollary 5.1.10, Theorem 3.3.8 and the fact that the corresponding frames may or may not be p-morphic image of the frames of models $Ch_\lambda(k)$ for the logic considered.

5.2   Quasi-characteristic Inference Rules

In the previous section a semantic description of structurally complete modal and superintuitionistic logics with the finite model property was presented. The same result can also be obtained by employing another technique which oriented with superintuitionistic and modal logics. The basic technique of this approach was discovered by A.Citkin in [29] where quasi-characteristic rules for superintuitionistic logics were offered. It is not difficult to extend this approach to modal logics and to use it to provide a semantic description of the structurally complete logics. Although the final semantic description of structurally complete logics will be the same as before, we prefer to give a proof of this description using this new mentioned above approach since the technique is very elegant and involves new algebraic tools. As a matter of fact, the notion of quasi-characteristic rules is a generalization of Jankov's [69] characteristic formulas. In order for the reader to compare them, we, in the beginning, will state the basic definitions and results concerning Jankov's characteristic formulas. All the research of this section will dealt only with modal logics extending $K4$ and superintuitionistic logics. First we need a description of finite subdirectly irreducible pseudo-boolean and modal $K4$-algebras. With respect to pseudo-boolean algebras, we know (see Theorem 2.1.23) that a finite pseudo-boolean algebra $\mathfrak{A}$ is subdirectly irreducible iff there is greatest element ω in $\{a \mid a \in |\mathfrak{A}|, a \neq \top\}$. For modal $K4$-algebras there is a similar description. Recall that we say that an element a of a modal algebra $\mathfrak{A}$ is *up-stable* if $\Box a \geq a$. , and by Lemma 2.2.21 a finite modal algebra $\mathfrak{A}$ from $Var(K4)$ is subdirectly irreducible iff there is a greatest up-stable element ω in $(|\mathfrak{A}| - \{\top\})$.

We introduce certain characteristic formulas by finite subdirectly irreducible modal algebras due to Jankov [69] (where characteristic formulas for finite subdirectly irreducible pseudo-boolean algebras were introduced). Let $\mathfrak{A}$ be a finite subdirectly irreducible modal $K4$-algebra. By Lemma 2.2.21 $\mathfrak{A}$ has the greatest up-stable element ω. The characteristic formula for $\mathfrak{A}$ is the formula $\chi(\mathfrak{A})$, where

$$\chi(\mathfrak{A}) := \Box\beta(\mathfrak{A}) \wedge \beta(\mathfrak{A}) \to p_\omega, \text{ where}$$

$$\beta(\mathfrak{A}) := C(\mathfrak{A}) \wedge D(\mathfrak{A}) \wedge I(\mathfrak{A}) \wedge N(\mathfrak{A}) \wedge B(\mathfrak{A}),$$

$$C(\mathfrak{A}) := \bigwedge_{x,y,z \in |\mathfrak{A}|, x \wedge y = z} (p_x \wedge p_y \equiv p_z),$$

$$D(\mathfrak{A}) := \bigwedge_{x,y,z \in |\mathfrak{A}|, x \vee y = z} (p_x \vee p_y \equiv p_z),$$

$$I(\mathfrak{A}) := \bigwedge_{x,y,z \in |\mathfrak{A}|, x \to y = z} (p_x \to p_y \equiv p_z),$$

$$N(\mathfrak{A}) := \bigwedge_{x,y \in |\mathfrak{A}|, \neg x = y} (\neg p_x \equiv p_y),$$

$$B(\mathfrak{A}) := \bigwedge_{x,y \in |\mathfrak{A}|, \Box x = y} (\Box p_x \equiv p_y).$$

These formulas characterize the properties of modal algebras as follows.

Lemma 5.2.1 *Let $\mathfrak{A}$ be a finite subdirectly irreducible modal $K4$-algebra, also let $\mathfrak{B}$ be a modal $K4$-algebra. Then $\mathfrak{B} \not\models \chi(\mathfrak{A})$ iff $\mathfrak{A}$ is a subalgebra of a homomorphic image of $\mathfrak{B}$, i.e. $\mathfrak{A} \in SH(\mathfrak{B})$.*

Proof. Suppose that $\mathfrak{A} \in SH(\mathfrak{B})$, i.e. $\mathfrak{A}$ is a subalgebra of $\varphi(\mathfrak{B})$, where φ is a homomorphism. We take the valuation of p_x, $x \in |\mathfrak{A}|$ as follows: $f(p_x)$ is a fixed element in $\varphi^{-1}(x)$. Then $\varphi(\beta(\mathfrak{A})(f(p_x))) = \top$ and consequently we obtain that $\varphi(\Box\beta(\mathfrak{A})(f(p_x)) \wedge \beta(\mathfrak{A})(f(p_x))) = \top$. At the same time $\varphi(f(p_\omega)) = \omega \neq \top$. Therefore

$$\mathfrak{B} \models \Box\beta(\mathfrak{A})(f(p_x)) \wedge \beta(\mathfrak{A})(f(p_x)) \not\leq f(p_\omega),$$

i.e. $\mathfrak{B} \not\models \chi(\alpha)$.

Conversely, suppose $\mathfrak{B} \not\models \chi(\alpha)$, i.e. there is a valuation f of p_x, $x \in |\mathfrak{A}|$ in $\mathfrak{B}$ such that

$$\mathfrak{B} \models \Box\beta(\mathfrak{A})(f(p_x)) \wedge \beta(\mathfrak{A})(f(p_x)) \not\leq f(p_\omega).$$

Let $c := \Box\beta(\mathfrak{A})(f(p_x)) \wedge \beta(\mathfrak{A})(f(p_x))$. We take the principal filter $\nabla(c)$ generated by c. Clearly, $\nabla(c)$ is a $\Box$-filter. By Lemma 2.2.13 there is the natural homomorphism φ from $\mathfrak{B}$ onto the quotient-algebra $\mathfrak{B}/\nabla(c)$. We define the mapping g from $\mathfrak{A}$ into $\mathfrak{B}/\nabla(c)$ as follows: $\forall x \in |\mathfrak{A}|(g(x) := \varphi f(p_x))$.

First we show that g is a homomorphism. In fact, consider, for example, the operation $\rightarrow$. Let $x, y \in |\mathfrak{A}|$ and $(x \rightarrow y) = z$. Then $g(x \rightarrow y) = \varphi f(p_z)$. The element $d := (f(p_z) \equiv (f(p_x) \rightarrow f(p_y)))$ is a member of $\nabla(c)$. Therefore $\varphi(d) = \top$, i.e. $\varphi(f(p_z)) \equiv (\varphi(f(p_x)) \rightarrow \varphi(f(p_y))) = \top$. That is

$$\varphi(f(p_z)) = (\varphi(f(p_x)) \rightarrow \varphi(f(p_y))) = g(x) \rightarrow g(y).$$

That g commutes with other operations can be shown in the same way. Thus g is a homomorphism.

To show g is one-to-one mapping suppose $g(x) = g(y)$ and $x \neq y$. Then $x \equiv y \neq \top$ and $g(x \equiv y) = \top$. By Theorem 2.2.22 the set $\nabla := g^{-1}(\top)$ is a $\Box$-filter in $\mathfrak{A}$, and we have $\nabla \neq \{\top\}$. Since $\mathfrak{A}$ is finite, ∇ is a principal finite filter and its smallest element a is up-stable. Since $\nabla \neq \{\top\}$, we obtain $a \neq \top$. The fact that ω is greatest up-stable element of $\mathfrak{A}$ among distinct with $\top$ elements entails that $a \leq \omega$. Then $g(a) \leq g(\omega) = \varphi(f(p_\omega))$. Since $c \not\leq f(p_\omega)$, we infer that $\varphi(f(p_\omega)) \neq \top$ which contradicts $g(a) = \top$. ∎

Thus the characteristic formula $\chi(\mathfrak{A})$ describes the property of the subdirectly irreducible modal $K4$-algebra $\mathfrak{A}$ *not being a subalgebra of homomorphic images*. The characteristic formulas are very useful instrument. For instance, first proofs that there are continuously many modal logics over $S4$ and continuously many superintuitionistic logics (Jankov [69]) have been obtained namely by using characteristic formulas.

Exercise 5.2.2 *Show that there are continuously many modal logics of depth 3 over $S4$.*

Hint: to use the characteristic formulas of subdirectly irreducible modal algebras associated to the posets of depth 3 from the sequence:

$$\mathcal{F}_n := \langle F_n, \leq \rangle, n \in N, n > 3, \text{ where}$$
$$b_n := \{\{0, 1, 2, ..., n\}, \forall i \in b_n(a_i := (b_n - \{i\})),$$
$$F_n := \{b_n\} \cup \bigcup_{i \in b_n, i \neq 0}\{a_i\},$$
$$x \leq y \Leftrightarrow (x = y) \vee (x \in y) \vee (x = 0),$$

Theorem 5.2.1 and the Birkhoff's representation theorem for varieties (Theorem 1.2.24).

For completeness and comparing, we present also the original characteristic Jankov's formulas for finite subdirectly irreducible pseudo-boolean algebras. Let $\mathfrak{A}$ be a finite subdirectly irreducible pseudo-boolean algebra. According to Theorem 2.1.23 $\mathfrak{A}$ has a greatest element ω among elements distinct from $\top$. The characteristic formula for $\mathfrak{A}$ is the formula $\chi(\mathfrak{A})$, where

$$\chi(\mathfrak{A}) := \beta(\mathfrak{A}) \to p_\omega, \text{ where}$$

$$\beta(\mathfrak{A}) := C(\mathfrak{A}) \wedge D(\mathfrak{A}) \wedge I(\mathfrak{A}) \wedge N(\mathfrak{A}),$$

$$C(\mathfrak{A}) := \bigwedge_{x,y,z \in |\mathfrak{A}|, x \wedge y = z} (p_x \wedge p_y \equiv p_z),$$

$$D(\mathfrak{A}) := \bigwedge_{x,y,z \in |\mathfrak{A}|, x \vee y = z} (p_x \vee p_y \equiv p_z),$$

$$I(\mathfrak{A}) := \bigwedge_{x,y,z \in |\mathfrak{A}|, x \to y = z} (p_x \to p_y \equiv p_z),$$

$$N(\mathfrak{A}) := \bigwedge_{x,y \in |\mathfrak{A}|, \neg x = y} (\neg p_x \equiv p_y).$$

Lemma 5.2.3 *Suppose $\mathfrak{A}$ is a finite subdirectly irreducible pseudo-boolean algebra and $\mathfrak{B}$ is an arbitrary pseudo-boolean algebra. Then $\mathfrak{B} \not\models \chi(\mathfrak{A})$ iff $\mathfrak{A}$ is a subalgebra of a homomorphic image of $\mathfrak{B}$, i.e. $\mathfrak{A} \in SH(\mathfrak{B})$.*

Proof. The proof is parallel to the proof of Lemma 5.2.1.

Now we turn to the so-called quasi-characteristic rules. First, for simplicity, we describe these rules only for finite subdirectly irreducible algebras, although there is very little difference between this and the general case. Second argument for consideration of this restricted case is the following. We will need only quasi-characteristic rules of finite subdirectly irreducible algebras for description of structurally complete logics with the fmp. After giving above mentioned description we will return to consider the general case of quasi-characteristic rules.

Borrowing an idea from A.Citkin [29], we introduce the *quasi-characteristic inference rules* for finite subdirectly irreducible algebras in the following way. Suppose $\mathfrak{A}$ is a certain finite subdirectly irreducible pseudo-boolean or a certain finite and subdirectly-irreducible modal $K4$-algebra. We denote by ω the greatest element among those distinct from $\top$ for pseudo-boolean algebras and the greatest up-stable element among those distinct from $\top$ for modal algebras.

Definition 5.2.4 *Suppose $\mathfrak{A}$ consists of the elements $\{a_i \mid 1 \le i \le n\}$. The quasi-characteristic inference rule for λ is the rule $r(\mathfrak{A})$, where*

$$r(\mathfrak{A}) := \frac{\{x_a * x_b \equiv x_{a*b} \mid a, b \in \mathfrak{A}\} \cup \{\circ x_a \equiv x_{\circ a} \mid a \in \mathfrak{A}\}}{x_\omega},$$

where $*, \circ$ are all possible binary and unary signature operations in $\mathfrak{A}$.

Recall that we say an algebra $\mathfrak{B}$ *invalidates* an inference rule r if there is a valuation of the variables from r in $\mathfrak{B}$ such that all the formulas in the premise of r receive under this valuation the value $\top$ but the conclusion of r receives another value. The most important property of quasi-characteristic rules is described in the following theorem.

Theorem 5.2.5 *Suppose $\mathfrak{A}$ is a finite subdirectly irreducible pseudo-boolean or modal $K4$-algebra. Then, for any algebra $\mathfrak{B}$, where $\mathfrak{B} \in Var(H)$ if $\mathfrak{A}$ is a pseudo-boolean algebra and $\mathfrak{B} \in Var(K4)$ if $\mathfrak{A}$ is a modal algebra, the inference rule $r(\mathfrak{A})$ is invalid in $\mathfrak{B}$) iff $\mathfrak{A}$ is isomorphically embeddable in $\mathfrak{B}$.*

Proof. It is clear that the inference rule $r(\mathfrak{A})$ is invalid in $\mathfrak{A}$ under the valuation $x_a \to a$. Therefore if $\mathfrak{A}$ is embeddable in $\mathfrak{B}$ then $r(\mathfrak{A})$ is invalid in $\mathfrak{B}$. Suppose $r(\mathfrak{A})$ is invalid in $\mathfrak{B}$. Then there is a valuation $x_d \to c_d \in \mathfrak{B}$ of all variables x_d from $r(\mathfrak{A})$ in $\mathfrak{B}$ such that

$$(\forall a, b \in \mathfrak{A})(c_a * c_b = c_{a*b}), \quad (\forall a \in \mathfrak{A})(\circ c_a = c_{\circ a})$$

but $c_\omega \neq \top$. Then, taking into account the equalities presented above, we conclude that the mapping $h : a \to c_a$ is a homomorphism from $\mathfrak{A}$ into $\mathfrak{B}$. If the homomorphism h is not one-to-one then there is an element $c \in \mathfrak{A}$ such that $c \neq 1$ but $h(c) = \top$. If $\mathfrak{A}$ is a pseudo-boolean algebra then $h(\omega) = \top$, which contradicts $h(\omega) = c_\omega \neq \top$. If $\mathfrak{A}$ is a modal $K4$-algebra then since $\mathfrak{A}$ is finite we can assume that c is up-stable. Then $c \leq \omega$ and it follows that $h(\omega) = \top$ which contradicts $h(\omega) = c_\omega = \neq \top$. Thus h is an isomorphism. ■

Using this theorem we can give another proof for Corollary 5.1.10 from the previous section which describes structurally complete modal and superintuitionistic logics with the fmp. The proof of sufficiency is the same as in the corresponding part of the proof of Theorem 5.1.8.

Necessity: Suppose λ is a structurally complete modal logic over $K4$ or a superintuitionistic logic. And suppose that λ has the fmp and there is a finite subdirectly irreducible algebra $\mathfrak{A}$ from $Var(\lambda)$ which is not a subalgebra of $\mathfrak{F}_\lambda(q)$ for all q. According to Theorem 5.2.5 this entails $r(\mathfrak{A})$ is valid in the algebra $\mathfrak{F}_\lambda(\omega)$; that is, $r(\mathfrak{A})$ is admissible in λ (see Theorem 1.4.5). At the same time, the formula $D(r(\mathfrak{A}))$ describing $r(\mathfrak{A})$ is false in $\mathfrak{A}$ since $\mathfrak{A}$ invalidates $r(\mathfrak{A})$. Thus $r(\mathfrak{A})$ is not derivable in λ, a contradiction. ■

Now we are going to describe the concept of quasi-characteristic rules in general not only for finite subdirectly irreducible algebras, following the original definition of A.Citkin [29]. For this we need the definition of *finite presentation* for algebras, which is the well-known notion of universal algebra.

Definition 5.2.6 *An algebra $\mathfrak{B}$ is called* finitely presented *in a variety Var if the following hold. Algebra $\mathfrak{B}$ is generated by a finite set of elements $\{a_1, ..., a_n\}$ and there is a finite list of terms $\gamma_i, \delta_i, 1 \leq i \leq m$ in the signature of $\mathfrak{A}$ depending on variables $\{x_1, ..., x_n\}$ such that*

   (a) $\mathfrak{B} \models \gamma_i(a_1, ..., a_n) = \delta_i(a_1, ..., a_n), 1 \leq i \leq m;$

   (b) *for every algebra $\mathfrak{A} \in Var$, if for some $\{b_1, ..., b_n\} \subseteq |\mathfrak{A}|$*

$$\mathfrak{A} \models \gamma_i(b_1, ..., b_n) = \delta_i(b_1, ..., b_n), 1 \leq i \leq m$$

   then the mapping $a_i \to b_i$ can be extended to a homomorphism from $\mathfrak{B}$ into $\mathfrak{A}$.

Suppose $\mathfrak{A}$ is a pseudo-boolean algebra or a modal $K4$-algebra. Suppose that if $\mathfrak{A}$ is a pseudo-boolean algebra then $\mathfrak{A}$ has a greatest element ω among its elements which are distinct from $\top$ and if $\mathfrak{A}$ is a modal algebra then $\mathfrak{A}$ has a greatest up-stable element ω among its up-stable elements distinct from $\top$. Suppose that $\mathfrak{A}$ is generated by $\{a_1, ..., a_n\}$, and the formulas $\gamma_i, \delta_i, 1 \leq i \leq m$ depending on the variables $\{x_1, ..., x_n\}$ are a finite presentation of $\mathfrak{A}$ (in $Var(H)$ or $Var(K4)$ respectively). And suppose $\chi(x_1, ..., x_n)$ is a formula such that $\chi(a_1, ..., a_n) = \omega$.

Definition 5.2.7 *The quasi-characteristic rule for* $\mathfrak{A}$ *is the rule* $r(\mathfrak{A})$, *where*

$$r(\mathfrak{A}) := \frac{\{\gamma_i \equiv \delta_i \mid 1 \leq i \leq m\}}{\chi}.$$

It is clear that all finite algebras (of finite signature) have certain finite presentations. Also note that quasi-characteristic rules of above-mentioned kind are not uniquely determined by generating algebra $\mathfrak{A}$, since any given $\mathfrak{A}$ may have distinct finite presentations and the choice of formula χ is also not deterministic. However, it will be shown below that all the quasi-characteristic rules of the same algebra $\mathfrak{A}$ are pairwise equivalent in a semantic sense. But first we present the most important property of quasi-characteristic rules.

Theorem 5.2.8 *(A. Citkin [29]). A quasi-characteristic rule* $r(\mathfrak{A})$ *of an algebra* $\mathfrak{A}$ *is invalid in an algebra* $\mathfrak{B}$ *(* $\mathfrak{B} \in Var(H)$ *or* $\mathfrak{B} \in Var(K4)$, *respectively) if and only if* $\mathfrak{A}$ *is isomorphically embeddable in* $\mathfrak{B}$.

Proof. It is clear that $r(\mathfrak{A})$ is invalid in $\mathfrak{A}$ under the valuation $x_i \rightarrow a_i$. Thus if $\mathfrak{A}$ is embeddable into $\mathfrak{B}$ then $r(\mathfrak{A})$ is invalid in $\mathfrak{B}$ as well. Now suppose $r(\mathfrak{A})$ is invalid in $\mathfrak{B}$. Then there is a valuation $x_i \rightarrow c_i \in \mathfrak{B}$ of all the variables $x_1, ..., x_n$ from $r(\mathfrak{A})$ in $\mathfrak{B}$ such that

$$\gamma_i(c_1, ...c_n) = \delta_i(c_1, ...c_n), 1 \leq i \leq m,$$
$$\chi(c_1, ...c_n) \neq \top.$$

Since formulas $\gamma_i, \delta_i, 1 \leq i \leq m$ form a finite presentation for $\mathfrak{A}$, there is a homomorphism of $\mathfrak{A}$ into $\mathfrak{B}$ such that $h(a_i) := c_i$. Suppose that h is not one-to-one mapping. Then there is an element $b \neq \top$ such that $h(b) = \top$. If $\mathfrak{A}$ is a pseudo-boolean algebra then $h(\omega) = \top$, which contradicts

$$h(\omega) = h(\chi(a_1, ...a_n)) = \chi(c_1, ...c_n) \neq \top.$$

If $\mathfrak{A}$ is a modal $K4$-algebra then for some $b \neq \top$, $h(\Box b) = \top$ holds. If $\Box b = \top$ then b is up-stable element, i.e. $b \leq \omega$, which entails $h(\omega) = \top$. If $\Box b \neq \top$ then $\Box b \leq \Box\Box b$, i.e. $\Box b$ is up-stable and $\Box b \leq \omega$. Thus again $h(\omega) = \top$. So in any case we have $h(\omega) = \top$. But similar to above we have $h(\omega) = h(\chi(a_1, ...a_n)) = \chi(c_1, ...c_n) \neq \top$, a contradiction. Thus h is an isomorphism into. $\blacksquare$

Corollary 5.2.9 *Let* $r(\mathfrak{A})$ *be a quasi-characteristic rule for* $\mathfrak{A}$ *and* r_1 *be a rule in the same language. Then* $r(\mathfrak{A})$ *is a semantic corollary of* r_1 *(in* $Var(H)$, *or in* $Var(K4)$, *respectively) iff* r_1 *is invalid in* $\mathfrak{A}$.

Proof. The rule $r(\mathfrak{A})$ is a semantic corollary of r_1 iff, for every algebra $\mathfrak{B}$ (of corresponding type), if $r(\mathfrak{A})$ is invalid in $\mathfrak{B}$ then r_1 is invalid in $\mathfrak{B}$. By Theorem 5.2.8 the latter is equivalent to the following: for every $\mathfrak{B}$, the fact that $\mathfrak{A}$

is a subalgebra of $\mathfrak{B}$ implies that r_1 is invalid in $\mathfrak{B}$ which is equivalent to r_1 is disprovable in $\mathfrak{A}$. ∎

From this corollary we immediately derive

Corollary 5.2.10 *If r_1 and r_2 are quasi-characteristic rules of the same algebra $\mathfrak{A}$ then r_1 and r_2 are equivalent in the semantic sense.*

Corollary 5.2.11 *The rule $r(\mathfrak{A})$ is a semantic corollary of $\mathcal{R}$ iff there is a rule $r_1 \in \mathcal{R}$ such that $r(\mathfrak{A})$ is a semantic corollary of r_1.*

Proof. For suppose $\mathcal{R} \models r(\mathfrak{A})$. Then since $r(\mathfrak{A})$ is false in $\mathfrak{A}$, there is a rule $r_1 \in \mathcal{R}$ such that r_1 is false in $\mathfrak{A}$. Then by Corollary 5.2.9 $r(\mathfrak{A})$ is a semantic corollary of r_1. ∎

Corollary 5.2.12 *Let $r(\mathfrak{A})$ be a quasi-characteristic rule of an algebra $\mathfrak{A}$. The rule $r(\mathfrak{A})$ is admissible in a logic λ, where $\lambda \subseteq \lambda(\mathfrak{A})$ (and λ is a superintuitionistic logic, or is a modal logic over $K4$, according to the type of $\mathfrak{A}$), iff $\mathfrak{A}$ is not embeddable isomorphically in the free algebra $\mathfrak{F}_\lambda(\omega)$.*

Proof. Indeed, if $r(\mathfrak{A})$ is admissible in λ then (Theorem 1.4.5) $r(\mathfrak{A})$ is valid in $\mathfrak{F}_\lambda(\omega)$. Consequently $\mathfrak{A}$ is not a subalgebra of $\mathfrak{F}_\lambda(\omega)$ since $r(\mathfrak{A})$ is invalid in $\mathfrak{A}$. Conversely, if $\mathfrak{A}$ is not a subalgebra of $\mathfrak{F}_\lambda(\omega)$ then by Theorem 5.2.8 $r(\mathfrak{A})$ is valid in $\mathfrak{F}_\lambda(\omega)$ and consequently (see Theorem 1.4.5) $r(\mathfrak{A})$ is admissible in λ. ∎

Now, using this consequence, we are going to describe the admissible quasi-characteristic rules for some popular modal and superintuitionistic logics. First we demonstrate this approach for Heyting's intuitionistic logic H. For this we need an admissible in H rule, which was discovered by G.Mints [104]:

$$r_M := \frac{(x \rightarrow y) \rightarrow x \vee z}{((x \rightarrow y) \rightarrow x) \vee ((x \rightarrow y) \rightarrow z)}.$$

As we know (see Example 3.5.29) the rule r_M is admissible but not derivable in H.

Definition 5.2.13 *Suppose $\mathcal{F}_1 := \langle F_1, R_1 \rangle$ and $\mathcal{F}_2 := \langle F_2, R_2 \rangle$ are certain frames. The consequent assembling of $\mathcal{F}_1$ and $\mathcal{F}_2$ is the frame $\mathcal{F}_1 \oplus \mathcal{F}_2$ which is based upon disjoint union of F_1 and F_2 and has the accessibility relation including R_1 and R_2 and the relation consisting of all pairs with a first element from F_1 and a second element from F_2.*

Definition 5.2.14 *A cluster C of a frame $\mathcal{F} := \langle F, R \rangle$ is a node if $\mathcal{F} = \mathcal{F}_1 \oplus C^{R \leq}$, where $\mathcal{F}_1$ is the frame obtained from $\mathcal{F}$ by removing all elements from $C^{R \leq}$, and the frame $C^{R \leq}$ has the generated from $\mathcal{F}$ accessibility relation.*

Theorem 5.2.15 *(A.Citkin [29]) Let $\mathfrak{A}$ be a finite pseudo-boolean algebra with a greatest element ω among elements distinct from $\top$. The following statements are equivalent:*

(i) *$\mathfrak{A}$ is isomorphically embeddable in $\mathfrak{F}_H(\omega)$;*

(ii) *Mints rule r_m is valid in $\mathfrak{A}$;*

(iii) *$\mathfrak{A}^+$ is a rooted poset which has the form $\mathcal{F}_1 \oplus \ldots \oplus \mathcal{F}_k$, where any $\mathcal{F}_i$ is a rooted poset of width not more than 2 and $\|\mathcal{F}_i\| \leq 3$ or $\mathcal{F}_i$ is 4-element distributive lattice, or $\mathcal{F}_i$ is two-element antichain.*

Proof. Since r_M is admissible, r_m is valid in $\mathfrak{F}_H(\omega)$ (see Theorem 1.4.5), therefore (i) implies (ii). (ii) $\Rightarrow$ (iii). Since $\mathfrak{A}$ has the element ω, $\mathfrak{A}^+$ is a rooted poset. Suppose $\mathfrak{A}^+$ has a node d with branching more than 2. Consider any three incomparable immediate successors a, b, c for d in $\mathfrak{A}^+$. Choose the valuation of the variables x, y, z from r_m in $\mathfrak{A}^{++}$ which is as follows: $V(x) := \{s \mid s \not\leq a\}$, $V(y) := \{s \mid s \not\leq b\}$, $V(z) := \{s \mid s \not\leq c\}$. Then all above mentioned sets are nonempty, in particular, $b \in V(x)$, $c \in V(y)$, $a \in V(z)$. And $(x \to y) \to x$ is false under V on a, and $(x \to y) \to z$ is false under V on c. Hence $((x \to y) \to x) \vee ((x \to y) \to z)$ is invalid under V in the root of $\mathfrak{B}^+$. Suppose $s \not\Vdash_V x \vee z$. Then $s \leq a$ and $s \leq c$. Since d is a node of $\mathfrak{B}^+$, we get $s \leq d$ and $s \leq b$. But $b \not\Vdash_V (x \to y)$. Thus $(x \to y) \to x \vee z$ is valid under V in any element. Hence r_M is invalid in $\mathfrak{A}^{++}$ and in $\mathfrak{A}$, a contradiction. Thus the branching of nodes in three or more elements is impossible.

Suppose d is a node of $\mathfrak{A}^+$ and d has branching 2. Assume a, b are its immediate successors. Assume that there is an immediate successor c for b such that $c \notin \{a\}^{\leq\leq}$. We take the valuation of variables x, y, z from r_m in $\mathfrak{A}^{++}$ as follows: $V(x) := \{s \mid s \not\leq a\}$, $V(y) := \{s \mid s \not\leq b, s \not\leq a\}$, $V(z) := \{s \mid s \not\leq c\}$. Then all above mentioned sets are nonempty, in particular, $b, c \in V(x)$, $c \in V(y)$, $a \in V(z)$. Furthermore, $(x \to y) \to x$ is invalid under V on a since any proper successor of a does not see b and a. Also $(x \to y) \to z$ is invalid under V on c because $c \Vdash_V y$. Thus $((x \to y) \to x) \vee ((x \to y) \to z)$ is false under V in the root of $\mathfrak{B}^+$. Suppose that $s \not\Vdash_V x \vee z$. Then $s \leq a$ and $s \leq c$. Because d is a node of $\mathfrak{B}^+$, we get $s \leq d$ and $s \leq b$. At the same time $b \not\Vdash_V (x \to y)$. Hence $(x \to y) \to x \vee z$ is valid under V on any element. We obtain r_M is invalid in $\mathfrak{A}^{++}$ and in $\mathfrak{A}$, a contradiction. Since above obtained limitations of the branching and $\mathfrak{B}^+$ is a rooted poset, we infer that $\mathfrak{B}^+$ has the required structure.

(iii) $\Rightarrow$ (i). It is easy to see that $\mathfrak{A}^+$ is a p-morphic image of the frame $Ch_H(1)$. Therefore $\mathfrak{A}$ is a subalgebra of $Ch_H(1)^+$ (Lemma 2.5.20). Therefore, since $r(\mathfrak{A})$ is invalid in $\mathfrak{A}$ (Theorem 5.2.5), it follows that $r(\mathfrak{A})$ is invalid in $Ch_H(1)^+$. By Theorem 3.5.8 and Lemma 3.4.2 and Theorem 1.4.5 we infer from this that $r(\mathfrak{A})$ is invalid in $\mathfrak{F}_H(\omega)$. Applying Theorem 5.2.5 we derive immediately that $\mathfrak{A}$ is a subalgebra of $\mathfrak{F}_H(\omega)$ ∎

From this theorem, Corollaries 5.2.9 and 5.2.12, we immediately derive

Corollary 5.2.16 *(A.Citkin [29]) A quasi-characteristic rule $r(\mathfrak{A})$ of a finite pseudo-boolean algebra $\mathfrak{B}$ is admissible in H iff $r(\mathfrak{A})$ is a semantic consequence of Mint's rule r_M, i.e. r_M is a basis in H for all admissible in H quasi-characteristic rules of all finite subdirectly irreducible pseudo-boolean algebras.*

Now it is not difficult to describe all the finite pseudo-boolean algebras such that all rules admissible in H are valid in these algebras.

Corollary 5.2.17 *(A.Citkin [29]) All inference rules admissible in H are valid in a finite pseudo-boolean algebra $\mathfrak{A}$ if and only if the following rule is valid in $\mathfrak{A}$:*

$$r_{M,g} := \frac{[(x \to y) \to x \vee z] \vee u}{([(x \to y) \to x) \vee ((x \to y) \to x)] \vee u} \quad \textit{Mint's generalized rule}$$

Proof. If all rules admissible in H are valid in $\mathfrak{A}$ then $r_{M,g}$ is also valid in $\mathfrak{A}$ since $r_{M,g}$ is admissible in H (because r_M is admissible in H and H has the disjunction property, see Theorem 2.4.31). Conversely, suppose that $r_{M,g}$ is valid in $\mathfrak{A}$. Consider any maximal (by capacity) rooted sharp open subframe $\mathcal{F}_i$ of $\mathfrak{A}^+$. If r_M is invalid in $\mathcal{F}_i$ it would follow that $r_{M,g}$ is invalid in $\mathfrak{A}^+$ and $\mathfrak{A}$, a contradiction. Therefore r_M is valid in any $\mathcal{F}_i$ and in any $\mathcal{F}_i^+$. Then by Theorem 5.2.15 all rules admissible in H are valid in any algebra $\mathcal{F}_i^+$ and any in frame $\mathcal{F}_i$, and consequently all of them are valid in the frame $\mathfrak{A}^+$ and, consequently, in the algebra $\mathfrak{A}$ also. ∎

Using Corollary 5.2.12 and Theorems 5.2.15, 5.2.15, and applying Corollaries 5.2.9 and 5.2.11, it could be shown that there are infinitely many not equivalent quasi-characteristic admissible in H rules, and none of them are derivable. We will now transfer our results concerning the description of the quasi-characteristic admissible in H rules to the modal system $S4$. For this we need the Gödel-McKinsey-Tarski translation of Mint's rule:

$$T(r_M) := \frac{[\Box(\Box x \to \Box y) \to \Box x] \vee \Box z}{([\Box(\Box x \to \Box y) \to \Box x) \vee (\Box(\Box x \to \Box y) \to \Box x)] \vee \Box z}.$$

and the following two rules:

$$r_{1,1} := \frac{\Diamond x \wedge \Diamond \neg x}{y},$$

$$r_{2,2} := \frac{\alpha_1 \wedge \alpha_2 \to \Box(\alpha_1 \wedge \alpha_2 \to \Diamond(\alpha_1 \wedge \alpha_2 \wedge y) \wedge \Diamond(\alpha_1 \wedge \alpha_2 \wedge \neg y))}{\neg(\alpha_1 \wedge \alpha_2)},$$

$$\alpha_1 := \Diamond(z_1 \wedge \Box \neg z_2), \quad \alpha_2 := \Diamond(z_2 \wedge \Box \neg z_1).$$

Exercise 5.2.18 *Show that* $T(r_M), r_{1,1}$ *and* $r_{2,2}$ *are admissible but not derivable in Lewis's modal system* $S4$.

Hint: The admissibility of $T(r_M)$ in $S4$ follows from the admissibility of r_M in H and Theorem 3.5.31, or directly from Theorem 3.5.1. Using Theorem 3.5.1 it is also easily to show that $r_{1,1}$ and $r_{2,2}$ are admissible in $S4$. In order to show that $T(r_M)$ is not derivable in $S4$ it is sufficient to use the frame $\mathcal{F} := \langle \{1, 2, 3, 4\}, \leq \rangle$, where $1 \leq 2, 1 \leq 3, 3 \leq 4$. To prove $r_{1,1}$ is non-derivable in $S4$, use the two element cluster. Finally, in order to show that $r_{2,2}$ is not derivable in $S4$, use the frame $\mathcal{F}_1$ obtained from $\mathcal{F}$ by removing 4 and replacing 1 by a two-element cluster. ∎

Definition 5.2.19 *Let* $\mathcal{F}$ *be a transitive reflexive frame. The skeleton of* $\mathcal{F}$ *is the frame obtained from* $\mathcal{F}$ *by contracting any cluster into a single reflexive element.*

Theorem 5.2.20 *Let* $\mathfrak{B}$ *be a finite modal subdirectly-irreducible algebra from* $Var(S4)$, *i.e. let* $\mathfrak{B}$ *have a greatest up-stable element* ω *among its up-stable elements which differ from* $\top$. *Then the following statements are equivalent:*

(i) $\mathfrak{B}$ *is isomorphically embeddable in* $\mathfrak{F}_{S4}(\omega)$;

(ii) rules $T(r_M), r_{1,1}$ *and* $r_{2,2}$ *are valid in* $\mathfrak{B}$;

(iii) (a): The skeleton of $\mathfrak{B}^+$ *is a rooted poset which has the form* $\mathcal{F}_1 \oplus ... \oplus \mathcal{F}_k$, *where any* $\mathcal{F}_i$ *is a rooted poset of width not more than 2 and* $||\mathcal{F}_i|| \leq 3$ *or* $\mathcal{F}_i$ *is 4-element distributive lattice, or* $\mathcal{F}_i$ *is the two-element antichain;*
(b): Some cluster of $\mathfrak{B}^+$ *maximal by the accessibility relation is single-element;*
(c): For any two incomparable elements a, b *from* $\mathfrak{B}^+$, *there is a single-element cluster* c *in* $\mathfrak{B}^+$ *which is a co-cover for* a *and* b.

Proof. Since $T(r_M), r_{1,1}$ and $r_{n,1}$ are admissible in $S4$, these rules are valid in $\mathfrak{F}_{S4}(\omega)$ (Theorem 1.4.5), therefore (i) implies (ii). (ii) $\Rightarrow$ (iii). Since $\mathfrak{B}$ has the element ω, $\mathfrak{B}^+$ is a finite rooted reflexive transitive frame. Since $T(r_M)$ is valid in $\mathfrak{B}$ and $\mathfrak{B}^+$, it follows that r_M is valid in the skeleton of $\mathfrak{B}^+$ (see Lemma 2.7.7). Therefore by Theorem 5.2.15 the skeleton of $\mathfrak{B}^+$ has the form required in (a) of (iii). If all clusters C of $\mathfrak{B}^+$ maximal by the accessibility relation are proper then it is easy to see that $r_{1,1}$ is invalid in $\mathfrak{B}^+$ and in $\mathfrak{B}$, a contradiction. Thus (b) from (iii) holds. If a and b are elements incomparable in $\mathfrak{B}^+$ then there are co-cover clusters for a and b since $\mathfrak{B}^+$ has a width of not more than 2 and (a). If all these co-cover clusters C_i are proper then it is not hard to check that $r_{2,2}$ is invalid in $\mathfrak{B}^+$ and $\mathfrak{B}$, a contradiction. Thus (c) from (iii) also holds.

(iii) $\Rightarrow$ (i). It is easy to see that if $\mathfrak{B}^+$ has the form described in (iii) then $\mathfrak{B}^+$ is a p-morphic image of the frame $Ch_{S4}(k)$ for some k. Therefore $\mathfrak{B}$ is a subalgebra of $Ch_{S4}(k)^+$ (Lemma 2.5.20). Then because $r(\mathfrak{B})$ is invalid in $\mathfrak{B}$ (Theorem 5.2.5), it follows that $r(\mathfrak{B})$ is invalid in $Ch_{S4}(k)^+$. Using Lemma 3.4.2 and Theorem 3.5.1, and then Theorem 1.4.5, we derive that $r(\mathfrak{B})$ is invalid in $\mathfrak{F}_{S4}(\omega)$. Applying Theorem 5.2.8 it follows that $\mathfrak{A}$ is a subalgebra of $\mathfrak{F}_{S4}(\omega)$ ∎

Using this theorem, Corollaries 5.2.9 5.2.11 and 5.2.12 we immediately derive

Corollary 5.2.21 *A quasi-characteristic rule $r(\mathfrak{B})$ of a finite modal algebra $\mathfrak{B}$ is admissible in $S4$ iff $r(\mathfrak{B})$ is a semantic consequence of rules $T(r_M), r_{1,1}$ and $r_{2,2}$, in other words, rules $T(r_M), r_{1,1}$ and $r_{2,2}$ form a basis in $S4$ for all admissible in $S4$ quasi-characteristic rules of all finite subdirectly irreducible modal $S4$-algebras.*

We also can (similarly to the case of pseudo-boolean algebras) describe all finite modal $S4$-algebras such that all rules admissible in $S4$ are valid in these algebras.

Corollary 5.2.22 *All rules admissible in $S4$ are valid in a finite modal $S4$-algebra $\mathfrak{B}$ if and only if the following rules $\Box\alpha \vee \Box u/\Box\beta \vee \Box u$, where u is a new variable which does not occur in α, β, and α/β is any rule from the list $T(r_M), r_{1,1}, r_{2,2}$, are valid in $\mathfrak{B}$*

Proof. If all admissible in $S4$ rules are valid in $\mathfrak{B}$ then the above mentioned rules also are valid in $\mathfrak{B}$ because $T(r_M), r_{1,1}$ and $r_{2,2}$ are admissible in $S4$, and consequently all the rules $\Box\alpha\vee\Box u/\Box\beta\vee\Box u$ are admissible in $S4$ as well (because $S4$ has the disjunction property, see Theorem 2.3.33). Conversely, suppose that all the rules $\Box\alpha \vee \Box u/\Box\beta \vee \Box u$ are valid in $\mathfrak{B}$. We take any maximal rooted sharp open subframe $\mathcal{F}_i$ of $\mathfrak{B}^+$. If some α/β would be false in $\mathcal{F}_i$ we would have $\Box\alpha\vee\Box u/\Box\beta\vee\Box u$ is false in $\mathfrak{B}^+$ and in $\mathfrak{B}$, a contradiction. Therefore all the rules α/β are valid in any $\mathcal{F}_i$ and in any $\mathcal{F}_i^+$. Then by Theorem 5.2.20 all admissible in $S4$ rules are valid in any algebra $\mathcal{F}_i^+$ and any frame $\mathcal{F}_i$, consequently all of them also are valid in the frame $\mathfrak{B}^+$ and in the modal algebra $\mathfrak{B}$. ∎

Now it remains only to consider the situation concerning admissibility quasi-characteristic rules in modal system $K4$.

Theorem 5.2.23 *Let $\mathfrak{B}$ be a finite modal subdirectly-irreducible algebra from $Var(K4)$, i.e. let $\mathfrak{B}$ have the greatest up-stable element ω among up-stable elements which differ from $\top$. Let $||\mathfrak{B}|| > 2$. Then $\mathfrak{B}$ is not embeddable in $\mathfrak{F}_{K4}(\omega)$ as a subalgebra.*

Proof. It is easy to see that $\mathfrak{B}^+$ is a rooted transitive frame and $||\mathfrak{B}|| > 1$. It is not difficult also to see that if $\mathfrak{B}^+$ has no two R-maximal elements one of which is reflexive and the other one is irreflexive then $\mathfrak{B}^+$ is not a p-morphic image of the frame of $Ch_{K4}(k)$ for any k. In particular, $\mathfrak{B}$ is not embeddable in $\mathfrak{F}_{K4}(\omega)$ (see Theorem 3.3.8). Suppose that two such R-maximal elements exist. Then again it is impossible to arrange a p-morphism from some $Ch_{K4}(k)$ onto $\mathfrak{B}^+$ due to the presence of irreflexive and reflexive co-covers for any two R-incomparable elements in $Ch_{K4}(k)$. Thus $\mathfrak{B}^+$ is not a p-morphic image of any $Ch_{K4}(k)$. In particular, $\mathfrak{B}$ is not a subalgebra of $Ch_{K4}(k)^+$ (see Corollary 2.5.24) for any k. Therefore by Theorem 5.2.8 $r(\mathfrak{B})$ is valid in $Ch_{K4}(k)^+$. Therefore applying Theorem 3.3.8 we conclude that $r(\mathfrak{B})$ is valid in the algebra $\mathfrak{F}_{K4}(k)$ for any k. Hence $r(\mathfrak{B})$ is valid in $\mathfrak{F}_{K4}(\omega)$. Using Theorem 5.2.8 we obtain $\mathfrak{B}$ is not a subalgebra of $\mathfrak{F}_{S4}(\omega)$ ∎

Applying this theorem, Corollary 5.2.12 and Theorem 1.4.5 we immediately derive

Corollary 5.2.24 *Any quasi-characteristic rule $r(\mathfrak{B})$ for any finite modal algebra $\mathfrak{B}$ with $\|\mathfrak{B}\| > 1$ is admissible in $K4$.*

Using this corollary it is not hard again to show that there are infinitely many independent quasi-characteristic admissible in $K4$ rules, and none of them are derivable in $K4$. Hence the logic $K4$ is entirely structurally incomplete with respect to quasi-characteristic rules: any such rule is admissible but not derivable.

5.3 Some Preliminary Technical Results

In this section we will investigate the particular properties of certain individual finite rooted frames. These properties are vital to our complete description in the next section of hereditarily structurally complete modal and superintuitionistic logics. The reader only interested in results concerning structural completeness can skip this section and return back to these special results only when they are used or referred to. Nevertheless this section contains several results which could also be of interest from a general point of view. For example, we find a sufficient condition for modal logics to have the fmp.

The following transitive frames of the kinds $F_1, ..., F_{14}$ which are depicted below in Fig. 5.1 will play a special role in our considerations. We will take • to mean a cluster consisting of a single reflexive element; ○ will denote an irreflexive element; we will use ⊙ to denote an improper cluster, i.e. cluster consisting of a single element; to denote the two element cluster we will use sign ⊙⊙ and □ will denote in pictures an arbitrary finite cluster.

Lemma 5.3.1 *Let $\mathcal{M} := \langle M, R, V \rangle$ be a transitive Kripke model. Suppose the logic $\lambda := \lambda(\mathcal{M})$ is not included in any logic of the kinds $\lambda(F_{11})$ or $\lambda(F_2)$. Then either $\mathcal{M}$ has no R-maximal irreflexive elements or the frame of $\mathcal{M}$ forms an antichain of irreflexive elements.*

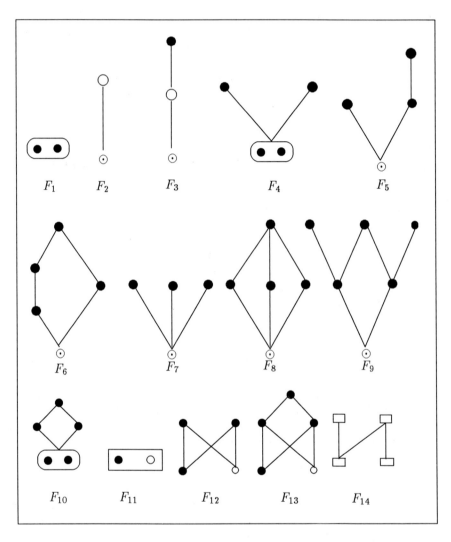

Figure 5.1: The frames of kinds F_1 - F_{14}.

Proof. Note that since $\lambda(\mathcal{M}) \not\subseteq \lambda(F_{11})$ and $\lambda(\mathcal{M}) \not\subseteq \lambda(F_2)$, for every definable variant $\mathcal{D}$ of $\mathcal{M}$ and for any open generated submodel $\mathcal{D}$ of $\mathcal{M}$,

$$\lambda(\mathcal{D}) \not\subseteq \lambda(F_{11}), \quad \lambda(\mathcal{D}) \not\subseteq \lambda(F_2).$$

We will actively use this observation further. Assume that $\mathcal{M}$ is not an antichain of irreflexive elements and that the frame of $\mathcal{M}$ has certain maximal irreflexive elements. Then $V(\Box\bot) \neq \emptyset$ in the model $\mathcal{M}$. By our supposition $V(\Box\bot) \neq M$ in $\mathcal{M}$ as well. If $\mathcal{M}$ has a certain R-maximal reflexive element then the generated open submodel $\mathcal{M}_1$ of $\mathcal{M}$, which consists of all R-maximal irreflexive elements and all R-maximal reflexive elements, is non-empty. In the model $\mathcal{M}_1$ the set of all R-maximal irreflexive elements and the set of all R-maximal reflexive elements are definable. This entails that λ is contained in $\lambda(F_{11})$, a contradiction.

Hence in this case $\mathcal{M}$ has no R-maximal reflexive elements. We choose the definable variant $\mathcal{D}$ of the given model $\mathcal{M}$ with valuation $S(p) := \Box\bot$. Then again $\lambda(\mathcal{D}) \not\subseteq \lambda(F_{11})$ and $\lambda(\mathcal{D}) \not\subseteq \lambda(F_2)$. We can suppose (applying if necessary an identifying p-morphism) that $\mathcal{D}$ has only one R-maximal cluster C and C is an irreflexive element. At the same time, $\mathcal{D}$ has other elements (because $\mathcal{M}$ is not an antichain of irreflexive elements). If there are certain elements a which do not R-see C then they do not R-see any R-maximal cluster. Therefore we can identify all such elements a in a single reflexive element by a corresponding contracting p-morphism. Then the obtained p-morphic image $\mathcal{D}_1$ of the model $\mathcal{D}$ has the same property:

$$\lambda(\mathcal{D}_1) \not\subseteq \lambda(F_{11}), \quad \lambda(\mathcal{D}) \not\subseteq \lambda(F_2).$$

But $\mathcal{D}_1$ has a R-maximal reflexive element, which gives, as we have seen above (the case of $\mathcal{M}$) a contradiction.

Thus all the elements of $\mathcal{D}$ R-see the maximal irreflexive cluster C if they differ from C. Thus if the set $(|\mathcal{D}| - C)$ has some R-maximal cluster C_1 then C_1 and C form a generated open submodel $\mathcal{D}_{C_1}$ of $\mathcal{D}$ with the frame of the kind F_2. Since C and C_1 are evidently definable in $\mathcal{D}_{C_1}$, we conclude

$$\lambda(\mathcal{D}) \subseteq \lambda(\mathcal{D}_{C_1}) \subseteq \lambda(F_2),$$

a contradiction. Hence the only remaining case is $(|\mathcal{D}| - C)$ has no maximal clusters and $(|\mathcal{D}| - C)$ is non-empty. Then we identify all elements of $(|\mathcal{D}| - C)$ in a new single reflexive element. It is easy to see that this contracting is a p-morphism f defined on $\mathcal{D}$. Then the corresponding p-morphic image of $\mathcal{D}$ under f is isomorphic to a frame of the kind F_2. But $\lambda(\mathcal{D}) \subseteq \lambda(f(\mathcal{D})) = \lambda(F_2)$, a contradiction. ∎

Lemma 5.3.2 *Let $\mathcal{M} := \langle M, R, V \rangle$ be a transitive Kripke model. Suppose that the logic $\lambda := \lambda(\mathcal{M})$ is not included in any logic of the kinds $\lambda(F_{11})$,*

$\lambda(F_2)$ *or* $\lambda(F_3)$. *Let* α *be a formula and* $Var(\alpha) \subseteq Dom(V)$. *Suppose that there is an irreflexive R-maximal element* b *in the set* $V(\alpha)$. *Then there are no elements* a *in* $\mathcal{M}$ *such that* aRb.

Proof. Assume that the set $\mathcal{B} := (V(\alpha) - V(\Diamond\alpha))$ (which consists of all R-maximal irreflexive elements of $V(A)$) is nonempty. Assume that there exist elements $a \in |\mathcal{M}|$ such that aRb and $b \in \mathcal{B}$. We take the definable variant $\mathcal{D}$ of $\mathcal{M}$ choosing the valuation $S(p) := \mathcal{B}$. Since $\mathcal{D}$ is a definable variant of $\mathcal{D}$ we conclude $\lambda(\mathcal{D}) \not\subseteq \lambda(F_{11})$, $\lambda(\mathcal{D}) \not\subseteq \lambda(F_2)$ and $\lambda(\mathcal{D}) \not\subseteq \lambda(F_3)$.

Let $X := (S(\neg\Diamond p) - S(p))$. If X is empty then we put $\mathcal{Q} := \mathcal{D}$. In this case each $b \in \mathcal{B}$ has no proper R-successors in $\mathcal{Q}$ and every element of $|\mathcal{Q}| - \mathcal{B}$ R-sees an element from $\mathcal{B}$. Assume that X is non-empty. By Lemma 5.3.1, $\mathcal{M}$ has no maximal irreflexive elements (since there are elements a which R-see irreflexive elements from $\mathcal{B}$). Therefore we can identify (contract) all the elements of the set X into a single reflexive element d by the corresponding p-morphism of the model $\mathcal{D}$. Denote the corresponding p-morphic image of the model $\mathcal{D}$ by $\mathcal{Q}$. Then it is clear that d is maximal in this model $\mathcal{Q}$. In particular, every $b \in \mathcal{B}$ either has a single proper successor in $\mathcal{Q}$, namely d, or has no successors at all. Therefore all elements $b \in \mathcal{B}$ in the model $\mathcal{Q}$ R-see the element d and only d, or all of them have no successors at all (otherwise λ would be contained in $\lambda(F_{11})$). Moreover, because X was nonempty and we have defined d, only the first case is possible (otherwise we will have $\lambda \subseteq \lambda(F_{11})$, a contradiction again). Hence in any considered case $\mathcal{Q}$ has the following form:

$$S_1(\mathcal{Q}) = \{d\}, \ \ Sl_2(\mathcal{Q}) = \mathcal{B}, \ \ (|\mathcal{Q}| - S_2(\mathcal{Q})) \neq \emptyset,$$

or $\mathcal{Q}$ is a model of this kind with d removed. Recall that $(|\mathcal{Q}| - S_2(\mathcal{Q})) \neq \emptyset$ since by our assumption there are elements a which R-see elements of $\mathcal{B}$. We can identify all irreflexive elements from $\mathcal{B}$ into a single irreflexive element u by a contracting p-morphism. Therefore we can suppose that $\mathcal{B}$ consists of only one irreflexive element, u say. Now, using the appropriate p-morphism (contraction), we can suppose that all clusters of $\mathcal{Q}$ are improper (i.e. each of them has only one-element). For $Y := (\Diamond\{u\} - \{u\})$, we have $Y \neq \emptyset$ since there are elements of $\mathcal{Q}$ which R-see u. If Y has a maximal cluster C then the open submodel $\mathcal{G}$ of $\mathcal{Q}$ generated by C has the frame isomorphic to a frame of the kind F_2 or of the F_3; and any cluster of $\mathcal{G}$ is definable in $\mathcal{G}$ (since p is valid under S only on u). Therefore

$$\lambda(\mathcal{Q}) \subseteq \lambda(\mathcal{G}) = \lambda(F_i),$$

where $i \in 2, 3$, a contradiction. Suppose that Y has no maximal clusters. Then we can identify (contract) all the elements of Y into a single reflexive element, and this gives us a p-morphism from $\mathcal{Q}$. And we obtain by this p-morphism a model $\mathcal{G}_1$ each element of which is definable in $\mathcal{G}_1$, and with a frame of the kind F_2 or F_3. Then we again get $\lambda(\mathcal{Q}) \subseteq \lambda(\mathcal{G}) = \lambda(F_i)$, where $i \in \{2,3\}$, a contradiction. Thus there are no elements $a \in |\mathcal{M}|$ such that aRb and $b \in \mathcal{B}$. ∎

The next lemma is, in a sense, a crucial result for further research and results concerning hereditary structural completeness. This lemma and its proof, in some ways resemble certain results of M.Kracht from [75] (where it is shown that the smallest logic of width 2 above $S4$ is a splitting of $S4$ by certain finite frames).

Lemma 5.3.3 *If a logic λ over $K4$ is not included in any logic from the list $\lambda(F_{11})$, $\lambda(F_2)$, $\lambda(F_3)$, $\lambda(F_5)$, $\lambda(F_6)$, $\lambda(F_7)$, $\lambda(F_8)$ and $\lambda(F_9)$ then λ has a width of not more than 2.*

Proof. Suppose that λ is a logic with a width of more than 2. Then the formula

$$\mathcal{W}_2 := \bigwedge_{i=0,1,2} \Diamond p_i \rightarrow \bigvee_{i,j=0,1,2,i\neq j} \Diamond(p_i \wedge (p_j \vee \Diamond p_j))$$

does not belong to λ. Consider the weak canonical model $\mathcal{M} := \langle M, R, V \rangle$ for λ where $Dom(V)$ is exactly the set of all propositional letters from the formula $\mathcal{W}_2$. This model is constructed like usual canonical models only the propositional language contains only propositional letters from $Dom(V)$. So M is the set of all complete consistent for λ theories all formulas of which are constructed from variables of $Dom(V)$. This model has all usual properties of canonical models. In particular, $\mathcal{W}_2$ is false in $\mathcal{M} := \langle M, R, V \rangle$ (see comment before Lemma 2.10.3). This means that there is an $a \in M$ such that

$$a \Vdash_V \bigwedge_{i=0,1,2} \Diamond p_i, \quad a \nVdash_V \bigvee_{i,j=0,1,2,i\neq j} \Diamond(p_i \wedge (p_j \vee \Diamond p_j)) \tag{5.1}$$

Let $\mathcal{F}_a$ be the open submodel generated by a in the model $\mathcal{M}$. It follows that $\mathcal{F}_a$ is not an antichain of irreflexive elements. Therefore, according to Lemma 5.3.1, the model $\mathcal{F}_a$ has no R-maximal irreflexive elements. Moreover, by Lemma 5.3.2 the model $\mathcal{F}_a$ has no R-maximal irreflexive elements in any its subset definable by formulas, which differ from $\{a\}$. We choose the definable variant $\mathcal{W}_1$ of $\mathcal{F}_a$ with the valuation

$$S(r_i) := V((p_i \vee \Diamond p_i) \wedge \bigwedge_{j \neq i} \neg \Diamond p_j), i = 0, 1, 2.$$

The contraction of all elements from the set $\Phi := S(\neg \Diamond r_0 \wedge \neg \Diamond r_1 \wedge \neg \Diamond r_2)$ into a single reflexive element m is a p-morphism from the model $\mathcal{W}_1$ (provided this set is nonempty). We denote the corresponding p-morphic image of $\mathcal{W}_1$ by $\mathcal{Q}_1$ (if Φ is empty then $\mathcal{Q}_1 := \mathcal{W}_1$). A sketch of the structure of the frame $\mathcal{Q}_1$ is displayed in Fig. 5.2. Then, in particular, we get $\lambda(\mathcal{Q}_1) \not\subseteq \lambda(F_i)$, for any frame of a kind

m

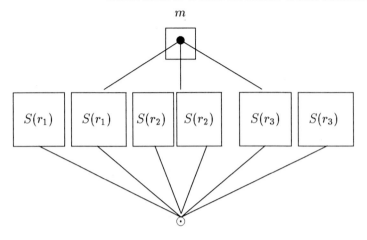

Figure 5.2: Fragment of the model $\mathcal{G}_1$.

F_i, where $i \in \{11, 2, 3, 5 - 9\}$. Consider the sets $\mathcal{X}_i := S(r_i)$. By (5.1) each $\mathcal{X}_i$ is nonempty and each element of $\mathcal{Q}_1$ which is R-accessible from an element of $\mathcal{X}_i$ is an element of $\mathcal{X}_i$ or is equal to m. Therefore the contraction of all the elements of the set $\mathcal{X}_i$ which R-see no m into a single reflexive element x_i is a p-morphism from $\mathcal{Q}_1$ (in fact, $\mathcal{X}_i$ has no maximal irreflexive elements). We denote the p-morphic image of $\mathcal{Q}_1$ with respect to this p-morphism by $\mathcal{Q}_2$ (if there are no elements for contraction we put $\mathcal{Q}_2 := \mathcal{Q}_1$).

Further, the contraction of all elements of the form $\mathcal{X}_i$ which differ from x_i and do not see x_i (in the model $\mathcal{Q}_2$) into a single reflexive element y_i which is R-incomparable with x_i is also a p-morphism. We denote by $\mathcal{Q}_3$ the p-morphic image of $\mathcal{Q}_2$ with respect to this p-morphism. Note that the p-morphic image of $\mathcal{X}_i$ in $\mathcal{Q}_3$ is a subset of $\{x_i, y_i\}$. Indeed, otherwise x_i and y_i exist and $Y_i := \mathcal{X}_i - \{x_i, y_i\}$ is a non-empty set which consists of elements such that both elements x_i and y_i are R-accessible from these elements. Because Y_i is definable in $\mathcal{Q}_3$, employing Lemma 5.3.2, we infer that Y_i has no maximal irreflexive elements. Then we can contract all the elements of Y_i into a single reflexive element b, and this contraction will be a p-morphism f from $\mathcal{Q}_3$. Then the open submodel of $f(\mathcal{Q}_3)$ generated by b will be based up on a frame of the kind F_5, and each element of this frame will be definable. This contradicts $\lambda(\mathcal{Q}) \not\subseteq \lambda(F_5)$. Thus each set $\mathcal{X}_i$ (in the model $\mathcal{Q}_3$) is a non-empty subset of $\{x_i, y_i\}$. Moreover, these sets are R-incomparable in $\mathcal{Q}_3$. Thus $\mathcal{Q}_3$ has a structure which is sketched in Fig. 5.3.

Consider the set Y of all elements from Q_3 which are not contained in the following set $\bigcup_i \mathcal{X}_i \cup \{m\}$. If $y \in Y$ then there is an element z from $\mathcal{X}_i$ such that yRz for some i. Because $y \notin \mathcal{X}_i$ and (5.1) holds, and because all fulfilled above

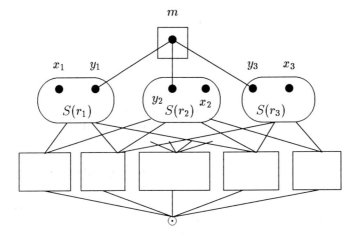

Figure 5.3: Fragment of the structure of Q_3.

contracting, we have $yRu, u \in \mathcal{X}_j, j \neq i$. Thus

$$\forall y \in Y \; \exists i, j \; (j \neq j) \wedge (yR\mathcal{X}_i, yR\mathcal{X}_j), \tag{5.2}$$

where $yR\mathcal{X}_i$ means y R-sees some element from $\mathcal{X}_i$. In particular, no one of variables r_i are valid on $y \in Y$ under S. Let $\Delta := \{m\}$ if m was defined, otherwise Δ is $\emptyset$. Any p-morphism ϕ from the frame $\bigcup_i \mathcal{X}_i \cup \Delta$ can be extended till a p-morphism ψ from the frame Q_3 (define ψ on elements from Y as one-to-one mapping). Every element of $\bigcup_i \mathcal{X}_i \cup \Delta$ is definable in Q_3. Therefore every such p-morphism ψ can be considered as a p-morphism of the definable variant of the model Q_3 in which the mentioned elements a definable. Using this observation we take the contracting p-morphism from the mentioned definable variant of the model Q_3 as follows:

(a) if m is defined and every $\mathcal{X}_i = \{x_i\}$ then we contract nothing;

(b) if m is defined and there is $\mathcal{X}_i = \{x_i\}$ but not all $\mathcal{X}_j$ have such representation then we contract every x_j such that $\mathcal{X}_j = \{x_j, y_j\}$ with x_i;

(c) if m is defined but every $\mathcal{X}_j \supseteq \{y_j\}$ then we contract all existing x_i with m;

(d) if m is not defined then every $X_i = \{x_i\}$ and we contract nothing.

The resulting p-morphic image Q_4 of Q_3 has exactly three incomparable reflexive elements $g_i, i = 0, 1, 2$ which are p-morphic images of the elements from

$\bigcup_i \mathcal{X}_i$. It is possible that $\mathcal{Q}_4$ has the p-morphic image h of the element m which is a cover for a collection of elements g_i. Since we know that $\mathcal{Q}_4$ has the set $\{g_i \mid i = 0, 1, 2\} \cup \Delta$ as the top part, and by (5.2)

$$\forall y \in (Q_4 - \{g_i \mid i = 0, 1, 2\} \cup \Delta)\, \exists i, j\, (j \neq j) \wedge (yRg_i, yRg_j), \qquad (5.3)$$

all the elements from $\{g_i \mid i = 0, 1, 2\} \cup \Delta$ are definable in $\mathcal{Q}_4$. We take now the new definable valuation in $\mathcal{Q}_4$ which evaluates exactly four (or three) new letters each of which is valid only on one corresponding element from the following set $\{g_i \mid i = 0, 1, 2\} \cup \Delta$.

Consider now the set $\mathcal{Z}$ of all elements from Q_4 which are not contained in $\{g_i \mid i = 0, 1, 2\} \cup \Delta$. By (5.3) any element of $\mathcal{Z}$ R-sees at least two incomparable elements from the list $g_i, i = 0, 1, 2$. We denote by $\mathcal{Z}_{u,v}$ the subset of $\mathcal{Z}$ consisting of all the elements each of which R-sees exactly two incomparable elements u, v from the list $g_i, i = 0, 1, 2$. If for the root a of $\mathcal{Q}_4$, $a \in \mathcal{Z}_{u,v}$ holds then we have that a R-sees only u, v, and does not R-see the other element of depth 2, which contradicts the fact that a is the root. Hence any element of any set $\mathcal{Z}_{u,v}$ must be accessible from a and a is not a member of these sets. Clear that the set $\mathcal{Z}_{u,v}$ is definable in $\mathcal{Q}_4$. Therefore, that $a \notin \mathcal{Z}_{u,v}$, a R-sees the set $\mathcal{Z}_{u,v}$, together with Lemma 5.3.2 imply that $\mathcal{Z}_{u,v}$ has no maximal irreflexive elements. If $\mathcal{Z}_{u,v}$ has no maximal irreflexive elements then, using (5.2), we can contract all elements from $\mathcal{Z}_{u,v}$ into a single reflexive element $z_{u,v}$ by the corresponding p-morphism. We denote the p-morphic image of $\mathcal{Q}_4$ under the composition of all such p-morphisms by $\mathcal{Q}_5$. The set $\mathcal{D} := (\mathcal{Q}_5 - (\{g_i \mid i = 0, 1, 2\} \cup \Delta \cup \bigcup_{u,v} \{z_{u,v}\})$ is nonempty. Moreover

$$\forall y \in \mathcal{D}\, \forall i = 0, 1, 2\, (yRg_i). \qquad (5.4)$$

Consider elements c from the set $\mathcal{D}$ which do not R-see any element of the kind $z_{u,v}$. Consider first possible case: there is a certain such element c which is accessible from a. Then we take the open submodel $\mathcal{W}_2$ of $\mathcal{Q}_5$ generated by c. The set $\mathcal{D}$ is definable in $\mathcal{Q}_5$, therefore $\mathcal{D}$ has no maximal irreflexive elements by Lemma 5.3.2. Because (5.2) and (5.4) hold, we can take the p-morphism of $\mathcal{W}_2$ contracting all the elements of $\mathcal{W}_2 \cap \mathcal{D}$ into a single reflexive element t. We denote by $\mathcal{E}$ the corresponding p-morphic image of the frame $\mathcal{W}_2$. Suppose that $\mathcal{H}$ is the subframe of $\mathcal{E}$ generated by t. It is easy to see that $\mathcal{H}$ is a frame of the kind F_7 or of the kind F_8 or $\mathcal{H}$ has among its p-morphic images a frame of the kind F_5. All elements of $\mathcal{H}$ are definable, therefore all the conclusions above concerning $\mathcal{H}$ contradict

$$\lambda(\mathcal{Q}_3) \nsubseteq \lambda(F_i)$$

for frames of kinds F_i, where $i \in \{5, 7, 8\}$. Consider second possible case: only $a \in \mathcal{Q}$ R-sees no any element of kind $z_{u,v}$. Then if aRb and b is not an element of the set $\{g_i \mid i = 0, 1, 2\} \cup \Delta$ then b must be equal to a according to (5.2). Thus, in this case, Q_5 itself looks as the frame $\mathcal{H}$ which gives a contradiction as it has been shown above.

Thus we have proved that every element of $\mathcal{D}$ must R-see at lest one element of the kind $z_{u,v}$. Consider the subsets $\mathcal{D}_{u,v}$ of $\mathcal{D}$ which consist of elements which R-see only one element of kind $z_{l,f}$, - namely $z_{u,v}$. Assume that a certain set $\mathcal{D}_{u,v}$ is nonempty. For every element y of $\mathcal{D}_{u,v}$, we have $yRz_{u,v}$ and, according to (5.4), yRg_i and $g_i \neq u, v$. Assume that

$$\exists y \in \mathcal{D}_{u,v}(aRy).$$

We take the open submodel $\mathcal{W}_3$ of Q_5 generated by y. It can be easily seen that the set $\mathcal{D}_{u,v}$ is definable in Q_5. Therefore $\mathcal{D}_{u,v}$ has no maximal irreflexive elements by Lemma 5.3.2. If y R-sees an element $b \in \mathcal{D}$ then, as we have seen above, b R-sees $z_{u,v}$ as well. Using this and (5.4), we can take the p-morphism of $\mathcal{W}_3$ which contracts all the elements of $\mathcal{W}_3 \cap \mathcal{D}_{u,v}$ into a single reflexive element w. Suppose $\mathcal{M}_3$ is the corresponding p-morphic image of $\mathcal{W}_3$ with respect to this p-morphism. The frame $\mathcal{M}_3$ is finite (it has not more than 7 elements) and each its element is definable. It is easy to see that the frame of $\mathcal{M}_3$ has among p-morphic images certain frames which have some open subframes of the kind F_5 or of the kind F_6. This contradicts

$$\lambda(Q_3) \nsubseteq \lambda(F_i)$$

for frames of kinds F_i, where $i \in \{5, 6\}$. The remaining case is: there are no $y \in \mathcal{D}_{u,v}$ such that aRy and $\mathcal{D}_{u,v} = \{a\}$. Then every b such that aRb and $b \in \mathcal{D}$ must be equal to a. Thus Q_5 itself is a finite model similar to $\mathcal{M}_3$ and as above we have a contradiction. Consequently, every $\mathcal{D}_{u,v}$ is empty.

Suppose we have already proved that, for every $\mathcal{E} \subseteq \{z_{u,v} \mid u, v = 0, 1, 2\}$ such that $1 \le \|\mathcal{E}\| \le n < 3$, each subset $Y(\mathcal{E}) \subseteq \mathcal{D}$ elements of which R-see only elements of $\mathcal{E}$ among elements of kind $z_{u,v}$ is empty. Consider a subset $\mathcal{E}$ of the set $\{z_{u,v} \mid u, v = 0, 1, 2\}$ which has $n + 1$ elements, and let $Y(\mathcal{E})$ be a subset of $\mathcal{D}$ defined as above. We will prove that this $Y(\mathcal{E})$ is empty as well. Assume that $Y(\mathcal{E})$ has an element y such that aRy. Suppose $\mathcal{W}_4$ is the open submodel of Q_5 generated by y. The set $Y(\mathcal{E})$ is definable in Q_5 and $Y(\mathcal{E})$ has no maximal irreflexive elements by Lemma 5.3.2. In particular, $\mathcal{W}_4 \cap Y(\mathcal{E})$ has no maximal irreflexive elements. Therefore if yRb and $b \in \mathcal{D}$ then by our assumption above b is an element of $\mathcal{W}_4 \cap Y(\mathcal{E})$. Therefore we can take the p-morphism from $\mathcal{W}_4$ contracting all elements from $\mathcal{W}_4 \cap Y(\mathcal{E})$ into a single reflexive element r. We denote by $\mathcal{M}_4$ the corresponding p-morphic image of $\mathcal{W}_4$. The frame of $\mathcal{M}_4$ is finite (has not more than 8 elements), and each element of $\mathcal{M}_4$ is definable in $\mathcal{M}_4$. It is not hard to see that the frame $\mathcal{M}_4$ has among its p-morphic images some frame of a kind F_9, F_8, F_5, F_6 or F_7, which contradicts

$$\lambda(Q_3) \nsubseteq \lambda(F_i)$$

for frames of kinds F_i, where $i \in \{9, 8, 5, 6, 7\}$.

Thus there are no elements $y \in Y(\mathcal{E})$ such that aRy and only the case $Y(\mathcal{E}) = \{a\}$ is possible. Then the model $\mathcal{Q}_5$ itself has not more than 8 elements and looks as the model $\mathcal{M}_4$. This implies as above a contradiction. Hence we proved that $Y(\mathcal{E})$ is empty. From the inductive procedure we derive that every set $Y(\mathcal{E})$ is empty for all possible $\mathcal{E}$. At the same time, $\mathcal{D}$ is nonempty, and, for the root a, $a \in \mathcal{D}$, and every element of $\mathcal{D}$ sees at least one element of the kind $z_{u,v}$. These two assertions are inconsistent. ∎

We need also

Lemma 5.3.4 *Let $\mathcal{F} := \langle F, R \rangle$ be a rooted transitive frame of width not more than 2, which has no infinite ascending chains of clusters and has no irreflexive elements except maybe the root. Suppose $\lambda(\mathcal{F})$ is not included in any logic of the kind $\lambda(F_6)$. Suppose C_1, C_2 and D are clusters of $\mathcal{F}$, and C_2 and D are not R-maximal, and still that C_1 R-sees the cluster C_2, and C_1 and D R-see no each other. Then D must R-see C_2.*

Proof. Assume that D does not R-see C_2. We can assume that all clusters C_1, C_2 and D are single-element and consist of reflexive elements c_1, c_2 and d, respectively (making if necessary corresponding contractions). Let

$$X := \{a \mid a \in F, \neg(aRc_2) \& \neg(aRd)\}.$$

The set X is nonempty because c_2 and d are not maximal. We contract X in a single reflexive element v by the corresponding p-morphism and obtain the frame which we denote by $\mathcal{E}$. Clearly that v is the R-greatest element of $\mathcal{E}$. Now we contract all the elements of $\mathcal{E}$ which are strictly R-accessible from c_1 with the element c_2. It is easy to see that this contraction is a p-morphism. Suppose $\mathcal{Q}$ is the p-morphic image of $\mathcal{E}$ with respect to this p-morphism. Now the element c_2 is the single immediate successor of c_1 in $\mathcal{Q}$. All elements from $\mathcal{Q}$ R-see d or c_1 or coincide with v or c_2. Let

$$Y := \{a \mid a \in |\mathcal{Q}|, aRc_1, \neg(aRd)\}.$$

We contract now all elements from Y with c_1. It is easy to see that again this contraction is a p-morphism from $\mathcal{Q}$. We denote the corresponding p-morphic image of $\mathcal{Q}$ by $\mathcal{W}$. It is obviously that the set $\mathcal{Z} := \{a \mid a \in |\mathcal{W}|, aRc_1, a \neq c_1\}$ is nonempty.

We pick a R-maximal element u from $\mathcal{Z}$ (this is possible because $\mathcal{F}$ has no infinite R-ascending chains of clusters). Then uRd, and any element b strictly R-accessible from u is an element of $c_1^{R\leq} \cup d^{R\leq}$ or b R-sees d and does not R-see c_1 (by the choice of Y and the fact that u is maximal in $\mathcal{Z}$). If u R-sees only elements of $c_1^{R\leq} \cup d^{\leq}$ then it follows that $\lambda(\mathcal{F})$ is included in the logic of a frame of the kind F_6 which contradicts the condition of this lemma.

If all elements of $\mathcal{W}$ which are strictly R-accessible from u and are not elements of the set $c_1^{R\leq} \cup d^\leq$ R-see no c_2 then we can contract all of them with the element d by the p-morphism contracting only these elements. This p-morphism maps $u^{R\leq}$ into a frame of the kind F_6 which contradicts the condition of this lemma again.

Assume that the set $H := \{s \mid s \in |\mathcal{W}|, uRs, s \notin c_1^{R\leq} \cup d^{R\leq}, sRc_2\}$ is nonempty. Then we make the following transformations. First we contract all the elements of the set

$$U := \{g \mid g \in |\mathcal{W}|, uRg, g \notin c_1^{R\leq} \cup d^{R\leq}, \neg(gRc_2)\}$$

with the element d by the corresponding p-morphism, and denote the corresponding p-morphic image by $\mathcal{L}$. The image H in $\mathcal{L}$ (which we denote by the same letter) is nonempty as well. We pick a maximal element c in H (as a subframe of the frame $\mathcal{L}$, $H \subseteq |\mathcal{L}|$) and contract all the elements of H with the element c. This contraction is a p-morphism. We denote the corresponding p-morphic image of $u^{R\leq}$ by $\mathcal{G}$. Contracting in the set $\mathcal{G}$ all the elements of $c_1^{R<}$ with the element v we obtain a frame of the kind F_6, which contradicts again the condition of this lemma. ∎

We need to recall the definition of the sequential composition of any two given frames. Let $\mathcal{Q}_1 := \langle Q_1, R_1 \rangle$ and $\mathcal{Q}_2 := \langle Q_2, R_2 \rangle$ be certain transitive frames. Then the *sequential composition* of $\mathcal{Q}_1$ and $\mathcal{Q}_2$ is the frame $\mathcal{Q}_1 \oplus \mathcal{Q}_2$ which is based up on the set $Q_1 \cup Q_2$ and has the accessibility relation

$$R := R_1 \cup R_2 \cup \{(a, b) \mid a \in Q_1, b \in Q_2\}.$$

We also fix the following denotation: for every frame $\langle F, R \rangle$ and any subset X of F, $X^\downarrow := \{a \in F, a \notin X, \exists b \in X(aRb)\}$. Let $\mathcal{F} := \langle F, R \rangle$ be a rooted transitive frame of width not more than 2. Suppose that $\mathcal{F}$ has no infinite R-ascending chains of clusters and has no irreflexive elements except maybe the root.

Lemma 5.3.5 *Suppose $\lambda(\mathcal{F})$ is not included in any logic which is a logic of a kind $\lambda(F_5)$ or $\lambda(F_6)$. Then, for every maximal (by the number of members) antichain X of clusters from $\mathcal{F}$ each of which has depth not less than 2, the following holds. The frame $\mathcal{F}$ is the sequential composition of the frame $X^\downarrow$ (with transferred from $\mathcal{F}$ accessibility relation R) and the open subframe $X^{R\leq}$ of $\mathcal{F}$.*

Proof. Consider the set of all maximal clusters of the set $|X^\downarrow|$ (this set is nonempty since $\mathcal{F}$ has no infinite R-ascending chains of clusters). Let C be a such maximal cluster. Assume that X has two clusters - C_1 and C_2 (maximal possibility). Then, by Lemma 5.3.4, C R-sees both C_1 and C_2. Thus we have showed that every element of the set $(C_1 \cup C_2)^\downarrow$ R-sees every element of $C_1^{R\leq} \cup C_2^{R\leq}$, which is what we needed.

Assume now that X has only one cluster C_1. Then any element of F either is R-comparable with C_1 or is R-incomparable with C_1 and has depth 1. The latter is impossible because, in this case, we would be able to produce a p-morphism from some generated subframe of $\mathcal{F}$ onto some frame of the kind F_5. Thus any element of F is R-comparable with C_1. ∎

Lemma 5.3.6 *Let $\mathcal{F}$ be a transitive finite rooted frame of width not more than 2. Suppose that only the root of $\mathcal{F}$ can be irreflexive. Suppose that $\lambda(\mathcal{F})$ is not included in any logic which is a logic of a kind $\lambda(F_5)$ or $\lambda(F_6)$. Then $\mathcal{F}$ is a frame of the form:*

$$Q_1 \oplus Q_2 \oplus ... \oplus Q_k \oplus \mathcal{W}_{14},$$

where $\mathcal{W}_{14}$ is either $\emptyset$ or a frame of the kind F_{14}, each Q_i, $1 \leq i \leq k$, is either

  (a) a single cluster, or

  (b) an antichain of two clusters.

Proof. The frame $\mathcal{F}$ has not more than 2 maximal clusters. If there is only single maximal cluster then the frame $S_2(\mathcal{F})$ has the required form. Assume that there are exactly two maximal clusters - C_1 and C_2. Then there are not more than 2 immediate R-predecessors for C_1 and C_2. If there is only one immediate R-predecessor, it is a R-predecessor for both maximal clusters, and the frame $S_2(F)$ again has the required form. Assume that there are two immediate predecessors D_1 and D_2 for C_1 and C_2. Then some D_i R-sees C_1 and C_2 simultaneously. For otherwise, we can obtain a frame of the kind F_5 from $\mathcal{F}$ by using p-morphisms and generated open subframes. Therefore $S_2(\mathcal{F})$ also has the required form in this case. Now we simply apply Lemma 5.3.5 moving from the elements of $\mathcal{F}$ of depth 3 to the root, and obtain the assertion of this lemma. ∎

Corollary 5.3.7 *Let $\mathcal{F}$ be a finite transitive rooted frame having the properties given in Lemma 5.3.6. Suppose $\lambda(\mathcal{F})$ is not included in any logic which is a logic of a kind $\lambda(F_1)$, $\lambda(F_4)$, or $\lambda(F_{10})$. Then $\mathcal{F}$ has a form given in Lemma 5.3.6, and every cluster of $\mathcal{F}$ having two different immediate successors must be improper, any maximal cluster of $\mathcal{F}$ is improper and reflexive.*

Corollary 5.3.8 *Let $\mathcal{F}$ be a finite transitive rooted frame. Suppose $\lambda(\mathcal{F})$ is not included in any logic of a kind $\lambda(F_i)$, $i = 1, ..., 13$. Then $\mathcal{F}$ is a sequential composition (in arbitrary order) of frames of the following kinds:*

  (α) rooted frames which are a sequential compositions of the single-element root and a collection of two-cluster antichains, where only maximal clusters can be proper;

  (β) clusters;

(γ) possibly a single frame of the kind F_{14}, as the terminal frame in this sequential composition, when (i): immediate R-predecessors of F_{14} in $\mathcal{F}$ are improper clusters, and (ii): only the cluster of depth 2 from F_{14} which R-sees only one element of depth 1 can be proper.

Moreover, the maximal clusters of $\mathcal{F}$ must be improper, reflexive and only the root of $\mathcal{F}$ can be irreflexive element. If the root of $\mathcal{F}$ is irreflexive (in this case we say that $\mathcal{F}$ is i-type, otherwise we say that $\mathcal{F}$ is r-type) and the root has exactly two immediate successors then $\mathcal{F}$ has the following form (where the sequential compositions mentioned below must be in the order given): $\mathcal{F}$ is either

(i) a sequential composition of the root and the some frame of the kind F_{14}), or

(ii) a sequential composition of the root and two frames each of which is a two-element antichain, or

(iii) a sequential composition of the root, its two incomparable successors, a certain (possible empty) chain of elements, and, possibly, a rooted 3-element reflexive frame with two maximal elements.

Proof. This follows immediately from Lemmas 5.3.1, 5.3.2 and 5.3.3, and Corollary 5.3.7 and the fact that the logic $\lambda(\mathcal{F})$ is not included in the logics $\lambda(F_{12})$ and $\lambda(F_{13})$. ∎

Now we are in a position to prove the following lemma, which is a second crucial point for establishing results of the next section concerning hereditary structural completeness.

Lemma 5.3.9 *Let λ be a modal logic extending $K4$ with a width of not more than 2. Suppose that λ is not included in any logic which is a logic of a kind $\lambda(F_i), i = 2, ..., 6, 10, 11$, then λ has the fmp.*

Proof. According to K.Fine's completeness theorem (Theorem 2.10.26), being a logic of a finite width above $K4$, λ is Kripke complete with respect to frames without infinite ascending chains of clusters. Let δ be a formula which is not a theorem of λ. Then there is a λ-frame $\mathcal{F} := \langle F, R \rangle$ rooted by a cluster C, and which has no infinite R-ascending chains of clusters, such that there is a valuation V of variables of δ in $\mathcal{F}$ which invalidates δ. Of course we can assume that every cluster of $\mathcal{F}$ has not more than 2^k elements, where k is the number of variables in δ.

By Lemma 5.3.2 the frame $\mathcal{F}$ has no irreflexive elements which differ from the root. According to Lemmas 5.3.5 and 5.3.6 the frame $\mathcal{F}$ has the following representation:

(1) either a frame $\mathcal{Q}$ obtained from a linear co-well ordered set $\langle S, \leq \rangle$ by replacing every element $x \in S$ by $g(x)$, where $g(x)$ is either a cluster or an antichain of two clusters; or

(2) a sequential composition of the described above frame $\mathcal{Q}$ and a frame of the kind F_{14} (only the root of $\mathcal{Q}$ can be irreflexive).

Hence, $\mathcal{F} = \mathcal{Q} \oplus \mathcal{D}$ where $\mathcal{D} = F_{14}$ or $\mathcal{Q} = \emptyset$. Now we will use a variant of the drop-point technique without real dropping elements but with certain redefining the valuation. Let X be the set of all subformulas of δ. For every $Y \subseteq X$, we define the set $\mathcal{E}(Y)$ as follows. $\mathcal{E}(Y)$ is the set of all maximal (with respect to the accessibility relation in $\mathcal{Q}$) clusters of $\mathcal{Q}$ having certain elements c_Y such that

$$\{\chi \mid \chi \in X, c_Y \Vdash_V \chi\} = Y$$

(if such clusters exist for given Y). Let

$$Max(\mathcal{F}, X) := \bigcup_{Y \subseteq X} \mathcal{E}(Y).$$

Suppose $\mathcal{M}(\mathcal{F})$ is the union of $Max(\mathcal{F}, X)$ and all the clusters from $\mathcal{D}$, and the root cluster of $\mathcal{Q}$. By assumption that there are no infinite ascending chains of clusters in $\mathcal{F}$, this set is well-defined. Since the width of the frame $\mathcal{F}$ is less or equal to 2, the set $\mathcal{M}(\mathcal{F})$ has finitely many elements, - not more than $2 * 2^{||X||} + 5$. Let

$$\mathcal{B} := \{g(x) \mid x \in S, \exists C \in \mathcal{M}(\mathcal{F})(C \in g(x))\}.$$

We will redefine the valuation V on the set F. Namely, the new valuation W of letters of the formula δ coincides with the valuation V on $\mathcal{D}$, and on all clusters from $\mathcal{B}$. Further, every cluster $G \subseteq (\mathcal{Q} - \mathcal{B})$ is placed exactly between some $g(y_G)$ and $g(z_G)$, where $g(y_G), g(z_G) \subseteq |\mathcal{B}|$, such that the following holds. Every element of $g(y_G)$ R-sees all the elements of G and each element of G R-sees all the elements of $g(z_G)$ (where y_G is a R-greatest, and z_G is R-smallest with such property).

If $g(z_G)$ has only one cluster then we choose the valuation W on all elements of G to be identical to the valuation V on some fixed element a_G of $g(z_G)$. If z_G has two clusters then we take an immediate predecessor e_G for the set z_G in S. Because λ is not contained in any logic of the kinds $\lambda(F_4)$ or $\lambda(F_{10})$, all the clusters of $g(e_X)$ must be single-element. We set the valuation W on elements from G as identical to the valuation V on some fixed element a_G of $g(e_G)$.

It can be verified directly by a simple induction on the length of formulas that the following holds. Truth values under V and W of all subformulas of δ on elements from $\mathcal{D}$ and $\mathcal{B}$, and elements of the clusters from $G \subseteq (|\mathcal{Q}| - |\mathcal{B}|)$

and the corresponding elements a_G coincide. In particular, W falsifies δ in $\mathcal{F}$. Note that it is possible to construct a p-morphism from the model $\langle F, R, W \rangle$ onto a finite frame. This is the p-morphism which contracts all elements from the clusters $G \subseteq (|\mathcal{Q}| - |\mathcal{B}|)$ with the corresponding elements a_G, and which is a one-to-one mapping on the other elements. The corresponding p-morphic image $\mathcal{Z}$ is a finite λ-frame which falsifies δ. Thus λ has the finite model property. ∎

5.4 Hereditary Structural Completeness

We have already investigated structural completeness and we now have certain semantic descriptions of what it is for a logic to be structurally complete. In this section we consider the structure of the class of structurally complete logics. For this we distinguish a class of logics which have a very strong kind of structural completeness.

Definition 5.4.1 *A logic λ is hereditarily structurally complete if every logic λ_1 extending λ (and having the same postulated inference rules) is structurally complete.*

We are now in a position to give a complete description for hereditarily structurally complete modal logics extending $K4$ and for hereditarily structurally complete superintuitionistic logics. This is the main result of this section. The following lemma gives us an infinite class of modal logics over $K4$ of this kind, and provides a sufficient condition for a logic to be structurally complete.

Lemma 5.4.2 *Let λ be a logic extending $K4$ which is not included in any logic of the kinds $\lambda(F_i)$, $i = 1, ..., 13$. Then λ is hereditarily structurally complete.*

Proof. It is sufficient to show that any such logic λ is simply structurally complete. By Lemmas 5.3.3 and 5.3.9 the logic λ has the finite model property. Therefore it is sufficient to prove (by Theorem 5.1.8 and Lemma 5.1.9 that any finite subdirectly irreducible modal algebra from $Var(\lambda)$ is a subalgebra of a certain free algebra of finite rank from $Var(\lambda)$.

Consider a finite subdirectly irreducible modal algebra $\mathfrak{B}$ from $Var(\lambda)$. Applying Lemma 2.2.21 and Theorem 2.5.6 it follows that $\mathfrak{B}$ has the form $\mathcal{Q}^+$, where $\mathcal{Q}$ is a finite transitive rooted λ-frame. By Theorem 3.3.8 the free algebra $\mathfrak{F}_\lambda(k)$ from $Var(\lambda)$ is isomorphic to $Ch_\lambda(k)^+(V(p_1), ..., V(p_n))$, where $Ch_\lambda(k)$ is the k-characterizing Kripke model for λ.

Suppose $\mathcal{Q}$ has k elements and the depth m. According to the construction of $Ch_\lambda(k)$, $\mathcal{Q}$ is an open generated subframe of $Ch_\lambda(k)$, and we supposed that $\mathcal{Q}$ has the depth m. Any element of $S_m(Ch_\lambda(k))$ is definable (expressible) by a formula according to Theorem 3.3.7. Therefore the new valuation S in $Ch_\lambda(k)$, where

$$Dom(S) := \{p_x \mid x \in S_m(Ch_\lambda(k))\}, \quad S(p_x) := \{x\},$$

is a definable valuation. Therefore the model $\mathcal{M} := \langle Ch_\lambda(k), S \rangle$ is a definable variant of $Ch_\lambda(k)$. Thus using Lemma 2.5.20, in order to prove our lemma, it is sufficient to show that there is a finite sequence of models having the following properties:

(i) this sequence begins with the model $\mathcal{M}$;

(ii) every not initial model of this sequence is a p-morphic image of a definable variant of the immediately preceding model;

(iii) the terminate model of this sequence is a model which is based upon the set $\mathcal{Q}$ and such that every element from $\mathcal{Q}$ is definable in this model.

By Lemma 5.3.6 the model $\mathcal{Q}$ has the form described in that lemma. Using this observation, we contract all the elements of $S_m(M)$ which see no maximal clusters of $\mathcal{Q}$ with an element of some maximal cluster of $\mathcal{Q}$ and obtain a frame $\mathcal{Q}_0$. Then we contract all the elements of $S_m(\mathcal{Q}_0)$ which R-see only single maximal cluster of $\mathcal{Q}_0$ and R-see no other clusters of $\mathcal{Q}$ with a fixed element of this maximal cluster, and obtain the frame $\mathcal{Q}_1$. Thus the elements of $[\mathcal{Q}_1 - \mathcal{Q}]$o R-see both two maximal clusters of $\mathcal{Q}$ (and they must exist then) or only single maximal cluster and some not maximal cluster of $\mathcal{Q}$. It is clear that these contractions give us a p-morphism from a definable variant of the model $\mathcal{M}$ onto $\mathcal{Q}_1$ and that $S_1(\mathcal{Q}_1) = S_1(\mathcal{Q})$. It can be easily seen that each element of $S_1(\mathcal{Q}_1)$ is definable in $\mathcal{Q}_1$.

Now consider the set $\mathcal{X}$ of all elements from $(|\mathcal{Q}_1| - |\mathcal{Q}|)$ which R-see no elements of depth 2 from $\mathcal{Q}$. Suppose that $\mathcal{X}$ is non-empty. Then every $y \in \mathcal{X}$ R-sees two maximal clusters of $\mathcal{Q}$, and every element accessible from y is also from $\mathcal{X}$ or is placed in a maximal cluster of $\mathcal{Q}$. Because, in this case, there are two maximal clusters in $\mathcal{Q}_1$, $S_2(\mathcal{Q})$ has at least one cluster C of depth 2 which R-sees all maximal elements of $\mathcal{Q}$, see Lemma 5.3.6. This cluster C must be a single-element because $\lambda \not\subseteq \lambda(F_4), \lambda(F_{10})$. Consider the following two cases. First case is: C is an irreflexive element. Then C is the root of $\mathcal{Q}$ (see Lemma 5.3.2), and all the elements of $\mathcal{X}$ also are irreflexive and do not R-see each other, which holds since $\lambda \not\subseteq \lambda(F_{12}), \lambda(F_{13})$ and Lemma 5.3.2. Then we contract all the elements from $\mathcal{X}$ with C, the obtained frame is denoted by $\mathcal{Q}_2$ (it follows, in particular, that $\mathcal{Q}_2 = \mathcal{Q}$ holds).

Second case: C is a reflexive single-element cluster. Than all maximal clusters from $\mathcal{X}$ are not irreflexive elements (because $\lambda \not\subseteq \lambda(F_{12}), \lambda(F_{13})$). Moreover, by Lemma 5.3.6 $\mathcal{Q}$ has a single-element cluster $\{u\}$ of depth 2 which R-sees all maximal clusters of $\mathcal{Q}$. We contract all the elements from $\mathcal{X}$ with the element u. In this case, the result of the contracting is also denoted by $\mathcal{Q}_2$. If $\mathcal{X} = \emptyset$ then $\mathcal{Q}_2 := \mathcal{Q}_1$. In all cases $\mathcal{Q}_2$ has the following properties:

(i) $\mathcal{Q}_2$ is a p-morphic image of a definable variant of $\mathcal{Q}_1$;

(ii) $S_2(Q_2) = S_2(Q)$;

(iii) each element of $S_2(Q_2)$ is expressible in the corresponding definable variant of Q_2;

(iv) all elements of Q differ with respect to the valuation in Q_2;

(v) every element of $|Q_2| - |Q|$ R-sees either two clusters of depth 2 from Q, or only a single cluster C from Q of depth 2 and only clusters from Q which are accessible from C.

To continue we now contract all the elements from $|Q_2| - |Q|$, which R-see only a single cluster C of depth 2 from Q and R-see no deeper clusters of Q, with a fixed element of C (contracting is permissible because C is not an irreflexive element). This contracting is a p-morphism of frames. We denote the result of these contractions by $Q_{2,1}$ and consider this model below.

If there are two clusters U, H of the depth 2 in Q then there is a singe-element cluster C of depth 3 in Q having them as immediate R-successors (since $\lambda \not\subseteq \lambda(F_{10})$ and Lemma 5.3.5). Consider the set Y of all clusters from $|Q_{2,1}| - |Q|$ which R-see U and H but R-see no deeper clusters of $|Q|$. If Y is empty we put $Q_3 := Q_{2,1}$. Suppose Y is not empty. Assume that C is an irreflexive element. Then all the elements of Y must be irreflexive and they do not R-see each other (since $\lambda \not\subseteq \lambda(F_{12})$, $\lambda(F_{13})$ and Lemma 5.3.2). Then we contract all the elements from Y with C. The result of these contractions is a p-morphic image of $Q_{2,1}$ which we denote by Q_3 (in particular, C must be the root of Q).

If C is a reflexive single-element cluster (the only remaining possibility) then all the maximal elements of Y must be single-element reflexive clusters (the properties $\lambda \not\subseteq \lambda(F_{13})$, $\lambda(F_{10})$ guarantee this). Using this observation we can contract all the elements of Y with C. The image of Q_2 after all these contractions is denoted by Q_3. Thus we defined Q_3 for all possible cases and Q_3 has the following properties: Q_3 is a p-morphic image of a definable variant of Q_2; $S_3(Q_3) = S_3(Q)$; each element of $S_3(Q_3)$ is expressible in Q_3; all elements of Q differ by valuation in Q_3; every element of $|Q_3| - |Q|$ R-sees two clusters of depth 3 from Q, or R-sees only a single cluster C of depth 3 from Q and only other clusters from Q which are accessible from C (see Lemma 5.3.5).

The next step in the construction of Q_4 from Q_3 and other following steps are quite similar to the construction of Q_3 from Q_2 above. Currying out this procedure, we obtain as a result a frame Q_m, where m is the depth of Q, and $Q_m = Q$. ∎

Lemma 5.4.3 *None of the logics $\lambda(F_i)$, for frames of the kinds F_i, $i = 1, ..., 13$, is structurally complete.*

Proof. By Theorem 5.1.8, Lemma 5.1.9 Theorem 3.3.8, and Lemma 2.5.20 it is sufficient to show that, for every frame F_i from the list above, there is no p-morphism from the frame of $Ch_{\lambda(F_i)}(k)$ (for every k) onto a rooted open generated subframe $\mathcal{E}$ of the frame F_i. It is not hard to see that, in cases where

$i := 4, ..., 10$ this rooted subframe $\mathcal{E}$ is the frame F_i itself. In fact, in these cases, assume that there is a p-morphism f from $Ch_{\lambda(F_i)}(k)$ onto F_i for some k. Then f maps an element b into the root a of F_i. Because F_i is a rooted $\lambda(F_i)$-frame with a maximal number of elements, the frame $b^{R\leq}$ is isomorphic to F_i. Then, for $i = 4, 5, 7, 9$, there is no element in F_i which can be an image of the single-element co-cover in $Ch_{\lambda(F_i)}(k)$ for some two element antichain in F_i containing of maximal elements. For the cases where $i = 6, 8, 10$, there is no element in F_i which can be an image of the single-element co-cover in $Ch_{\lambda(F_i)}(k)$ for a certain two-element antichain containing an element of depth 2 in $F_i \subseteq Ch_{\lambda(F_i)}(k)$.

In the cases where $i = 2, 3$ we choose $\mathcal{E}$ as the open subframe of F_i rooted by the maximal irreflexive element a of F_i. In this case $\mathcal{E}$ has no element which could be an image under a p-morphism of the co-cover of the irreflexive elements in $Ch_{\lambda(F_i)}(k)$ which have a as their image. For $i := 1$ there is no p-morphism onto F_1 from $Ch_{\lambda(F_i)}(k)$ because the latter frame has single-element reflexive maximal elements. For $i = 12, 13$, $\mathcal{E}$ is an open subframe of F_i generated by the irreflexive element a of F_i. Clearly, for $i = 12$, $\mathcal{E}$ has no elements which could be an image under a p-morphism of a reflexive single-element co-cover for distinct maximal elements of $Ch_{\lambda(F_{12})}(k)$ which should be mapped into distinct maximal elements of $\mathcal{E}$. For $i = 13$, $\mathcal{E}$ has no elements which could be an image under a p-morphism of a reflexive single-element co-cover for distinct elements of depth 2 from $Ch_{\lambda(F_{13})}(k)$ which should be mapped into the two-element antichain of elements of depth 2 from $\mathcal{E}$. ■

From Lemmas 5.4.2 and 5.4.3 we directly derive the next theorem which is the main result of this section.

Theorem 5.4.4 *(Rybakov, [156])* In order for a modal logic λ extending $K4$ to be hereditarily structurally complete, it is necessary and sufficient that λ not be included in any of the logics $\lambda(F_i), i = 1, ..., 13$.

Thus none of the well known (and in a sense basic) modal logics $K4$, $K4.1$, $K4.2$, $K4.3$, $S4$, $S4.1$, $S4.2$, $S4.3$, $S5$, Grz and $Grz.2$ hereditarily structurally complete. But $Grz.3$, for example, is hereditarily structurally complete.

Definition 5.4.5 *A logic λ is said to be* structurally precomplete *if this logic is itself structurally incomplete but any logic strictly including λ is structurally complete (in other words, if λ is a maximal structurally incomplete logic).*

It is not difficult to see that all the logics $\lambda(F_i)$, $i \in 1, 2, ..., 13$ are incomparable. Therefore we immediately derive

Corollary 5.4.6 *(Rybakov, [156])* There exist exactly twenty structurally precomplete modal logics over $K4$, and they are all tabular.

The content of this corollary is noticeable, since now we know better the position of structurally complete logics in the lattice of all modal logics, in the sense that any structurally incomplete logic must be included in one of the tabular logics $\lambda(F_i)$, $i \in 1, 2, ..., 13$.

Now on the basis of the result obtained we are going to give a description of hereditarily structurally complete and structurally precomplete superintuitionistic logics (a description of this kind was obtained in A.Citkin [30] by employing another technique). For this, we first need to clarify the connection between the structural completeness in superintuitionistic logics and their modal counterparts.

Generally speaking, the modal counterparts of superintuitionistic logics do not inherit structural completeness. For instance, we know (see Theorem 5.4.4) that the smallest modal counterpart $S5$ for the structurally complete classical propositional calculus PC (see Exercise 5.1.11) is structurally incomplete. Indeed, the rule $\Diamond x \wedge \Diamond \neg x / y$ is admissible but not derivable in $S5$. However the greatest modal counterparts $\sigma(\lambda)$ of a superintuitionistic logics λ preserve structural completeness, as it turns out, and we show this below. Recall that T is the Gödel-Mckinsey-Tarski translation of formulas in the language of intuitionistic logic into modal propositional formulas.

Theorem 5.4.7 *In order for a superintuitionistic logic λ to be structurally complete, it is necessary and sufficient that $\sigma(\lambda)$ be structurally complete.*

Proof. Assume that $\sigma(\lambda)$ is structurally complete. Let $r := \alpha_1, ..., \alpha_n/\beta$ be a rule admissible in λ. Then by Theorem 3.2.2 $T(r)$ is admissible in $\sigma(\lambda)$. By our assumption concerning the structural completeness of $\sigma(\lambda)$ and the deduction theorem for $\sigma(\lambda)$ we have

$$T(\alpha_1) \wedge .... \wedge T(\alpha_n) \to T(\beta) \in \sigma(\lambda).$$

Then $\alpha_1 \wedge ... \wedge \alpha_n \to \beta$ is a certain theorem of the logic λ (since $\sigma(\lambda)$ is a modal counterpart for λ).

Conversely, let λ be a structurally complete superintuitionistic logic. Let $\alpha_1, ..., \alpha_n/\beta$ be an inference rule admissible in $\sigma(\lambda)$. And assume that the formula $\Box\alpha_1 \wedge ... \wedge \Box\alpha_n \to \beta$ is not a theorem of $\sigma(\lambda)$. Then, by the definition of $\sigma(\lambda)$, there is a modal algebra $\mathfrak{A}$ from $Var(\sigma(\lambda))$ generated by elements of the form $\Box a$, and such that: (i) the pseudo-boolean algebra $G(\mathfrak{A})$ of all open elements of $\mathfrak{A}$ belongs to $Var(\lambda)$, and (ii) $\mathfrak{A}$ invalidates the formula $\Box\alpha_1 \wedge ... \wedge \Box\alpha_n \to \beta$. Then $\mathfrak{A}$ invalidates the formula $\Box\alpha_1 \wedge ... \wedge \Box\alpha_n \to \beta$ as well. This means that there are terms $t_i(x_1, ..., x_m)$ built up out of $\Box x_1, ..., \Box x_m$ (we abbreviate this by $t_i(x_1, ..., x_n) = g_i(\Box x_j)$) such that

$$\Box(\bigwedge_{1 \leq k \leq n} \Box\alpha_k(g_i(\Box a_j)) \to \Box\beta(g_i(\Box a_j))) \neq \top$$

in the algebra $\mathfrak{A}$. By Lemma 2.7.14 the formulas $\Box\alpha_k(g_i(\Box x_j))$ for any k and $\Box\beta(g_i(\Box x_j))$ are equivalent in $S4$ to certain formulas $T(\gamma_k)$ and $T(\delta)$ respectively, where γ_k and δ are propositional formulas in the language of intuitionistic logic. Using this fact we conclude that the formula $\bigwedge_{1<k\le n}\gamma_k \to \beta$ is not a theorem of λ. λ is assumed to be structurally complete. Therefore the rule

$$\frac{\gamma_1,\,...,\,\gamma_n}{\delta}$$

in not admissible in λ. Then by Theorem 3.2.2 the rule

$$\frac{\Box\alpha_1(g_i(\Box x_j)),\,...,\,\Box\alpha_n(g_i(\Box x_j))}{\Box\beta(g_i(\Box a_j))}$$

is not admissible in $\sigma(\lambda)$. At the same time this rule is obtained as a substitution instance of the rule $\Box\alpha_1,\,...,\,\Box\alpha_n/\beta$ which is admissible in $\sigma(\lambda)$, a contradiction. ∎

Theorem 5.4.8 *(Citkin [30]) For any superintuitionistic logic λ, λ is hereditarily structurally complete if and only if λ is not included in any logic from the list $\lambda(\mathcal{Q}_i)$, where $i = 5,...,9$, where each $\mathcal{Q}_i$ is a partial ordered frame of the kind F_i.*

Proof. If λ is a superintuitionistic logic which is not contained in each logic from the list above then $\sigma(\lambda)$ is hereditarily structurally complete by Theorem 5.3.5. Therefore Theorem 5.4.7 implies that λ is also hereditarily structurally complete. Conversely, suppose that λ is contained in a $\lambda(\mathcal{Q}_i)$. Then the modal logic λ_1 of the frame $\mathcal{Q}_i$ is not structurally complete by Lemma 5.4.3. At the same time $\lambda_1 = \sigma(\lambda(\mathcal{Q}_i))$ (see Lemma 2.7.1, 2.7.17). Consequently, by Theorem 5.4.7, the logic $\lambda(\mathcal{Q}_i)$ is structurally incomplete. ∎

Corollary 5.4.9 *(Citkin [30]) There are exactly five structurally precomplete superintuitionistic logics (namely $\lambda(\mathcal{Q}_i)$, $i = 5,...,9$), and they are all tubular.*

Note that the limitations discovered in the structure of λ-frames for hereditarily structurally complete logics λ also allow us to extract various other positive properties for logics of this kind.

Theorem 5.4.10 *(Rybakov [156]) Let λ be a modal logic extending $K4$ or a superintuitionistic logic, and let λ be hereditarily structurally complete. Then*

  (a) λ has the fmp and is complete with respect to a class of finite frames of the kind described in Corollary 5.3.8;

  (b) λ is finitely axiomatizable and decidable.

In particular, there are only countably many logics of this kind.

Proof. First we consider the case when λ is a modal logic. Then, in fact, λ has the fmp and is complete with respect to a class of finite frames, which is pointed above in (a), according to Theorem 5.4.4, and Lemmas 5.3.3, 5.3.5, 5.3.6 and Corollary 5.3.8. Using the structure of frames of the kind described in Corollary 5.3.8 we can prove that all hereditarily structurally complete logics above $K4$ are finitely axiomatizable (i.e. (b) holds). To show this we need to the point to modify only the proof of Theorem 4.3.11 which implies that all extensions of $S4.3$ are finitely axiomatizable. All frames mentioned above in (a) have either r-type or i-type (see Corollary 5.3.8). In order to prove finite axiomatizability of λ it is sufficient to show that any infinite sequence $\mathcal{S}$ of finite frames of these kinds has a subsequence $\mathcal{S}_1$ such that every it's frame $\mathcal{F}$ is a p-morphic image of every successor of $\mathcal{F}$ in $\mathcal{S}_1$.

Of course we may suppose that our sequence $\mathcal{S}$ consists of only i-frames, or only r-frames (see Corollary 5.3.8). Now we can pick up an infinite subsequence of $\mathcal{S}$ consisting of frames which either (1) have no terminal frames of kind F_{14} (in theirs decomposition into a sequential composition), or (2) all they have this kind terminal frame. First we investigate the case (1) under assumption that all frames of $\mathcal{S}$ are r-frames. We consider a description of such frames as finite lists of elements from the union of two sets: the set of natural numbers N and the set $A := N^3$. In this lists, the elements of A indicates the numbers of clusters in frames of kind α (see Corollary 5.3.8) and the numbers of elements in their maximal clusters. And numbers of N indicate the numbers of elements in nodal clusters of kind β. Now it is not difficult to modify the proof of Theorem 4.3.11 for our case: we simply consider finite sequences of elements from A and N in these lists separately and consequently. This reasoning gives us that there exists a subsequence $\mathcal{S}_1$ of $\mathcal{S}$ in which any frame is p-morphic image of all its successors in $\mathcal{S}_1$, which is what we needed.

Suppose that (1) holds and all the frames of $\mathcal{S}$ are i-frames. If there is an infinite subsequence of $\mathcal{S}$ all frames of which have only single immediate successor of the root then we simply ignore the root in this subsequence and carry out, for such restricted frames, the reasoning as above, which gives us required result. If there is an infinite subsequence $\mathcal{S}_1$ of $\mathcal{S}$ all frames of which has two immediate successors for the root then these frames have the kind described in (i), (ii) and (iii) of Corollary 5.3.8. Therefore we can remove from $\mathcal{S}_1$ all frames of kind (i) and (ii), this gives us an infinite subsequence which we denote $\mathcal{S}_2$.

In $\mathcal{S}_2$ there is an infinite subsequence $\mathcal{S}_3$ all frames of which either have a single maximal element or have two maximal elements. Now, in any frame from $\mathcal{S}_3$ we remove both maximal elements (if they exist) and the root, two immediate successors of the root, and the nodes immediate successors for them. Now we apply to the remaining parts of frames the reasoning as in the case (1) when all frames are r-frames and obtain the required result.

Consider now the case (2). First we remove the frame of the kind F_{14} from all frames of $\mathcal{S}$ and obtain a sequence $\mathcal{S}_1$. And then we carry out the reasoning as in

(1). We get that all modified frames of some infinite subsequence $\mathcal{S}_2$ of the modified sequence $\mathcal{S}_1$ have the following property: any frame is a p-morphic image of any its successor in $\mathcal{S}_2$. Now we consider all the frames of $\mathcal{S}$ corresponding to modified frames from $\mathcal{S}_2$. These frames form a subsequence $\mathcal{S}_3$. Now we take a subsequence $\mathcal{S}_4$ of $\mathcal{S}_3$ in which the number of elements in proper cluster of F_{14} increase (not strictly). It can be easily seen that $\mathcal{S}_4$ has the required property. Thus in modal case we have proved (a) and (b) and our theorem is complete (note that, like to Theorem 4.3.11, we also can use here the Kruskal's Tree Theorem [79] to proof this step).

In order to complete our theorem for the case of superintuitionistic logics we need the following simple lemma.

Lemma 5.4.11 *A superintuitionistic logic λ is hereditarily structurally complete iff $\sigma(\lambda)$ is hereditarily structurally complete.*

Proof. If a superintuitionistic logic λ is hereditarily structurally complete then by Theorem 5.4.8 λ is not included in any logic from the list in Theorem 5.4.8. Then $\sigma(\lambda)$ is not included in any logic from the list from Theorem 5.4.4 and consequently $\sigma(\lambda)$ is also hereditarily structurally complete. Conversely, let $\sigma(\lambda)$ be a hereditarily structurally complete logic. Let λ_1 be a superintuitionistic logic extending λ. Then $\sigma(\lambda) \subseteq \sigma(\lambda_1)$ and $\sigma(\lambda_1)$ is structurally complete. Using Theorem 5.4.7 λ_1 is also structurally complete. ∎

To continue the proof of our theorem consider a hereditarily structurally complete superintuitionistic logic λ. By Lemma 5.4.11 $\sigma(\lambda)$ is hereditarily structurally complete. Therefore by proved already part of our theorem for the case of modal logics $\sigma(\lambda)$ has the fmp and complete with respect to described in (a) finite frames. Then by Corollary 2.7.21 λ also has the fmp and by Theorem 2.7.20 and Corollary 2.7.21 and Lemma 2.7.6 λ is also complete with respect to frames of the kind described in (a). Thus (a) holds for superintuitionistic logics also. In order to prove (b) we can apply the same and even simplified proof as for the proof of (b) for the case of modal logics. ∎

Note that (a) in our theorem above gives an exhaustive semantic description of all hereditarily structurally complete modal and superintuitionistic logics. We shall close this section by considering structural completeness for infinite inference rules. The principal reason for this consists of the fact that the main questions concerning structural completeness for infinite rules can be very easy reduced to the questions concerning structural completeness for usual finite inference rules, as it was shown by W.Dziobiak [36].

Definition 5.4.12 *An infinite inference rule is an expression r of the form $\mathcal{X}/\beta$ where β is a formula and $\mathcal{X}$ is an infinite set of formulas.*

The notions of admissibility and derivability can be extended to infinite rules in the obvious way.

Definition 5.4.13 *We say that a logic* λ *is* structurally complete in the infinitary sense *if any infinite inference rule admissible in* λ *(in particular, every finite rule) is derivable in* λ.

A logic λ is hereditarily structurally complete in the infinitary sense if any extension of λ is structurally complete in the infinitary sense.

Lemma 5.4.14 *If a modal logic* λ_1 *extending S4 (or a superintuitionistic logic* λ_1*) is structurally complete in the infinitary sense then* λ_1 *is tabular.*

Proof. Let λ_1 be a non-tabular logic. Then $\lambda_1 \subseteq \lambda$, where λ is a pretabular logic (see Proposition 4.5.3). Using Theorems 4.3.31 and 4.3.32 we conclude that λ has the fmp. In particular, there are rooted finite λ-frames with more then n elements for each given n. From this observation it is easy to see that the infinite inference rule

$$\{(p_i \equiv p_j) \to p_0 \mid 0 < i < j < \infty\}/p_0$$

is not derivable in λ. Now it remains only to note that this inference rule is admissible in λ, which follows directly from the fact that λ has the fmp. Thus λ_1 is not structurally complete. ∎

Theorem 5.4.15 *(W.Dziobiak [36]) A modal logic over S4 (a superintuitionistic logic)* λ_1 *is hereditarily structurally complete in the infinitary sense iff* λ_1 *is hereditarily structurally complete in the finitary sense and is tabular.*

Proof. In fact, by Lemma 5.4.14 we only need to establish the sufficiency of the condition. Let a tabular logic λ_1 be a hereditarily structurally complete logic in the finitary sense. Let λ be a logic extending λ_1. Then λ is tabular and structurally complete in the finitary sense. Let $r := \mathcal{X}/\alpha$ be an infinite inference rule admissible in λ. We claim that there is a finite $\mathcal{Y} \subset \mathcal{X}$ such that $\mathcal{Y}/\alpha$ is admissible in λ.

Suppose otherwise: let for every finite $\mathcal{Y}$, the rule $\mathcal{Y}/\alpha$ be not admissible. Clearly, every rule $\mathcal{Y}/\alpha$ is admissible in λ iff the corresponding quasi-identity is valid in $\mathfrak{F}_\lambda(n)$, where n is the number of all subsets of a finite frame $\mathcal{Q}$, where $\lambda = \lambda(\mathcal{Q})$ (see Lemma 4.1.10 and Theorem 1.4.5). Therefore each finite set consisting of the complete diagram Φ of $\mathfrak{F}_\lambda(n)$, a certain finite set of formulas $\{\mathcal{B} = \top \mid \mathcal{B} \in \mathcal{Y}\}$ and $\alpha \neq \top$ is consistent. By the compactness theorem for first order logic, this means that the set $\{\mathcal{B} = \top \mid \mathcal{B} \in \mathcal{X}\} \cup \{\mathcal{A} \neq \top, \Phi\}$ is consistent as well. And this implies the rule r is not admissible in λ. Thus, for some $\mathcal{Y}$, $\mathcal{Y}/\alpha$ is admissible. λ is structurally complete in the finitary sense, therefore $\mathcal{Y}/\alpha$ is derivable and, consequently, r is also derivable in λ. ∎

5.5   Structurally Complete Fragments

As we have seen above many important and popular logical systems are structurally incomplete. However, certain inference rules of a particular form can be derived in such systems provided these rules are admissible. We might expect that such rules will have a rather homogeneous structure and that the classes of such rules will be representative. This leads us to properly investigate such classes and to specify inference rules which are admissible iff they are derivable. The main content of this section is a description of G.Mints [104] for a representative class of inference rules which are equivalent with respect to admissibility and derivability in the intuitionistic logic H.

Definition 5.5.1 *Let λ be a logic, and let C be the set of all logical connectives of the language of λ. Suppose $\mathcal{D}$ is a subset of C. The $\mathcal{D}$-fragment $\mathcal{R}_{\mathcal{D}}$ of any set $\mathcal{R}$ of inference rules in the language of λ is the set of all rules from $\mathcal{R}$ which only contain occurrences of logical connectives from $\mathcal{D}$. We call the $\mathcal{D}$-fragment of all rules in the language of λ by $\mathcal{D}$-rules.*

Definition 5.5.2 *A rule r is called stable in a logic λ if r is admissible in λ if and only if r is derivable in λ.*

We begin with a simple observation concerning the semantic properties of derivable rules.

Lemma 5.5.3 *A rule $r := \alpha_1, ..., \alpha_n/\beta$ in the language of an algebraic logic λ which possesses a general deduction theorem with n-deduction formula β, is derivable in λ if $\beta(\alpha_1, ..., \alpha_n, \beta) \in \lambda$ or, equivalently, if $\beta(\alpha_1, ..., \alpha_n, \beta)$ is valid in every algebra from a class $\mathcal{K}$ generating $Var(\lambda)$.*

Since the logic H which we will consider below satisfies the condition described above, we can recognize the derivability of rules in H applying this lemma.

Definition 5.5.4 *We say a rule r is reducible to a rule r_1 in a logic λ if*

> *(i) the admissibility of r in λ implies r_1 is admissible in λ;*
>
> *(ii) the derivability of r_1 in λ implies r is derivable in λ.*

A rule r is reducible to a pair of rules r_1, r_2 in a logic λ if

> *(iii) the admissibility of r in λ implies r_1 and r_2 are also admissible in*
> *the logic λ;*
>
> *(ii) the derivability of both rules r_1 and r_2 in λ implies*
> *r is derivable in λ.*

We will now investigate the intuitionistic logic H, which is itself structurally incomplete (see Examples 3.5.19, 3.5.21 and Corollary 3.5.27). We will often use the Glivenko's Theorem (see Corollary 2.8.3) below. Recall this theorem says that any formula α is provable in PC iff $\neg\neg\alpha$ is provable in H. In particular, if $\alpha \equiv \beta \in PC$ then $\neg\alpha \equiv \neg\beta \in H$. Thus we can make classically equivalent substitutions in formulas of kind $\neg\gamma$ preserving their equivalence in H. The possibility of using such substitutions will be very often used below without additional references.

Lemma 5.5.5 *The following hold*

(1) *Any rule of the form* $\alpha \wedge \beta, \Gamma/\gamma$ *is reducible in* H *to* $\alpha, \beta, \Gamma/\gamma$,

(2) *Any rule of the form* $(\alpha \vee \beta), \Gamma/\gamma$ *is reducible in* H *to the pair* $[\alpha, \Gamma/\gamma]$, $[\beta, \Gamma/\gamma]$,

(3) *Any rule of the form* $\Gamma/\alpha \wedge \beta$ *is reducible in* H *to the pair* $[\Gamma/\alpha]$, $[\Gamma/\beta]$,

(4) *Any rule of the form* $\Gamma/\alpha \to \beta$ *is reducible in* H *to* $\alpha, \Gamma/\beta$,

(5) *Any rule admissible in* H *and having the form* $\Gamma/\neg\gamma$ *is derivable in* H,

(6) *If formulas* $(\bigwedge_{1 \leq i \leq n} \delta_i) \to ((\bigwedge_{1 \leq i \leq m} \alpha_i) \equiv (\bigwedge_{1 \leq i \leq k} \beta_i))$ *and* $(\bigwedge_{1 \leq i \leq n} \delta_i) \to (\alpha \equiv \beta)$ *are theorems of* H *then the rule* $\bigwedge_{1 \leq i \leq n} \delta_i, \bigwedge_{1 \leq i \leq m} \alpha_i/\alpha$ *is reducible in* H *to the rule* $\bigwedge_{1 \leq i \leq n} \delta_i, \bigwedge_{1 \leq i \leq k} \beta_i/\beta$

(7) *Any rule of the form* $(y \equiv \alpha), \Gamma/\beta$, *where* y *is a propositional letter, is reducible in* H *to the rule* $\Gamma(\frac{y}{\alpha})/\beta(\frac{y}{\alpha})$,

(9) *Any rule of the form* $y, \Gamma/\beta$, *where* y *is a propositional letter, is reducible in* H *to the rule* $\Gamma(\frac{y}{\top})/\beta(\frac{y}{\top})$.

Proof. (1) is evident. (2): the part concerning the derivability follows from the deduction theorem for H (see Theorem 1.3.9) and the fact that

$$(\alpha \to \gamma) \to ((\beta \to \gamma) \to (\alpha \vee \beta \to \gamma))$$

is a theorem of H. The part concerning the admissibility is evident. (3) and (4) follow easy from deduction theorem for H. (5): Suppose $\Gamma/\neg\gamma$ is not derivable in H. Then $\chi := \bigwedge \Gamma \to \neg\gamma \notin H$. Since H has the fmp (Corollary 2.6.14), the formula χ is falsified in a finite poset $\mathcal{F}$, which implies that γ is false in the single-element frame $\mathcal{A}$. Therefore there is a substitution of formulas $\top$ and $\bot$ in place of all variables of γ, such that the value of $\bigwedge \Gamma$ in $\mathcal{A}$ is $\top$ and the value of γ is $\bot$. Since the value of any formula built up out of $\top$ and $\bot$ in any pseudoboolean algebra and the two-element boolean algebra are the same, we get $\Gamma/\neg\gamma$ is not admissible in H. (6) and (7) are obvious. (8) follows easy from (7), (6) and $(y \equiv \top) \equiv y \in H$. ∎

Theorem 5.5.6 *(Mints [104]) Every* $\wedge, \vee, \neg$-*rule is stable in* H.

Proof. Suppose $r := \Gamma/\alpha$ is a rule which does not include occurrences of the connective $\rightarrow$. Suppose r is admissible in H. By (1), (2) and (8) from Lemma 5.5.5 we may assume that all formulas from Γ begin with $\neg$. By (3) and (6) from Lemma 5.5.5 we may assume that if α is not $\top$ then $\alpha = x_1 \vee ... \vee x_k \vee \neg \delta_1 \vee ... \vee \neg \delta_v$, where $x_1, ..., x_k$, $k \geq 0$, are variables, $\delta_1, ... \delta_v$ are certain formulas and $k + v > 0$. It may also be assumed that $\alpha \neq \top$ and that the formulas of Γ form a set which is consistent in H (otherwise the result is obvious). Assume that a rule r (in the form obtained) is not derivable in H.

Then since H has the fmp, there is a finite rooted poset model $\mathfrak{M}$ which invalidates the formula $\bigwedge \Gamma \rightarrow \alpha$. And the frame of $\mathfrak{M}$ is an open rooted subframe $\mathcal{F}$ of the frame $Ch_H(m)$ for some m. We take the following valuation S of variables from r in $Ch_H(m)$. S coincides on the maximal elements of $\mathcal{F}$ with the valuation of $\mathfrak{M}$ on the corresponding elements. On the maximal elements of $Ch_H(m)$ which are outside $\mathcal{F}$ the valuation S of variables from r is the same as that of on some fixed maximal element from $\mathcal{F}$. For the other elements of $Ch_H(m)$ we stipulate that no variable is true under S. Clearly S invalidates the rule r in $Ch_H(m)$. By Theorem 3.5.8 it follows that r is not admissible in H, a contradiction. Therefore r is derivable in H and, since we reduced the original inference rule to the final one, it follows that the original rule r is derivable in H as well. ∎

For simplicity of notation, for any finite set of formulas $\mathcal{S}$, we let $\mathcal{S}$ stand for the conjunction of all its members $\bigwedge \mathcal{S}$, And we use capital letters to denote subformulas which are conjunctions of some formulas.

Lemma 5.5.7 *Any rule of the form* $r := \Delta, \Gamma/x$, *where*

$$\Delta := \{\mathcal{S}_1 \rightarrow x_1, ..., \mathcal{S}_m \rightarrow x_m, \mathcal{S}_{m+1} \rightarrow \\ \neg\neg x_{m+1}, ..., \mathcal{S}_{m+n} \rightarrow \neg\neg x_{m+n}, \},$$

and $x_1, ..., x_{m+n}$ *are variables distinct from the variable* x, *is reducible in* H *to the rule* $\Gamma(^{x_1,...,x_{m+n}}_{\beta_1,...,\beta_{m+n}})/x$, *where* $\beta_i := \Delta \rightarrow x_i$, $i = 1, 2, ..., m + n$.

Proof. For simplicity we will denote the formula $\gamma(^{x_1,...,x_{m+n}}_{\beta_1,...,\beta_{m+n}})$ by γ^* and, for any set of formulas S, we let S^* denote the set $\{\gamma^* \mid \gamma \in S\}$. If the rule r is admissible in H then the rule $\Delta^*, \Gamma^*/x$ is also admissible in H. We will show that any formula from the list Δ^* is derivable in H. Indeed, for $i \leq m$, the i-th member χ_i of this list has the form $\chi_i := \mathcal{S}_i^* \rightarrow (\Delta \rightarrow x_i)$. Since $\Delta \rightarrow (\beta_i \equiv x_i)$ is derivable in H, we replace all β_i in χ_i by the corresponding x_i and obtain the formula $\epsilon_i := \mathcal{S}_i \rightarrow (\Delta \rightarrow x_i)$, which is equivalent to χ_i modulo derivability in H (because $(p \rightarrow (q \rightarrow u)) \equiv (q \rightarrow (p \rightarrow u)) \in H$). But ϵ_i is evidently derivable in H since the set Δ includes the formula $\mathcal{S}_i \rightarrow x_i$. Therefore χ_i is derivable in H as well. For $i > m$, the i-th member of Δ^* has the form $\chi_i := \mathcal{S}_i^* \rightarrow \neg\neg(\Delta \rightarrow x_i)$.

Again since $(p \to \neg\neg(q \to u)) \equiv (q \to \neg\neg(p \to u)) \in H$ and $\Delta \to (\beta_i \equiv x_i)$ is derivable in H, we again replace all β_i in χ_i by the corresponding x_i, and we obtain the formula $\epsilon_i := \mathcal{S}_i \to \neg\neg(\Delta \to x_i)$, which is equivalent to χ_i modulo derivability in H. But once again ϵ_i is derivable in H since the set Δ includes the formula $\mathcal{S}_i \to x_i$, and $\chi_i \in H$. Thus all the formulas from Δ^* are derivable in H and Γ^*/x is admissible in H.

Suppose Δ^*/x is derivable in H, i.e. $\Delta^* \to x \in H$. We replace everywhere x by the formula $\Delta \to x$ in $\Delta^* \to x$. Then using (i) the fact that $\Delta \to (\beta_i \equiv x_i) \in H$ and $\Delta \to ((\Delta \to x) \equiv x) \in H$ and (ii) the fact that we may replace formulas $\Delta \to x$ by x and formulas β_i by x_i $(i = 1, 2, ...m+n)$ in the corresponding parts, we obtain that the formula $\Gamma \to (\Delta \to x)$ is derivable in H. From this it follows immediately that the rule r is derivable in H. $\blacksquare$

Lemma 5.5.8 *Any admissible in H rule r of the form*

$$\frac{\mathcal{S}_1 \to x, ..., \mathcal{S}_m \to x, \mathcal{S}_{m+1} \to \neg\neg x, ..., \mathcal{S}_{m+n} \to \neg\neg x}{x}$$

is derivable in H.

Proof. Assume that r is admissible in H. For short we will denote the set of all premises of the rule r by Δ. We substitute the formula $\Delta \to x$ in place of x in r and as result we obtain a rule r_1 which is also admissible in H. Clearly all premises of r_1 are derivable in H, consequently the conclusion of r_1, namely $\Delta \to x$, is also derivable in H, hence the rule r is derivable in H. $\blacksquare$

Theorem 5.5.9 *(Mints [104]) Every $\{\wedge, \to, \neg\}$-rule is stable in the intuitionistic logic H.*

Proof. Suppose r is an $\{\wedge, \to, \neg\}$-rule. Using (3), (4) and (5) from Lemma 5.5.5 we may suppose that the conclusion of r is a variable x. Using (1), (6) and (8) from Lemma 5.5.5 and

$$(p \to (q \to u)) \equiv (p \wedge q \to u) \in H,$$

$$(p \to q \wedge u)) \equiv (p \to q) \wedge (p \to u)) \in H,$$

$$(p \to \neg q) \equiv \neg(p \wedge q) \in H$$

we can assume also that all premises of r either have a form $\mathcal{S} \to z$, where z is a variable, or begins with $\neg$. All premises of the form $\mathcal{S} \to z$, where $z \neq x$, can be rejected by Lemma 5.5.7. Finally, any pair of formulas $\neg\beta, \neg\gamma$ in the premise can be replaced by the formula $\neg\neg(\neg\beta \wedge \neg\gamma)$ according to Glivenko's Theorem and (6) from Lemma 5.5.5. After all these reductions, the rule r is reduced to a rule of the form

$$\neg\alpha, \mathcal{S}_1 \to x, ..., \mathcal{S}_k \to x/x. \tag{5.5}$$

We will show now that the formula $\Gamma \to \neg\neg x$, where Γ is the list of premises in (5.5), is derivable in H provided the rule in (5.5) is admissible in H. First we replace x in (5.5) by the formula $\neg\neg x$, and then, using (6) from Lemma 5.5.5 and applying Glivenko's Theorem and $(p \to \neg\neg q) \equiv \neg(p \wedge \neg q) \in H$, we replace $\neg\neg x$ by x everywhere within $\neg\alpha$ and the formulas of premises $\mathcal{S}_1, ..., \mathcal{S}_k$. This way we conclude that the following rule is admissible in H:

$$\neg\alpha, \mathcal{S}_1 \to \neg\neg x, ..., \mathcal{S}_k \to \neg\neg x / \neg\neg x.$$

By (5) from Lemma 5.5.5 the formula $\neg\alpha, \mathcal{S}_1 \to \neg\neg x, ..., \mathcal{S}_k \to \neg\neg x \to \neg\neg x$ (recall we mean here the conjunction of formulas) is derivable in H which implies that the required formula $\neg\alpha, \mathcal{S}_1 \to x, ..., \mathcal{S}_k \to\to \neg\neg x$ is also derivable in H. Thus we can add $\neg\neg x$ to the list of premises of the rule in (5.5) and then, applying (6) from Lemma 5.5.5 we can exclude all occurrences of x in the formula $\neg\alpha$, using that $\neg\neg x \wedge \neg\alpha \equiv (\neg\neg x \wedge \neg\alpha(\frac{x}{\top})) \in H$. The next step concerns a reducing of the number of variables in $\neg\alpha$. We pick a variable p occurring in $\neg\alpha$ and decompose $\neg\alpha$ with respect to p:

$$\neg\alpha \equiv ((\beta \to \neg\neg p) \wedge (p \to \neg\delta)) \in PC,$$

where formulas β, δ do not include p (the fact of existence of such decomposition can be easily obtained from the presentation of $\neg\alpha$ within PC in the conjunctive normal form and the observation that we may assume that formulas β and δ have only connectives $\wedge, \neg$). Then by Glivenko's Theorem

$$\neg\neg\neg\alpha \equiv \neg\neg((\beta \to \neg\neg p) \wedge (p \to \neg\delta)) \in H,$$

but $\neg\neg\neg\alpha \equiv \neg\alpha \in H$ and

$$\neg\neg((\beta \to \neg\neg p) \wedge (p \to \neg\delta)) \equiv ((\beta \to \neg\neg p) \wedge (p \to \neg\delta)) \in H$$

(this can be easy verified by using the fmp of H). Hence

$$\neg\alpha \equiv ((\beta \to \neg\neg p) \wedge (p \to \neg\delta)) \in H.$$

Now we replace $\neg\alpha$ in the premise of the rule by the following formula $(\beta \to \neg\neg p) \wedge (p \to \neg\delta)$ and then we exclude $\beta \to \neg\neg p$ by means of Lemma 5.5.7, substituting $(\beta \to \neg\neg p) \to p$ for p, and as a result we obtain the rule

$$((\beta \to \neg\neg p) \to p) \to \neg\delta), \mathcal{S}_1^* \to x, ..., \mathcal{S}_k^* \to x, \neg, \neg x / x,$$

where staring means the results of corresponding substitutions. The formula $(((\beta \to \neg\neg p) \to p) \to \neg\delta)$ is equivalent in PC to the formula $\beta \vee p \to \neg\delta$. Consequently $(((\beta \to \neg\neg p) \to p) \to \neg\delta)$ is equivalent in H to $\beta \vee p \to \neg\delta$ and to $(\beta \to \neg\delta) \wedge (p \to \neg\delta)$. Therefore the first premise of our reduced rule can be replaced by the pair $\beta \to \neg\delta, p \to \neg\delta$ the first member of which no longer include p and is equivalent in H to the formula $\neg(\beta\wedge\delta)$. Since the second member of this pair is equivalent in H to the formula $p \equiv (p\wedge\neg\delta)$, we finally exclude $p \to \neg\delta$ by (7) from Lemma 5.5.5. Carrying on this procedure we finally reject all formulas in the premise beginning with $\neg$ except $\neg\neg x$ and complete the proof applying Lemma 5.5.8. ∎

Theorems 5.5.6 and 5.5.9 give us descriptions of representative classes of stable in H rules, i.e. rules for which admissibility and derivability in H are equipotent. Also using Theorems 5.5.6 and 5.5.9 we immediately infer

Theorem 5.5.10 *(Mints [104]) If a rule r is not stable in H then r has occurrences of connectives $\vee$ and $\to$.*

We know (cf. Example 3.5.29) that the Mints rule

$$\frac{(x \to y) \to x \vee z}{((x \to y) \to x) \vee ((x \to y) \to z)}$$

is admissible but not derivable in H. Therefore the presence of only connectives $\vee$ and $\to$ is already sufficient in order to produce an example non-stable in H rule.

Chapter 6

Related Questions

6.1 Rules with Meta-Variables

The considerations of the previous sections have showed us that there is a close connection between the admissibility of inference rules and the validness of the corresponding quasi-identities in the free algebras of the corresponding varieties. Using studied various tools we can recognize the admissibility of inference rules and the truth of quasi-identities in free algebras. In this section we will study generalized inference rules - rules which have meta-variables, and investigate their admissibility. The motivation for this comes from two distinct sources. First of these is the so-called substitution problem and the problem of solvability of logical equations for non-standard logics. Second source is purely algebraic: it is the problem of decidability of equations in free algebras corresponding to algebraic logics. We begin by explaining the substitution problem. For a given logic λ, the *substitution problem* for that logic is the following question: given an arbitrary formula of the form $\alpha(x_1, ..., x_n, p_1, ..., p_k)$, where $x_1, ..., x_n$ are variables and $p_1, ..., p_n$ are propositional letters, exists there a tuple of formulas $\beta_1, ..., \beta_n$ such that

$$\alpha(\beta_1, ..., \beta_n, p_1, ..., p_k) \in \lambda.$$

Of course all the x_i and all the p_j above are merely propositional letters, but we divide them into two classes considering the x_i as variables for formulas and the p_j as meta-variables (or parameters). For instance, if a formula has no variables and has only certain meta-variables, i.e. has the form $\alpha(p_1, ..., p_k)$, then the substitution question is simply reduced to the question of whether $\alpha(p_1, ..., p_k)$ is a theorem of λ. This shows why we call the p_j by metavariables. In essence we can now consider p_j as variables for arbitrary formulas in the sense that if $\alpha(x_1, ..., x_n, p_1, ..., p_k) \in \lambda$ then we can conclude that $\alpha(\gamma_1, ..., \gamma_n, p_1, ..., p_k) \in \lambda$ for any formulas $\gamma_1, ..., \gamma_n$.

Definition 6.1.1 *The substitution problem for a logic λ is decidable if there is an algorithm which decides for any given formula $\alpha(x_1, ..., x_n, p_1, ..., p_k)$ whether there exist formulas $\beta_1, ..., \beta_n$ such that $\alpha(\beta_1, ..., \beta_n, p_1, ..., p_k) \in \lambda$.*

For algebraic logics, the substitution problem is a particular case of a more general problem: the problem of decidability for logical equations and the problem of finding their solutions.

Definition 6.1.2 *A logical equation for an algebraic logic λ is an expression*

$$\alpha(x_1, ..., x_n, p_1, ..., p_k) \equiv \beta(x_1, ..., x_n, p_1, ..., p_k),$$

where $\alpha(x_1, ..., x_n, p_1, ..., p_k)$ and $\beta(x_1, ..., x_n, p_1, ..., p_k)$ are formulas in the language of λ built up out of the variables $x_1, ..., x_n, p_1, ..., p_k$, where the letters x_i are the variables (for formulas) and p_j are the meta-variables (or *parameters). A solution to this equation in λ is a tuple of formulas $\beta_1, ..., \beta_n$ such that*

$$\alpha(\beta_1, ..., \beta_n, p_1, ..., p_k) \equiv \beta(\beta_1, ..., \beta_n, p_1, ..., p_k) \in \lambda.$$

We pause briefly to comment on this definition. There is an evident resemblance between the usual notion of algebraic equations and the definition of logical equations given above. In fact, a general algebraic equation in n variables is an algebraic expression $f(x_1, ..., x_n, p_1, ..., p_k) = g(x_1, ..., x_n, p_1, ..., p_k)$, where $x_1, ..., x_n$ are variables, and $p_1, ..., p_k$ are coefficients (or parameters). Its solution is a tuple of elements $a_1, ..., a_n$ such that

$$f(a_1, ..., a_n, p_1, ..., p_k) = g(a_1, ..., a_n, p_1, ..., p_k).$$

So, in a sense, there is no difference between logical and algebraic equations, we just consider the equivalence $\equiv$ instead of equality $=$, but $\equiv$ has a similar meaning to $=$.

The reduction of the substitution problem to that concerning the decidability of logical equations is evident:

Lemma 6.1.3 *The logical equation $\alpha(x_1, ..., x_n, p_1, ..., p_k) \equiv \top$ is decidable in an algebraic logic λ iff there are formulas $\beta_1, ..., \beta_n$ such that $\alpha(\beta_1, ..., \beta_n, p_1, ..., p_k) \in \lambda$.*

Hence there is the required substitution iff the corresponding logical equation is decidable. For the classical propositional calculus PC, as well as for any uniformly locally finite algebraic logic λ, the substitution problem and the problem of decidability of logical equations can be easily answered.

Lemma 6.1.4 *If λ is an algebraic uniformly locally finite decidable logic then the problem of recognizing whether a given logical equation is decidable, and the problem of finding its solutions, have effective decision.*

Proof. Indeed, for formulas $\gamma_1, ..., \gamma_n$,

$$\alpha(\gamma_1, ..., \gamma_n, p_1, ..., p_k) \equiv \beta(\gamma_1, ..., \gamma_n, p_1, ..., p_k) \in \lambda \Leftrightarrow$$
$$\alpha(\delta_1, ..., \delta_n, p_1, ..., p_k) \equiv \beta(\delta_1, ..., \delta_n, p_1, ..., p_k) \in \lambda,$$

where $\delta_1, ..., \delta_n$ are built up out of variables $p_1, ..., p_k$. Since λ is uniformly locally finite we may consider only finitely many formulas $\delta_1, ..., \delta_n$ from a finite effectively bounded set of formulas. Since λ is decidable, this gives us the required algorithm for recognizing whether a given logical equation is decidable and an algorithm for obtaining solutions of logical equations. ∎

For logics which are not locally finite the problem of the decidability of their logical equations and the substitution problem may become very difficult. For instance, the substitution problem for the intuitionistic logic H was an open question from 1950's and was much discussed by the Moscow's school of logic (which was then headed by P.Novikov) and then also by the Sankt-Peterburg logical school. Below we will solve this problem and a number related problems for H and other logics by means of an approach which uses inference rules with meta-variables and their admissibility (we will, of course, use the tools for deciding the admissibility of ordinary inference rules which were developed in the previous chapters). First we describe how the question concerning the decidability of logical equations can be reduced to the problem of the admissibility of certain inference rules with meta-variables.

Definition 6.1.5 *An inference rule with meta-variables (or parameters) in a language $\mathcal{L}$ is an expression*

$$r := \frac{\alpha_1(x_1, ..., x_n, p_1, ..., p_k), ..., \alpha_m(x_1, ..., x_n, p_1, ..., p_k)}{\beta(x_1, ..., x_n, p_1, ..., p_k)},$$

where all $\alpha_j(x_1, ..., x_n, p_1, ..., p_k)$ and $\beta(x_1, ..., x_n, p_1, ..., p_k)$ are formulas in the language $\mathcal{L}$ built up out of the letters $x_1, ..., x_n, p_1, ..., p_k$, and all the x_q are variables of r, and all the p_t are meta-variables of r.

Definition 6.1.6 *A rule with metavariables*

$$r := \frac{\alpha_1(x_1, ..., x_n, p_1, ..., p_k)\alpha_m(x_1, ..., x_n, p_1, ..., p_k), ...,}{\beta(x_1, ..., x_n, p_1, ..., p_k)}$$

is admissible in a logic λ iff for any formulas $\gamma_1, ..., \gamma_n$, $\beta(\gamma_1, ..., \gamma_n, p_1, ..., p_k)$ is a theorem of λ provided $\forall j, \alpha_j(\gamma_1, ..., \gamma_n, p_1, ..., p_k) \in \lambda$.

Thus the only difference between our definition of admissibility for rules with meta-variables and our definition of admissibility for ordinary rules is that we cannot substitute formulas in place of meta-variables.

Definition 6.1.7 *A tuple of formulas $\gamma_1, ..., \gamma_n$ form an* obstacle *for a rule r with metavariables (in described above form) in a logic λ if*

$$\forall j, \alpha_j(\gamma_1, ..., \gamma_n, p_1, ..., p_k) \in \lambda, \quad but \quad \beta(\gamma_1, ..., \gamma_n, p_1, ..., p_k) \notin \lambda.$$

The questions of determining whether a logical equation has a solution in a logic can be evidently reduced to the question whether the appropriate inference rule with meta-variables is admissible in this logic.

Lemma 6.1.8 *Let λ be an algebraic logic. And let δ be a formula without variables (say $\delta = \bot$) and $\delta \notin \lambda$. A logical equation*

$$\alpha(x_1, ..., x_n, p_1, ..., p_k) \equiv \gamma(x_1, ..., x_n, p_1, ..., p_k)$$

has a solution $\beta_1, ..., \beta_1$ in λ iff the rule with metavariables

$$r := \frac{\alpha(x_1, ..., x_n, p_1, ..., p_k) \equiv \gamma(x_1, ..., x_n, p_1, ..., p_k)}{\delta}$$

is not admissible in λ and $\beta_1, ..., \beta_n$ form an obstacle for r in λ.

Thus we can answer the problem of determining the decidability of logical equations and the substitution problem in a logic λ provided we have an algorithm for deciding whether inference rules with meta-variables are admissible in λ. And this fact leads us to investigate such rules, it is the first motivation mentioned above. The second motivation comes from algebra or applied model theory. We explain this motivation below simultaneously describing a certain approach to handle rules with meta-variables.

Theorem 6.1.9 *An inference rule with meta-variables*

$$r := \frac{\alpha_1(x_1, ..., x_n, p_1, ..., p_k), ..., \alpha_m(x_1, ..., x_n, p_1, .., p_k)}{\alpha(x_1, ..., x_n, p_1, ..., p_k)},$$

is admissible in an algebraic logic λ iff the quasi-identity

$$q(r) := \alpha_1 = \top, ..., \alpha_m = \top \Rightarrow \alpha = \top$$

is valid in the free algebra of countable rank $\mathfrak{F}_\lambda(\omega)$ from the variety $Var(\lambda)$ under any valuation which interprets $p_1, ..., p_k$ as distinct free generators of $\mathfrak{F}_\lambda(\omega)$.

Proof. The proof is parallel to the proof of Theorem 1.4.5. We simply take additional care in the interpretation of $p_1, ..., p_k$. ■

Conversely, it is also possible to describe the truth of quasi-identities in the free algebra $\mathfrak{F}_\lambda(\omega)$ of the signature enriched with constants, which are interpreted as free generators of $\mathfrak{F}_\lambda(\omega)$, in terms of admissibility in λ of the corresponding inference rules with metavariables.

Theorem 6.1.10 *Suppose λ is an algebraic logic. Any given quasi-identity $q := g_1 = f_1, ..., g_m = f_m \Rightarrow g = f$ in the signature of $\mathfrak{F}_\lambda(\omega)$, enriched by constants p_i for the free generators, is valid in $\mathfrak{F}_\lambda(\omega)$ iff all the inference rules with meta-variables p_i from the collection of all inference rules*

$$r(q) := \frac{(g_1 \equiv f_1), ..., (g_m \equiv f_m)}{g \equiv f}$$

(we set above that any formula in the tuple of formulas $g \equiv f$ produces the single individual rule) are admissible in λ.

Proof. We follow the proof of Theorem 1.4.6 bearing in ming that the constants p_i for free generators have to be treated as meta-variables everywhere in rules. ∎.

We can now explain more precisely why we are interested in rules with meta-variables from algebraic point of view . In the light of Theorem 6.1.10 the question whether there is a solution to algebraic equations in the free algebra $\mathfrak{F}_\lambda(\omega)$ can be reduced to the question of whether the corresponding rules with meta-variables are admissible in the corresponding logics λ. The question concerning the decidability of equations in various free algebraic systems was under active investigating (especially for a number classic algebraic objects). For instance, investigation of equations in free groups goes back to Lyndon [86, 87]. The algorithms for finding solutions to equations in free semi-groups were discovered in Makanin [88], [89]. The interest of solutions to equations in free algebras is evident: if an equation has a solution in a free algebra then this equation has solutions in any algebra from the corresponding variety, when coefficients corresponding to free generators interpreted in arbitrary way. Whether equations in free algebras of varieties corresponding to non-standard algebraic logics have solutions is also very interesting since, besides being a pure algebraic property, this expresses the logical essence of some permissible theorem-schemes for logics, and solves the substitution problem.

Of course the theorems mentioned above are merely simple reductions and translations. If we are interested deeper results we need to develop a theory in fact. We could study inference rules with meta-variables earlier in the chapter where we developed criteria for the admissibility of ordinary rules. But in doing this we would have had some additional combinatorial difficulties and our proofs would become more complexity. Therefore the study of rules with meta-variables was postponed until this section. We will follow our proofs for the ordinary rules, proving only the new parts which concern the presence of meta-variables. Therefore we will use all the results and their proofs from Sections 3, 4 and 5 of Chapter 3, and close acquaintance with those results is presupposed.

In a similar way to the case of ordinary inference rules, by considering n-characterizing models we can give a very simple criterion for the admissibility of inference rules with meta-variables.

Theorem 6.1.11 *Let* $\mathcal{K}_n, n \in N$ *be a sequence of n-characterizing Kripke models for a modal or superintuitionistic logic* λ. *And let* $r := \alpha_1, ..., \alpha_m/\beta$ *be a rule with k variables and n meta-variables. Then r is admissible for* λ *iff the following hold: for every* $m \in N$, $m \geq n$, *for each valuation* S *of k variables and n meta-variables of r in* $\mathcal{K}_m$ *such that*

(i) for any variable x from r, $S(x)$ is definable in $\mathcal{K}_m$, and

(ii) for any meta-variable p_i from r, $S(p_i) = V(p_i)$, where V is the original valuation $\mathcal{K}_m$ of the letters $p_1, ..., p_n$,

the following holds: if $S(\alpha_1) = |\mathcal{K}_m|, ..., S(\alpha_m) = |\mathcal{K}_m|$ then $S(\beta) = |\mathcal{K}_m|$.

Proof. The proof is parallel to the proof of Lemma 3.3.3 and is left as an easy exercise.

As in the case of the ordinary rules this theorem has a straightforward application through canonical models. But, again, this theorem does not convenient criterion for evaluating the admissibility of rules because it is not easy to verify necessary properties for all expressible valuations. In order to overcome this difficulty we first will clarify the structure of invalidating valuations in a similar way to Lemmas 3.4.1 and 3.4.2. We can simply use fragments of their proofs preserving the corresponding denotations and the scheme. First we need to extend the condensing lemma for n-characterizing Kripke models to the case of inference rules with meta-variables.

Lemma 6.1.12 *Let λ be a modal logic over $K4$ or a superintuitionistic logic, and let λ have the fmp. Let r be a rule with k variables and m meta-variables. Suppose that there is a valuation S of the variables and the meta-variables of r in the frame of an n-characterizing model $Ch_\lambda(n)$ for $n \geq m$ which has the following properties. S invalidates r, and S coincides with the original valuation V of $Ch_\lambda(n)$ on the meta-variables $p_1, ..., p_n$. Then there is a valuation S_1 of all variables and meta-variables of r in the frame of $Ch_\lambda(k+m)$ which invalidates r and coincides with the original valuation of the model $Ch_\lambda(k+m)$ on letters $p_1, ..., p_n$.*

Proof. The proof is similar to the proof of Lemma 3.4.2. We simply take care of the correspondence of the truth value on the meta-variables. ∎

We need a simple generalization of the notion of the property of branching below m.

Definition 6.1.13 *We say that a logic λ with the fmp has the* generalized property of branching below m *if the following hold.*

(i) If λ admits the irreflexive d-branching below m, and $\mathcal{W}$ is a finite generated subframe with depth of not less m of a component $Ch_\lambda(k)$, and $\mathcal{W}$ has n roots, where $n \leq d$, then the frame $\mathcal{W}_1$, obtained from $\mathcal{W}$ by adding a new irreflexive root, is a λ-frame.

(ii) If λ has a finite rooted λ-frame with a root consisting of k-element cluster C which has depth strictly more m and has d immediate successor clusters (in this case we say that λ admits reflexive k, d branching below m) then the following holds. If $\mathcal{W}$ is a finite generated subframe of a component of $Ch_\lambda(k)$, which has depth not less m and has n roots, where $n \leq d$, then the frame $\mathcal{W}_1$, which is obtained from $\mathcal{W}$ by adding the root consisting of the cluster C is a λ-frame.

We also need a simple generalization of the view-realizing property. Let $\mathcal{M}$ be a model, and V be a valuation of this model. Let x be an element of $|\mathcal{M}|$, and let Y be a set of formulas closed with respect to subformulas. Suppose all the propositional letters of formulas from Y are included in $Dom(V)$. As before we fixe the following denotation:

$$\Diamond(x, Y) := \{\alpha \mid \alpha \in Y, \Diamond\alpha \in Y, a \Vdash_V \Diamond\alpha\},$$
$$V(x, Y) := \{\alpha \mid \alpha \in Y, a \Vdash_V \alpha\}.$$

Let λ be a modal logic with the fmp extending logic $K4$. Let $\mathcal{M} := \langle M, R, S \rangle = \bigsqcup M_j$ be a λ-model. Suppose that $Dom(S)$ is finite and Y is a finite set of formulas built up out of propositional letters from $Dom(S)$. Let $Var(Y)$ be the set of all letters from Y. We fix a subset $\mathcal{Q}$ of $Var(Y)$. Recall that for any frame $\mathcal{F}$, $d(\mathcal{F})$ is the depth of $\mathcal{F}$, and for any $x \in |\mathcal{F}|$, $d(x)$ is the depth of x in $\mathcal{F}$.

Definition 6.1.14 *We say that $\mathcal{M}$ has the generalized view-realizing proper-ty for $Y, \mathcal{Q}$ below m with respect to λ if the following hold.*

Let $\mathcal{F}$ be a finite generated subframe of a component M_j and let $d(\mathcal{F}) \geq m$. Suppose the set E of all minimal clusters from $\mathcal{F}$ has d clusters. Then the following hold.

(i) If λ admits the irreflexive d-branching below m then, for every $\mathcal{Z} \subseteq \mathcal{Q}$, there exists an element $x_{E,\mathcal{Z},i}$ in the model M_j such that $m \leq d(x_{E,\mathcal{Z},i})$, $S(x_{E,\mathcal{Z},i}, Y) \cap \mathcal{Q} = \mathcal{Z}$ and

$$\Diamond(x_{E,\mathcal{Z},i}, Y) = \bigcup\{\Diamond(y, Y) \cup S(y, Y) \mid y \in E\}.$$

(ii) If λ admits the reflexive k, d-branching below m, $\mathcal{R} \subseteq \{\mathcal{Z} \mid \mathcal{Z} \subseteq \mathcal{Q}\}$ and $\|\mathcal{R}\| \leq k$ then, for any $\mathcal{Z} \in \mathcal{R}$, there exist el-ements $x_{E,\mathcal{R},\mathcal{Z},r}$ in the model M_j such that $m \leq d(x_{E,\mathcal{R},\mathcal{Z},r})$, $S(x_{E,\mathcal{R},\mathcal{Z},r}, Y) \cap \mathcal{R} = \mathcal{Z} \in \mathcal{R}$ and

$$\Diamond(x_{E,\mathcal{R},\mathcal{Z},r}, Y) = \bigcup\{\Diamond(y, Y) \cup S(y, Y) \mid y \in E\}\cup$$

$$\cup\{S(x_{E,\mathcal{Z}_1,r}, Y) \mid \mathcal{Z}_1 \in \mathcal{R}\}.$$

Now we are in a position to extend Lemma 3.4.9 to inference rules with meta-variables. Let λ be a modal logic with the fmp extending $K4$ and let λ have the generalized property of branching below m and the effective m-drop points prop-erty. Suppose g is the effective computable function for λ from the definition of the effective m-drop points property. Let $r = \alpha_1, ..., \alpha_n/\beta$ be an inference rule with meta-variables, which has k variables and k_1 meta-variables. Let $SV(r)$ be the set of all variables and meta-variables of our rule r. $MV(r)$ denotes the set of all meta-variables of r and $Sub(r)$ is the set of all subformulas of r.

Lemma 6.1.15 *Suppose that there is a valuation S of the set $SV(r)$ in the frame $\langle T, R \rangle$ of a n-characterizing model $Ch_\lambda(n)$ for λ, where $n \geq k_1$, such that*

(i) $\langle T, R, S \rangle$ invalidates r;

(ii) S coincides on the set of metavariables $p_1, \ldots, p_{k_1}$ of the rule r with the original valuation V of $Ch_\lambda(n)$ on the letters $p_1, \ldots, p_{k_1}$.

Then there is a λ-model $\mathcal{M} := \langle M, R, S \rangle = \bigsqcup \mathcal{M}_j$ with a valuation S of the variables and the meta-variables of r such that:

(a) the model $S_m(\mathcal{M})$ is isomorphic to the model $S_m(Ch_\lambda(k+k_1))$, where the letters $p_1, \ldots p_{k_1}$ correspond to the meta-variables $p_1, \ldots, p_{k_1}$;

(b) $\mathcal{M}$ has at most $g(e, q) * 2^{2^{k+k_1+1}}$ elements, where q is the number of subformulas of the rule r and e is the number of elements in $S_m(Ch_{K4}(k + k_1))$;

(c) the model $\mathcal{M}$ invalidates the rule r;

(d) the model $\mathcal{M}$ has the generalized view-realizing property below m for $Sub(r), MV(r)$ with respect to λ;

(e) each $\mathcal{M}_j$ is a generated submodel of $Ch_\lambda(k + k_1)$.

Proof. We will follow very close the proof of Lemma 3.5.7. By Theorem 6.1.11 and Lemma 6.1.12 there is a valuation S of $k + k_1$ variables and meta-variables of r in the model $Ch_\lambda(k + k_1)$ which invalidates the rule r. The model $Ch_\lambda(k + k_1)$ has a decomposition of the kind

$$Ch_\lambda(k + k_1) = \bigsqcup \{H_j \mid j \in J\}.$$

Also $\|S_m(H_j)\| \leq \|S_m(Ch_{K4}(k + k_1))\|$, and the number of the components H_j in $Ch_\lambda(k + k_1)$ is not more than $2^{2^{k+k_1+1}}$. We pick in every H_j some finite submodel K_j as follows. For every $Z \subseteq Sub(r)$, we choose and fix in H_j some single element x_Z of depth strictly more than m such that $S(x_Z, Sub(r)) = Z$ (if such elements exist for given Z). Then we take the open submodel K_j of H_j generated by $S_m(H_j)$ and all x_Z. It is clear that every K_j and $\bigsqcup K_j$ are finite, and the valuation S invalidates the rule r in the model $\bigsqcup K_j$. Note that

$$S_m(\bigsqcup \{K_j \mid j \in J\}) = S_m(Ch_\lambda(k + k_1))$$

(it is important for our proof now). Since our logic λ has the effective m-drop points property, for each j, there exist a submodel G_j of the model K_j, which contains the set $max(Sub(r), K_j, m) \cup S_m(K_j)$, and a homomorphism f_j from G_j onto a λ-model D_j with all the properties from the definition of effective m-drop point property. We fix certain G_j and D_j for each j and we set

$$\mathcal{M}_j := D_j, \qquad \mathcal{W}_1 := \bigsqcup \{\mathcal{M}_j \mid j \in J\}.$$

By (iii) from the definition of effective m-drop points property it follows that

$$S_m(D_j) = S_m(H_j) = S_m(K_j)$$

and property (a) from the assertion of our lemma holds for $\mathcal{W}_1$. (b) holds for $\mathcal{W}_1$ by (i) from the definition of effective m-drop points property. (c) holds for the model $\mathcal{W}_1$ by the choice of $\mathcal{W}_1$ and (ii) from the definition of effective m-drop points property.

In the prove of (d) there is a difference with Lemma 3.4.9 since meta-variables now present. Suppose that $\mathcal{F}$ is a generated open subframe of D_j, and that the depth $d(\mathcal{F})$ of $\mathcal{F}$ satisfies $m \leq d(\mathcal{F})$. Suppose also that $\mathcal{F}$ has d minimal clusters. The antichain of all minimal clusters in $\mathcal{F}$ is denoted by E. For each cluster C from E, we pick a single representative z from C. We denote by U the set which consists of all these representatives z. For every $z \in U$ we pick a single cluster from G_j which contains some element from $f_j^{-1}(z)$, and the set of all such clusters is denote by $U(f_j^{-1})$. It is clear that $U(f_j^{-1})$ is an antichain with d clusters (since f_j preserves the accessibility relation). Some such cluster has depth at least m by property (iii) from the definition of the effective m-drop points property. We now take the open generated subframe $U(f_j^{-1})^{R<} \cup U(f_j^{-1})$ of the frame H_j.

Suppose that λ admits irreflexive d-branching below m and $\mathcal{Z} \subseteq MV(r)$. Since λ has the generalized property of branching below m, there is a λ-frame $\mathcal{A}$ with an irreflexive root c such that the frame $c^{R<}$ is isomorphic to the frame $U(f_j^{-1})^{R<} \cup U(f_j^{-1})$. Hence according to the construction of models $Ch_\lambda(k)$, there exists an irreflexive element $y \in H_j$ such that $y^{R<}$ is the frame $U(f_j^{-1})^{R<} \cup U(f_j^{-1})$ and $S(y, Sub(r)) \cap MV(r) = \mathcal{Z}$. Then y has depth at least $m+1$ (more than the depth of the clusters in $U(f_j^{-1})$). Also it is easy to verify directly that

$$\Diamond(y, Sub(r)) = \bigcup \{\Diamond(f_j^{-1}(z), Sub(r)) \mid z \in U\} \cup$$

$$\bigcup \{S(f_j^{-1}(z), Sub(r)) \mid z \in U\} \qquad (6.1)$$

By the choice of K_j, there exists an y_1 in K_j which has depth of more m such that $S(y_1, Sub(r)) = S(y, Sub(r))$. Therefore the model G_j has an element y_2 of depth more m such that $S(y_2, Sub(r)) = S(y, Sub(r))$. Besides we have that $S(f_j(y_2), Sub(r)) = S(y, Sub(r))$ by (ii) from the definition of effective m-drop points property. Also, for each $z \in U$, $S(f_j^{-1}(z), Sub(r)) = S(z, Sub(r))$ (also by (ii) from the definition of the effective m-drop points property). Now we define $x_{E,Z,i} := f_j(y_2)$. Then by (iii) from the definition of e-m-dpp, $x_{E,Z,i}$ has depth at least $m + 1$, and using (6.1), we infer that $x_{E,Z,i} := f_j(y_2)$ satisfies the necessary property from the definition of the generalized view-realizing property for $Sub(r)$, $MV(r)$ from (d).

Suppose now that λ admits a reflexive u, d-branching below m. Suppose $\mathcal{R} \subseteq \{Z \mid Z \subseteq MV(r)\}$, $\|\mathcal{R}\| \leq u$. Since λ has the generalized property of branching below m, there is a λ-frame $\mathcal{A}$ rooted by a cluster C such that:

(1) C has exactly u reflexive elements and

(2) the frame $C^{R<}$ is isomorphic to the frame $U(f_j^{-1})^{R<} \cup U(f_j^{-1})$.

Then by the construction of models $Ch_\lambda(k)$, there exists a cluster C_1 in H_j such that

(3) $C_1^{R<}$ is the frame $U(f_j^{-1})^{R<} \cup U(f_j^{-1})$ and

(4) $\{S(y, Sub(r)) \cap MV(r) \mid y \in C_1\} = \mathcal{R}$.

The cluster C_1 has depth at least $m + 1$ (more than the depth of the clusters in $U(f_j^{-1})$). It can be easily seen that for any $y \in C_1$,

$$\Diamond(y, Sub(r)) = \bigcup\{\Diamond(f_j^{-1}(z), Sub(r)) \mid z \in U\} \cup$$

$$\bigcup\{S(f_j^{-1}(z), Sub(r)) \mid z \in U\} \cup \{S(v, Sub(r)) \mid v \in C_1\}. \tag{6.2}$$

By the choice of K_j, for every $v_i \in C_1$, there is an element z_i in K_j of depth more than m such that $S(z_i, Sub(r)) = S(v_i, Sub(r))$. Then the model G_j has an element w_i of depth more than m such that $S(w_i, Sub(r)) = S(z_i, Sub(r))$. Besides $S(f_j(w_i), Sub(r)) = S(w_i, Sub(r))$ by (ii) from the definition of *effective m-drop points property*. For each $z \in U$, $S(f_j^{-1}(z), Sub(r)) = S(z, Sub(r))$ (again by (ii) from the definition of e-m-dpp). Now we define, for every $Z \in \mathcal{R}$, $x_{E,\mathcal{R},Z,r} := f_j(w_i)$, where $S(f_j(w_i), Sub(r)) \cap \mathcal{R} = Z$ (w_i exists by the choice of C_1 above). Then by (iii) from the definition of e-m-dpp, $x_{E,\mathcal{R},Z,i}$ has a depth of at least $m + 1$, and by (6.2) it follows that the set of elements $x_{E,\mathcal{R},Z,r} := f_j(w_i)$ satisfies the necessary property from the definition of generalized view-realizing property for $Sub(r)$, $MV(r)$ from (d).

It remains now only to make a corresponding condensing p-morphism of $\mathcal{W}_1$ starting with a depth of $m + 1$, and to obtain a model $Cg(\mathcal{W}_1)$ which satisfies (e). We put $\mathcal{M} := Cg(\mathcal{W}_1)$, and then properties (a) and (b) hold evidently for $\mathcal{M}$, and properties (c) and (d) are true by Lemma 3.4.1. ∎

The proof of the analog of Lemma 3.4.10 depends upon the presence of meta-variables in much more degree than previous lemma, meta-variables have to be treated in special way, and for this reason the proof of the following lemma is more complicated and is divided in several intermediate lemmas.

Lemma 6.1.16 *Let λ be a modal logic with the fmp extending $K4$ and having the generalized property of branching below m and the effective m-drop points property. Suppose there exists a λ-model $\mathcal{M}$ with properties (a) - (e) from Lemma 6.1.15. Then there is an effectively constractable definable valuation S of the k variables and k_1 metavariables of the rule r in the frame of the model $Ch_\lambda(k + k_1)$ such that*

(i) S invalidates r in $Ch_\lambda(k + k_1)$ and

(ii) S coincides on the meta-variables $p_1, ..., p_{k_1}$ of r with the original valuation V of the model $Ch_\lambda(k + k_1)$ on the letters $p_1, ..., p_{k_1}$.

Proof. We follow the proof of Lemma 3.4.10 making certain alterations and introducing certain new constructions due to presence of meta-variables. Hence, again, given a λ-model $\mathcal{M} = \langle M, R, S \rangle = \bigsqcup \mathcal{M}_j$ which has properties (a) - (e) from Lemma 6.1.15 and the model $Ch_\lambda(k + k_1)$ has a decomposition of the kind $Ch_\lambda(k + k_1) := \bigsqcup H_j$. By (a) from Lemma 6.1.15, we may assume $S_m(H_j) = S_m(\mathcal{M}_j)$ (as Kripke models) for all corresponding j and that

$$S_m(\mathcal{M}) = S_m(Ch_\lambda(k + k_1)).$$

By (e) from Lemma 6.1.15 the model $\mathcal{M}_j$ is a generated submodel of the model H_j. Now we consider all the elements of $|\mathcal{M}|$ as the corresponding elements of $Ch_\lambda(k + k_1)$.

By Theorem 3.3.7 each element of the model $Ch_\lambda(k)$ is definable, therefore we can assume that S is a valuation in the frame of $\mathcal{M}$, as an open generated subframe of $Ch_\lambda(k + k_1)$, which is definable valuation in $Ch_\lambda(k + k_1)$. Also we can effectively write out formulas defining S. In particular, S coincides with the original valuation V from $Ch_\lambda(k + k_1)$ on letters $p_1, ..., p_{k_1}$ on all elements from $\mathcal{M}$. By (c) from Lemma 6.1.15 S invalidates the rule r in $\mathcal{M}$. Now we, as with the proof of Lemma 3.4.10, will extend the valuation S from $\mathcal{M}$ onto whole the frame of $Ch_\lambda(k + k_1)$ in a such a way that the premises of r will be true in $Ch_\lambda(k)$.

Any component H_j has the corresponding component $\mathcal{M}_j$ in the model $\mathcal{M}$, and the frame of $\mathcal{M}_j$ is a open subframe of H_j. We pick any j, put $Q = H_j$, and extend S from $\mathcal{M}_j$ onto the whole frame Q (recall j is fixed) by means of the construction described below. As with the proof of Lemma 3.4.10, we construct a sequence of subsets $\Sigma(x, t)$ of the frame Q, where x are all elements of $\mathcal{M}_j$, $0 \le t \le m_1$, where m_1 is the number of elements in $\mathcal{M}_j$. The sequence $\Sigma(x, t), x \in \mathcal{M}_j, 0 \le t \le m_1$ will have the following properties:

(a1) $\forall x_i \in \mathcal{M}_j(\Sigma(x, t) \subseteq \Sigma(x, t + 1))$,
$\forall x_i, x_j(x_i \ne x_j \Rightarrow \Sigma(x_i, t) \cap \Sigma(x_j, t) = \emptyset)$;

(b1) $\forall \Sigma(x, t)$ there exists a modal formula $\alpha(x, t)$ such that under the original valuation V in the model $Q = H_j$
$$V(\alpha(x, t)) := \{y \mid y \in |Q|, \ y \Vdash_V \alpha(x, t)\} = \Sigma(x, t);$$

(c1) $\bigcup_{x_i \in M_j} \Sigma(x_i, t)$ forms a generated subframe of the frame Q for any given t;

(d1) We set the valuation S of propositional letters from r in the frame $\bigcup\{\Sigma(x, t) \mid x \in M_j\}$ as follows $\forall p \in Var(r)$, $S(p) := \bigcup\{\Sigma(x, t) \mid x \in M_j, x \Vdash_S p$ in the model $M_j\}$. Then $\forall a \in \Sigma(x, t), \forall \alpha \in Sub(r) \ (a \Vdash_S \alpha) \Leftrightarrow (x \Vdash_S \alpha$ in the model $M_j)$;

(d2) For each $x \in \Sigma(x, t)$ and any metavariable p_i of the rule r, $x \Vdash_S p_i \Leftrightarrow x \Vdash_V p_i$ in the model $Ch_\lambda(k + k_1)$;

(e1) For any t, $\forall y \in Q$, that $(y \notin \bigcup\{\Sigma(x, t) \mid x \in M_j\}$ implies $\exists$ $t+1$-element subset W of $|M_j|$ such that
$$\forall x \in W(y^{R<} \cap \Sigma(x, t) \neq \emptyset);$$

(f1) If $y \in (\Sigma(x, t) - \Sigma(x, 0))$ then the depth of x in M_j is at least m.

First, we choose, for all $x \in |M_j|$, $\Sigma(x, 0) := \{x\}$. It can be easily seen that all properties (a1) - (e1) mentioned above are valid for $t = 0$. Suppose that all the sets $\Sigma(x, \mu), x \in M_j$, where $1 \leq \mu \leq t$, were already constructed, and that all properties (a1) - (f1) hold. Consider all sets D, where $D \subseteq |M_j|$ and D has exactly $t+1$ elements, and has at least one element with depth of not less m. For each such D, we denote by $C(D)$ the set of all the clusters from the frame of M_j which contain elements of D. It is not difficult to see that there are antichains E of elements (i.e. sets of elements incomparable by the accessibility relation) of these clusters, which have the following property (recall that the model M_j has the valuation S of letters from $SV(r)$). E has an element with depth of at least m and

$$\bigcup\{\Diamond(y, Sub(r)) \cup S(y, Sub(r)) \mid y \in E\} = \tag{6.3}$$

$$\bigcup\{\Diamond(y, Sub(r)) \cup S(y, Sub(r)) \mid y \in D\}.$$

Indeed, E, which consists of some representative elements of the clusters the antichain of all R-minimal clusters of $C(D)$, satisfies (6.3). For every mentioned above D, we fix an antichain E which satisfies (6.3) and has minimal number of elements. Now we consider every chosen E. Assume that a given E has d elements.

Suppose λ admits the irreflexive d-branching below m. Let $\mathcal{Z} \subseteq MV(r)$. By (d) from Lemma 6.1.15, there is an $x_{E, \mathcal{Z}, i} \in |M_j|$ with the properties concerning the generalized view-realizing property described in Lemma 6.1.15:

$$S(x_{E, \mathcal{Z}, i}, Sub(r)) \cap MV(r) = \mathcal{Z},$$

$$\Diamond(x_{E,\mathcal{Z},i}, Sub(r)) = \bigcup\{\Diamond(y, Sub(r)) \cup S(y, Sub(r)) \mid y \in E\}, \qquad (6.4)$$

For any given E, we fix a single $x_{E,\mathcal{Z},i}$ for which all required above hold.

Assume that λ admits a u, d-reflexive branching below m and $||u|| \leq 2^{MV(r)}$. Suppose $\mathcal{R} \subseteq 2^{MV(r)}$ and $||\mathcal{R}|| \leq u$. Again by (d) of Lemma 6.1.15 there are elements $x_{E,\mathcal{R},\mathcal{Z},r} \in |\mathcal{M}_j|$ for $\mathcal{Z} \in \mathcal{R}$ such that:

$$S(x_{E,\mathcal{R},\mathcal{Z},r}, Sub(r)) \cap MV(r) = \mathcal{Z},$$

$$\Diamond(x_{E,\mathcal{R},\mathcal{Z},r}, Sub(r)) = \bigcup\{\Diamond(y, Sub(r)) \cup S(y, Sub(r)) \mid y \in E\} \cup \quad (6.5)$$

$$\cup \{S(x_{E,\mathcal{R},\mathcal{Z}_1,r}, Sub(r)) \mid \mathcal{Z}_1 \in \mathcal{R}\}.$$

We fix a tuple $x_{E,\mathcal{R},\mathcal{Z},r}$ for any such $\mathcal{R}$ and any $\mathcal{Z} \in \mathcal{R}$.

Now we introduce special formulas which are necessary to enlarge the sets of the kind $\Sigma(x,t)$ to the sets of the kind $\Sigma(x, t+1)$. Let

$$q(D) := \bigwedge_{x \in D} \Diamond\alpha(x,t) \wedge \bigwedge_{x \notin D, x \in \mathcal{M}_j} \neg\Diamond\alpha(x,t) \wedge \neg(\bigvee_{x \in \mathcal{M}_j} \alpha(x,t)).$$

$$\forall \mathcal{Z} \subseteq MV(r),$$

$$p(\mathcal{Z}) := \bigwedge\{p_i \mid p_i \in \mathcal{Z}\} \wedge \bigwedge\{\neg p_i \mid p_i \in (MV(r) - \mathcal{Z})\}).$$

Recall that, for any element x of a frame, $d(x)$ is the depth of x. A special role in our definition will play the fact whether the following property holds: $\exists g \in D$ such that $d(g) \geq m$) and

$$\Diamond(g, Sub(r) = \bigcup\{\Diamond(x, Sub(r)) \cup S(x, Sub(r)) \mid x \in D\}). \qquad (6.6)$$

Suppose first that (6.6) does not hold. If for given D, E and $\mathcal{R} \subseteq MV(r)$, where $1 \leq ||\mathcal{R}||$, all the elements $x_{E,\mathcal{R},\mathcal{Z},r}$ from $\mathcal{M}_j$ for all $\mathcal{Z} \in \mathcal{R}$ were fixed for D, E and $\mathcal{R}, \mathcal{Z}$ then

$$\chi_1(D, \mathcal{Z}, \mathcal{R}) := p(\mathcal{Z}) \wedge q(D) \wedge$$

$$\bigwedge_{\mathcal{Z}_1 \in \mathcal{R}} \Diamond[p(\mathcal{Z}_1) \wedge q(D) \wedge \delta(\mathcal{R}, D)] \wedge \beta(\mathcal{R}, D), \text{ where}$$

$$\delta(\mathcal{R}, D) := \bigwedge_{\mathcal{Z}_1 \in \mathcal{R}} \Diamond[p(\mathcal{Z}_1) \wedge q(D)],$$

$$\beta(D, \mathcal{R}) := \Box(q(D) \rightarrow \bigvee_{\mathcal{Z}_1 \in \mathcal{R}} [p(\mathcal{Z}_1) \wedge (\bigwedge_{\mathcal{Z}_2 \in \mathcal{R}} \Diamond(p(\mathcal{Z}_2) \wedge q(D) \wedge$$

$\wedge \delta(\mathcal{R}, \mathcal{Z}))]).$

If for given D, E and $\mathcal{Z} \subseteq MV(r)$ the element $x_{E,\mathcal{Z},i}$ in $\mathcal{M}_j$ was fixed by D, E and $\mathcal{Z}$ then

$$\varphi_1(D, \mathcal{Z}) := p(\mathcal{Z}) \wedge q(D) \wedge \Box(\bigvee_{x \in D} \alpha(x,t)).$$

Suppose now that (6.6) holds. Consider the set

$$P(D) := \{S(x, Sub(r)) \cap MV(r)) \mid x \in D, d(x) \geq m, x \text{ satisfy } (6.6)\}.$$

We fix a subset $\mathcal{G}(D) \subseteq D$ which consists of some elements from D with depth of at least m and satisfying (6.6), and such that the following holds: $\|\mathcal{G}(D)\| = \|P(D)\|$ and $P(\mathcal{G}(D)) = P(D)$, where $P(\mathcal{G}(D)) := \{S(g, Sub(r)) \cap MV(r) \mid g \in \mathcal{G}(D)\}$. Then we introduce necessary formulas in the following way.

Consider any $\mathcal{Z} \in P(\mathcal{G}(D)))$ and any $g \in \mathcal{G}(D)$, where $S(g, Sub(r) \cap MV(r) \cap = \mathcal{Z}$. Let

$$\psi(D, \mathcal{Z}, g) := p(\mathcal{Z}) \wedge q(D) \wedge \Box(q(D) \rightarrow \bigvee_{\mathcal{Z}_1 \in P(\mathcal{G}(D))} p(\mathcal{Z}_1)), \tag{6.7}$$

If for given D, E and $\mathcal{Z} \subseteq MV(r)$ the element $x_{E,\mathcal{Z},i}$ in $\mathcal{M}_j$ was fixed by D, E and $\mathcal{Z}$ and $\mathcal{Z} \notin P(\mathcal{G}(D))$ then

$$\varphi_2(D, \mathcal{Z}) := p(\mathcal{Z}) \wedge q(D) \wedge$$

$$\wedge \Box(\bigvee_{x \in D} \alpha(x,t) \vee \bigvee_{g \in \mathcal{G}(D), \mathcal{Z}_1 \in P(\mathcal{G}(D))} \psi(D, \mathcal{Z}_1, g)).$$

If for given D, E and $\mathcal{R} \subseteq MV(r)$, $1 \leq \|\mathcal{R}\|$, the elements $x_{E,\mathcal{R},\mathcal{Z},r}$ from $\mathcal{M}_j$ for all $\mathcal{Z} \in \mathcal{R}$ were fixed for D, E and $\mathcal{R}, \mathcal{Z}$ and there is some $\mathcal{Z}_1 \in \mathcal{R}$ such that $\mathcal{Z}_1 \notin P(\mathcal{G}(D))$ then we put

$$\chi_2(D, \mathcal{Z}, \mathcal{R}) := p(\mathcal{Z}) \wedge q(D) \wedge$$

$$\wedge \bigwedge_{\mathcal{Z}_1 \in \mathcal{R}} \Diamond[p(\mathcal{Z}_1) \wedge q(D) \wedge \bigwedge_{\mathcal{Z}_2 \in \mathcal{R}} \Diamond[p(\mathcal{Z}_2) \wedge q(D)] \wedge \beta_1(\mathcal{R}, D), \text{ where}$$

$$\beta_1(D, \mathcal{R}) := \Box(q(D) \rightarrow \bigvee_{\mathcal{Z}_1 \in \mathcal{R}} [p(\mathcal{Z}_1) \wedge$$

$$\wedge \bigwedge_{\mathcal{Z}_2 \in \mathcal{R}} \Diamond(p(\mathcal{Z}_2) \wedge q(D) \wedge \delta(\mathcal{R}, D))] \vee \Theta),$$

$$\delta(\mathcal{R}, D) := \bigwedge_{\mathcal{Z}_1 \in \mathcal{R}} \Diamond[p(\mathcal{Z}_1) \wedge q(D)],$$

where the formula Θ is the disjunction of all formulas of the kind $\psi(D, \mathcal{Z}_1, g)$, for fixed D and arbitrary possible $\mathcal{Z}_1$ and g; and still we stipulate, for all the cases in the above definition when formulas of required type were no introduced for some D, $\mathcal{Z}$, $\mathcal{R}$ or g, that these formulas are equal to $\perp$.

Now we define the sets $\Sigma(x, t+1)$, for $x \in |\mathcal{M}_j|$, as follows (note that varying below elements $x_{E,\mathcal{Z},i}$ and $x_{E,\mathcal{R},\mathcal{Z},r}$ we also vary D, and in the last $\bigvee_D$ D and x themselves stipulate $\mathcal{Z}$).

$$\Sigma(x, t+1) := \Sigma(x, t) \cup \bigcup_{x_{E,\mathcal{Z},i}=x} V(\varphi_1(D, \mathcal{Z}) \vee \varphi_2(D, \mathcal{Z})) \cup$$

$$\cup \bigcup_{x_{E,\mathcal{R},\mathcal{Z},r}=x} V(\chi_1(D, \mathcal{Z}, \mathcal{R}) \vee \chi_2(D, \mathcal{Z}, \mathcal{R})) \cup V(\bigvee_{D,\mathcal{Z}} \psi(D, \mathcal{Z}, x)).$$

It is clear that $\Sigma(x, t+1) = V(\alpha(x, t+1))$, where

$$\alpha(x, t+1) := \alpha(x, t) \vee \bigvee_{x_{E,\mathcal{Z},i}=x} (\varphi_1(D, \mathcal{Z}) \vee \varphi_2(D, \mathcal{Z})) \vee$$

$$\vee \bigvee_{x_{E,\mathcal{R},\mathcal{Z},r}=x} (\chi_1(D, \mathcal{Z}, \mathcal{R}) \vee \chi_2(D, \mathcal{Z}, \mathcal{R})) \vee \bigvee_{D,\mathcal{Z}} \psi(D, \mathcal{Z}, x).$$

Thus (b1) holds for the indexes $t+1$.

Lemma 6.1.17 *The sets $\Sigma(x, t+1)$ satisfy (a1).*

Proof. It is clear that $\Sigma(x, t) \supseteq \Sigma(x, t+1)$. Moreover, the all sets of kind $V(\varphi_\alpha(D, \mathcal{Z}))$ for $\alpha = 1, 2$, and the sets $V(\chi_\alpha(D, \mathcal{Z}, \mathcal{R}))$ for $\alpha = 1, 2$, and the set $V(\psi(D, \mathcal{Z}, g))$ joined to the old sets are disjoint with the old sets due to presence of the formula $q(D)$ as a conjunct member in formulas defining the new mentioned above sets.

If $D_1 \neq D_2$ then the all sets of kind $V(\varphi_\alpha(D_1, \mathcal{Z}_1))$, where $\alpha := 1, 2$ or all the sets $V(\chi_\alpha(D_1, \mathcal{Z}_1, \mathcal{R}_1))$, where $\alpha := 1, 2$, or sets $V(\psi(D_1, \mathcal{Z}_1, g_1))$ and the sets of the kind $V(\varphi_\beta(D_2, \mathcal{Z}_2))$, where $\beta := 1, 2$, or the sets $V(\chi_\beta(D_2, \mathcal{Z}_2, \mathcal{R}_2))$, where $\beta := 1, 2$ or the sets $V(\psi(D_2, \mathcal{Z}_2, g_2))$ are disjoint since formulas $q(D_1)$ and $q(D_2)$ are the corresponding conjuncts in the mentioned above formulas. If $\mathcal{Z}_1 \neq \mathcal{Z}_2$ and $D_1 = D_2$ then we have all the mentioned above sets corresponding to these distinct $\mathcal{Z}_1$ and $\mathcal{Z}_2$ are disjoint since formulas $p(\mathcal{Z}_1)$ and $p(\mathcal{Z}_2)$ are the corresponding conjuncts in the corresponding formulas. Thus it remains only to consider the new joined sets corresponding the same D and $\mathcal{Z}$.

All the sets of the kind $V(\chi_1(D, \mathcal{Z}, \mathcal{R}))$ or $V(\varphi_1(D, \mathcal{Z}))$ are disjoint with all the sets of the kind $V(\varphi_2(D, \mathcal{Z}))$ or $V(\chi_2(D, \mathcal{Z}, \mathcal{R}_1))$ since D does not satisfy

(6.6) for the sets of the first kind, and does satisfy (6.6) for the sets of the second kind. Thus D cannot be the same in that formulas.

The sets $V(\chi_1(D, \mathcal{Z}, \mathcal{R}))$ and $V(\varphi_1(D, \mathcal{Z}))$ are disjoint due to presence the third conjuncts in the corresponding formulas. The sets of kind $V(\varphi_2(D, \mathcal{Z}))$, and the sets $V(\chi_2(D, \mathcal{Z}, \mathcal{R}))$ are disjoint by presence the third conjuncts in the corresponding formulas, since there is $\mathcal{Z}_1 \in (\mathcal{R} - (\mathcal{G}(D))$ for this case.

The sets of the kind $V(\psi(D, \mathcal{Z}, g))$ are disjoint with the all sets of the kind $V(\varphi_\alpha(D, \mathcal{Z}))$, $\alpha = 1, 2$ and $V(\chi_\alpha(D, \mathcal{Z}, \mathcal{R}))$, where $\alpha = 1, 2$, due to presence first, second and third conjunct members in $\psi(D, \mathcal{Z}, g)$ since (6.6) holds. ∎

Lemma 6.1.18 $\bigcup_{x \in \mathcal{M}_j} \Sigma(x, t + 1)$ *is an open subframe of Q, i.e. (c1) holds for $t + 1$.*

Proof. If $x \in \bigcup \Sigma(x, t)$ and xRy then y is a member of the set $\bigcup \Sigma(x, t)$ by (c1) for the index t, and consequently y is included in the new sets. Now let xRy and $x \in (\bigcup \Sigma(x, t + 1) - \bigcup \Sigma(x, t))$. Then

$$x \Vdash_V \varphi_1(D, \mathcal{Z}) \vee \chi_1(D, \mathcal{Z}, \mathcal{R}) \vee$$
$$\vee \varphi_2(D, \mathcal{Z}) \vee \chi_2(D, \mathcal{Z}, \mathcal{R}) \vee \psi(D, \mathcal{Z}, g)$$

for some $D, \mathcal{Z}, \mathcal{R}, g$. Suppose $y \notin \bigcup_{x \in \mathcal{M}_j} \Sigma(x, t)$. Then it is easy to see by (e1) for t that $y \Vdash_V q(D)$.

If $x \Vdash_V \varphi_1(D, \mathcal{Z})$ we have a contradiction with the fact $y \Vdash_V \bigvee_{x \in D} \alpha(x, t)$. If $x \Vdash_V \chi_1(D, \mathcal{Z}, \mathcal{R})$ then it is not hard to see that $y \Vdash_V \chi_1(D, \mathcal{Z}_1, \mathcal{R})$ for some $\mathcal{Z}_1$ and consequently $y \in \bigcup_{x \in \mathcal{M}_j} \Sigma(x, t + 1)$. If $x \Vdash_V \psi(D, \mathcal{Z}, g)$ then it can be easily seen that $y \Vdash_V \psi(D, \mathcal{Z}_1, g_1)$ for some $\mathcal{Z}_1, g_1$ and we again have $y \in \bigcup_{x \in \mathcal{M}_j} \Sigma(x, t + 1)$. If $x \Vdash_V \varphi_2(D, \mathcal{Z})$ then it is not hard to see that $y \Vdash_V \psi(D, \mathcal{Z}_1, g_1)$ which yields $y \in \bigcup_{x \in \mathcal{M}_j} \Sigma(x, t + 1)$. Finally if $x \Vdash_V \chi_2(D, \mathcal{Z}, \mathcal{R})$ then we immediately infer, for some $\mathcal{Z}_1, g_1$,

$$[y \Vdash_V \chi_2(D, \mathcal{Z}_1, \mathcal{R})] \text{ or } [y \Vdash_V \psi(D, \mathcal{Z}_1, g_1)].$$

which again yields $y \in \bigcup_{x \in \mathcal{M}_j} \Sigma(x, t + 1)$. ∎

Now we turn to prove that (d1) holds for the sets of kind $\Sigma(x, t + 1)$.

Lemma 6.1.19 *If we let the valuation S of all the propositional letters from r in $\bigcup \{\Sigma(x, t + 1) \mid x \in |\mathcal{M}_j|\}$ be the following: $\forall p \in Var(r)$,*

$$S(p) := \bigcup \{\Sigma(x, t + 1) \mid x \in |\mathcal{M}_j|, x \Vdash_S p \text{ in the model } \mathcal{M}_j \}.$$

Then $\forall a \in \Sigma(x, t + 1)$,

$$\forall \alpha \in Sub(r)[(a \Vdash_S \alpha) \Leftrightarrow (x \Vdash_S \alpha \text{ in the model } \mathcal{M}_j)].$$

Proof. If $a \in \Sigma(x,t)$ it follows that the condition holds by the inductive hypothesis for t. Suppose $a \in (\Sigma(x,t+1) - \bigcup \Sigma(x_1,t))$. Then for some $D, \mathcal{Z}, \mathcal{R}, g$,

$$a \Vdash_V \varphi_1(D, \mathcal{Z}) \vee \chi_1(D, \mathcal{Z}, \mathcal{R}) \vee \qquad (6.8)$$
$$\vee \varphi_2(D, \mathcal{Z}) \vee \chi_2(D, \mathcal{Z}, \mathcal{R}) \vee \psi(D, \mathcal{Z}, g).$$

We carry out the proof by induction on the length of the subformulas α from $Sub(r)$. If a formula α is a propositional letter (i.e. either a variable of r or a meta-variable of r) then this lemma holds immediately by the definition of the valuation S. The inductive steps for all the nonmodal logical connectives are trivial and routine, so it remains only to consider the connective $\Diamond$. Suppose first that $a \Vdash_S \Diamond\alpha$ in the model which is based up on $\bigcup\{\Sigma(x,t+1) \mid x \in |\mathcal{M}_j|\}$. Then there is an b such that aRb and $b \Vdash_S \alpha$.

Consider the first case, when $b \in \bigcup \Sigma(x_1,t)$. From (6.8) and the fact that the formula $q(D)$ is a conjunct in all the formulas mentioned in (6.8) it follows that $b \Vdash_V \alpha(x_l,t)$ for some $x_l \in D$. Then by the inductive hypothesis $x_l \Vdash_S \alpha$ holds. It can be easily seen from (6.3) that in this case $\alpha \in \bigcup\{\Diamond(y, Sub(r)) \cup S(y, Sub(r)) \mid y \in E\}$. Because $x = x_{E,\mathcal{Z},i}$ or $x = x_{E,\mathcal{Z},\mathcal{R},r}$ or $x = g \in \mathcal{G}(D)$ by our fixing x, applying (6.4) or (6.5), or, (6.6) respectively we obtain $x \Vdash_S \Diamond\alpha$ in $\mathcal{M}_j$.

Now we turn to the second case, when $b \notin \bigcup \Sigma(x_1,t)$. Then by (6.8) and (e1) for t and the presence of the formula $q(D)$ as a conjunct in all formulas mentioned in (6.8), it follows that

$$b \Vdash_V \chi_1(D, \mathcal{Z}_1, \mathcal{R}) \vee \vee \chi_2(D, \mathcal{Z}_1, \mathcal{R}) \vee \psi(D, \mathcal{Z}_1, g_1).$$

for the same $D, \mathcal{R}$ and certain $\mathcal{Z}_1, g_1$. In particular, b is included in a set $\Sigma(y, t+1)$ for some y, where $y = x_{E,\mathcal{Z}_1,\mathcal{R}}$ or $y = g_1 \in \mathcal{G}(D)$. Then by inductive assumption we infer $y \Vdash_S \alpha$ in $\mathcal{M}_j$. It this case we have (a): $x = x_{E,\mathcal{Z},\mathcal{R},r}$, and (6.6) is fail, and $y = x_{E,\mathcal{Z}_1,\mathcal{R}}$, or (b): (6.6) holds and either (b.1): $x = x_{E,\mathcal{Z},i}$ or $x = x_{E,\mathcal{Z},\mathcal{R},r}$, or $x = g \in \mathcal{G}(D)$ and $y = g_1 \in \mathcal{G}(D)$, or (b.2): $x = x_{E,\mathcal{Z},\mathcal{R},r}$ and $y = x_{E,\mathcal{Z}_1,\mathcal{R},r}$. If (a) holds then applying (6.5) we obtain $x \Vdash_S \Diamond\alpha$ in $\mathcal{M}_j$. If (b.1) holds then using (6.3) and (6.4) or (6.5), or (6.6) and the choice of $\mathcal{G}(D)$ it follows $x \Vdash_S \Diamond\alpha$ in $\mathcal{M}_j$. Finally if (b.2) holds then we conclude $x \Vdash_S \Diamond\alpha$ using (6.5). Thus we have proved that $a \Vdash_S \Diamond\alpha \Rightarrow x \Vdash_S \Diamond\alpha$ in the model $\mathcal{M}_j$.

Conversely, suppose $x \Vdash_S \Diamond\alpha$ in the model $\mathcal{M}_j$ and $x = x_{E,\mathcal{Z},i}$ or $x = x_{E,\mathcal{Z},\mathcal{R},r}$, or $x = g \in \mathcal{G}(D)$. Then applying (6.3) and (6.4) or (6.5) or (6.6), it follows that there exists $z \in E$ such that (i): $z \Vdash_S \Diamond\alpha$ or $z \Vdash_S \alpha$, or (ii): $x = x_{E,\mathcal{Z},\mathcal{R},r}$ and for some $\mathcal{Z}_1 \in \mathcal{R}$ and $x_{E,\mathcal{Z}_1,\mathcal{R},r} \Vdash_S \alpha$.

If (ii) holds then using the structure of formulas $\chi_\gamma(E, \mathcal{Z}, \mathcal{R})$, $\gamma \in \{1,2\}$, we immediately obtain aRb, where $b \Vdash_V \chi_\gamma(E, \mathcal{Z}_1, \mathcal{R})$ and by inductive hypothesis it follows that $b \Vdash_S \alpha$. Hence $a \Vdash_V \Diamond\alpha$. If (i) holds then using formula $q(D)$ it follows that there exists b such that aRb and $b \in \alpha(z,t)$. Then by inductive hypothesis for t we infer $b \Vdash_S \Diamond\alpha$ or $b \Vdash_S \alpha$, i.e. $a \Vdash_V \Diamond\alpha$. ∎

Thus by Lemma 6.1.19 the property (d1) holds for all the sets $\Sigma(x, t+1)$ of level $t+1$. Note that (d2) holds directly by definition of S in all the sets $\Sigma(x, t+1)$. Also it is easy to see that if $a \in [\Sigma(x, t+1) - \Sigma(x, t)]$ then x has a depth in $\mathcal{M}_j$ of at least m. Indeed, this follows from the fact that all elements $x_{E,\mathcal{Z},i}$ and $x_{E,\mathcal{R},\mathcal{Z},r}$, and elements x corresponding to formulas $\psi(D, \mathcal{Z}, x)$, in definitions of sets $\Sigma(x, t+1)$ are certain elements of depth at least m. Thus (f1) holds for sets $\Sigma(x, t+1)$.

Lemma 6.1.20 *The sets* $\Sigma(x, t+1)$, $x \in \mathcal{M}_j$ *satisfy (e1).*

Proof. Consider any element y such that

$$y \in (Q - \bigcup\{\Sigma(x, t+1) \mid x \in \mathcal{M}_j\}). \tag{6.9}$$

Then, in particular, y does not belong to the sets which were constructed while step t. By (e1) for t, this implies that there exists a (t+1)-element set D of elements from $\mathcal{M}_j$ such that

$$\forall x \in D[y^{R\leq} \cap \Sigma(x, t) \neq \emptyset].$$

Thus if there exists a $x_u \in (\mathcal{M}_j - D)$ such that $y^{R\leq} \cap \Sigma(x_u, t+1) \neq \emptyset$ then the assertion (e1) holds for y and $t+1$. Suppose that there are no such elements x_u. We may suppose that y is a R-maximal element not included in the union of sets of the kind $\Sigma(x, t+1)$. Then it can be easily seen that $y \Vdash_V q(D)$. Consider any element v which is R-seen from y and also is not included in the sets of the kind $\Sigma(x, t)$ and any element v of the cluster containing y. Then for every v considered, $v \Vdash_V q(D)$ by (e1) for the case t. Note that any D stipulates some E. For any v the following holds.

Lemma 6.1.21 *If* v *is irreflexive then for given* D *there exist elements* $x_{E,\mathcal{Z},i}$ *for any* $\mathcal{Z} \subseteq MV(r)$. *If* v *is reflexive then there exist elements* $x_{E,\mathcal{Z},\mathcal{R},r}$ *for any* $\mathcal{R} \subseteq MV(r)$, *and* $\mathcal{Z} \in \mathcal{R}$, *where* $\|\mathcal{R}\| \leq \|C(v)\|$, *where* $C(v)$ *is the cluster containing* v.

Proof. Let $G = \{z_1, ..., z_l\}$ be a set of representatives of all clusters which are R-immediate successors of $C(v)$. Note that v R-sees the same sets of kind $\Sigma(x_0, t)$ which are R-seen from y. The frame $\{v\}^{R<} - C(v)$ has a depth of at least m since the sets of the kind $\Sigma(x_0, t)$ cover the frame $S_m(Q)$. Hence, if v is irreflexive then the logic λ admits the irreflexive l-branching below m, and if v is reflexive than λ admits a $\|C(v)\|$, l-reflexive branching below m. By the choice of v, it follows that each $z \in G$ is an element of a set $\Sigma(x_z, t+1)$ for some element $x_z \in |\mathcal{M}_j|$. If any element from G is included in a certain $\Sigma(x_z, 0)$ then, since $\Sigma(x_z, 0) := \{x\}$, some x_z has in $\mathcal{M}_j$ depth of m or more. Otherwise some $z \in G$ is included in $(\Sigma(x_z, t+1) - \Sigma(x_z, 0))$. Then by (f1) for $t+1$, x_z has a depth of at least m. Thus, in any case, there is an x_z which has in $\mathcal{M}_j$ depth of at least m.

Let H be the set of all R-minimal in $\mathcal{M}_j$ elements of kind x_z considered. Recall that, by choice of y, all x_z must be from D. Then using that $v \Vdash_V q(D)$, property (d1) for the sets of the level $t + 1$, and that all proper R-successors of v belong to sets of kind $\Sigma(w, t + 1)$, where $w \in D$, we conclude that H has property of the set E from (6.3) with respect to D. The set H has at most l elements, thus the set E has at most l elements as well. We have showed above that if v is irreflexive then the logic λ admits irreflexive l-branching below m, and if v is reflexive then λ admits $||C(v)||$, l-reflexive branching below m Therefore by (d) from Lemma 6.1.15 we derive that all elements $x_{E,\mathcal{Z},i}$, where $\mathcal{Z} \subseteq MV(r)$, exist when v is irreflexive, and all elements $x_{E,\mathcal{Z},\mathcal{R},r}$, for any $\mathcal{Z} \in \mathcal{R}$, $||\mathcal{R}|| \leq ||C(v)||$, exist when v is reflexive. ∎

To continue the proof of Lemma 6.1.20 we suppose first that (6.6) for D does not hold. By Lemma 6.1.21 for any R-maximal v considered $v \Vdash_V \chi_1(D, \mathcal{Z}, \mathcal{R}) \vee \varphi_1(D, \mathcal{Z})$ for some $\mathcal{Z}$ and $\mathcal{R}$. That is $v \in \Sigma(x_0, t + 1)$, where $x_0 := x_{E,\mathcal{Z},i}$ or $x_0 := x_{E,\mathcal{R},\mathcal{Z},r}$, in particular x_0 has depth m or more, and by supposition that (6.6) is fail for D, $x_0 \notin D$. Hence the case $y = v$ contradicts the choice y, otherwise yRv for some R-maximal v again contradicts the choice y.

Suppose now that (6.6) holds for D. If y do not R-see any element of kind v considered then y is irreflexive, and then by Lemma 6.1.21 $y \Vdash_V \psi(D, \mathcal{Z}, g) \vee \varphi_2(D, \mathcal{Z})$ for some $\mathcal{Z}$ and g. That is $y \in \Sigma(x_0, t+1)$ for some x_0, a contradiction. Thus this case is impossible and y R-sees some v. By assumption that y do not belong to sets $\Sigma(x_0, t + 1)$ for any x_0 we conclude $y \nVdash_V \psi(D, \mathcal{Z}, g)$ for any $\mathcal{Z}$ and g. Therefore either $y \Vdash_V p(\mathcal{Z}_1)$ and $y \Vdash_V q(D)$, where $\mathcal{Z}_1 \notin P(\mathcal{G}(D))$, or for some z, yRz and $z \Vdash_V p(\mathcal{Z}_1)$, $\mathcal{Z}_1 \notin P(\mathcal{G}(D))$ and $z \Vdash_V q(D)$. We take an R-maximal z with properties described above or, if there is no such z, then y is irreflexive and has itself these properties and we take $z = y$.

Note that in the last case we have $y \Vdash_V p(\mathcal{Z}_1)$, where $\mathcal{Z}_1 \notin P(\mathcal{G}(D))$ and if yRy_1 and $y_1 \Vdash_V q(D)$ then $y_1 \Vdash_V p(\mathcal{Z}_2)$ where $\mathcal{Z}_2 \in P(\mathcal{G}(D))$. Since y is irreflexive, by Lemma 6.1.21 elements $x_{E,\mathcal{Z}_0,i}$ exist for all $\mathcal{Z}_0$, and $y \Vdash_V \varphi_2(D, \mathcal{Z}_1)$ for our $\mathcal{Z}_1$. Thus $y \in \Sigma(x_{E,\mathcal{Z}_1,i}, t + 1)$ which contradicts with the choice of y. Consider the first case. Then z is chosen as R-maximal that yRz and $z \Vdash_V p(\mathcal{Z}_1)$, $\mathcal{Z}_1 \notin P(\mathcal{G}(D))$ and $z \Vdash_V q(D)$

If z is irreflexive then by Lemma 6.1.21 elements $x_{E,\mathcal{Z}_0,i}$ with depth of m or more exist for all $\mathcal{Z}_0$ and we conclude $z \Vdash_V \varphi_2(D, \mathcal{Z})$ for some $\mathcal{Z}$ and $z \in \Sigma(x_{E,\mathcal{Z},i}, t+1)$. If $x_{E,\mathcal{Z},i} \in D$ then by (6.4) $x_{E,\mathcal{Z},i}$ satisfies (6.6). Consequently $S(x_{E,\mathcal{Z},i}, Sub(r)) \cap MV(r) \in P(D)$. Therefore we obtain $S(x_{E,\mathcal{Z},i}, Sub(r)) \cap MV(r) = \mathcal{Z} \in P(\mathcal{G}(D))$ and by (d2) for $t + 1$ we get that $z \Vdash_V p(\mathcal{Z}_1)$ implies $\mathcal{Z} \in P(\mathcal{G}(D))$, a contradiction. Hence $x_{E,\mathcal{Z},i} \notin D$ and $z \in \Sigma(x_{E,\mathcal{Z},i}, t + 1)$, yRz, which again contradicts the choice y.

Suppose finally that z is reflexive. Then by Lemma 6.1.21 elements $x_{E,\mathcal{R},\mathcal{Z},r}$ with depth of at least m , with $\mathcal{R} := \{S(w, Sub(r)) \cap MV(r) \mid w \in C(z)\}$ and any $\mathcal{Z} \in \mathcal{R}$ exists. And it is not hard to see that $z \Vdash_V \chi_2(D, \mathcal{Z}_1, \mathcal{R})$, where $\mathcal{R}$ is described above and cz_1 is the fixed above tuple, i.e. $z \Vdash_V p(\mathcal{Z}_1)$, $\mathcal{Z}_1 \notin P(\mathcal{G}(D))$. Thus $z \in \Sigma(x_{E,\mathcal{Z}_1,\mathcal{R},r}, t+1)$. Suppose that $x_{E,\mathcal{Z}_1,\mathcal{R},r} \in D$, then (6.5) yields $x_{E,\mathcal{Z},i}$ satisfies (6.6). Hence $S(x_{E,\mathcal{Z},\mathcal{R},r}, Sub(r)) \cap MV(r) \in P(D)$ which implies $S(x_{E,\mathcal{Z},\mathcal{R},r}, Sub(r)) \cap MV(r) \in P(\mathcal{G}(D))$. Using (d2) for $t+1$ we infer from this that $z \Vdash_V p(\mathcal{Z}_1)$ implies $\mathcal{Z}_1 \in P(\mathcal{G}(D))$, a contradiction. Consequently we arrive at $x_{E,\mathcal{Z},\mathcal{R},r} \notin D$ and $z \in \Sigma(x_{E,\mathcal{Z},\mathcal{R},r}, t+1)$ and yRz, which again contradicts with the choice y. Hence by exhausting all possibilities we showed that the property (e1) holds for $t+1$. ∎

In such way we construct all the sequence of the sets $\Sigma(x,t)$, $x \in |\mathcal{M}_j|$ and $1 \leq t \leq m_1$. Note now that if an element $v \in Q$ was not included in the constructed sets of the kind $\Sigma(x,t)$ until the step $m_1 - 1$, where m_1 is the number of elements in $\mathcal{M}_j$, then v will be included in one of the sets constructed in step m_1.

Indeed, if $y \notin \bigcup \Sigma(x, m_1 - 1)$ then $y \Vdash_V q(D)$, where $D := |\mathcal{M}_j|$, by (e1) for the index $m_1 - 1$. Suppose that $y \notin \bigcup \Sigma(x, m_1)$. We may assume that y is R-maximal element with this property. Consider all the elements v which are not included in the sets of the kind $\Sigma(x, m_1 - 1)$ and are R-accessible from y or do belong to the cluster containing y. Then Lemma 6.1.21 holds for all corresponding v and $t := m_1 - 1$. If (6.6) for $D := |\mathcal{M}_j|$ does not hold then existing by Lemma 6.1.21 elements $x_{E\mathcal{Z},i}$ or $x_{E,\mathcal{Z},\mathcal{R},r}$ do not belong to $D = |\mathcal{M}_j|$, a contradiction.

Consequently (6.6) for $D := |\mathcal{M}_j|$ holds. Consider any v. If v is irreflexive then by Lemma 6.1.21 there exists an element $x_{E,\mathcal{Z},i} \in |\mathcal{M}_j| = D$ with depth of at least m, where $S(v, Sub(r)) \cap MV(r) = \mathcal{Z}$ and $S(x_{E,\mathcal{Z},i}, Sub(r)) \cap MV(r)) = \mathcal{Z} \in P(\mathcal{G}(\mathcal{M}_j))$. If v is reflexive than by Lemma 6.1.21 there exists an element $x_{E,\mathcal{Z},\mathcal{R},r} \in |\mathcal{M}_j| = D$ with depth of at least m, where $\mathcal{R} := \{S(w, Sub(r) \cap MV(r) \mid w \in C(v)\}$ and $S(v, Sub(r)) \cap MV(r) = \mathcal{Z}$ and $S(x_{E,\mathcal{Z},i}, Sub(r)) \cap MV(r)) = \mathcal{Z} \in P(\mathcal{G}(\mathcal{M}_j))$.

Thus for any v, $v \Vdash_V p(\mathcal{Z})$, for some $\mathcal{Z} \in P(\mathcal{G}(\mathcal{M}_j))$, and for all such $\mathcal{Z}$ elements $g \in \mathcal{G}(\mathcal{M}_j)$ with $S(g, Sub(r)) \cap MV(r) = \mathcal{Z}$ exists. Therefore it follows that $y \Vdash_V \psi(D, \mathcal{Z}, g)$ for some $g \in D = |\mathcal{M}_j|$, i.e. $y \in \Sigma(g, m_1)$. Hence, finally, we arrive at $\bigcup \Sigma(x, m_1) = Q$, what we need.

Thus, in the way described above we have extended the valuation S onto whole the frame Q. We make such extending for all H_j and $\mathcal{M}_j$ in the described above way. By (d1) for the sets of level m_1 and (c) from Lemma 6.1.15 we conclude that the valuation S invalidates the inference rule r in some H_j, and that the premise of r is true under S in every H_j. Denoting the definable valuation S on H_j by S_j, we have $S_j(x) := V_j(\beta_{j,x})$, where $x \in (Var(r) - MV(r))$, $\beta_{j,x}$ is the corresponding defining formula. If $H_l \neq H_j$ then these models have no common maximal clusters. By Theorem 3.3.7 each element of $Ch_\lambda(k + k_1)$ is definable. Therefore the set $\mathfrak{M}_l$ of all the elements of all the maximal clusters from the model H_l is definable by a formula δ_l for each l. Finally, we choose the valuation S in $Ch_\lambda(k + k_1)$ as follows $\forall x \in (Var(r) - MV(r))$

$$S(x) := V(\bigvee_j (\beta_{j,x} \wedge \bigwedge_{l \neq j} \neg(\delta_l \vee \Diamond\delta_l))).$$

Then it can be easily seen that this definable valuation invalidates the inference rule r in $Ch_\lambda(k)$ and coincides with the original valuation of $Ch_\lambda(k + k_1)$ on meta-variables of r. $\blacksquare$.

Now combining Lemmas 6.1.15, 6.1.16 and Theorem 6.1.11 we obtain the following theorem.

Theorem 6.1.22 Semantic criterion for admissibility. *Let λ be a modal logic with the fmp extending $K4$ and having the generalized property of branching below m and the effective m-drop points property. An inference rule r with k variables and k_1 metavariables is admissible for λ if and only if r is valid in the frame of the model $Ch_\lambda(k + k_1)$ with respect to any valuation coinciding on $MV(r)$ with the original valuation V of $Ch_\lambda(k + k_1)$.*

We are now also in a position to prove our main theorem concerning effective recognizing of the admissibility for inference rules with meta-variables in modal logics.

Theorem 6.1.23 *Let λ be a modal logic extending $K4$ with the fmp and the following properties:*

(i) *λ has the generalized property of branching below m;*

(ii) *λ has the effective m-drop points property;*

(iii) *there is an algorithm for recognizing finite λ-frames (which holds, in particular, if λ is finitely axiomatizable).*

Then

(a) *There is an algorithm which determines admissibility in λ for inference rules with meta-variables.*

(b) *There is an algorithm which determines whether any given logical equation has solutions in λ and which, if so, provides a solution to that equation.*

Proof.   Consider an inference rule r with k variables and k_1 meta-variables. We will investigate whether there is an λ-model $\langle M, R, S \rangle$ satisfying properties (a) - (e) from Lemma 6.1.15 with at most $g(e, q) * 2^{2^{k+k_1+1}}$ elements (where g is the recursive function occurring in the definition of the effective m-drop points property; e is the number of elements in $|S_m(Ch_{K4}(k + k_1))|$; and q is the number of subformulas in r). If such model exists then by Lemma 6.1.16 there exists an expressible valuation S with effectively constructed defining formulas which invalidates r in $Ch_\lambda(k)$. Then, by Theorem 6.1.11 the inference rule r is not admissible in λ. If we discovered that such λ-model $\mathcal{M}$ does not exist then the inference rule r is admissible for λ. In fact, let r be not admissible. Then by Theorem 6.1.11, Lemma 6.1.12 and Lemma 6.1.15 we obtain that a model $\mathcal{M}$ with properties described above exists - a contradiction. Hence we have proved (a). In order to show (b) we use Lemma 6.1.8 and (a), i.e. the algorithm recognizing admissibility of rules with metavariables in λ, and the fact that this algorithm works on the basis of the Lemma 6.1.16, where we have effectively constructed the disproving definable valuation. ■

Immediately from this theorem, Theorem 6.1.9 and Theorem 6.1.10 we obtain

Corollary 6.1.24 *Let λ be a finitely axiomatizable modal logic with the fmp extending $K4$ and having the generalized property of branching below m and the effective m-drop points property. Then*

(a) The quasi-equational theory of the free modal algebra $\mathfrak{F}_\lambda(\omega)$ in the signature enriched with constants for the free generators is decidable.

(b) There is an algorithm examining decidability of equations in the free modal algebra $\mathfrak{F}_\lambda(\omega)$, which constructs certain solutions to solvable equations.

(c) If λ has the disjunction property then the universal theory of the free modal algebra $\mathfrak{F}_\lambda(\omega)$ in the signature enriched with constants for the free generators is decidable as well.

As with the case of ordinary inference rules (without meta-variables) we will now transform the obtained results concerning modal logics onto superintuitionistic logics. First we note that the proof of the Theorem 3.2.2 can be immediately transferred to inference rules with meta-variables, and we obtain the following result.

Theorem 6.1.25 *An inference rule r with meta-variables $p_1, ..., p_k$ is admissible in a superintuitionistic logic λ iff its Gödel-McKinsey-Tarski translation $T(r)$, which is a modal rule with the same meta-variables $p_1, ..., p_k$, is admissible in the modal logic $\sigma(\lambda)$.*

We can now present the central theorem for superintuitionistic logics.

Theorem 6.1.26 *If a superintuitionistic logic λ (i) has the strong effective m-drop points property, (ii) the property of branching below m and (iii) the finite λ-frames are effectively recognizable (which holds, in particular, when λ is finitely axiomatizable) then*

(a) there is an algorithm recognizing the admissibility in λ of inference rules with meta-variables.

(b) there is an algorithm examining the decidability of logical equations in λ, which constructs certain solutions to solvable equations.

Proof. By Theorem 6.1.25 an inference rule r with meta-variables is admissible in our superintuitionistic logic λ if and only if the rule $T(r)$ is admissible in $\sigma(\lambda)$. Using this, Lemma 3.5.6, Lemma 3.5.5 and Theorem 6.1.23 we obtain an algorithm recognizing the admissibility of rules with meta-variables in λ. (b) follows from (a), Lemma 6.1.8 and the fact that the algorithm from Theorem 6.1.23 recognizing admissibility of $T(r)$ works on basis of Lemma 6.1.16, where we have effectively constructed the invalidating valuations. ∎

Using this theorem, Theorem 6.1.9 and Theorem 6.1.10 we immediately obtain an algebraic counterpart of this theorem:

Corollary 6.1.27 *Suppose λ is a finitely axiomatizable superintuitionistic logic with the fmp, the property of branching below m and the strong effective m-drop points property. Then*

(a) The quasi-equational theory of the free pseudo-boolean algebra $\mathfrak{F}_\lambda(\omega)$ in the signature enriched with constants for the free generators is decidable.

(b) There is an algorithm examining decidability of the equations in the free pseudo-boolean algebra $\mathfrak{F}_\lambda(omega)$, which constructs certain solutions to solvable equations.

(c) If λ has the disjunction property then the universal theory of the free pseudo-boolean algebra $\mathfrak{F}_\lambda(\omega)$ in the signature enriched with constants for the free generators is decidable.

We can also infer a parallel to the modal case semantics criterion concerning superintuitionistic logics:

Theorem 6.1.28 *Let λ be a superintuitionistic logic with the fmp, the property of branching below m, and the strong effective m-drop point property. Then an inference rule r with k variables and k_1 meta-variables is admissible in λ iff r is valid in the frame of the model $Ch_\lambda(k + k_1)$ with respect to any valuation coinciding on meta-variables of r with the original valuation of this model.*

Proof. If r is not admissible in λ then by Theorem 6.1.11 there is a valuation S invalidating r in a model $Ch_\lambda(k + k_1 + m)$, and coinciding with the original valuation on meta-variables. Applying Lemma 6.1.12 we obtain the required result. Conversely, suppose that there exists a valuation S of variables and meta-variables from r in the frame $Ch_\lambda(k + k_1)$ invalidating r and coinciding on metavariables with the original valuation of that model. Then S invalidates the rule $T(r)$ also. And there is a p-morphism from $Ch_{\sigma(\lambda)}(k + k_1)$ onto $Ch_\lambda(k + k_1)$ which preserves truth of formulas p_i, where p_i are meta-variables of r. Therefore $T(r)$ is invalidated in $Ch_{\sigma(\lambda)}$ by a valuation which coincides with the original valuation of the model $Ch_{\sigma(\lambda)}$ on the meta-variables of r. Therefore by Theorem 6.1.22 the rule $T(r)$ is not admissible in $\sigma(\lambda)$. This and Theorem 6.1.25 immediately yield that r is not admissible in λ. ∎

Now we demonstrate some applications the general theorems obtained above to certain individual logics and certain classes of logics. We know (see Section 5 of Chapter 3) that modal logics $K4$, $K4.1$, $K4.2$, $K4.3$, $S4$, $S4.1$, $S4.2$, $S4.3$, Grz, $Grz.2$, $Grz.3$ and GL have the property of branching below 1, and it is easy to see that all they have also generalized property of branching below 1. Also all they have the strong effective 1-drop points property. Therefore we immediately infer from Theorem 6.1.23 the following corollary.

Corollary 6.1.29 *For any modal logic λ from the list $K4$, $K4.1$, $K4.2$, $K4.3$, $S4$, $S4.1$, $S4.2$, $S4.3$, Grz, $Grz.2$, $Grz.3$ and GL,*

(i) *There is an algorithm which determines admissibility of inference rules with meta-variables in λ.*

(ii) *There is an algorithm examining solvability of logical equations in λ, which constructs certain solutions to equations which are solvable.*

(iii) *The quasi-equational theory of the free modal algebra $\mathfrak{F}_\lambda(\omega)$ in the signature enriched with constants for the free generators is decidable.*

(iv) *There is an algorithm examining solvability of equations in the free modal algebra $\mathfrak{F}_\lambda(\omega)$, which constructs certain solutions to equations if they are solvable.*

(v) *For $\lambda = K4$, $K4.1$ $S4$, $S4.1$, Grz, GL, the universal theory of the free modal algebra $\mathfrak{F}_\lambda(\omega)$ in the signature enriched with constants for the free generators is decidable.*

Similarly we can apply Theorems 6.1.26, 6.1.28 and Corollary 6.1.27 to superintuitionistic logics H and KC since they have the fmp, the property of branching below 1 and the strong effective 1-drop points property (cf. Corollary 3.5.10). In particular we have

Corollary 6.1.30 *Intuitionistic propositional logic H and superintuitionistic logic $KC := H \oplus \neg p \vee \neg\neg p$ are*

(i) *decidable by admissibility of inference rules with metavariables,*

(ii) *the problem of solvability for logical equations is decidable for these logics,*

(iii) *the substitution problem is decidable for H and KC.*

We know (see Exercise 3.5.13) that the Gabbay-De-Jongh superintuitionistic logics D_n and modal logics $K4_{b,n}$, $S4_{b,n}$, $Grz_{b,n}$, $GL_{b,n}$ have the disjunction property and (see Corollary 3.5.14 and Lemma 3.5.15) all of them have the property of branching below 1, and it can be easily seen that they also have the

generalized property of branching below 1, and the strong effective 1-drop point
property. Therefore from Theorems 6.1.23 and 6.1.26 and Corollary 6.1.27 we
immediately derive

Corollary 6.1.31 *For any logic λ from the list D_n, $K4_{b,n}$, $S4_{b,n}$, $Grz_{b,n}$,
$GL_{b,n}$,*

- *(i) There is an algorithm which recognizes the admissibility of
  inference rules with meta-variables in λ;*

- *(ii) There is an algorithm which determines whether a logical
  equations has a solution in λ, which, if so, provides a solution.*

- *(iii) The quasi-equational theory of the free algebra $\mathfrak{F}_\lambda(\omega)$ in the
  signature enriched with constants for the free generators is
  decidable.*

- *(iv) There is an algorithm which determines whether an equa-
  tion has a solution in the free modal algebra $\mathfrak{F}_\lambda(\omega)$ and if so
  provides a solution.*

These results are applicable to many logics more, we conclude this section
with the following

Corollary 6.1.32 *For any finitely axiomatizable modal logic $\lambda \subseteq K4$ of finite
depth m, and for any finitely axiomatizable superintuitionistic logic λ of finite
depth m, for any m, the following hold:*

- *(i) There is an algorithm which determines the admissibility of inference
  rules with meta-variables in λ;*

- *(ii) There is an algorithm which recognizes whether any given logical equation
  has a solution in λ, and which, if so, provides a solution.*

- *(iii) The quasi-equational theory of the free algebra $\mathfrak{F}_\lambda(\omega)$ in the signature
  enriched with constants for the free generators is decidable.*

Proof. The proof follows immediately from Theorem 6.1.23 and Theorem 6.1.26
in view of lack elements with depth more m in λ-frames. ∎

6.2   The Preservation of Admissible for S4 Ru-
les

It is not hard to see that the extensions of an axiomatic system sometimes do not
inherit the admissible inference rules of the original axiomatic system. Therefore for certain important axiomatic systems $\mathcal{AS}$ it would be desirable to describe those extensions which inherit, or preserve, the inference rules admissible

in $\mathcal{AS}$, and those which do not. In this section we investigate those modal logics which preserve the admissible rules of the Lewis modal system $S4$. We begin with a precise definition of the preservation of admissibility for inference rules.

Definition 6.2.1 *Suppose λ_1 and λ_2 are algebraic logics. We say that logic λ_2 preserves all inference rules admissible in λ_1 if every rule admissible in λ_1 also is admissible in λ_2.*

Having dealt here with modal logics, we now introduce a semantic notion which reflects the preservation of admissibility.

Definition 6.2.2 *Let $\mathcal{F} := \langle F, R \rangle$ be a frame and let X be an antichain of clusters from $\mathcal{F}$. We say that an element c of $\mathcal{F}$ is a co-cover for X if $c^{R<} = X^{R\leq}$ and $\{c\}$ forms a cluster.*

Definition 6.2.3 *Let $\mathcal{F}$ be a finite rooted transitive and reflexive frame. A frame $\mathcal{M}$ is said to be a co-covers poster for $\mathcal{F}$ if $\mathcal{M}$ is obtained from F by adding a finite number of elements in the following way. Let $\mathcal{F}_0 := \mathcal{F}$. In every step $i, i \in N$ we add to the frame $\mathcal{F}_i$ a single new reflexive element c as follows. We choose some non-trivial antichain of clusters X (having at least two clusters) in $\mathcal{F}_i$ which has no co-cover in $\mathcal{F}_i$ and add to $\mathcal{F}_i$ a new element c which is co-cover for X. After a finite number of steps $i = 1, 2, ..., n$ this procedure terminates and the resulting frame is $\mathcal{M}$.*

Definition 6.2.4 *We say that a modal logic λ extending $S4$ has the co-cover property if, for any rooted finite λ-frame $\mathcal{F}$, every co-covers poster of $\mathcal{F}$ is a λ-frame.*

Recall that we say a rule r is invalidated in a frame $\mathcal{F}$ by valuation V if all premises of r are valid in $\mathcal{F}$ with respect to V but the conclusion of r is false in F under V.

Lemma 6.2.5 *If a modal logic $\lambda \supseteq S4$ has the co-cover property and has the finite model property then λ preserves all admissible in $S4$ inference rules.*

Proof. It is clear that we may consider only inference rules which have single premise. Suppose α/β is an admissible in $S4$ inference rule which is not admissible in λ. Then there exist formulas γ_i such that $\alpha(\gamma_i) \in \lambda$ but $\beta(\gamma_i) \notin \lambda$. Formulas γ_i contain a finite number of propositional letters, say n letters. Consider the n-characterizing model $Ch_\lambda(n)$ for λ (see Theorem 3.3.6), since this modal characterizes theorems λ in n variables, we have

$$Ch_\lambda(n) \Vdash \alpha(\gamma_i), \quad Ch_\lambda(n) \nVdash \beta(\gamma_i).$$

Thus the valuation $S(x_i) := V(\gamma_i)$ invalidates the inference rule α/β in a sub-frame $a^{R\leq}$ of $Ch_\lambda(n)$ generated by some a. Moreover we can assume that the cluster containing a is R-maximal with this property.

We stipulate $\mathcal{F}_0 := a^{R\leq}$ and $\mathcal{F}_{m+1}$ is obtained from $\mathcal{F}_m$ as follows. We take every non-trivial antichain X of clusters from $\mathcal{F}_m$ which has no co-cover in $\mathcal{F}_m$ and add to $\mathcal{F}_m$ a single reflexive element which is co-cover for X. By our construction $\mathcal{F}_{m+1}$ is a co-covers poster for $\mathcal{F}_m$. The logic λ has the co-cover property. Therefore each $\mathcal{F}_m$ can be considered as a generated subframe of $Ch_\lambda(n)$ and still $\mathcal{F}_m$ is a generated subframe of $\mathcal{F}_{m+1}$.

Thus $\mathcal{F}_\infty := \bigcup_{m<\infty} \mathcal{F}_m$ is a generated subframe of $Ch_\lambda(n)$ and the rule α/β is false in $\mathcal{F}_\infty$ with respect to the valuation S. Moreover, according to its construction, $\mathcal{F}_\infty$ has a co-cover for every finite non-trivial antichain of arbitrary clusters. Note that according to its construction, the model $Ch_{S4}(n)$ has co-covers for any non-trivial antichain of clusters, and, as it can be easily seen, $Ch_\lambda(n)$ is an open submodel of $Ch_{S4}(n)$. From this together with the property of $\mathcal{F}_\infty$ to have co-covers for all non-trivial finite antichains of clusters, we directly conclude that there exists a p-morphism of the frame $Ch_{S4}(n)$ onto the frame $\mathcal{F}$. Using this p-morphism we can transfer the valuation S to the frame $Ch_{S4}(n)$. Because p-morphisms preserve truth of formulas, it follows that S invalidates the rule α/β in $Ch_{S4}(n)$. Theorem 3.5.1 and Lemma 3.4.2 entail that an inference rule r is admissible in $S4$ iff r is valid in the frame of $Ch_{S4}(n)$ for every n and every valuation. Therefore it follows that α/β is not admissible in $S4$, a contradiction. Thus λ preserves all admissible in $S4$ inference rules. ∎

Lemma 6.2.6 *If a modal logic λ extending $S4$ preserves all admissible in $S4$ inference rules and has the finite model property then λ has the co-cover property.*

Proof. Let $\lambda \supseteq S4$ be a logic with the fmp. Suppose λ does not have the co-cover property. This entails that there is a subframe $\mathcal{F} := a^{R\leq}$ of $Ch_\lambda(n)$ generated by a cluster a and there exists a sequence $\mathcal{F}_0, ..., \mathcal{F}_k$ of generated finite subframes of $Ch_\lambda(n)$ which have the following properties:

a) $\mathcal{F}_0 := \mathcal{F}$.

b) All antichains of clusters from $\mathcal{F}_i$ of depth less than i have in $\mathcal{F}_i$ some co-covers.

c) The frame $\mathcal{F}_{i+1}$ is obtained from the frame $\mathcal{F}_i$ when $i < k$ as follows: We consider the set A_i consisting of all non-trivial antichains of clusters from $\mathcal{F}_i$ having at least one cluster of depth i. If all antichains from A_i have co-covers in $\mathcal{F}_i$ then we set $\mathcal{F}_{i+1} := \mathcal{F}_i$. Otherwise, for every antichain X from A_i which has no co-cover, we add to $\mathcal{F}_i$ a single reflexive co-cover for X.

  d) There exists an antichain Δ of clusters from $\mathcal{F}_k$ which has no
    co-cover in $Ch_\lambda(n)$ and has a cluster with depth k.

Then, in particular, all antichains of $\mathcal{F}_k$ with depth of elements of not more than
$k - 1$ have co-covers in $\mathcal{F}_k$. Moreover, we can choose a sequence $\mathcal{F}_1, ..., \mathcal{F}_k$ with
smallest number k of members and pointed above properties. Now we need
special modal formulas in order to express necessary properties of the model
$Ch_\lambda(n)$. Let $M := S_k(Ch_\lambda(n)) \cup \mathcal{F}_k$ and $M_1 := Sl_{k+1}(Ch_\lambda(n)) \cup M$.

Any cluster from $Ch_\lambda(n)$ is a subset of M_1 or M or has empty intersection
with M_1 and M. For each cluster $i \subseteq M_1$ (now it will be convenient for us to
denote clusters by small letters like indexes), we introduce a new propositional
variable p_i (*new* means that p_i is not a variable from the domain of the valuation
of $Ch_\lambda(n)$). The special formulas $f(i)$ for clusters i from M_1 are introduced by
induction on the depth of i as follows. If i and j are clusters then iRj means that
all elements of j are accessible by R from all elements of i. If a cluster i belongs
to $S_1(M_1)$ then $f(i) := \Box p_i$. If the formulas $f(i)$ are already introduced for all
clusters i from $S_t(M_1)$, and j is a cluster from $Sl_{t+1}(M_1)$, then we put

$$f(j) := \Box p_j \wedge \bigwedge \{\neg \Box p_i \mid i \neq j, i \subseteq Sl_{t+1}(M_1)\} \wedge \bigwedge_{j\,Ri, \neg(iRj)} \Diamond f(i) \wedge \bigwedge$$

$$\{\neg \Diamond f(i) \mid \neg(jRi), i \subseteq S_t(M_1)\} \wedge \bigwedge_{\neg(iRj)} \neg \Box p_i \wedge u(j), \text{ where}$$

$$u(j) := \Box([\Box p_i \wedge \bigwedge \{\neg \Box p_i \mid i \neq j, i \subseteq Sl_{t+1}(M_1)\} \wedge \bigwedge_{j\,Ri, \neg(iRj)} \Diamond f(i) \wedge$$

$$\bigwedge_{j\,Ri, \neg(iRj)} \neg f(i)] \vee \bigvee_{j\,Ri, \neg(iRj)} f(i))$$

The inference rule r has the following definition:

$$g := \bigwedge_{i \in M_1} \neg f(i) \wedge \bigvee_{i \in Sl_{k+1}(M_1)} \Diamond f(i), \quad r := \frac{\bigvee \{f(i) \mid i \subseteq M_1\} \vee g}{\neg f(a)},$$

where a is the root of the frame of $\mathcal{F}_0$. We choose the valuation W of variables
p_i from r in $Ch_\lambda(n)$ as follows: $W(p_i) := \{j \mid iRj\}$. It can be verified directly
by induction on the depth of clusters, that, for each cluster i from M_1 and any
$a \in i$, $a \Vdash_W f(i)$ holds. and that, for each cluster i from $(Ch_\lambda(n) - M_1)$ and
every $a \in i$, $a \Vdash_W g$ holds. Thus r is invalidated in $Ch_\lambda(n)$ by the valuation W,
that is

$$W(\bigvee_{i \subseteq M_1} f(i) \vee g) = |Ch_\lambda(n)|, \quad W(\neg(f(a)) \neq |Ch_\lambda(n)|. \tag{6.10}$$

According to Theorem 3.3.7, every element of $Ch_\lambda(n)$ is definable. Therefore the valuation W assigns some definable subsets of $Ch_\lambda(n)$ to propositional letters p_i. Therefore by Theorem 3.3.3 inference rule r is not admissible in λ

Our aim now is to show that r is admissible in $S4$. Suppose that r is not admissible in $S4$. Then, according to Theorem 3.3.3, there is a definable valuation W in some n-characterizing model $Ch_{S4}(n)$ for $S4$ such that

$$W(\bigvee_{i \in M_1} f(i) \wedge g)) = |Ch_{S4}(n)| \& W(\Box \neg f(a)) \neq |Ch_{S4}(n)|. \tag{6.11}$$

Then there is $b_a \in |Ch_{S4}(n)|$ such that $b_a \Vdash_W f(a)$. It is easy to see from the definition of $f(i)$ that

$$\forall b_i \in |Ch_{S4}(n)|(b_i \Vdash_W f(i) \& j \subseteq M_1 \&$$

$$\& (iRj) \& \neg (jRi) \implies \exists b_j (b_i Rb_j \& b_j \Vdash_W f(j))) \tag{6.12}$$

Suppose that $b_i \Vdash_W f(i), b_j \Vdash_W f(j)$ and $b_i Rb_j$. If $\neg(iRj)$ then $b_j \Vdash_W \neg\Box p_i$. But $b_i \Vdash_W f(i)$ therefore $b_i \Vdash_W \Box p_i$. This and the fact that $b_i Rb_j$ imply $b_j \Vdash_W \Box p_i$, a contradiction. Hence

$$\forall b_i, b_j \in |Ch_{S4}(n)|(b_i \Vdash_W f(i) \& b_j \Vdash_W f(j) \tag{6.13}$$

$$\& \neg(iRj) \Rightarrow \neg(b_i Rb_j))$$

From (6.12) and (6.13) we obtain immediately

$$b_i \Vdash_W f(i) \& (iRj) \& \neg(jRi) \Rightarrow \tag{6.14}$$

$$\exists b_j ((b_i Rb_j) \& \neg (b_j Rb_i) \& (b_j \Vdash_W f(j))).$$

From (6.13) it follows that

$$b_i \Vdash_W f(i) \& b_j \Vdash_W f(j) \& b_i Rb_j \implies iRj. \tag{6.15}$$

Again using (6.13) we infer that

$$[b_i \Vdash_W f(i)] \& [b_j \Vdash_W f(j)] \& \neg(iRj) \& \neg(jRi) \Rightarrow \tag{6.16}$$

$$\neg(b_i Rb_j) \& \neg(b_j Rb_i)$$

Also we immediately derive from the definition of $f(i)$ that

$$(b_i \Vdash_W f(i)) \& (b_i Rd) \implies \exists j((iRj) \& (d \Vdash_W f(j))) \tag{6.17}$$

Moreover, from this and (6.14) it follows that

$$(\forall j)(aRj) \Rightarrow \exists b_j ((b_a Rb_j) \wedge (b_j \Vdash_W f(j))) \tag{6.18}$$

Let Y be a non-trivial antichain of clusters from M which have depth not more than k. Suppose that there exists a collection of elements $b_i \in |Ch_{S4}(n)|$ such that $b_i \Vdash_W f(i)$ for all $i \in Y$. Then by (6.16) all the clusters C_i including elements $b_i, i \in Y$ form an antichain of clusters in $Ch_{S4}(n)$. According to the construction of the model $Ch_{S4}(n)$, there is a co-cover v in this model for the antichain consisting of all C_i. The relation (6.11) implies that $v \Vdash_W f(j)$ for some $j \in M$ or $v \Vdash_W g$. We claim that

$$\exists j \in M_1 (v \Vdash_W f(j)). \tag{6.19}$$

Indeed, suppose otherwise. Then we have $v \Vdash_W g$ and $v \Vdash_W \Diamond f(x)$, where $x \subseteq Sl_{k+1}(M_1)$ and $v \Vdash_W \neg f(j)$ for all $j \subseteq M_1$. Hence there exists b_x such that vRb_x and $\neg b_x Rv$, and $b_x \Vdash_W f(x)$. Then $b_i R b_x$ for some i from Y. From this and (6.18) we infer that iRx, which contradicts with $x \in Sl_{k+1}(M_1)$. Hence (6.19) holds. If $j \subseteq M_1$ is a cluster and $v \Vdash_W f(j)$ then

$$j \text{ is a co-cover for } Y \text{ in } M_1 \tag{6.20}$$

Indeed, by our assumption we have that $\forall i \in Y (b_i \Vdash_W f(i) \& vRb_i) \& \neg (b_i Rv)$. This together with (6.15) yield $\forall i \in Y (jRi)$. Suppose a cluster l is a strict successor of j in M_1. Then by (6.14) there exists a b_l from $|Ch_{S4}(n)|$ such that $(b_l \Vdash_W f(l)) \& (vRb_l \& \neg (b_l Rv))$. Hence $b_i Rb_l$ for some $i \in Y$. This implies iRl by (6.15). Therefore we have that j is a co-cover for the antichain Y and (6.20) is proved. Applying (6.12), (6.18), (6.19) and (6.20) we obtain that, for any cluster x,

$$(x \subseteq S_k(\mathcal{F}_k)) \Rightarrow \exists b_x \in |Ch_{S4}(n)| (b_x \Vdash_W f(x)) \tag{6.21}$$

Let Δ be an antichain of clusters from $\mathcal{F}_k$ (having clusters of depth k) which has no co-cover in $Ch_\lambda(n)$. Let $D := \{b_x \mid x \in \Delta\}$, where $b_x \Vdash_W f(x)$ (we use (6.21)). According to (6.16) the clusters containing elements from D form an antichain in $Ch_{S4}(n)$. Each antichain from $Ch_{S4}(n)(n)$ has a co-cover in this model by construction of $Ch_{S4}(n)$. Let w be a co-cover for D. By (6.19) and (6.20) we have that there exists j which is a co-cover for Δ in M_1 and in $Ch_\lambda(n)$ which contradicts our assumption. ∎

From Lemmas 6.2.5 and 6.2.6 we immediately derive

Theorem 6.2.7 *[156] In order for a modal logic λ with the fmp to preserve all admissible in S4 inference rules it is necessary and sufficient that λ to have the co-cover property.*

Looking at the structure of approximating finite frames for modal logics with the fmp and the co-cover property, it seems that all these classes of frames are very similar and uniform. So at first glance it looks very likely that the class of all such logics has not many representatives. But in real it does not a case.

Theorem 6.2.8 *[156] There are continuously many modal logics over modal system Grz which preserve all admissible in S4 inference rules and have the fmp.*

Proof. We introduce a sequence of partially ordered frames $A_n, n \in N$ as follows. Let A_1 be the three-elements antichain. Every A_{n+1} is obtained from A_n by adding a single co-cover for each non-trivial antichain from A_n having elements of depth n. We define $\mathcal{F}_n$ to be the rooted by a_n partially ordered frame, where $(\mathcal{F}_n - \{a_n\}) = A_n$, and we set G_n is the rooted by b_n partially ordered frame, where $(G_n - \{b_n\}) = \mathcal{F}_n$. For any $Y \subseteq N$ we set $\lambda(Y)$ is the logic $\lambda(\{\mathcal{F}_n \mid n \in N\} \cup \{G_n \mid n \in Y\})$. Then obviously all logics $\lambda(Y)$ are extensions of the modal logic Grz. Moreover all logics $\lambda(Y)$ have the co-cover property. In order to show this we take any finite rooted λ-frame $\mathcal{B}$ (i.e. $\mathcal{B} \Vdash \lambda(Y)$). First we note that $\mathcal{B}$ must be a poset.

Furthermore, the frame $\mathcal{B}$ has not more than three maximal elements because all frames G_n and all frames $\mathcal{F}_n$ have exactly three maximal elements each, and the property to have not more than three maximal accessible elements can be expressed by a modal formula. If $\mathcal{G}$ is a co-covers poster of $\mathcal{B}$ then $\mathcal{G}$ is a generated subframe of a p-morphic image of an appropriate finite disjoint union $\mathcal{F}_n \sqcup \ldots \sqcup \mathcal{F}_n$ of the frame $\mathcal{F}_n$ for some n. This can be easily shown bearing in mind the structure of frames $\mathcal{F}_n$. Because taking of disjoint unions, p-morphic images and generated subframes preserves the truth of formulas, it follows that $\mathcal{G} \Vdash \lambda(Y)$ which we needed. Thus every $\lambda(Y)$ has the co-cover property.

Every logic $\lambda(Y)$ has the finite model property by the definition. Therefore all $\lambda(Y)$ preserve all admissible in $S4$ inference rules according to Theorem 6.2.7. It remains only to show that all $\lambda(Y)$ are pairwise distinct. It is a consequence of the following proposition,

$$\forall n(\lambda(N - \{n\}) \nsubseteq \lambda(G_n)). \tag{6.22}$$

In order to prove (6.22) we suppose that $\exists n(\lambda(N - \{n\}) \subseteq \lambda(G_n))$. Then according to the correspondence between varieties of modal algebras and modal logics, using Birkhoff's Representation Theorem for varieties, it follows that

$$G_n^+ = HSP(\{G_m^+ \mid m \in N - \{n\}\} \cup \{\mathcal{F}_k^+ \mid k \in N\}).$$

It can be easily seen that

$$G_n^+ = HS(G_{m_1}^+ \times \ldots \times G_{m_k}^+ \times \mathcal{F}_{m_1}^+ \times \ldots \times \mathcal{F}_{m_d}^+)$$

because algebra G_n^+ is finite. Now using the correspondence between homomorphic images, subalgebras and directs products of finite modal algebras, and generated subframes, p-morphic images and disjoint unions of theirs corresponding frames, we have that G_n is a generated subframe of a frame $\mathcal{D}$, where

$$\mathcal{D} = f(G_{m_1} \sqcup \ldots \sqcup G_{m_k} \sqcup \mathcal{F}_{m_1} \sqcup \ldots \sqcup \mathcal{F}_{m_d}),$$

and f is a corresponding p-morphism. The frame G_n is a rooted frame, therefore the equality above implies that there is p-morphism f from a rooted by a subframe $F(a)$ of the frame G, where $G = G_{m_w}$ or $G = \mathcal{F}_{m_w}$ for some w, onto the frame G_n. Let for short $l := m_w$. Any p-morphism does not increase the depth of elements, therefore $n \leq l$ and a has the depth of not less than n. Moreover, f maps $S_1(F(a))$ one-to-one and f is a mapping onto $S_1(G_n)$. If S is an antichain from $S_1(F(a))$ and a_S is its co-cover from $F(a)$ then f maps a_S into some co-cover in G_n for f-image of S. Conversely, let b_D be a co-cover in G_n for some antichain D from $S_1(G_n)$. Then there exists a co-cover c_n for $f^{-1}(D)$ in $F(a)$ such that f maps c_n in b_D. Thus f is an isomorphism from $S_2(H)$ onto $S_2(G_n)$. By the same reasoning, we show that f is an isomorphism from $S_i(G)$ onto $S_i(G_n)$, for each $i \in \{1, ..., n\}$. Then every co-cover a_E in G for $Sl_n(G)$ is mapped by f into a_n.

Suppose that $G = \mathcal{F}_n$. Then there are no elements in G which can be mapped by f into the generating element b_n of G_n, a contradiction. Suppose that $G = \mathcal{F}_l$, where $n < l$. Then the co-cover of a two-elements antichain (for instance) from $Sl_n(H)$ should be mapped by f in some co-cover in G_n for f-image of this antichain, but G_n has no such co-covers, a contradiction. The remaining case $G = G_l, n < l$ can be considered quite similar and we will have a contradiction again. Thus (6.22) holds and the theorem is proved. ∎

Because we are interested also in applications of our description of modal logics preserving inference rules admissible in $S4$, we will now focus our attention on tabular modal logics, those that are logics of finite Kripke frames. We will apply Theorem 6.2.7 to tabular logics in order to determine which tabular modal logics preserve all those rules admissible for $S4$.

Recall that a frame $\mathcal{F}$ has width n if $\mathcal{F}$ has no antichains of clusters with more than n clusters but has an antichain with n clusters. We know that the property of having width of not more than n can be expressed for transitive rooted frames by a certain modal propositional formula. More precisely, the formula $\mathcal{W}_n$, where

$$\mathcal{W}_n := \bigwedge_{0 \leq i \leq n} \Diamond p_i \rightarrow \bigvee_{i,j=0,1,...,n,i \neq j} \Diamond(p_i \wedge (p_j \vee \Diamond p_j))$$

has the following property: for any transitive rooted frame $\mathcal{F}$, $\mathcal{F} \Vdash \mathcal{W}_n \iff \mathcal{F}$ has a width of not more than n. Recall also that we say that a modal logic λ over $K4$ has a width of not more than n if $\mathcal{W}_n \in \lambda$. We say that λ has a width of more than n if $\mathcal{W}_n \notin \lambda$.

We need a special collection of finite reflexive transitive frames which are described below:

$$M_1 := \langle \{a, b_1, b_2, c_1, d\}, \leq \rangle, a \leq b_1, a \leq c_1, b_1 \leq b_2, c_1 \leq d, b_2 \leq d.$$

$$M_2 := \langle \{a, b_1, b_2, c_1\}, \leq \rangle, a \leq b_1, a \leq c_1, b_1 \leq b_2.$$

$M_3 := \langle \{a, b_1, b_2, c_1, c_2\}, \leq \rangle, a \leq b_1, a \leq c_1, b_1 \leq b_2, c_1 \leq c_2, b_1 \leq c_1.$

$M_4 := \langle \{a, b_1, b_2, c_1, c_2, d\}, R \rangle, aRb_1, aRb_2,$
$\qquad b_1 Rb_2, b_2 Rb_1, b_1 Rc_1, b_1 Rc_2, b_2 Rc_1, b_2 Rc_2, c_1 Rd, c_2 Rd.$

$M_5 := \langle \{a, b_1, b_2, c_1, c_2\}, R \rangle, aRb_1, aRb_2,$
$\qquad b_1 Rb_2, b_2 Rb_1, b_1 Rc_1, b_1 Rc_2, b_2 Rc_1, b_2 Rc_2.$

The structure of these frames is displayed in Figure 6.1 below.

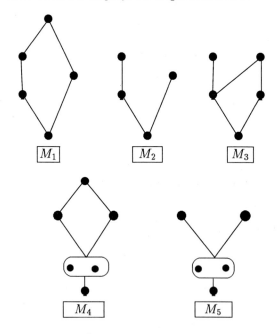

Figure 6.1: Structure of Frames M_1 - M_5

Any tabular modal logic λ has a representation: $\lambda := \lambda(\mathcal{F}_1 \sqcup ... \sqcup \mathcal{F}_n)$, where $\mathcal{F}_1 \sqcup ... \sqcup \mathcal{F}_n$ is the disjoint union of certain rooted finite frames such that the logics $\lambda(\mathcal{F}_i)$ are incomparable. We fix some such representation for any given tabular logic.

Lemma 6.2.9 *A tabular modal logic extending S4 $\lambda := \lambda(\mathcal{F}_1 \sqcup ... \sqcup \mathcal{F}_n)$ preserves all the inference rules admissible for S4 iff each logic $\lambda(\mathcal{F}_i)$ does.*

Proof. If some logic $\lambda(\mathcal{F}_i)$ does not preserve an inference rule α/β admissible in $S4$ then the logic λ does not preserve the rule $\Box \alpha \lor \Box z / \Box \beta \lor \Box z$ which is admissible in $S4$ in virtue of the disjunction property of $S4$. The converse assertion of our lemma is evident. ∎

Lemma 6.2.10 *If a tabular logic λ has the co-cover property and a finite frame F is a co-covers poster for a rooted λ-frame then F has a width of not more than 2.*

Proof. Assume that F has a 3-element antichain of clusters. According to the co-cover property, taking consequently co-covers posters of F, we obtain that, for any $n \in N$, there is a frame $\mathcal{F}_n$ with depth n such that all theorems of λ are valid in $\mathcal{F}_n$. But λ is tabular, i.e., $\lambda = \lambda(Q)$ for some finite frame Q which has, say, a depth k. Then $\mathcal{W}_k \in \lambda$, where $\varphi(k)$ is our formula expressing the property of frames to have depth of not more than k. This contradicts the truth of λ in the above mentioned frames $\mathcal{F}_n$ for each n. $\blacksquare$

As we have seen above, for any tabular logic λ, $\lambda := \lambda(\mathcal{F}_1 \sqcup ... \sqcup \mathcal{F}_n)$, where $\mathcal{F}_1 \sqcup ... \sqcup \mathcal{F}_n$ is a disjoint union of certain rooted finite frames, and all logics $\lambda(\mathcal{F}_i)$ are incomparable.

Definition 6.2.11 *We say that a rooted frame $\mathcal{F}_k$ is properly rooted if the root has only single immediate successor cluster or the root is single-element cluster.*

Now we are in a position to prove the concluding description theorem for tabular logics preserving inference rules admissible in $S4$.

Theorem 6.2.12 *A tabular logic $\lambda := \lambda(\mathcal{F}_1 \sqcup ... \sqcup \mathcal{F}_n)$, preserves all inference rules admissible in $S4$ iff each $\mathcal{F}_j$ is a properly rooted frame with width of not more than 2, and for all $i, 1 \leq i \leq 5, \lambda \not\subseteq \lambda(M_i)$.*

Proof. Suppose λ preserves all inference rules admissible in $S4$. Then by Theorem 6.2.7 and Lemma 6.2.9 every $\lambda(\mathcal{F}_j)$ has the co-cover property. According to Lemma 6.2.10 λ has width of 2 or smaller. Suppose λ has width of not more than 2 but λ is included in some logic $\lambda(M_i)$, where $1 \leq i \leq 3$. We fix some i such that $\lambda \subseteq \lambda(M_i), i \in \{1, 2, 3\}$. Then there exists a p-morphism from a rooted by a cluster C generated subframe Q_j of some frame $\mathcal{F}_j$ onto M_i. Moreover, we can suppose that C is a maximal cluster with respect to accessibility relation R in $\mathcal{F}_j$ such that there is a p-morphism f from $C^{R\leq}$ onto M_i. Then, in particular, all the strict successors of C are not mapped by f into the root of M_i. Hence there are two clusters C_1 and C_2 which are immediate successor clusters for C. Then f maps C_1 and C_2 into immediate successors a_1 and a_2 (respectively) of the root of M_i. By definition of M_i there exists an immediate successor b of a_1 (or a_2) such that b is not a successor of a_2 (a_1 respectively). Therefore there is an immediate successor cluster B for the cluster C_1 which is mapped by f into b and which is not a successor for the cluster C_2.

Thus the set of two clusters antichains $\{C_1, C_2\}$ from $\mathcal{F}_j$ such that one cluster, say C_1, has an immediate successor cluster C_g which is not a successor for another cluster C_2 is nonempty. We choose a such X with clusters C_1 and C_2 such that X is minimal with respect to R (i.e. with minimal with respect to R possible clusters). There is no co-cover for G_g, C_2 in $\mathcal{F}_j$, otherwise X would be not minimal. Therefore the co-covers poster F obtained from $\mathcal{F}_j$ by adding a co-cover c for the antichain of clusters C_g, C_2 differs from $\mathcal{F}_j$. Now we take the co-covers poster Q of F obtained from F by adding a co-cover d for the root of $\mathcal{F}_j$ and c. Because $\lambda(\mathcal{F}_j)$ has the co-cover property, Q must be a $\lambda(\mathcal{F}_j)$-frame, i.e. $\lambda(\mathcal{F}_j) \subseteq \lambda(Q)$. Moreover, $\mathcal{F}_j$ is a generated subframe of Q, hence $\lambda(Q) \subseteq \lambda(\mathcal{F}_j)$. Therefore

$$\lambda(Q) \subset \lambda(\mathcal{F}_j) \subseteq (Q),$$

where the first inclusion is proper because the depth of Q is strictly more than the depth of $\mathcal{F}_j$. Thus $\lambda(Q) \subset \lambda(Q)$, a contradiction.

Suppose now every $\mathcal{F}_j$ is properly rooted and λ is included in $\lambda(M_i)$, where $i = 4$ or $i = 5$, and that, for $i \in \{1, 2, 3\}$, λ is not included in $\lambda(M_i)$. We fix i such that $\lambda \subseteq \lambda(M_i)$ and $i \in \{4, 5\}$. Then again there exists a p-morphism f from a certain rooted generated subframe Q_j of some frame $\mathcal{F}_j$ onto M_i. The set X of clusters from G_j, which f maps into immediate successors c_1, c_2 of the two-element cluster from M_i, do not have in $\mathcal{F}_j$ co-covers (because, for $k = 1, 2, 3$, $\lambda \not\subseteq \lambda(M_k)$). We take the co-covers posters for M_j by adding a co-cover c for X, and then, by adding a co-cover for c and the root of M_i, and obtain a contradiction in the same way as above. Finally, suppose that some $\mathcal{F}_j$ is not properly rooted. Then this frame has the root cluster with at least two elements and with two immediate successor clusters C_1 and C_2 . In this case we immediately obtain a contradiction by adding a co-cover c for C_1 and C_2 and then by adding a co-cover for c and the root.

Now we turn to prove the sufficiency of the condition. Suppose that λ has a width of not more than 2, and that $\lambda \not\subseteq \lambda(M_i)$, $1 \le i \le 5$, and that all $\mathcal{F}_j$ are properly rooted. Then every logic $\lambda(\mathcal{F}_j)$ has the same properties. In order to show that λ preserves all inference rules admissible in $S4$ by Lemma 6.2.9 it is sufficient to verify that all $\lambda(\mathcal{F}_i)$ preserve inference rules admissible in $S4$. For this according to Theorem 6.2.7 it is sufficient to prove that all $\lambda(\mathcal{F}_j)$ have the co-cover property. We show this below. The root of $\mathcal{F}_j$ will be a co-cover for its immediate successors because $\mathcal{F}_j$ is properly rooted. From this together with $\lambda \not\subseteq \lambda(M_i)$ for every $i, 1 \le i \le 5$ it is not hard to derive that every non-trivial antichain of clusters from $\mathcal{F}_j$ has some co-cover inside the frame $\mathcal{F}_i$. Therefore if a Q is a rooted finite $\lambda(\mathcal{F}_j)$-frame which has no co-covers for some non-trivial antichain Y inside itself then the following holds. There exists a p-morphic image Q_1 of a rooted generated subframe D_j of $\mathcal{F}_j$ such that Q is a generated subframe of Q_1 and Q_1 has co-covers for all non-trivial antichains of clusters. Therefore the logic $\lambda(\mathcal{F}_j)$ has the co-cover property, as it was required. ∎

6.3 The Preservation of Admissibility for H

In this section we will investigate superintuitionistic logics which preserve inference rules admissible in Heyting intuitionistic logic H. We will use here the same approach and ideas as in the case concerning the preservation of rules admissible in the modal logic $S4$. As a matter of fact, the proofs are so similar to those case that it would be possible to consider also superintuitionistic logics in previous section and to carry out all the proofs in parallel way distinguishing only certain places, where it is important to make clear whether we are dealing with a modal or a superintuitionistic logic. But this would increase the length and complexity of the proofs, and we decided to present the proofs for superintuitionistic logics separately.

The notions of a co-cover in a frame and a co-covers poster of a frame are directly transferable to posets from the case of transitive reflexive frames. We recall here that, for any poset $\mathcal{F}$ and any antichain X from $\mathcal{F}$, an element y from $\mathcal{F}$ is a co-cover for X if $y^< = X^\le$. A finite poset $\mathcal{F}_1$ is a co-cover generated from $\mathcal{F}$ if $\mathcal{F}_1$ is obtained from $\mathcal{F}$ by adding to $\mathcal{F}$ a single co-cover for some non-trivial antichain from $\mathcal{F}$ which has no co-covers in $\mathcal{F}$. We say that a finite poset $\mathcal{G}$ is a co-cover's poster of a finite poset $\mathcal{F}$ if there is a finite sequence of finite posets such that $\mathcal{F}$ is its first element, $\mathcal{G}$ is its last element, and each poset of this sequence is a co-cover generated from the previous member.

Definition 6.3.1 *We say that a superintuitionistic logic λ has the co-cover property if for any rooted (sharp) finite poset $\mathcal{F}$ if $\mathcal{F} \Vdash \lambda$ and $\mathcal{F}_1$ is a co-cover's poster of $\mathcal{F}$ then $\mathcal{F}_1 \Vdash \lambda$.*

Lemma 6.3.2 *If a superintuitionistic logic λ has the co-cover property and the finite model property then λ preserves all inference rules admissible in intuitionistic logic H.*

Proof. Investigating admissibility in superintuitionistic logics we can have dealt only with rules which have only one formula in the premise. Suppose that there is an inference rule α/β admissible in H but not admissible in λ. Then there is a tuple of formulas γ_i such that $\alpha(\gamma_i) \in \lambda$ but $\beta(\gamma_i) \notin \lambda$. The mentioned above formulas have a finite number of propositional letters, say, n letters. The model $Ch_\lambda(n)$ is n-characterizing for λ according to Theorem 3.3.11, therefore we have $Ch_\lambda(n) \Vdash \alpha(\gamma_i)$ and $Ch_\lambda(n) \not\Vdash \beta(\gamma_i)$.

Then the inference rule α/β is invalidated in the frame $a^\le$ where $a \in |Ch_\lambda(n)|$ under the valuation $S(x_i) := V(\gamma_i)$. Let $F_0 := a^\le$. We introduce the frames F_n by induction on n. Let F_{m+1} be the co-cover's poster of F_m obtained by adding co-covers for all non-trivial antichains of F_m which have no co-covers in F_n. The logic λ has the co-cover property. Therefore each F_m can be considered as a generated subframe of $Ch_\lambda(n)$; also any F_m is a generated subframe of F_{m+1}.

Thus $F := \bigcup_{m<\infty} F_m$ is a generated subframe of $Ch_\lambda(n)$ and the rule α/β is invalidated in F by the valuation S (since it was invalid in $Ch_\lambda(n)$ and $a \in |F|$).

Now note that all antichains of F have certain co-covers inside of F itself according to the construction of F. Using this fact and the fact that F is an open subframe of $Ch_H(n)$, it is not hard to conclude that there is a p-morphism f from the frame of $Ch_H(n)$ onto the frame F. Using f we can transfer back the valuation S from F to $Ch_H(n)$. Since p-morphisms preserve truth of formulas, we obtain that this transferred by f valuation (pro-image of S) will invalidate the rule α/β in the frame of $Ch_H(n)$. Applying Theorem 3.5.8 we derive from this fact that the rule α/β) is not admissible in H, a contradiction. ∎

Lemma 6.3.3 *If a superintuitionistic logic λ preserves all inference rules admissible in H and has the finite model property then λ has the co-cover property.*

Proof. Suppose that a superintuitionistic logic λ has the fmp and does not have the co-cover property. It means that there is a generated subframe $F := a^{\leq}$ of $Ch_\lambda(n)$ and a sequence $F_0, ..., F_k$ of generated subframes of the frame of $Ch_\lambda(n)$ with the following properties. (i) $F_0 := F$. (ii) If all antichains having elements of depth i from F_i have co-covers in F_i then $F_{i+1} := F_i$, otherwise F_{i+1} $(i < k)$ is the co-cover's poster obtained from F_i by adding co-covers for all antichains of F_i with elements of depth i which do not have co-covers. (iii) All antichains of F_i of depth less than i have co-covers in F_i. (iv) For some antichain Δ with elements of depth k in F_k, there is no a co-cover for Δ in $Ch_\lambda(n)$. In particular, all antichains of F_k with depth of elements not more than $k - 1$ have co-covers in F_k. Moreover, we can suppose that k is the smallest possible number with properties concerning frames F_i described above.

We introduce special modal formulas in order to describe certain particular properties of the model $Ch_\lambda(n)$. We set $M := S_k(Ch_\lambda(n)) \cup F_k$ and $M_1 := Sl_{k+1}(Ch_\lambda(n)) \cup M$. For each $i \in M_1$ we introduce a new propositional variable p_i. Using these letters with these indexes, we introduce special formulas $f(i)$ for $i \in M_1$ by induction on the depth of i. If $i \in S_1(M_1)$ then $f(i) := \Box p_i$. If formulas $f(i)$ are introduced for all $i \in S_t(M_1)$ and $j \in Sl_{t+1}(M_1)$ then let

$$f(j) := \Box p_j \wedge \bigwedge_{i \neq j, i \in Sl_{t+1}(M_1)} \neg \Box p_i \wedge \bigwedge_{j < i} \Diamond f(i) \wedge$$

$$\wedge \bigwedge_{\neg(j < i), i \in S_t(M_1)} \neg \Diamond f(i) \wedge \bigwedge_{\neg(i \leq j)} \neg(\Box p_i) \wedge \Box([\Box p_i \wedge$$

$$\bigwedge_{i \neq j, i \in Sl_{t+1}(M_1)} \neg \Box p_i \wedge \bigwedge_{j < i} \Diamond f(i) \wedge \bigwedge_{j < i} \neg f(i)] \vee \bigvee_{j < i} f(i)).$$

Now let

$$g := \bigwedge_{i \in M} \neg f(i) \wedge \bigvee_{i \in Sl_{k+1}(M_1)} \Diamond f(i).$$

Using described above formulas we introduce the following modal inference rule:

$$r := \Box(\bigvee_{i \in M} f(i) \vee g)/\Box \neg f(a),$$

where a is the root of F_0. The valuation W of the variables occurring in r in the frame of $Ch_\lambda(n)$ is defined as follows:

$$W(p_i) := \{j \mid i \leq j\}.$$

It can be verified directly that, for each $i \in M_1$, $i \Vdash_W f(i)$ and that, for each $i \in (Ch_\lambda(n) - M)$, $i \Vdash_W g$. Hence the inference rule r is invalidated in $Ch_\lambda(n)$ by the valuation W.

According to Lemma 2.7.14 there are formulas α and β of intuitionistic propositional calculus H such that

$$T(\alpha) \equiv_{S4} \Box(\bigvee_{i \in M} f(i) \vee g), \quad T(\beta) \equiv_{S4} \Box \neg f(a), \tag{6.23}$$

where T is the Gödel-McKinsey-Tarski translation of intuitionistic formulas into modal formulas. Thus the inference rule $T(\alpha)/T(\beta)$ is invalidated in $Ch_\lambda(n)$ by the valuation W. It means that the intuitionistic inference rule α/β is invalidated in $Ch_\lambda(n)$ by the intuitionistic valuation W because, for any propositional non-modal formula γ,

$$\forall x \ (x \Vdash_W T(\gamma) \Longleftrightarrow x \Vdash_W \gamma). \tag{6.24}$$

Recall that each rooted generated subframe of the n-characterizing H Kripke model $Ch_H(n)$ is definable by an intuitionistic formula according to Theorem 3.3.13. Thus each rooted generated subframe of $Ch_\lambda(n)$ is definable as well because $Ch_\lambda(n)$ is an open submodel of $Ch_H(n)$. Thus the valuation W is definable by formulas. Therefore the fact that α/β is invalidated in $Ch_\lambda(n)$ under W implies that α/β is not admissible in λ (see Theorem 3.5.8), because the model $Ch_\lambda(n)$ is n-characterizing for λ.

We proceed now to proof that α/β is admissible in H which will give us a contradiction. Suppose that α/β is not admissible in H. Then according to Theorem 3.5.8 there exists a n-characterizing model $Ch_H(n)$ of H and a definable valuation W on $Ch_H(n)$ such that $W(\alpha) = |Ch_H(n)|$ and $W(\beta) \neq |Ch_H(n)|$. Then according to (6.23), we infer

$$W(\bigvee_{i \in M} f(i) \wedge g)) = |Ch_H(n)| \ \& \ W(\Box \neg f(a)) \neq |Ch_H(n)| \tag{6.25}$$

Furthermore, there is $b_a \in Ch_H(n)$ such that $b_a \Vdash_W f(a)$. It is easy to see that

$$\forall b_i \in Ch_H(n)(b_i \Vdash_W f(i) \& j \in M_1 \& (i < j) \Rightarrow \qquad (6.26)$$

$$\exists b_j(b_i \leq b_j \& b_j \Vdash_W f(j)))$$

bearing in mind the third conjunction member of $f(i)$. Suppose that $b_i \Vdash_W f(i)$, $b_j \Vdash_W f(j)$ and $b_i \leq b_j$. If $i \not\leq j$ then $b_j \Vdash_W \neg\Box p_i$. But $b_i \Vdash_W f(i)$ therefore $b_i \Vdash_W \Box p_i$. This fact and $b_i \leq b_j$ imply $b_j \Vdash_W \Box p_i$, a contradiction. Thus

$$\forall b_i, b_j \in Ch_H(n)(b_i \Vdash_W f(i) \ \& \ b_j \Vdash_W f(j) \ \& \ \neg(i \leq j) \Rightarrow \qquad (6.27)$$

$$\neg(b_i \leq b_j))$$

From (6.26) and (6.27) we obtain immediately

$$b_i \Vdash_W f(i) \ \& \ (i < j) \implies \exists b_j((b_i < b_j) \ \& \ (b_j \Vdash_W f(j))) \qquad (6.28)$$

Another consequence of (6.27) is

$$b_i \Vdash_W f(i) \ \& \ b_j \Vdash_W f(j) \ \& \ b_i \leq b_j \implies i \leq j \qquad (6.29)$$

From (6.29) we immediately derive

$$(b_i \Vdash_W f(i)) \ \& \ (b_j \Vdash_W f(j)) \ \& \ (i \not\leq j) \ \&$$

$$\& \ (j \not\leq i) \implies \neg(b_i \leq b_j) \ \& \ \neg(b_j \leq b_i) \qquad (6.30)$$

Also, the following holds:

$$(b_i \Vdash_W f(i)) \& (b_i < d) \implies \exists j((i \leq j) \& (d \Vdash_W f(j))), \qquad (6.31)$$

which can be easily shown using the structure of the formula $f(i)$. Moreover, by (6.27)

$$(\forall j > a) \exists b_j((b_a < b_j) \wedge (b_j \Vdash_W f(j))) \qquad (6.32)$$

Let Y be an antichain in M with depth of elements not more than k and $b_i \in Ch_H(n)$ and $b_i \Vdash_W f(i)$, where $i \in Y$. Then by (6.30) b_i, $i \in Y$ form an antichain in $Ch_H(n)$. According to the construction of $Ch_H(n)$, there is a co-cover v for the antichain $b_i, i \in Y$. The relation (6.25) implies that $v \Vdash_W f(j)$ for some $j \in M$ or $v \Vdash_W g$. We claim that

$$\exists j \in M(v \Vdash_W f(j)) \qquad (6.33)$$

Indeed, let $v \Vdash_W g$. Then $v \Vdash_W \Diamond f(x)$, where $x \in Sl_{k+1}(M_1)$ and $v \Vdash_W \neg f(j)$ for all $j \in M$. So there exists a b_x such that $v < b_x$ and $b_x \Vdash_W f(x)$. Then $b_i \leq$

b_x for some i from Y. From this and (6.27) we obtain $i \leq x$, which contradicts $x \in Sl_{k+1}(M_1)$. Hence (6.33) holds.

If $j \in M$ and $v \Vdash_W f(j)$ then

$$j \text{ is a co-cover for } Y \text{ in } M. \tag{6.34}$$

Indeed, by assumption we have $\forall i \in Y(b_i \Vdash_W f(i) \ \& \ v < b_i)$. This, by (6.29), gives us that $\forall i \in Y(j < i)$. Suppose that $(j < l)$ for a certain l from M. Then, by (6.28), there exists a b_l from $Ch_H(n)$ such that $(b_l \Vdash_W f(l)) \& (v < b_l)$. Since, for some $i \in Y$, $(b_i \leq b_l)$ which gives us $i \leq l$ by (6.29). Therefore it follows that j is a co-cover for the antichain Y and (6.34) is proved.

By using (6.26), (6.32), (6.33) and (6.34) we obtain

$$\forall x \in S_k(F_k) \ \exists b_x \in Ch_H(n)(b_x \Vdash_W f(x)) \tag{6.35}$$

Let Δ be an antichain of depth k from F_k which has no co-covers in $Ch_\lambda(n)$ and let $D := \{b_x \mid x \in \Delta\}$, where $b_x \Vdash_W f(x)$ (we use (6.35)). According to (6.29) D forms an antichain in $Ch_H(n)$. Each antichain from $Ch_H(n)$ has a co-cover by the construction of the model $Ch_H(n)$. Let w be a co-cover for D. Then, by (6.25) we have that either $w \Vdash_W f(j)$ for some $j \in M$ or $w \Vdash_W g$, but not both simultaneously. By (6.33) only the first case hold. Then by (6.34) the corresponding j should be a co-cover for Δ, a contradiction. ∎

Now we are able to give a description of the superintuitionistic logics with the finite model property which preserve all admissible inference rules of the intuitionistic logic.

Theorem 6.3.4 *A superintuitionistic logic λ with the finite model property preserves all inference rules admissible in intuitionistic logic H if and only if λ has the co-cover property.*

Proof follows immediately from Lemmas 6.3.2 and 6.3.3.

Now we have a semantic description of those superintuitionistic logics with the fmp which preserve the inference rules admissible in H. Using this description we proceed to studying this class in more detail. First we will determine how many logics there are in this class. As in the case of modal logics, at first sight there seems little difference between the non-tabular logics with the fmp from this class (since finite frames for these logics have homogeneous construction). But as with the case of modal logics, this class is rather big.

Theorem 6.3.5 *There are continuously many superintuitionistic logics with the finite model property preserving all inference rules admissible in H.*

Proof. We need the two following sequences of finite posets F_n and G_n, $n \in N$. First we introduce by induction the sequence of frames A_n, $n \in N$. Let A_1 be the three-elements antichain, and if A_n has already been constructed then A_{n+1} is obtained from A_n by adding a single co-cover for each non-trivial antichain of A_n which has elements with depth of n and which has no a co-cover. Let $F_n := \{a_n\}^{\leq}$ be the rooted frame, where $\{a_n\}^{<} = A_n$ and $G_n := \{b_n\}^{\leq}$, where $\{b_n\}^{<} = F_n$. For any $Y \subseteq N$, let

$$\lambda(Y) := \lambda(\{F_n \mid n \in N\} \cup \{G_n \mid n \in Y\})$$

First we check that all the logics $\lambda(Y)$ have the co-cover property. Indeed, let $\mathcal{B}$ be a finite rooted $\lambda(Y)$-frame, that is $\mathcal{B} \Vdash \lambda(Y)$. Then $\mathcal{B}$ has not more than three maximal elements, because all the frames from Y and all the frames F_n have only three maximal elements each, and this property is expressible by an intuitionistic propositional formula for rooted posets. If M is a co-cover's poster of $\mathcal{B}$ then M is a generated subframe of a p-morphic image of a finite disjoint union $F_n \sqcup ... \sqcup F_n$ of the frame F_n for some n. This can be easily shown taking into account the structure of the frames F_n. Because all the constructions given above preserve the truth of formulas, we have $M \Vdash \lambda(Y)$ which is what we needed. Thus any logic $\lambda(Y)$ has the co-cover property.

By definition $\lambda(Y)$ has the finite model property therefore any $\lambda(Y)$ preserves all inference rules admissible in H by Theorem 6.3.4.

Now we will prove that all the logics $\lambda(Y)$ are distinct. This is a consequence of the proposition:

$$\forall n (\lambda(N - \{n\}) \not\subseteq \lambda(G_n)) \tag{6.36}$$

To prove (6.36) we suppose that $\exists n (\lambda(N - \{n\}) \subseteq \lambda(G_n))$. Then from the correspondence between varieties of pseudo-boolean algebras and superintuitionistic logics and Birkhoff's Representation Theorem for varieties it follows that

$$G_n^+ = HSP(\{G_m^+ \mid m \in N - \{n\}\} \cup \{F_k^+ \mid k \in N\})$$

It can be easily seen that

$$G_n^+ = HS(G_{m_1}^+ \times ... \times G_{m_k}^+ \times F_{m_1}^+ \times ... \times F_{m_d}^+)$$

because G_n^+ is finite. Then from the correspondence between homomorphic images, subalgebras and directs products of finite pseudo-boolean algebras, and generated subframes, p-morphic images and disjoint unions of the corresponding frames, it follows that G_n is a generated subframe of a frame W, where

$$W = f(G_{m_1} \sqcup ... \sqcup G_{m_k} \sqcup F_{m_1} \sqcup ... \sqcup F_{m_d}),$$

and f is a p-morphism. The frame G_n is rooted therefore the previous fact implies that there is p-morphism f from a rooted subframe $a^{\leq}$ of the frame H onto

G_n, where $H = G_{m_v}$ or $H = F_{m_v}$ for some v. Let $l = m_v$. Since p-morphisms cannot increase depth, it follows that $n \leq l$ and a has a depth of not less than n. Moreover, f maps the set $S_1(a^{\leq})$ one-to-one and onto $S_1(G_n)$. If S is an antichain from $S_1(a^{\leq})$ and a_S is its co-cover from $a^{\leq}$ then f maps a_S into a co-cover in G_n of the f-image of S. Conversely, if b_D is a co-cover in G_n of an antichain D from $S_1(G_n)$ then a certain co-cover for $f^{-1}(D)$ should be in $a^{\leq}$, and f maps this co-cover into b_D. Thus f is an isomorphism of $S_2(H)$ onto $S_2(G_n)$. In a same way, we can show that f is an isomorphism of $S_i(H)$ onto $S_i(G_n)$, for each $1 \leq i \leq n$. Then the co-cover a_E in H for $Sl_n(H)$ is mapped by f into a_n.

Suppose that $H = F_n$. Then there are no elements in H which can be mapped by f into the generating element b_n of G_n, a contradiction. Suppose that $H = F_l$, where $m < l$. Then the co-cover of a two-elements antichain (for instance) in $Sl_n(H)$ should be mapped by f in some co-cover in H for the f-image of this antichain. But G_n has no such co-cover, a contradiction. The remaining case is $H = G_l, n < l$. But in this case we can repeat the above reasoning for $H = F_l, n < l$. Thus our supposition is wrong and (6.36) holds. which completes thee proof of our theorem. ∎

Thus the class of superintuitionistic logics which preserve all inference rules admissible in H and have the fmp is rather rich. In particular, it contains undecidable logics without finite axiomatizations. Now we can answer which tabular superintuitionistic logics preserve all inference rules admissible in H. Note that an answer follows simply from Corollary 5.2.17: they are exactly all those superintuitionistic tabular logics for which the Mints's generalized rule

$$\frac{((x \to y) \to (x \to z)) \vee w}{((x \to y) \to x) \vee ((x \to y) \to z) \vee w}$$

is admissible. Below, we give a semantic description of these tabular logics using Theorem 6.3.4. In fact the result described above follows immediately from this our description.

We again need the three finite rooted frames M_i, $i = 1, 2, 3$ which were introduced in Section 2 of this chapter. Recall that M_1 is the poset isomorphic to the 5-element lattice which invalidates the modularity law (*Pentagon*), M_2 is the frame obtained from M_1 by removing the greatest element, and, finally, M_3 is obtained from M_2 by adding a cover for the elements of M_2 with depth of 2 (see Fig. 6.1). Recall also that the property of a rooted poset to have the width of not more than n can be expressed by an intuitionistic propositional formula. In other words, there is a formula $W_{i,n}$ such that, for each rooted frame F, $F \Vdash W_{i,n} \Longleftrightarrow F$ has a width of not more than n. We say that a superintuitionistic logic λ has a width of not more than n if $W_{i,n} \in \lambda$, and that λ has a width of more than n if $W_{i,n} \notin \lambda$.

Theorem 6.3.6 *A tabular superintuitionistic logic λ preserves all inference rules admissible in H iff λ has a width of not more than 2 and, $\forall i (1 \leq i \leq 3), \lambda \not\subseteq \lambda(M_i)$.*

Proof. Let $\lambda := \lambda(F_1 \sqcup ... \sqcup F_n)$, where the logics of finite rooted posets F_i are incomparable.

Lemma 6.3.7 *Logic λ preserves all inference rules admissible in H iff each logic $\lambda(F_i)$ does.*

Proof. Indeed, if a logic $\lambda(F_i)$ does not preserve an inference rule α/β admissible in H then the logic λ does not preserve the rule $\alpha \lor z \;/\; \beta \lor z$ which is admissible in H according to the disjunction property of H. The converse assertion is evident. ∎

Lemma 6.3.8 *If a superintuitionistic tabular logic λ has the co-cover property and a finite frame F is a co-cover's poster for a rooted λ-frame then F has a width of not more than 2.*

Proof. Assume that the frame F has a 3-element antichain. According to the co-cover property, taking consequently co-cover's posters of F we obtain that all theorems of λ are valid in some frame $\mathcal{F}_n$ with a depth of n for each n. But λ is a tabular logic, that is $\lambda = \lambda(Q)$ for a finite frame Q which have (say) a depth of k. Then $D(k) \in \lambda$, where D_k is the formula which expresses the property of frame having a depth of not more than k. This contradicts the truth of all theorems of the logic λ in the mentioned above frames $\mathcal{F}_n$ for each number n. ∎

Lemma 6.3.9 *Let λ be a superintuitionistic logic, and let $\lambda = \lambda(Q)$, where Q is a rooted finite poset. Then λ preserves all inference rules admissible in H iff Q has a width of not more then 2 and, for any two-element antichain $\{x, y\}$ from Q, any immediate successor z for x also is an immediate successor for y*

Proof. If, for two-elements antichains, Q has the property pointed above then it can be easily seen that λ has the co-cover property. Therefore by Theorem 6.3.4 λ preserves all inference rules admissible in H.

Conversely, if the logic λ preserves all inference rules admissible in H then this logic has the co-cover property by Theorem 6.3.4, and any generated subframe of the frame Q has a width of not more than 2 by Lemma 6.3.8. Let x, z be a minimal (by the partial order of Q) pair of incomparable elements from Q and z is an immediate successor of x. Suppose that y and z are incomparable. If they have no a co-cover in Q then we add a co-cover for they, and, after that, we add a co-cover for the smallest element of Q and this added co-cover, and obtain a λ-frame which is deeper than Q, a contradiction. Thus if y and z are incomparable then there is a co-cover in Q for the pair $\{y, z\}$, which contradicts the choice of x and y. Thus $y < z$.

If z is not an immediate successor of y then we consider an immediate successor d of y in the chain (y, z) from x to z. Reasoning in a way similar to above, and changing x, y, z on y, x, d, we obtain that $x < d < z$ which contradicts the fact that z is an immediate successor for x. Thus z is an immediate successor of y. Continuing this reasoning, and moving to the top of Q, we complete the proof of this lemma. ∎

To continue the proof of our theorem assume that λ preserves all inference rules admissible in H, and that $\lambda \subseteq \lambda(M_i)$, for some $i = 1, 2, 3$. We fix a such i. Then $M_i = \langle \{a, b, c, d\} \cup X, \leq \rangle$ where $a < b, a < c, b < d$ and X is the empty set or a one-element set which consists of a cover for the antichain $\{c, d\}$ or the antichain $\{c, b\}$. We have $M_i \Vdash \lambda$. Therefore there is a p-morphism f from a generated subframe $a_1^{\leq}$ of a frame F_j onto the frame M_i. This is simple corollary of Birkhoff's Representation Theorem, and also can be verified in a way similar to the part of the proof of Theorem c6s38 placed after (6.36). We can suppose that a_1 is a maximal element from Q which is mapped into a by f. Then there are two immediate successors b_1 and c_1 of a_1 (which are exactly all successors of a_1 because F_i has a width of not mote then 2) which are mapped onto b and c by f respectively. According to the definition of a p-morphism, there is $d_1 \in b_1^{\leq}$ such that $f(d_1) = d$. Because c and d are incomparable, c_1 and d_1 are incomparable as well. This contradicts the assertion of Lemma 6.3.9, which holds for F_i by Lemma 6.3.7.

Conversely, assume that the logic λ is not included in the logics $\lambda(M_i)$, $1 \leq i \leq 3$ and λ has a width of not more than 2. Then each logic $\lambda(F_i)$ is not included in these logics as well. We proceed to the proof that each $\lambda(F_i)$ preserves all inference e rules admissible in H. To show this it is sufficient to prove that every F_i has the property from Lemma 6.3.9. Suppose that this property does not hold. Then, since the frame F_i has a width of not more than 2, it is easy to see that there is a p-morphism from F_i onto a frame M_i, where $1 \leq i \leq 3$. This contradicts $\lambda(F_i) \not\subseteq \lambda(M_i)$. Hence all the logics $\lambda(F_i)$ preserve all inference rules admissible in H. Therefore λ also preserves all inference rules admissible in H. ∎

Lemmas 6.3.7 and 6.3.9 give, in particular, a semantic description of the tabular superintuitionistic logics which preserve admissibility in H. They are all the logics which are generated by frames of the following structure: a sequential finite union (in any order) of some finite chains of elements and some frames of kind E_n, where $0 \leq n$,

$$E_n := \langle \{a, b_1, ..., b_n, c_1, ..., c_n, d\}, \leq \rangle,$$

$$a < b_1, a < c_1, b_i < b_{i+1}, c_i < c_{i+1}, b_i < c_{i+1}, c_i < b_{i+1}, b_n < d, c_n < d$$

and some frames D_n which are obtained from E_n by removing the top or bottom (note that this description precisely corresponds to the description in [29] and our description in Section 2 of Chapter 5 of subdirectly irreducible finite pseudo-boolean algebras which are embeddable into $\mathfrak{F}_H(\infty)$). It is easy to see that tabular logics mentioned above are precisely all the tabular superintuitionistic logics for which Mints's generalized rule is admissible.

Corollary 6.3.10 *All the tabular superintuitionistic logics preserving all inference rules admissible in H form an infinite filter in the lattice of all superintuitionistic logics (in particular, they form a sublattice of this lattice).*

Proof. This follows immediately from Theorem 6.3.6.

Recall that we say a logic λ has finite model property with respect to admissibility if, for any not admissible in λ rule r, there is a finite algebra from $Var(\lambda)$ such that r in invalid in $\mathfrak{A}$ but any admissible in λ rule is valid in that algebra. From Theorem 6.3.6 we can immediately derive also the following corollary.

Corollary 6.3.11 *The intuitionistic propositional logic H fails to have the finite model property with respect to admissibility.*

Proof. Indeed the inference rule $x \equiv x/W_{i,2}$ is not admissible in H but cannot be detached from the inference rules admissible in H by finite frames (pseudo-boolean algebras) according to Theorem 6.3.6 ∎

As we have seen above, the tabular superintuitionistic logics, which preserve all inference rules admissible in H, form an infinite sublattice of the lattice of all superintuitionistic logics. Is it true for all superintuitionistic logics with the fmp? In other words, do form all superintuitionistic logics which have the fmp and which preserve all inference rules admissible in H a sublattice of the lattice of all superintuitionistic logics? Of course the intersection of arbitrary family of superintuitionistic logics which preserve all admissible rules of H does preserve the admissibility also. Is this true for the union (at least, finite union) of logics? Tools which were developed above basically work for logics with the finite model property. And we can very easily give a positive answer in the case where the union of logics has the finite model property.

Lemma 6.3.12 *If λ_1 and λ_2 are superintuitionistic logics with the fmp, and both preserve all inference rules admissible in H, and if $\lambda_1 \cup \lambda_2$ has the fmp, then the logic $\lambda_1 \cup \lambda_2$ preserves all inference rules admissible in H as well.*

Proof. By Theorem 6.3.4 it is sufficient to establish that the logic $\lambda_1 \cup \lambda_2$ has the co-cover property. Let F be a rooted finite frame for $\lambda_1 \cup \lambda_2$. Then it is a frame for the logics λ_i, $i = 1, 2$. Both these logics have the co-cover property by Theorem 6.3.4. Therefore, for each co-cover's poster M of F, we have $M \Vdash \lambda_i$,

$i = 1, 2$. Thus $M \Vdash \lambda_1 \cup \lambda_2$. Hence the logic $\lambda_1 \cup \lambda_2$ has the co-cover property and by our assumption this logic has the fmp, and Theorem 6.3.4 implies that $\lambda_1 \cup \lambda_2$ preserves all inference rules admissible in H. ∎

6.4 Non-Compact Modal Logics

Kripke semantics plays an important role in the study of non-standard logics by many reasons, in particular, since this semantics allows us to involve actively our geometric intuition in research. At the same time not all modal and super-intuitionistic logics are Kripke complete. First examples of Kripke incomplete modal logics were independently found by K.Fine [45] and S.K.Thomason [179]. Examples of Kripke incomplete modal logic with simple axiomatic system were given by J. van Benthem [190], we described these examples in Section 9 of Chapter 2. The first example of a Kripke incomplete superintuitionistic logic was discovered by V.Shehtman [170]. An innovative research program investigating the effect of Kripke incompleteness has been developed, in particular, in works of W.Block [12] and A.Chagrov and Zakharyaschev [27, 28]. In this section we touch on only one phenomenon concerning Kripke incompleteness and inference rules simultaneously. It is phenomenon of Kripke non-compactness which was introduced into consideration in the article of S.K.Thomason [178]. The concept of Kripke non-compactness generalizes the concept of Kripke incompleteness. Before giving the explicit definition of non-compactness, we briefly explain the result of this effect and the motivation of giving rise to this definition.

Since some logics are Kripke incomplete, it is natural to consider the Kripke completions of such logics. In fact, we can extend any logic λ to a Kripke complete logic, it is sufficient to take the intersection $KC(\lambda)$ of all Kripke complete logics which extend the logic λ. Clearly $KC(\lambda)$ is the smallest Kripke complete logic which contains λ. The question naturally arises how we can obtain the Kripke completion of a given logic λ in a syntactic way, by altering the axiomatic system for λ. The obvious way is just to add all the new formulas from $KC(\lambda)$ as new axioms. But this does not clarify the essence of Kripke completion very much. Another approach is to fix rigidly the original set of axioms for λ and simply to enlarge the original set of inference rules for λ by adding new rules which preserve the truth of formulas in Kripke frames. If this approach could always be realized then we would be able in effect specify new rules which precisely reflect the semantic consequence in Kripke frames. However it turns out that this approach cannot be carried out for a fundamental reason, which we describe below, connected with semantic consequence in Kripke frames. We consider first all logics λ for which the notion of Kripke completeness is meaningful.

Definition 6.4.1 *A logic λ is said to be* compact *(or Kripke compact) if the following holds. If for any finite set X of formulas from λ there is a frame $\mathcal{F}$*

which separates a formula α from $\mathcal{X}$ (which means $\mathcal{F} \Vdash \mathcal{X}$ and $\mathcal{F} \nVdash \alpha$) then there is a frame $\mathcal{F}$ which separates α from whole λ.

Consequently, a logic λ is Kripke non-compact if there is a formula α which cannot be separated by a frame from λ, but can be separated by an appropriate frame from any finite subset of λ. The following lemma is an evident observation concerning non-compact logics.

Lemma 6.4.2 *If a logic λ is Kripke non-compact then λ is Kripke incomplete and cannot have a finite axiomatic system.*

Proof. In fact, since λ is Kripke non-compact then there is a formula α such that (i) for any finite subset $\mathcal{X}$ of λ, there is a frame $\mathcal{F}_{\mathcal{X}}$ such that $\mathcal{F}_{\mathcal{X}} \Vdash \mathcal{X}$ but $\mathcal{F}_{\mathcal{X}} \nVdash \alpha$, and (ii) $\forall \mathcal{F}(\mathcal{F} \Vdash \lambda \Rightarrow \mathcal{F} \Vdash \alpha)$. Because (i), α is not a theorem of λ. Therefore (ii) entails λ is Kripke incomplete. Suppose λ has a finite set of axioms, say a set $\mathcal{A}x$. (i) yields that there is a frame $\mathcal{F}$ such that $\mathcal{F} \Vdash \mathcal{A}x$ and $\mathcal{F} \nVdash \alpha$. Since $\mathcal{A}x$ is a set of axioms for λ, $\mathcal{F}_{\mathcal{A}x} \Vdash \lambda$ holds, which contradicts (ii). ∎.

It turns out that there are Kripke non-compact superintuitionistic and modal logics. Such examples will be given below. Note that existence of Kripke non-compact logics shows that it is impossible at all to obtain the Kripke completions of logics by means of adding any collection of new inference rules which preserve the validness by Kripke (i.e. the validness in Kripke frames), to the axiomatic systems of that logics.

Lemma 6.4.3 *If λ is a Kripke non-compact logic then the set of theorems of the logic $\lambda \oplus \mathcal{R}$, where $\mathcal{R}$ is the set of all inference rules which preserve validness of formulas in frames, is strictly less than $KC(\lambda)$.*

Proof. Indeed, since λ is Kripke non-compact, it follows that there is a formula α which is separable by an appropriate frame from any finite subset $\mathcal{Z}$ of λ, but which is valid in any λ-frame. Suppose α is a theorem of $\lambda \oplus \mathcal{R}$. Then there is a derivation $\mathcal{S}$ of α from a finite set $\mathcal{X}$ of axioms of λ by means of postulated inference rules of λ and certain rules from $\mathcal{R}$. Since all these rules preserve the validness in frames, α is valid in any frame $\mathcal{F}$ such that $\mathcal{X}$ is valid in $\mathcal{F}$. But the formula α is separable from $\mathcal{X}$ by a frame $\mathcal{F}$, a contradiction. ∎

This lemma is a reflection of the fact that modal and superintuitionistic logics are rather strong fragments of the monadic second order logic with respect to Kripke semantics, and the fact that, for the second order monadic logic, the analog of the compactness theorem of the first-order logic does not hold. Now we turn to our construction of Kripke non-compact logics. We give below examples of Kripke non-compact modal logics extending $S4$. For this we need the following modal formulas:

$$\beta_0 := q_0, \ \beta_1 := q_1, \gamma_0 := r_0, \gamma_1 := r_1,$$
$$\beta_{m+2} := \Diamond \beta_{m+1} \wedge \Diamond \gamma_m \wedge \neg \Diamond \gamma_{m+1},$$

$$\gamma_{m+2} := \Diamond\gamma_{m+1} \wedge \Diamond\beta_m \wedge \neg\Diamond\beta_{m+1},$$
$$\alpha_m := \Diamond\beta_{m+1} \wedge \Diamond\gamma_{m+1} \wedge \neg\Diamond\beta_{m+2} \wedge \neg\Diamond\beta_{m+2},$$

$$\psi_1 := \Box(\beta_1 \to \Diamond\beta_0 \wedge \neg\Diamond\gamma_0),$$
$$\psi_2 := \Box(\gamma_1 \to \Diamond\gamma_0 \wedge \neg\Diamond\beta_0),$$

$$\psi_3 := \Box(\beta_0 \to \neg\Diamond(\beta_1 \vee \gamma_1)),$$
$$\psi_4 := \Box(\gamma_0 \to \neg\Diamond(\beta_1 \vee \gamma_1)),$$

$$\delta := \psi_1 \wedge \psi_2 \wedge \psi_3 \wedge \psi_4,$$
$$\chi_1 := \neg(p \wedge \Box(p \to \Diamond(\neg p \wedge \Diamond p))),$$
$$\chi_2 := \neg(p_0 \wedge \Box(p_0 \to \Diamond(\neg p_0 \wedge p_1 \wedge \Diamond(\neg p_0 \wedge \neg p_1 \wedge \Diamond p_0)))),$$
$$\phi_m := \neg\Diamond\alpha_{n-1} \wedge \Diamond\alpha_n \wedge \Diamond\alpha_{n+1},$$
$$\mu := \neg\chi_1 \wedge \delta \wedge \Diamond\phi_1 \wedge \Diamond\phi_2 \to \Box(\phi_1 \to \Diamond\phi_2),$$
$$\psi := \delta \wedge \Diamond\psi_1 \wedge \Diamond\psi_2 \to \Box(\psi_1 \to \psi_2),$$
$$\epsilon := \neg\chi_1 \wedge \delta \wedge \Diamond\alpha_0 \wedge \Box(\beta_0 \to \Box\beta_0) \wedge \Box(\gamma_0 \to \Box\gamma_0),$$
$$\varphi := \delta \wedge \Diamond\alpha_0 \wedge \Diamond\alpha_1 \wedge \Box(\beta_0 \to \Box\beta_0) \wedge \Box(\gamma_0 \to \Box\gamma_0),$$
$$\rho_n := \epsilon \to \Diamond\phi_n,$$
$$\xi_n := \varphi \to \Diamond\phi_n.$$

It is not hard to see that $Grz = S4 \oplus \chi_1$ (see Theorem 2.6.25). The main content of this section is the proof of the fact that the modal logics introduced below are Kripke non-compact. Let

$$\lambda_1 := S4 \oplus \{\chi_2, \mu\} \cup \{\rho_m \mid m \in N\}.$$

$$\lambda_2 := S4 \oplus \{\chi_1, \psi\} \cup \{\xi_m \mid m \in N\}.$$

Theorem 6.4.4 *The logic λ_1 is Kripke non-compact and $S4 \subset \lambda_1 \subset Grz$.*

Theorem 6.4.5 *The logic λ_2 is Kripke non-compact and $Grz \subset \lambda_2$.*

In order to prove these theorems we need a number of technical lemmas concerning properties of formulas introduced above.

Lemma 6.4.6 *Let $\mathcal{F} := \langle F, R \rangle$ be a reflexive transitive frame. Then for every $x \in F$,*

(i) there exists a valuation V such that $x \nVdash_V \chi_1$ iff x R-sees an infinite ascending chain of clusters or a certain proper cluster from $\mathcal{F}$.

(ii) there exists a valuation V such that $x \nVdash_V \chi_2$ iff x R-sees an infinite ascending chain of clusters or a proper cluster from $\mathcal{F}$ with not less than 3 elements.

Proof. It can be shown by straightforward verification, details are left to the reader as an exercise. ■.

For any formula α, the n-power α^n of α is the substitution of formulas β_n, β_{n+1}, γ_n and γ_{n+1} into α in place of $\beta_0, \beta_1, \gamma_0$ and γ_1 respectively.

Lemma 6.4.7 *For any m and n, $\beta_n^m \equiv \beta_{n+m} \in S4$, $\gamma_n^m \equiv \gamma_{n+m} \in S4$, $\alpha_n^m \equiv \alpha_{n+m} \in S4$ and $\phi_n^m \equiv \phi_{n+m} \in S4$.*

Proof. The proof can simply be effected by induction on n and m, we leave details to the reader as an exercise. ■

Lemma 6.4.8 *$\delta^m \to \delta^{m+1} \in S4$ for any $m \geq 1$, in particular $\delta \to \delta^n \in S4$.*

Proof. The proof consists of simple induction on m using Lemma 6.4.7. ■

Suppose a formula α has no occurrences of q_0, q_1 and r_0, r_1 and that β is an arbitrary formula. Then

Lemma 6.4.9 *Let $\mathcal{F} := \langle F, R \rangle$ be an arbitrary reflexive transitive frame. If $\forall n(\mathcal{F} \Vdash \alpha \wedge \beta \wedge \delta \to \Diamond \phi_n)$ and $\mathcal{F} \Vdash \alpha \wedge \delta \wedge \Diamond \phi_1 \wedge \Diamond \phi_2 \to \Box(\phi_1 \to \phi_2)$, and $(\exists x \in F)(\forall V)(x \Vdash_V \alpha \wedge \beta \wedge \delta)$ then $\mathcal{F}$ contains an infinite ascending chain of clusters.*

Proof. Suppose that $a \in F$ and there exist a valuation V such that $a \Vdash_V \alpha \wedge \beta \wedge \delta$. We will show that $\mathcal{F}$ contains an infinite ascending chain $u_n, n \in N$ of elements such that

$$aRu_n, u Ru_{n+1}, \neg(u_n Ru_m) \text{ for } m < n, \text{ and } u_n \Vdash_V \phi_n. \tag{6.37}$$

We have $a \Vdash_V \alpha \wedge \beta \wedge \delta$ and $(\mathcal{F} \Vdash \alpha \wedge \beta \wedge \delta \to \Diamond \phi_1)$, hence $a \Vdash_V \Diamond \phi_1$, and there exists u_1 such that aRu_1 and $u_1 \Vdash_V \phi_1$. Assume that elements $u_1, u_2, ..., u_n$ which satisfy (6.37) have already been found. Since $a \Vdash_v \delta$ we infer from Lemma 6.4.8 that $a \Vdash_V \delta_{n-1}$. By the hypothesis of our lemma, $\mathcal{F} \Vdash \alpha \wedge \beta \wedge \delta \to \Diamond \phi_n$ and $\mathcal{F} \Vdash \alpha \wedge \beta \wedge \delta \to \Diamond \phi_{n+1}$. Also $a \Vdash_V \alpha \wedge \beta \wedge \delta$, thus $a \Vdash_V \Diamond \phi_n$ and $a \Vdash_V \Diamond \phi_{n+1}$. By Lemma 6.4.7 $\phi_n \equiv \phi_1^{n-1} \in S4$ and $\phi_{n+1} \equiv \phi_2^{n-1} \in S4$. Therefore $a \Vdash_V \Diamond \phi_1^{n-1}, a \Vdash_V \Diamond \phi_2^{n-1}$. Since $a \Vdash_V \alpha$ and $\alpha^n = \alpha$ it follows that $a \Vdash_V \alpha_{n-1} \wedge \delta_{n-1} \wedge \Diamond \phi_1^{n-1} \wedge \Diamond \phi_2^{n-1}$. But

$$\mathcal{F} \Vdash \left(\alpha \wedge \delta \wedge \Diamond \phi_1 \wedge \Diamond \phi_2 \to \Box(\phi_1 \to \phi_2)\right)^{n-1}.$$

Therefore $a \Vdash_V \Box(\phi_1^{n-1} \to \Diamond \phi_2^{n-1})$. Using Lemma 6.4.7 we conclude $a \Vdash_V \Box(\phi_n \to \phi_{n+1})$. By the inductive hypothesis it follows that $u_n \Vdash_V \phi_n$ and aRu_n, hence there exists u_{n+1} such that $u_n Ru_{n+1}$ and $u_{n+1} \Vdash_v \phi_{n+1}$. It remains only to show that $\neg(u_{n+1}Ru_k)$ for $k \leq n$. Indeed, $u_n \Vdash_V \phi_n$ holds, and this yields

$u_n \Vdash_v \Diamond \alpha_n$ and $u_k \Vdash_V \Diamond \alpha_n$. Since $u_{n+1} \Vdash_V \phi_{n+1}$, it follows that $u_{n+1} \nVdash_V \neq \Diamond \alpha_n$. Consequently, $\neg(u_{n+1} R u_k)$. Thus $u_n, n \in N$ with (6.37) exists. ∎

We need also some facts concerning the truth value of certain formulas in special frames. We describe the structure of these frames below. For $n \geq 0$, $\mathcal{W}_n := \langle W_n, R \rangle$, were

$$W_n := \{x_0, ..., x_{n+2}, y_0, ..., y_{n+2}, t_{0,0}, t_{0,1}, t_1, ..., t_{n+1}, z_0, ..., z_{n+1}\},$$

R is reflexive and transitive relation and

$$x_i R x_j (j \leq i \leq n+2), y_i R y_j (j \leq i \leq n+2), t_i R t_j (j \leq i \leq n+1),$$
$$x_{i+2} R y_i (i \leq n), y_{i+2} R x_i (i \leq n), z_i R x_{i+1} (i \leq n+1),$$
$$z_i R y_{i+1} (i \leq n+1), t_i R z_i (i \leq n+1),$$
$$t_{0,0} R t_{0,1}, t_{0,1} R t_{0,0}, t_{0,0} R t_1.$$

For vividity, a fragment of the frame $\mathcal{W}_n$ is depicted below in Fig. 6.2. Note that these frames are merely certain modifications of the frame actively employed in K.Fine's [45]. The frame $\mathcal{U}_n$ is obtained from the frame $\mathcal{W}_n$ by removing the element $t_{0,1}$. We also fixe the following 5-element poset $\mathcal{H}$ depicted in Fig. 6.3 below.

Lemma 6.4.10 *For any placement of $\mathcal{H}$ as a subframe of the frame $\mathcal{W}_n$ or the frame $\mathcal{U}_n$, the following hold. For some i, either (a) $u_o = x_i$, $v_o = y_i$, $u_1 = x_{i+1}$, $v_1 = y_{i+1}$, or (b) $u_0 = y_i$, $v_0 = x_i$, $u_1 = y_{i+1}$, $v_1 = x_{i+1}$.*

Proof. Note that $u_0 \neq t_{0,0}, t_{0,1}$ since these elements are R-comparable to all elements of $\mathcal{W}_n$ ($\mathcal{U}_n$). Suppose $u_0 = t_i$, $i \geq 1$. Then the only possibility is $v_0 = z_k$, hence $v_1 = t_l$ for some l. Still u_0 and v_1 are R-comparable, a contradiction. Furthermore, in the frame $\mathcal{W}_n$ ($\mathcal{U}_n$), for all i, $u_0 \neq z_i$. In fact, if $u_0 = z_i$ then $u_1 = t_j$ for some j, consequently $v_0 = z_k$, which implies $v_1 = t_l$, i.e. v_1 is R-comparable to u_1, a contradiction. Therefore obligatory $u_0 = x_i$ (or $u_0 = y_i$) holds. By the same reasoning it follows that $y_0 = y_j (x_j)$. Since u_0 is R-incomparable to v_0, it follows that $j \in \{i_1, i, i+1\}$. If $j = i - 1$ then $u_1 R v_0$. If $j = i + 1$ then $v_1 R u_0$. Consequently, we have $j = i$. If $u_1 \neq x_{i+1}(y_{i+1})$ then $u_1 R v_0 (v_1 R u_0)$, hence $u_1 = x_{i+1}(y_{i+1})$. In a similar way we show that $v_1 = y_{i+1}(x_{i+1})$. ∎

The proofs of the following three lemmas depend on which case, (a) or (b) holds in our lemma above. We will assume that case (a) holds always, since in case (b) the lemmas and their proofs remain the same, except we must interchange always all x_m and all y_m.

Lemma 6.4.11 *For any placement the frame $\mathcal{H} := \langle \{a, u_1, v_1, u_0, v_0\}, R \rangle$ as a subframe of the frame $\mathcal{W}_n$ ($\mathcal{U}_n$) and any valuation V in $\mathcal{W}_n$ ($\mathcal{U}_n$), the following hold. If $a \Vdash_V \delta$, $u_0 \Vdash_V \beta_0$, $v_0 \Vdash_v \gamma_0$, $v_0 \Vdash_V \beta_1$, $v_0 \Vdash_v \gamma_1$ and i is the number from Lemma 6.4.10 then*

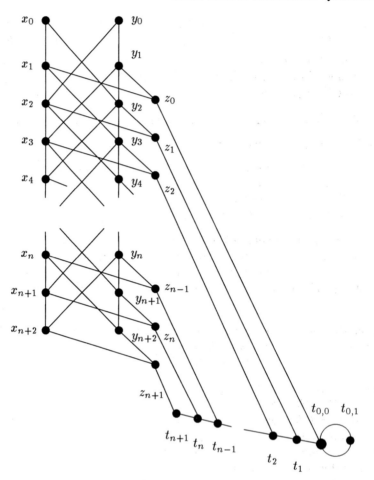

Figure 6.2: Fragment of the frame $\mathcal{W}_n$.

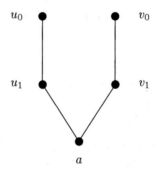

Figure 6.3: The 5-element frame $\mathcal{H}$.

(1) $\{x_i\} \subseteq V(\beta_0) \cap a^{R\leq} \subseteq \{x_i, z_{i_1}, ..., z_0\}$;

(2) $\{y_i\} \subseteq V(\gamma_0) \cap a^{R\leq} \subseteq \{y_i, z_{i_1}, ..., z_0\}$.

Proof. By Lemma 6.4.10 it follows that $u_0 = x_i$, $v_0 = y_i$, $u_1 = x_{i+1}$ and $v_1 = y_{i+1}$. Therefore $\{x_i\} \subseteq V(\beta_0) \cap a^{R\leq}$ and $\{y_i\} \subseteq V(\gamma_0) \cap a^{R\leq}$. Note that $\forall t_\rho(t_\rho \in a^{R\leq} \Rightarrow t_\rho \nVdash_V \beta_0)$. Indeed, if aRt_ρ and $t_\rho \Vdash_V \beta_0$ then $t_\rho Rx_{i+1}$ and $x_{i+1} \Vdash_V \beta_1$. Hence $t_k \nVdash_V \psi_3$ which contradicts $a \Vdash_V \delta$. We also have $\forall k \geq 0$ $(z_{i+k} \in a^{R\leq} \Rightarrow z_{i+k} \nVdash_V \beta_0)$. In fact if aRz_{i+k} and $z_{i+k} \Vdash_V \beta_0$ then, since $z_{i+k} Rx_{i+1}$ and $x_{i+1} \Vdash_V \beta_1$ we again obtain a contradiction the fact that $a \Vdash_V \delta$.

Furthermore, $(\forall k \geq 0)(x_{i-1-k} \in a^{R\leq} \Rightarrow x_{i-1-k} \nVdash_V \beta_0)$. Indeed, if we have that aRx_{i-1-k} and $x_{i-1-k} \Vdash_v \beta_0$ then $y_{i+1} Rx_{i-1-k}$ and $y_{i+1} \Vdash_V \gamma_1$, contrary to $a \Vdash_V \psi_2$. Note also that $(\forall k \geq 0)(x_{i+1+k} \in a^{R\leq} \Rightarrow x_{i+1+k} \nVdash_V \beta_0)$. In fact, if aRx_{i+1+k} and $x_{i+1+k} \Vdash_V \beta_0$ then $x_{i+1+k} Rx_{i+1}$ and $x_{i+1} \Vdash_V \beta_1$, contrary to $a \Vdash_V \psi_3$. Also, $(\forall k \geq 0)(y_{i-k} \in a^{R\leq} \Rightarrow y_{i-k} \nVdash_V \beta_0)$. Indeed, if aRy_{i-k} and $y_{i_k} \Vdash_V \beta_0$ then, since $y_{i+1} Ry_{i-k}$ and $y_{i+1} \Vdash_V \gamma_1$, we obtain a contradiction $a \Vdash_V \delta$. Still we have $(\forall k \geq 0)(y_{i+1+k} \in a^{R\leq} \Rightarrow y_{i+1+k} \nVdash_V \beta_0)$. Indeed, suppose aRy_{i+1+k} and $y_{n+1+k} \Vdash_V \beta_0$. Then $y_{i+1+k} Ry_{i+1}$ and $y_{i+1} \Vdash_V \gamma_1$ which contradicts $a \Vdash_v \delta$. Thus we have proved that $\{x_i\} \subseteq V(\beta_0) \cap a^{R\leq} \subseteq \{x_i, z_{i_1}, ..., z_0\}$, i.e. (1) holds. The situation with γ_0 is completely symmetric and, arguing in a similar way (replacing β_0 onto γ_0 and interchanging x_m and y_m) we prove (2). ∎

Lemma 6.4.12 *If the hypothesis of Lemma 6.4.11 are satisfied and $k \geq 1$, then*

(1) $(\forall b \in a^{R\leq})(b \Vdash_V \beta_k \Leftrightarrow b = x_{i+k})$ *for $i + k \leq n + 2$;*

(2) $(\forall c \in a^{R\leq})(c \Vdash_V \gamma_k \Leftrightarrow c = y_{i+k})$ *for $i + k \leq n + 2$;*

(3) $V(\beta_k) \cap a^{R\leq} = \emptyset$, $V(\gamma_k) \cap a^{R\leq} = \emptyset$ *for $i + k > n + 2$.*

Proof. We carry out the proof by induction on k. Consider the basis: $k = 1$. By Lemma 6.4.10 $u_1 = x_{i+1}$, aRx_{i+1}, $x_{i+1} \Vdash_V \beta_1$, and $i + 1 \leq n + 2$. Now we must prove the following: $b \in a^{R\leq}$ and $b \Vdash_V \beta_1$ only if $b = x_{i+1}$. First note that $\forall \rho(aRt_\rho \Rightarrow t_\rho \nVdash_V \beta_1)$. Indeed, otherwise $t_\rho \Vdash_V \beta_1$, aRt_ρ, $t_\rho Ry_i$ and by Lemma 6.4.10 $y_i \Vdash_V \gamma_0$, which contradicts $a \Vdash_V \delta$. Also $\forall j \geq i - 1(aRz_j \Rightarrow z_j \nVdash_V \beta_1)$. Otherwise aRz_j, $z_j \Vdash_V \beta_1$, $z_j Ry_i$, and by Lemma 6.4.10 $y_i \Vdash_V \gamma_0$ which again contradicts $a \Vdash_V \delta$. Next, we need to show that $\forall j < i - 1(aRz_j \Rightarrow z_j \nVdash_V \beta_1)$. Again, otherwise aRz_j, $z_j \Vdash_V \beta_1$ and, since $a \Vdash_V \delta$, there is a certain s such that $z_j Rs$, $s \Vdash_V \beta_0$ and $\neg(sRz_j)$. Then $y_{i+1} Rs$ and $s \Vdash_V \beta_0$. By hypothesis of Lemma 6.4.10 it follows that $y_{i+1} \Vdash_V \gamma_1$. This contradicts $a \Vdash_V \delta$. Note that $\forall j < i + 1(aRx_j \Rightarrow x_j \nVdash_V \beta_1)$. In fact, otherwise aRx_j, $x_j \Vdash_V \beta_1$, and $x_i Rx_j$, and $x_i \Vdash_V \beta_0$ (Lemma 6.4.10), which contradicts $a \Vdash_V \delta$. Furthermore, $\forall j > i + 1(aRx_j \Rightarrow x_j \nVdash_V \beta_1)$. In fact, otherwise aRx_j, $x_j \Vdash_V \beta_1$, $x_j Ry_i$ and $y_i \Vdash_V \gamma_0$ (Lemma 6.4.10), which contradicts $a \Vdash_v \delta$. Moreover, $\forall j > i(aRy_j \Rightarrow y_j \nVdash_V \beta_1)$. In fact, if aRy_j, $y_j \Vdash_V \beta_1$ then, since $y_j Ry_i$ and

$y_i \Vdash_V \gamma_0$ (Lemma 6.4.10), we get a contradiction the fact that $a \Vdash_V \delta$. Note now that $\forall j \le i (aRy_j \Rightarrow y_j \nVdash_V \beta_1)$. Indeed, if aRy_j and $y_j \Vdash_V \beta_1$ then since $y_i Ry_j$ and $y_i \Vdash_V \gamma_0$ (Lemma 6.4.10) we again obtain contradiction $a \Vdash_V \delta$. Thus we showed that $(\forall b \in a^{R\le})(b \Vdash_V \beta_1 \Leftrightarrow b = x_{i+1})$. Similarly, we show that $(\forall c \in a^{R\le})(c \Vdash_V \gamma_1 \Leftrightarrow b = y_{i+1})$ Since by Lemma 6.4.10 $i + 1 \le n + 2$, the lemma is proved for $k = 1$.

Now we prove our lemma for $k = 2$ (the second inductive basis step). If $i+2 > n + 2$, i.e. $i + 1 = n + 2$, then $\forall b \in a^{R\le}(b \nVdash_V \beta_2)$. Indeed,

$$\beta_2 := \Diamond\beta_1 \wedge \Diamond\gamma_0 \wedge \neg\Diamond\gamma_1$$

and by Lemma 6.4.11 $V(\gamma_0) \cap a^{R\le} \subseteq \{y_{n+1}, z_n, ..., z_0\}$. Also using what has been proved above, it follows $V(\beta_1) \cap a^{R\le} = \{x_{n+2}\}$, hence $\forall b \in a^{R\le}(b \nVdash_V \beta_2)$. In a similar way we show that $\forall c \in a^{R\le}(c \nVdash_V \gamma_2)$. Consequently (3) is proved for $k > 2$. For $k = 2$, $V(\beta_k) \cap a^{R\le} = \emptyset$ holds and also it follows that $V(\gamma_k) \cap a^{R\le} = \emptyset$, and the assertion of our lemma is proved.

Suppose now that $i + 2 \le n + 2$. Assume that aRb and $b \Vdash_V \beta_2$. Since $\beta_2 = \Diamond\beta_1 \wedge \Diamond\gamma_0 \wedge \neg\Diamond\gamma_1$, by Lemma 6.4.11 and the case $k = 1$ of our lemma, it follows that bRx_{i+1}, $\neg(bRy_{i+1})$ and some element of the set $\{y_i, z_{i-1}, ..., z_0\}$ is R-accessible from b. This is possible if and only if $b = x_{i+2}$. Thus (1) is proved for $k + 2$. The proof of (2) for $k + 2$ is completely analogous.

Now assume that our lemma has been proved for $k = m - 1$ and $k = m$ and that $i + k \le n + 2$. We have to prove the assertion of our lemma for $k = m + 1$. If we assume that $i + m > n + 2$ then $i + m - 1 = n + 2$. By the inductive assumption, $V(\beta_m) \cap a^{R\le} = \{x_{n+2}\}$, $V(\gamma_m) \cap a^{R\le} = \{y_{n+2}\}$ and $V(\gamma_{m-1}) \cap a^{R\le} = \{y_{n+1}\}$. Therefore $\forall b \in a^{R\le}(b \nVdash_V \beta_{m+1})$ since $\beta_{m+1} = \Diamond\beta_m \wedge \Diamond\gamma_{m-1} \wedge \neg\Diamond\gamma_m$. In a similar way we show that $\forall c \in a^{R\le}(c \nVdash_V \gamma_{m+1})$. Hence it follows that if $k \ge m+1$, then $V(\beta_k) \cap a^{R\le} = V(\gamma_k) \cap a^{R\le} = \emptyset$ and the lemma is proved. Assume now that $i + m \le n + 2$. Suppose aRb and $b \Vdash_V \beta_{m+1}$. Since $\beta_{m+1} = \Diamond\beta_m \wedge \Diamond\gamma_{m-1} \wedge \neg\Diamond\gamma_m$ and our lemma has been proved for $k = m$ and $k = m - 1$, we obtain bRx_{i+m}, bRy_{i+m-1} and $\neg(bRy_{i+m})$, which holds if and only if $b = x_{i+m+1}$. Thus (1) is shown for $k = m + 1$. A similar arguing shows that (2) also holds for $i + m + 1 \le n + 2$. (3) follows immediately from (1) and (2). ∎

Lemma 6.4.13 *If the hypothesis of Lemma 6.4.11 are satisfied, then*

(1) $(\forall g \in a^{R\le})(g \Vdash_V \alpha_k \Leftrightarrow g = z_{i+k}), i + k \le n$,

(2) $V(\alpha_k) \cap a^{R\le} = \emptyset$ *for* $i + k > n + 1$,

(3) $V(\alpha_k) \cap a^{R\le} = \{z_{n+1}, t_{0,\rho}, t_1, ..., t_{n+1}\} \cap a^{R\le}$ *for* $i+k = n+1$,

(4) $(\forall f \in a^{R\le})(f \Vdash_V \phi_k \Leftrightarrow f = t_{i+k}), i + k \le n$,

(5) $V(\phi_k) \cap a^{R\le} = \emptyset$ *for* $i + k > n$.

Proof. Suppose $i + k \leq n$. Assume that aRg and $g \Vdash_V \alpha_k$. Since

$$\alpha_k := \Diamond\beta_{k+1} \wedge \Diamond\gamma_{k+1} \wedge \neg\Diamond\beta_{k+2} \wedge \neg\Diamond\gamma_{k+2},$$

it follows from Lemma 6.4.12 that aRg and $g \Vdash_V \alpha_k$ are equivalent to gRx_{i+k+1}, gRy_{i+k+1} and $\neg(gRx_{i+k+2})$, $\neg(gRy_{i+k+2})$, which is equivalent to $y = z_{i+k}$. Thus (1) is proved. Let $i+k > n+1$. Since by Lemma 6.4.12, $V(\beta_{k+1}) \cap a^{R\leq} = \emptyset$, we conclude $V(\alpha_k) \cap a^{R\leq} = \emptyset$ and (2) is shown. Finally, suppose that $i + k = n + 1$. Then by Lemma 6.4.12 $V(\beta_{k+2}) \cap a^{R\leq} = \emptyset$ and $V(\gamma_{k+2}) \cap a^{R\leq} = \emptyset$, and $V(\beta_{k+1}) \cap a^{R\leq} = \{x_{n+2}\}$, $V(\gamma_{k+1}) \cap a^{R\leq} = \{y_{n+2}\}$. Therefore

$$V(\alpha_k) \cap a^{R\leq} = \{z_{n+1}, t_{0,\rho}, t_1, ..., t_{n+1}\} \cap a^{R\leq}$$

and (3) is proved. Now we turn to the proof the assertions concerning ϕ_k. Suppose $i + k \leq n$. Assume that $f \in a^{R\leq}$ and $f \Vdash_V \phi_k$. Since

$$\phi_k := \neg\Diamond\alpha_{k-1} \wedge \Diamond\alpha_k\Diamond\alpha_{k+1},$$

it follows from (1) and (3) that fRz_{i+k}, $\neg(fRz_{i+k-1})$ and an element of the set $\{z_{i+k+1}, t_{0,\rho}, t_1, ..., t_{n+1}\}$ is R-accessible from f, which is possible iff $f = t_{i+k}$. Thus (4) is shown. If $i + k > n$ then according to (2) $V(\alpha_k) \cap a^{R\leq} = \emptyset$, hence $V(\phi_k) \cap a^{R\leq} = \emptyset$, and consequently (5) holds. ∎

In order to show Kripke non-compactness of the modal logic λ_1 we need also the following three lemmas.

Lemma 6.4.14 $\mathcal{W}_n \Vdash \chi_2 \wedge \mu \wedge \rho_1 \wedge ... \wedge \rho_n$ *and* $\mathcal{W}_n \nVdash \neg\epsilon$ *for any* $n \geq 0$.

Proof. We define a valuation V in $\mathcal{W}_n$ as follows:

$$V(p) := \{t_{0,0}\}, \quad V(q_j) := \{x_j\}, \quad V(r_j) := \{y_j\}, \quad j = 0, 1.$$

Obviously that than we obtain $t_{0,0} \Vdash_V \neg\chi_1 \wedge \delta \wedge \Box(\beta_0 \to \Box\beta_0) \wedge \Box(\gamma_o \to \Box\gamma_0)$. Moreover, we are in the situation of Lemma 6.4.12 with $i = 0$ since $t_{0,0} \Vdash_V \Box(\beta_0 \to \Box\beta_0) \wedge \Box(\gamma_o \to \Box\gamma_0)$. By Lemma 6.4.13 we derive $z_0 \Vdash_V \alpha_0$ and $t_{0,0}Rz_0$. Hence $t_{0,0} \Vdash_V \Diamond\alpha_0$, which implies $t_{0,0} \Vdash_V \epsilon$ and $\mathcal{W}_n \nVdash \neg\epsilon$.

By Lemma 6.4.6 $\mathcal{W}_n \Vdash \chi_2$. Now we will show that $\mathcal{W}_n \Vdash \mu$. If $a \in |\mathcal{W}_n|$ and $a \Vdash_V \neg\chi_1 \wedge \delta \wedge \Diamond\phi_1 \wedge \Diamond\phi_2$ then $a \Vdash_v \Diamond\alpha_1$. This yields $a \Vdash_V \Diamond\beta_1 \wedge \Diamond\gamma_1$, i.e. there are u_1 and v_1 such that aRu_1, aRv_1 and $u_1 \Vdash_V \beta_1$, $v_1 \Vdash_V \gamma_1$. Since $a \Vdash_V \delta$, there exist u_0 and v_0 such that u_1Ru_0, $u_0 \Vdash_V \beta_0$, v_1Rv_0, $v_0 \Vdash_V \gamma_0$, $\neg(u_1Rv_0)$ and $\neg(v_1Ru_0)$. Thus we are in the situation of Lemma 6.4.11. Since $a \Vdash_V \Diamond\phi_1$, by Lemma 6.4.13 it follows that $i+2 \leq n$ and $V(\phi_1) \cap a^{R\leq} = \{t_{i+1}\}$ and $V(\phi_2) \cap a^{R\leq} = \{t_{i+2}\}$. Therefore $a \Vdash_V \Box(\phi_1 \to \Diamond\phi_2)$. Thus we showed that $\mathcal{W}_n \Vdash \mu$.

Now we turn to prove $\mathcal{W}_n \Vdash \rho_1 \wedge ... \wedge \rho_n$. Assume that $a \Vdash_V \epsilon$. Since $a \Vdash_V \neg\chi_1$, by Lemma 6.4.6 it follows that $a = t_{0,0}$ (or $a = t_{0,1}$, in this case the reasoning must be the same). Since $a \Vdash_V \delta \wedge \Diamond \alpha_0$, there are u_1 and v_1 such that aRu_1, aRv_1, $u_1 \Vdash_V \beta_1$ and $v_1 \Vdash_v \gamma_1$. Since $a \Vdash_V \delta$, there exist u_0 and v_0 such that $u_1 R u_0$, $u_0 \Vdash_V \beta_0$, $v_1 R v_0$, $v_0 \Vdash_V \gamma_0$, $\neg (u_1 R v_0)$ and $\neg (v_1 R u_0)$. Thus we are again in the situation of Lemma 6.4.11. Moreover, since $a \Vdash_V \Box(\beta_0 \to \Box \beta_0)$, $a \Vdash_V \Box(\gamma_0 \to \Box \gamma_0)$, we have $i = 0$ in the condition of Lemma 6.4.11. By Lemma 6.4.13, $V(\phi_k) = \{t_k\}$, where $k \leq n$. But $a = t_{0,0}$, hence $a \Vdash_V \Diamond \phi_1 \wedge ... \wedge \Diamond \phi_n$. Thus $\mathcal{W}_n \Vdash_V \rho_1 \wedge ... \wedge \rho_n$. ∎

Lemma 6.4.15 *If $\mathcal{F}$ is a transitive reflexive frame, $\mathcal{F} \Vdash_V \chi_2 \wedge \mu$ and $\mathcal{F} \Vdash_V \rho_n$ for any n, then $\mathcal{F} \Vdash_V \neg \epsilon$.*

Proof. Suppose there is a valuation V in $\mathcal{F}$ and there is an element u such that $u \Vdash_V \epsilon$. Using Lemma 6.4.9 and taking as α the formula $\neg \chi_1$, as β the formula $\Diamond \alpha_0 \wedge \Box(\beta_0 \to \Box \beta_0) \wedge \Box(\gamma_0 \to \Box \gamma_0)$, we conclude that $\mathcal{F}$ contains an infinite R-ascending chain, which in view of Lemma 6.4.6 contradicts $\mathcal{F} \Vdash \chi_2$. ∎

Lemma 6.4.16 $S4 \subseteq \lambda_1 \subseteq Grz$.

Proof. We know (see Theorem 2.6.25) that Grz is the modal logic generated by the class of all finite posets. Let $\mathcal{F}$ be a finite poset. Then by Lemma 6.4.6 $\mathcal{F} \Vdash \chi_2$. Moreover $\mathcal{F} \Vdash_V \mu$ and $\mathcal{F} \Vdash_V \rho_n$ for any n since the premises of these formulas are invalidated in $\mathcal{F}$ under any valuation (using again Lemma 6.4.6). Thus all theorems of λ_1 are valid in $\mathcal{F}$ and $S4 \subseteq \lambda_1 \subseteq Grz$. ∎

Using Lemmas 6.4.14, 6.4.15 and Lemma 6.4.16 we immediately obtain the assertion of Theorem 6.4.4. ∎

The next corollary immediately follows from Theorem 6.4.4.

Corollary 6.4.17 *There are certain modal logics λ extending $S4$ such that (i): λ is Kripke non-compact, and consequently Kripke incomplete, having no the fmp, and having no finite axiomatizations, but (ii) their intuitionistic fragment $\rho(\lambda)$ coincides with H, in particular, $\rho(\lambda)$ has the fmp and is finitely axiomatized.*

Now we turn to show that the modal logic λ_2 is Kripke non-compact also. This example is especially important for our further research when we will show that there are Kripke non-compact but decidable logics.

Lemma 6.4.18 $\mathcal{U}_n \Vdash_V \chi_1 \wedge \psi \wedge \xi_1 \wedge ... \wedge \xi_n$ *and* $\mathcal{U}_n \nVdash \neg \phi$ *for any* $n \geq 0$.

Proof. We define a valuation V in $\mathcal{U}_n$ as follows:

$$V(q_j) := \{x_j\}, \quad V(r_j) := \{y_j\}, \quad j = 0, 1.$$

It is not hard to see that $t_{0,0} \Vdash_V \neg \delta \wedge \Box(\beta_0 \to \Box\beta_0) \wedge \Box(\gamma_0 \to \Box\gamma_0)$. Now we are in the situation of Lemma 6.4.12 and $i = 0$ for this case. Using Lemma 6.4.13 we conclude $z_0 \Vdash_V \alpha_0$ and $z_1 \Vdash_V \alpha_1$. Hence

$$t_{0,0} \Vdash_V \delta \wedge \Diamond\alpha_0 \wedge \Diamond\alpha_1 \wedge \Box(\beta_0 \to \Box\beta_0) \wedge \Box(\gamma_0 \to \Box\gamma_0).$$

Thus $t_{0,0} \Vdash_V \phi$ and $\mathcal{U}_n \not\Vdash \neg\phi$.

We will show that $\mathcal{U}_n \Vdash \psi$. Let $a \in |\mathcal{U}_n|$ and $a \Vdash_V \delta \wedge \Diamond\phi_1 \wedge \Diamond\phi_2$. Then $a \Vdash_V \Diamond\alpha_1$ and $a \Vdash_V \Diamond\beta_1 \wedge \Diamond\gamma_1$. Consequently there are u_1 and v_1 such that aRu_1 and aRv_1, and $u_1 \Vdash_V \beta_1$, $v_1 \Vdash_V \gamma_1$. Since $a \Vdash_V \delta$, there exist u_0 and v_0 such that $u_1 R u_0$, $u_0 \Vdash_V \beta_0$, $v_1 R v_0$, $v_0 \Vdash_V \gamma_0$, $\neg(u_1 R v_0)$ and $\neg(v_1 R u_0)$. Thus we are in the situation of Lemma 6.4.11. Since $a \Vdash_V \Diamond\phi_1$, by Lemma 6.4.13 we get $i + 2 \leq n$ and $V(\phi_1) \cap a^{R\leq} = \{t_{i+1}\}$ and $V(\phi_2) \cap a^{R\leq} = \{t_{i+2}\}$. Therefore $a \Vdash_V \Box(\phi_1 \to \Diamond\phi_2)$. Hence $\mathcal{U}_n \Vdash \psi$.

Now we turn to prove that $\mathcal{U}_n \Vdash \xi_1 \wedge \ldots \wedge \xi_n$. Let $a \Vdash_V \phi$. Then $a \Vdash_V \delta \wedge \Diamond\alpha_0 \wedge \Diamond\alpha_1$. This yields there are u_1 and v_1 such that aRu_1, aRv_1, $u_1 \Vdash_V \beta_1$ and $v_1 \Vdash_V \gamma_1$. Using $a \Vdash_V \delta$ we conclude that there exist u_0 and v_0 such that $u_1 R u_0$, $u_0 \Vdash_V \beta_0$, $v_1 R v_0$, $v_0 \Vdash_V \gamma_0$, $\neg(u_1 R v_0)$ and $\neg(v_1 R u_0)$. Thus we are again in the situation of Lemma 6.4.11 and $i = 0$.

By Lemma 6.4.13, since $a \Vdash \Diamond\alpha_0 \wedge \Diamond\alpha_1$, it follows that $V(\alpha_0) \cup a^{R\leq} = \{z_0\}$, $V(\alpha_1) \cup a^{R\leq} = \{z_1, t_{0,0}, t_1\}$ or $V(\alpha_1) \cup a^{R\leq} = \{z_1\}$. Since $a \Vdash \Diamond\alpha_0 \wedge \Diamond\alpha_1$, we have aRz_0 and some element of $\{z_1, t_{0,0}, t_1\}$ is R-accessible from a, hence $a = t_{0,0}$. Also by Lemma 6.4.13, it follows that $t_1 \Vdash_V \phi_1, \ldots, t_n \Vdash_V \phi_n$. Therefore $a \Vdash_V \Diamond\phi_1 \wedge \ldots \wedge \Diamond\phi_n$. Thus, $\mathcal{U}_n \Vdash \xi_1 \wedge \ldots \wedge \xi_n$. ∎

Lemma 6.4.19 *Let $\mathcal{F}$ be a reflexive transitive frame and $\mathcal{F} \Vdash \chi_1 \wedge \phi$. Suppose that $\mathcal{F} \Vdash \xi_n$ for any n. Then $\mathcal{F} \Vdash \neg\phi$.*

Proof. Suppose that there is a valuation V in $\mathcal{F}$ such that, for a certain element u, $u \Vdash_V \phi$. Using Lemma 6.4.9 and taking as α the formula $\top$, as β the formula $\Diamond\alpha_0 \wedge \Diamond\alpha_1 \wedge \Box(\beta_0 \to \Box\beta_0) \wedge \Box(\gamma_0 \to \Box\gamma_0)$, we obtain that $\mathcal{F}$ contains an infinite R-accessible ascending chain of clusters. This contradicts $\mathcal{F} \Vdash \chi_1$ in view of Lemma 6.4.6. ∎

Using Lemmas 6.4.18 and 6.4.19 we immediately obtain the assertion of Theorem 6.4.5. ∎

Note that, using the technique developed above, it is possible to construct continuously many Kripke non-compact modal logics in the interval $[S4, Grz]$ and continuously many Kripke non-compact modal logics extending Grz (see Rybakov [130]). Note also that, using constructed above Kripke non-compact logics, it is possible to show due to S.K.Thomason [178] that the logical consequence relation in Kripke transitive reflexive frames, as well as posets, is not recursively enumerable.

Definition 6.4.20 *Let X be a recursively enumerable set of formulas and α be a formula. We say that α is a semantic Kripke consequence of X, written $X \Vdash \alpha$, iff, for any frame $\mathcal{F}$, $\mathcal{F} \Vdash \alpha$ if $\mathcal{F} \Vdash X$.*

Corollary 6.4.21 *The semantic Kripke consequence relation $\Vdash$ on posets is not recursively enumerable.*

Proof. Consider the standard enumeration $W_0, W_1, W_2, \ldots$ of the recursively enumerable sets of natural numbers. We also assume the standard Gödel numbering of the formulas, we mean α_n is the formula with Gödel number n. Consider a recursive function $f(x)$ such that $f(n)$ is Gödel number of the formula $\chi_1 \wedge \psi \wedge \xi_0 \wedge \ldots \wedge \xi_n$. Suppose $\Vdash$ is recursively enumerable. Then it is not hard to see that the set

$$S := \{e \mid e \in N, \{\alpha_{f(x)} \mid x \in W_e\} \Vdash \neg\phi\}\}$$

is also recursively enumerable. By Lemmas 6.4.18 and 6.4.19, $e \in S \Leftrightarrow W_e$ is infinite. It is well known that $\{e \mid W_e$ is infinite $\}$ is not recursively enumerable. ∎

Therefore it is not possible to offer a finite axiomatic system for $\Vdash$ but moreover, no recursive axiomatization for $\Vdash$ exists.

6.5 Decidable Kripke Non-compact Logics

The aim of this section is to clarify how deeply the decidability of a logic can be combined with certain negative properties. It is known that there are decidable logics which do not have the fmp and which are Kripke incomplete (cf., for instance, [31]). Using the technique developed in previous section we are now in a position to construct an example of a Kripke non-compact modal logic (consequently a logic without the fmp, Kripke incomplete, and lacking even a finite axiomatization) which is nevertheless decidable. The content of this section is a construction of a such logic and proving it has all necessary properties.

We will need a simple modification of the algebraic tools which we previously applied to modal logics.

Definition 6.5.1 *We say that a model (not an algebra) $\mathfrak{B}$ with partially-defined operations in the signature of modal algebras is a partial S4-modal algebra if*

      (i) $\mathfrak{B}$ is a boolean algebra (i.e. all boolean operations are always defined on $\mathfrak{B}$ and satisfy the identities of boolean algebras);

(ii) the value of the modal operator $\Diamond$ is not necessary defined for all elements of $\mathfrak{B}$, but $\Diamond$ satisfies the identities

$$\Diamond\Diamond x = \Diamond x,$$
$$\Diamond x \vee x = \Diamond x,$$
$$\Diamond x \vee \Diamond y = \Diamond(x \vee y),$$
$$\Diamond\top = \top.$$

provided all the values in the corresponding identities are defined.

Note that above we correctly employ the term modal algebra since the required identities in general define modal $S4$-algebras (see Lemmas 2.2.3 and 2.2.4).

Definition 6.5.2 *If $\mathfrak{B}_1$ and $\mathfrak{B}_2$ are partial $S4$-modal algebras then we call $\mathfrak{B}_1$ a subalgebra of $\mathfrak{B}_2$ (denotation $\mathfrak{B}_1 < \mathfrak{B}_2$) if $\mathfrak{B}_1$ is isomorphically embeddable in $\mathfrak{B}_2$ as a boolean algebra by a mapping f and $\Diamond f(x)$ is defined and $f(\Diamond x) = \Diamond f(x)$ whenever $\Diamond x$ is defined.*

Definition 6.5.3 *Let $\mathfrak{B}$ be a partial $S4$-modal algebra and $\alpha(x_1, ..., x_n)$ be a modal formula built up out of variables $x_1, ..., x_n$. We say α is invalid in $\mathfrak{B}$ (denotation $\mathfrak{B} \not\models \alpha$) if there are some $a_1, ..., a_n \in |\mathfrak{B}|$ such that the following hold. In the term obtained from $\alpha(x_1, ..., x_n)$ by replacing the connectives by the corresponding operations and $x_1, ..., x_n$ by $a_1, ..., a_n$ respectively, the value of all the subterms is defined in $\mathfrak{B}$ and $\alpha(a_1, ..., a_n) \neq \top$.*

Lemma 6.5.4 *If $\mathfrak{B}$ is a modal $S4$-algebra, α is a modal formula and $\mathfrak{B} \not\models \alpha$ then there exists a partial $S4$-modal algebra $\mathfrak{B}_1$ such that $\mathfrak{B}_1 < \mathfrak{B}$, $\mathfrak{B}_1 \not\models \alpha$ and $||\mathfrak{B}|| \leq 2^{2^k}$, where k is the number of subformulas in α.*

Proof. The proof follows immediately from our definitions and local finititude of boolean algebras.

The following lemma is also evident.

Lemma 6.5.5 *If $\mathfrak{B}_1$ and $\mathfrak{B}_2$ are partial $S4$-modal algebras, $\mathfrak{B}_1 < \mathfrak{B}_2$ and $\mathfrak{B}_1 \not\models \alpha$ then $\mathfrak{B}_2 \not\models \alpha$ as well.*

Now we will study certain special properties of partial $S4$-modal subalgebras of modal algebras $\mathcal{U}_n^+$, $n \in N$, where posets $\mathcal{U}_n$ were defined in the previous section. First, we fix the polynomial $P(x, y) := 8(x^2 + 4x + y + 3)$.

Lemma 6.5.6 *If $\mathfrak{B}$ is a partial $S4$-modal subalgebra of the modal algebra $\mathcal{U}_n^+$ and d is a number smaller n and $||\mathfrak{B}|| = 2^k$ then there is some m such that $\mathfrak{B} < \mathcal{U}_m^+$ and $d \leq m \leq P(2^k, d)$.*

Proof. For any subset X of the basis set of the frame $\mathcal{U}_n$, we define $s(X)$, assuming that $sup(\emptyset) = \emptyset)$, as follows:

$$s(X) := \{sup\{x_i \mid x_i \in X\}\} \cup \{sup\{y_i \mid y_i \in X\}\} \cup$$

$$\cup\{sup\{t_i \mid t_i \in X\}\} \cup \{z_j \mid j = max\{i \mid a_i \in X\}\}.$$

It is easy to see that $s(X) \subseteq X$ and $\Diamond_{\mathcal{U}_n} X = X \cup \Diamond_{\mathcal{U}_n} s(X)$, where the index $\mathcal{U}_n$ for $\Diamond$ means the taking of $\Diamond$ in the poset $\mathcal{U}_n$. By our assumption $|\mathfrak{B}| = \{X_1, ..., X_{2^k}\}$, where $X_i \subseteq |\mathcal{U}_n|$. If $X_i \neq X_j$, then there exists $\varepsilon_{i,j}$ such that $\varepsilon_{i,j} \in X_i$ but $\varepsilon_{i,j} \notin X_j$ (or conversely). Consider the following set $\mathcal{Z}$ of elements from $\mathcal{U}_n$:

$$\mathcal{Z} := \{\varepsilon_{i,j} \mid 1 \leq i < j \leq 2^k\} \cup (\bigcup\{s(X) \mid X \in |\mathfrak{B}|\}) \cup$$

$$\cup\{x_0, x_1, ..., x_{d+2}\}.$$

Note that $d \leq ||\mathcal{Z}|| \leq 2^{2*k} + 4 * 2^k + d + 3$. By *a bundle* we will mean a quadruple of elements from $|\mathcal{U}_n|$ of the form $\langle t_j, z_j, x_{j+1}, y_{j+1}\rangle$. We denote the set of all bundles formed from elements of $\mathcal{U}_n$ by $b(\mathcal{U}_n)$. We also put

$$\mathcal{Z}^b := \bigcup\{\mathcal{D} \mid \mathcal{D} \in b(\mathcal{U}_n) \wedge \exists x(x \in \mathcal{D} \wedge x \in \mathcal{Z}))\} \cup \{x_0, y_0\}.$$

Continuing our construction, we define the set

$$\mathcal{Z}^{b+} := \mathcal{Z}^b \cup \{\bigcup\{\mathcal{D} \mid \mathcal{D} \in b(\mathcal{U}_n) \wedge$$

$$\wedge \exists i((x_i \in (\mathcal{D} - \mathcal{Z}^b) \wedge (x_{i-1} \in \mathcal{Z}^b)) \wedge \exists j(j > i \& (x_j \in \mathcal{Z}^b)))\}.$$

It is not hard to see that

$$d \leq ||\mathcal{Z}|| \leq ||\mathcal{Z}^{b+}|| \leq 8 * ||\mathcal{Z}|| \leq 8(2^{2*k} + 4 * 2^k + d + 3) = P(2^k, d).$$

In the set $\mathcal{Z}^{b+}$ we define a relation R_1 as follows:

$$R_2 := (R \cap (\mathcal{Z}^{b+})^2 - T_1 \cup T_2), \text{ where}$$

$$T_1 := \{\langle y_j, x_i\rangle \mid y_j R x_i \& x_i \in (\mathcal{Z}^{b+} - \mathcal{Z}^b)$$
$$\& y_j \in \mathcal{Z}^b \& x_{i+1} \notin \mathcal{Z}^b \& \forall l(y_l \in \mathcal{Z}^b \& y_l R x_i \Rightarrow j \leq l)\},$$
$$T_2 := \{\langle x_j, y_i\rangle \mid x_j R y_i \& y_i \in (\mathcal{Z}^{b+} - \mathcal{Z}^b)$$
$$\& x_j \in \mathcal{Z}^b \& y_{i+1} \notin \mathcal{Z}^b \& \forall l(x_l \in \mathcal{Z}^b \& x_l R y_i \Rightarrow j \leq l)\}.$$

and R_1 is the reflexive transitive closure of R_2. Now it is not hard to see that the frame $\langle \mathcal{Z}^{b+}, R_1\rangle$ is isomorphic to the frame $\mathcal{U}_m$, where $d \leq m \leq P(2^k, d)$. Now we define a mapping $\varphi : \mathfrak{B} \mapsto \mathcal{U}_m^+$ as follows:

$$\varphi(X_i) := X_i \cap \mathcal{Z}^{b+}.$$

Obviously that φ is a boolean homomorphism from $\mathfrak{B}$ into $\mathcal{U}_m^+$. Since $\mathcal{Z}^{b+}$ contains elements $\varepsilon_{i,j}$ for any $X_i, X_j \in |\mathfrak{B}|$, φ is an one-to-one mapping. Thus it remains only to show that $\varphi(\Diamond X_i) = \Diamond \varphi(X_i)$ if $\Diamond X_i$ is defined in $\mathfrak{B}$. Suppose $\Diamond X_i$ is defined, we have to prove that $\Diamond X_i \cap \mathcal{Z}^{b+} = \Diamond_{\mathcal{U}_m}(X_i \cap \mathcal{Z}^{b+})$.

First we will establish that $\Diamond_{\mathcal{U}_m}(X_i \cap \mathcal{Z}^{b+}) \subseteq \Diamond X_i \cap \mathcal{Z}^{b+}$. Suppose that $x \in \Diamond_{\mathcal{U}_m}(X_i \cap \mathcal{Z}^{b+})$. Then $xR_1 y$ and $y \in X_i \cap \mathcal{Z}^{b+}$ for some y. Since $R_1 \subseteq R \cap (\mathcal{Z}^{b+})^2$ and $x, y \in \mathcal{Z}^{b+}$ we have xRy. Consequently $x \in \Diamond X_i$ and $x \in \Diamond X_i \cap \mathcal{Z}^{b+}$.

Now we will show the converse inclusion: $\Diamond X_i \cap \mathcal{Z}^{b+} \subseteq \Diamond_{\mathcal{U}_m}(X_i \cap \mathcal{Z}^{b+})$. If $x \in X_i \cap \mathcal{Z}^{b+}$ then $x \in \Diamond_{\mathcal{U}_m}(X_i \cap \mathcal{Z}^{b+})$. Assume that $x \in \Diamond X_i \cap \mathcal{Z}^{b+}$ and $x \notin X_i$. Then as we have seen above xRy for some $y \in s(X_i)$. If $y = z_j \in s(X_i)$ then $a_j \in \mathcal{Z} \subseteq \mathcal{Z}^{b+}$ and $xR_1 z_j$ since $x, z_j \in \mathcal{Z}^{b+}$, xRz_j and R_1 is obtained from R by removing pairs among which there are none of the form $\langle x, z_j \rangle$. Thus $x \in \Diamond_{\mathcal{U}_m}(X_i \cap \mathcal{Z}^{p+})$. The case $y = t_j \in S(X_i)$ is handled analogously. Suppose $y = x_j \in s(X_i)$. Then $x_j \in \mathcal{Z} \subseteq \mathcal{Z}^b \subseteq \mathcal{Z}^{b+}$. We have xRx_j and the pairs of the form $\langle x, x_j \rangle$ are removed from R in the definition of R_1 only when $x_j \notin \mathcal{Z}_b$. Thus $xR_1 x_j$ and $x \in \Diamond_{\mathcal{U}_m}(X_i \cap \mathcal{A}^{b+})$. The last case when $y = y_j \in S(X_i)$ is handled like the preceding one. Hence provided $\Diamond X_i$ is defined in $\mathfrak{B}$, $\varphi(\Diamond X_i) = \Diamond_{\mathcal{U}_m}(\varphi(X_i))$ holds. Hence we have showed $\mathfrak{B} < \mathcal{U}_m^+$ and $d \leq m \leq P(2^k, d)$ as it has been required. ∎

The following lemma together with the one just proved, will serve the basis of the decision algorithm for the Kripke non-compact logic to be constructed below.

Lemma 6.5.7 *Suppose $\mathfrak{B}$ is a partial S4-modal algebra and $\mathfrak{B} < \mathcal{U}_n^+$, $\|\mathfrak{B}\| = 2^k$ and $n \geq 3k$. Then $\mathfrak{B} < \mathcal{U}_m^+$ for any $m \geq n + 3$.*

Proof. We may consider elements of $\mathfrak{B}$ as subsets of $|\mathcal{U}_n|$. Suppose $X_1, ..., X_k$ are all the atoms of $\mathfrak{B}$. Since $3k \leq n$, there exists an atom X_l such that $x_\alpha, x_\beta, x_\gamma, x_\delta \in X_l$, where $\alpha < \beta < \gamma < \delta$.

We construct the frame $\mathcal{A}_m := \langle A_m, R_1 \rangle$ as follows:

$$A_m := |\mathcal{U}_n| \cup \left(\bigcup \{ \{a_{j-1}, b_j, c_j) \mid 1 \leq j \leq m - n \} \right),$$

$R_0 := R \cup \left(\bigcup \{ \{\langle a_{j-1}, b_j \rangle, \{\langle a_{j-1}, c_j \rangle \} \mid 1 \leq j \leq m - n \} \right) \cup$
$\cup \{ \langle b_{j+1}, b_j \rangle \mid 1 \leq j \leq m - n - 1 \}$
$\cup \{ \langle c_{j+1}, c_j \rangle \mid 1 \leq j \leq m - n - 1 \}$
$\cup \{ \langle b_{j+2}, c_j \rangle \mid 1 \leq j \leq m - n - 2 \}$
$\cup \{ \langle c_{j+2}, b_j \rangle \mid 1 \leq j \leq m - n - 2 \}$
$\cup \{ \langle b_1, x_\beta \rangle, \langle c_1, y_\beta \rangle, \langle b_1, y_{\beta-1} \rangle, \langle c_1, x_{\beta-1} \rangle, \}$
$\cup \{ \langle b_2, y_\beta \rangle, \langle c_2, x_\beta \rangle, \langle x_{\beta+1}, b_{m-n} \rangle, \langle y_{\beta+1}, c_{m-n} \rangle, \} \cup$
$\cup \{ \langle y_{\beta+2}, b_{m-n} \rangle, \langle x_{\beta+2}, c_{m-n} \rangle, \langle x_{\beta+1}, c_{m-n-1} \rangle, \langle y_{\beta+1}, b_{m-n-1} \rangle \},$

and R_1 is the reflexive transitive closure of R_0. Thus $R_1 \cap |\mathcal{U}_n|^2 \supseteq R$. It is also easy to see that

$$R_1 \cap |\mathcal{U}_n|^2 = R \cup \{\langle y_{\beta+1}, x_\beta \rangle, \langle x_{\beta+1}, y_\beta \rangle\}.$$

It is also not hard to see that $\mathcal{A}_m$ differs from $\mathcal{U}_m$ only by the absence of the elements $t_{\beta+1}, ..., t_{\beta+m_n}$.

We introduce a mapping $\varphi : \mathfrak{B} \mapsto \mathcal{A}_m^+$ as follows:

$$\varphi(X) := \begin{cases} X \cup D & \text{if } X_l \subseteq X, \\ X & \text{if } X_l \not\subseteq X \end{cases}$$

where $D := \bigcup \{\{a_{j-1}, b_j, c_j\} \mid 1 \le j \le m_n\}$. Since $D \cap |\mathcal{U}_n| = \emptyset$, it follows that φ is a one-to-one mapping. Since $\{X \mid X \in \mathfrak{B} \& X_l \subseteq X\}$ is an ultrafilter in $\mathfrak{B}$, the mapping φ is a boolean isomorphism of $\mathfrak{B}$ into $\mathcal{A}_m^+$. We will prove that $\varphi(\Diamond X) = \Diamond_{\mathcal{A}_m}(\varphi(X))$ if $\Diamond X$ is defined in $\mathfrak{B}$. We consider the following three cases.

(a) $X_l \subseteq X$. In this case $X_l \subseteq \Diamond X$. So we have $\varphi(\Diamond X) = \Diamond X \cup D$, $\varphi(X) = X \cap D$. We will show that $\Diamond X \cup D = \Diamond_{\mathcal{A}_m}(X \cup D)$. Since $R_1 \cap |\mathcal{U}_n|^2 \supseteq R$, it follows that $\Diamond X \cup D \subseteq \Diamond_{\mathcal{A}_m}(X \cup D)$. We will prove the converse inclusion. Suppose $x \in \Diamond_{\mathcal{A}_m}(X \cup D)$ and $x \notin X \cup D$. Then $xR_1y, y \in X \cup D$. If $y \in D$ then xR_1b_1 or xR_1c_1. Therefore xR_1x_β or $xR_1x_{\beta-1}$. Hence xR_1x_α, but $x_\alpha \in X_l$ and $X_l \subseteq X$. Consequently $x \in \Diamond X \cup D$ since in $(R_1 \cap |\mathcal{U}_n|^2 - R)$ there are no pairs of the form $\langle u, b_\alpha \rangle$. Now assume that $y \in (X - D)$. If $\langle x, y \rangle \notin R$ then $y = x_\beta$ or $y = y_\beta$ and correspondingly $x = y_{\beta+1}$ or $x = x_{\beta+1}$. Moreover, $x_\alpha \in X_l \subseteq X$ and $x_{\beta+1}Rx_\alpha, y_{\beta+1}Rx_\alpha$. Therefore $x \in \Diamond X \cup D$. Thus $\Diamond X \cup D = \Diamond_{\mathcal{A}_m}(X \cup D)$.

(b) $X_l \not\subseteq X$ and $X_l \subseteq \Diamond X$. Then $\varphi(X) := X$, $\varphi(\Diamond X) = \Diamond X \cup D$. We have to show that $\Diamond X \cup D = \Diamond_{\mathcal{A}_m}(X)$. Indeed, $\Diamond X \subseteq \Diamond_{\mathcal{A}_m}(X)$. Moreover $D \subseteq \Diamond_{\mathcal{A}_m}(X)$ since $x_\alpha \in \Diamond X$ (because $X_l \subseteq \Diamond X$) and, for any $g \in D$, it follows gR_1x_α. Thus $\Diamond X \cup D \subseteq \Diamond_{\mathcal{A}_m}(X)$. To prove the converse inclusion suppose xR_1y and $y \in X$. If $x \in D$ then $x \in \Diamond X \cup D$. Assume that $x \notin D$. Then $x, y \in |\mathcal{U}_n|$ holds and xR_1y. If $\langle x, x \rangle \notin R$ then $x = x_{\beta+1}$ or $x = y_{\beta+1}$. But $x_\alpha \in \Diamond X$ and $x_{\beta+1}Rx_\alpha, y_{\beta+1}Rx_\alpha$. Consequently $x \in \Diamond X \cup D$.

(c) $X_l \not\subseteq \Diamond X$. We have $\varphi(X) = X$, $\varphi(\Diamond X) = \Diamond X$. Since $R_1 \cup |\mathcal{U}_n|^2 \supseteq R$, it follows that $\Diamond X \subseteq \Diamond_{\mathcal{A}_m}(X)$. We need also to show the converse inclusion. First we show that $D \cap \Diamond_{\mathcal{A}_m}(X) = \emptyset$. Indeed, if xR_1y and $x \in D, y \in X$ then $b_{m-n}R_1y$ or $c_{m_n}R_1y$. But then $x_\delta R_1y$ and $\langle x_\delta, u \rangle \notin (R_1 \cap |\mathcal{U}_n|^+ - R)$ for any u, i.e. $x_\delta Ry$ and $y \in X$. Hence $x_\delta \in \Diamond X$, which contradicts $X_l \not\subseteq \Diamond X_l$, inasmuch as $\Diamond X$ is defined in $\mathfrak{B}$ and X_l is an atom of $\mathfrak{B}$.

Now suppose that xR_1y and $y \in X$, then, since $D \cap \Diamond_{\mathcal{A}_m}(X) = \emptyset$, it follows $x \notin D$. If $\langle x, y \rangle \notin R$ then $y = x_\beta$ or $y = y_\beta$. But then $x_\delta Ry$, hence $x_\beta \in \Diamond X$ which, as before, contradicts $X_l \not\subseteq \Diamond X$. Thus xRy and $y \in \Diamond X$. Hence $\Diamond_{\mathcal{A}_m}(X) = \Diamond X$.

Thus we have proved that φ is an isomorphic embedding of the partial $S4$-algebra $\mathfrak{B}$ into $\mathcal{A}_m^+$.

To continue our proof we introduce the frame $\mathcal{B}_m$ as follows: $\mathcal{B}_m := \langle B_m, R_2 \rangle$, where

$$B_m := A_m \cup \{t_1^1, ..., t_{m-n}^1\},$$

$$R_3 := R_1 \cup \{\langle t_{i-1}^1, t_i^1 \rangle \mid 2 \le i \le m_n\} \cup \{\langle t_\beta, t_1^1 \rangle\} \cup \{\langle t_{m-n}^1, t_{\beta+1} \rangle\} \cup \{\langle t_i^1, a_{i-1} \rangle \mid 1 \le i \le m - n\},$$

and R_2 is the reflexive transitive closure of R_3.

It is easy to see that $\mathcal{B}_m$ is isomorphic to the poset $\mathcal{U}_m$ and that $R_2 \cap A_m^2 = R_1$. We define the mapping f of the partial $S4$ algebra $\varphi(\mathfrak{B}) < \mathcal{A}_m^+$ into the modal algebra $\mathcal{B}_m^+$ as follows:

$$f(X) := \begin{cases} X \cup F & \text{if } t_{\beta+1} \in X \\ X & \text{if } t_{\beta+1} \notin X \end{cases}$$

where $F := \{t_1^1, ..., t_{m-n}^1\}$. It is easy to see that f is a boolean isomorphism from the partial algebra $\varphi(\mathfrak{B})$ into the modal algebra $\mathcal{B}_m^+$. We will show that $f(\Diamond_{\mathcal{A}_m}(X)) = \Diamond_{\mathcal{B}_m}(f(X))$ for any X if the value of $\Diamond_{\mathcal{A}_m}(X)$ is defined in $\varphi(\mathfrak{B})$. The proof is divided into three cases.

(d) $t_{\beta+1} \in X$. In this case we have $f(X) = X \cup F$ and, also, since $t_{\beta+1} \in X \subseteq \Diamond_{\mathcal{A}_m}(X)$, it follows $f(\Diamond_{\mathcal{A}_m}(X)) = \Diamond_{\mathcal{A}_m}(X) \cup F$. We have to show that $\Diamond_{\mathcal{B}_m}(X \cup F) = \Diamond_{\mathcal{A}_m}(X) \cup F$. Obviously, $\Diamond_{\mathcal{A}_m}(X) \cup F \subseteq \Diamond_{\mathcal{B}_m}(X \cup F)$. To show the converse inclusion suppose $x R_2 y$, $y \in X \cup F$. If $x \in F$ then $x \in \Diamond_{\mathcal{A}_m}(X) \cup F$. If $x \notin F$ then $x \in A_m$. If $y \in F$ then $x = t_l$, where $l \le \beta$. Hence $x R_1 t_{\beta+1}$, which, in view of $t_{\beta+1} \in X$ implies $x \in \Diamond_{\mathcal{A}_m}(X) \cup F$. If $y \notin F$ then $x R_2 y$ implies $x R_1 y$. Thus again $x \in \Diamond_{\mathcal{A}_m}(X) \cup F$. Consequently $\Diamond_{\mathcal{B}_m}(X \cup F) = \Diamond_{\mathcal{A}_m}(X) \cup F$.

(e) $t_{\beta+1} \notin X$, $t_{\beta+1} \in \Diamond_{\mathcal{A}_m}(X)$. In this case the following holds

$$f(X) = X, f(\Diamond_{\mathcal{A}_m}(X)) = \Diamond_{\mathcal{A}_m}(X) \cup F.$$

We must prove that $\Diamond_{\mathcal{B}_m}(X) = \Diamond_{\mathcal{A}_m}(X) \cup F$. Since $t_{\beta+1} \in \Diamond_{\mathcal{A}_m}(X)$, it follows $t_{\beta+1} \in \Diamond_{\mathcal{B}_m}(X)$, which implies $F \subseteq \Diamond_{\mathcal{B}_m}(X)$. Moreover, $\Diamond_{\mathcal{A}_m}(X) \subseteq \Diamond_{\mathcal{B}_m}(X)$. Thus $\Diamond_{\mathcal{A}_m}(X) \cup F \subseteq \Diamond_{\mathcal{B}_m}(X)$. To prove the converse inclusion suppose that $x \in \Diamond_{\mathcal{B}_m}(X)$. If $x \in F$ then $x \in \Diamond_{\mathcal{A}_m}(X) \cup F$. If $x \notin F$ and $x R_2 y$, where $y \in X$, then $x R_1 y$, hence $x \in \Diamond_{\mathcal{A}_m}(X) \cup F$. Thus $\Diamond_{\mathcal{B}_m}(X) = \Diamond_{\mathcal{A}_m}(X) \cup F$.

(f) $t_{\beta+1} \notin X$, $t_{\beta+1} \notin \Diamond_{\mathcal{A}_m}(X)$. In this case $f(X) = X$, $f(\Diamond_{\mathcal{A}_m}(X)) = \Diamond_{\mathcal{A}_m}(X)$ holds. We have to prove that $\Diamond_{\mathcal{B}_m}(X) = \Diamond_{\mathcal{A}_m}(X)$. By definition of R_2 it follows that $\Diamond_{\mathcal{A}_m}(X) \subseteq \Diamond_{\mathcal{B}_m}(X)$. We must show the converse inclusion. We claim that

$$\Diamond_{\mathcal{B}_m}(X) \cap F = \emptyset. \tag{6.38}$$

Indeed, otherwise $x R_2 y$, $y \in X$, $x \in F$. Since $t_{\beta+1} \notin \Diamond_{\mathcal{A}_m}(X)$, it follows that $\{t_{\beta+1}, ..., t_{n+1}\} \cap X = \emptyset$ and $\{z_{\beta+1}, ..., z_{n+1}\} \cap X = \emptyset$, and also, for all i, j,

$x_i, b_j, y_i, c_j \notin X$, consequently (as $x R_2 y$ and $x \in F$) $y = a_i$. Therefore $X \cap D \neq \emptyset$. But $X = \varphi(Y)$ for some $Y \subseteq |\mathcal{U}_n|$, and by the definition of φ, either $\varphi(Y) = Y \cup D$ or $\varphi(Y) = Y$, where $Y \cap D = \emptyset$. Therefore $D \subseteq X$ which yields $b_1 \in X$ and $t_{\beta+1} \in \Diamond_{\mathcal{A}_m}(X)$ which contradicts the hypothesis of the case (f). Thus (6.38) holds.

Now suppose that $x R_2 y$ and $y \in X$. In view of (6.38), it follows that $x \in A_m$ and since $R_2 \cap A_m^2 = R_1$ we arrive at $x R_1 y$. Hence $x \in \Diamond_{\mathcal{A}_m}(X)$ and we showed $\Diamond_{\mathcal{B}_m}(X) = \Diamond_{\mathcal{A}_m}(X)$.

Thus we have proved that f is an isomorphism of the partial $S4$-modal algebra $\varphi(\mathfrak{B})$ into $\mathcal{B}_m^+$. Finally, summarizing the shown above inclusions we conclude

$$\mathfrak{B} \cong \varphi(\mathfrak{B}) < \mathcal{A}_m^+,$$

$$f(\varphi(\mathfrak{B})) < \mathcal{B}_m^+ \cong \mathcal{U}_m^+,$$

i.e. $\mathfrak{B} < \mathcal{U}_m^+$ as it was required. ∎

We introduce the modal logic λ_3 as follows:

$$\lambda_{3,n} := \lambda(\{\mathcal{U}_n \mid m \geq n\}) \text{ and}$$

$$\lambda_3 := \bigcup_{n \geq 1} \lambda_{3,n}.$$

Lemma 6.5.8 *The modal logic λ_3 is Kripke non-compact and extends the modal logic Grz.*

Proof. In the previous section we introduced formulas $\xi_n, m \geq 0, \chi_1, \psi, \phi$. In Lemma 6.4.18 it has been shown that for any n,

$$\mathcal{U}_n \Vdash \chi_1 \wedge \psi \wedge \bigwedge_{1 \leq i \leq n} \xi_i \text{ and } \mathcal{U}_n \nVdash_V \neg\phi.$$

Consequently $\lambda_3 \supseteq Grz = S4 \oplus \chi_1$, and, for any n, $\neg\phi \notin \lambda_{3,n}$. Therefore $\neg\phi \notin \lambda_3$. Furthermore, $\chi_1 \wedge \psi \wedge \bigwedge_{1 \leq i \leq n} \xi_i \in \lambda_3$ for any n. By Lemma 6.4.19, $\neg\phi$ cannot be separated from the set $\{\chi_1, \psi\} \cup \{\xi_i \mid i \in N\}$ by a Kripke frame. In particular, we conclude that $\neg\phi$ cannot be separated from λ_3 by a Kripke frame (hence λ_3 is Kripke incomplete).

Assume that $X \subseteq \lambda_3$ is a finite set. Since $\lambda_3 = \bigcup_{1 \leq n} \lambda_{3,n}$ and $\lambda_{3,m} \subseteq \lambda_{3,m+1}$, it follows that there exists a certain m such that $X \subseteq \lambda_{3,m}$. Thus $\mathcal{U}_m \Vdash X$ and, as it has been shown, $\mathcal{U}_n \nVdash \neg\phi$. Thus $\neg\phi$ can be separated by Kripke frames from any finite subset X of the logic λ_3 but cannot be separated by frames from λ_3 itself. Thus λ_3 is Kripke non-compact ∎

Lemma 6.5.9 *The logic λ_3 is decidable.*

Proof. Suppose α is a formula which has k subformulas. We claim that

$$\alpha \notin \lambda_3 \Leftrightarrow \exists i \in N[(3 * 2^k \leq i \leq P(2^{2^k}, 3 * 2^k)) \& (U_i \nVdash \alpha)]. \tag{6.39}$$

To prove this relation, we first suppose that $\alpha \notin \lambda_3$. Then, for any n, there is some $m \geq n$ such that $U_m \nVdash \alpha$. In particular, there is a certain $m \geq 3 * 2^k$ such that $U_m \nVdash \alpha$. Then $U_m^+ \nvDash \alpha$ and by Lemma 6.5.4 there exists a partial $S4$-modal algebra $\mathfrak{B}$ such that $\mathfrak{B} < U_m^+$, $\mathfrak{B} \nvDash \alpha$ and $||\mathfrak{B}|| \leq 2^{2^k}$. Using Lemma 6.5.6 it follows that there exists i such that $\mathfrak{B} < U_i^+$ and $3 * 2^k \leq i \leq P(2^{2^k}, 3 * 2^k)$. By Lemma 6.5.5 we derive $U_i^+ \nvDash \alpha$ and $U_i \nVdash \alpha$. Thus our assertion is proved in the one direction.

Now suppose $3 * 2^k \leq i \leq P(2^{2^k}, 3 * 2^k)$ and $U_i \nVdash \alpha$. Then $U_i^+ \nvDash \alpha$ and by Lemma 6.5.4 there exists a partial $S4$-modal algebra $\mathfrak{B}$ such that $\mathfrak{B} < U_i^+$, $\mathfrak{B} \nvDash \alpha$ and $||\mathfrak{B}|| \leq 2^{2^k}$. By Lemma 6.5.7 $\mathfrak{B} < U_m^+$ for any $m \geq i+3$. Therefore, for any $m \geq i+3, U_m^+ \nvDash \alpha$ by Lemma 6.5.5. Thus, for any $m \geq i+3, U_m \nVdash_V \alpha$. Therefore $\alpha \notin \lambda_{3,n}$ for any n, and consequently $\alpha \notin \lambda_3$. Thus (6.39) holds, and this equivalence yields that λ_3 is decidable. ∎

It is easily seen that the offered in the lemma above deciding algorithm for λ_3 is primitively recursive, i.e. it has the effective upper bound for the number of steps by the length of a given for the test formula. Immediately from Lemmas 6.5.8 and 6.5.9 we derive the main result of this section:

Theorem 6.5.10 *(V.V.Rybakov, 1978, [131]) The modal logic λ_3 is Kripke non-compact but decidable.*

Thus, in particular, λ_3 do not have the fmp, is Kripke incomplete, and has no a finite axiomatization, but nevertheless is decidable. Our construction of the logic λ_3 shows also that the union of a certain strictly ascending chain of modal logics with the fmp can have negative properties: lack of the fmp, lack of Kripke completeness, and even lack of Kripke compactness.

6.6 Non-Compact Superintuitionistic Logics

In this section we describe an example of Kripke non-compact (and, consequently, Kripke incomplete) superintuitionistic logic. This example was found by V.Shehtman [173]. By induction we introduce the following sequence of propositional formulas:

$$\beta_0 := \neg(p \wedge q), \quad \gamma_0 := \neg(\neg p \wedge q),$$

$$\beta_1 := \gamma_0 \to \beta_0 \vee q, \quad \gamma_1 := \beta_1 \to \gamma_0 \vee q,$$

$$\beta_{n+1} := \gamma_n \to \beta_n \vee \gamma_{n-1},$$

$$\gamma_{n+1} := \beta_n \to \gamma_n \vee \beta_{n-1},$$

$$\alpha_n := \beta_{n+2} \wedge \gamma_{n+2} \to \beta_{n+1} \vee \gamma_{n+1},$$

$$\mu_n := \alpha_n \vee \beta_{n+2}, \quad \delta_n := (\alpha_n \to \mu_{n+1}) \to \mu_n.$$

Let

$$\beta r_2 := \bigwedge_{0 \leq i \leq 2} ((p_i \to \bigvee_{j \neq i} p_j) \to \bigvee_{j \neq i} p_j) \to \bigvee_{0 \leq i \leq 2} p_i.$$

Lemma 6.6.1 *Let $\mathcal{F}$ is a poset such that, for any $a \in |\mathcal{F}|$, the open subframe generated by a is finite. Then $\mathcal{F} \Vdash \beta r_2$ if any element from $\mathcal{F}$ has a branching of not more than* 2.

Proof. The proof is a straightforward verification and is left to the reader as a simple exercise.

We define the superintuitionistic logic λ_1 as follows:

$$\lambda_1 := H \oplus \{\beta r_2\} \cup \{\delta_n \mid 0 \leq n\}.$$

Theorem 6.6.2 *The superintuitionistic logic λ_1 is Kripke non-compact.*

Proof. We need several preliminary lemmas.

Lemma 6.6.3 *For any λ_1-frame $\mathcal{F}$, $\mathcal{F} \Vdash_V \mu_0$.*

Proof. We pick any poset $\mathcal{F}$ such that $\mathcal{F} \Vdash \delta_n$ for all $n \leq 0$. Suppose that there is a valuation V which invalidates μ_0 on some element v_0. In order to prove our lemma it is sufficient to show that $\mathcal{F} \nVdash \beta r_2$. We consider the following sets

$$V_n := |\mathcal{F}| - V(\bigwedge_{m < n} \alpha_m \to \mu_n)$$

$$T_n := |\mathcal{G}| - V(\bigwedge_{m < n} \alpha_m \wedge \beta_{n+2}\gamma_{n+2} \to \beta_{n+1} \vee \gamma_{n+1}),$$

where any empty conjunction is understood as usually, that is as $\top$. We need to show that

$$V_n \subseteq \Diamond V_{n+1} := \{a \mid a \in |\mathcal{F}|, \exists b \in V_{n+1}(a \leq b)\}. \tag{6.40}$$

Indeed if $x \in V_n$ then $x \nVdash_V \mu_n$ and, since $\mathcal{F} \Vdash \delta_n$, it follows that $x \nVdash_V \alpha_n \to \mu_{n+1}$, i.e. there is an element y such that $x \leq y$ and $y \Vdash_V \alpha_n$ and $y \nVdash_V \mu_{n+1}$. Since $x \in V_n$, we have $y \Vdash_V \bigwedge_{m<n} \alpha_n$. Therefore $y \in V_{n+1}$.

$$V_n \subseteq \Diamond T_n. \tag{6.41}$$

In fact, if $x \in V_n$ then $x \nVdash_V \alpha_n$ which entails that there is some y such that $x \leq y$ and $v \Vdash_V \beta_{n+1} \wedge \gamma_{n+1}$, and $y \nVdash_V \beta_{n+1} \vee \gamma_{n+1}$. However we have $x \Vdash_V \bigwedge_{m<n} \alpha_m$, therefore $y \in T_n$.

For any $x \in |\mathcal{F}|$, we define $N(x) := \{n \mid x \in \Diamond T_n\}$.

$$\forall m, k((m, k \in N(x)) \& (k < m) \Rightarrow \exists n((n \leq k) \& x \in V_n)). \tag{6.42}$$

In fact, if the premise holds then the set $\{l \mid x \nVdash_V \alpha_l\}$ is non-empty since $x \nVdash_V \alpha_k$. Let n be the smallest element of the set $\{l \mid x \nVdash_V \alpha_l\}$. Then $x \Vdash_V \bigwedge_{i<n} \alpha_i$ and $x \nVdash_V \alpha_n$. Besides $x \nVdash_V \beta_{m+1}$, and since $n+2 \leq k+2 \leq m+1$, we conclude $x \nVdash_V \beta_{n+2}$, consequently $x \in V_n$.

$$k \leq n \Rightarrow T_n \cap V_k = \emptyset. \tag{6.43}$$

Indeed, by (6.40) $V_n \subseteq \Diamond V_{n+1}$. Moreover, we have $T_n \cap \Diamond V_{n+1} = \emptyset$ since $x \in \Diamond V_{n+1}$ implies $x \nVdash_V \beta_{n+3}$.

Let $T_n^+ := \{x \mid x \in |\mathcal{F}|, N(x) := \{n\}\}$.

$$T_n \subseteq T_n^+. \tag{6.44}$$

To show this suppose $x \in T_n$. Then $n \in N(x)$ and either $N(x) := \{x\}$ or $x \in V_k$ for some $k \leq n$ by (6.42). However the latter is impossible since (6.43).

$$\Diamond T_n = \Diamond T_n^+. \tag{6.45}$$

This follows directly from $T_n \subseteq T_n^+ \subseteq \Diamond T_n$.

$$m \neq n \Rightarrow \Diamond T_n^+ \cap T_m^+ = \emptyset \tag{6.46}$$

follows immediately from (6.45). We need also to show that

$$\forall m, n(V_m \cap T_n^+ = \emptyset). \tag{6.47}$$

In fact, by (6.41) $V_{m+1} \subseteq \Diamond T_{m+1}$, and by (6.40) $V_m \subseteq \Diamond V_{m+1}$. Consequently $V_m \subseteq \Diamond V_{m+1} \subseteq \Diamond T_{m+1}$. Using (6.45) we derive $\Diamond T_{m+1} = \Diamond T_{m+1}^+$. If $n \neq m+1$ it follows that $V_m \cap T_n^+ \subseteq \Diamond T_{m+1}^+ \cap T_n^+ = \emptyset$ using (6.46). If $n = m + 1$ then $V_m \subseteq \Diamond T_m$ using (6.41) and (6.45) $\Diamond T_m = \Diamond T_m^+$. Therefore $V_m \cap T_{m+1}^+ \subseteq \Diamond T_m^+ \cap T_{m+1}^+ = \emptyset$ applying (6.46).

$$V_m \cap \Diamond V_n = \emptyset \text{ for } m < n. \tag{6.48}$$

Indeed, if $x \in V_m$ then $x \nVdash_V \alpha_m$, and if $x \in V_n$ then $x \Vdash_V \alpha_m$.

Now consider the set $\mathcal{Z} := \{x \mid x \in |\mathcal{F}|, N(x) = \emptyset.\}$. Using proved above relations (6.40) - (6.48) it follows directly that V is a disjoint union of sets $\mathcal{Z}$, V_n and T_n^+, $n \in N$ (some of them can be empty). Note also that all the sets $T_n^+ \cup \mathcal{Z}$

and $\mathcal{Z}$ are open (generated) subsets of $\mathcal{F}$. Using this observation we can define the following intuitionistic valuation H on $\mathcal{F}$:

$$H(p_0) := \mathcal{Z} \cup \bigcup_{0 \leq n} \mathcal{Z} \cup T_{3n}^+,$$
$$H(p_1) := \mathcal{Z} \cup \bigcup_{0 \leq n} \mathcal{Z} \cup T_{3n+1}^+,$$
$$H(p_2) := \mathcal{Z} \cup \bigcup_{0 \leq n} \mathcal{Z} \cup T_{3n+2}^+.$$

We will show that $v_0 \in (|\mathcal{F}| - H(\beta r_2))$. Indeed, by (6.40) and (6.41) it follows that $v_0 \in V_0 \subseteq \Diamond T_0 \cup \Diamond T_1 \cup \Diamond T_2$. Consequently $v_0 \not\Vdash_V \bigvee_{0 \leq i \leq 2} p_i$. To show that

$$\forall x \in |\mathcal{F}|, x \Vdash_H \bigwedge_{0 \leq i \leq 2} ((p_i \to \bigvee_{j \neq i} p_j) \to \bigvee_{j \neq i} p_j)$$

consider the case when $i = 0$. If $x \not\Vdash_H p_1 \vee p_2$ it follows that

$$x \notin \mathcal{Z} \cup \bigcup_{0 \leq n} T_{3n+1}^+ \cup \bigcup_{0 \leq n} T_{3n+2}^+.$$

Consequently $x \in \bigcup_{0 \leq n}(T_{3n}^+ \cup V_n)$. If $x \in T_{3n+}$ it follows that $x \not\Vdash_H (p_0 \to p_1 \vee p_2)$. If $x \in V_n$ we pick a certain k, where $3k \geq n$. Then $x \in \Diamond T_{3k}^+$ (by (6.40), (6.41) and (6.45)), and again $x \not\Vdash_V p_0 \to p_1 \vee p_2$. Cases $i := 1, 2$ can be considered in a similar way. Thus $v_0 \not\Vdash_H \beta r_2$, which completes the proof. ∎

We fix the Fine's frame $\mathcal{F}_f$ (see K.Fine [45]), which is the following poset: $\mathcal{F}_f := \langle W, \leq \rangle$,

$$W := \bigcup_{0 \leq n} \{a_n, b_n, c_n, d_n\},$$

$$b_1 \leq b_0, c_1 \leq c_0, b_{n+1} \leq b_n, b_{n+1} \leq c_{n-1},$$
$$c_{n+1} \leq c_n, c_{n+1} \leq b_{n-1}, a_n \leq b_{n+1}, a_n \leq c_{n+1},$$
$$d_n \leq a_n, d_n \leq d_{n+1}.$$

For convenience a fragment of the frame $\mathcal{F}_f$ is depicted in Fig. 6.4 below. Using the frame $\mathcal{F}_f$ we define a sequence of frames $\mathcal{F}_{f,n}$, $n \in N$ as follows: $\mathcal{F}_{f,n}$ is the poset obtained from $\mathcal{F}_f$ by erasing of all elements d_m for $m \geq n+2$.

Lemma 6.6.4 $\forall m, m \leq n \Rightarrow \mathcal{F}_{f,n} \Vdash \delta_m$.

Proof. We pick out an arbitrary valuation V in $\mathcal{F}_{f,n}$. Suppose that there is an element x of $\mathcal{F}_{f,n}$ such that $x \not\Vdash_V \mu_m$. Then it is not hard to see that there are certain elements of this frame which disprove formulas $\beta_0, ..., \beta_{m+1}$ and $\gamma_0, ..., \gamma_{m+1}$ under V. Furthermore, for any formula from the list $\beta_0, ..., \beta_{m+1}$ or the list $\gamma_0, ..., \gamma_{m+1}$, there is an element on which this formula is valid under V.

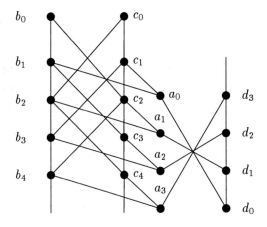

Figure 6.4: A fragment of K.Fine's frame $\mathcal{F}_f$.

In particular, for some x, $x \Vdash_V \beta_0$. Then $x \leq y$, $y \Vdash_V p \wedge q$. Consequently either $b_0 \Vdash_V p \wedge q$ or $c_0 \Vdash_V p \wedge q$. Similarly, starting with γ_0, we show that $b_0 \Vdash_V \neg p \wedge q$ or $c_0 \Vdash_V \neg p \wedge q$. Therefore only two following possibilities can take place. First one is:

(i) $b_0 \Vdash_V p \wedge q$, $c_0 \Vdash_V \neg p \wedge q$.

Then by induction on k we will show that, for any k,

$$V(\beta_k) = (|\mathcal{F}_{f,n}| - \Diamond\{b_k\}),$$

$$V(\gamma_k) = (|\mathcal{F}_{f,n}| - \Diamond\{c_k\}).$$

The basis of induction: clearly $V(\beta_0) = \{c_0, c_1\}$, what we needed. Similarly, we infer $V(\gamma_0) = \{b_0, b_1\}$, which is again what we needed. Now consider the case $k = 1$. Since $x \nVdash \mu_m$ we get $x \nVdash_V \beta_{m+2}$, and consequently there exists a z such that $z \nVdash_V \beta_1$. Suppose that $x \nVdash_V \beta_1 := \gamma_0 \rightarrow \beta_0 \wedge q$. Then $x \leq y$ with $y \Vdash_V \gamma_0$ and $y \nVdash_V \beta_0 \wedge q$. Then by the inductive hypothesis we have $y \not\leq c_0$ and $y \leq b_0$, but $y \nVdash_V q$, i.e. $y \neq b_0$, consequently $y = b_1$. Thus $V(\beta_1) = (|\mathcal{F}_{f,n}| - \Diamond\{b_1\})$. That $x \nVdash \mu_m$ yields $x \nVdash_V \alpha_m$ which implies there exists a z such that $z \nVdash_V \gamma_1$. Suppose that $x \nVdash_V \gamma_1 := \beta_0 \rightarrow \gamma_0 \wedge q$. This entails $x \leq y$ with $y \Vdash_V \beta_0$ and $y \nVdash_V \gamma_0 \wedge q$ and by inductive hypothesis $y \not\leq b_0$ and $y \leq c_0$, but $y \nVdash_V q$, i.e. $y \neq c_0$, consequently $y = c_1$. Thus $V(\gamma_1) = (|\mathcal{F}_{f,n}| - \Diamond\{c_1\})$. The inductive steps are quite similar to the proof above of the case $k = 1$, and by means of such reasoning we obtain the required result.

Second case is:

(ii) $c_0 \Vdash_V p \wedge q$, $b_0 \Vdash_V \neg p \wedge q$.

In this case, quite parallel to the case (i) we show that in the case (ii) the equalities similar to the case (i) hold for any k.

Using the observations concerning $V(\beta_k)$ and $V(\gamma_k)$ in cases (i) and (ii), it is not hard to see that $V(\alpha_k) = (|\mathcal{F}_{f,n}| - \Diamond\{a_k\})$ in both cases (i) and (ii) for any k. Thus, for any $x \in |\mathcal{F}_{f,n}|$, the relation $x \nVdash_V \mu_m$ implies that $x \leq d_m$. Since $d_{m+1} \in \mathcal{F}_{f,n}$ and $d_m \leq d_{m+1}$ it follows that $x \nVdash_V (\alpha_m \to \mu_{m+1})$. ∎

Lemma 6.6.5 *For any n, there exists a poset $\mathcal{F}$ which detaches the formula μ_0 from the logic $H \oplus \{\delta_m \mid m \leq n\} \cup \{\beta r_2\}$.*

Proof. By Lemma 6.6.4 it is sufficient to show that $\mathcal{F}_{f,n} \nVdash_V \mu_0$ for any n. We take the valuation $V(p) := \{b_0\}$, $V(q) := \{b_0, c_0\}$. Quite similarly to the proof of Lemma 6.6.4 we derive that, for any k, $V(\beta_k) = (|\mathcal{F}_{f,n}| - \Diamond\{b_k\})$, $V(\gamma_k) = (|\mathcal{F}_{f,n}| - \Diamond\{c_k\})$, $V(\alpha_k) = (|\mathcal{F}_{f,n}| - \Diamond\{a_k\})$. Therefore we conclude that $d_0 \nVdash_V \mu_0$. ∎

Immediately from this lemma together with Lemma 6.6.2 we conclude that λ_1 is Kripke non-compact superintuitionistic logic. ∎

Index

Bibliography

[1] ARTEMOV S.N. Modal Logics Axiomatizing Provability. *Izvestiya Akad. Nauk SSSR Ser. Math.*, V.49, 1985, 1123-1154 (in Russian, English transl. in *Math. USSR Izvestiya*, 27(1986)).

[2] ARTEMOV. S.N., DZHAPARIDZE G. Finite Kripke Models and Predicate Logics of Provability. *J. of Symbolic Logic*, V.55, 1990, No. 3, 1090-1098.

[3] ARTEMOV S., BEKLEMISHEV L. On Propositional Quantifiers in Provability Logic. Steklov Mathematical Institute, Moscow, 1992, Preprint, LCS-92-1.

[4] ARTEMOV S.N. Logic of Proofs. *Annals of Pure and Applied Logic*, V.69, 1994, 29-59.

[5] BAKER K. Finite Equational Basis for Finite Algebras a Congruence-Distributive Equational Class. *Advanced Mathematics*, V.24, 1977, No.3, 207-243.

[6] BALBES R., HORN A. Injective and Projective Heyting Algebras I, *Trans. Amer. Math. Society*, V.148, 1970, No.2, 549-559.

[7] BEKLEMISHEV L.D. Provability Logics for Natural Turing Progressions of Arithmetical Theories, *Studia Logica*, V.50, 1991, 107-128.

[8] BEKLEMISHEV L.D. On Bimodal Logics of Provability, *Annals of Pure and Applied Logic*, V.68, 1994, 115-159.

[9] BELLISSIMA F. Finitely Generated Free Heyting Algebras. *Journal of Symbolic Logic*, V.51, 1986, No.1, 152-165.

[10] BELLISSIMA F. On the Lattice of Extensions of the Modal Logic $K.Alt_n$. *Arch. Math. Logic*, V.27, 1988, 107-114.

[11] BELLISSIMA F. Atoms of Tense Algebras. *Algebra Universalis*, V.28, 1991, 52-78.

[12] BLOK W.J. Varieties of Interior Algebras. *Ph.D Dissertation*, University of Amsterdam, 1976.

[13] BLOK W.J. The Free Closure Algebra on Finitely Many Generators. *Indagationes Mathematicae*, V.39, 1977, No.5, 362 - 379.

[14] BLOK W.J. Pretabular Varieties of Modal Algebras. *Studia Logica*, V.39, 1980, 101-124.

[15] BLOK W.J. The Lattice of Modal Logics: an Algebraic Investigation. *J. of Symbolic Logic*, V.45, 1980, 221-236.

[16] BLOK W.J., KÖHLER. P. Algebraic Semantics for Quasi-Classical Modal Logics. *J.of Symbolic Logic*, V.48, 1983, 941-964.

[17] BLOK W.J., PIGOZZI D. A Finite Basis Theorem for Quasivarieties. *Algebra Universalis*, V.22, 1986, 1-13.

[18] BLOK W.J., PIGOZZI D. Algebraizable Logics. *Memoirs of the Amer. Math, Soc.*, V.77, 1989, No.396.

[19] BLOOM S.L. A Representation Theorem for the Lattice of Standard Consequence Operator. *Studia Logica*, V.34, 1975, 235-237.

[20] BOOLOS G. The Unprovability of Consistency. *Cambridge: Cambridge University Press*, 1979.

[21] BULL R. That All Normal Extension of *S*4.3 Have the Finite Model Property. *Z. für Math. Log. und Grundl. der Math.*, V.12, 1967, 325 - 329.

[22] BULL R., SEGERBERG K. Basic Modal Logic. In Dov Gabbay and Franz Guenthner, editors, *Handbook of Philosophical Logic*, Vol. 2, 1-88. Reidel, 1984.

[23] BURGESS J.P. Basic Tense Logic. In Dov Gabbay and Franz Guenthner, editors, *Handbook of Philosophical Logic*, Vol. 2, 89-133. Reidel, 1984.

[24] BURRIS S., SANKAPPANAVAR H.P. A Course in Universal Algebra. *Springer*, No. 78 in Graduate Texts in Mathematics, 1981.

[25] CHANG L., KEISLER S. Model Theory, North-Holland, Amsterdam, 3-d edition, 1990.

[26] CHAGROV A. Decidable Modal Logic with Undecidable Admissibility Problem. *Algebra and Logic*, V.31, 1992, No.1, 53-61 (in Russian).

[27] CHAGROV A., ZAKHARYASCHEV M. The Undecidability of the Disjunction Property and Other Related Problems. J. OF SYMBOLIC LOGIC, V.58, 1993, 967-1002.

[28] CHAGROV A., ZAKHARYASCHEV M. Modal Logics. Manuscript of Book, 1996, will appear in *Cambridge Press.*

[29] CITKIN A.I. On Admissible Rules of Intuitionistic Propositional Logic. *Math. USSR Sbornik*, V.31, 1977, No.2, 279 - 288.

[30] CITKIN A.I. On Structurally Complete Superintuitionistic Logics, *Soviet Math. Dokl.* V.19, 1978), No. 4, 816 - 819.

[31] CRESWELL M. An Incomplete Decidable Logic. *J. of Symbolic Logic*, V.49, 1984, 520-527.

[32] CZELAKOWSKI J. Equivalential Logics (I), (ii). *Studia Logica*, V. 40, 1981, 227-236, 335-372.

[33] DOSEN K. Sequent Systems and Groupoid Models. *Studia Logica*, V. 47, 1988, 353 -385.

[34] DUMMETT M., LEMMON E. Modal Logics Between S4 and S5. *Z. für Math. Log. und Grundl. der Math.*, V.5, 1959, 260-264.

[35] DZIOBIAK W. Concerning Axiomatizability of the Quasivariety Generated by a Finite Heyting or Topological Boolean Algebra. *Studia Logica*, V.41, 1982, No. 4, 415-428.

[36] DZIOBIAK W. Structural Completeness of Modal Logics Containing $K4$. *Bull. Sect. Logic Pol. Acad. Sci.*, V.12, 1983, No.1, 32-36.

[37] EHRENFEUCHT A. Decidability of the Theory of Linear Ordering Relation. *Notices of Amer. Math. Soc.*, V.6, 1959, 268 - 269.

[38] ERSHOV YU,L., LAVROV I.A., TAIMANOV A.D., TAISLIN M.A. Elementary Theories. *Uspechi Matematicheskich Nauk*, V.20, 1965, No.4, 37 - 108 (in Russian).

[39] ESAKIA L., MESKHI V. Christmas Trees. On Free Cyclic Algebras in Some Varieties of Closure Algebras. *Bull. Sec. Logic Pol. Acad. Sci.*, V.4, 1975, No 3, 95-102.

[40] ESAKIA L., MESKHI V. The Critical Modal Systems. *Theoria*, V.43, 1977, 52-60.

[41] FAGIN R., HALPERN J.Y., VARDI M.Y. What is an Inference Rule. *J. of Symbolic Logic*, V.57, 1992, No 3, 1018 - 1045.

[42] FEFERMAN S., VAUGHT R. The First-Order Properties of Algebraic Systems. *Fund. Math.*, V.47, 1959, 57 - 103.

[43] FINE K. Logics Containing S4.3.*Z. für Math. Logic and Grundl. Math.*, V.17, 1971, 371 - 376.

[44] FINE K. An Ascending Chain of S4 Logics. *Theoria*, V.40, 1974, 110 - 116.

[45] FINE K. An Incomplete Logic Containing S4. *Theoria*, V.40, 1974, 23 - 29.

[46] FINE K. Logics Containing K4, Part I. *J. of Symbolic Logic*, V.39, 1974, 229 - 237.

[47] FINE K. Logics Containing K4. Part II. *J. of Symbolic Logic*, V.50, 1985, No. 3, 619-651.

[48] FRIEDMAN H. One Hundred and Two Problems in Mathematical Logic. *J. of Symb. Logic*, V.40, 1975, No.3 113 - 130.

[49] M.FITTING Intuitionistic Logic, Model Theory and Forcing. *North-Holland*, Amsterdam-New-York, 1969.

[50] DE JONGH D., TROELSTRA S. On the Connection of Partially Ordered Sets with Some Pseudo-boolean Algebras. *Indagationes Mathematicae*, V.28, 1966, 317-329.

[51] DE JONGH D., VISSER A. Explicit Fixed Points in Interpretability Logic. *Studia Logica*, V.50, 1991, 39-49.

[52] DE JONGH D., VISSER A. Embeddings of Heyting Algebras (revised Version), *Preprint*, Department of Philosophy - Utrecht University, 1994, No. 115.

[53] GABBAY D.M. Selective Filtration in Modal Logic. *Theoria*, V.30, 1970, 323-330.

[54] GABBAY D. M., DE JONGH D. A Sequence of Decidable Finitely Axiomatizable Intermediate Logics with the Disjunction Property. *J. of Symbolic Logic*, V.39, 1974), No 1, 67 - 78.

[55] GABBAY D.M., GUENTHNER F., editors. Handbook of Philosophical Logic. vol. 2, *Reidel*, 1984.

[56] GLIVENKO V. Sur quelques points de la logique de M.Brouwer. *Academie Royale de Belgique, Bulletins de la Classe des Sciences*, ser. 5, V.15 (1929), 183-188.

[57] GÖDEL K. Eine Interpretation der Intuitionistisher Aussagenkalculus, *Ergebnisse math. Kolloquiums*, V.4, 1933, 39-40.

[58] GOLDBLATT R.I. Metamathematics of Modal Logics. *Reports on Mathematical Logic*, V.6, 1976, 41 - 78 (Part 1), and V.7, 1976, 21 - 52 (Part 2).

[59] GOLDBLATT R.I. The McKinsey Axiom is not Canonical. *J. of Symbolic Logic*, V.56, 1991), No. 2, 554-562.

[60] GORANKO V. Refutation Systems in Modal Logic. *Studia Logica*, V. 53, 1994, 299-324.

[61] GRÄTZER G. Universal Algebra. *van Nostrand*, New-York, 1968.

[62] GRIGOLIA R.SH. Free S4.3-Algebras of Finite Rank. In book: *Investigation in Non-Classical Logics and Formal Theories*, Moscow, Nauka, 1983, 281-286 (in Russian)

[63] GRZEGORCZYK A. Some Relational Systems and Associated Topological Spaces. *Fund. Math.*, V.60, 1967, 223-231.

[64] HARROP R. Concerning Formulas of the Types $A \to B \vee C$, $A \to \exists x B(x)$ in Intuitionistic Formal System. *J. of Symbolic Logic*, V. 25, 1960, 27-32.

[65] HENKIN L., MONK J.D., TARSKI A. Cylindric Algebras. Parts 1 and 2,*North-Holland Pub. Co.*, Amsterdam, 1971, 1985.

[66] HERRMANN B., RAUTENBERG W. Finite Replacement and the Finite Hilbert-Style Axiomatizability. *Z. für Math. Log. und Grundl. der Math.*, V. 38, 1992, No. 4, 327 -344.

[67] HORN A. Free S5-Algebras. *Notre Dame J. of Formal Logic*, V.19, 1978, 189-191.

[68] HOSOI T. On Intermediate Logics I. *J. Fac. Sci. Univ. Tokyo, sec. I*, V. 14, 1967, 293-312.

[69] JANKOV V.A. The Construction of a Sequence of Strongly Independent Superintuitionistic Propositional Calculi. *Soviet Mathematics*, V.9, 1968, 806-807.

[70] JONSSON B. Algebras Whose Congruence Lattices are Distributive. *Math. Scan.*, V.21, 1967, 110-121.

[71] JONSSON B., TARSKI A. Boolean Algebras with Operators. *American J. of Mathematics*, V.73, 1951, 891 - 939.

[72] KOZEN D., PARIKH R. An Elementary Proof of the Completeness of PDL. THEORETICAL COMPUTER SCIENCE, V.14, 1981, 113-118.

[73] KRACHT M. An Almost General Splitting Theorem for Modal Logic. *Studia Logica*, V.49, 1990, 455-470.

[74] KRACHT M. Prefinitely Axiomatizable Modal and Intermediate Logics. *Mathematical Logic Quarterly*, V.39, 1993, 301-322.

[75] KRACHT M. Splittings and Finite Model Property. *J. of Symbolic Logic*, V.58, 1993, 139-157.

[76] KRACHT M. Highway to the Danger Zone. *J. of Logic and Computation*, V.5, 1995, 93-109.

[77] KRACHT M., WOLTER F. Properties of Independently Axiomatizable Bimodal Logics. *J. of Symbolic Logic*, V.56, 1991, No. 4, 1469-1485.

[78] KRIPKE S. Semantic Analysis of Modal Logic. *Z. für Math. Log. und Grundl. der Math.*, V.9, 1963, 67 - 96.

[79] KRUSKAL J.B. Well-quasi-ordering, the Tree Theorem and Vazsonyi's Conjecture. *Trans. Amer. Math. Soc.*, V.95, 1960, 210 -225.

[80] KUZNETSOV A.V. On Superintuitionistic Logics. *Mathematical Investigations*, V.10, 1975, No.2, 150-157 (in Russian).

[81] KUZNETSOV A.V. On Methods of Recognizing of Nondeducible and Inexpressibility. In Book: *Logical Derivation*, Moscow, Nauka, 1979, 5-33 (in Russian).

[82] LEWIS C., LANGFORD G. Symbolic Logic, New-York, 1932.

[83] LEMMON Algebraic Semantics for Modal Logics I, II. *J. of Symbolic Logic*, V. 31, 1966, No. 1, 46 - 65, No.2, 191 - 218.

[84] LORENZEN P. Einfüng in Operative Logik und Mathematik. Berlin - Göttingen - Heidelberg, 1955.

[85] LOS J., SUSZKO R. Remarks on Sentential Logic. *Indagationes Mathematicae*, V.20, 1958, 177-183.

[86] LYNDON R.C. Equations in Free Groups. *Trans. Amer. Math. Soc.*, V.96, 1960, 445-457.

[87] LYNDON R.C. Equations in Free Metabelian Groups. *Proc. Amer. Math. Soc.*, V.17, 1966, 728-730.

[88] MAKANIN G.S. Problem of Solvability for Equations in Free Semigroup. *Mathematical Sbornik* V.103, 1977, No. 2, 147 - 236 (in Russian).

[89] MAKANIN G.S. Decidability of Universal and Positive Theories of Free Group. *Izvestiya Acad. of Sci. USSR, ser. mathem.*, V.48, 1984, No. 4, 735 - 749 (in Russian).

[90] MAKINSON D.C. Some Embedding Theorems for Modal Logic. *Notre Dame J. of Formal Logic*, V.12, 1971, 252-254.

[91] MAKSIMOVA L.L. Pretabular Superintuitionistic Logics. *Algebra and Logic*, V.11, 1972, No.5, 558-570 (in Russian, Engl. transl. in: Amer. Math. Soc. Transl., *Algebra and Logic*, 1972).

[92] MAKSIMOVA L.L. Pretabular Extensions of Lewis's Logic S4. *Algebra and Logic*, V.14, 1975, No.1, 28-55 (in Russian, Engl. transl. in: Amer. Math. Soc. Transl., *Algebra and Logic*, 1975).

[93] MAKSIMOVA L.L. Finite-Level Modal Logics. *Algebra and Logic*, V.14, 1975, 304 - 319 (in Russian, Engl. transl. in: Amer. Math. Soc. Transl., *Algebra and Logic*, 1975).

[94] MAKSIMOVA L.L. Definability and Interpolation in Classical Modal Logics. *Contemporary Mathematics*, V.131, 1992, 583-599.

[95] MAKSIMOVA L.L. The Beth Properties, Interpolation, and Amalgamability in Varieties of Modal Algebras. *Soviet Math. Dokl.* V. 44, 1992, No. 1, 327 - 331.

[96] MAKSIMOVA L.L., SKVORTSOV D.P., SHEHTMAN V.B., The Impossibility of a Finite Axiomatization of Medvedev's Logic of Finite Problems. *Soviet Math. Dokl.*, V.20, 1979, No. 2, 394 - 398.

[97] MAKSIMOVA L.L, RYBAKOV V.V. A Lattice of Normal Modal Logics. *Algebra and Logic*, V.13, 1974, No.2, 105 - 122.

[98] MALT'SEV A.I Algebraic Systems. *Springer-Verlag*, Berlin, Heidelberg, New York, 1973.

[99] McKAY C.C. A Class of Decidable Intermediate Propositional Logics. *J. of Symbolic Logic*, V.36, 1971, No.1, 127 - 128.

[100] McKINSEY J.C.C, TARSKI A. The Algebra of Topology. *Ann. Math.*, V.45, 1944, 141-191.

[101] McKINSEY J.C.C, TARSKI A. Some Theorems about the Sentential Calculi of Lewis and Heyting. *J. of Symbolic Logic*, V.13, 1948, 1 - 15.

[102] MENDELSON E. Introduction to Mathematical Logic. *D. Van Nostrand Company, Inc.*, Princeton, New Jersey, Toronto, New York, London, 1964.

[103] MONK J.D. Substitutionless Predicate Logic with Identity. *Arch. Math. Logic*, V.7, 1965, 353 - 358.

[104] MINTS G.E. Derivability of Admissible Rules. *J. of Soviet Mathematics*, V. 6, 1976, No. 4, 417 - 421.

[105] NEMETI I. On Varieties of Cylindric Algebras with Applications to Logic. *Annals of Pure and Applied Logic*, V.36, 1987, 235-277.

[106] NISHIMURA I. On Formulas of One Variable in Intuitionistic Propositional Calculus. *J. of Symb. Logic*, V.25, 1960, 327-331.

[107] ONO H. A Study of Intermediate Predicate Logics. *Publ. RIMS, Kyoto Univ*, V. 8, 1973, 113 - 130.

[108] ONO H. On Finite Linear Intermediate Predicate Logics. *Studia Logica*, V. XLVII, 1988, No. 4, 391 - 399.

[109] ONO H. Semantics for Substructural Logics. *Substructural Logics*, *K.Dozen and P.Schroeder-Heister eds.*, Clarendon Press, Oxford, 1993, 259 - 291.

[110] ONO H., KOMORI Y. Logics without the Contraction Rule. *J. of Symbolic Logic*, V.50, 1985, 169 - 201.

[111] PIECZKOWSKI A. Undecidability of the Homogeneous Formulas of Degree 3 of the Predicate Calculus. *Studia Logica*, V.XXII, 1968, 7-14.

[112] PIGOZZI D. Finite Basis Theorem for Relatively Congruence-Distributive Quasivarieties. *Transactions Amer. Math. Soc.*, V.310, 1988, No.2, 499-533.

[113] PIXLEY A.F. Distributivity and permutability of congruence relations in equational classes of algebras. *Proc. Amer. Math. Soc.*, V. 14, 1963, No.1, 105 - 109.

[114] POGORZELSKI W.A. Structural Completeness of the Propositional Calculi. *Bull. de l'Acad. Polonaise des Sciences*, V.19, 1971, No. 5, 345-351.

[115] PORT J. The Deducibilities of S5. *J. of Phil. Logic*, V.10, 1981, 409-422.

[116] PORT J. Axiomatization and Indepedence in S4 and S5. *Reports on Mathematical Logic*, V.16, 1983, 23 - 33.

[117] PRUCNAL T. The Structural Completeness the Medvedev's Logic. *Reports on Math. Logic*, V.6, 1976, 103-105.

[118] PRUCNAL T. On Two Problems of Harvey Friedman. *Studia Logica*, V.38, 1979, No.3, 247-262.

[119] PRUCNAL T. Structural Completeness of Some Fragments of Intermediate Logics. *Bull. Sec. Log. Pol. Acad. Sci.* V.12, 1983, No.1, 18-21.

[120] RASIOWA H. An Algebraic Approach to Non-Classical Logics. *Studies in Logic and Foundation of Mathematics*, V.78, American Elsevier Publishing Company, Amsterdam - London - New-York, 1974.

[121] RASIOWA H., SIKORSKI R. The Mathematics of Metamathematics. *Polska Akedemia Nauk*, V. 41, Warszawa, 1963.

[122] RAUTENBERG W. Klassische und nichtklassische Aussagenlogik. Braunschweig - Wiesbaden, Deutschland, 1979.

[123] RAUTENBERG W. Splitting Lattices of Logics. *Archive für Math. Logik*, V. 20, 1980, 155 -159.

[124] RAUTENBERG W. 2-Element Matrices. *Studia Logica*, V.40, 1981, No. 4, 315-353.

[125] RAUTENBERG W. Modal Tableau Calculi and Interpolation. *J. Phil. Logic*, V.12, 1983, 403-423.

[126] RAUTENBERG W. Applications of Weak Kripke Semantics to Intermediate Consequences. *Studia Logica*, V.XLV, 1984, No.1, 119- 134.

[127] RAUTENBERG W. Strongly Finitely Based Equational Theories. *Algebra Universalis*, V.28, 1991, 549-558.

[128] ROZIERE P. Admissible and Derivable Rules. *Math. Struct. in Comp. Science*, V.3, 1993, 129 - 136. Cambridge University Press.

[129] ROGERS H. Theory of Recursive Functions and Effective Computability. McGraw-Hill Book Company. New-York-St.Louis,etc. 1967.

[130] RYBAKOV V.V. Noncompact Extensions of Logic *S4*. *Algebra and Logic*, V.16, 1977, No. 4, 472-490. (in Russian, Engl. Transl. in Amer. Math. Soc. Transl. *Algebra and Logic*, 1977).

[131] RYBAKOV V.V. Decidable Noncompact Extension of Logic *S4*. *Algebra and Logic*, V.17, 1978, No.2, 210-219. (in Russian, Engl. Transl. in Amer. Math. Soc. Transl. *Algebra and Logic*, 1978).

[132] RYBAKOV V.V. Modal Logics with LM-Axioms. *Algebra and Logic*, V.17, 1978, No. 4, 455-467. (in Russian, Engl. Transl. in Amer. Math. Soc. Transl. *Algebra and Logic*, 1978).

[133] RYBAKOV V.V. Admissible Rules of Pretabular Modal Logics. *Algebra and Logic*, V.20, 1981, 291-307 (English translation).

[134] RYBAKOV V.V. Bases for Quasi-identities of Finite Modal Algebras. *Algebra and Logic*, V.21, 1982, No. 2, 219-228 (in Russian, Engl. Transl. in Amer. Math Soc. Transl. *Algebra and Logic*, 1982).

[135] RYBAKOV V.V. A Criterion for Admissibility of Rules in the Modal System $S4$ and the Intuitionistic Logic. *Algebra and Logic*, V.23 (1984), No 5, 369 - 384 (Engl. Translation).

[136] RYBAKOV V.V. Admissible Rules for Logics Containing S4.3. *Sibirski Math. Journal*, V.25, 1984, No 5, 141 - 145 (in Russian, Engl. transl. in Siberian Math. Journal, 1984).

[137] RYBAKOV V.V. Elementary Theories of Free Topo-Boolean and Pseudo-boolean Algebras. *Matematicheskie Zametki*, V.37, 1985, No.6, 797-802 (in Russian, English Transl. in USSR Math. Zametki, 1985).

[138] RYBAKOV V.V. A Criterion for Admissibility of Rules of Inference in Modal and Intuitionistic Logic. *Soviet Math. Dokl.*, V.32, 1985, No.2, 452 - 455.

[139] RYBAKOV V.V. The Bases for Admissible Rules of Logics S4 and Int. *Algebra and Logic*, V.24, 1985, 55-68 (English translation).

[140] RYBAKOV V.V Decidability by Admissibility of the Modal System Grz and Intuitionistic Logic. *Izvestiya Acad. Nauk SSSR, Ser. Math.*, V.50, 1986, No.3 (in Russian, Engl.transl. in *Math. USSR Izvestiya*, V.28, 1987, No. 3, 589 - 608).

[141] RYBAKOV V.V. Bases of Admissible Rules of the Modal System Grz and of Intuitionistic Logic. *Math. Sbornik*, V.128(170), 1985, No.3 (in Russian, Engl.Transl. in *Math USSR Sbornik*, Vol.56, 1987, No. 2, 311 - 331).

[142] RYBAKOV V.V. Equations in Free Topoboolean Algebra and the Substitution Problem. *Soviet Math. Dokl.*, V.33, 1986, No.2, 428 - 431.

[143] RYBAKOV V.V. Equations in Free Topoboolean Algebra. *Algebra and Logic*, V.25, 1986, No. 2, 172-204 (in Russian, Engl.Transl. in Amer. Math Soc. Transl. *Algebra and Logic*, 1986)

[144] RYBAKOV V.V. Problems of Admissibility and Substitution, Logical Equations and Restricted Theories of Free Algebras. In book:
Logic Methodology and Philosophy of Science VIII, Studies in Logic and Foundations of Mathematics, V. 126, Eds. J.E.Fienstad, I.T.Frolov and R.Hilpinen, North. Holland, Amsterdam, 1989, 121 - 139.

[145] RYBAKOV V.V. Problems of Substitution and Admissibility in the Modal System Grz and Intuitionistic Calculus. *Annals of Pure and Applied Logic*, V.50 , 1990, 71-106.

[146] RYBAKOV V.V. Metatheories of First-order Theories. In: *Proc. of the Fourth Asian Logic Conference,* CSK Educational Center, Tokyo, Japan, 1990, 16 - 17.

[147] RYBAKOV V.V. Logical Equations and Admissible Rules of Inference with Parameters in Modal Provability Logics. *Studia Logica,* V. XLIX, 1990, No 2, 215 - 239.

[148] RYBAKOV V.V. Admissibility of Rules of Inference, and Logical Equations in Modal Logics Axiomatizing Provability. *Izvestiya Akad. Nauk SSSR, Ser. Math.,* V.54, 1990, No.2, 357 -377 (in Russian, Engl. transl. in *Math. USSR Izvestiya,* V.36, 1991, No.2, 369 - 390).

[149] RYBAKOV V.V. Poly-modal Logic as Metatheory of Pure Predicate Calculus. *Abstracts of 9-th Intern. Congress of Log., Method. and Phil. of Sci.,* Sections 1 - 5, Uppsala, Sweden, 1991, p.158.

[150] RYBAKOV V.V. Rules of Inference with Parameters for Intuitionistic Logic. *J. of Symbolic Logic,* V.57, 1992, No 3, 912 - 923.

[151] RYBAKOV V.V. A Modal Analog for Glivenko's Theorem and its Applications. *Notre Dame J. of Formal Logic,* V.33, 1992, No 2, pp 244-248.

[152] RYBAKOV V.V. Intermediate Logics Preserving Admissible Inference Rules of Heyting Calculus. *Mathematical Logic Quarterly,* V.39, 1993, 403 - 415.

[153] RYBAKOV V.V. Elementary Theories of Free Algebras for Varieties Corresponding to Non-Classical Logics. *In book: Algebra, Proc. of the III-rd International Conference on Algebra, Krasnoyarsk, Russia, 1993.* Ed. *Yu.L.Ershov, E.I.Khukhro etc.,* de Gruyter, Berlin-New York, 199-208.

[154] RYBAKOV V.V. Criteria for Admissibility of Inference Rules. Modal and Intermediate Logics with the Branching Property. *Studia Logica,* V.53, 1994, No.2, 203-225.

[155] RYBAKOV V.V. Preserving of Admissible Inference Rules in Modal Logic. In Book: *Logical Foundations of Computer Science, Lecture Notes in Computer Science,* Eds.: A.Nerode, Yu.V.Matiyasevich, V.813, 1994, Springer-Verlag, 304-316.

[156] RYBAKOV V.V. Hereditarily Structurally Complete Modal Logics. *J. of Symbolic Logic,* V.60, 1995, No.1, 266-288

[157] RYBAKOV V.V Modal Logics Preserving Admissible for S4 Inference Rules. In book: *Computer Science Logic, Lecture Notices in Computer Science,* Eds. L.Pacholski, J.Tiurin, V.993, 1995, Springer-Verlag, 512 - 526.

[158] RYBAKOV V.V. Even Tabular Modal Logics Sometimes Do Not Have Independent Base for Admissible Rules. *Bulletin of the Section of Logic*, V.24, 1995, No.1, 37 - 40.

[159] RYBAKOV V.V. Scheme-Logics and Admissible Rules for First-Order Theories and Poly-Modal Propositional Logic. *Notre Dame J. of Formal Logic*, 1996, accepted, will appear.

[160] SAHLQVIST H. First and Second-order Semantics for Modal Logic. In book: Ed. Stig Kanger, *Proc. of the Third Scandinavian Logic Symposium*, North-Holland, Amsterdam, 1975, 110-143.

[161] SAMBIN G. An Effective Fixed-Point Theorem in Intuitionistic Diagonalizable Algebras. *Studia Logica*, V.35, 1975, 345-361.

[162] SAMBIN G., VACCARO V., Topology and Duality in Modal Logic. *Annals of Pure and Applied Logic*, V.37, 1988, 249-296.

[163] SAMBIN G., VACCARO V. A Topological Proof of Sahlqvist Theorem. *J. of Symbolic Logic*, V.54, 1989, 992-999.

[164] SHAVRUKOV V.JU. Two extensions of modal provability logic. *Mathematical Sbornik*, V.181, 1990, N0. 2, 240 - 255. (in Russian)

[165] SEGERBERG K. An Essay in Classical Modal Logic. Filosofiska Studier, Mimeograph, Vol. 1 - 3, Uppsala, 1971.

[166] SEGERBERG K. That All Extension of S4.2 are Normal. In book: Ed. Stig Kanger, *Proc. of the Third Scandinavian Logic Symposium*, North-Holland, Amsterdam, 1975, 194-196.

[167] SEGERBERG K. A Completeness Theorem in the Modal Logic of Programs. *Notes of the Amer. Math. Soc.*, V.24(6), 1977, A-522.

[168] SEGERBERG K. Modal Logics with Functional Alternative Relations. *Notre Dame J. of Formal Logic*, V.27, 1986, 504-522.

[169] SELMAN A. Completeness of Calculi for Axiomatically Defined Classes of Algebras. *Algebra Universalis*, V.2, 1972, No.1, 20-32.

[170] SHEHTMAN V.B. On Incomplete Propositional Logics. *Doklady AN SSSR*, V.235, 1977, No. 3, 542-545 (in Russian, Engl. Transl. in *Soviet Math. Dokl.*).

[171] SHEHTMAN V.B. Undecidable Superintuitionistic Propositional Calculus. *Doklady AN SSSR*, V.240, 1978, No. 3, 549-552 (in Russian, Engl. Transl. in *Soviet Math. Dokl.*).

[172] SHEHTMAN V.B. Riger-Nishimura's Ladders. *Doklady AN SSSR*, V.241, 1978, No. 6, 1288-1291 (in Russian, Engl.Transl. in *Soviet Math. Dokl.*).

[173] SHEHTMAN V.B. Topological Models of Propositional Logics. *Semiotika i Informatika*, V.15, 1980, 74-98. (in Russian)

[174] SOLOVAY R. Provability Interpretations of Modal Logic. *Israel J. Math.*, V.25, 1976, 287 - 304.

[175] SURMA S.J. The Deduction Theorems Valid in Certain Fragment of the Lewis's System S2 and the System *T* of Feys-von Wright. *Studia Logica*, V.31, 1972, 127-136.

[176] SUSZKO R. Concerning the Method of Logical Schemes, the Notion of Logicae Calculus and Role of Consequence Relations, *Studia Logica*, V.11, 1961, 185-214.

[177] THOMASON S.K. Semantic Analysis of Tense Logic. *J. of Symbolic Logic*, V.37, 1972, 150-158.

[178] THOMASON S.K. Noncompactness in Propositional Modal Logics. *J. of Symbolic Logic*, V.37, 1972, No. 4, 716-720.

[179] THOMASON S.K. An Incompleteness Theorem in Modal Logic. *Theoria*, V.40, 1974, No. 1, 30-34.

[180] THOMASON S.K. Reduction of Tense Logic to Modal Logic. *J. of Symbolic Logic*, V.39, 1974, 549-551.

[181] THOMASON S.K. Reduction of Tense Logic to Modal Logic II. *Theoria*, V.41, 1975, 154-169.

[182] TOKARZ. M. On Structural Completeness of Lukasiewicz logics. In book: *Selected Papers on Lukasiewicz Sentential Calculi*, Wroclow, 1977, 171-176.

[183] URQUHART A., Decidability and the Finite Model Property. *J. of Phil. Logic*, V.10, 1981, 367-370.

[184] URQUHART A., The Undecidability of Entailment and Relevant Implication *J. of Symbolic Logic*, V.49, 1984, 1059-1073.

[185] URQUHART A., The Complexity of Gentzen Systems for Propositional Logic. *J. Theoretical Computer Science*, V.66, 1989, 87 - 97.

[186] URQUHART A., Failure of Interpolation in Relevant Logics. *J. of Phil. Logic*, V.22, 1993, 449 - 479.

[187] VAKARELOV D. Filtration Theorem for Dynamic Algebras with Tests and Inverse Operator. In book: *Logics of Programs and Their Applications, Lecture Notes in Computer Science*, Ed.: A. Salwicki, No.148, 1983, Springer, 314-324.

[188] VAKARELOV D. A Modal Logic for Similarity Relations in Pawlak Knowledge Representation Systems. *Fundamenta Informaticae*, V. XV, 1991, 61 -79.

[189] VAN BENTHEM J.F.A.K. Modal Reduction Principles. *J.of Symbolic Logic*, vol. 41, 1976, 301-312.

[190] VAN BENTHEM J.F.A.K. Two simple Incomplete Modal Logics. *Theoria*, V. 44, 1978, No.1, 25 - 37.

[191] VAN BENTHEM J.F.A.K. Syntactic Aspects of Modal Incompleteness Theorems. THEORIA, V.45, 1979, 63-77.

[192] VAN BENTHEM J.F.A.K. Modal and Classical Logic. Bibliopolis, 1983.

[193] VAN BENTHEM J.F.A.K. Correspondence Theory. In book: Eds.: D.Gabbay and F.Guenthner, *Handbook of Philosophical Logic*, V.2, 1984, Reidel, 167-247.

[194] VAN BENTHEM J.F.A.K. Notes on Modal Definability. *Notre Dame J. of Formal Logic*, V.39, 1989, 20-39.

[195] VENEMA Y. Many-Dimensional Modal Logic. *Ph.D. Dissertation*, Amsterdam University, Faculty of Mathematics and Information Science, 1992, 177pp.

[196] VENEMA Y. Derivation Rules as Anti-axioms in Modal logic. *J. of Symbolic Logic*, V.59, 1993, No.3, 1003-1034.

[197] VISSER A. The Provability Logics of Recursively Enumerable Theories. *J. of Phil. Logic*, V.13, 1984, 97 -113.

[198] VISSER A. The Formalization of Interpretability. *Studia Logica*, V.LI, 1991, 81 - 105.

[199] VISSER A. Interpretations over Heyting's Arithmetic. Preprint,*Logic Group Preprint Series*, Utrecht University, Utrecht, The Netherlands, 1995.

[200] WILLIAMSON T. Some Admissible Rules in Non-normal Modal Systems. *Notre Dame J. of Formal Logic*, V.34, 1993, No. 3, 378 - 400.

[201] WOJCICKI R. Some Remarks on the Consequence Operation in Sentential Logic. *Fund. Math.*, V.58, 1970, 269-279.

[202] WOJCICKI R. Note on Decidability and Many-validness. *J. of Symbolic Logic*, V.39, 1974, No. 3, 563 - 566.

[203] WOJCICKI R. Theory of Logical Calculi. *Kluwer Press*, Dordrecht, 1988.

[204] WOJTYLAK P. On Structural Completeness of Many-valued Logics. *Studia Logica*, V.37, 1978, No.2, 139-147.

[205] WOJTYLAK P. Independent Axiomatizability of Sets of Sentences. *Annals of Pure and Applied. Logic*, V.44, 1989, 259 - 299.

[206] WOLTER FR. Lattice of Modal Logics. *Ph.D. Thesis*, Free University, Berlin, 1993.

[207] WOLTER FR. Solution to a Problem of Goranko and Passy. *J. of Logic and Computation*, V.4, 1994, 21-22.

[208] WOLTER FR. What is the Upper Part of the Lattice of Bimodal Logics ? *Studia Logica*, V.53, 1994, 235-242.

[209] WOLTER FR. The Finite Model Property in Tense Logics. *J. of Symbolic Logic*, V.60, 1995, No. 3, 757 - 774.

[210] WRONSKI A. On Cardinalities of Matrices Strongly Adequate for the Intuitionistic Propositional Logic. *Reports on Math. Logic*, 1974, No. 3, 67-72.

[211] WRONSKI A. On Finitely Based Consequence Operations. *Studia Logica*, V.35, 1976, No.4, 452-458.

[212] ZAKHARYASCHEV M.V. Syntax and Semantics of Modal Logics Containing $S4$. *Algebra and Logic*, V. 27, 1987, 659-689 (in Russian, English Transl. in Amer. Math. Soc. Transl., *Algebra and Logic*, 1987).

[213] ZAKHARYASCHEV M.V. Modal Companions of Intermediate Logics: Syntax, Semantics and Preservation Theorems. *Mathematical Sbornik*, V. 180, 1989, 1415 - 1427 (in Russian).

[214] ZAKHARYASCHEV M.V. Canonical Formulas for $K4$. Part I. Basic Results. *J. of Symbolic Logic*, V.57, 1992, 1377-1402.

[215] ZYGMUNT J. Direct Product of Consequence Operations. *Bull. Sec. Logic Pol. Acad. Sci.*, V.1, 1972, No.4, 61-64.